The Philosophy of Science

An Encyclopedia

EDITORIAL ADVISORY BOARD

Justin Garson
University of Texas at Austin

Paul Griffiths
University of Queensland

Cory Juhl
University of Texas at Austin

James Justus
University of Texas at Austin

Phillip Kitcher
Columbia University

Ian Nyberg
University of Texas at Austin

Anya Plutyinski
University of Utah

Sherrilyn Roush
Rice University

Laura Ruetsche
University of Pittsburgh

John Stachel
Boston University

William Wimsatt
University of Chicago

The Philosophy of Science

An Encyclopedia

Volume 1
A–M

Sahotra Sarkar
Jessica Pfeifer
EDITORS

EDITORIAL ASSISTANTS
Justin Garson, James Justus, and Ian Nyberg
University of Texas

Published in 2006 by
Routledge
Taylor & Francis Group
270 Madison Avenue
New York, NY 10016

Published in Great Britain by
Routledge
Taylor & Francis Group
2 Park Square
Milton Park, Abingdon
Oxon OX14 4RN

© 2006 by Taylor & Francis Group, LLC
Routledge is an imprint of Taylor & Francis Group

Printed in the United States of America on acid-free paper
10 9 8 7 6 5 4 3 2 1

International Standard Book Number-10: 0-415-93927-5 (Hardcover)
International Standard Book Number-13: 978-0-415-93927-0 (Hardcover)
Library of Congress Card Number 2005044344

No part of this book may be reprinted, reproduced, transmitted, or utilized in any form by any electronic, mechanical, or other means, now known or hereafter invented, including photocopying, microfilming, and recording, or in any information storage or retrieval system, without written permission from the publishers.

Trademark Notice: Product or corporate names may be trademarks or registered trademarks, and are used only for identification and explanation without intent to infringe.

Library of Congress Cataloging-in-Publication Data

The philosophy of science : an encyclopedia / Sahotra Sarkar, Jessica Pfeifer, editors.
   p. cm.
   Includes bibliographical references and index.
   ISBN 0-415-93927-5 (set : alk. paper)--ISBN 0-415-97709-6 (v. 1 : alk. paper) -- ISBN 0-415-97710-X (v. 2 : alk. paper)
   1. Science--Philosophy--Encyclopedias. I. Sarkar, Sahotra. II. Pfeifer, Jessica.

Q174.7.P55 2005
501'.03--dc22
                                              2005044344

Taylor & Francis Group is the Academic Division of T&F Informa plc.

Visit the Taylor & Francis Web site at
http://www.taylorandfrancis.com

and the Routledge Web site at
http://www.routledge-ny.com

Dedicated to the memory of Bob Nozick, who initiated this project.

TABLE OF CONTENTS

Introduction — xi

List of Contributors — xxvii

A to Z List of Entries — xxxiii

Thematic List of Entries — xxxvii

Entries A–M — 1

THE PHILOSOPHY OF SCIENCE: AN INTRODUCTION

Philosophy of science emerged as a recognizable sub-discipline within philosophy only in the twentieth century. The possibility of such a sub-discipline is a result of the post-Enlightenment disciplinary and institutional separation of philosophy from the sciences. Before that separation, philosophical reflection formed part of scientific research—as, indeed, it must—and philosophy was usually guided by a sound knowledge of science, a practice that gradually lost currency after the separation. In the nineteenth century, philosophical reflection on science resulted in a tradition of natural philosophy, particularly in Britain (with the work of Mill, Pearson, Whewell, and others), but also in continental Europe, especially in Austria (with Bolzano, Mach, and others). What is called philosophy of science today has its roots in both the British and the Austrian traditions, although with many other influences, as several entries in this Encyclopedia record (see, for instance, Duhem Thesis; Poincaré, Henri).

This Encyclopedia is intended to cover contemporary philosophy of science. It is restricted to conceptual developments since the turn of the twentieth century. Its treatment of major figures in the field is restricted to philosophers (excluding scientists, no matter what the extent of their philosophical influence has been) and, with very few exceptions (notably Chomsky, Noam; Putnam, Hilary; and Searle, John), to those whose work is distant enough to allow "historical" appraisal. Conceptual issues in the general philosophy of science (including its epistemology and metaphysics) as well as in the special sciences are included; those in mathematics have been left for a different work. This Introduction will provide a guided tour of these conceptual issues; individual figures will only be mentioned in passing.

Historically, the themes treated in the Encyclopedia are those that have emerged starting with the period of the Vienna Circle (see Vienna Circle), including the figures and developments that influenced it (see Bridgman, Percy Williams; Duhem Thesis; Mach, Ernest; Poincaré, Jules Henri). The work of the members of the Vienna Circle provide a link between the older natural philosophy, especially in its Austrian version, and the later philosophy of science, which borrowed heavily from the concepts and techniques of the mathematical logic that was being created in the first three decades of the last century (see Hilbert, David; Ramsey, Frank Plumpton; Russell, Bertrand; see also Ayer [1959] and Sarkar [1996a]). The new set of doctrines—or, more accurately, methods—came to be called "logical positivism" and, later, "logical empiricism" (see Logical Empiricism; see also Sarkar [1996b]). By the 1930s these views had spread beyond the confines of Vienna and had attracted allegiance from many other similarly-minded philosophers (see Ayer, A. J.; Quine, Willard Van; Reichenbach, Hans). Two attitudes were widely shared within this group: a belief that good philosophy must be conversant with the newest developments within the sciences (see Rational Reconstruction), and a rejection of traditional metaphysics imbued with discussions with no empirical significance (see Cognitive Significance; Verifiability).

Some members of the Vienna Circle also took the so-called linguistic turn (see Carnap, Rudolf) and viewed scientific theories as systems formalized in artificial languages (Sarkar 1996c). Arguably, at least, this work lost the prized contact with the practice of science, and this development contributed to the eventual rejection of logical empiricism by most philosophers of science in the late twentieth century. However, a number of the original logical empiricists, along with many others, rejected the linguistic turn, or at least did not fully endorse it (see Neurath, Otto; Popper, Karl Raimund; Reichenbach, Hans). The tensions between the two views were never fully articulated during this period, let alone resolved, because the Vienna Circle as an institution and logical empiricism as a movement both came under political attack in Europe with the advent of Nazism. Most of the figures involved in the movement migrated to the United Kingdom and the United States. In the United States, many of the logical empiricists also later fell afoul of McCarthyism (see Logical Empiricism).

In the United States, Nagel probably best exemplifies what philosophy of science became in the period of the dominance of logical empiricism. The discussions of Nagel's (1961) *Structure of Science* typically include careful formal accounts of conceptual issues, but these are supplemented by detailed "nonformal" discussions in the spirit of the tradition of natural philosophy—this book may be viewed as a summary of where logical empiricism stood at its peak (see Nagel, Ernest). However, starting in the late 1940s, many of the theses adopted by the logical empiricists came under increasing attack even by those committed to keeping philosophy in contact with the sciences (Sarkar 1996e). (The logical empiricists had explicitly advocated and practiced intense self-criticism, and many of these attacks came from within their ranks—see Hempel, Carl Gustav.) Some of this criticism concerned whether cherished doctrines could be successfully formulated with the degree of rigor desired by the logical empiricists (see Analyticity; Cognitive Significance).

However, the most serious criticism came from those who held that the logical empiricists had failed to give an account of scientific confirmation and scientific change (see "Confirmation," "Scientific Discovery," and "Scientific Change," below). Feyerabend, for one, argued that the logical empiricists had placed science under an inadmissible rational straitjacket (see Feyerabend, Paul). As philosophy of science took a distinctly historical turn, analyzing the development of science in increasing historical detail, many felt that the logical empiricists had misinterpreted the historical processes of scientific change (see Hanson, Norwood Russell; Kuhn, Thomas). Kuhn's (1962) *Structure of Scientific Revolutions*, originally written for an encyclopedia sponsored by the logical empiricists, was particularly influential. By the mid-1960s logical empiricism was no longer the dominant view in the philosophy of science; rather, it came to be regarded as a "received view" against which philosophers of science defined themselves (Suppe 1974). However, this interpretation of logical empiricism ignores the disputes and diversity of viewpoints within the tradition (see, especially, Logical Empiricism), arguably resulting in a caricature rather than a responsible intellectual characterization.

Nevertheless, for expository ease, the term "received view" will be used in this Introduction to indicate what may, at least loosely, be taken to be the majority view among the logical empiricists.

Scientific realism and various forms of naturalism, sometimes under the rubric of "evolutionary epistemology," have emerged as alternatives to the logical empiricist interpretations of science (see Evolutionary Epistemology; Scientific Realism). Meanwhile, science has also been subject to feminist and other social critiques (see Feminist Philosophy of Science). Kuhn's work has also been used as an inspiration for interpretations of science that regard it as having no more epistemological authority than "knowledge" generated by other cultural practices (see Social Constructionism). However, whether such work belongs to the philosophy of science, rather than its sociology, remains controversial. While no single dominant interpretation of science has emerged since the decline of logical empiricism, the ensuing decades have seen many innovative analyses of conceptual issues that were central to logical empiricism. There has also been considerable progress in the philosophical analyses of the individual sciences. The rest of this Introduction will briefly mention these with pointers to the relevant entries in this work.

Theories

The analysis of scientific theories—both their form and content—has been a central theme within the philosophy of science. According to what has become known as "the received view," which was developed in various versions by the logical empiricists between the 1920s and 1950s, theories are a conjunction of axioms (the laws of nature) and correspondence rules specified in a formalized ideal language. The ideal language was supposed to consist of three parts: logical terms, observational terms, and theoretical terms. Logical claims were treated as analytic truths (see Analyticity), and were thought by many to be accepted as a matter of convention (see Conventionalism). Observational claims were also thought to be unproblematic, initially understood as referring to incorrigible sense-data and later to publicly available physical objects (see Phenomenalism; Physicalism; Protocol Sentences). The correspondence rules were supposed to allow the logical empiricists to give cognitive significance (see Cognitive Significance; Verifiability) to the theoretical portion of the language, by specifying rules for connecting theoretical and observational claims. In their extreme version, these correspondence rules took the form of operational definitions (see Bridgeman, Percy Williams). One goal of such attempts was to distinguish science from non-science, especially what the logical empiricists derided as "metaphysics" (see Demarcation, Problem of).

Starting in the 1960s, the received view encountered a number of problems. Even earlier, difficulties had arisen for the correspondence rules, which took various forms over the years as a result of these problems. Initially understood as explicit definitions, they were later treated as partial definitions, and in the end the theoretical terms were merely required to make a difference to the observational consequences of the theory. One central focus of the criticism was on the observation-theory distinction (see Observation). It was argued that the theoretical and observational portions of language are not distinct (Putnam 1962; Achinstein 1968; see also Putnam, Hilary), that the distinction between entities that are observable and those that are not is vague (Maxwell 1962), and that observations are theory-laden (Hanson 1958; see also Hanson, Norwood Russell; Observation). In addition, there were problems ruling out unintended models of theories, which became a source of counterexamples. In hindsight, it is also clear that the problem of demarcating science from non-science was never fully solved.

More recently, a number of philosophers have questioned the important place given to laws of nature on this view, arguing that there are scientific theories in which laws do not appear to play a significant role (see Biology, Philosophy of; Laws of Nature). Others have questioned not the occurrence of laws within theories, but whether any of these entities should be conceptualized as linguistic entities (which is quite foreign to the practice of science). Still others have wondered whether the focus on theories has been an artifact of the received view being based primarily on physics, to the detriment of other sciences. As the received view fell out of favor, starting in the 1960s, a number of philosophers developed various versions of what is known as the semantic view of theories, which understands theories as classes of models, rather than as linguistic entities specifiable in an axiomatic system. While not without its problems, the semantic view seemed to bring philosophical accounts of theories more in line with the practices of scientists and has become the generally accepted view of theories (see Scientific Models; Theories). Nevertheless, there is at present no consensus within the discipline as to how theories should be philosophically characterized.

Scientific Models

Models are central to the practice of science and come in a bewildering variety of forms, from the double helix model of DNA to mathematical models of economic change (see Scientific Models). Scientific models were regarded as being of peripheral philosophical interest by the received view. Little philosophical work was done on them until the 1970s, with Hesse's (1963) *Models and Analogies in Science* being a notable exception. That situation has changed drastically, with models probably now being the locus of even more philosophical attention than theories.

Two developments have contributed to the burgeoning philosophical interest in models:

(i) *The Semantic Interpretation of Theories.* The development of various versions of the semantic interpretation of theories has put models at the center of theoretical work in science (see Theories). For many proponents of the semantic view, the received view provided a syntactic interpretation of theories, regarding theories as formalized structures. Scientific models are then supposed to be construed in analogy with models in formal logic, providing semantic interpretations of syntactic structures. The semantic view inverts this scheme to claim that models are epistemologically privileged and that theories should be regarded as classes of models. The various semantic views have made many contributions to the understanding of science, bringing philosophical analysis closer to the practice of science than the received view. Nevertheless, almost all versions of the semantic view are at least partly based on a dubious assumption of similarity between models in logic and what are called "models" in science.

(ii) *Historical Case Studies.* How dubious that presumed similarity has been underscored by the second development that helped generate the current focus on scientific models: the detailed studies of the role of models in science that has been part of the historical turn in the philosophy of science since the 1960s. That turn necessitated a focus on models because much of scientific research consists of the construction and manipulation of models (Wimsatt 1987). These studies have revealed that there are many different types of models and they have a variety of dissimilar functions (see Scientific Models for a taxonomy). At one end are models of data and representational material models such as the double helix. At the other are highly idealized models (see Approximation), including many of the mathematical models in the different sciences. Some models, such as the Bohr model of the atom (see Quantum Mechanics) or the Pauling models of chemical bonds (see Chemistry, Philosophy of), are both mathematical and accompanied by a visual picture that help their understanding and use (see also Visual Representation).

At present, no unified treatment of the various types and functions of scientific models seems possible. At the very least, the rich tapestry of models in science cannot entirely be accommodated to the role assigned to them by the semantic interpretation of theories or any other account that views models as having only explanatory and predictive functions. The ways in which models also function as tools of exploration and discovery remain a topic of active philosophical interest (Wimsatt 1987).

Realism

A central concern of philosophers of science has long been whether scientists have good reason to believe that the entities (in particular the unobservable entities) referred to by their theories exist and that what their theories say about these entities is true or approximately true (see Realism). In order for theories to refer to or be true about unobservable entities, they must actually be claims about these entities. This was denied by many logical empiricists, building on concerns raised by Mach, Duhem, and Poincaré (see Mach, Ernest; Poincaré, Henri). As noted above, the logical empiricists were interested in providing cognitive significance to theoretical terms by attempting to reduce theoretical claims to claims in the observation language. Even when this proved impossible, many nevertheless argued that theoretical terms are simply convenient instruments for making predictions about observable entities, rather than claims about unobservable entities (see Instrumentalism).

Because of the difficulties with theory-observation distinction discussed above (see Observation; Theories), this view fell out of favor and was replaced with a milder version of anti-realism. Van Fraassen (1980), for example, argues that while claims about unobservables might have a truth–value, scientists only have good reason to believe in their empirical adequacy, not their truth. Such a view might broadly be understood as instrumentalist in the sense that the truth of theories does not underwrite the functions they serve. There are two main arguments provided in support this version of anti-realism. First, given the problem of underdetermination raised by Duhem and Quine, there will always be more than one rival hypothesis compatible with any body of evidence (see Duhem Thesis; Underdetermination of Theories). Therefore, since these hypotheses are incompatible, the evidence cannot provide adequate reason to believe that one or the other theory is true. Second, some have argued that history provides evidence against believing in the truth of scientific theories. Given the large number of theories once thought true in the past that have since been rejected as false, history provides inductive evidence that science's current theories are likely to be false as well (see Laudan 1981).

There have been a number of responses to these arguments, including attempts to show that the problem of underdetermination can be solved, that anti-realism depends on a distinction between observable and unobservable entities that cannot be sustained, and that the realist need only claim that theories are approximately true or are getting closer to the truth (see Verisimilitude). In addition, arguments have been provided in support of realism about theories, the most influential of which is Putnam's miracle argument (see Putnam, Hilary). There are various versions of this argument, but the central premise is that science is successful (what this success amounts to varies). The contention is that the only way this success can be explained is if scientific theories are approximately true (see Abduction); otherwise the success of science would be a miracle.

This argument has been criticized in three central ways. First, Fine (1986) criticizes the miracle argument for being viciously circular. Second, some have argued that science is in fact not very successful, for reasons outlined above. Third, it is argued that the success of science does not depend on its truth, or perhaps does not even require an explanation. Van Fraassen (1980), for example, has argued that it is not surprising that scientific theories are predictively successful, since they are chosen for their predictive success. Therefore, the success of theories can be explained without supposing their truth. Others have responded that this would not, however, explain the predictive success of theories in novel situations (e.g., Leplin 1997).

Due to these problems, other forms of realism have been defended. Hacking (1983), for example, defends entity realism. He argues that, while scientists do not have good reason to believe their theories are true, they do have good reason to believe that the entities referred to in the theories exist, since scientists are able to manipulate the entities. Others have attempted to defend a more radical form of anti-realism, according to which the entities scientists talk about and the theories they invent to discuss them are merely social constructs (see Social Constructionism).

Explanation

In an attempt to avoid metaphysically and epistemically suspect notions such as causation (see Causality), Hempel and Oppenheim (1948) developed a covering law model of explanation: the deductive-nomological (D-N) account (see Explanation; Hempel, Carl). Rather than relying on causes, they argued that scientific explanations cite the law or laws that cover the phenomena to be explained. According to the D-N model, explanations are deductive arguments, where the conclusion is a statement expressing what is to be explained (the *explanandum*), and the premises (the *explanans*) include at least one law-statement. Often statements about particular antecedent conditions from which the *explanandum* can be derived. Initially developed only to cover explanations of particular facts, the D-N model was expanded to include explanations of laws, such as the explanation of Kepler's laws by deriving them from Newton's laws of motion (along with particular facts about the planets). To account for explanations of particular events and laws governed by *statistical* laws, the inductive-statistical (I-S) and deductive-statistical (D-S) models were developed (Hempel 1965). According to the D-S model, statistical laws are explained by deductively deriving them from other statistical laws. However, statements describing particular facts cannot be deduced from statistical laws. Instead, according to the I-S model, the explanans containing statistical laws must confer a high inductive probability to the particular event to be explained. In this way, the covering law model of explanation was able to link explanation with predictability (see Prediction) and also make clear why the reduction of, say, Kepler's laws to Newton's laws of motion could be explanatory (see Reductionism).

In the ensuing years, these accounts ran into a number of problems. The covering law model seemed unable to account for cases where scientists and non-scientists appear to be giving perfectly good explanations without citing laws (see Biology, Philosophy of; Function; Mechanism; Social Sciences, Philosophy of). Several counterexamples were developed against the D-N model, including the purported explanation of events by citing irrelevant factors, such as the explanation of Joe's failure to get pregnant by citing the fact that he took birth-control pills, and the explanation of causes by citing their effects, such as the explanation of the height of a flagpole by citing the length of its shadow. Deductive relations, unlike explanatory relations, can include irrelevant factors and need not respect temporal asymmetries. The I-S model also encountered difficulties. According to the I-S model, improbable events cannot be explained, which runs counter to many philosophers' intuitions about such cases as the explanation of paresis by citing the fact that a person had untreated syphilis. Moreover, developing an account of inductive probability proved difficult (see Inductive Logic; Probability). Attempts to provide an adequate account of laws within an empiricist framework also encountered problems. According to Hempel and Oppenheim, laws are expressed by universal generalizations of unlimited scope, with purely qualitative predicates, and they do not refer to particular entities. The problem is that there are accidental generalizations, such as 'All pieces of gold have a mass of less than 10,000 kg,' that satisfy these conditions. Laws appear to involve the modal features that Hume and the logical empiricists were intent on avoiding; unlike accidental generalization, laws seem to involve some sort of natural necessity. The difficulty is to develop an account of laws that makes sense of this necessity in a way that does not make knowledge of laws problematic (see Laws of Nature).

In response to these problems, some have attempted to rescue the covering-law model by supplementing it with additional conditions, as in unificationist accounts of explanation. According to these accounts, whether an argument is explanatory depends not just on the argument itself, but on how it fits into a unified theory (see Unity and Disunity of Science). Scientists explain by reducing the number of brute facts (Friedman 1974) or argument patterns (Kitcher 1989) needed to derive the largest number of consequences. Others have developed alternatives to the covering law model. Van Fraassen (1980) has defended a pragmatic account of explanation, according to which what counts as a good explanation depends on context. Others have developed various causal accounts of explanation. Salmon (1971) and others have argued that explanatory and causal relations can be understood in terms of statistical relevance; scientists explain by showing that the *explanans* (a causal factor) is statistically relevant for the event to be explained. Salmon (1984) eventually rejected this view in favor of a causal mechanical model, according to which explanations appeal to the mechanisms of causal propagation and causal interactions (see Mechanism). Along with the development of various causal accounts of explanation have come numerous accounts of causation, as well as attempts to develop a better epistemology for causal claims through, for example, causal modeling (see Causality).

Prediction

Traditionally, prediction has been regarded as being as central to science as explanation (see Prediction). At the formal level, the received view does not distinguish between explanation and prediction. For instance, in the D-N model, the conclusion derived from the laws and other assumptions can be regarded as predictions in the same way that they can be regarded as explanations. While prediction is generally taken to refer to the future—one predicts future events—philosophically, the category includes retrodiction, or prediction of past events, for instance the past positions of planets from Newton's laws and their present positions and momenta. (On some accounts of hypothesis confirmation, retrodiction is even more important than forward prediction—see Bayesianism.)

The D-N model assumes that the laws in question are deterministic (see Determinism). Statistical explanations are also predictive, but the predictions are weaker: they hold probabilistically and can only be confirmed by observing an ensemble of events rather than individual events (see Confirmation Theory). Interest in statistical explanation and prediction initially arose in the social sciences in the nineteenth century (Stigler 1986; see also Social Sciences, Philosophy of the). In this case, as well as in the case of prediction in classical statistical physics, the inability to predict with certainty arises because of ignorance of the details of the system and computational limitations. A different type of limitation of prediction is seen when predictions must be made about finite samples drawn from an ensemble, for instance, biological populations (see Evolution; Population Genetics). Finally, if the laws are themselves indeterministic, as in the case of quantum mechanics, prediction can only be statistical (see Quantum Mechanics). The last case has generated the most philosophical interest because, until the advent of quantum mechanics, the failure to predict exactly was taken to reflect epistemological limitations rather than an ontological feature of the world. That the models of statistical explanation discussed earlier do not distinguish between these various cases suggests that there remains much philosophical work to be done. Meanwhile, the failure of determinism in quantum mechanics has led to much re-examination of the concept of causality in attempts to retain the causal nature of physical laws even in a probabilistic context (see Causality).

Prediction, although not determinism, has also been recently challenged by the discovery that there exist many systems that display sensitivity to initial conditions, the so-called chaotic systems. Determinism has usually been interpreted as an ontological thesis: for deterministic systems, if two systems are identical at one instant of time, they remain so at every other instant (Earman 1986; see Determinism). However, satisfying this criterion does not ensure that the available—and, in some cases, all obtainable—knowledge of the system allows prediction of the future. Some physical theories may prevent the collection of the required information for prediction (Geroch 1977; see also Space-Time). Even if the information can be collected, pragmatic limitations become relevant. The precision of any information is typically limited by measurement methods (including the instruments). If the dynamical behavior of systems is exceedingly sensitive to the initial conditions, small uncertainties in the initial data may lead to large changes in predicted behavior—chaotic systems exemplify this problem (see Prediction).

Confirmation

Hume's problem—how experience generates rational confidence in a theory—has been central to philosophy of science in the twentieth century and continues to be an important motivation for contemporary research (see Induction, Problem of). Many of the logical empiricists initially doubted that there is a logical canon of confirmation. Breaking with earlier logical traditions, for many of which inductive logic was of central importance, these logical empiricists largely regarded confirmation as a pragmatic issue not subject to useful theoretical analyses. That assessment changed in the 1940s with the work of Carnap, Hempel, and Reichenbach, besides Popper (see Carnap, Rudolf; Hempel, Carl Gustav; Popper, Karl Raimund; Reichenbach, Hans). Carnap, in particular, began an ambitious project of the construction of a logic of confirmation, which he took to be part of semantics, in the process reviving Keynes' logical interpretation of probability. Early versions of this project were distant from the practice of science, being restricted to formal languages of excessively simplified structures incapable of expressing most scientific claims. Later versions came closer to scientific practice, but only to a limited extent (see Carnap, Rudolf). Whether or not the project has any hope remains controversial among philosophers. Although the relevant entries in this Encyclopedia record some progress, there is as yet no quantitative philosophical theory of confirmation (see Confirmation Theory; Inductive Logic; Probability).

Meanwhile, within the sciences, the problem of confirmation was studied as that of statistical inference, bringing standard statistical methods to bear on the problem of deciding how well a hypothesis is supported by the data. Most of these methods were only invented during the first half of the twentieth century. There are two approaches to statistics, so-called orthodox statistics (sometimes called "frequentist" statistics) and Bayesian statistics (which interprets some probabilities as degrees of belief). The former includes two approaches to inference, one involving confidence intervals and largely due to Neyman and E. S. Pearson and the other due to Fisher. These have received some attention from philosophers but, perhaps, not as much as they deserve (Hacking 1965; see Statistics, Philosophy of). In sharp contrast, Bayesian inference has been at the center of philosophical attention since the middle of the twentieth century. Interesting work points to common ground between traditional confirmation theory and Bayesian methodology. Meanwhile, within the sciences, newer computational methods have made Bayesian statistics increasingly popular (see Statistics, Philosophy of), for instance, in the computation of phylogenies in evolutionary biology (see Evolution). Bayesian inference methods also have the advantage of merging seamlessly with contemporary decision theory (see Decision Theory), even though most of the methods within decision theory were invented in an orthodox context.

Philosophically, the differences between orthodox and Bayesian methods remain sharply defined. Orthodox methods do not permit the assignment of a probability to a hypothesis, which, from the perspective of most Bayesians, makes them epistemologically impotent. (Bayesians also usually argue that orthodox inferential recipes are *ad hoc*—see Bayesianism.) Meanwhile Bayesian methods require an assignment of *prior* probabilities to hypotheses before the collection of data; for the orthodox such assignments are arbitrary. However, in the special sciences, the trend seems to be one of eclecticism, when orthodox and Bayesian methods are both used with little concern for whether consistency is lost in the process. This situation calls for much more philosophical analysis.

Experimentation

The logical empiricists' focus on the formal relations between theory and evidence resulted in Anglo-American philosophers neglecting the role of experimentation in science. Experimentation did receive some philosophical treatment in the late nineteenth and early twentieth centuries, in particular by Mill, Mach, and Bernard (see Mach, Ernest). In twentieth century Germany, two traditions developed around the work of Dingler and Habermas. It is only in the past three decades that experimentation has received more attention from Anglo-American philosophers, historians, and sociologists. Since then, there have been a number of careful analyses of the use of experiments by practicing scientists, with historians and sociologists focusing largely on the social and material context of experiments and philosophers focusing on their epistemic utility.

From a philosophical perspective, the neglect of experimentation was particularly problematic, since experimentation seems to affect the very evidential relations empiricists were interested in formalizing. Whether experimental results are good evidence for or against a hypothesis depends on how the results are produced—whether the data are reliably produced or a mere artifact of the experimental procedure. Moreover, this reliability often comes in degrees, thereby affecting the degree to which the data confirms or disconfirms a hypothesis. In addition, how data are produced affects what sorts of inferences can be drawn from the data and how these inferences might be drawn. As Mill argues, "Observations, in short, without experiment . . . can ascertain sequences and coexistences, but cannot prove causation" (1874, 386). How experimental results are obtained can also affect whether replication is necessary and how statistical methods are used. In some cases, statistics is used to analyze the data, while in others, it is involved in the very production of the data itself (see Experimentation; Statistics, Philosophy of).

One of the central issues in the philosophy of experimentation is what experiments are. Experiments are often distinguished from observations in that the former involve active intervention in the world, whereas the latter are thought to be passive. However, it is unclear what counts as an intervention. For example, are the use of sampling methods or microscopes interventions? There are also questions about whether thought experiments or computer simulations are "real" experiments or if they merely function as arguments. Moreover, it is not always clear how to individuate experiments—whether it is possible, especially with the increasing use of computers as integral parts of the experimental set-up, to disambiguate the experiment from the analysis of the data.

Another fundamental issue is whether and what epistemic roles experiments can play (Rheinberger 1997). They are purportedly used in the testing of theories, in garnering evidence for the existence of entities referred to by our theories (see Realism), in the creation (and thereby discovery) of new phenomena, in the articulation of theories, in the development of new theories, in allowing scientists to "observe" phenomena otherwise unobservable (see Observation), and in the development and refinement of technologies.

Whether experiments can reliably serve these epistemic functions has been called into question in a number of ways. First, sociologists and historians have argued that social factors affect or even determine whether an experiment "confirms" or "disconfirms" a theory (see Social Constructionism). It is also argued that experiments are theory-laden, since experiments require interpretation and these interpretations rely on theories (Duhem 1954). Whether this is a problem depends in part on what use is made of the experiment and what sorts of theories are needed—the theory being tested, theories of the phenomena being studied but not being tested, or theories about the experimental apparatus being used. As Hacking (1983) and Galison (1987) both argue, experiments and experimental traditions can have a life of their own independent of higher-level theories.

The theory-ladenness of experimentation also raises questions about whether experiments can be used to test hypotheses in any straightforward way no matter which level of theory is used, since predictions about experimental results rely on auxiliary hypotheses that might be called into question (see Duhem Thesis). Experiments are also purported to be "practice-laden," relying on tacit knowledge that cannot be fully articulated (Collins 1985; see also Polanyi 1958). According to Collins, this leads to problems with replication. The reliability of experiments is often judged by the ability of scientists to replicate their results. However, what counts as replication of the "same" experiment is often at issue in scientific disputes. Since, according to Collins, tacit knowledge (which cannot be made explicit) is involved in the replication of experiments and even in judgments about what constitutes the "same" experiment, adjudicating these disputes on rational grounds is problematic. Collins, in addition, questions whether there can be independent grounds for judging whether an experiment is reliable, which he calls "the experimenters' regress." Whether an experimental procedure is reliable depends on whether it consistently yields correct results, but what counts as a correct result depends on what experimental procedures are deemed reliable, and so on (Collins 1985; for a reply, see Franklin 1994). Experiments also typically involve manipulation of the world, often creating things that are not naturally occurring, which has led some to question whether experiments represent the world as it naturally is. At one extreme are those who argue that experimentation actually constructs entities and facts (Latour and Woolgar 1979; Pickering 1984; Rheinberger 1997; see also Social Constructionism). Others argue that experiments can produce artifacts, but that these can be reliably distinguished from valid results (Franklin 1986). A milder version of this worry is whether laboratory settings can accurately reproduce the complexities of the natural world, which is exemplified in debates between field and experimental biologists. The effect of interventions on experimental outcomes is even more problematic in quantum physics (see Quantum Measurement Problem).

Scientific Change

Scientific change occurs in many forms. There are changes in theory, technology, methodology, data, institutional and social structures, and so on. The focus in the philosophy of science has largely been on theory change and whether such changes are progressive (see Scientific Change; Scientific Progress). The primary concern has also been with how scientific theories are justified and/or become accepted in the scientific community, rather than how they are discovered or introduced into the community in the first place. Over the years, there have been various notions of progress correlated with the different goals scientific theories are purported to have: truth, systematization, explanation, empirical adequacy, problem solving capacity, and so on. (Notice that if the focus were on, say, technological or institutional changes, the goals attended to might be very different; for example, does the technology have greater practical utility or is the institutional change just?)

Traditionally, scientific change has been thought of as governed by rational procedures that incrementally help science achieve its goals. For the logical empiricists, the aim of scientific theories was to systematize knowledge in a way that yields true predictions in the observational language (see Theories). As such, science progresses through the collection of additional confirming data, through the elimination of error, and through unification, typically by reducing one theory to another of greater scope. To make sense of these sorts of changes, the logical empiricists developed accounts of reduction, explanation, and inductive logic or confirmation theory (see Confirmation Theory; Explanation; Inductive Logic; Reductionism; Unity and Disunity of Science). Others, such as Popper, offered a different account of theory change. Popper defended an eliminativist account much like Mill's, whereby science attempts to eliminate or falsify theories. Only those theories that pass severe tests ought to be provisionally accepted (see Corroboration). This was also one of the earliest versions of evolutionary epistemology (see Evolutionary Epistemology; Popper, Karl Raimund).

As discussed in the previous sections, these accounts ran into difficulties: Quine extended Duhem's concerns about falsification, criticized the analytic/synthetic distinction, and raised questions about the determinacy of translation (see Duhem Thesis; Quine, Willard Van; Underdetermination); Popper and Hanson argued that observations are theory-laden (see Hanson, Norwood Russell; Observation; Popper, Karl Raimund); there were problems with Carnap's inductive logic; and so on. Partly influenced by these difficulties and partly motivated by a concern that philosopher's theories about science actually fit the practices of science, Kuhn's *The Structure of Scientific Revolutions* (1962) challenged the way philosophers, historians, sociologists, and scientists thought about scientific change (see Kuhn, Thomas). He argued that scientific change is not in general cumulative and progressive, but develops through a series of distinct stages: immature science (when there is no generally accepted paradigm), normal science (when there is an agreed upon paradigm), and revolutionary science (when there is a shift between paradigms). Kuhn's notion of paradigms also expanded the focus of scientific change beyond theories, since paradigms consisted, not just of theories, but of any exemplary bit of science that guides research. While the development of normal science might in some sense be incremental, Kuhn argued that the choice between paradigms during a revolution involves something like a *Gestalt* shift. There are no independent methods and standards, since these are paradigm-laden; there is no independent data, since observations are paradigm-laden; and the paradigms may not even be commensurable (see Incommensurability). Consequently, paradigm shifts seemed to occur in an irrational manner.

The responses to Kuhn's influential work took two very different paths. On the one hand, strongly influenced by Kuhn, members of the Strong Programme argued that scientific change ought to be explained sociologically—that the same social causes explain both "good" and "bad" science. Others (e.g. Latour and Woolgar 1979) went further, arguing that scientists in some sense construct facts (see Social Constructionism). Focus on the social aspects of scientific research also led to developments in feminist philosophy of science, both in the close analysis of the gender and racial biases of particular sciences and in the development of more abstract feminist theories about science (see Feminist Philosophy of Science).

The other, a very different sort of response, involved a defense of the rationality and progress of science. There were attempts to show that competing scientific theories and paradigms are not incommensurable in the sense of being untranslatable. Davidson (1974) argues the very idea of a radically different, incommensurable paradigm does not make sense; others (e.g., Scheffler 1967) argued that sameness of reference is sufficient to ensure translatability, which was later buttressed by referential accounts of meaning (see Incommensurability). The rationality of scientific change was also defended on other grounds. Lakatos developed Popper's ideas in light of Kuhn into his methodology of scientific research programs (see Lakatos, Imre; Research Programmes); and Laudan (1977) argued that progress can be made sense of in terms of problem solving capacity. Another approach to showing that scientific change is progressive can be found in realism. Rather than arguing that each change involves a rational choice, defenses of realism can be seen as attempts to establish that science is approaching its goal of getting closer to the truth (see Realism). Of course, anti-realists might also argue that science is progressing, not toward truth, but toward greater empirical adequacy.

More recently, there have been attempts to develop formal methods of theory choice beyond confirmation theory and inductive logic (see Bayesianism;

Statistics, Philosophy of). There have also been attempts to model discovery computationally, which had been thought not to be rule governed or formalizable. Some of these try to model the way humans discover; others were developed in order to make discoveries (e.g., data mining), whether or not humans actually reason in this way. As a normative enterprise, such modeling can also be used as a defense of the rationality of scientific discovery and, therefore, scientific change (see Scientific Change).

Perhaps the longest-lasting influence in the philosophy of science of Kuhn's influential work has been to encourage philosophers to look more closely at the actual practices of the various sciences. This has resulted in a proliferation of philosophies of the special sciences.

Foundations of the Special Sciences

The logical empiricists believed in the unity of science (see Unity of Science Movement). However, the theme was interpreted in multiple ways. At one extreme were views according to which unification was to be achieved through hierarchical reduction (see Reductionism) of sociology to individual psychology (see Methodological Individualism), psychology to biology (see Psychology, Philosophy of), biology to physics and chemistry (see Biology, Philosophy of), and chemistry to physics (see, Chemistry, Philosophy of); for an influential defense of this view, see Oppenhiem and Putnam (1958). At the other extreme were those who believed that unification required no more than to be able to talk of the subjects of science in an interpersonal (that is, non-solipsistic) language—this was Carnap's (1963) final version of physicalism. Somewhere in between were stronger versions of physicalism, which, for most logical empiricists and almost all philosophers of science since them, provides some vision of the unity of science (see Physicalism).

Perhaps with the exception of the most extreme reductionist vision of the unity of science, all other views leave open the possibility of exploring the foundations and interpretations of the special sciences individually. During the first few decades of the twentieth century, most philosophical attention to the special sciences was limited to physics; subsequently, psychology, biology, and the social sciences have also been systematically explored by philosophers. In many of these sciences, most notably biology and cognitive science, philosophical analyses have played a demonstrable role in the further development of scientific work (see Biology, Philosophy of; Cognitive Science; Intentionality).

Physical Sciences

The first three decades of the twentieth century saw the replacement of classical physics by relativity theory and quantum mechanics, both of which abandoned cherished classical metaphysical principals (see Quantum Mechanics; Space-Time). It is therefore not surprising that many philosophers interested in "scientific philosophy" (see Logical Empiricism) did significant work in this field. In particular, Popper and Reichenbach made important contributions to the interpretation of quantum mechanics; Reichenbach and, to a lesser extent, Carnap also contributed to the philosophy of space-time (see Carnap, Rudolf; Popper, Karl Raimund; Reichenbach, Hans). In both quantum mechanics and relativity, philosophers have paid considerable attention to issues connected with causality and determinism, which became problematic as the classical world-view collapsed (see Causality; Determinism). Arguably, Reichenbach's work on space-time, especially his arguments for the conventionality of the metric, set the framework for work in the philosophy of space-time until the last few decades (see Conventionalism). Reichenbach also produced important work on the direction of time.

Several philosophers contributed to the clarification of the quantum measurement problem (see Quantum Measurement Problem), the concept of locality in quantum mechanics (see Locality), and the nature and role of quantum logic (see Putnam, Hilary; Quantum Logic). Meanwhile, many physicists, including Bohr, Einstein, Heisenberg, and Schrödinger, also produced seminal philosophical work on the foundations of physics (see also Bridgman, Percy Williams; Duhem Thesis). The only consensus that has emerged from all this work is that, whereas the foundations of relativity theory (both special and general) are relatively clear, even after eighty years, quantum mechanics continues to be poorly understood, especially at the macroscopic level (see Complementarity).

Perhaps because of the tradition of interest in quantum mechanics, philosophers of physics, starting mainly in the 1980s, also began to explore the conceptual structure of quantum field theory and particle physics (see Particle Physics; Quantum Field Theory). However, one unfortunate effect of the early focus on quantum mechanics and relativity is that other areas of physics that also deserve philosophical scrutiny did not receive adequate attention, as Shimony (1987) and others have emphasized. (See the list of questions in the entry, Physical Sciences, Philosophy of.) Only in recent years have philosophers begun to pay attention to questions such as reductionism and irreversibility in kinetic theory (see Irreversibility; Kinetic Theory) and condensed matter physics (see Batterman [2002] and Reductionism). One interesting result has been that the question of reductionism within physics is now believed to be far more contentious than what was traditionally thought (when it was assumed that biology, rather than the physics of relatively large objects, presented a challenge to the program of physical reductionism—see Emergence).

Finally, beyond physics, some philosophical attention is now being directed at chemistry (see Chemistry, Philosophy of) and, so far to a lesser extent, astronomy (see Astronomy, Philosophy of). As in the case of macroscopic physics, the question of the reduction of chemistry to physics has turned out to be unexpectedly complicated with approximations and heuristics playing roles that make orthodox philosophers uncomfortable (see Approximation). It is likely that the future will see even more work on these neglected fields and further broadening of philosophical interest in the physical sciences.

Biology

Professional philosophers paid very little attention to biology during the first few decades of the twentieth century, even though the advent of genetics (both population genetics and what came to be called classical genetics [see Genetics]) was transforming biology in ways as profound as what was happening in physics. Professional biologists—including Driesch, J. B. S. Haldane, J. S. Haldane, and Hogben—wrote philosophical works of some importance. However, the only philosopher who tried to interpret developments in biology during this period was Woodger (1929, 1937), better known among philosophers as the translator of Tarski's papers into English. Philosophers paid so little attention to biology that not only the evolutionary "synthesis" (see Evolution), but even the formulation of the double helix model for DNA (see Reduction), went unnoticed by philosophers of those generations (Sarkar 2005).

All that changed in the 1960s, when the philosophy of biology emerged as a recognizable entity within the philosophy of science. The first question that occupied philosophers was whether molecular biology was reducing classical biology (see Molecular Biology; Reductionism). Initial enthusiasm for reductionism gave place to a skeptical consensus as philosophers began to question both the standard theory-based account of reductionism (due to Nagel 1961; see Nagel, Ernest) and whether molecular biology had laws or theories at all (Sarkar 1998). In the 1970s and 1980s, attention shifted almost entirely to evolutionary theory (see Evolution), to the definitions of "fitness" (see Fitness) and "function" (see Function), the nature of individuals and species (see Individual; Species), the significance of adaptation and selection (see Adaptation and Adaptationism; Population Genetics), and, especially, the units and levels of selection. Philosophical work has contributed significantly to scientific discussions of problems connected to units of selection, although no consensus has been reached (see Altruism; Natural Selection). Besides evolution, there was some philosophical work in genetics (see Genetics; Heredity and Heritability).

As in the case of the philosophy of physics, the last two decades have seen a broadening of interest within the philosophy of biology. Some of the new work has been driven by the realization that molecular biology, which has become most of contemporary biology, is not simply the study of properties of matter at lower levels of organization, but has a conceptual framework of its own. This framework has largely been based on a concept of information that philosophers have found highly problematic (see Biological Information). Formulating an adequate concept of biological information—if there is one—remains a task to which philosophers may have much to contribute (see Molecular Biology).

There has also been some attention paid to biodiversity (see Conservation Biology), ecology (see Ecology), immunology (see Immunology), and developmental biology, especially in the molecular era (see Molecular Biology). Neurobiology has sometimes been approached from the perspective of the philosophy of biology, although philosophical work in that area typically has more continuity with psychology (see "Psychology" below and Neurobiology). Philosophers have also argued on both sides of attempts to use biology to establish naturalism in other philosophical areas, especially epistemology and ethics—this remains one of the most contested areas within the philosophy of biology (see Evolutionary Epistemology; Evolutionary Psychology). Some philosophers of science have also interpreted the philosophy of medicine as belonging within the conceptual terrain of the philosophy of biology (Schaffner 1993). Finally, work in the philosophy of biology has also led to challenges to many of the traditional epistemological and metaphysical assumptions about science, about the nature of explanations, laws, theories, and so on (see Biology, Philosophy of; Mechanism).

Psychology

Philosophy and psychology have an intimate historical connection, becoming distinct disciplines only in the late nineteenth and early twentieth centuries. Even since then, many of the topics covered by psychology have remained of interest to philosophers of mind and language, although the route taken to address these questions might be very different. However, while philosophers of science did address concerns about the human sciences more generally (see "Social Sciences" below), it is only in the last twenty years or so that philosophy of psychology has developed as a distinct area of philosophy of science.

The intimate connection between philosophy and psychology can be seen throughout the history of psychology and the cognitive sciences more broadly. In an attempt to make psychology scientific, Watson (1913), a philosopher, founded behaviorism, which dominated the field of psychology for the first half of the twentieth century (see Behaviorism). This view fit well with empiricist attempts to reduce theoretical claims to those in the observational language by providing operational definitions (see Hempel 1949; see also Bridgeman, Percy Williams; Theories; Verificationism). However, the combined weight of objections from philosophers, linguists, and psychologists led to the demise of behaviorism. These criticisms, along with developments in mathematical computation (see Artificial Intelligence; Turing, Alan) and the influential work of Chomsky (see Chomsky, Noam; Linguistics, Philosophy of), resulted in the cognitive revolution in psychology; it became generally agreed upon that psychological theories must make reference to internal representations (see Intentionality; Searle, John). These developments also led to the creation of the interdisciplinary field of cognitive science, which included psychology, linguistics, computer science, neuroscience, and philosophy (see Cognitive Science).

Philosophers of psychology have been broadly interested in foundational issues related to the cognitive sciences. Among the topics of concern are the content of representation, the structure of thought, psychological laws and theories, and consciousness, each of which is briefly discussed below:

(i) *The Content of Representations.* One central question is what fixes the content of representations—is content determined by internal features of the agent (e.g., conceptual role semantics), features of the external physical environment (e.g., causal and teleological theories), or features of the external social environment? There are also debates about whether the representations are propositional in form, whether they require language (see Linguistics, Philosophy of), whether some are innate (see Empiricism; Innate/acquired Distinction), and whether representations are local or distributed (see Connectionism).

(ii) *The Structure of Thought.* The nature of cognition has also been a topic of dispute. Some argue that human cognition takes the form of classical computation (see Artificial Intelligence; Cognitive Science); connectionists argue that it is more similar to parallel distributed processing (see Connectionism); and more recently other accounts have been proposed, such as dynamical and embodied approaches to cognition. Also at issue is whether the cognitive structures in the mind/brain are modular (see Evolutionary Psychology), whether cognition is rule-governed, and whether some of the rules are innate (see Chomsky, Noam; Innate/Acquired Distinction).

(iii) *Theories and Laws.* Questions have been raised about the nature of theories in the cognitive sciences (see Neurobiology), about whether there are psychological or psychophysical laws (see Laws of Nature), and about how the theories and laws in different areas of the cognitive sciences relate, such as whether psychology is reducible to neurobiology (see Neurobiology; Physicalism; Reductionism; Supervenience). In addition, there is disagreement about how to interpret theories in the cognitive sciences—whether to interpret them realistically, as an attempt to represent how the mind/brain actually works, or merely instrumentally, as a means of saving the phenomena or making predictions (see Instrumentalism; Realism). Moreover, the problems of reflexivity and the intentional circle discussed below, along with difficulties peculiar to the various areas of the cognitive sciences, raise questions about the testability of psychological theories (see Neurobiology; Psychology, Philosophy of).

(iv) *Consciousness*. There has been a resurgence of interest in consciousness (see Consciousness; Searle, John). There have been attempts to clarify what "consciousness" involves in its various senses, as well as debates about how to explain consciousness. To this end, a number of theories of consciousness have been proposed, including higher-order theories, neurological theories, representational theories, and various non-physical theories.

Social Sciences

Philosophical interest in the foundations of the social sciences has a long history, dating back at least to Mill's influential work on the social sciences. Some foundational issues have also been systematically discussed by social scientists themselves, such as Durkheim (1895/1966) and Weber (1903/1949). Around the middle of the twentieth century, the social sciences again received serious philosophical attention. The focus was largely on their being *human* sciences and the philosophical issues this raised. More recently, philosophers have directed their attention to the different social sciences in their own right, especially economics (see Economics, Philosophy of).

A central focus of discussion is whether the social sciences are fundamentally different from the natural sciences. Logical empiricists attempted to incorporate the social sciences into their models for the natural sciences (see Unity of Science Movement). Others have argued that the social sciences are unique. This has framed many of the debates within the philosophy of the social sciences, a number of which are briefly discussed in what follows (see Social Sciences, The Philosophy of):

(i) *Are There Social Science Laws?* Laws played important roles in empiricist accounts of explanation, theories, confirmation, and prediction, but it is unclear whether there are laws of the social sciences (see Laws of Nature). Social phenomena are complex, involve reference to social kinds, and require idealizations. As a result, many argue that generalizations of the social sciences, if they are laws at all, require ineliminable ceteris paribus clauses. Others argue that the social sciences ought not even attempt to create generalizations or grand theories, as social phenomena are essentially historical and local.

(ii) *Do Social Scientific Theories Yield Testable Predictions?* Because of the complexity of social systems, social scientific theories require idealizations. Given the nature of these idealizations, deriving empirical predictions from social scientific theories is difficult at best (see Prediction). As a result, many argue that social scientific theories are not testable. This is exacerbated by the reflexive nature of social science theories: the very act of theorizing can change the behavior one is theorizing about. Moreover, if human action is explained by agents' desires and beliefs, social scientists seem to be caught in an intentional circle, making it difficult to derive any testable claims (see Rosenberg 1988).

(iii) *Is the Methodology of the Social Sciences Distinct?* Given that social sciences involve humans and human behavior on a large scale, experimentation has not played a significant role in the social sciences (see Experimentation). There are also many who question whether the social sciences can be naturalized. Some argue that understanding social action is essentially a hermeneutic enterprise, distinctly different from the natural sciences.

(iv) *What Are the Ontological Commitments of Scientific Theories?* Beginning with Mill and, subsequently, Durkheim and Weber, there have been debates as to whether social scientific theories are reducible to theories about individual behavior (see Methodological Individualism). Moreover, after Nagel's influential account of intertheoretic reduction, it has been argued that social phenomena are multiply realizable, and therefore, social science theories are not reducible to lower-level theories (see Emergence; Reductionism; Supervenience). Additionally, given that social scientific theories involve idealizations, there are questions about whether these theories ought to be interpreted realistically or instrumentally (see Instrumentalism; Realism).

(v) *What Is the Nature of Social Scientific Explanations?* Some, such as Hempel (1962), have argued that social scientific explanations are no different than in the physical sciences. Others, however, have questioned this. If there are no social scientific laws, then social scientific explanation cannot be captured by the covering law model (see Explanation). Social sciences also often rely on functional explanations, which, while similar to biology, seem to be different from explanations in physics (see Function). Others, following Winch (1958), have argued that social sciences explain action, not behavior, which requires understanding the meaning of the action (not its causes), and therefore must include the actors' intentions and social norms. Moreover, some have argued that actions are governed by reasons, and are therefore not susceptible to causal explanation, a view that was later convincingly refuted by Davidson (1963). An alternative account of how beliefs and desires can explain actions has been formalized in rational choice theory (see Decision Theory), although there are questions about whether such explanations capture how people actually behave, rather than how they ought to behave.

(vi) *What Is the Relationship Between Social Science and Social Values?* There has also been concern with the connection between social values and the social sciences. Taylor (1971), for example, argues that social theory is inherently value-laden, and Habermas (1971) argues that social theory *ought* to engage in social criticism.

Concluding Remarks

Philosophy of science remains a vibrant sub-discipline within philosophy today. As this introduction has documented, many of the traditional questions in epistemology and metaphysics have been brought into sharper profile by a focus on scientific knowledge. Moreover, philosophical engagement with the special sciences has occasionally contributed to the development of those sciences and, as philosophers become more immersed in the practice of science, the number and level of such contributions can be expected to increase. The trend that philosophers of science engage all of the special sciences—not just physics—will also help produce a more complete picture of the growth of science, if not all knowledge, in the future.

With few exceptions (e.g., Demarcation, Problem of and Feminist Philosophy of Science) the entries in the Encyclopedia are not concerned with the social role of science. But, as science and technology continue to play dominant roles in shaping human and other life in the near future, philosophers may also contribute to understanding the role of science in society. Moreover, in some areas, such as the environmental sciences and evolutionary biology, science is increasingly under ill-motivated attacks in some societies, such as the United States. This situation puts philosophers of science, because of their professional expertise, under an obligation to explain science to society, and, where ethically and politically appropriate, to defend the scientific enterprise. How such defenses should be organized without invoking a suspect criterion of demarcation between science and non-science remains a task of critical social relevance. The Encyclopedia should encourage and help such efforts.

JESSICA PFEIFER
SAHOTRA SARKAR

References

Achinstein, P. (1968), *Concepts of Science: A Philosophical Analysis.* Baltimore: Johns Hopkins University Press.

Ayer, A. J. 1959. Ed. Logical Positivism. New York: Free Press.

Batterman, R. W. 2002. *The Devil in the Details.* Oxford: Oxford University Press.

Carnap, R. 1963. "Replies and Systematic Expositions." In Schilpp, P. A. Ed. *The Philosophy of Rudolf Carnap.* La Salle: Open Court, pp. 859–1013.

Collins, Harry (1985), *Changing Order: Replication and Induction in Scientific Practice.* London: Sage.

Davidson, D. (1963), "Actions, Reasons, and Causes", *Journal of Philosophy*, 60, reprinted in Davidson (1980), *Essays on Actions and Events*, Oxford: Clarendon Press.

——— (1974), *Inquiries into Truth and Interpretation.* Oxford: Oxford University Press.

Duhem, Pierre (1954), *The Aim and Structure of Physical Theory.* Princeton: Princeton University Press.

Durkheim, E. (1895/1966), *The Rules of Sociological Method, 8th edition.* Solovay and Mueller (transl.) and Catlin (ed.), New York: Free Press.

Earman, J. 1986. *A Primer on Determinism.* Dordrecht: Reidel.

Fine, A. (1986), *The Shaky Game: Einstein, Realism, and the Quantum Theory.* Chicago: University of Chicago Press.

Franklin, Allan (1994), "How to Avoid the Experimenters' Regress." *Studies in the History and Philosophy of Science* 25: 97–121.

——— (1986), *The Neglect of Experiment.* Cambridge: Cambridge University Press.

——— (1997), "Calibration," *Perspectives on Science* 5: 31–80.

Friedman, M. (1974), "Explanation and Scientific Understanding," *Journal of Philosophy* 71: 5–19.

Galison, P. (1987), *How Experiments End.* Chicago: University of Chicago Press.

Geroch, R. (1977), "Prediction in General Relativity." *Minnesota Studies in the Philosophy of Science* 8: 81–93.

Habermas, J. (1971), *Knowledge and Human Interests.* McCarthy (transl.), Boston: Beacon Press.

Hacking, I. (1965), *Logic of Statistical Inference*, Cambridge: Cambridge University Press.

Hacking, Ian (1983), *Representing and Intervening*. Cambridge: Cambridge University Press.

Hanson, N. R. (1958), *Patterns of Discovery*. Cambridge: Cambridge University Press.

Hempel, C. (1949), "The Logical Analysis of Psychology," in H. Feigl and W. Sellars (eds.), *Readings in Philosophical Analysis*. New York: Appleton-Century-Crofts, 373–384.

—— (1962), "Explanation is Science and in History," in Colodny (ed.) *Frontiers of Science and Philosophy*. Pittsburgh: University of Pittsburgh Press.

—— (1965), *Aspects of Scientific Explanation and Other Essays in the Philosophy of Science*. New York: The Free Press.

Hempel, C. and P. Oppenheim (1948), "Studies in the Logic of Explanation," *Philosophy of Science* 15: 135–175.

Hesse, M. (1963), *Models and Analogies in Science*. London: Sheed and Ward.

Kitcher, P. (1989), "Explanatory Unification and the Causal Structure of the World," in Kitcher and Salmon (eds.), *Scientific Explanation. Minnesota Studies in the Philosophy of Science*, Vol. XIII. Minneapolis: University of Minnesota Press, 410–505.

Kuhn, T. (1962), *The Structure of Scientific Revolutions*. Chicago: University of Chicago Press.

Kuhn, Thomas (1977), *The Essential Tension*. Chicago: Chicago University Press.

Latour, Bruno and Steve Woolgar (1979), *Laboratory Life: The Social Construction of Scientific Facts*. London: Sage.

Laudan, L. (1977), *Progress and its Problems: Towards a Theory of Scientific Growth*. Berkeley: University of California Press.

—— (1981), "A Confutation of Convergent Realism," *Philosophy of Science* 48: 19–50.

Leplin, J. (1997). *A Novel Defense of Scientific Realism*. New York: Oxford University Press.

Maxwell, G. (1962), "The Ontological Status of Theoretical Entities," in Feigl and Maxwell (eds.) *Minnesota Studies in the Philosophy of Science, Vol III*. Minneapolis: University of Minnesota Press, 3–27.

Mayo, Deborah (1996), *Error and the Growth of Experimental Knowledge*. Chicago: University of Chicago Press.

Mill, John Stuart (1874), *A System of Logic: Ratiocinative and Inductive, 8th edition*. Reprinted in *Collected Works of John Stuart Mill, Vols. VII-VIII*, edited by J. M. Robson. London and Toronto: Routledge and Kegan Paul and University of Toronto Press, 1963–1991.

Nagel, E. (1961), *The Structure of Science*. New York: Harcourt, Brace and World.

Norton, John (1996) "Are Thought Experiments Just What You Thought?" *Canadian Journal of Philosophy*, 26, pp. 333–66.

Oppenheim, P. and Putnam, H. (1958), "The Unity of Science as a Working Hypothesis." In Feigl, H., Scriven, M., and Maxwell, G. Eds. *Concepts, Theories, and the Mind-Body Problem*. Minneapolis: University of Minnesota Press, pp. 3–36.

Pickering, Andrew (1984), *Constructing Quarks*. Chicago: University of Chicago Press.

Polanyi, Michael (1958), *Personal Knowledge*. Chicago: University of Chicago Press.

Putnam, H. (1962), "What Theories are Not," in Nagel, Suppes, Tarski (eds.) *Logic, Methodology, and Philosophy of Science: Proceedings of the Nineteenth International Congress*. Stanford: Stanford University Press, 240–251.

Rheinberger, H. J. (1997), *Toward a History of Epistemic Things: Synthesizing Proteins in the Test Tube*. Stanford: Stanford University Press.

Rosenberg, A. (1988), *Philosophy of Social Science*. Boulder: Westview Press.

Salmon, W. (1971), "Statistical Explanation and Statistical Relevance", in Salmon, Greeno, and Jeffrey (eds.), *Statistical Explanation and Statistical Relevance*. Pittsburgh, University of Pittsburgh Press, 29–87.

—— (1984), *Scientific Explanation and the Causal Structure of the World*. Princeton: Princeton University Press.

Sarkar, S. Ed. (1996a), *Science and Philosophy in the Twentieth Century: Basic Works of Logical Empiricism. Vol. 1. The Emergence of Logical Empiricism: From 1900 to the Vienna Circle*. New York: Garland.

—— Ed. (1996b), *Science and Philosophy in the Twentieth Century: Basic Works of Logical Empiricism. Vol. 2. Logical Empiricism at Its Peak: Schlick, Carnap, and Neurath*. New York: Garland.

—— Ed. (1996c), *Science and Philosophy in the Twentieth Century: Basic Works of Logical Empiricism. Vol. 3. Logic, Probability, and Epistemology: The Power of Semantics*. New York: Garland.

—— Ed. (1996d), *Science and Philosophy in the Twentieth Century: Basic Works of Logical Empiricism. Vol. 4. Logical Empiricism and the Special Sciences: Reichenbach, Feigl, and Nagel*. New York: Garland.

—— Ed. (1996e), *Science and Philosophy in the Twentieth Century: Basic Works of Logical Empiricism. Vol. 5. Decline and Obsolescence of Logical Empiricism: Carnap vs. Quine and the Critics*. New York: Garland.

—— (1998), *Genetics and Reductionism*. New York: Cambridge University Press.

—— (2005), *Molecular Models of Life: Philosophical Papers on Molecular Biology*. Cambridge, MA: MIT Press.

Schaffner, K. F. (1993), *Discovery and Explanation in Biology and Medicine*. Chicago: University of Chicago Press.

Scheffler, I. (1967), *Science and Subjectivity*. Indianapolis: Bobbs-Merrill.

Shimony, A. (1987), "The Methodology of Synthesis: Parts and Wholes in Low-Energy Physics." In Kargon, R. and Achinstein, P. Eds. *Kelvin's Baltimore Lectures and Modern Theoretical Physics*. Cambridge, MA: MIT Press, pp. 399–423.

Stigler, S. M. (1986), *The History of Statistics: The Measurement of Uncertainty before 1900*. Cambridge: Harvard University Press.

Suppe, F. Ed. (1974), *The Structure of Scientific Theories*. Urbana: University of Illinois Press.

Taylor, C. (1971), "Interpretation and the Sciences of Man," *Review of Metaphysics* 25: 3–51.

Van Fraasen, Bas. *The Scientific Image*. Oxford: Clarendon Press, 1980.

Watson, John B. (1913), "Psychology as a Behaviorist Views It," *Psychological Review* 20: 158–177.

Weber, Max (1903/1949) *The Methodology of the Social Sciences*. Shils and Finch (eds.), New York: Free Press.

Wimsatt, W. (1987), "False Models as Means to Truer Theories." In Nitecki, N. and Hoffman, A. Eds. *Neutral Models in Biology*. Oxford: Oxford University Press, pp. 23–55.

Winch, P. (1958), *The Idea of Social Science and its Relation to Philosophy*. London: Routledge and Kegan Paul.

Woodger, J. H. (1929), *Biological Principles*. Cambridge: Cambridge University Press.

——— (1937), *The Axiomatic Method in Biology*. Cambridge: Cambridge University Press.

LIST OF CONTRIBUTORS

Alexander, Jason London School of Economics, United Kingdom

Armendt, Brad Arizona State University

Backe, Andrew Independent Scholar

Barrett, Jeff University of California at Irvine

Bechtel, William Washington University in St. Louis

Bouchard, Frederic University of Montreal, Canada

Bradie, Michael Bowling Green State University

Brown, Joshua University of Michigan

Byrne, Alex Massachusetts Institute of Technology

Cat, Jordi Indiana University

Craver, Carl Washington University in St. Louis

de Regt, Henk W. Vrije Universiteit, The Netherlands

Dickson, Michael Indiana University

DiSalle, Robert The University of Western Ontario, Canada

Downes, Stephen University of Utah

Eckardt, Barbara von Rhode Island School of Design

Eells, Ellery University of Wisconsin-Madison

Elga, Adam Princeton University

Ewens, Warren J. University of Pennsylvania

Falk, Raphael The Hebrew University, Israel

LIST OF CONTRIBUTORS

Fetzer, James University of Minnesota

Fitelson, Branden University of California at Berkeley

Folina, Janet Macalester College

Frigg, Roman London School of Economics, United Kingdom

Garson, Justin University of Texas at Austin

Gillies, Anthony University of Texas at Austin

Glennan, Stuart Butler University

Grandy, Richard Rice University

Gregory, Paul Washington and Lee University

Griffiths, Paul University of Queensland, Australia

Hájek, Alan California Institute of Technology

Hall, Ned Massachusetts Institute of Technology

Halvorson, Hans Princeton University

Hankinson-Nelson, Lynn University of Missouri at St. Louis

Hardcastle, Gary Bloomsburg University of Pennsylvania

Hardcastle, Valerie Virginia Polytechnic Institute and State University

Hartmann, Stephan London School of Economics and Political Science, United Kingdom

Hochberg, Herbert University of Texas at Austin

Hodges, Andrew University of Oxford

Hooker, Cliff A. University of Newcastle, Australia

Hull, David Northwestern University

Irvine, Andrew University of British Columbia, Canada

Joyce, Jim University of Michigan

Juhl, Cory University of Texas at Austin

Justus, James University of Texas at Austin

LIST OF CONTRIBUTORS

Kamlah, Andreas University of Osnabrück, Germany

Kincaid, Harold Center for Ethics and Values in the Sciences

Koertge, Noretta Indiana University

Larvor, Brendan University of Hertfordshire, United Kingdom

Laubichler, Manfred Arizona State University

Leplin, Jarrett University of North Carolina

Lipton, Peter University of Cambridge, United Kingdom

Little, Daniel University of Michigan-Dearborn

Lloyd, Elisabeth Indiana University

Loomis, Eric University of South Alabama

Ludlow, Peter University of Michigan

Lynch, Michael Cornell University

Lyre, Holger University of Bonn, Germany

MacLaurin, James University of Otago, New Zealand

Magnus, P. D. State University of New York at Albany

Majer, Ulrich University of Goettingen, Germany

Martinich, A. P. University of Texas at Austin

Motterlini, Matteo University of Trento, Italy

Nagel, Jennifer University of Toronto, Canada

Nickles, Thomas University of Nevada, Reno

Niiniluoto, Ilkka University of Helsinki

Noe, Alva University of California at Berkeley

Nyberg, Ian University of Texas at Austin

Odenbaugh, Jay University of California at San Diego

Okasha, Samir University of Bristol, United Kingdom

LIST OF CONTRIBUTORS

Perini, Laura Virginia Polytechnic Institute and State University

Pfeifer, Jessica University of Maryland, Baltimore County

Piccinini, Gualtiero Washington University in St. Louis

Plutynski, Anya University of Utah

Pojman, Paul University of Utah

Radder, Hannes Vrije Universiteit Amsterdam

Ramsey, Jeffry Smith College

Ratcliffe, Matthew University of Durham, United Kingdom

Richardson, Alan University of British Columbia

Roberts, John University of North Carolina

Rosenkrantz, Roger Independent Scholar

Roush, Sherrilyn Rice University

Ruetsche, Laura University of Pittsburgh

Sandell, Michelle Independent Scholar

Sankey, Howard University of Melbourne, Australia

Sarkar, Sahotra University of Texas at Austin

Shapere, Dudley Wake Forest University

Sigmund, Karl University of Vienna, Austria

Simchen, Ori University of British Columbia, Canada

Sober, Elliott University of Wisconsin-Madison

Stachel, John Boston University

Stadler, Friedrich K. University of Vienna, Austria

Stanford, P. Kyle University of California, Irvine

Stoljar, Daniel Australian National University

Stöltzner, Michael Institute Vienna Circle, Austria

LIST OF CONTRIBUTORS

Sundell, Tim University of Michigan

Suppes, Patrick Stanford University

Tauber, Alfred I. Boston University

Thornton, Stephen P. University of Limerick, Ireland

Vineberg, Susan Wayne State University

Wayne, Andrew Concordia University, Canada

Wilson, Jessica University of Toronto, Canada

Wilson, Robert A. University of Alberta, Canada

Wimsatt, William The University of Chicago

Witmer, D. Gene University of Florida

A TO Z LIST OF ENTRIES

A

Abduction
Adaptation and Adaptationism
Altruism
Analyticity
Anthropic Principle
Approximation
Artificial Intelligence
Astronomy, Philosophy of
Ayer, A. J.

B

Bayesianism
Behaviorism
Biological Information
Biology, Philosophy of
Bridgman, Percy Williams

C

Carnap, Rudolf
Causality
Chemistry, Philosophy of
Chomsky, Noam
Classical Mechanics
Cognitive Science
Cognitive Significance
Complementarity
Confirmation Theory
Connectionism
Consciousness
Conservation Biology
Conventionalism
Corroboration

D

Decision Theory
Demarcation, Problem of
Determinism
Duhem Thesis
Dutch Book Argument

E

Ecology
Economics, Philosophy of
Emergence
Empiricism
Epistemology
Evolution
Evolutionary Epistemology
Evolutionary Psychology
Experiment
Explanation
Explication

F

Feminist Philosophy of Science
Feyerabend, Paul Karl
Fitness
Function

G

Game Theory
Genetics

A TO Z LIST OF ENTRIES

H

Hahn, Hans
Hanson, Norwood Russell
Hempel, Carl Gustav
Heritability
Hilbert, David

I

Immunology
Incommensurability
Individuality
Induction, Problem of
Inductive Logic
Innate/Acquired Distinction
Instrumentalism
Intentionality
Irreversibility

K

Kinetic Theory
Kuhn, Thomas

L

Lakatos, Imre
Laws of Nature
Linguistics, Philosophy of
Locality
Logical Empiricism

M

Mach, Ernest
Mechanism
Methodological Individualism
Molecular Biology

N

Nagel, Ernest
Natural Selection
Neumann, John von
Neurath, Otto
Neurobiology

O

Observation

P

Parsimony
Particle Physics
Perception
Phenomenalism
Physical Sciences, Philosophy of
Physicalism
Poincaré, Jules Henri
Popper, Karl Raimund
Population Genetics
Prediction
Probability
Protocol Sentences
Psychology, Philosophy of
Putnam, Hilary

Q

Quantum Field Theory
Quantum Logic
Quantum Measurement Problem
Quantum Mechanics
Quine, Willard Van

R

Ramsey, Frank Plumpton
Rational Reconstruction
Realism
Reductionism
Reichenbach, Hans
Research Programs
Russell, Bertrand

S

Schlick, Moritz
Scientific Change
Scientific Domains
Scientific Metaphors
Scientific Models
Scientific Progress
Scientific Revolutions
Scientific Style

Searle, John
Social Constructionism
Social Sciences, Philosophy of the
Space-Time
Species
Statistics, Philosophy of
Supervenience

T

Theories
Time
Turing, Alan

U

Underdetermination of Theories
Unity and Disunity of Science
Unity of Science Movement

V

Verifiability
Verisimilitude
Vienna Circle
Visual Representation

THEMATIC LIST OF ENTRIES

Biology

Adaptation and Adaptationism
Altruism
Biological Information
Biology, Philosophy of
Conservation Biology
Ecology
Evolution
Fitness
Genetics
Heritability
Immunology
Individuality
Innate/Acquired Distinction
Molecular Biology
Natural Selection
Population Genetics
Species

Epistemology and Metaphysics

Abduction
Analyticity
Approximation
Bayesianism
Causality
Cognitive Significance
Confirmation Theory
Conventionalism
Corroboration
Decision Theory
Demarcation, Problem of
Determinism
Duhem Thesis
Dutch Book Argument
Emergence
Empiricism
Epistemology
Evolutionary Epistemology
Experiment
Explanation
Explication
Feminist Philosophy of Science
Function
Incommensurability
Induction, Problem of
Inductive Logic
Instrumentalism
Laws of Nature
Logical Empiricism
Mechanisms
Observation
Parsimony
Perception
Phenomenalism
Physicalism
Prediction
Probability
Protocol Sentences
Rational Reconstruction
Realism
Reductionism
Research Programs
Scientific Change
Scientific Domains
Scientific Metaphors
Scientific Models
Scientific Progress
Scientific Revolutions
Scientific Style
Social Constructionism
Supervenience
Theories
Underdetermination of Theories
Unity and Disunity of Science
Unity of Science Movement
Verifiability
Verisimilitude
Vienna Circle
Visual Representation

THEMATIC LIST OF ENTRIES

Physical Sciences

Anthropic Principle
Astronomy, Philosophy of
Chemistry, Philosophy of
Classical Mechanics
Complementarity
Irreversibility
Kinetic Theory
Locality
Particle Physics
Physical Sciences, Philosophy of
Quantum Field Theory
Quantum Logic
Quantum Measurement Problem
Quantum Mechanics
Space-Time
Time

Principal Figures

Ayer, A. J.
Bridgman, Percy Williams
Carnap, Rudolf
Chomsky, Noam
Feyerabend, Paul Karl
Hahn, Hans
Hanson, Norwood Russell
Hempel, Carl Gustav
Hilbert, David
Kuhn, Thomas
Lakatos, Imre
Mach, Ernest
Nagel, Ernest
Neumann, John von
Neurath, Otto
Poincaré, Jules Henri
Popper, Karl Raimund
Putnam, Hilary
Quine, Willard Van
Ramsey, Frank Plumpton
Reichenbach, Hans
Russell, Bertrand
Schlick, Moritz
Searle, John
Turing, Alan

Psychology and Mind

Artificial Intelligence
Cognitive Science
Connectionism
Consciousness
Evolutionary Psychology
Intentionality
Neurobiology
Psychology, Philosophy of

Social Sciences

Behaviorism
Economics, Philosophy of
Game Theory
Linguistics, Philosophy of
Methodological Individualism
Social Sciences, Philosophy of the

Statistics

Statistics, Philosophy of

ABDUCTION

Scientific hypotheses cannot be deduced from the empirical evidence, but the evidence may support a hypothesis, providing a reason to accept it. One of the central projects in the philosophy of science is to account for this nondemonstrative inductive relation. The justification of induction has been a sore point since the eighteenth century, when David Hume ([1777] 1975) gave a devastating skeptical argument against the possibility of any reason to believe that nondemonstrative reasoning will reliably yield true conclusions (see Induction, Problem of). Even the more modest goal of giving a principled description of our inductive practices has turned out to be extremely difficult. Scientists may be very good at weighing evidence and making inferences; but nobody is very good at saying how they do it.

The nineteenth-century American pragmatist Charles Sanders Peirce (1931, cf. 5.180–5.212) coined the term *abduction* for an account, also now known as "Inference to the Best Explanation," that addresses both the justification and the description of induction (Harman 1965; Lipton 1991, 2001; Day and Kincaid 1994; Barnes 1995). The governing idea is that explanatory considerations are a guide to inference, that the hypothesis that would, if correct, best explain the evidence is the hypothesis that is most likely to be correct. Many inferences are naturally described in this way (Thagard 1978). Darwin inferred the hypothesis of natural selection because although it was not entailed by his biological evidence, natural selection would provide the best explanation of that evidence. When astronomers infer that a galaxy is receding from the Earth with a specified velocity, they do so because the recession would be the best explanation of the observed redshift of the galaxy's spectrum.

On the justificatory question, the most common use of abduction has been in the miracle argument for scientific realism. Hilary Putnam (1978, 18–22) and others have argued that one is entitled to believe that empirically highly successful hypotheses are at least approximately true—and hence that the inductive methods scientists use are reliable routes to the truth—on the grounds that the truth of those hypotheses would be the best explanation of their empirical success (see Putnam, Hilary; Realism) because it would be a miracle if a hypothesis that is fundamentally mistaken were found to have so many precisely correct empirical consequences. Such an outcome is logically possible, but the correctness of the hypothesis is a far better explanation of its success. Thus the miracle argument is itself an abduction—an inference to the best explanation—from the predictive success of a hypothesis to its correctness.

Like all attempts to justify induction, the miracle argument has suffered many objections. The miracle argument is itself an abduction, intended as a justification of abduction. One objection is that this is just the sort of vicious circularity that Hume argued against. Moreover, is the truth of a hypothesis really the best explanation of its empirical successes? The feeling that it would require a miracle for a false hypothesis to do so well is considerably attenuated by focusing on the history of science, which is full of hypotheses that were successful for a time but were eventually replaced as their successes waned (see Instrumentalism). The intuition underlying the miracle argument may also be misleading in another way, since it may rest solely on the belief that most possible hypotheses would be empirically unsuccessful, a belief that is correct but arguably irrelevant, since it may also be the case that most successful hypotheses would be false, which is what counts.

Abduction seems considerably more promising as an answer to the descriptive question. In addition to the psychologically plausible account it gives of many particular scientific inferences, it may avoid weaknesses of other descriptive accounts, such as enumerative induction, hypothetico-deductivism, and Bayesianism. Enumerative inferences run from the premise that observed Fs are G to the conclusion that all Fs are G, a scheme that does not cover the common scientific case where hypotheses appeal to entities and processes not mentioned in the evidence that supports them. Since those unobservables are often introduced precisely because they would explain the evidence, abduction has no difficulty allowing for such "vertical" inferences (see Confirmation theory).

If the enumerative approach provides too narrow an account of induction, hypothetico-deductive models are too broad, and here again abduction does better. According to hypothetico-deductivism, induction runs precisely in the opposite direction from deduction, so that the evidence supports the hypotheses that entail it. Since, however, a deductively valid argument remains so whatever additional premises are inserted, hypothetico-deductivism runs the risk of yielding the absurd result that any observation supports every hypothesis, since any hypothesis is a member of a premise set that entails that observation. Abduction avoids this pitfall, since explanation is a more selective relationship than entailment: Not all valid arguments are good explanations.

The relationship between abduction and Bayesian approaches is less clear. Like abduction, Bayesianism avoids some of the obvious weaknesses of both enumerative and hypothetico-deductive accounts. But the Bayesian dynamic of generating judgments of prior probability and likelihood and then using Bayes's theorem to transform these into judgments of posterior probability appears distant from the psychological mechanisms of inference (see Bayesianism). It may be that what abduction can offer here is not a replacement for Bayesianism, but rather a way of understanding how the Bayesian mechanism is "realized" in real inferential judgment. For example, consideration of the explanatory roles of hypotheses may be what helps scientists to determine the values of the elements in the Bayesian formula. Bayesianism and abduction may thus be complementary.

One objection to an abductive description of induction is that it is itself only a poor explanation of our inductive practices, because it is uninformative. The trouble is that philosophers of science have had as much difficulty in saying what makes for a good explanation as in saying what makes for a good inference. What makes one explanation better than another? If "better" just means more probable on the evidence, then the abductive account has left open the very judgment it was supposed to describe. So "better" had better mean something like "more explanatory." However, it is not easy to say what makes one hypothesis more explanatory than another. And even if the explanatory virtues can be identified and articulated, it still needs to be shown that they are scientists' guides to inference. But these are challenges the advocates of abduction are eager to address.

PETER LIPTON

References

Barnes, Eric (1995), "Inference to the Loveliest Explanation," *Synthese* 103: 251–277.
Day, Timothy, and Harold Kincaid (1994), "Putting Inference to the Best Explanation in Its Place," *Synthese* 98: 271–295.
Harman, Gilbert (1965), "The Inference to the Best Explanation," *Philosophical Review* 74: 88–95.
Hume, David ([1777] 1975), *An Enquiry Concerning Human Understanding*. Edited by L. A. Selby-Bigge and P. H. Nidditch. Oxford: Oxford University Press.
Lipton, Peter (1991), *Inference to the Best Explanation*. London: Routledge.
——— (2001), "Is Explanation a Guide to Inference?" in Giora Hon and Sam S. Rakover (eds.), *Explanation: Theoretical Approaches and Application*. Dordrecht, Netherlands: Kluwer Academic Publishers, 93–120.

Peirce, Charles Sanders (1931), *Collected Papers*. Edited by C. Hartshorn and P. Weiss. Cambridge, MA: Harvard University Press.

Putnam, Hilary (1978), *Meaning and the Moral Sciences*. London: Hutchinson.

Thagard, Paul (1978), "The Best Explanation: Criteria for Theory Choice," *Journal of Philosophy* 75: 76–92.

See also **Bayesianism; Confirmation Theory; Induction, Problem of; Instrumentalism; Realism**

ADAPTATION AND ADAPTATIONISM

In evolutionary biology, a phenotypic trait is said to be an *adaptation* if the trait's existence, or its prevalence in a given population, is the result of natural selection. So for example, the opposable thumb is almost certainly an adaptation: Modern primates possess opposable thumbs because of the selective advantage that such thumbs conferred on their ancestors, which led to the retention and gradual modification of the trait in the lineage leading to modern primates. Usually, biologists will describe a trait not as an adaptation per se but rather as an adaptation *for* a given task, where the task refers to the environmental "problem" that the trait helps the organism to solve. Thus the opposable thumb is an adaptation *for* grasping branches; the ability of cacti to store water is an adaptation *for* living in arid deserts; the brightly adorned tail of the peacock is an adaptation *for* attracting mates; and so on. Each of these statements implies that the trait in question was favored by natural selection *because* it conferred on its bearer the ability to perform the task. In general, if a trait *T* is an adaptation for task *X*, this means that *T* evolved because it enabled its bearers to perform *X*, which enhanced their Darwinian fitness. This can also be expressed by saying that the *function* of the trait *T* is to perform *X*. Thus there is a close link between the concepts of adaptation and evolutionary function (Sterelny and Griffiths 1999; Ariew, Cummins, and Perlman 2002; Buller 1999).

Many authors have emphasized the distinction between a trait that is an adaptation and a trait that is *adaptive*. To describe a trait as adaptive is to say that it is *currently* beneficial to the organisms that possess it, in their current environment. This is a statement solely about the present—it says nothing about evolutionary history. If it turned out that Darwinism were wholly untrue and that God created the universe in seven days, many phenotypic traits would still qualify as adaptive in this sense, for they undeniably benefit their current possessors. By contrast, to describe a trait as an adaptation *is* to say something about evolutionary history, namely that natural selection is responsible for the trait's evolution. If Darwinism turned out to be false, it would follow that the opposable thumb is *not* an adaptation for grasping branches, though it would still be adaptive for primates in their current environment. So the adaptive/adaptation distinction corresponds to the distinction between a trait's *current utility* and its *selective history*.

In general, most traits that are adaptations are also adaptive and vice versa. But the two concepts do not always coincide. The human gastrointestinal appendix is not adaptive for contemporary human beings—which is why it can be removed without loss of physiological function. But the appendix is nonetheless an adaptation, for it evolved to help its bearers break down cellulose in their diet. The fact that the appendix no longer serves this function in contemporary humans does not alter the (presumed) fact that this is why it originally evolved. In general, when a species is subject to rapid environmental change, traits that it evolved in response to previous environmental demands, which thus count as adaptations, may cease to be adaptive in the new environment. Given sufficient time, evolution may eventually lead such traits to disappear, but until this happens these traits are examples of adaptations that are not currently adaptive.

It is also possible for a trait to be adaptive without being an adaptation, though examples falling into this category tend to be controversial. Some linguists and biologists believe that the capacity of humans to use language was not directly selected for, but emerged as a side effect of natural selection for larger brains. According to this theory, there was a direct selective advantage to having a large brain, and the emergence of language was simply an incidental by-product of the resulting increase in brain size among proto-humans. *If* this theory is correct, then human linguistic ability does not qualify as an adaptation and has no evolutionary function; thus it would be a mistake to look for a specific environmental demand to which it is an evolved response. But the ability to use language is presumably adaptive for humans in their *current* environment, so this would be an example of an adaptive trait that is not an adaptation. It should be noted, however, that many biologists and linguists are highly suspicious of the idea that human linguistic capacity was not directly shaped by natural selection. (See Pinker and Bloom 1990 and Fodor 2000 for opposing views on this issue.)

It sometimes happens that a trait evolves to perform one function and is later co-opted by evolution for a quite different task. For example, it is thought that birds originally evolved feathers as a way of staying warm, and only later used them to assist with flight. This is an interesting evolutionary phenomenon, but it creates a potential ambiguity. Should birds' feathers be regarded as an adaptation for thermoregulation or for efficient flight? Or perhaps for both? There is no simple answer to this question, particularly since feathers underwent considerable evolutionary modification after they first began to be used as a flying aid. Gould and Vrba (1982) coined the term "exaptation" to help resolve this ambiguity. An exaptation is any trait that originally evolves for one use (or arises for nonadaptive reasons) and is later co-opted by evolution for a different use.

How is it possible to tell which traits are adaptations and which are not? And if a particular trait is thought to be an adaptation, how is it possible to discover what the trait is an adaptation *for*, that is, its evolutionary function? These are pressing questions because evolutionary history is obviously not directly observable, so can be known only via inference. Broadly speaking, there are two main types of evidence for a trait's being an adaptation, both of which were identified by Darwin (1859) in *On the Origin of Species*. First, if a trait contributes in an obvious way to the "fit" between organism and environment, this is a prima facie reason for thinking it has been fashioned by natural selection. The organism/environment fit refers to the fact that organisms often possess a suite of traits that seem specifically tailored for life in the environments they inhabit. Consider for example the astonishing resemblance between stick insects and the foliage they inhabit. It seems most unlikely that this resemblance is a coincidence or the result of purely chance processes (Dawkins 1986, 1996). Much more plausibly, the resemblance is the result of many rounds of natural selection, continually favoring those insects who most closely resembled their host plants, thus gradually bringing about the insect/plant match. It is obvious *why* insects would have benefited from resembling their host plants—they would have been less visible to predators—so it seems safe to conclude that the resemblance is an adaptation for reducing visibility to predators. Biologists repeatedly employ this type of reasoning to infer a trait's evolutionary function.

Second, if a phenotypic trait is highly *complex*, then many biologists believe it is safe to infer that it is an adaptation, even if the trait's evolutionary function is not initially known. Bodily organs such as eyes, kidneys, hearts, and livers are examples of complex traits: Each involves a large number of component parts working together in a coordinated way, resulting in a mechanism as intricate as the most sophisticated man-made device. The inference from complexity to adaptation rests on the assumption that natural selection is the only serious scientific explanation for how organic complexity can evolve. (Appealing to an intelligent deity, though intellectually respectable in pre-Darwinian days, no longer counts as a serious explanation.) Again, inferences of this sort do not strictly amount to proof, but in practice biologists routinely assume that complex organismic traits are adaptations and thus have evolutionary functions waiting to be discovered.

The definition of an adaptation given above— any trait that has evolved by natural selection—is standard in contemporary discussions. In this sense, all biologists would agree that every extant organism possesses countless adaptations. However, the term has sometimes been understood slightly differently. R. A. Fisher, one of the founders of modern Darwinism, wrote that an organism

"is regarded as adapted to a particular situation ... only in so far as we can imagine an assemblage of slightly different situations, or environments, to which the animal would on the whole be less well adapted; and equally only in so far as we can imagine an assemblage of slightly different organic forms, which would be less well adapted to that environment." (1930, 41)

It is easy to see that Fisher's notion of adaptation is more demanding than the notion employed above. Fisher requires a very high degree of fit between organism and environment before the concept of adaptation applies, such that any small modification of either the organism or the environment would lead to a reduction in fitness. In modern parlance, this would normally be expressed by saying that the organism is *optimally* adapted to its environment.

It is quite possible for an organism to possess many adaptations in the above sense, i.e., traits that are the result of natural selection, without being optimally adapted in the way Fisher describes. There are a number of reasons why this is so. First, natural selection is a gradual process: Many generations are required in order to produce a close adaptive fit between organism and environment. Suboptimality may result simply because selection has yet to run its full course. Second, unless there is considerable environmental constancy over time, it is unlikely that organisms will evolve traits that adapt them optimally to any particular environment, given the number of generations required. So suboptimality may result from insufficient environmental constancy. Third, there may be evolutionary trade-offs. For example, the long necks of giraffes enable them to graze on high foliage, but the price of a long neck might be too high a center of gravity and thus a suboptimal degree of stability. Evolution cannot always modify an organism's phenotypic traits independently of each other: Adjusting one trait to its optimal state may inevitably bring suboptimality elsewhere. Finally, as Lewontin (1985) and others have stressed, natural selection can drive a species from one point in phenotypic space to another only if *each intermediate stage* is fitness enhancing. So, suboptimality may result because the optimal phenotypic state cannot be accessed from the actual state by a series of incremental changes, each of which increases fitness. For all these reasons, it is an open question whether natural selection will produce optimally adapted organisms.

It is worth noting that the Fisherian concept of optimal adaptation employed above is not *totally* precise, and probably could not be made so, for it hinges on the idea of a small or slight modification to either the organism or the environment, leading to a reduction in fitness. But "small" and "slight" are vague terms. How large can a modification be before it counts as too big to be relevant to assessing whether a given organism is optimally adapted? Questions such as this do not have principled answers. However, any workable concept of optimality is likely to face a similar problem. It is unacceptable to say that an organism is optimally adapted if there is no *possible* modification that would raise its fitness, for by that token no organism would qualify as optimally adapted. With sufficient imagination, it is always possible to think of phenotypic changes that would boost an organism's fitness—for example, doubling its fecundity while leaving everything else unchanged. (As John Maynard Smith wrote, "it is clearly impossible to say what is the "best" phenotype unless one knows the range of possibilities" [Maynard Smith 1978, 32]). So to avoid trivializing the concept of optimality altogether, some restriction must be placed on the class of possible modifications whose effects on organismic fitness are relevant to judging how well adapted the organism is in its current state. Spelling out the necessary restriction will lead to a concept similar to Fisher's, with its attendant vagueness.

The constraints on optimality noted in the earlier discussion of suboptimality show that natural selection *may* fail to produce organisms that are optimally adapted. But how important these constraints are in practice is a matter of considerable controversy. Some biologists think it is reasonable to assume that most extant organisms are optimally or nearly optimally adapted to their current environment. On this view, any phenotypic trait of an organism can be studied on the assumption that selection has finetuned the trait very precisely, so that there is an evolutionary reason for the character being exactly the way it is. Other biologists have less confidence in the power of natural selection. While not denying that selection *has* shaped extant phenotypes, they see the constraints on optimality as sufficiently important to invalidate the assumption that what has actually evolved is optimal in Fisher's sense. They would not seek adaptive significance in every last detail of an organism's phenotype. (See Maynard Smith 1978 and Maynard Smith et al. 1985 for a good discussion of this issue.)

The optimality question is just one aspect of an important and sometimes heated debate concerning the legitimacy of what is called "adaptationism" in evolutionary biology (Sober and Orzack 2001; Dupre 1987). Adaptationism encompasses both an empirical thesis about the world and a methodology for doing evolutionary research (Godfrey-Smith 2001). Empirically, the main claim is that natural selection has been by far the most important determinant of organismic phenotypes in evolutionary history—all or most traits have been directly fashioned by natural selection. Typically, adaptationists will also show some sympathy for the view that extant organisms are optimally adapted to their environments, in at least certain respects. Methodologically, adaptationists believe that the best way to study the living world is to search for the evolutionary function of organisms' phenotypic traits. Thus, for example, if an adaptationist observes an unusual pattern of behavior in a species of insect, the adaptationist will immediately assume that the behavior has an evolutionary function and will devote effort to trying to discover that function. Opponents of adaptationism reject both the empirical thesis and the methodological strategy. They emphasize the constraints on optimality noted above, as well as others; additionally, they point out that natural selection is not the only cause of evolutionary change and that organisms possess certain features that are nonadaptive and even maladaptive. Thus, it is a mistake to view the living world through an exclusively adaptationist lens, they argue.

The basic contours of the adaptationism debate have been in place for a long time, and indeed trace right back to Darwin. But the modern debate was instigated by Stephen Jay Gould and Richard Lewontin's famous article "The Spandrels of San Marco and the Panglossian Paradigm" (Gould and Lewontin 1979). These authors launched a forthright attack on what they saw as the extreme adaptationism prevalent in many evolutionary circles. They accused adaptationists of (a) uncritically assuming that every organismic trait *must* have an evolutionary function, (b) failing to accord a proper role to forces other than natural selection in evolution, and (c) paying insufficient heed to the constraining factors that limit selection's power to modify phenotypes at will. Unusually for a scientific article, "Spandrels" contains two striking literary allusions. Firstly, adaptationists are compared to Dr. Pangloss, a protagonist in Voltaire's satirical novel *Candide*, who despite suffering terrible misfortunes continues to believe that he inhabits the "best of all possible worlds." Gould and Lewontin's suggestion is that adaptationists commit a similar absurdity by viewing every aspect of an organism's phenotype as optimized by selection. Secondly, adaptationists are accused of inventing "Just So Stories" in their relentless search for evolutionary functions, that is, devising speculative hypotheses about traits' adaptive significance that owe more to their ingenuity than to empirical evidence. The reference here is to Rudyard Kipling's famous collection of children's stories, which include "How the Leopard Got Its Spots" and "How the Camel Got His Hump."

The title of Gould and Lewontin's paper illustrates what is perhaps their central complaint against adaptationist logic: the assumption, in advance of specific empirical evidence, that every trait has adaptive significance of its own. *Spandrel* is an architectural term that refers to the roughly triangular space between two adjacent arches and the horizontal above them; they are necessary by-products of placing a dome (or a flat roof) on arches. The spandrels beneath the great dome of St. Mark's Cathedral in Venice are decorated with elaborate mosaics of the four evangelists. Gould and Lewontin's point is that despite their ornate design, the spandrels are obviously not the raison d'être of the whole construction: Rather, they are inevitable by-products of the architectural design. Similarly, they suggest, certain anatomical and morphological traits of modern organisms may be inevitable by-products of their overall design, rather than directly shaped by selection. If so, such traits would not be adaptations, and it would be inappropriate to search for their evolutionary function. The human chin is a commonly cited example of a spandrel.

Gould and Lewontin's attack on adaptationism provoked an array of different reactions. Some of their opponents accused them of caricaturing adaptationism and thus attacking a strawman, on the grounds that no evolutionist had ever claimed every phenotypic trait of every organism to be adaptation, less still an optimal adaptation. There is certainly an element of truth to this charge. Nonetheless, Gould and Lewontin were writing at the height of the controversy over human sociobiology; and it is also true that *some* of the early proponents of that discipline advanced highly speculative hypotheses about the supposed evolutionary function of various behavioral patterns in humans, often on the basis of flimsy and anecdotal evidence. (This was not true of the best work in human sociobiology.) Gould and Lewontin's critique, even if overstated, was a useful corrective to this sort of naive adaptationism and led to a greater degree of methodological self-awareness among evolutionary biologists.

With hindsight, it seems that Gould and Lewontin's article has tempered, but not altogether eliminated, the enthusiasm felt by evolutionary biologists for adaptationism (cf. Walsh, "Spandrels," forthcoming). Many biologists continue to believe that cumulative natural selection over a large number of generations is the most plausible way of explaining complex adaptive traits, and such traits are abundant in nature. And despite the potential methodological pitfalls that the "Spandrels" paper warns against, the adaptationist research program continues to be highly fruitful, yielding rich insights into how nature works, and it has no serious rivals. Moreover, it *is* possible to test hypotheses about the adaptive significance of particular traits, in a variety of different ways. The *comparative method*, which involves comparing closely related species and trying to correlate phenotypic differences among them with ecological differences among their habitats, is one of the most common (cf. Harvey and Pagel 1991); it was employed by Darwin himself in his discussion of the Galapagos finches' beaks. Experimentally altering a trait, e.g., painting the plumage of a bird, and then carefully observing the effect on the organism's survival and reproductive success is another way of learning about a trait's adaptive significance. The most sophisticated work in evolutionary biology routinely uses these and other tests to adjudicate hypotheses about evolutionary function, and they bear little relation to the crude storytelling that Gould and Lewontin criticize. (See Endler 1986 for a good discussion of these tests.)

On the other hand, there is a grain of truth to Gould and Lewontin's charge that when a particular hypothesis about a trait's adaptive function is falsified, biologists will normally invent another adaptationist hypothesis rather than conclude that the trait is not an adaptation at all. However, not everyone agrees that reasoning in this way is methodologically suspect. Daniel Dennett agrees that adaptationists like himself offer "purely theory-driven explanations, argued a priori from the assumption that natural selection tells the true story—some true story or other—about every curious feature of the biosphere," but he regards this as perfectly reasonable, given the overall success of Darwinian theory (1995, 245). It is doubtful whether what Dennett says is literally true, however. There are many "curious features" of the biosphere for which it is not known whether there is an adaptationist story to be told or not. Take for example the prevalence of repeat sequences of noncoding "junk" DNA in the eukaryotic genome. This certainly qualifies as a curious feature—it took molecular biologists greatly by surprise when it was first discovered in the 1970s. But junk DNA has no known function—hence its name—and many people suspect that it has no function at all (though the current evidence on this point is equivocal; see Bejerano et al. 2004 for a recent assessment). So although Dennett is right that there is a general presumption in favor of adaptationist explanations among biologists, it is not true that every trait is *automatically* assumed to be an adaptation.

SAMIR OKASHA

References

Ariew, Andre, Robert Cummins, and Mark Perlman (eds.) (2002), *Functions: New Essays in the Philosophy of Psychology and Biology*. Oxford: Oxford University Press.

Bejerano, Gill, Michael Pheasant, Igor Makunin, Stuart Stephen, W. James Kent, John S. Mattick, and David Haussler (2004), "Ultraconserved Elements in the Human Genome," *Science* 304: 1321–1325.

Buller, David J. (ed.) (1999), *Function, Selection and Design*. Albany: State University of New York Press.

Darwin, Charles (1859), *On the Origin of Species by Means of Natural Selection*. London: John Murray.

Dawkins, Richard (1986), *The Blind Watchmaker*. New York: W. W. Norton.

——— (1996), *Climbing Mount Improbable*. New York: W. W. Norton.

Dennett, Daniel (1995), *Darwin's Dangerous Idea*. New York: Simon and Schuster.

Dupre, John (ed.) (1987), *The Latest on the Best: Essays on Evolution and Optimality*. Cambridge, MA: MIT Press.

Endler, John A. (1986), *Natural Selection in the Wild*. Princeton, NJ: Princeton University Press.

Fisher, Ronald A. (1930), *The Genetical Theory of Natural Selection*. Oxford: Clarendon Press.

Fodor, Jerry (2000), *The Mind Doesn't Work That Way*. Cambridge MA: MIT Press.

Godfrey-Smith, Peter (2001), "Three Kinds of Adaptationism." In Elliott Sober and Steven Hecht Orzack (eds.), *Adaptationism and Optimality*. Cambridge: Cambridge University Press, 335–357.

Gould, Stephen Jay, and Elizabeth Vrba (1982), "Exaptation: A Missing Term in the Science of Form," *Paleobiology* 8: 4–15.

Gould, Stephen Jay, and Richard Lewontin (1979), "The Spandrels of San Marco and the Panglossian Paradigm: A Critique of the Adaptationist Programme," *Proceedings of the Royal Society of London*, Series B 205: 581–598.

Harvey, Paul, and Mark Pagel (1991), *The Comparative Method in Evolutionary Biology*. Oxford: Oxford University Press.

Lewontin, Richard (1985), "Adaptation." In Richard Levins and Richard Lewontin (eds.), *The Dialectical Biologist*. Cambridge MA: Harvard University Press, 65–84.

Maynard Smith, John (1978), "Optimization Theory in Evolution," *Annual Review of Ecology and Systematics* 9: 31–56.

Maynard Smith, John, Richard Burian, Stuart Kaufman, Pere Alberch, James Campbell, Brian Goodwin, Russell Lande, David Raup, and Lewis Wolpert (1985), "Developmental Constraints and Evolution," *Quarterly Review of Biology* 60: 265–287.

Pinker, Steven, and Paul Bloom (1990), "Natural Language and Natural Selection," *Behavioral and Brain Sciences* 13: 707–784.

Sober, Elliott, and Steven Hecht Orzack (eds.) (2001), *Adaptationism and Optimality*. Cambridge: Cambridge University Press.

Sterelny, Kim, and Paul Griffiths (1999), *Sex and Death: An Introduction to Philosophy of Biology*. Chicago: Chicago University Press.

Walsh, Denis (ed.) (forthcoming). *Spandrels of San Marco: 25 Years Later*. Cambridge: Cambridge University Press.

See also **Evolution; Fitness; Natural Selection**

ALTRUISM

The concept of altruism has led a double life. In ordinary discourse as well as in psychology and the social sciences, behavior is called *altruistic* when it is caused by a certain sort of motive. In evolutionary biology, the concept applies to traits (of morphology, physiology, and behavior) that enhance the fitness of others at some cost to the self.

A behavior can be altruistic in the evolutionary sense without being an example of psychological altruism. Even if a honeybee lacks a mind, it nonetheless counts as an evolutionary altruist when it uses its barbed stinger to attack an intruder to the hive: The barb disembowels the bee but allows the stinger to keep pumping venom into the intruder even after the bee has died, thus benefiting the hive.

Symmetrically, a behavior can be altruistic in the psychological sense without being an example of evolutionary altruism. If one gives another a volume of piano sonatas out of the goodness of one's heart, one's behavior may be psychologically altruistic. However, the gift will not be an example of evolutionary altruism if it does not improve the other's prospects for survival and reproductive success or does not diminish one's own fitness.

Both types of altruism have given rise to controversy. According to psychological egoism, all our motives are ultimately selfish, and psychological altruism is merely a comforting illusion. Egoism was the dominant position in all major schools of twentieth-century psychology (Batson 1991). Within evolutionary biology, there has been considerable hostility to the idea that altruistic traits evolve because they benefit the group; according to one influential alternative viewpoint, the gene—not the group or even the individual organism—is the unit of selection (Dawkins 1976; Williams 1966).

Evolutionary Altruism

Evolutionary altruism poses a puzzle—it appears to be a trait that natural selection will stamp out rather than promote. If altruists and selfish individuals live in the same group, altruists will donate "fitness benefits" to others, whereas selfish individuals will not. Altruists receive benefits from the donations of other altruists, but so do selfish individuals. It follows that altruists will be less fit than selfish individuals in the same group. Natural selection is a process that causes fitter traits to increase in frequency and less fit traits to decline. How, then, can natural selection explain the existence of evolutionary altruism?

Darwin's answer was the hypothesis of group selection. Although altruists are less fit than selfish individuals in the same group, groups of altruists will be fitter than groups of selfish individuals. Altruistic traits evolve because they benefit the group and in spite of the fact that they are deleterious to altruistic individuals. Darwin (1859, 202) applied this idea to explain the barbed stinger of the honeybee; he also invoked the hypothesis to explain why men in a tribe feel morally obliged to defend the tribe and even sacrifice their lives in time of war (1871, 163–165).

With regard to natural selection, Darwin was a "pluralist": He held that some traits evolve because they are good for the individual, while others evolve because they are good for the group (Ruse 1980). This pluralism became a standard part of the evolutionary biology practiced in the period 1930–1960, when the "modern synthesis" was created. The idea of group adaptation was often applied uncritically during this period; however, the same can be said of the idea of individual adaptation. The situation changed in the 1960s, when group selection was vigorously attacked (Hamilton 1964; Maynard Smith 1964; Williams 1966). Its exile from evolutionary theory was hailed as one of the major advances of twentieth-century biology. Since then, the hypothesis of group selection has been making a comeback; many biologists now think that group selection is well grounded theoretically and well supported empirically as the explanation of some—though by no means all—traits (Sober and Wilson 1998). However, many other biologists continue to reject it.

The arguments mustered against group selection during the 1960s were oddly heterogeneous. Some were straightforwardly empirical; for example, Williams (1966) argued that individual selection predicts that sex ratios should be close to even, whereas group selection predicts that organisms should facultatively adjust the mix of daughters and sons they produce so as to maximize group productivity. Williams thought (mistakenly) that sex ratios were almost always close to even and concluded that group selection was not involved in the evolution of this trait. Other arguments were sweeping in their generality and almost a priori in character; for example, it was argued that genes, not groups, were the units of selection, because genes provide the mechanism of heredity (Dawkins 1976; Williams 1966). A third type of argument involved proposing alternatives to group selection. One reason why group selection was largely abandoned during this period was that inclusive fitness theory (Hamilton 1964), the theory of reciprocal altruism (Trivers 1971), and game theory (Maynard Smith and Price 1973) were widely viewed as alternatives. One reason the controversy continues is that it is disputed whether these theories are alternatives to or implementations of the idea of group selection (Sober and Wilson 1998).

Psychological Altruism

Although the concept of psychological altruism is applied to people and to actions, the best place to begin is to think of psychological altruism as a property of motives, desires, or preferences. Eve's desire that Adam have the apple is other directed; the proposition that she wants to come true—that Adam has the apple—mentions another person, but not Eve herself. In contrast, Adam's desire that he have the apple is self-directed. In addition to purely self-directed and purely other-directed desires, there are mixed desires: People desire that they and specific others be related in a certain way. Had Eve wanted to share the apple with Adam, her desire would have been mixed.

An altruistic desire is an other-directed desire in which what one wants is that another person do well. Altruistic desires, understood in this way, obviously exist. The controversy about psychological altruism concerns whether these desires are ever ultimate or are always merely instrumental. When one wishes others well, does one ever have this desire as an end in itself, or does one care about others only because one thinks that how they do will affect one's own welfare? According to psychological egoism, all ultimate motives are self-directed. When Eve wants Adam to have the apple, she has this other-directed desire only because she thinks that his having the apple will benefit her.

Psychological hedonism is one variety of egoistic theory. Its proponents claim that the only ultimate motives people have are the attainment of pleasure and the avoidance of pain. The only things one cares about as ends in themselves are states of one's own consciousness. This special form of egoism is the hardest one to refute. It is easy enough to see from human behavior that people do not always try to maximize their wealth. However, when someone chooses a job with a lower salary over a job that pays more, the psychological hedonist can say that this choice was motivated by the desire to feel good and to avoid feeling bad. Indeed, hedonists think they can explain even the most harrowing acts of self-sacrifice—for example, the proverbial soldier in a foxhole who throws himself on a live grenade to protect his comrades. The soldier supposedly does this because he prefers not existing at all over living with the tormenting knowledge that he allowed his friends to die. This hedonistic explanation may sound strained, but that does not mean it must be false.

Since hedonism is difficult to refute, egoism is also difficult to refute. However, that does not mean it is true. Human behavior also seems to be consistent with a view called *motivational pluralism;* this is the claim that people have both self-directed and other-directed ultimate aims. This theory does not assert that there are human actions driven solely by other-directed ultimate desires. Perhaps one consideration lurking behind everything one does is a concern for self. However, since actions may be caused by several interacting desires, pluralism is best understood as a claim about the character of our desires, not about the purity of our actions.

It is an interesting fact about our culture that many people are certain that egoism is true and many others are certain it is false. An extraterrestrial anthropologist might find this rather curious, in view of the fact that the behaviors one observes in everyday life seem to be consistent with both egoism and pluralism. One's convictions evidently outrun the evidence one has at hand. Is the popularity of egoism due to living in a culture that emphasizes individuality and economic competition? Is the popularity of pluralism due to the fact that people find it comforting to think of benevolence as an irreducible part of human nature? These questions are as fascinating as they are difficult to answer.

Social psychologists have tried to gather experimental evidence to decide between egoism and motivational pluralism. Egoism comes in a variety of forms; if each form could be refuted, this would refute egoism itself. According to one version of egoism, one helps needy others only because witnessing their suffering makes one uncomfortable; one helps help them for the same reason that one adjusts the thermostat when the room becomes too

ALTRUISM

hot. According to a second version of egoism, people help only because they want to avoid censure from others or self-censure. According to a third version, people help only because helping provides them with a mood-enhancing reward. Batson (1991) argues that the experimental evidence disconfirms all the versions of egoism formulated so far (but see Sober and Wilson 1998).

The Prisoners' Dilemma

Game theory was developed by economists and mathematicians as a methodology for modeling the process of rational deliberation when what is best for one agent depends on what other agents do (von Neumann and Morgenstern 1947). These ideas were subsequently modified and extended, the result being evolutionary game theory, in which organisms (that may or may not have minds) interact with each other in ways that affect their fitness (Maynard Smith 1982). The prisoners' dilemma began as a problem in game theory, but it also has been much discussed in evolutionary game theory. It is central to understanding the conditions that must be satisfied for evolutionary altruism to evolve. Economists and other social scientists have studied the prisoners' dilemma because they think that people's behavior in such situations throws light on whether their motivation is altruistic or selfish.

Two actors come together; each must decide independently which of two actions to perform: cooperate or defect. The payoffs to each player are shown in the illustration.

Prisoners' dilemma: payoffs	Column player	
	Cooperate	Defect
Row player Cooperate	Both get 3.	Row gets 1. Column gets 4.
Defect	Row gets 4. Column gets 1.	Both get 2.

A simple dominance argument shows that each player should defect. Suppose you are the row player. If the column player cooperates, you are better off defecting (since $4 > 3$); and if the column player defects, you are better off defecting (since $2 > 1$). The column player is in exactly the same position. However, the resulting solution—defection by both—means that both players are worse off than they would have been if both had chosen to cooperate (since $3 > 2$). The lesson is that rational decision making, with full information about the consequences of one's choices, can make one worse off.

How might the players avoid this dispiriting outcome? One possibility is for them to become less selfish. If they care about payoffs to self and payoffs to others in equal measure, each player will choose to cooperate (since $6 > 5$ and $5 > 4$). Another possibility is for the players to be forced to make the same decision, either by a mutually binding agreement or by a third party (a "leviathan"). Each of these changes puts the players into a new game, since the defining features of the prisoners' dilemma have now been violated.

In an evolutionary setting, the utilities described in the table are reinterpreted as "fitnesses"—that is, numbers of offspring. Suppose there are two types of individuals in a population: cooperators and defectors. Individuals form pairs, the members of each pair interact, and then each reproduces asexually, with the number of an organism's offspring dictated by the table of payoffs. Offspring exactly resemble their parents (there are no mutations). The older generation then dies, and the new generation of cooperators and defectors form pairs and play the game again. If this process is iterated over many generations, what will be the final configuration of the population? Since mutual defection is the solution to the prisoners' dilemma when the problem involves rational deliberation, can we infer that the solution for the evolutionary problem is universal defection?

The answer is no. Everything depends on how the pairs are formed. If they form at random, then universal defection will evolve. But suppose that like always pairs with like. In this instance cooperators receive a payoff of 3 and defectors receive a payoff of 2, which means that cooperation will increase in frequency. The dominance argument that settles the deliberational problem has no force in the evolutionary problem (Skyrms 1996). All one can tell from the table of payoffs is that the average fitness of cooperation must be somewhere between 3 and 1, and the average fitness of defection must be somewhere between 4 and 2. Cooperation and defection in evolutionary game theory are nothing other than evolutionary altruism and selfishness. The evolution of altruism thus crucially depends on how much correlation there is between interacting individuals (Skyrms 1996; Sober 1992).

When economists run experiments in which subjects play the prisoners' dilemma game, they find that people often behave "irrationally"—they do not always defect. What could explain this? One obvious possibility is that the payoff used in the experiment (usually money) doesn't represent everything that the subjects care about. It is true that defection can be reduced in frequency by manipulating the magnitude of monetary payoffs, but this does not show that the impulse to cooperate is always nonexistent; it shows only that this impulse is sometimes weaker than the impulse of self-interest. Furthermore, defenders of psychological egoism can easily bring cooperative behavior within the orbit of their theory. If people cooperate because and only because cooperation brings them pleasure, the behavior is consistent with psychological hedonism. Indeed, there is evidence from neuroscience that subjects do experience pleasure when they cooperate in the prisoners' dilemma (Billing et al. 2002). Note, however, that motivational pluralism is consistent with this finding. This raises the question whether behavior in such experiments throws light on the dispute between psychological egoism and motivational pluralism.

ELLIOTT SOBER

References

Batson, C. Daniel (1991), *The Altruism Question: Toward a Social-Psychological Answer*. Hillsdale, NJ: Lawrence Erlbaum Associates.

Billing, J. K., D. A. Gutman, T. R. Zeh, G. Pagnoni, G. S. Berns, and C. D. Kitts (2002), "A Neural Basis for Social Cooperation," *Neuron* 35: 395–405.

Darwin, C. (1859), *On the Origin of Species*. London: Murray.

——— (1871), *The Descent of Man and Selection in Relation to Sex*. London: Murray.

Dawkins, R. (1976), *The Selfish Gene*. New York: Oxford University Press.

Hamilton, William D. (1964), "The Genetical Evolution of Social Behavior, I and II," *Journal of Theoretical Biology* 7: 1–16, 17–52.

Maynard Smith, John (1964), "Group Selection and Kin Selection," *Nature* 201: 1145–1146.

——— (1982), *Evolution and the Theory of Games*. Cambridge: Cambridge University Press.

Maynard Smith, John, and George Price (1973), "The Logic of Animal Conflict," *Nature* 246: 15–18.

Ruse, Michael (1980), "Charles Darwin and Group Selection," *Annals of Science* 37: 615–630.

Skyrms, Brian (1996), *The Evolution of the Social Contract*. New York: Cambridge University Press.

Sober, Elliott (1992), "The Evolution of Altruism: Correlation, Cost, and Benefit," *Biology and Philosophy* 7: 177–188.

Sober, Elliott, and David Sloan Wilson (1998), *Unto Others: The Evolution and Psychology of Unselfish Behavior*. Cambridge, MA: Harvard University Press.

Trivers, R. L. (1971), "The Evolution of Reciprocal Altruism," *Quarterly Review of Biology* 46: 35–57.

von Neumann, John, and Oscar Morgenstern (1947), *Theory of Games and Economic Competition*. Princeton, NJ: Princeton University Press.

Williams, George C. (1966), *Adaptation and Natural Selection*. Princeton, NJ: Princeton University Press.

See also **Evolution; Game Theory; Natural Selection**

ANALYTICITY

Analyticity can be characterized vaguely as follows. Analytic sentences, statements, or propositions are either true or known to be true "in virtue of" the meanings of the terms or concepts contained within them. Although a variety of notions resembling analyticity appear in writings of the late seventeenth century, notably in the work of Gottfried Leibniz (who uses "analytic" as a synonym for "practical" [Leibniz 1981, 624]), the more contemporary use of the notion of analyticity first appears in the work of Immanuel Kant. The account below will survey several historically significant notions of analyticity and then turn to more recent developments stemming from important objections to the notion of analyticity developed by Willard Van Quine and others. A strategy in defense of analyticity in response to Quine's challenge will then be sketched.

Historical Background

In his *Critique of Pure Reason*, Kant characterized an *analytic* judgment in two ways. Firstly, he posited that

> the relation of a subject to the predicate is . . . possible in two different ways. Either the predicate *B* belongs to the subject *A*, as something which is (covertly) contained in this concept *A*; or *B* lies outside *A*, although it does indeed stand in connection with it. In the one case I entitle the judgment analytic, in the other synthetic. (Kant 1965, A6–7/B10)

ANALYTICITY

Later Kant stated:

> All analytic judgments rest wholly on the principle of contradiction.... For because the predicate of an affirmative analytic judgment has already been thought in the concept of the subject, it cannot be denied of the subject without contradiction. (1965, A150/B189)

Subsequent philosophers such as Ayer (1946, 78) have questioned the equivalence of these two definitions, arguing that the former appears to provide a psychological criterion for analyticity, while the latter is purely logical.

It is tempting to impute to Kant the view that analytic judgments are vacuous (Ayer 1946, 77) or that they are "true by virtue of meanings and independently of fact" only (Quine 1953a, 21), but it is not clear that such attributions are correct. Subject–predicate judgments that Kant regarded as synthetic a priori, such as "A straight line between two points is the shortest," are plausibly regarded as being both necessary and true by virtue of meaning, since the predicate is conjoined to the subject through a kind of construction in intuition, and this construction conforms to what Kant called a "necessity inherent in the concepts themselves" (1965, B16–17; cf. Proust 1986; Hintikka 1974). Moreover, Kant at one point rejected the supposition that "tautologous" or vacuous judgments such as identity statements could be analytic, precisely in virtue of their vacuity:

> Propositions which explain *idem per idem* advance cognition neither analytically nor synthetically. By them I have neither an increase in distinctness nor a growth in cognition. (1965, XXIV, 667)

What then does distinguish analytic truths for Kant? A specification of the exact nature of analytic truths in Kant's work is complicated by his rather detailed theory of concepts (and his at times imprecise formulations), but if one attributes to him a common eighteenth-century doctrine of complex concepts according to which such concepts are built up by composition from simpler ones (such as a concept's *genus* and *differentia*), then a judgment of the form "*S* is *P*" could be said to be analytic if and only if the concept *S* has an analysis in which the concept *P* appears as a composite element (cf. De Jong 1995). Kant thought that "All bodies are extended" was of this form. In contrast, synthetic a priori subject–predicate judgments forge a connection between their subject and predicate through a construction in intuition by which the predicate is somehow "attached necessarily" to the subject but not antecedently contained within it (1965, B16–17). However, Kant believed that both forms of judgment advanced cognition and that both could eventuate in necessary truth: analytic judgments through clarification of a previously given but unanalyzed content, synthetic a priori judgments through construction of the predicate in intuition.

While the Kantian notion of analyticity thus bears only a loose resemblance to more contemporary ones, certain features of the contemporary notion were anticipated soon after Kant in the *Wissenshaftslehre* of Bernhard Bolzano. Bolzano thought that Kant's characterization of analyticity was too vague because the notion of the "containment" of the predicate by the subject was merely figurative (Bolzano 1973, 201). Furthermore, Kant's characterization allowed a proposition like "The father of Alexander, King of Macedon, was King of Macedon" to be classed as analytic even though, Bolzano claimed, no one would want to describe it as such (1973, 201). In brief, Bolzano proposed characterizing analytic propositions as those propositions that contained some referring idea such that, given any arbitrary substitution of that idea with another referring idea, the truth value of the original proposition would be preserved (provided that both ideas shared the same "range of variation"). Thus, the proposition "A depraved man does not deserve respect" is regarded as analytic because it contains a referring idea (man) that can be replaced with any other within a certain range while preserving truth (1973, 198). By contrast, the proposition "God is omniscient" is synthetic, since no constituent referring idea of this proposition could be arbitrarily replaced without altering the truth of the proposition.

Bolzano's account allowed him to introduce a distinction within the class of analytic propositions between those propositions recognizable as analytic solely by virtue of "logical knowledge" (such as "An *A* that is *B* is *A*") and those that required "extra-logical knowledge." His account also anticipated much later developments, both in making propositions (as opposed to judgments) the basis of the designation *analytic* and in relating logically analytic propositions to his notions of satisfaction and logical validity (Bolzano 1973, §147f).

Despite its originality, Bolzano's account of analytic propositions has certain counterintuitive consequences. For instance, the conjunction of any true proposition like "Grass is green" with an analytic proposition (e.g., "Grass is green and (*A* is *A*)") would appear to be analytic, since *A* occurs in the second conjunct vacuously and so can be arbitrarily substituted while preserving the truth value. But by the same criterion, the proposition "A bachelor is an unmarried adult male" would be synthetic, since for example "A widower is an unmarried adult male" is false and we can produce false sentences by substituting "married," "infant," and "female" for the terms "unmarried," "adult," and "male," respectively. Moreover, Bolzano's theory of propositions remained limited by his insistence that every statement be regarded as having the simple subject–copula–predicate form "*S* has *P*" (1973, 173ff).

The latter limitation was largely overcome with Gottlob Frege's development of the first-order predicate calculus, which, through the use of variable-binding generality operators ("quantifiers"), allowed logicians to deal with statements not obviously of the subject–predicate form, such as multiply general statements, or such statements as the claim that 1 plus 1 equals 2 or that any two distinct points determine a unique line. In his *Foundations of Arithmetic,* Frege defined analytic truths as all and only those truths that were provable from general logical laws and definitions (1974, §4). Since Frege believed that he could establish that arithmetic rested upon logic alone, his characterization of analyticity within the context of his new logic allowed him to assert, *contra* Kant, that mathematical truths were analytic. While this significantly improved logical apparatus thus allowed Frege to capture a notion of logical consequence that encompassed many new intuitively valid inferences, his restriction of analytic statements to those provable from general logical laws meant that statements such as "If *A* is longer than *B*, and *B* is longer than *C*, then *A* is longer than *C*" did not count as analytic.

This problem was addressed by Frege's student, Rudolf Carnap. Carnap attempted to "explicate" analyticity by giving the concept a precise characterization within formally specified languages (see Explication). In his *Logical Syntax of Language,* Carnap identified analytic statements as those statements that were consequences—in a given specified language such as his "Language II"—of logic and classical mathematics (1937, 100–1), and he later introduced a distinction between analytic truths that depend upon the meaning of nonlogical terms and those that do not (1956, 223–4). Unlike Bolzano and Frege, who saw in nonlogical analytic truths the need for the introduction of concepts "wholly alien to logic" (cf. Bolzano 1973, 198), Carnap thought that nonlogical analytic truths could nonetheless be precisely and unproblematically accommodated through the introduction of "meaning postulates" that stipulated relations of implication or incompatibility for the nonlogical predicates of an analytic statement (Carnap 1990, 68f). Since such postulates were stipulated, knowledge of them was for Carnap no different in kind than knowledge of the laws of logic or mathematics, which he also regarded as stipulated (see Conventionalism). It is apparently only through the introduction of something like Carnap's meaning postulates or allied devices that a reasonably precise characterization of analyticity in terms of truth by virtue of meaning is possible (see Carnap, Rudolf).

Besides Carnap, other members of the Vienna Circle (see Logical Empiricism; Vienna Circle) had seen in the notion of analytic truth the possibility of explaining how certain statements, such as "Either some ants are parasitic or none are" or the statements of the laws and theorems of logic, could be regarded as knowable a priori within an empiricist epistemology. The idea was that analytic statements, understood as statements that are true by virtue of meaning alone, would be true independently of matters of fact and known a priori to be true by anyone who understood the words used in them. As mere consequences of word meaning, such statements would not express genuine knowledge about the world and so would pose no threat to an epistemology that regarded all genuine knowledge as empirical (see Ayer 1946, 71–80). It should be noted, however, that although Carnap was allied with the Vienna Circle, the extent to which he viewed analyticity as serving in this role has been questioned (cf. Friedman 1987).

Quine's Two Dogmas of Empiricism

In "Two Dogmas of Empiricism," Quine (1953a) identified two "dogmas" that he claimed had been widely accepted by logical empiricists such as Carnap and other members of the Vienna Circle. One dogma was the view that there is a distinction between analytic statements, which are true by virtue of meaning alone, and synthetic ones, which have empirical content. The other dogma was the view that individual statements have empirical content, in the sense that each such statement can be confirmed or disconfirmed "taken in isolation from its fellows" (41). Quine suggested, in opposition to this dogma of "reductionism," that "our statements about the external world face the tribunal of sense experience not individually but only as a corporate body" (41) (see Verifiability).

Against the first dogma, Quine argued that the notion of analyticity is suspect, in that it cannot be given a satisfactory explanation in independent terms. For example, he argued that one can define analytic statements as those that become truths of logic via a sequence of replacements of expressions by synonymous ones. However, Quine then noted that synonymy can be defined in terms of analyticity, in that two terms are synonymous just in case a statement of their equivalence is analytic. Thus what might have seemed to be an explication of the notion of analyticity turns out, in Quine's view, to cast doubt on the legitimacy of the notion of synonymy instead. Quine considered other attempts at defining analyticity, including appeals to the notion of a definition. For most terms in use, there is no record of anyone introducing the term into our language by explicit stipulation. Interestingly, Quine allowed that in cases of "the explicitly conventional introduction of novel notations for purposes of sheer abbreviation ... the definiendum becomes synonymous with the definiens simply because it has been created expressly for the purpose of being synonymous with the definiens" (1953a, 26). However, asserting that stipulative definitions create synonymies is of doubtful coherence with the rest of what Quine advocates, as will be discussed further below.

Carnap agreed that expressions within natural languages are so vague and ambiguous that it is often unclear which should count as analytic. One can, however, create more precisely defined artificial languages in which the question of which sentences are analytic is perfectly well defined, in Carnap's view. Carnap called some basic sentences *meaning postulates* and essentially defined analytic statements for such a language L to be those sentences that follow, via the logical rules of L, from the meaning postulates alone (cf. Carnap 1956, 222f). Quine objected that Carnap's appeal to artificial languages is unhelpful, since Carnap does not explain what is special about meaning postulates of a language other than that they are listed under the heading "Meaning Postulates of L" in Carnap's books (Quine 1953a, 34). What Quine appeared to want was a language-transcendent notion of analyticity that would enable one (given an arbitrary language) to pick out the analytic sentences, and he denied that Carnap provided this.

The second dogma is related to the first, according to Quine, since if reductionism (confirmational localism, as opposed to holism) were true, then one could consider individual sentences "in isolation" and see whether they were confirmed and remained so under all possible observational circumstances. If so, they would count as analytic in this stipulated sense. However, Quine believed that whether any sentence is confirmed on a given occasion depends on its connections to other sentences and to further sentences in an immense web of belief. He thought that because of this holism, given any sentence S, there would always be some imaginable circumstances under which S would be disconfirmed. Thus, there are no sentences that would be confirmed in all circumstances, and so no analytic statements.

During the next decade or two, a large number of counterattacks and further objections were raised, culminating in Harman's summary of the current state of the argument as of about 1967. After that, most philosophers either were converted to Quineanism or, "however mutinously" (cf. Harman 1967), stopped claiming that anything was analytic.

At the turn of the twenty-first century, the arguments commonly raised against analyticity are essentially the same as those of Quine and Harman. The very few papers one finds in defense of analyticity, while they might contain an objection to one or another Quinean argument, nevertheless admit an inability actually to present a positive view concerning the distinction. Yet at the same time, most contemporary philosophers, even most who agree with Quine on analyticity, doubt Quine's conclusions about the "indeterminacy of translation." Most seem to think that if analyticity has to go, it will be for reasons much more straightforward and less contentious than those supporting the indeterminacy of translation (and of the indeterminacy of meaning generally). In what follows, the indeterminacy-based arguments against analyticity will be ignored (for these, see Boghossian 1996). The legitimacy of modal (or "intensional") notions such as possibility, which Quine rejected as part of his rejection of analyticity (Quine 1953b, 139ff), will also not be disputed. Again, most contemporary philosophers, including those who reject analyticity, are quite happy to employ intensional notions.

Early Responses to Quine: Carnap, Grice and Strawson, and Others

Carnap himself responded to Quine in a number of places (see especially 1956, 233f; 1990, 427ff) and also noted responses to Quine by other philosophers. One is by Martin (1952), who objects that although Quine finds the notions of synonymy and analyticity to be problematic on the grounds that no one, including Carnap, has given a language-transcendent concept of synonymy, Quine nonetheless accepts the notion of truth as unproblematic. Yet, neither has anyone provided a language-transcendent notion of truth. Parity of reasoning would thus seem to require that Quine abandon this latter notion also. Benson Mates (1951) argues that in fact there are behavioristically detectable differences between our treatment of analytic statements and merely well-confirmed synthetic ones, so that even if one adopts Quine's strict behaviorist methodology, one can draw the distinction.

More recently, John Winnie (1975) has argued that Carnap's later attempt at defining analyticity overcomes most of Quine's objections. The basic idea is that one takes a theory T, "Ramsifies" it (essentially, existentially generalize on the theoretical predicate terms by using second-order variables and quantifiers) to yield R, and then constructs the sentence

$$R \rightarrow T.$$

This conditional is taken to be the analytic content of the theory. While Winnie's proposal will not be addressed, it is mentioned as a potentially fruitful avenue for the defender of one form of analyticity.

Among the earliest and best-known responses to the Quinean offensive against analyticity is Grice and Strawson's "In Defense of a Dogma" (1956). The argument in effect states that since there is broad agreement in a number of central cases of analytic and synthetic sentences/propositions, there must be some substantive distinction that is being drawn, even if it is difficult to spell it out clearly. Harman has a response to this argument that must be confronted, however (see below).

In their paper, Grice and Strawson further suggest that Quine's position on stipulative definition for the purposes of abbreviation is incoherent. If, as Quine argues at length, the concept of analyticity is unintelligible, then how can it be intelligible to stipulate that some sentence has this unintelligible feature? Some important later defenders of Quine seem to agree that Quine made a slip here. Harman (1967), for instance, rejects Quine's concession on behalf of analyticity. Even later supporters of Quine's view of analyticity such as William Lycan (1991) continue to try to make a case against stipulative definition as a source of analyticity. It will be taken for granted here, along with Grice and Strawson, Harman, and Lycan, that the opponent of analyticity should disavow stipulative varieties. Furthermore, their main arguments all work against the case of stipulation if they work at all, so that good responses on behalf of stipulation will undercut the Quinean threat as a whole. For these reasons the discussion below will focus on the crucial issue of stipulative definitions when suggesting possible responses to Quinean arguments.

Harman's Synthesis of the Case Against Analyticity

Gilbert Harman neatly summarized the state of the dialectic as of the mid-1960s (Harman 1967), and similar arguments have continued to define terms of the debate into the present. Harman presents a broadly Quinean case against analyticity, incorporating Quinean responses to a number of "first wave" criticisms. Harman's arguments can be summarized as follows:

1. *Circularity.* Analyticity can be defined or explained only in terms of a family of concepts interdefinable with it (1967, 128f).
2. *Empty extension.* There are in fact no analytic sentences, just as there are no witches (125).
   a. Any (assent disposition toward a) sentence is revisable, and since analytic sentences would have to be unrevisable, there are no such sentences.
3. *Nonexplanatoriness.* The analytic/synthetic distinction does not explain anything; it is part of a "bad empirical theory" (136ff).
4. Analyticity does not account for a priori knowledge in the way logical empiricists claimed (130f):
   a. "Truth in virtue of meaning" makes no sense.
   b. Knowledge of truth by virtue of knowledge of meanings is impossible.
   c. Stipulation does not guarantee truth, in part because there is no way to draw a distinction between "substantive postulates" and stipulative definitions (Harman differs on this point from Quine [1953a]).

Depending upon the views of a given defender of analyticity, these objections might be faced in a variety of ways. Rather than present a panoply of possible responses and counterobjections, one strand of the dialectic will be emphasized, exhibiting one possible avenue of defense against these objections. Keeping in mind that this is only one possible view concerning analyticity, one can distinguish between statements that no empirical evidence is allowed to count for or against and those that are empirically defeasible. Call the former statements analytic and the latter synthetic. Examples of analytic statements might be that bachelors are unmarried men, that trilaterals have three sides, and that for any A and B, if A murdered B, then B died. Stipulative definitions of newly introduced terms, for purposes of abbreviation, are in the present view paradigmatic cases of analytic statements. A necessary condition for being a *statement*,

ANALYTICITY

in the present use, is that a sentence used on an occasion be in accord with certain norms of use. The notion of a norm may well appear suspect to a Quinean (cf. Quine 1953b, 32f), and this raises a general question as to whether and to what extent a defender of the notion of analyticity is required to use only notions that Quine himself regards as legitimate. This important issue will be discussed below.

Many authors have responded to the first argument presented by Harman above, that analyticity has no noncircular characterization. A particularly clear response is presented by Hans-Johann Glock (1992), who examines Quine's criteria of adequacy for a satisfactory definition of the analytic. Quine's objections to various proposed definitions invariably involved his claiming that concepts used in the definienda are themselves definable in terms of analyticity. But here Glock points out that it appears that Quine has unfairly stacked the rules against his opponents:

> [Quine] rejects Carnap's definition—via "meaning postulates"—of metalinguistic predicate "analytic in L_0" for formal languages on the grounds that this only explains "analytic in L_0" but not "analytic." ... And this objection must surely amount to the requirement that an explanation of "analytic" give its *intension,* and not merely its *extension.* ... It follows that Quine's challenge itself depends essentially upon the use of an intensional concept ... whilst he forbids any response which grants itself the same license. Yet, it is simply inconsistent to demand that the meaning of "analytic" be explained, but to reject putative definitions solely on the grounds that they depend on the use of intensional concepts.

There is another inconsistency between Quine's insistence that the notion of analyticity be reduced to extensional ones and his requests for an explanation of the meaning of analyticity. For the latter demand cannot be fulfilled without using those concepts to which, as Quine himself has shown, the explanandum is *synonymous* or *conceptually* related, that is, the notions he prohibits. Consequently Quine's circularity charge comes down to the absurd complaint that the analytic can be explained only via synonyms or notions to which it is conceptually related and not via notions to which it is conceptually unrelated. (Glock 1992, 159). In other words, it appears that Quine is requiring an explanation of the meaning of the term *analytic* (as opposed to simply a specification of its extension), while simultaneously denying the intelligibility of the distinction. Furthermore, if the circularity worry is that analyticity cannot be defined except in terms that are interdefinable with it, it is unclear that analyticity fares worse in this respect than any concept whatsoever. The Quinean may retort that there are more or less illuminating explanations of meaning, and those given by defenders involve such small circles that they are not helpful at all. Even granting this, however, the concern remains that Quine's demands are inconsistent with his own expressed views.

Harman's second argument states that just as scientists have discovered that there are no witches and no phlogiston, Quine and others have discovered that there are no analytic sentences. A possible reply to this objection would be to suggest that the cases of witches and phlogiston are crucially disanalogous to the case of analyticity. Witches, for instance, are alleged to possess special causal powers. It is plausible to think that scientists can and have discovered that there are no beings with such causal powers. Furthermore, whether someone is a witch is logically or conceptually independent of whether anyone believes that that person is a witch, or treats that person as a witch. By contrast, analytic sentences are arguably more like chess bishops. In such a view, what makes something a chess bishop is that it is treated as one. Similar accounts might be given for presidents or tribal chieftains, laws, and a host of other familiar items.

Suppose, for the purposes of analogy, that some gamers invent a game at one of the weekly meetings of the Game-Definers' Club and call it Chess. They stipulate that bishops move only diagonally and begin at a square on which there appears a small sculpture resembling a Roman Catholic bishop. Suppose now that someone who has observed the proceedings and agrees to the stipulation says that it has been discovered that chess bishops move horizontally, or alternatively that there are no chess bishops. What could the gamers say to such a bizarre bishop skeptic? It seems as though skepticism is beside the point in such cases, or is even unintelligible. Similarly, one might respond to Quine's and Harman's second objection by noting that their talk of "discovering that there are no analytic sentences" makes no sense, in a way akin to that in which talk of discovering that there are no chess bishops, or that chess bishops move horizontally rather than diagonally, makes no sense. Someone who denies a stipulated rule of a game simply fails to understand the proceedings.

The second argument presented by Harman above goes on to state that there are in fact no analytic sentences, since scientists do (or would, in some conceivable circumstance) count some evidence against any sentence. There are a number of issues raised by this objection, to which several replies are possible. First, even if one decided that, as a matter of empirical fact, no one has ever used a sentence in an "evidentially isolated" way, this would be irrelevant to the question whether someone *could* do so, or whether scientists or philosophers can do so now, for instance in an attempt to "rationally reconstruct" or explicate scientific practice. Second, no advocate of analyticity ever denied that any given *sentence* could be used in a variety of ways in different contexts, for instance as evidentially isolated in one context or to express a synthetic claim in another. Here the analogy with game pieces can be extended. The fact that a miniature piece of sculpture could be used as a chess piece, and later as a paperweight, need not undercut the claim that it is a chess bishop as employed on a particular occasion. Finally, the defender of analyticity need not deny that it may be unclear in particular cases whether one should count a sentence as it was used on a particular occasion as analytic or as synthetic. One might grant that the language "game" often proceeds without completely determinate rules governing the uses of sentences while at the same time maintaining that some applications of expressions are evidence independent in the sense required for analyticity.

Harman's third major claim is that the notion of analyticity is part of a bad empirical theory. In reply, one might propose that analyticity need not be conceived as part of an empirical theory at all, any more than chess bishophood is. Carnap and some other logical empiricists treated metalinguistic notions as ultimately part of a (behavioristic) psychology, thereby ultimately to be evaluated in terms of their causal-explanatory virtues. The advocate of such a position would presumably have to address Harman's objection by pointing out ways in which standard behavior with respect to analytic sentences could be distinguished from standard behavior with respect to nonanalytic ones (cf. Mates 1951). However, a defender of analyticity might suggest instead that analyticity is a concept that is not typically employed within a predictive theory. For instance, when one says that a certain piece on a board is a bishop, one is typically not making a predictive claim about how it will move (although on some occasions one might be using the sentence in this way), but rather specifying what rules govern the movements of a certain object within a certain game. Likewise, when someone is described as a local chieftain, it is often simply to note how the chieftain is to be treated, what counts as appropriate responses to the chieftain's commands, and so on. So too, when one stipulatively defines terms, one may assign evidentially isolated roles to the definitions in the practice of description. This is not to deny that one can make predictions about how people within some group will move a bishop, treat a chieftain, or hold on to a definition sentence in the face of evidence. But stipulations are not empirical hypotheses about future behavior. They are prescriptions about what constitutes appropriate uses of expressions in a language. Moving a bishop in chess along the rank and file, for instance, does not falsify an empirical prediction but rather indicates a failure to understand or play by the rules. That such a prescriptive, evidentially isolated function exists for some statements is evident from an examination of common activities of playing games, exhibiting deferential attitudes toward leaders, and other practices. If Harman's third argument is to go through, then it appears that he must show how it is that all such apparently prescriptive statements are actually elements of predictive empirical theories.

His fourth objection includes the caveat that truth by virtue of meaning is an incoherent notion. Harman notes that it is difficult to see why the statements "Copper is a metal" and even "Copper is copper" are not made true by the fact that copper is a metal, and that copper is copper (or a more general fact concerning self-identities), respectively. Harman is correct to note that logical empiricists often thought that analytic sentences did not "answer to facts" for their truth in the way that nonanalytic sentences do. But a defender of analyticity need not make such claims. One might grant that the statement "Copper is copper" is made true by the fact that copper *is* copper. Furthermore, one can agree that copper would be copper whether or not anyone stipulated anything, although the fact that the sentence "Copper is copper" expresses what it does depends on human convention.

Perhaps of most importance in Harman's fourth argument is the claim that knowledge of truth by virtue of knowledge of meanings is impossible. The reason for the importance of this objection is that empiricists and others employed the notion of analyticity chiefly in order to solve an epistemological problem concerning the apparent a priori nature of our knowledge of logic and mathematics. If analyticity does not solve the basic problem for which it was invoked, then any further appeal to the notion threatens to be pointless, even if the notion proves to be coherent. Harman's objection can be put as follows. An empiricist might claim that what is distinctive about an analytic truth is that one cannot understand the claim without "seeing" that it is true. But "seeing" is a success term that begs the question. Even if it turns out that people by and large cannot fail to believe a sentence S as soon as they understand S, this could be due to a mere psychological compulsion, and so would have no epistemic (justificatory) relevance.

A possible reply on behalf of a defender of analyticity may be presented by returning to the meeting of the Game-Definers' Club. At this meeting, someone specifies a game for the first time and calls it Chess. Part of the specification includes the rule that there are to be bishops on the board at the beginning of any game-play sequence, that four miniature sculptures he points to are to be used as the bishops in the game to be played on that night, and that bishops move only diagonally if at all. Everyone agrees—well, almost everyone. After hearing the specification of the new game, the most recent Quinean infiltrator of the club, Skep, complains that no one in the room knows that bishops move diagonally, or that the particular piece pointed to is a chess bishop. Just saying that something is true does not make it so, Skep asserts, even if all of the club members assent to it. Further, Skep "shows us how to falsify" the claim that bishops move only diagonally by moving one of the pieces that club members agreed to treat as bishops along a file rather than diagonally. Club members might refrain (at first) from violent methods for the elimination of Quinean skepticism at such meetings and try to explain that the troublesome club member fails to understand the practice of stipulative definition. As for the concern that saying that something is so does not make it so, the appropriate response might be, "Yes it does, in the case of stipulative definition of novel terms." A Quinean may simply insist that stipulation does not guarantee truth and that no one has shown that it does. In response, one might grant that it has not been demonstrated that coherent stipulative definitions are true, any more than it has been proved that in chess, bishops move only diagonally, or that it is illegal to drive more than 70 miles per hour in certain areas. But it looks as though the Quinean must either deny that *any feature at all* can be guaranteed by conventional stipulation, including features such as legality, or explain why truth is relevantly different from these other features. The former denial seems hopelessly counterintuitive, whereas the latter seems of doubtful coherence, since if, e.g., the legality of X can be stipulated, the truth that X is legal seems guaranteed. Nevertheless, interesting issues may be raised here concerning what features of sentences can be said to be up to us, for even if one grants that one can make the case that no empirical evidence counts against a sentence S (and thereby nothing counts against the truth of the proposition expressed by S), this does not directly yield the truth of S.

What sorts of features of sentences can be determined by "arbitrary" decision? If one can decide what counts as evidence for or against a sentence, why not also grant that one can decide that a sentence expresses a truth? But can one even decide what counts for or against an arbitrary sentence? A completely satisfying account of analyticity will require plausible answers to these and related questions.

Does one *know* that stipulative definitions are true? Suppose, for simplicity, that one knows that the stipulative definition is not logically inconsistent with sets of other purportedly analytic statements and that the collection of stipulative definitions does not entail any paradigmatically empirical claim. There are several questions that could be considered, but let us focus on two: whether it is known that the sentence S is true and whether it is known that p is true, where p is the proposition expressed by S.

As regards S, there can be intelligible and even appropriate skepticism concerning, say, whether people within a certain group have all agreed to treat S as evidentially isolated, and thus skepticism as to whether S expresses an analytic proposition within the language of that group. Harman, Lycan, and other Quineans, however, apparently believe that even in a setting in which everyone agrees that S is evidentially isolated and in which it is common knowledge that everyone agrees, S still might, so far as anyone knows, be false. In the view suggested here on behalf of the advocate of analyticity, there is no intelligible room for skepticism concerning the truth of S for those present, any more than there is room for skepticism about how bishops move in the game of Chess at the Game-Definers' Club meeting.

When Quineans are asked for an account in which such sentences are falsified by evidence, they often provide an account that, to their opponents, looks like a story in which a different game is eventually called Chess, or in which a different concept is meant by the term *bachelor*. The Quineans then complain that the criteria employed to individuate games, concepts, or languages beg the question. Criteria of individuation, they claim, are themselves further claims open to empirical falsification, whereas the advocate of analyticity appeals to constitutive rules in the case of Chess and meaning postulates in the case of concepts or languages. So the advocate of analyticity who adopts the strategy suggested should present an account of individuation that coheres with his or her overall position. A natural proposal here would be to say that criteria of individuation are themselves analytic and stipulated (at least in the case of some terms, including new terms stipulatively defined, artificially extended languages, and some games). To be sure, the Quinean will object that this response assumes the notion being defended. But defenders of analyticity might reply that they need only defend the coherence of the notion of analyticity, and that it is the Quinean who has not supplied an independent argument against analyticity that a defender should feel any compulsion to accept. A related issue, a disagreement over whether the introduction of a new concept (a term as it is employed) requires justification, whether epistemic or pragmatic, is discussed a bit further in the final section.

As for knowledge that p is true, one might think that talk of having such knowledge makes sense in contrast to "mere" true belief, or perhaps mere justified true belief. But in the case of stipulative definitions, there is arguably no such contrast to be drawn. In such a conception of what is central to knowledge, one should either deny that one knows that p is true, where p is something stipulated, or warn that such claims are misleading. On the other hand, if one thinks that what is fundamental to having knowledge that p is true is the impossibility of being wrong about p, then if stipulation guarantees truth (of the sentence S stipulated, by guaranteeing that S expresses a true proposition), one might want to say that one knows that p is true, where p is the proposition expressed by S and where it is common knowledge that this is how S is being employed.

A number of the most important objections to the concept of analyticity (or the claim that some sentences express analytic truths) have been discussed thus far, along with one possible line of defense against the Quineans. Many other lines of objection and reply can be given (see Mates 1951; Martin 1952; Grice and Strawson 1956; Winnie 1975; and Boghossian 1996). The following discussion turns to some related questions.

Quine's Web of Belief

Quine grants that one can draw a distinction between sentences based upon their proximity to the center of a so-called web of belief. According to Quine's metaphor, altering one's assent behavior in regard to some sentences will "force" more drastic revisions in the rest of the web, whereas revising assent behaviors toward other sentences "near the periphery" will not have much effect on the rest of the web (see, e.g., Quine 1953a, 42–43). Thus for Quine, evidential isolation is a matter of degree, whereas according to the picture presented by the advocate of analyticity, it seems to be an all-or-nothing affair. Quineans think that the web metaphor captures everything sensible in the notion of analyticity while avoiding its pitfalls.

This picture has met with some significant challenges. In order to see why, one must look closely at the web metaphor that Quine employs. One problem noted by Laurence BonJour and others (BonJour 1998, 63ff; see also Wright 1980) is that Quine's talk of "connections" between elements of the web is misleading, in that in Quine's own view these connections are not normative or conceptual, but are instead causal or empirical. Further, for Quine, logical principles are themselves simply elements of the web. But if this is the case, it is unclear why any revision of the web is at any time required. As BonJour puts it:

> Thus the basis for any supposed incompatibility within any set of sentences (such as that which supposedly creates a need for revision in the face of experience) can apparently only be some further sentence in the system. . . . The upshot is that even the revision of one's epistemic or logical principles . . . turns out not to be necessary, since at some level there will inevitably fail to be a further sentence saying that the total set of sentences that includes those principles and that seems intuitively inconsistent really is inconsistent. This means that *any* non-observational sentence or set of sentences can always be retained. (BonJour 1998, 94)

It is not clear that the problem that BonJour raises can be restricted to nonobservational sentences. BonJour's argument is analogous to the famous "Achilles and the tortoise" argument of Lewis Carroll (1895). The tortoise asks Achilles why B should be inferred from A and $A \bigcup B$, and Achilles simply adds further sentences to the premise set at each stage, thereby failing to make any progress toward his goal. Similarly, BonJour argues that showing why a revision is required at some stage of inquiry cannot be accomplished by merely adding further sentences to the set supposedly requiring revision.

Another problem concerns the fact that Quine's metaphor of a web of belief allows for a loosely defined notion of greater or lesser degrees of the revisability of assent dispositions toward sentences. There are at least two notions of proximity to the center of the web to consider here. One is in terms of the pragmatic difficulties involved in giving up assent to a sentence S. The other is in terms of the "responsiveness" (or probabilities of change), to nerve firings, of assent dispositions toward S. One is a pragmatic notion and the other a ("merely") causal notion. Quine's view is that the former feature explains the latter. However, consider the classic philosophical example, "Bachelors are unmarried." Both Quine and Carnap will agree that this has a high degree of evidential isolation. As even supporters of Quine have acknowledged, it is false to say that giving up this sentence would require drastic revisions in our web of belief. Thus, the pragmatic difficulty of giving up such a sentence seems irrelevant to its low probability of being revised.

Putnam (1975), for example, thinks that Quine is dead wrong when he takes isolation to be a result of proximity to the center of the web. Putnam thinks that the only unrevisable sentences are near the periphery, at least in the pragmatic sense of not being connected with much else in the web. Putnam suggests that one call analytic those terms that have only a single criterion for their application, since there cannot arise any empirical pressure to revise the definition of a term if it is a "one-criterion" term, whereas if a term has two distinct criteria of application, then the possibility of an incompatibility between these criteria, and hence a breakdown in the application of the term, can arise. Putnam's notion of a one-criterion term as an account of analyticity, however, has been objected to on the basis that whether a term is one-criterion or not presupposes an analytic/synthetic distinction. For instance, should one count "Bachelors are unmarried men" and "Bachelors are unwed men" as two criteria or as two synonymous expressions of the same criterion? Clearly the latter seems more natural, and yet it presupposes a form of synonymy from which analyticity is definable, and thus does not provide a conceptually independent characterization of analyticity.

A complaint made against a number of responses to Quine's and Harman's arguments is that some, if not all, of them presuppose the distinction between analytic and synthetic propositions and for that reason are inefficacious. It is indeed difficult to respond to someone who denies the notions of meaning and synonymy without appealing to such notions, at least implicitly. However, in response to this worry the defender of analyticity can reasonably try to shift the onus of proof. One might suggest that the Quinean fails to correctly assess the status of an attack on a distinction. One sort of objection to a concept is that it is useless. Another is that the concept is incoherent or self-contradictory. From the point of view of a nonpragmatist, the Quinean claim that distinctions that are not explanatory or are otherwise useless are therefore nonexistent is puzzling. One may stipulate that "redchelors" are red-haired bachelors, for instance. Is there no such thing as a redchelor, or no distinction between redchelors and others, just because no one should have any interest in who is a redchelor, or in how many redchelors there are? It is one thing to claim that the distinction is useless for any explanatory purposes. It is another to say that the distinction is ill-defined, or alternatively that it is unreal. So far, it might be argued, the concept of analyticity has not been shown to be incoherent. Whether it is useless is also doubtful, especially in the clear case of stipulative definitions of new terms for the purposes of abbreviation. In response, the Quinean might grant much of this and yet propose that Quine's criticism has at the very least cast into doubt the pretensions of the logical empiricists, who thought that they could capture a wide variety of statements, such as those of logic and mathematics, with a single criterion.

It remains unclear whether adopting a Quinean practice (which disallows setting aside sentences as evidentially isolated in the absence of pragmatic justification) is somehow pragmatically deleterious compared with the practice of counting some sentences as isolated. To the extent that all that one is interested in is predicting future nerve-ending states from past ones, it may be that a Quinean practice works at least as well as a practice that allows for stipulative definitions to be permanently isolated from empirical refutation. On the other hand, if one can show that the practice of evidential isolation is even coherent, then this leaves open the possibility of so-called analytic theories of mathematical and other apparently a priori knowledge. This in itself might count as a pragmatic (albeit "theoretical") advantage of adopting a practice that allows for analyticity. Whether such a program can be carried through in detail, however, remains to be seen.

CORY JUHL
ERIC LOOMIS

The authors acknowledge the helpful input of Samet Bagce, Hans-Johan Glock, Sahotra Sarkar, and David Sosa.

References

Ayer, Alfred J. (1946), *Language, Truth and Logic.* New York: Dover Publications.
Boghossian, Paul (1996), "Analyticity Reconsidered," *Noûs* 30: 360–391.

Bolzano, Bernhard (1973), *Theory of Science.* Translated by B. Terrell. Dordrecht, Netherlands: D. Reidel.

BonJour, Laurence (1998), *In Defense of Pure Reason.* Cambridge: Cambridge University Press.

Carroll, Lewis (1895), "What the Tortoise Said to Achilles," *Mind* 4: 278–280.

Carnap, Rudolf (1937), *The Logical Syntax of Language.* Translated by Amethe Smeaton. London: Routledge and Kegan Paul.

——— (1956), *Meaning and Necessity: A Study in Semantics and Modal Logic.* Chicago: University of Chicago Press.

——— (1990), "Quine on Analyticity," in R. Creath (ed.), *Dear Carnap, Dear Van: The Quine–Carnap Correspondence and Related Work.* Berkeley and Los Angeles: University of California Press, 427–432.

De Jong, Willem R. (1995), "Kant's Analytic Judgments and the Traditional Theory of Concepts," *Journal of History of Philosophy* 33: 613–641.

Frege, Gottlob (1974), *The Foundations of Arithmetic: A Logico-Mathematical Inquiry into the Concept of Number.* Translated by J. L. Austin. Oxford: Blackwell.

Friedman, Michael (1987), "Carnap's *Aufbau* Reconsidered," *Noûs* 21: 521–545.

Glock, Hans-Johann (1992), "Wittgenstein vs. Quine on Logical Necessity," in Souren Tehgrarian and Anthony Serafini (eds.), *Wittgenstein and Contemporary Philosophy.* Wakefield, NH: Longwood Academic, 154–186.

Grice, H. P., and P. F. Strawson (1956), "In Defense of a Dogma," *Philosophical Review* LXV: 141–158.

Harman, Gilbert (1967), "Quine on Meaning and Existence I," *Review of Metaphysics* 21: 124–151.

——— (1996), "Analyticity Regained?" *Nous* 30: 392–400.

Hintikka, Jaakko (1974), *Knowledge and the Known: Historical Perspectives in Epistemology.* Dordrecht, Netherlands: D. Reidel.

Kant, Immanuel (1965), *Critique of Pure Reason.* Translated by N. K. Smith. New York: St. Martin's Press.

——— (1974), *Logic.* Translated by R. S. Harriman and W. Schwartz. Indianapolis: Bobbs-Merrill.

Leibniz, Gottfried W. F. (1981), *New Essays Concerning Human Understanding.* Translated by P. Remnant and J. Bennett. New York: Cambridge University Press.

Lycan, William (1991), "Definition in a Quinean World," in J. Fetzer and D. Shatz (eds.), *Definitions and Definability: Philosophical Perspectives.* Dordrecht, Netherlands: Kluwer Academic Publishers, 111–131.

Martin, R. M. (1952), "On 'Analytic,'" *Philosophical Studies* 3: 65–73.

Mates, Benson (1951), "Analytic Sentences," *Philosophical Review* 60: 525–534.

Proust, Joelle (1986), *Questions of Form: Logic and the Analytic Proposition from Kant to Carnap.* Translated by A. A. Brenner. Minneapolis: University of Minnesota Press.

——— (1975), "The Analytic and the Synthetic," in *Philosophical Papers II: Mind, Language, and Reality.* Cambridge: Cambridge University Press, 33–69.

Quine, Willard Van (1953a), "Two Dogmas of Empiricism," in *From a Logical Point of View.* Cambridge: Harvard University Press, 21–46.

——— (1953b), "Reference and Modality," in *From a Logical Point of View.* Cambridge: Harvard University Press, 139–159.

Winnie, John (1975), "Theoretical Analyticity," in Jaakko Hintikka (ed.), *Rudolph Carnap, Logical Empiricist.* Dordrecht, Netherlands: D. Reidel, 149–159.

Wright, Crispin (1980), *Wittgenstein on the Foundations of Mathematics.* London: Duckworth.

See also **Carnap, Rudolf; Conventionalism; Explication; Logical Empiricism; Quine, Willard Van; Verificationism; Vienna Circle**

ANTHROPIC PRINCIPLE

An anthropic principle is a statement that there is a relation between the existence and nature of human life and the character of the physical universe. In the early 1970s, cosmologist Brandon Carter gave this name to a new family of principles, some of which he was also the first to formulate (Carter 1974; McMullin 1993). Cosmologists discussed these principles in response to the fine-tuning for life discovered in describing the evolution of the universe, and later also in the parameters of the Standard Model of particle physics (Collins and Hawking 1973; Barrow and Tipler 1986; Roush 2003). A universe is fine-tuned for X whenever (1) its parameters have values allowing for X in that universe, (2) that universe would not have (or be) X if the values of those parameters were slightly different, and (3) X is a significant, gross, or qualitative feature. (The values of the parameters in most cases currently have no explanation.) To say that a universe is fine-tuned in this way implies nothing about how the universe got the way it is—that an intelligent being tuned it would be an additional claim. That a universe is fine-tuned means only that key properties depend sensitively on the values of parameters.

The myriad examples of fine-tuning are often expressed in the form of counterfactual statements, for example:

1. If the ratio of the mass of the electron and the mass of the neutron were slightly different, then there would be no stable nuclei.
2. If the ratio of the strong to the electrical force were slightly different, then there would be few stable nuclei.
3. If any of the parameters affecting gravity were different, then the universe would have either lasted too short a time for galaxies to evolve or expanded too fast to allow clumping of matter into objects.

Since if there were no stable nuclei or material objects there would not be human life, such statements as these imply that if the universe were slightly otherwise than it is physically in any of a number of ways, then there would be no human beings to contemplate the fact.

Some, even if they would not put it in these terms, see this relationship as a spooky indication that human life was "meant to be," that the universe came to have the parameter values it has in order to make possible the existence of human beings, that the existence of human beings was a necessity antecedent to the determination of the physical features of the universe. These are the ideas behind the strong anthropic principle (SAP), which says that the universe must be such as to contain life (or human beings) in it at some point in its history. One must wonder, though, whether one should draw the same conclusion from the fact that human life depends on the existence of green plants, and the evolution of green plants depended on numerous contingencies. This analogy highlights the fact that the SAP class of responses to fine-tuning has involved arguments analogous to those of eighteenth-century models for the existence of God. In those arguments, the observation in the natural and especially biological world of an improbably well functioning phenomenon was supposed to make the existence of a designer-God more probable.

It is true that the physical universe is improbable on the assumption that its parameters were determined by direct chance, since fine-tuning implies that only a tiny proportion of possible universes possess the basic features in question; chance was unlikely to find them. (It is a matter of philosophical debate whether that makes the existence of God, or in some views multiple universes, more probable. See e.g., Earman 1987; Hacking 1987; White 2000.) However, direct chance is not the only possible hypothesis for explaining the values of parameters. Just as Darwin's concept of natural selection showed how it was possible for well-adapted organisms to arise without antecedent design, the next better physical theory may explain the values of the parameters in a way that makes no reference to a goal of providing for the existence of life (see Natural Selection). In the history of physics, there is precedent for a newer, deeper theory explaining previously unexplained values of parameters, as when, for example, the ideal gas constant, determined merely from observation in classical thermodynamics, got derived and explained in statistical thermodynamics. Indeed, physicists intend the next better theory to explain the values of the currently adjustable parameters of the Standard Model, and most see fine-tuning as an indication that a better physical theory than the Standard Model is needed. This is in keeping with a tradition in physics for eschewing frankly teleological explanation.

While SAP responses to fine-tuning try to use the fact that some basic physical characteristics of the universe are improbably fine-tuned to support a positive (and rather grand) thesis about the existence of a designer or of many universes, the weak anthropic principle (WAP) draws from the fact that for human-type observers, certain features of the physical universe are necessary in order for there to be a negative, or cautionary, lesson about what is presently known. The WAP says that what is observed may be restricted by the conditions necessary for the presence of humans as observers: For example, since humans are beings who could not have evolved in any place that lacked galaxies, there is reason to think that casual observation of the neighborhood humans find themselves in is not going to yield a representative sample of the cosmos with respect to the existence of galaxies. If there were regions of the universe that lacked galaxies, then humans probably would not have found them with this method of observation. Thus, instances of galaxies observed in this way would support the generalization that there are galaxies everywhere in the universe to a lesser degree than one would have expected if the instances had been a fair sample.

This type of argument, which found its first use by R. H. Dicke (1961) against the evidence P. A. M. Dirac (1937, 1938, 1961) had presented for a speculative cosmological hypothesis, is obviously an inference about bias and is generally endorsed by physicists and philosophers alike. There are disputes, however, about how best to model inferences about bias in evidence that is introduced by the method or process of obtaining the evidence. Some Bayesians think that the process through which evidence is produced is irrelevant to the degree to which the evidence supports a hypothesis and so, presumably, must reject the WAP style of argument (see Bayesianism). An error statistician will entirely approve of inferences about bias that exploit counterfactual dependencies that reveal the presence of error in the procedures used. Some Bayesians think that the bias introduced by the process of gathering evidence should be acknowledged and that this is best done by including an account of the process in the total evidence, a strategy whose rationale would be the principle of total evidence. Other Bayesian strategies for modeling WAP argumentation as legitimate may founder on the fact that the only analogy between WAP cases and ordinary cases of bias produced by evidence-gathering procedures lies in counterfactual statements that Bayesians resist incorporating into reasoning about evidence, such as: If there were regions without galaxies, then this method of observing would not have discovered them. These counterfactuals are not supported by the causal process of gathering the evidence but by background assumptions concerning the general possibility of that process.

Carter (1974) and others have claimed that anthropic principles represent a reaction against overwrought submission to the lesson learned from Copernicus when he proposed that the Earth was not at the center of the cosmos, or in other words that human beings were not special in the universe. While this diagnosis is clearly correct for the SAP, which does contemplate the specialness of human beings, it is misleading about the WAP. Indeed, reasoning that is associated with the WAP could be said to follow Copernicus, since it is analogous to the inference Copernicus made when he considered that the heavenly appearances might be the same regardless of whether the Earth were stationary and the stars moving or the Earth moving and the stars at rest. He inferred that evidence—generated by standing on the Earth—was inconclusive and therefore that the hypothesis that the Earth moves should not be scorned (Roush 2003).

SAP and WAP are misnamed with respect to each other, since WAP is not a weak form of SAP. WAP is not "weakly" anti-Copernican where SAP is "strongly" anti-Copernican, because WAP is not anti-Copernican at all. Also, WAP does not provide a way of weakly using the facts of fine-tuning to infer a speculative hypothesis about the universe, nor even does it use the facts of fine-tuning to infer a weaker speculative hypothesis; it provides a way of blocking such inferences. For these reasons, it would be better to refer to the SAP as the *metaphysical anthropic principle* (MAP), and to the WAP as the *epistemic anthropic principle* (EAP).

SHERRILYN ROUSH

References

Barrow, J. D., and F. J. Tipler (1986), *The Anthropic Cosmological Principle.* Oxford: Oxford University Press.
Carter, B. (1974), "Large Number Coincidences and the Anthropic Principle in Cosmology," in M. S. Longair (ed.), *Confrontation of Cosmological Theories with Observational Data.* Boston: D. Reidel Co., 291–298.
Collins, C. B., and S. W. Hawking (1973), "Why Is the Universe Isotropic?" *Astrophysical Journal* 180: 317–334.
Dicke, R. H. (1961), "Dirac's Cosmology and Mach's Principle," *Nature* 192: 440–441.
Dirac, P. A. M. (1937), "The Cosmological Constants," *Nature* 139: 323.
——— (1938), "A New Basis for Cosmology," *Proceedings of the Royal Society of London,* Series A 165: 199–208.
——— (1961), Response to "Dirac's Cosmology and Mach's Principle," *Nature* 192: 441.
Earman, John (1987), "The SAP Also Rises: A Critical Examination of the Anthropic Principle," *American Philosophical Quarterly* 24: 307–317.
Hacking, Ian (1987), "The Inverse Gambler's Fallacy: The Argument from Design: The Anthropic Principle Applied to Wheeler Universes," *Mind* 96: 331–340.
McMullin, Ernan (1993), "Indifference Principle and Anthropic Principle in Cosmology," *Studies in the History and Philosophy of Science* 24: 359–389.
Roush, Sherrilyn (2003), "Copernicus, Kant, and the Anthropic Cosmological Principles," *Studies in the History and Philosophy of Modern Physics* 34: 5–36.
White, Roger (2000), "Fine-Tuning and Multiple Universes," *Nous* 34: 260–276.

See also **Particle Physics**

ANTI-REALISM

See **Instrumentalism; Realism; Social Constructionism**

APPROXIMATION

An approximation refers to either a *state* of being a nearly correct representation or the *process* of producing a nearly correct value. In the former sense a planet's orbit approximates an ellipse, while in the latter sense one approximates by ignoring terms or factors in order to facilitate computation. Examples of approximations in both senses are plentiful in the mathematically oriented physical, biological, and engineering sciences, including:

- Attempts to solve the general relativistic field equations without the Schwarzchild assumption of a nonrotating, spherically symmetric body of matter;
- The generalized three-body problem;
- Many problems in fluid mechanics;
- Attempts to solve the Schrödinger wave equation for atoms and molecules other than hydrogen;
- Multilocus/multiallele problems in population genetics and evolutionary biology; and
- Problems associated with the description of ecosystems and their processes.

Scientists often employ approximations to bypass the computational intractability or analytical complexity involved in the description of a problem. This can occur in two ways, which are often combined in multistep problem solutions. For example, in solving a three-body problem, one can write down the complete, 18-variable equation first and make approximations as the solution is generated. Here, an approximate solution to an exact equation is generated. Alternatively, one can assume that two of the three bodies dominate the interaction, treating the third body as a small perturbation on the other two bodies' motion. Here, the process of making the approximation is completed before the solution is attempted in order to generate an exact solution to an approximate equation. In this second sense, approximations resemble the processes of idealization and abstraction quite strongly. It is probably pointless to insist on a strict division among the three strategies, since many approximations are justified not by purely mathematical arguments but by arguments that some variables are unimportant for the particular class of problems being investigated. However, roughly speaking, one can say that approximations (of both types) are invoked upon contemplation of the complete description of the problem, whereas idealizations and abstractions are deliberate attempts to study only part of the complete problem.

Until relatively recently, philosophers have tended to view approximations as uninteresting or ephemeral categories of scientific activity. John Stuart Mill (1874, 416) claimed that "approximate generalizations" (Most *A*s are *B*s) were of little use in science "except as a stage on the road to something better." Further, philosophers have focused almost entirely on the *state* sense defined above. For Mill as well as the twentieth-century logical empiricists, this was a result of their focus on logic to the exclusion of the mathematical difficulties of solving equations. Discussions of "nearly correct" results are missing in Mill's analysis and also in those of many of the early logical positivists. Eventually, when it was pointed out that, for example, Kepler's laws were not deducible from Newtonian mechanics and gravitational theory (since Kepler's laws assert that the orbit of a planet is elliptical, but according to Newtonian theory a planet's orbit around the Sun will not be exactly elliptical due to the influence of the other planets), Hempel (1966) responded that there is a law entailed by Newtonian theory that Kepler's laws approximate. Similar remarks were made regarding the explanation of individual events (as opposed to laws). Schaffner (1967) provided a number of examples of approximate explanation in the context of the reduction of one theory or law to another. Subsequently, structuralist philosophers of science have developed a theory that assesses the amount of approximation between any two statements in a given language (Balzer, Ulises-Moulines, and Sneed 1987, Chap. 6). However, while scientifically important and interesting, discussion of the state sense of approximation appears philosophically limited. One can say that a value is or is not "close enough" to count as a correct prediction or explanation, but the question as to how close is close enough must be decided on pragmatic grounds supplied by the science available at the time. More importantly, it is difficult to address whether the process that produced the "nearly correct" value is justifiable.

Recently, more attention has been directed toward the process of making approximations. Analyses have shown that approximations are tied to a number of methodological and interpretive issues. Perhaps most basically, approximations raise the question of whether a theory actually accounts for a given datum, phenomenon, law, or other theory. For example, "the fact remains that the exponential decay law, for which there is so much empirical support in radioactive processes, is not a rigorous consequence of quantum mechanics but the result of somewhat delicate approximations" (Merzbacher, quoted in Cartwright 1983, 113). In the absence of more robust derivations from the underlying theory, one can quite seriously ask what kind of understanding one has of the phenomenon in question.

In addition, approximations sometimes produce intertheoretic relations. Philosophers have traditionally assumed that approximations frustrate intertheoretic reduction because they produce gaps in the derivation of one theory, law, or concept from another. Woody (2000) has analyzed the approximations involved in producing molecular orbital theory as a case of the approximations producing the intertheoretic connections. She notes that the complete, nonapproximate *ab initio* solution of the Schrödinger wave equation for molecules is rarely ever solved for because it leads to a solution in which the semantic information prized by chemists cannot be recovered. (Importantly, even this nonapproximate solution is based on an approximate representation of the molecule in the idealized Hamiltonian.) That is, it leads to a solution in which the information about bonds and other measurable properties is distributed throughout the terms of the solution and is not localized so that it can be recognized and connected with any measurable properties. In order to recover the chemical properties, only one step in the complete set of calculations is performed. Specifically, the spatial portion of the basis sets of the calculation are integrated to give atomic orbitals, the spherical and lobed graphical representations familiar from chemistry texts. So this involves an approximation to the full solution. The upshot is that the quantum mechanical description (or at least some portion of it) is connected with familiar chemical ideas via only an approximation. Without the approximation, the intertheoretic connection just does not appear.

The process of making approximations also sometimes creates terms that are given subsequent physical interpretations and sometimes eliminates terms that are later recovered and given interpretations. Questions then arise whether the entities that are created and eliminated are artifactual or not. The case of the atomic orbitals illustrates the issues surrounding the creation of entities. If one does not make the approximations, there is some justification for saying that such orbitals do not "exist" outside the model. This is because the complete solution produces orbitals that are not localized in the way the approximate solution pictures (cf. Scerri 1991). Other examples include (1) the construction of rigid, three-dimensional molecules via the Born-Oppenheimer approximation that treats nuclear and electronic motions separately; (2) potential energy surfaces in various physical sciences that arise due to the assumption that various kinds of energy functions are isolable from each other; and (3) pictures of fitness landscapes in evolutionary biology that give the impression that fitness values are static across wide ranges of environments.

Cartwright's (1983) discussion of the derivation of the exponential decay law and the discovery of the Lamb shift is a classic example of how approximations can eliminate terms that should be given a physical interpretation in some circumstances.

APPROXIMATION

Typically, one is considering an excited atom in an electromagnetic field. The question is how to describe the process of decay from an excited to an unexcited state. The experimental data speak for an exponential law; the issue is to try to reproduce this law solely on theoretic grounds. (As yet, it cannot be done.) In one derivation of the decay law, scientists integrate certain slowly varying frequencies of the field first and the amplitude of the excited state second. The result is an equation that relates the amplitude of the excited state to the magnitude of the line broadening only. However, if the integrations are reversed (which can be done because the amplitude of the excited state is slowly varying compared with its rapid oscillations with the field), the resulting equation relates the amplitude to the line broadening and an additional term that describes a small displacement in the energy levels. That additional term is the Lamb shift, discovered by Willis Lamb and R. C. Retherford (1947). Thus, in this case the process of approximation affects what is believed to be going on in the world.

These interpretive issues give rise to a number of epistemological and methodological questions, especially with regard to the justifiability and the purpose of the approximation. As indicated by Merzbacher, noted above, approximations produce tenuous connections, so that the claim that a theory accounts for a given law or phenomenon is to some degree questionable. In such circumstances, how should one decide that the approximations are justified? Clearly, one important step toward answering this question is to delineate the features of the approximation that is being made. To this end, one can distinguish whether the approximation (1) is implicit or explicit; (2) is corrigible, incorrigible in practice, or incorrigible in principle; (3) has effects that may be estimable, not estimable in practice, or not estimable in principle; (4) is justified mathematically, with respect to some more foundational theory, a combination of both, or neither; and (5) is context dependent or independent and involves counterfactual assumptions (as when Galileo assumed what would happen to falling bodies in a vacuum). (For further discussion of these issues within the specific context of the reduction of one theory to another, see Sarkar 1998, chap. 3.) Once it is known how the approximation falls out with respect to these distinctions, one can begin to assess whether the approximation is warranted.

In addition to assessing whether an approximation is mathematically justified or justified with respect to some more fundamental theory, scientists typically assess approximations according to whether they track the set of causal distinctions drawn in the theory, whether the approximation works across a wide range of cases, and whether the approximation allows a plausible or coherent interpretation of the available data to be constructed (Laymon 1989, 1985; Ramsey 1990). Whether these considerations are exhaustive and whether any one or more of them is more basic remains unclear.

As a final methodological issue, consider the question of the purpose of the approximation. The examples thus far illustrate the purpose of producing a prediction or explanation that is superior to the one that can be given without the approximation. Yet, some analyses of approximations are post hoc justifications of idealizations already in use. Often, this happens by embedding the approximate analysis within a more complex theory so that the idealization can be interpreted as what results when certain terms are eliminated (Laymon 1991). Given different purposes for approximations, it is reasonable to expect that different kinds of justifications will be acceptable in the two situations. Whether this is so remains unanswered at present.

What has been said in this article probably represents only a small subset of the philosophical issues surrounding the use of approximations in the sciences. Given the ubiquity of approximations in many sciences and their varied modes of justification, much philosophical work remains to be completed.

JEFFRY L. RAMSEY

References

Balzer, W., C. Ulises-Moulines, and J. Sneed (1987), *An Architectonic for Science: The Structuralist Program*. Boston: D. Reidel.
Cartwright, N. (1983), *How the Laws of Physics Lie*. New York: Oxford University Press.
Hempel, C. (1966), *Philosophy of Natural Science*. Englewood Cliffs, NJ: Prentice-Hall.
Lamb, W., and Retherford, R. C. (1947), "Fine Structure of the Hydrogen Atom," *Physical Review* 72: 241.
Mill, J. S. (1874), *System of Logic,* 8th ed. New York: Harper and Brothers.
Laymon, R. (1989), "Applying Idealized Scientific Theories to Engineering," *Synthese* 81: 353–371.
——— (1985), "Idealizations and the Testing of Theories by Experimentation," in P. Achinstein and O. Hannaway (eds.), *Observation, Experiment and Hypothesis in Modern Physical Science*. Cambridge, MA: MIT Press, 147–173.
——— (1991), "Computer Simulations, Idealizations and Approximations," *Philosophy of Science* 2 (Proceedings): 519–534.
Ramsey (1990), "Beyond Numerical and Causal Accuracy: Expanding the Set of Justificational Criteria," *Philosophy of Science* 1 (Proceedings): 485–499.
Sarkar (1998), *Genetics and Reductionism*. New York: Cambridge University Press.

Schaffner, K. (1967), "Approaches to Reduction," *Philosophy of Science* 34: 137–147.

Scerri, E. (1991), "The Electronic Configuration Model, Quantum Mechanics and Reduction," *British Journal for the Philosophy of Science* 42: 309–325.

Woody, A. (2000), "Putting Quantum Mechanics to Work in Chemistry: The Power of Diagrammatic Representation," *Philosophy of Science* 67: S612–S627.

ARTIFICIAL INTELLIGENCE

Artificial intelligence (AI) aims at building machines that exhibit intelligent behaviors, understood as behaviors that normally require intelligence in humans. Examples include using language, solving puzzles, and playing board games. AI is also concerned with reproducing the prerequisites for those activities, such as pattern recognition and motor control. AI has typically been pursued using only computing machines. It has been influenced by the sciences of mind and brain as well as philosophical discussions about them, which it has influenced in turn. AI is a successful discipline, with many applications, ranging from guiding robots to teaching.

Origins

In 1936, Alan Turing offered a rigorous definition of computable functions in terms of a class of computing machines—now called Turing Machines—and argued persuasively that his machines could perform any computation (Turing [1936–7] 1965) (see Turing, Alan). In 1943, Warren McCulloch and Walter Pitts published their computational theory of mind, according to which mental processes are realized by neural computations (McCulloch and Pitts 1943). In the late 1940s, the first stored-program computers were built (von Neumann 1945). If Turing's thesis is correct, stored-program computers can perform any computation (until they run out of memory) and can reproduce mental processes. And whether or not mental processes are computational, it might be possible to program computers to exhibit intelligent behavior (Turing 1950).

Starting in the late 1940s, researchers attempted to put those ideas into practice. Some designed computing machines formed by networks of interactive units, which were intended to reproduce mental processes by mimicking neural mechanisms (Anderson and Rosenfeld 1988). This was the beginning of connectionism. Other researchers wrote computer programs that performed intelligent activities, such as playing checkers and proving theorems (Feigenbaum and Feldman 1963). Finally, some researchers built whole artificial agents, or robots. Sometimes the label "AI" is reserved for the writing of intelligent computer programs, but usually it also applies to connectionism as well as the construction of robots that exhibit intelligent behaviors (see Connectionism).

Artificial Intelligence and Computation

Although there is no room in this article for an overview of the field, this section introduces some fundamental concepts in AI (for further reference, standard AI textbooks include Russell and Norvig 1995 and Nilsson 1998).

Computation deals with strings of symbols. A symbol is a particular (e.g., *a*, *b*, *c*) that falls under a type (e.g., a letter of the English alphabet) and can be concatenated with other symbols to form strings (e.g., *abc*, *cba*, *bccaac*). Symbols are discrete in the sense that they fall under only finitely many types.

A computation is a mechanical process that generates new strings of symbols (outputs) from old strings of symbols (inputs plus strings held in memory) according to a fixed rule that applies to all strings and depends only on the old strings for its application. For example, a computational problem may be, for any string, to put all of the string's symbols in alphabetical order. The initial string of symbols is the input, whereas the alphabetized string of symbols is the desired output. The computation is the process by which any input is manipulated (sometimes together with strings in memory) to solve the computational problem for that input, yielding the desired output as a result. This is digital computation. There is also analog computation, in which real variables are manipulated. Real variables are magnitudes that are assumed to change over real time and take values of an uncountable number of types. In digital computation, accuracy can be increased by using longer strings of symbols, which are designed to be measured reliably. By contrast, accuracy in analog computation can be increased only by measuring real variables more precisely. But measuring real variables is always subject to error. Therefore, analog computation is not as flexible and precise as digital computation, and it has not found many AI applications.

Ordinary computations—like arithmetic additions—are sequences of formal operations on strings of symbols, i.e., operations that depend on only the symbols' types and how the symbols are concatenated. Examples of formal operations include deleting the first symbol in a string, appending a symbol to a string, and making a copy of a string. In order to define a computation that solves a computational problem (e.g., to put any string in alphabetical order), one must specify a sequence of formal operations that are guaranteed to generate the appropriate outcome (the alphabetized string) no matter what the input is. Such a specification is called an *algorithm*. (For an alternative computing scheme, see Connectionism. Connectionist computations manipulate strings of symbols based on which units of a connectionist network are connected to which, as well as their connection strengths.)

Executing algorithms takes both a number of steps and some storage space to hold intermediate results. The time and space—the resources—required to solve a computational problem with a specific algorithm grows with the size of the input and of the strings in memory: The same computation on a bigger input requires more steps and more storage space than on a smaller input. The rate at which needed resources grow is called computational *complexity*. Some computations are relatively simple: The required time and storage space grow linearly with the size of the input and the strings in memory. Many interesting computations are more complex: The required resources may grow as fast as the size of the input and strings in memory raised to some power. Other computations, including many AI computations, are prohibitively complex: Needed resources grow exponentially with the size of the input and strings in memory, so that even very large and fast computing machines may not have enough resources to complete computations on inputs and stored strings of moderate size.

When all the known algorithms for a computational problem have high complexity, or when no algorithm is known, it may be possible to specify sequences of operations that are not guaranteed to solve that computational problem for every input but will use a feasible amount of resources. These procedures, called *heuristics,* search for the desired output but they may or may not find it, or they may find an output that only approximates the desired result. Since most AI computations are very complex, AI makes heavy use of heuristics (Newell and Simon 1976).

Symbols have their name because they can be interpreted. For instance, the symbol '/' may represent the letter *a* or the number 1. A string of symbols may be interpreted in a way that depends on its component symbols and their concatenation. For example, assume that '/' represents 1. Then, under different interpretations, the string '///' may represent the number 3 (in unary notation), 7 (in binary notation), 111 (in decimal notation), etc. Interpreted strings of symbols are often called *representations.*

Much of AI is concerned with finding effective representations for the information—AI researchers call it "knowledge"—that intelligent agents (whether natural or artificial) are presumed to possess. If an artificial system has to behave intelligently, it needs to respond to its environment in an adaptive way. Because of this, most AI systems are guided by internal states that are interpreted as representations of their environment. Finding efficient ways to represent the environment is a difficult problem, but an even harder problem—known as the frame problem—is finding efficient ways to update representations in the face of environmental change. Some authors have argued that the frame problem is part of the more general problem of getting machines to learn from their environment, which many see as something that needs to be done in order to build intelligent machines. Much AI research and philosophical discussion have been devoted to these problems (Ford and Pylyshyn 1996).

Given an interpretation, a string of symbols may represent a formal operation on strings, in which case the string is called an *instruction* (e.g., "write an *a*"). A sequence of instructions, representing a sequence of operations, is called a *program* (e.g., "[1] append an *a* to the string in register 0025; [2] if register 0034 contains *ccccc*, stop computing; [3] append a *c* to the string in register 0034; [4] go back to instruction [1]"). Given a domain of objects, such as numbers, it may be useful to derive some new information about those objects, for instance their square roots. Given inputs representing numbers under some notation, there may be an algorithm or heuristic that generates outputs representing the square roots of the numbers represented by the inputs. That algorithm or heuristic can then be encoded into a program, and the computation generated by the program can be interpreted as operating on numbers. So, relative to a task defined over a domain of objects encoded by a notation, a program operating on that notation may represent an algorithm, a heuristic, or just any (perhaps useless) sequence of formal operations.

A stored-program computer is a machine designed to respond to instructions held in its memory by executing certain primitive operations on inputs and other strings held in memory, with the effect that certain outputs are produced. Because of this, a computer's instructions are naturally interpreted as representing the operations performed by the computer in response to them. Most stored-program computers execute only one instruction at a time (serial computers), but some execute many instructions at a time (parallel computers). All stored-program computers execute their instructions in one step, which requires the processing of many symbols at the same time. In this sense, most computers are parallel (Turing Machines are an exception), which is also the sense in which connectionist systems are parallel. Therefore, contrary to what is often implied, parallelism is not what distinguishes connectionist systems from stored-program computers.

Given that stored-program computers can compute any computable function (until they run out of memory) and that computations can be defined over interpreted strings of symbols, stored-program computers are very flexible tools for scientific investigation. Given strings of symbols interpreted as representations of a phenomenon, and a series of formal operations for manipulating those representations, computers can be used to model the phenomenon under investigation by computing its representations. In this way, computers have become a powerful tool for scientific modeling.

Typically, AI is the programming of stored-program computers to execute computations whose outputs are interpretable as intelligent behavior. Usually, these AI computations involve the manipulation of strings of symbols held in memory, which are interpreted as representations of the environment. Some AI research is devoted to building complete artificial agents (robots), which are usually guided by an internal computing machine hooked up to sensors and motor mechanisms. Some roboticists have downplayed the importance of representation, attempting to develop a nonrepresentational framework for robotics (Brooks 1997).

Philosophical Issues

The remainder of this article describes some of the debates within and about AI that are likely to interest philosophers of science, and some of the relations between the issues that arise therein.

Engineering Versus Science

Some say that AI is a mathematical discipline, concerned with formalisms for representing information and techniques for processing it. Others say it is a branch of engineering, aimed at constructing intelligent machines. Still others say it is a *bona fide* empirical science, whose subject matter is intelligent behavior by natural and artificial agents (Newell and Simon 1976) and whose experimental method consists of building, testing, and analyzing AI artifacts (Buchanan 1988). A few have argued that AI is a form of philosophical investigation that turns philosophical explications into computer programs (Glymour 1988) and whose main method is a priori task analysis (Dennett 1978).

These views need not be incompatible. As in any other science, work in AI can be more theoretical, more experimental, or more driven by applications, and it can be conducted following different styles. At least three kinds of AI research can be usefully distinguished (Bundy 1990):

- *Applied AI* builds products that display intelligent behavior. It is a form of software engineering that uses AI techniques.
- *Cognitive simulation* develops and applies AI techniques to model human or animal intelligence. It is constrained by the experimental results of psychology and neuroscience and is part of the science of mind.
- *Basic AI* develops and studies techniques with the potential to simulate intelligent behavior. In developing these techniques, different researchers follow different styles. Some proceed more a priori, by task analysis; others proceed more *a posteriori*, by looking at how humans solve problems. All are constrained by mathematical results in computability and complexity theory. Basic AI is a largely formal or mathematical discipline, but it often proceeds by exploring the capabilities of artifacts that embody certain formalisms and techniques. In obtaining results, it might be unfeasible to prove results about AI systems and techniques mathematically or by a priori argument. The only practical way to evaluate a design might be to build the system and see how it performs under various measures. In

this respect, basic AI is experimental, though more in the sense in which engineering is experimental than in the sense in which experimental physics is.

To the extent that AI theories say what intelligent behaviors can be obtained by what means, they are relevant to our scientific and philosophical understanding of mind and intelligent behavior.

Strong Versus Weak

Strong AI holds that a computing machine with the appropriate functional organization (e.g., a stored-program computer with the appropriate program) has a mind that perceives, thinks, and intends like a human mind. Strong AI is often predicated on the computational theory of mind (CTM), which says that mental processes are computations (McCulloch and Pitts 1943; Putnam 1967; Newell and Simon 1976). Alternatively, strong AI can be grounded in instrumentalism about mentalistic language, according to which ascribing mental features to agents is not a matter of discovering some internal process but of using mentalistic predicates in a convenient way (Dennett 1978; McCarthy 1979) (see Instrumentalism).

Those who believe that ascribing genuine mental features is a matter of discovery, as opposed to interpretation, and that mental processes are not computations reject strong AI (Searle 1980). They agree that AI has useful applications, including computational models of the mind and brain, but they submit that these models have no genuine mental features, any more than computational models of other natural phenomena (e.g., the weather) have the properties of the systems they model (e.g., being humid or windy). Their view is called weak AI.

Some supporters of strong AI have replied that although performing computations may be insufficient for having a mind, performing the appropriate computations plus some other condition, such as being hooked up to the environment in appropriate ways, is enough for having at least some important mental features, such as intentionality (Pylyshyn 1984) and consciousness (Lycan 1987) (see Consciousness, Intentionality).

Hard Versus Soft

Not all attempts at writing intelligent computer programs purport to mimic human behavior. Some are based on techniques specifically developed by AI researchers in order to perform tasks that require intelligence in humans. Research in this tradition starts with a task, for example playing chess, and analyzes it into subtasks for which computational techniques can be developed. This approach is sometimes called hard AI, because it attempts to achieve results without regard for how humans and animals generate their behavior (McCarthy 1960).

Another tradition, sometimes called soft AI, attempts to build intelligent machines by mimicking the way humans and animals perform tasks. Soft AI—based on either computer programs or connectionist networks—is adopted by many cognitive psychologists as a modeling tool and is often seen as the basis for the interdisciplinary field of cognitive science (see Cognitive Science). In this guise, called *cognitive simulation,* it aims at mimicking human behavior as closely as possible. In contrast, hard AI does not aim at mimicking human behavior but at approaching, matching, and eventually outperforming humans at tasks that require intelligence.

Weak Versus Strong Equivalence

When two agents exhibit the same behavior under the same circumstances, they are weakly equivalent. When their identical behaviors are generated by the same internal processes, they are strongly equivalent (Fodor 1968). The distinction between weak and strong equivalence should not be confused with that between weak and strong AI. The latter distinction is about whether or not a machine weakly equivalent to a human has genuine mental features. If the answer is yes, then one may ask whether the machine is also strongly equivalent to a human. If the answer is no, then one may ask what the human has that the machine lacks, and whether a different kind of machine can have it too. The issue of strong equivalence takes a different form depending on the answer to the strong vs. weak AI question, but it arises in either case.

Those who originally developed the methodology of cognitive simulation endorsed CTM. They saw mental processes as computations and attempted to discover the computations performed by minds, sometimes by engaging in psychological research. When writing AI programs, they aimed at reproducing mental computations, that is, building machines that were strongly equivalent to humans.

Some authors think that strong equivalence is too much to hope for, because two agents whose behavior is empirically indistinguishable may nevertheless be performing different computations, and there is no empirical criterion for distinguishing between their internal processes (Anderson 1978). These authors still aim at cognitive simulation but see its purpose as to mimic intelligent behavior, without attempting to reproduce the internal processes generating that behavior in humans or animals. Their position is a form of instrumentalism about cognitive science theories, which aims at building machines weakly equivalent to humans and animals.

Those who take a realist position about cognitive scientific theories think that weak equivalence is an unsatisfactory goal for a science of mind. They argue that by reproducing in machines certain aspects of human behavior, such as the temporal duration of the process and the patterns of error, it should be possible to use cognitive simulation to study human internal computational processes, thereby striving for strong equivalence (Pylyshyn 1984).

Programs, Models, Theories, and Levels

Within cognitive simulation, there has been a dispute about the relationship between AI artifacts, models, and theories of intelligent behaviors. Computational models are common in many sciences, for instance in meteorology. Usually, these models are computations driven by programs and interpreted to represent some aspect of a phenomenon (e.g., the weather). The fact that the models perform computations is not modeling anything—computing is not a feature of the phenomenon but rather the means by which the model generates successive representations of the phenomenon.

When a cognitive simulation mimics human or animal behavior, there are two ways in which it can be seen as a model. Those who lean toward weak AI believe that appropriately interpreted AI artifacts are models in the same sense as are computational models of the weather: The model computes representations of the phenomenon without the phenomenon being computational. In contrast, those who incline toward strong AI, especially if they also believe CTM, hold that AI models are different from computational models in other disciplines in that when a model is correct, the computations themselves model the computational process by which humans and animals generate their behavior.

In AI (and cognitive science), a theory of an intelligent behavior should be specific enough that it can guide the design of machines exhibiting that behavior. It may be formulated at different levels of abstraction and detail:

1. A theory may specify a task, possibly by describing the characteristics that, given certain inputs, the outputs should have. For example, vision may consist of generating three-dimensional representations of physical objects from retinal stimuli. This is sometimes called the *computational level* (Marr 1982).
2. A theory may specify how to perform a task by giving a finite set of rules, regardless of whether those rules are part of the procedure through which an agent performs the task. For example, a theory of syntax may consist of rules for generating and recognizing grammatical sentences, regardless of how humans actually process linguistic items. This is sometimes called the *competence level* (Chomsky 1965).
3. A theory may specify a procedure by which an agent performs a task. This may be an algorithm or heuristic that operates on representations. This is sometimes called the *algorithmic level* (Marr 1982).
4. A theory may specify a mechanism (usually a program) that implements the representations and algorithm or heuristic of level 3. This corresponds to the building of a computational model and its interpretation. (In connectionist AI, levels 3 and 4 are the same, corresponding to the design and interpretation of a connectionist system for performing the task.)
5. A theory may specify the architectural constraints, such as the size of the memory, and how they affect the behavior of human and animal subjects and their errors in the task being performed. This is sometimes called the *performance level* (Chomsky 1965).
6. A theory may specify the physical components of the system and their mutual functional relations. This is sometimes called the *implementation level* (Marr 1982). It can be further subdivided into sublevels (Newell 1980).

The distinctions discussed in this aricle have arisen within computational approaches to AI and the mind, but they are largely independent of computational assumptions.

Although theories at levels 1–3 are usually proposed within a computational framework, they are not necessarily committed to CTM, because they do not specify whether the implementing mechanism is computational. Theories at level 4 may be interpreted either as describing the computations by which brains compute or as using computations to describe noncomputational cognitive processes (analogously to programs for weather forecasting). Levels 5 and 6 arise regardless of whether the mind is computational.

The distinction between strong and weak AI applies to any attempt to build intelligent machines. The question is whether a machine that appears intelligent has a genuine mind, and how to find evidence one way or the other. AI is committed to either dismissing this question as meaningless or answering it by forming hypotheses about the mind and testing them empirically. What constitutes empirical testing in this domain remains controversial.

The distinction between approaches that attempt to mimic natural agents (soft AI) and those that do not (hard AI) applies to any means of reproducing intelligence, whether computational or not.

Finally, within any attempt to mimic human intelligence, instrumentalists aim only at simulating behavior (weak equivalence), while realists attempt to reproduce internal processes (strong equivalence).

GUALTIERO PICCININI

References

Anderson, J. A., and E. Rosenfeld (eds.) (1988), *Neurocomputing: Foundations of Research.* Cambridge, MA: MIT Press.

Anderson, J. R. (1978), "Arguments Concerning Representations for Mental Imagery," *Psychological Review* 85: 249–277.

Brooks, R. A. (1997), "Intelligence without Representation," in J. Haugeland (ed.), *Mind Design II.* Cambridge, MA: MIT Press: 395–420.

Buchanan, B. G. (1988), "Artificial Intelligence as an Experimental Science," in J. H. Fetzer (ed.), *Aspects of Artificial Intelligence.* Dordrecht, Netherlands: Kluwer, 209–250.

Bundy, A. (1990), "What Kind of Field Is AI?" in D. Partridge and Y. Wilks (eds.), *The Foundations of Artificial Intelligence: A Sourcebook.* Cambridge: Cambridge University Press: 215–222.

Chomsky, N. (1965), *Aspects of a Theory of Syntax.* Cambridge, MA: MIT Press.

Dennett, D. C. (1978), *Brainstorms.* Cambridge, MA: MIT Press.

Glymour, C. (1988), "Artificial Intelligence Is Philosophy," in J. H. Fetzer (ed.), *Aspects of Artificial Intelligence.* Dordrecht, Netherlands: Kluwer: 195–207.

Feigenbaum, E. A., and J. Feldman (1963), *Computers and Thought.* New York: McGraw-Hill.

Fodor, J. A. (1968), *Psychological Explanation.* New York: Random House.

Ford, K. M., and Z. W. Pylyshyn (1996), *The Robot's Dilemma Revisited: The Frame Problem in Artificial Intelligence.* Norwood, NJ: Ablex.

Lycan, W. (1987), *Consciousness.* Cambridge, MA: MIT Press.

Marr, D. (1982), *Vision.* New York: Freeman.

McCarthy, J. (1960), "Programs with Common Sense," in *Mechanisation of Thought Processes* (Proceedings of the Symposium at the National Physical Laboratory) 1, 77–84.

——— (1979), "Ascribing Mental Qualities to Machines," in M. Ringle (ed.), *Philosophical Perspectives in Artificial Intelligence.* Atlantic Highlands, NJ: Humanities Press, 161–195.

McCulloch, W. S., and W. H. Pitts (1943), "A Logical Calculus of the Ideas Immanent in Nervous Nets," *Bulletin of Mathematical Biophysics* 7: 115–133.

Newell, A. (1980), "Physical Symbol Systems," *Cognitive Science* 4: 135–183.

Newell, A., and H. A. Simon (1976), "Computer Science as an Empirical Enquiry: Symbols and Search," *Communications of the ACM* 19: 113–126.

Nilsson, N. J. (1998), *Artificial Intelligence: A New Synthesis.* San Francisco: Morgan Kauffman.

Putnam, H. (1967), "Psychological Predicates," in W. H. Capitan and D. D. Merill (eds.), *Art, Mind, and Religion.* Pittsburgh, PA: University of Pittsburgh Press, 37–48.

Pylyshyn, Z. W. (1984), *Computation and Cognition.* Cambridge, MA: MIT Press.

Russell, S. J., and P. Norvig (eds.) (1995), *Artificial Intelligence: A Modern Approach.* Englewood Cliffs, NJ: Prentice Hall.

Searle, J. R. (1980), "Minds, Brains, and Programs," *Behavioral and Brain Sciences* 3: 417–457.

Turing, A. ([1936–7] 1965), "On Computable Numbers, with an Application to the Entscheidungs problem," in M. Davis (ed.), *The Undecidable: Basic Papers on Undecidable Propositions, Unsolvable Problems and Computable Functions.* Hewlett, NY: Raven Press.

——— (1950), "Computing Machinery and Intelligence," *Mind* 59: 433–460.

von Neumann, J. (1945), *First Draft of a Report on the EDVAC.* Philadelphia, PA: Moore School of Electrical Engineering, University of Pennsylvania.

See also **Chomsky, Noam; Cognitive Science; Connectionism; Consciousness; Instrumentalism; Putnam, Hilary; Realism; Searle, John; Turing, Alan**

PHILOSOPHY OF ASTRONOMY

It has been claimed that inside every astronomer lies the heart of a cosmologist (Hoskin 1997, 108). This may have well been true when the universe was thought to be relatively small and simple, but as the understanding of the size and complexity of the universe has enlarged, especially in the latter half of the twentieth century, so with it has grown the establishment of subject areas whose degree of specialization may touch nary at all on cosmology—a subject that too has matured into a field all its own. Therefore, a philosophy of astronomy cannot be content simply to track the historical development of cosmological thought (see Munitz 1957). One also finds astronomers currently who disavow any self-identification with cosmology, and at times also theoretical astrophysics, at least in part because of cosmology's highly complicated theoretical nature, and because of cosmology's relative isolation from technology. No comprehensive philosophical account of astronomy up to the present day yet exists, and such an analysis should distinguish astronomical work as such from cosmology and theoretical astrophysics, since many of the important developments in astronomy have occurred as a result of technological change, without a theory in place to understand the new phenomena astronomers and engineers uncovered. Consequently, this article will use an historical approach, distinguishing phases of astronomy's theoretical development through the technologies in use. (For more details of the historical material presented here, see Lankford 1997.)

The Naked Eye

Artifacts such as Ireland's 3000 B.C.E. Newgrange Passage Tomb and Mesoamerica's Monte Alban and Xochicalco zenith tubes are evidence that human beings have long been aware of regular celestial cycles. Records of the Sun's, moon's, and stars' movements were kept by the Babylonians (1800 B.C.E.–75 C.E.) and Egyptians (starting perhaps as early as 3000 B.C.E.), who recognized a connection between time and celestial phenomena. By the second century B.C.E. the Greeks, informed by these records, devised geometrical models to create a unified image of what the substance and motion of the heavenly bodies were like. Classical Western-style philosophizing about the nature of heavens starts with the works of Plato and Aristotle. By all appearances, the Earth is fixed, while the domed ceiling of the sky revolves around it. These basic impressions coupled with physical arguments on the nature of matter and motion (cf. particularly Aristotle's *De caelo* and *Metaphysics*) led to the long-standing view that the Earth sits unmoving at the center of the universe, while all the celestial bodies—the Sun, moon, planets, and fixed stars, perfectly spherical bodies composed of a special unearthly substance—travel about in circular motion, carried along by a system of revolving crystalline spheres. These ideas remained fundamental to astronomy for nearly 2,000 years, ever modified to accommodate data gathered by the naked eye using a set of instruments that dominated the science for that entire period. The armillary sphere, a device for measuring position, was used by Eratosthenes around 204 B.C.E. and Hipparchus of Rhodes circa 150–125 B.C.E. The cross-staff, used in Alexandria around 284 B.C.E., measures separation between celestial objects. The quadrant and astrolabe, invented by Ptolemy circa 150 C.E., also measure position. With a fair degree of accuracy, one could chart the meanderings of the wandering stars, or planets, as they rise and fall in elevation, growing brighter and dimmer as their motion appears to accelerate, decelerate, stop, and go in reverse. In an Aristotelian view, one would adjust the number of spheres, each contributing its own element of motion, to explain planetary motion retroactively. Alternatively, around 200 B.C.E., Hipparchus proposed that the Earth was in the middle of the universe but slightly off-center, so that the stars, moon, and Sun traveled around it in eccentric orbits. Thus, depending upon whether the stars and Sun were closer or farther away from the Earth, the planets would appear brighter or dimmer, and the seasons would be longer or shorter

Ptolemy's *Almagest* ([137 C.E.] 1984) built upon the model of the universe Hipparchus introduced. Using the epicycle, eccentric orbits, and the equant point (the off-center position across from the Earth from where the planets appear to have uniform motion), Ptolemy's model was quite fruitful in generating predictions, although the metaphysics it endorsed went against the deep-seated belief that the Earth was at the very center of everything. The *Almagest* generated controversy and attempts to revive a purer geocentrism for nearly 1,500 years. For instance, Martianus Capella (365–440) argued that Mercury and Venus are visible only near the Sun at dusk or dawn because they circle the Sun (rather than the Earth) as the Sun orbits the Earth in a circle. (See Neugebauer 1969 for details of the history of astronomy in antiquity.) Between the fifth and tenth centuries C.E., Muslim astronomers contributed the bulk of observational records and improved instrumentation. Western European astronomy owed much to their colleagues in the East for introducing them in the tenth century not only to Greek classics like Aristotle but also to Ptolemy's *Almagest*, novel terminology (e.g., terms like "azimuth" and "zenith" and the names of several stars such as

Betelgeuse and Algol), and tools such as the sextant (invented by the Arabic astronomer Abu Abdullah al Battanti around 1000 C.E.). For the next several centuries, newly found texts were translated into Latin and disseminated across European centers of learning.

Dissatisfaction with Ptolemy's eccentric orbits and equant points was raised an order of magnitude around 1543, when Copernicus released his theory of a Sun-centered universe, where the Earth not only revolves around the Sun but also rotates on its own axis. Although not as exact for predictions as Ptolemy's model, Copernicus offered a less complicated geometry that made his theory desirable, even if controversial. The Greeks and the Bible were revisited for arguments against the motion of the Earth, although, by now, several counterarguments existed. The absence of stellar parallax—change in a star's apparent position due to the motion of the Earth around the Sun—was also considered telling evidence against heliocentrism. But breaking the Earth free from the center of the universe opened up the way for new ideas.

Tycho Brahe famously disagreed with Copernicus's theory, even though as an observer Brahe was unquestionably the better astronomer. In the late seventeenth century he improved upon the measurements and instrumentation gained from the Muslims, motivated by his 1572 observation of a nova in Cassiopeia, a startling event because the fixed stars were supposed to be unchangeable. In 1577, he measured a comet with such accuracy that he could determine its orbit to be within the region of the planets, indicating that the planets could not be carried about on material spheres. Brahe asked Johannes Kepler to join him in organizing his extensive data. Brahe also improved resolvability—the capacity to distinguish between two objects in the sky—to one arcminute (where there are 60 arcminutes to a degree, and 60 arcseconds to an arcminute).

Kepler, who had been educated in and was not averse to Copernicus's work, devoted his time to finding the most efficient geometrical model for celestial motion. No combination of simple circular orbits matching planets' motions east to west was able simultaneously to accommodate their changes in altitude. The models were off by as much as 8 arcminutes, now unacceptable because Brahe's measurements were more precise than that. Kepler fashioned his own geometrical account for planetary motion, publicly voiced in *Mysterium cosmographicum* (1596) and in *Astronomia nova* ([1609] 1992): Planetary orbits are ellipses with the Sun at one focus, and planetary motion is not uniform. (For more detail, see Pannekoek 1961.)

The Telescope

As Kepler disseminated his theory of planetary motion, Galileo was experimenting with the next generation's defining icon of astronomy: the optical telescope. Originating in the Netherlands but spreading rapidly throughout Europe, this device single-handedly transformed the conception of the nature of heavenly bodies and their relation to Earth. With an eightfold magnification, Galileo saw things never before imagined: that the moon and Saturn were not perfectly smooth spherical bodies, that Jupiter was orbited by its own moons, and that the planets appeared proportionately enlarged through magnification but the stars did not, indicating that the stars were extraordinarily far away. Received with skepticism at first, telescopes increased in popularity as confirmation of Galileo's findings grew. In 1656, Christiaan Huygens declared that Saturn's earlike appendages were rings that encircled the planet; others saw dark patches on the Sun (sunspots) and that Venus exhibited phases much like the Earth's moon.

Long-held beliefs about celestial bodies were severely strained, but jettisoning an Aristotelian organization of matter meant being without a natural explanation for why the planets orbit and why objects on the Earth do not fly off into space as the Earth rotates. In 1687, Isaac Newton proposed just such an explanation in his foundational work in physics with a theory of gravitation (see Classical Mechanics). The mathematical sophistication of his *Principia* (Newton [1687] 1999) made it inaccessible to many, but through colleagues such as Samuel Clark, Newton's word gradually spread.

The eighteenth and nineteenth centuries witnessed the invention of new telescope designs such as Newtonian, Cassegrain, and Gregorian focuses, and the micrometer. Mirrors gradually replaced lenses, and mirrors first made with polished speculum were replaced with lighter and easier-to-construct glass with silver reflective surfaces. In 1826, 9.5 inches was the largest diameter with which a good mirror could be made; by 1897 it was 40 inches (Pannekoek 1961). Between the seventeenth and nineteenth centuries, astronomical announcements began appearing in such journals as the *Philosophical Transactions of the Royal*

Society of London, which began publishing in 1665; the monthly *Astronomische Nachrichten* in 1823; *Monthly Notices of the Royal Astronomical Society* in 1827; and the *Astronomical Journal* in 1849. International collaboration was also becoming practical as the ease and rapidity of communication improved (the telegraph was introduced in 1838, the telephone in 1876). In 1887, 56 scientists from 19 nations met in Paris to begin a collaborative photographic sky atlas, the *Carte du ciel*. Sapping the energies of some observatories' staffs for several decades, the *Carte* was never completed. Not trivially, changes in scientific publishing helped speed the process of transmission and exchange of information. In November 1782, the star Algol was seen to change regularly in brightness, one night passing from what appeared to be of fourth magnitude to second in a matter of mere hours; it kicked off a flurry of projects to discover variable stars. William Herschel discovered Uranus in 1781. An asteroid belt between Mars and Jupiter was detected in 1801–1807. By 1846, astronomers were fully aware of how subtly Mercury's perihelion moves faster in longitude than Newton's theory of gravitation predicted.

By 1838, resolution had improved to the point that astronomers could finally detect stellar parallax, giving long-anticipated evidence of the Earth's orbit around the Sun. In 1785, William Herschel, assisted by his sister Caroline, published a general star count, categorizing stars with others nearby of similar magnitude, and included a map of how the stars are distributed across the sky. In 1785, Herschel was also the first to detect star clusters and, later, in 1790, what came to be called planetary nebulae. In 1802, he released a catalog of 2,500 nebulae. John Herschel followed in his father's line of work and took a 20-inch telescope to the Cape of Good Hope. In 1847, he released an all-sky survey (Pannekoek 1961).

No one knew exactly what the nebulae were; they appeared as wispy light spots, but were they gas clouds inside the galaxy or something else entirely? In 1845, William Parsons was able to make out, with a 6-foot reflecting telescope of his own construction, that the nebula M51 was spiral shaped. In 1852, Stephen Alexander suggested that the solar system's galaxy was possibly spiral as well, and he began conducting an analysis of stars starting with the closest and working outward. By the turn of the century it was understood that most stars lay in a flat disc, 8–10 times greater in diameter than in thickness, and with a radius of approximate 10–20 light years. But astronomy would have never passed beyond the stage of merely recording positional and luminosity measurements without the incorporation of spectroscopy.

The Spectrometer

As early as 1670, Newton taught that light from the Sun, commonly thought to be simple, was composed of several colors. In 1802, William Hyde Wollaston looked more closely and found seven dark bands in the Sun's spectrum, which he assumed were natural spaces between the colors, but 15 years later, Joseph Fraunhofer looked at the Sun's light still more closely with the first spectrometer set up with a telescope and found several hundred absorption lines. Gustav Kirchhoff and Robert Bunsen began interpreting these dark lines in 1858 as being due to elements in excited states, with each element having its unique set of lines. Spectroscopy, especially after early twentieth-century improvements in physics, provides a wealth of information about objects in space, their constitution, their density, and their physical conditions. Earlier in the nineteenth century, August Comte had declared that the chemical constitution of the Sun was inherently unknowable. But by 1862, A. J. Ångstrom identified hydrogen within the Sun, and 50 more elements were identified by the end of the decade. Studies of the Sun during solar eclipses revealed spectral lines not yet identified on Earth. One of them, called "helium," from the Greek word for the Sun, was later, in 1895, isolated in laboratories as a product of radioactive decay. In 1864, Giovan Battista Donati observed that comets' emission—presumed to be no more than reflected sunlight—had its own unique composition, containing carbon monoxide, hydrogen, methane, and ethylene. In conjunction with photographic techniques introduced in the nineteenth century, and the longer exposure and integration times photographic plates allowed, astronomers obtained spectra from ever fainter sources. The star Vega was daguerreotyped in 1850. In 1882, Henry Draper took a 137-minute exposure of the Orion Nebula that showed the entire nebula and faint stars in it. In 1879, William Huggins demonstrated that Vega had a spectrum in the ultraviolet.

With spectroscopy, astronomers can determine velocity of a star along the line of sight and its distance. Spectral lines exhibit Doppler shifting, already understood in the case of sound waves in 1842 to be a measure of motion toward and away from an observer, and the first stellar radial velocities were measured as early as 1890. It is also found on the basis of spectral features that stars fall into groups. Widely used today is the Harvard classification of spectral types, which began in the late 1880s. In 1911 and 1913, Ejnar Hertzsprung and Henry Norris Russell plotted the relationship of spectral type against known absolute magnitudes (or luminosity), creating the first of what have come to be called *H-R diagrams*. Ever since their introduction, the distance to a star can be determined by locating it on an H-R diagram on the basis of its observed spectrum and a reading of its absolute magnitude from its location. Since brightness is a function of distance, the difference between a star's absolute magnitude and its observed magnitude gives its distance.

Instead of spectral lines, Harlow Shapley and Henrietta Leavitt probed galactic distances using Cepheid variable stars. Cepheids cycle regularly in brightness, with a period related to its absolute magnitude. Shapley calculated the distance to nearby clusters of stars containing Cepheids and, assuming that star clusters are similar, compared the apparent magnitudes of the brightest stars in distant and nearby star clusters. By 1917, Shapley concluded that the remotest clusters were on the order of 200,000 light years away, and that the clusters were symmetrically concentrated around the region of Sagittarius. Shapley's proposal that the center of the galaxy was some 200,000–300,000 light years away was received with skepticism. If the solar system was so off-center from the galaxy, observers should see systematically red- and blue-shifted spectra from neighborhood stars circling around with the Earth—evidence that Jan Oort provided 10 years later.

By the end of the nineteenth century, astronomers' comprehension of the solar system had become quite reliably systematic. Outside the solar system it was entirely another matter: Data vastly exceeded explanatory power. Astronomers had a start on figuring the constitution of the Sun and other stars, but no good idea of how they worked. Shapley gave a picture of the Sun's place in the galaxy, but no one knew what the galaxy overall was like. As early as 1844, Bessel recognized that Sirius had an invisible companion as massive as the Sun and with the size of the Earth, but nothing in physics accounted for such density. Explanations for such phenomena would not be forthcoming until the early decades of the 1900s.

Meanwhile astronomers began thinking that the Milky Way may not be the only galaxy in the universe. As early as 1755, Immanuel Kant proposed that faint nebulous patches of light were more likely far away stellar systems than nearby diffusions of glowing gas. In 1913, V. M. Silpher's spectral analysis of M31, the Andromeda nebula, showed its radial velocity to be 300 km/s—quite extreme considering that normal stellar velocities tended around 20 km/s. In 1917, the spectra of 25 nebulae showed that 4 had velocities greater than 1000 km/s, although this conclusion remained hotly contested. Adriaan van Maanen argued that spiral nebulae were observed to be rotating—indicating that they must be small and relatively nearby (that is, within the galaxy), and even Shapley, the ingenious observer that he was, did not believe they were anything like external galaxies.

In 1923, Edwin Hubble examined spiral nebulae for signs of their being composed of stars—novae and Cepheid variables. That he found the requisite signs gave highly supportive evidence for the claim that M31 was an external stellar system, starkly opening up the realization there were other galaxies in the universe besides Earth's. By 1929, Hubble had calculated red shifts for 24 galaxies, finding that the velocity of a galaxy was proportional to its distance—the constant of proportionality now known as *Hubble's constant*. Why galaxies should be moving apart with such high velocity, and why the elements that are observed came to exist at all, remained a matter of speculation in the early twentieth century. Some theories were put forward, such as Georges Lemaître's 1931 primeval atom hypothesis, a close relative to George Gamow's 1948 theory of the "Big Bang" (see Berger 1984). Hermann Bondi, Fred Hoyle, and Thomas Gold in 1948 alternatively supported the thesis that the universe was constantly creating matter and maintaining itself in a steady state (Bondi and Gold 1948). The conflict between the Big Bang and Steady State cosmologies has had a significant influence on theory and experiment in astronomy in the latter half of the twentieth century.

Note that for the entire period of astronomy's existence so far discussed, observers had concerned themselves with only the small fraction of the electromagnetic spectrum available to the human eye. The next generation of astronomical research, beginning in the 1930s, was marked by scientists, quite often not starting out as astronomers, looking to the universe at other wavelengths.

Beyond the Optical

Radio waves from space were first found by Karl Jansky in the 1930s, but the study of radio astronomy did not take off until physicists and engineers, primarily in Britain and Australia, refocused their radar antennas after World War II. To general surprise, including their own, they often found intense signals where nothing optically interesting was seen. For several years they had to battle the common (but not universal) professional preconception that radio wavelengths would not show anything interesting about the universe. By the 1960s, radio instruments had improved to such an extent that optical astronomers were able to get help with optical identification, and radio astronomy was shown to have discovered some very interesting things.

Because radio waves are so long (ranging from 1 mm to 100 m), they travel relatively undisturbed by dust and gas through space. This means that radio waves originating extremely far away in space and time can be detected, and consequently radio astronomy played a central role in adjudicating between the Big Bang and Steady State cosmologies. Atoms and molecules also emit radiation at radio wavelengths: Atomic hydrogen was the first detected through radio waves in 1951 and was found to exist abundantly between the stars. Since 1951, radio astronomers have detected over 100 molecules, many of which are organic and complex. Atomic and molecular spectroscopy has revolutionized the understanding of the constitution of interstellar space, which had long been thought to be either empty or too vulnerable to cosmic rays for molecules to exist. These developments spawned the now mature interdisciplinary research field of astrochemistry (see Sullivan 1984 for more historical detail).

Starting in the 1930s, astronomical and military interests combined to develop finer infrared technology. Similar to radio, infrared wavelengths are long enough to be quite immune to dust blockage, although parts of its spectrum are blocked by the Earth's atmosphere, making high-altitude or satellite observations often preferable and sometimes necessary. Infrared astronomers are able to look at the youngest of stars developing deep inside their cocoons of dust and gas. Some mature stars, like Vega, were found to radiate more in infrared than one would expect for its spectral type. At least 50 extrasolar planetary systems have been detected through various instruments. Hundreds of galaxies have been detected as radiating over 95% of their total luminosity in the infrared (whereas the Milky Way radiates approximately 50%). Some of the more crucial measurements of the cosmic microwave background radiation have occurred at infrared frequencies.

Ultraviolet, x-ray, and gamma ray astronomy also came of age after the 1930s. All of these areas of study must collect their data above the Earth's atmosphere. As a result, spectroscopic and photometric surveys have been done initially using V-2 rockets and, later, satellite conveyors. These studies have aided the understanding of what the universe is like in its more energetic states. Ultraviolet spectroscopy provides evidence of the existence of molecular hydrogen existence throughout the universe, and provides a lower limit ($10^6\,°K$) on the temperature of some of the gas in and around the galaxy. X-ray astronomy has recorded evidence of the high-energy features around black holes, and these were found to exist to an unexpected extent. At the shortest wavelengths, sudden bursts of extraordinarily intense gamma radiation have been detected, and with the assistance of instruments operating at other wavelengths, some of the sources have been isolated, although their nature is not yet completely understood. Astronomy, with the inclusion of techniques from physics, chemistry, and many types of technology, has radically transformed humanity's conception of itself and its place in the universe. It was once believed that Planet Earth was unique and at the center of everything. Now it is known that the Earth is far from the center and lies in only one of millions of galaxies in the universe, where neither solar systems nor carbon-rich molecules are unusual.

Astronomy has been dubbed a passive, observational enterprise. But the exploration of other wavelength regimes outside of the small fraction of the electromagnetic spectrum comprised by the optical range stretches the sense of "observation" far beyond any commonsense conception of the word (see Observation). Astronomers employ a complicated technological network on Earth and in space to interact with the causal nexus stemming from their objects of study. Designating astronomy as "passive" hardly seems a fit descriptor of the science. Although dated, it is philosophically common to construe scientific activity as a process of systematically testing theories against data, with an emphasis on theory. But much of the history of astronomy is marked by important periods when increasing quantities of data had no good theoretical explanation, and sometimes no explanation at all. By and large, theories of the universe and nature of the bodies occupying it have largely remained inert until new technologies powered change in sometimes highly unexpected directions. Consider astrobiology and dark matter as a couple of contemporary examples.

The issue of whether or not life exists elsewhere than Earth is on record for as long as any issue in philosophy, all decisions on the issue being able to rest on little more than a priori reasoning.

Recent technological developments have provided an unprecedented empirical basis for determining that there are indeed planets around other stars. Astrochemistry has revealed a universe replete with organic molecules. In 1986, what had been dubbed "exobiology" became the more institutionally organized disciple of astrobiology, focusing upon the detection of chemicals in space indicative of signatures of forms of life, such as with the detection of oxygen or methane from extrasolar planets, and perhaps the detection of extraterrestrial life itself, or its remains (on, e.g., Mars, Europa).

MICHELLE SANDELL

References

Berger, A. (ed.) (1984), *The Big Bang and Georges Lemaître* Dordrecht, Holland: Reidel.

Bondi, H., and Gold, T. (1948), "The Steady State Theory of the Expanding Universe," *Monthly Notices of the Royal Astronomical Society* 108: 252–270.

Hoskin, Michael (1997), *The Cambridge Illustrated History of Astronomy*. Cambridge, UK: Cambridge University Press.

Kepler, J. ([1609] 1992), *The New Astronomy*. Cambridge, UK: Cambridge University Press.

Lankford, J. (ed.) (1997), *History of Astronomy: An Encyclopedia*. New York: Garland.

Munitz, Milton K. (ed) (1957), *Theories of the Universe*. Glencoe, IL: Free Press.

Neugebauer, Otto (1969), *The Exact Sciences in Antiquity* (2nd ed.). New York: Dover Publications.

Newton, I. ([1687] 1999), *The Principia: Mathematical Principles of Natural Philosophy* Berkeley and Los Angeles: University of California Press.

Ptolemy ([137 C.E.] 1984), *Almagest*. Berlin: Springer.

Pannekoek, A. (1961), *A History of Astronomy*. New York: Dover Publications.

Sullivan, W. T. (1984), *The Early Years of Radio Astronomy*. Cambridge, UK: Cambridge University Press.

See also **Classical Mechanics**

A. J. AYER

(29 October 1910–27 June 1989)

Alfred Jules Ayer attended Eton College and then Oxford, taking a first in classics in 1932. Impressed by Ludwig Wittgenstein's *Tractatus,* which he studied in late 1931, Ayer embarked on a career in philosophy. He was Lecturer (and then Fellow) at Christ College, and subsequently at Wadham College, Oxford. In 1946 he was elected Grote Professor of the Philosophy of Mind and Logic at University College, London, and in 1959 he became Wykeham Professor of Logic at Oxford. He was the author of more than 100 articles and several books about knowledge, experience, and language, among them *Language, Truth, and Logic* ([1936] 1946), *The Foundations of Empirical Knowledge* (1940), and *The Problem of Knowledge* (1956) (Rogers 1999).

Language, Truth, and Logic

Ayer's significance to the philosophy of science comes primarily as author of *Language, Truth, and Logic* (LTL), published in January of 1936. LTL was Ayer's first book and the most widely read early English discussion of the *wissenschaftliche Weltauffassung* (scientific world-conception) of the logical positivists of the Vienna Circle, whose meetings Ayer attended from December 1932 through March 1933 (see Logical Empricism). LTL is directly concerned less with science and the philosophical issues it raises than with philosophy and the form philosophy takes in light of the *verifiability criterion of significance,* a condition of meaningfulness applied to propositions (see Verificationism and Cognitive Significance). Philosophy in this relatively new form involved the dismissal of a swath of traditional philosophical problems (deemed "metaphysical" or "nonsensical") and the identification of what remained with the logical analysis of scientific claims. LTL's significance to the philosophy of science therefore arises less from the answers it offered to questions *within* the philosophy of science than from the vision of philosophy it popularized, a vision that made philosophy dependent upon science. Thus while LTL begins, famously, with the declaration that the "traditional disputes of philosophers are, for the most part, as unwarranted as they are unfruitful" ([1936] 1946, 33), it ends with an admonition:

> [P]hilosophy must develop into the logic of science . . . [which is] the activity of displaying the logical relationship of . . . hypotheses and defining the symbols which occur in them. . . . What we must recognize is that it is necessary for a philosopher to become a scientist, in this sense, if he is to make any substantial contribution towards the growth of human knowledge. (153)

In 1936, this view of philosophy was hardly novel, nor did Ayer claim it was. In LTL's original preface, Ayer credited his views to "the doctrines of Bertrand Russell and Wittgenstein, . . . themselves the logical outcome of the empiricism of Berkeley and David Hume" ([1936] 1946, 31). And among the members of the Vienna Circle, Ayer singled out as influential Rudolf Carnap, whom he met in London in 1934 when Carnap lectured on his recently published *Logische Syntax der Sprache* (Carnap 1934; Rogers 1999, 115).

Though Ayer's message was familiar to many, the verve with which LTL advanced this vision attracted readers who had ignored scientific philosophy, and it allowed LTL to orient (particularly British) discussion of scientific philosophy. The result was considerable attention for LTL and (particularly in Britain) the near identification of Ayer with logical positivism. For a time, Ayer was a (if not the) leading English proponent of scientific philosophy. It was a curious mantle to fall to someone who, in contrast to the leading members of the Vienna Circle, lacked (and would continue to lack) scientific knowledge or training of any sort (Rogers 1999, 129–130).

Behind Ayer's scientific conception of philosophy was the verifiability criterion, according to which, as Ayer ([1936] 1946) formulated it early in LTL, the

> question that must be asked about any putative statement of fact is, Would any observations be relevant to the determination of its truth or falsehood? And it is only if a negative answer is given to this . . . question that we conclude that the statement under consideration is nonsensical. (38)

Otherwise, the "putative" statement of fact is meaningful.

This criterion did not originate with Ayer, nor did he claim originality. Moreover, Ayer followed the logical empiricists in applying the criterion only to matters of fact, not mathematical or logical propositions. These were not verifiable, but neither were they meaningless. They were true because they were *analytic,* that is, their truth was a consequence of the meanings of their terms. And it was, moreover, thus that analytic (and only analytic) propositions could be known a priori. Thus Ayer adopted the familiar division of propositions into either synthetic *a posteriori* propositions (subject to the verifiability criterion and informative) or analytic a priori propositions (tautologous and uninformative). Any purported proposition falling into neither category was a meaningless *pseudo*-proposition (Ayer [1936] 1946, 31; cf. e.g., Carnap [1932] 1959).

Ayer's "modified" verificationism required neither that verification be certain nor that a proposition's verifiability be immediately decidable. In Ayer's view, a proposition is meaningful if evidence can be brought to bear on it *in principle.* "There are mountains on the farther side of the moon" was Ayer's example (borrowed from Moritz Schlick); this claim could not be tested in 1936, nor its truth or falsity established with certainty, but it is nevertheless significant or meaningful (Ayer [1936] 1946, 36). Even in such modified form, the criterion ultimately met with insurmountable criticism and came to be recognized as inadequate (see Cognitive Significance for further detail).

In the remaining chapters of LTL, Ayer applied the verifiability criterion to traditional philosophical problems, illustrating more than arguing for LTL's vision of a scientific philosophy. The range of philosophical issues across which Ayer wielded the criterion, combined with his efficient and unflinching (if ham-fisted) manner, remains remarkable. It is perhaps in range and vigor that Ayer and LTL can claim to have contributed to, rather than just echoed, logical empiricism's antimetaphysical project. In just its first two chapters, Ayer addresses Cartesian skepticism about knowledge of the world, monism versus pluralism, and the problem of induction—finding all of these to be "fictitious." Most notably, verifiability led infamously to emotivism, the view that "in every case in which one would commonly be said to be making an ethical judgment, the function of the relevant ethical word is purely 'emotive.' It is used to express feeling about certain objects, but not to make any assertion about them" ([1936] 1946, 108).

It is instructive to compare LTL with Carnap's *Überwindung der Metaphysik durch Logische Analyse der Sprache* (Carnap [1932] 1959), an influential essay Ayer read and cited in LTL. Nearly all of LTL's conceptual apparatus—verifiability, analyticity, opposition to traditional philosophy—are found in the *Überwindung*. However, where Carnap ([1932] 1959, 77) characterized philosophy not as making statements but as applying a *method*, Ayer ([1936] 1946, 57) regarded philosophy as consisting of tautological statements clarifying our language. And this is no small difference, for it reflects Carnap's recognition of the *significance* of metaphysics as an "expression of an attitude toward life" rather than a theory about the world. Metaphysicians were often guilty, claimed Carnap, of trying to express as a theory an attitude to be expressed by poetry or music; but metaphysics *itself* was not useless. For Ayer, though, metaphysics had *no* authority over life; its pseudo-propositions were grammatical confusions, not expressions of a legitimate "attitude" (cf. Rogers 1999, 97–98). Ayer's antimetaphysical bent ran deeper than Carnap's and was less tolerant, a fact that may explain logical empiricism's subsequent reputation for intolerance.

After *Language, Truth, and Logic*

The revised 1946 edition of LTL gave Ayer occasion to respond to criticism. While asserting that "the point of view which [LTL] expresses is substantially correct," he ceded ground on several fronts, most notably in recognizing basic empirical propositions, which "refer solely to the content of a single experience" and can be "verified conclusively" by it (Ayer [1936] 1946, 10; [1940] 1959). Ayer's epistemic foundationalism remained characteristic of his views and provoked others, especially J. L. Austin (Rogers 1999, 146–147).

Such adjustments to verificationism are often regarded as evidence of logical positivism's decline, resulting in replacement by a holism defended by Carl Hempel, Willard Van Quine, and Thomas Kuhn. Verificationism did wane, but the scientific philosophy Ayer popularized did not depend on verificationism; and, moreover, Hempel, Quine, and Kuhn retained elements of scientific philosophy (Friedman 1999; Hardcastle and Richardson 2003). And it was in these terms that Ayer gauged the influence of LTL, for example lamenting (in 1959) that "among British philosophers" there was little "desire to connect philosophy with science" (Ayer 1959, 8), but noting with pride that "in the United States a number of philosophers . . . conduct logical analysis in a systematic scientific spirit . . . [close] to the . . . ideal of the Vienna Circle" (7–8). To the considerable extent to which Ayer and LTL caused this, Ayer's influence upon the philosophy of science is significant.

Gary L. Hardcastle

References

Ayer, Alfred Jules ([1936] 1946), *Language, Truth, and Logic*. London: Victor Gollancz.

——— ([1940] 1959), "Verification and Experience," in *Logical Positivism*. New York: The Free Press, 228–243. Originally published in *Proceedings of the Aristotelian Society* 37: 137–156.

——— (1940), *The Foundations of Empirical Knowledge*. London: Macmillan Press, 1940.

——— (1956), *The Problem of Knowledge*. New York: St. Martin's Press.

——— (1959), *Logical Positivism*. New York: Free Press.

Carnap, Rudolf ([1932] 1959), "The Elimination of Metaphysics through Logical Analysis of Language," in A. J. Ayer (ed.), *Logical Positivism*. New York: Free Press, 60–81. Originally published as "Überwindung der Metaphysik durch der Logische Analyse der Sprache," *Erkenntnis* 2: 219–241.

——— ([1934] 1937), *Logical Syntax of Language*. London: Kegan Paul. Originally published as *Logische Syntax der Sprache*. Vienna: Springer.

Friedman, Michael (1999), *Reconsidering Logical Positivism*. Cambridge: Cambridge University Press.

Hardcastle, Gary L., and Richardson, Alan (eds.) (2003), *Logical Empiricism in North America*. Minneapolis: University of Minnesota Press.

Rogers, Ben (1999), *A. J. Ayer*. New York: Grove Press.

See also **Carnap, Rudolf; Cognitive Significance; Hempel, Carl Gustav; Kuhn, Thomas; Logical Empiricism; Quine, Willard Van; Verifiability**

B

BASIC SENTENCES

See **Popper, Karl; Protocol Sentences**

BAYESIANISM

Thomas Bayes's *Essay* ([1763] 1958) initiated a penetrating mathematical analysis of inductive reasoning based on his famous rule for updating a probability assignment (see equation 1b below) (see Induction, Problem of; Inductive Logic). Rediscovered and generalized by Laplace a decade later (equation 1d), it found widespread applications in astronomy, geodesy, demographics, jurisprudence, and medicine. Laplace used it, for example, to estimate the masses of the planets from astronomical data and to quantify the uncertainty of such estimates. Laplace also advanced the idea that optimal estimation must be defined relative to an error or loss function—as minimizing the expected error (Hald 1998, sec. 5.3; Jaynes 2003, 172–174). This was an important source of decision theory, whose revitalization at the hands of Ramsey, von Neumann, Wald, and Savage helped launch the modern Bayesian revival (see Decision Theory; Ramsey, Frank Plumpton; von Neumann, John). Bayesian decision theory has been extended in recent years to game theory by Harsanyi, Skyrms, and others. The focus of the present entry, however, is the distinctive and influential philosophy of science extracted from Laplace's rule (equation 1d). A survey of the issues that divide the Bayesian from rival philosophies is followed by a sketch of the mathematical analysis of inductive and predictive inference initiated by Bayes and Laplace and its extensions. There follows a section on minimal belief change, then some remarks on the alleged subjectivity of the prior probability inputs needed for a Bayesian inference and various attempts to "objectify" these inputs.

Bayesian Logic and Methodology

Cox (1946) derived the usual rules of probability, the *product rule,*

$$P(A \wedge B \mid I) = P(A \mid B \wedge I) P(B \mid I), \quad (1a)$$

and the negation rule, $P(A \mid I) = 1 - P(A \mid I)$, from a requirement of agreement with qualitative common sense and a consistency requirement that *two ways of performing a calculation permitted by the rules must yield the same result* (Jaynes 2003, chaps. 1–2). Using the equivalence of the conjunctions $A \wedge B$ and $B \wedge A$ and equation 1a, this requirement yields the symmetric form of the product rule,

$$P(A \mid B \wedge I) P(B \mid I) = P(B \mid A \wedge I) P(A \mid I)$$

of which Bayes's rule is, in more suggestive notation, the trivial variant:

$$P(H \mid D \wedge I) = P(H \mid I) P(D \mid H \wedge I) / P(D \mid I), \quad (1b)$$

where H is hypothesis, D is data, and I is the assumed information, which is included to note that probabilities depend as much on the background information as on the data.

The probability $P(D \mid H \wedge I)$ is called the sampling distribution when considered as a function of the data D, and the likelihood function *qua* function of the (variable) hypothesis H. The "most likely" hypothesis (or parameter value) is thus the one that maximizes the likelihood function, that is, the one that accords D the highest probability. It follows, then, from equation 1b that the hypothesis of a pair comparison that accords D the higher probability is *confirmed* (made more probable) and its rival *disconfirmed.* In fact, Laplace was led to the odds form of equation 1b, which expresses the updated odds as the product of the initial odds, by the likelihood ratio (LR), thus

$$P(H \mid D \wedge I) : P(K \mid D \wedge I)$$
$$= [P(H \mid I) : P(K \mid I)] \times [P(D \mid H \wedge I) : P(D \mid K \wedge I)] \quad (1c)$$

as a quantitative sharpening of this qualitative condition. But he lacked a compelling reason for preferring equation 1c to alternative rules, for example, rules that multiply the initial odds by a positive power of the LR, and this led to much agonizing over the basis for Bayes's rule. It has even led some contemporary philosophers to claim that all such rules, agreeing as they do in their qualitative behavior, are on an equal footing. To appreciate the force of Cox's derivation of equation 1a is to recognize that *all these alternatives to Bayes's rule are inconsistent* (see Rosenkrantz 1992 for some illustrations). Moreover, the optimality theorem shows that Bayesian updating outperforms all rivals in maximizing one's expected score after sampling under any *proper scoring rule,* that is, a method of scoring forecasts that gives one no incentive to state degrees of prediction different from one's actual degrees of belief.

According to equation 1b, $P(H \mid D \wedge I)$ is directly proportional to $P(D \mid H \wedge I)$ and inversely proportional to $P(D \mid I)$. The latter is generally computed from the partitioning formula:

$$P(D \mid I) = P(D \mid H_1 \wedge I) P(H_1 \mid I) + \ldots \quad (2)$$
$$+ P(D \mid H_n \wedge I) P(H_n \mid I)$$

where $H_1, \ldots, H_n$ are mutually exclusive and jointly exhaustive in light of I. Given such a partition of hypotheses (the "live" possibilities from the perspective of I), Laplace recast equation 1b as

$$P(H_j \mid DI) = \frac{P(H_j \mid I) P(D \mid H_j \wedge I)}{\sum_{j=1}^{n} P(D \mid H_j \wedge I) P(H_j \mid I)}. \quad (1d)$$

Hence, H_j is confirmed, so that $P(H_j \mid DI) > P(H_j \mid I)$, when $P(D \mid H_j I) > P(D \mid I)$, or when the probability that H_j accords D (in light of I) exceeds the weighted average (equation 2) of the likelihoods. In particular, this holds when $P(D \mid H_j I) = 1$ provided $P(D \mid I) < 1$, or: *Hypotheses are confirmed by their consequences and the more so as these are unexpected on the considered alternative hypotheses.*

Seen here as well are two further implications for the scientific method: first, the dependence of D's import for a hypothesis on the considered alternatives and, second, the tenet that evidence cannot genuinely disconfirm, much less rule out, a hypothesis merely because it assigns a low probability to the data or observed outcome. Rather, disconfirmation of H requires that some alternative accord D a higher probability. Thus, Bayesian inference is inherently *comparative;* it appraises hypotheses relative to a set of considered alternatives. At the same time, there is nothing to stop one from enlarging the "hypothesis space" when none of the considered alternatives appears consonant with the data (see Jaynes 2003, 99) or when predictions based on a given hypothesis space fail.

These illustrations already hint at the many canons of induction and scientific method that follow *in a sharper quantitative form* from equation 1. While critics of Bayesianism deny it, there is an abundance of evidence that working scientists are guided by these norms, many of them illustrated in the extensive writings of Polya on induction in mathematics. Then, too, there is the growing army of Jaynesians, Bayesian followers of Jaynes and Jeffreys, who apply Bayes's rule and its maximum entropy extension to an ever-expanding range of scientific problems, expressly endorse its methodological implications, and condemn violations thereof (Loredo 1990).

The Testing Approach

To bring out salient features of the Bayesian approach, it will be useful to survey some of the issues that divide it from its major rivals, which may be lumped together, notwithstanding minor variations, as the *testing approach*. In its most developed form (Giere 1983; Mayo 1996), it holds that a hypothesis h is confirmed—rendered more *trustworthy* as well as more *testworthy*—when and only when it passes a *severe* test, that is, one with a low probability of passing a false hypothesis. This approach grew out of the writings of Peirce (see Mayo 1996, chap. 12) and of Popper and, above all, out of Neyman and Pearson's approach to testing statistical hypotheses, embraced by most "orthodox" statisticians (see de Groot 1986, Chap. 7; Hodges and Lehmann 1970, Chap. 13, for accessible introductions). This approach directly opposes Bayesianism in denying any distinctive evidential relation between data and hypotheses. It relies on only "direct" probabilities of outcomes conditional on hypotheses and the assumed model of the experiment.

Consider a medical diagnostic test with the conditional probabilities given in Table 1.

Given that a patient tests positive (+), a Bayesian multiplies the prior odds of infection by the LR,

$$P(+\mid h_0):P(+\mid h_1) = 94:2$$

to obtain the updated odds. (An LR of 49:1 is evidence of infection slightly stronger than the

Table 1 Conditional probabilities for infected and noninfected

	+	−
Infected (h_0)	0.94	0.06
Uninfected (h_1)	0.02	0.98

32:1 LR a run of 5 heads accords the hypothesis that a coin is two-headed rather than fair.) Writing $\alpha = P(-\mid h_0)$ and $\beta = P(+\mid h_1)$ for the probabilities of the two possible errors, *viz.*, false negatives and false positives, one could view the LR of $1 - \alpha:\beta$ as a plausible quantification of the (qualitative) characterization of a severe test as one with low probabilities, α and β, of passing a false hypothesis. The Bayesian approach based on LRs and the testing approach based on error probabilities do not differ appreciably in such cases. The differences show up when numerical outcomes like frequency or category counts come into play, such as in the cure rate of a new drug or phenotypic category counts in a genetic mating experiment.

Bayesians look at the likelihood function, $L(\theta) = f(x\mid\theta)$, of the outcome x *actually observed* to estimate the unknown parameter, θ, or to compare hypotheses about it. On the other hand, testing theorists look at *sets* of outcomes, rejecting h_0 in favor of h_1 just in case the outcome x falls in a critical region R of the outcome space. Labeling the hypotheses of a pair comparison so that erroneously rejecting h_0 is the more serious of the two errors, the recommended Neyman-Pearson (NP) procedure is to fix the probability,

$$\alpha = P(X \in R \mid h_0)$$

of this type I error at an acceptable level, $\alpha \leq \alpha_0$, called the *size* of the test, and then choose among all tests of that size, α_0, one of maximal *power*, $1 - \beta$, where

$$\beta = P(X \notin R \mid h_1)$$

is the probability of the less serious type II error of accepting h_0 when h_1 is true. Naturally, there are questions about what it means to accept (or reject) a hypothesis (for a good discussion of which see Smith 1959, 297), but these are left aside in deference to more serious objections to the NP procedure.

Consider a test of $h_0: P = 0.5$ versus $h_1: P = 0.3$, where P is the success rate in n Bernoulli trials. The test for the best 5 percent of $n = 20$ trials has $R = [X \leq 5]$, i.e., rejects h_0 when 5 or fewer successes are observed. Thus, h_0 is accepted when $X = 6$, even though that is exactly the number of successes expected when $P = 0.3$. At $n = 900$ trials, $R = [X \leq 425]$ even though the boundary point, $X = 425$, is a lot closer to the 450 successes expected when $P = 0.5$ than the 270 expected when $P = 0.3$. In fact, the LR in *favor* of h_0 when $X = 425$ is

$$P(X = 425 \mid h_0) : P(X = 425 \mid h_1)$$
$$= (5/3)^{425}(5/7)^{475} = 7.5 \times 10^{24}$$

and tends to ∞ at (or near) the boundary of R as $n \to \infty$. There is, then, a *recognizable* subset of R whose elements strongly favor the rejected hypothesis. The overall type I error rate of 5% is achieved by averaging the much higher than advertised 5% probability of being misled by outcomes near the boundary of R against the much lower probability of being misled by the elements of R farthest from the boundary. For testing theorists, however, the individual test result draws its meaning solely from the application of a generally reliable rule of rejection as attested by its *average* performance over many real or imagined repetitions of the experiment. This objection to the NP procedure was first lodged by a non-Bayesian, Fisher (1956, secs. 1 and 5 of chap. 6, especially 93).

Among the strongest methodological implications of equation 1 is the likelihood principle (LP), which counts two outcomes (of the same or different experiments) as *equivalent* if they give rise to the same likelihood function *up to a constant of proportionality*. Suppose, in the last example, that a second statistician elects to sample until the 6th success occurs and this happens, perchance, on the 20th trial. For this experiment, the likelihood function is:

$$L_1(P) = \binom{n-1}{5} P^6 (1 - P^{n-6})$$

which, for $n = 20$, is proportional to the likelihood function for $r = 6$ successes in $n = 20$ trials:

$$L_2(P) = \binom{20}{6} P^6 (1 - P)^{14}.$$

So the likelihood functions are proportional. In particular, both experiments yield the same LR of $(3/5)^6 (7/5)^{14} = 5.2$ in favor of h_1. But for the second experiment $R = [n \geq 19]$ is the best 5% test and so h_0 is rejected when $n = 20$. In a literal sense, both statisticians observe the same result, 6 successes in 20 Bernoulli trials, yet one accepts h_0 while the other rejects it. NP theory is thus charged with the most obvious inconsistency—allowing opposite conclusions to be drawn from the same data (Royall 1997, Chap. 3). Testing theorists are open to the same charge of inconsistency at a higher ("meta") level (Jaynes 1983, 185), in as much as they concede the validity of equation 1 and so, by implication, of the LP, when the needed prior probabilities are "known" from frequency data, as when one knows the incidence of a disease or of a rare recessive trait. Is one to base an evaluation of a methodological principle on such contingencies as whether given prior probabilities are known or only partially known?

Indeed, it seems perfectly reasonable for an experimenter to stop sampling as soon as the incoming data are deemed sufficiently decisive. That is, after all, the idea behind Wald's extension of NP theory to so-called sequential analysis. Could it make a difference whether one planned beforehand to stop when the sample proportion of defectives exceeded B or fell below A or so decided upon observing this event? Moreover, this issue of "optional stopping" has an ethical dimension in that failure to terminate an experiment when sufficiently strong evidence has accumulated can expose experimental subjects to needless risk or even death (see Royall 1997, Sec. 4.6, for a chilling real-life example).

But what about a fraud who determines to go on sampling until some targeted null hypothesis is rejected? This is indeed possible using significance tests, for as was seen, the power of such a test approaches unity as the sample size increases. But using the likelihood to assess evidence, the probability of such deception is slight. When h_0 holds, the probability of obtaining with any finite sample an LR, $L_1 : L_0 \geq k$, in favor of h_1 against h_0 is less than $1/k$, for if S is the subset of outcomes for which the LR exceeds k, then there is the Smith-Birnbaum inequality:

$$P(L_1 : L_0 \geq k \mid h_0) = \sum_{x \in S} P(x \mid h_0)$$
$$\leq k^{-1} \sum_{x \in S} P(x \mid h_1) \leq 1/k.$$

More generally, as Fisher also emphasized (1956, 96), "the infrequency with which ... decisive evidence is obtained should not be confused with the force, or cogency, of such evidence." In planning an experiment, one can compute *the probability of misleading evidence*, i.e., of an LR in favor of either hypothesis in excess of $L*$ when the other hypothesis is true, just as readily as one can compute probabilities (for an NP test) of rejecting h_0 when it is true or accepting it when it is false. These probabilities,

$$P(f(x \mid h_1) : f(x \mid h_0) > L^* \mid h_0)$$

and

$$P(f(x \mid h_1) : f(x \mid h_0) < 1/L^* \mid h_1),$$

which are governed by the Smith-Birnbaum inequality, tell the experimenter how large a sample to take in order to control adequately for the probability of misleading results. But this still leaves one free to break off sampling if sufficiently strong evidence turns up before many trials are observed. Thus, one can have the best of both worlds: (a) the use of likelihood to assess the import of the results of one's experiment and (b) control, in the planning stage, of the probability of obtaining weak or misleading evidence—the feature that made NP theory so attractive in the first place. Richard Royall (1997, Chap. 5) refers to this as the "new paradigm" of statistics.

One is free, in particular, to test new hypotheses against old data and to use the data as a guide to new models that can be viewed, in Bayesian terms, as supporting those data. Testing theorists take issue with this: first, in proscribing what Pearson branded "the dangerous practice of basing one's choice of a test . . . upon inspection of the observations"; and, second, in maintaining that "evidence predicted by a hypothesis counts more in its support than evidence that accords with a hypothesis constructed after the fact" (Mayo 1996, 251). The idea in both cases is that such tests are insufficiently severe.

Even orthodox statisticians routinely transgress, as when they check the assumptions of a model against the same data used to test the relevant null hypothesis (and, perhaps, base their choice of test on the departure from the relevant assumption indicated by those data), or when they test the hypothesis that two normal variances are equal before applying a t test (see Jaynes 1983, 157), or when they quote "critical" or "exact" significance levels (Lehmann 1959, 62) or test a complication of a model against the data that prompted that complication. Indeed, it is literally impossible to live within the confines of a strict predesignationism that requires that the sampling rule, the tested hypothesis, and the critical region (or rejection rule) all be stated prior to sampling; and the examples to show this come from the bible of orthodox testing (Lehmann 1959, 7). For Bayesians, the evidential relation is timeless, and no special virtue attaches to prediction. See Giere (1983, 274–276) and Mayo (1996, 252) for some of the history of this controversy in the philosophy of science and (new and old) references to such figures as Whewell, Mill, Jevons, Peirce, and Popper. Taking the extreme view, Popper insisted that a scientific theory cannot be genuinely confirmed ("corroborated") at all by fitting extant data or known effects but only by withstanding "sincere attempts" at refutation (see Corroboration; Popper, Karl Raimund). Critics were quick to point out the paradoxicality of Popper's attempt to confer greater objectivity on theory appraisal by appeal to a psychologistic notion. In responding to this criticism, Popper did what scientists often do (but he forbids): He amended his characterization of a severe test to read: "A theory is [severely] tested by applying it to . . . cases for which it yields results different from those we should have expected without that theory, or in the light of other theories" (1995, 112). The problem for testing theorists is to detach this criterion from its transparently Bayesian provenance.

For Giere and Mayo, a severe test is one in which the theory has a low probability of passing if false, but the question is how to compute that probability without reference to alternative hypotheses. One knows how to compute such error probabilities in statistical contexts where they are given by the assumed model of the experiment. But what meaning can be attached to such probabilities in scientific contexts, as in Giere's (1983) example of the white spot that Fresnel's wave theory of diffraction predicted would appear at the center of the shadow cast by a circular disk? Merely to label such shadowy probabilities "propensities" does not confer on them any objective reality (see Probability).

In a study of the original sources bearing on the acceptance of eight major theories, Stephen Brush (1994, 140) flatly declares that in no case was the theory accepted, "primarily because of its successful prediction of novel facts." About quantum mechanics, he writes (136) that "confirmation of novel predictions played actually no role in its acceptance," that instead its advocates "argued that quantum mechanics accounted at least as well for the facts explained by the old theory, explained several anomalies that its predecessor had failed to resolve, and gave a simple method for doing calculations in place of a collection of *ad hoc* rules" (137). Brush even turns the tables in contending that retrodiction often counts more in a theory's favor than novel predictions, that for example, Einstein's general theory of relativity was more strongly supported by the previously known advance of the perihelion of Mercury than by the bending of light in the gravitational field of the Sun (138). The main reason he offers is that Mercury's perihelion advance was a *recognized anomaly*. The failure (up to 1919) to account for it in Newtonian terms was strong indication that no such explanation would be forthcoming, and so, in Bayesian terms, the effect was essentially "inexplicable" in other theories, while the general theory of relativity was able to account for it with quantitative precision.

Problems of Induction

James Bernoulli recognized that if one is to discover a population proportion q of some trait Q

empirically—say, the proportion of 65-year-olds who survive 10 years or more—then the proportion q_n of Qs in a random sample of n (drawn with replacement) must approximate q arbitrarily well with "moral certainty" for sufficiently large samples. Bernoulli's elegant proof yields an upper bound on the least sample size, n_0, for which the sum C_n of the "central" binomial probabilities satisfying $|q_n - q| < \frac{1}{r+s}$ with $q:(1-q) = r:s$ exceeds $c(1 - C_n)$ whenever $n \geq n_0$. Notice, $\epsilon = \frac{1}{(r+s)}$ can be made arbitrarily small without changing the ratio, $r:s$ and that $C_n > c(1 - C_n)$ if and only if $C_n > \frac{c}{(1+c)}$. Bernoulli's rather loose bound was subsequently improved by his nephew, Nicholas, who interested de Moivre in the problem, leading the latter to his discovery of the normal approximation to the binomial.

The limitations of his approach notwithstanding, James Bernoulli had taken a major step toward quantifying uncertainty and deriving frequency implications from assumed probabilities. Yet, his actual goal of justifying and quantifying inductive inference eluded him and his followers, including de Moivre. Even though the statement that q_n lies within ϵ of q holds if and only if q lies within ϵ of q_n, it *does not follow* that these two statements have the same probability. The probability of the former is just a sum of binomial probabilities with q and n given, but there is no such simple way of computing the probability that an observed sample proportion lies within ϵ of the *unknown* population proportion. For all one knows, that sample may be quite deviant or unrepresentative.

While it is uncertain whether Bernoulli fell prey to this rather seductive fallacy, it is certain that Bayes, who had detected other subtle fallacies of this kind (Hald 1998, 134; Stigler 1986, 94–95), did not. He realized that a distinctively inductive inference is required. In fact, Bayes posed and solved a far more difficult problem than that of "inverting" Bernoulli's weak law of large numbers, *viz.*, given the number of Qs and non-Qs in a sample *of any size*, to find the probability that q "lies...between any two degrees of probability that can be named," i.e., in *any* subinterval of $[0,1]$, when nothing is known about the population before sampling. His solution appeared in his *Essay*, published posthumously by his friend Price in 1763.

Bayes offered a subtle argument. He equated the uninformed state of prior knowledge ("ignorance" of q) with one in which all outcomes of n Bernoulli trials were equiprobable, whatever the value of n. That condition holds if the prior density, $w(q)$ of q is the *uniform density*, $w(q) = 1$, which assigns equal probability to intervals of equal length. Bayes tacitly assumed that conversely, the *only* density of q meeting this condition is the uniform density. The central moments of the uniform density are:

$$E(q^n) = \int_0^1 q^n w(q) dq = \frac{1}{n+1}$$

when $w(q) = 1$. Then, because the central moments uniquely determine a density that is concentrated on a finite interval (see de Groot 1986, Sec. 4.4), Bayes's assumption is proved correct. Later critics, among them Boole, Venn, and Fisher, all overlooked Bayes's criterion of ignorance, which is immune to the charge of inconsistency they leveled at "Bayes's postulate."

Bayes's solution of the "inverse problem" now required one more step, the extension of the partitioning formula, equation 2, to continuous parameters. With $B = [k$ successes in n trials$]$ and $A = [t_0 \leq q \leq t_1]$, the probability Bayes sought was $P(A \mid B \wedge I_0)$ computed as $P(A \wedge B \mid I_0)/P(B \mid I_0)$ with I_0 being the assumed model of Bernoulli trials and the "empty" state of knowledge about the parameter. Then, replacing the sum in equation 2 with an integral, Bayes found for $w(q) = 1$:

$$P(B \mid I_0) = \int_0^1 \binom{n}{k} q^k (1-q)^{n-k} w(q) dq = \frac{1}{n+1}.$$

He then found that:

$$P(A \mid B \wedge I_0) = \frac{(n+1)!}{k!(n-k)!} \int_{t_0}^{t_1} q^k (1-q)^{n-k} dq. \quad (3)$$

This method works, however, only for small samples, and so Bayes devoted the remainder of the *Essay* to approximating the solution of the more general case—a formidable undertaking (see Stigler 1986, 130ff; Hald 1998, Sec. 8.6). His ongoing work on this problem was the main cause of the delay in publishing the *Essay*, and not, as some have alleged, misgivings about his formalization of ignorance (Stigler 1986, 129).

Bayes's solution thus incorporated three highly original extensions of the probability theory he inherited from the Bernoullis and de Moivre: *first*, his rule (equation 1a) for updating a probability assignment; *second*, his novel criterion of ignorance leading to a uniform prior density of the unknown population proportion; and *third*, the extension of equation 2 to continuous parameters.

Price was fully cognizant of the relevance of the *Essay*, not only to Bernoulli's problem of justifying induction, but to the skeptical arguments Hume mounted against that possibility in his *Treatise of Human Nature* (1739) and even more emphatically in *Enquiry Concerning Human Understanding* (1748) (see Empiricism; Induction, Problem of). Hume contended that one's expectation of future successes following an unbroken string of successes is merely the product of "custom" or habit and lacks any rational foundation. In his *Four Dissertations* of 1767, Price writes:

> In [Bayes's essay] a method [is] shown of determining the exact probability of all conclusions founded on induction.... So far is it from being true, that the understanding is not the faculty which teaches us to rely on experience, that it is capable of determining, in all cases, what conclusions ought to be drawn from it, and what precise degree of confidence should be placed in it.

Possibly, Bayes, too, was motivated in part by the need to answer Hume, but all that is known of his motivation is what Price says in the introduction to the *Essay*. Price also contributed an appendix to the *Essay*, in which he drew attention to the special case of equation 3 in which $k = n$ and found that the probability was

$$(n+1)\int_{.5}^{1} q^n dq = \frac{2^{n+1} - 1}{2^{n+1}}$$

that q lay between $t_0 = .5$ and $t_1 = 1$. He also showed that the probability tended to 1 and that q lay in a small interval around q_n. Price also moved into deeper waters with his suggestion that the inverse probability engine kicks in only after the first trial has revealed the existence of Qs (see Zabell 1997, 363–369, for more on Price's intriguing appendix).

It is curious that neither Bayes nor Price considered predictive probabilities per se. It was left to Laplace to take this next step, in 1774. He did this in a natural way by equating the probability of success on the next trial, following k successes in an observed sequence of n, with the mean value of the posterior density of q:

$$P(X_{n+1} = 1 \mid B_n = k, I_0)$$
$$= \frac{(n+1)!}{k!(n-k)!} \int_0^1 q^{k+1}(1-q)^{n-k} dq$$
$$= \frac{(n+1)!}{k!(n-k)!} \cdot \frac{(k+1)!(n-k)!}{(n+2)!}.$$

This simplifies to:

$$P(X_{n+1} = 1 \mid B_n = k, I_0) = \frac{k+1}{n+2}, \quad (4)$$

or Laplace's *law of succession*, which specializes to $P(X_{n+1} = 1 \mid B_n = n, I_0) = \frac{n+1}{n+2}$ when $k = n$. Notice that equation 4 does not equate the probability of success on the next trial with the observed relative frequency of success, the *maximum likelihood estimate* of q. That rule would accord probability 1 to success on the next trial following an observed run of n successes even when $n = 1$. In Bayesian terms, that is tantamount to prior knowledge that the population is *homogeneous*, with all Qs or all non-Qs.

Laplace went beyond the derivation of equation 4 in later work (see Hald 1998, Chaps. 9, 10, 15). His aim was to obtain predictive probabilities of all sorts of events by "expecting" their sampling probabilities against the posterior distribution based on an observed outcome sequence. In particular, the probability, $P(c \mid m, a, n)$, of c Qs and $d = m - c$ non-Qs in the next m trials given a Qs and $b = n - a$ non-Qs observed in $n = a + b$ trials is:

$$P(c \mid m, a, n) = \frac{(a+b+1)!}{a!b!} \binom{m}{c}$$
$$\int_0^1 Q^c(1-Q)^d Q^a(1-Q)^b dq. \quad (5)$$

Laplace approximated equation 5 by

$$P(c \mid m, a, n) \approx$$
$$\binom{m}{c} \frac{(a+c)^{a+c+\frac{1}{2}}(b+d)^{b+d+\frac{1}{2}} n^{n+\frac{1}{2}}}{a^{a+\frac{1}{2}} b^{b+\frac{1}{2}} (n+m)^{n+m+\frac{1}{2}}} \frac{n+1}{n+m+1}. \quad (5a)$$

When c and d are small compared with a and b, so that $m \ll n$, this simplifies further to

$$P(c \mid m, a, n) \approx \binom{m}{c} (a/n)^c (b/n)^d, \quad (5b)$$

the sampling probability, $\binom{m}{c} Q^c(1-Q)^d$, with the observed sample proportion, $Q_n = a/n$, in place of q.

He even showed that Q is asymptotically normally distributed about the observed sample proportion $h = q_n$, with variance hk/n, $k = 1 - h$ (Hald 1998, 169–170), hence that for any $\epsilon > 0$, the posterior probability

$$P_\epsilon = (n+1) \binom{n}{a} \int_{h-\epsilon}^{h+\epsilon} Q^a (1-Q)^b dQ \to 1$$

that $h - \epsilon \leq q \leq h + \epsilon$ approaches 1 as $n \to \infty$. This is the counterpart to the inverse probability of Bernoulli's weak law of large numbers. Finally, Laplace showed that the mode of equation 5 is the integer part, $\lfloor (m+1)Q_n \rfloor$, of $(m+1)Q_n$. Thus, the most probable sample frequency in a second sample is the one closest to that observed in the first sample. This is, arguably, the high point of the early Bayesian response to Hume, delivering a precise sense in which one can *reasonably* expect the future to "resemble" the past when nothing is known beyond what the observed random sample conveys.

It is astonishing that in two centuries of discussion of the problem of justifying induction, scarcely any mention has been made of these fundamental results of Bayes and Laplace. The subsequent development of inductive logic consists mainly of generalizations of Laplace's rule (equation 4) (see Inductive Logic). (For objections to this rule, most of them predicated on ignoring the conditions of its validity, *viz.*, independent trials and prior ignorance of q, see Fisher 1956, Chap. 2; Jaynes 2003, 563–85.)

Consider, first, random sampling *without* replacement. Jaynes (2003) has given this rather shopworn topic a new lease on life and illustrated, at the same time, how to teach probability and statistics as a unified whole (Chaps. 4 and 6). Does equation 4 generalize to this case? Given an urn containing R red and $W = N - R$ white balls, write R_j [red on jth trial] for the event of drawing a red ball on trial j, $j = 1, \ldots, N$. Then the probability of red on trial 1 is $P(R_1 \mid B) = R/N$, writing B for the assumed background knowledge, while

$$P(R_2 \mid B) = P(R_2 \mid R_1 \wedge B)P(R_1 \mid B) + P(R_2 \mid W_1 \wedge B)P(W_1 \mid B)$$
$$= \frac{R-1}{N-1}\frac{R}{N} + \frac{R}{N-1}\frac{N-R}{N} = \frac{R}{N},$$

the same as $P(R_1 \mid B)$. By mathematical induction, drawing red (respectively, white) has the same probability on *any* trial, prior, of course, to sampling.

Moreover, by the product rule, and using $P(R_k \mid B) = P(R_j \mid B)$:

$$P(R_j \mid R_k \wedge B) = P(R_k \mid R_j \wedge B) \qquad (6)$$

Thus, knowledge that a red ball was drawn on a later trial has the same effect on the probability of red as knowledge that a red ball was drawn on an earlier trial. (This also poses a barrier to any propensity interpretation of these conditional probabilities [Jaynes 2003, Sec. 3.2].) (see Probability)

Next:

$$P(R_1 \wedge W_2 \mid B) = \frac{R}{N}\frac{N-R}{N-1} = \frac{N-R}{N}\frac{R}{N-1}$$
$$= P(W_1 \wedge R_2 \mid B),$$

and an obvious extension of this shows that each sequence containing r red and w white has probability

$$\frac{R^{\underline{r}}(N-R)^{\underline{w}}}{N^{\underline{n}}} = \frac{R!(N-R)!(N-n)!}{(R-r)!(N-R-w)!N!} \qquad (7)$$

with $x^{\underline{k}} = x(x-1)\ldots(x-k+1)$, irrespective of the *order* in which the red and white balls are drawn. Such sequences of trials are called *exchangeable*. It follows, just as when sampling with replacement, that the probability $h(r \mid N, R, n)$ of drawing r red and w white balls in $n = r + w$ trials is obtained by multiplying equation 7 by $\binom{n}{r}$ to yield:

$$h(r \mid N, R, n) = \frac{n!}{(n-r)!r!}\frac{R!}{(R-r)!}\frac{(N-R)!}{(N-R-w)!}$$
$$\frac{(N-n)!}{N!} = \frac{\binom{R}{r}\binom{N-R}{n-r}}{\binom{N}{n}}. \qquad (8)$$

In this derivation of the familiar *hypergeometric* distribution, equation 8 brings out the many parallels between random sampling with and without replacement; above all, their common exchangeability.

Having dealt with the "direct" probabilities that arise in this connection, Jaynes (2003, ch. 6) then turns to the inverse problem, where $D = (n, r)$ is given (the data) and the population parameters (N, R) are both unknown. This is a far richer and more challenging problem than its binomial counterpart—Bayes's problem—since it involves two unknown parameters. (Indeed, it may well lie beyond the capabilities of orthodox statistics.) Here the import of the data is inextricable from the prior information, which may take many forms, and, in addition, if interest centers on R or the population proportion, R/N, N then enters as a well-named "nuisance" parameter, a real stumbling block for both orthodox and likelihood methods.

The joint posterior distribution, $P(N \wedge R \mid DI)$, may be written using the product rule as:

$$P(N \wedge R \mid DI) = P(N \mid I)p(R \mid N \wedge I)\frac{P(D \mid N \wedge R \wedge I)}{P(D \mid I)}.$$

Hence, the marginal distribution of N is given either by

$$P(N \mid D \wedge I) = \sum_{R=0}^{N} P(N \wedge R \mid D \wedge I)$$
$$= P(N \mid I)\sum P(R \mid NI)\frac{P(D \mid N \wedge R \wedge I)}{P(D \mid I)}$$

or directly from equation 1 by

$$P(N \mid D \wedge I) = P(N \mid I) \frac{P(D \mid N \wedge I)}{P(D \mid I)}.$$

Now it would be natural to assume that $D = (n, r)$ can do no more than eliminate values of N less than n, leaving the relative probabilities of those greater than n unchanged. The general condition on $p(R \mid N \wedge I)$ that the data say no more about N than to exclude values less than n is that

$$P(D \mid N \wedge I) = \sum_{R=0}^{N} P(D \mid N \wedge R \wedge I) P(R \mid N \wedge I) = f(n, r)$$

where $f(r, n)$ may depend on the data but not on N, or using equation 8, that

$$\sum \binom{R}{r}\binom{N-R}{n-r} P(R \mid N \wedge I) = f(n,r)\binom{N}{n}. \quad (9)$$

Thus, it took Bayes's rule to ferret out this condition, one that unaided intuition would never have discovered.

As the condition is commonly met, Jaynes (2003) next turns his attention to the factor $p(R \mid DNI)$ in the joint posterior distribution of N and R, $P(N \wedge R \mid D \wedge I) = p(N \mid D \wedge I) p(R \mid D \wedge N \wedge I)$. By Bayes's rule (equation 1),

$$P(R \mid D \wedge N \wedge I) = P(R \mid N \wedge I) \frac{P(D \mid N \wedge R \wedge I)}{P(D \mid I)}.$$

To begin with, assume that I_0 is the state in which nothing is known about R beyond $0 \leq R \leq N$. Then the prior is uniform over this range, that is $p(R \mid N \wedge I_0) = \frac{1}{N+1}$. The posterior distribution of R is then

$$P(R \mid D \wedge N \wedge I_0) = \binom{N+1}{n+1}^{-1} \binom{R}{r}\binom{N-R}{n-r}. \quad (10)$$

From this, it is easy to show that Jaynes's condition, equation 9, is satisfied. For predictive purposes, what is needed is the posterior mean:

$$\langle R \rangle = E(R \mid D \wedge N \wedge I_0) = \sum_{R=0}^{N} R P(R \mid D \wedge N \wedge I_0).$$

Using equation 10, after much algebraic manipulation:

$$\langle R \rangle + 1 = (N+2)\frac{r+1}{n+2}, \quad (11)$$

which for large N, n, and r is close to the mode, $R' = (N+1)r/n$, of equation 10. Moreover, the expected fraction $\langle F \rangle$ of red balls left after the sample (n, r) is

$$\langle F \rangle = \frac{\langle R \rangle - r}{N - n} = \frac{r+1}{n+2}. \quad (11a)$$

Finally, the probability of drawing red on the next trial, given (n, r), is obtained by averaging the probability, $(R-r)/(N-n)$, of drawing red from the depleted urn against the posterior distribution:

$$P(R \mid DNI_0) =$$
$$\sum_{R=0}^{N} \frac{R-r}{N-n} \binom{N+1}{n+1}^{-1} \binom{R}{r}\binom{N-R}{n-r}$$
$$= \frac{1}{N-n}[\langle R \rangle + 1 - (r+1)] = \langle F \rangle$$

whence

$$P(R_{n+1} \mid DNI_0) = \frac{r+1}{n+2}, \quad (11b)$$

the same result Laplace obtained for sampling with replacement. The rediscovery of equation 11b by Broad in 1918 and the surprise that it did not depend on N sparked a revival of interest in the mathematical analysis of inductive reasoning by Jeffreys, Johnson, Keynes, and Ramsey.

Jaynes goes on to consider other priors for sampling an urn, but the basic conclusion is already apparent. The import of the data depends on the prior information; the two are inextricable. Hence, different priors may or may not lead to different inferences from the same data.

A path to the treatment of more substantial prior knowledge begins by imagining that this prior information comes from a pilot study. In the binomial case Bayes treated, a pilot sample issuing in a successes and b failures could be pooled with a subsequent experiment issuing in r successes and $s = n - r$ failures to yield a posterior beta density,

$$f_\beta(q \mid a+r+1, b+s+1)$$
$$= B(a+r+1, b+s+1)^{-1} q^{a+r}(1-q)^{b+s}$$

starting from Bayes's uniform prior, which is $f_\beta(q \mid 1, 1)$. And this is, of course, the same posterior density obtained from the beta prior

$$f_\beta(q \mid a+1, b+1) = \frac{(a+b+1)!}{a!b!} q^a (1-q)^b,$$

which is the posterior density obtained from the pilot sample. (Jaynes has dubbed this property of Bayesian inference "chain consistency.")

Given, therefore, r successes in n Bernoulli trials, the posterior density, $f_\beta(q \mid a+r, b+n-r)$, for a beta prior, $f_\beta(q \mid a, b)$, yields a posterior mean or predictive probability of

$$\frac{a+r}{a+b+n}$$

and, in the case $a = b$ of a symmetric beta prior, with mean, mode, and median all equal to 1/2, this becomes

$$P(R_{n+1} \mid n, r) = E_\beta(Q \mid a+r, b+n-r) = \frac{a+r}{2a+n}$$

or, putting $\lambda = 2a$,

$$P(R_{n+1} \mid n, r) = \frac{r + \frac{\lambda}{2}}{n + \lambda}. \quad (12)$$

Again, one gets a weighted average,

$$\left[n\frac{r}{n} + \lambda\frac{1}{2}\right]/(n + \lambda)$$

of the sample proportion and the prior expectation of 1/2, the weights $n/(n + \lambda)$ and $\lambda / (n + \lambda)$ reflecting the relative weights of the sample information and the prior information. (Here the separation is clean.) Thus, one arrives at a whole new continuum of rules of succession, which Carnap (1952) dubbed the λ – *continuum*. Laplace's rule is included as the special case $\lambda = 2$. Unfortunately, Carnap, who sought the holy grail of a universally applicable rule of succession, wrote as if one must choose a single value of λ for life, as a function of logical and personal considerations, like risk averseness, while the derivation from symmetric beta priors shows that different choices of λ merely correspond to different states of prior knowledge of q, or to different beliefs about the uniformity of the relevant population.

The uniqueness of the beta prior, $f_\beta(q \mid a, a)$, that corresponds to the λ – *rule* with $\lambda = 2a$ follows from the uniqueness part of de Finetti's celebrated *representation theorem*. Recall, binary random variables, $X_1, \ldots, X_n$, are *exchangeable* if their joint distribution is permutation invariant:

$$P(X_1 = e_1, \ldots, X_n = e_n) = P(X_1 = e_{\sigma(1)}, \ldots, X_n = e_{\sigma(n)})$$

for any permutation σ of $\{1, 2, \ldots, n\}$. An infinite sequence $X_1, \ldots, X_n, \ldots$ is exchangeable if every finite subsequence of length n is exchangeable for all $n \geq 1$. This nails down the idea that the probability of any finite (binary) outcome sequence depends only on the number of 1's it contains and not the particular trial numbers on which they occur. This is manifestly true of binomial outcome sequences, as well as probability mixtures of exchangeable sequences. De Finetti's theorem is a strong converse, affirming that every infinite exchangeable sequence is a mixture of binomials, so that if $S_n = X_1 + \ldots + X_n$, then there is a *unique* distribution F of q such that for all n and k,

$$P(S_n = k) = \int_0^1 \binom{n}{k} q^k (1-q)^{n-k} dF(q). \quad (13)$$

Notice, one and the same F works for all n and k. If F is continuous (admits a density), equation 13 becomes:

$$P(S_n = k) = \int_0^1 \binom{n}{k} q^k (1-q)^{n-k} f(q) dq. \quad (13a)$$

An immediate corollary is that if $P(S_n = k) = \frac{1}{n+1}$ for all n and k, then the uniform density, for which

$$\frac{1}{n+1} = \int_0^1 \binom{n}{k} q^k (1-q)^{n-k} dq$$

holds for all n, and k must be the *only* one for which Bayes's criterion of ignorance holds. Next, assuming that only the trials to which the λ – *rules* (equation 12) apply are exchangeable, it follows that the corresponding beta density is the only one that yields the λ – *rule*, being the "mixer" in equation 12a. To see this, note first that for any exchangeable sequence satisfying equation 12, $P(R_1) = P(W_1) = 1/2$. Then if, say, $\lambda = 4$, so that $a = 2$, de Finetti's theorem affirms that for a unique distribution, F,

$$P(W_1 \wedge R_2) = \int_0^1 q(1-q) dF(q).$$

But

$$P(W_1 \wedge R_2) = P(W_1) P(R_2 \mid W_1)$$
$$= \frac{1}{2} \int_0^1 q f_\beta(q \mid 2+0, 2+1) dq,$$

which simplifies to

$$\int_0^1 q(1-q) f_\beta(q \mid 2, 2) dq.$$

An extension of this argument shows that for any n and k, a binary exchangeable sequence of length n and k 1's has probability

$$\int_0^1 q^k (1-q)^{n-k} f_\beta(q \mid 2, 2) dq.$$

Hence, the unique mixer of equation 13a can only be $f_\beta(q \mid 2, 2)$.

For $K > 2$ colors, the $\lambda - rules$ generalize to

$$P(X_{n+1} = i \mid (n_1, \ldots, n_K)) = \frac{n_i + \frac{\lambda}{K}}{n + \lambda} \quad (12a)$$

with n_i the number of trials on which color i is drawn and $\sum_{i=1}^{K} n_i = 1$.

The corresponding prior densities—or mixers in the extension of de Finetti's theorem to this case—are the symmetric Dirichlet priors:

$$\frac{\Gamma(Ka)}{\Gamma(a)} p_1^{a-1} p_2^{a-1} \cdots p_K^{a-1}$$

where p_i is the population proportion of color i and $\sum p_i = 1$.

Again, this correspondence is one to one.

Exchangeability in the multicolored case means that two outcome sequences with the same vector, $(n_1, \ldots, n_k)$, of category counts are equiprobable. Thus, it makes no difference in which trials the different colors are drawn. Johnson generalized Bayes's criterion for $K = 2$ colors to the requirement that every "ordered $K-$ partition" $(n_1, \ldots, n_k)$ of n is equiprobable, a much stronger condition than exchangeability. Later, he weakened it to require that the probability of color i in trial $n + 1$ depends only on n_i and n, not on the frequencies with which other colors are drawn, i.e.,

$$P(X_{n+1} = 1 \mid X_1 = i_1, \ldots, X_n = i_n) = f(n_i, n). \quad (14)$$

He then showed that $f(n_i, n)$ is given by equation 12a, provided $K > 2$. Carnap rediscovered all of this two decades later. Johnson's derivation of equation 12a was a milestone, for it casts into sharp relief the question of when and whether a symmetric Dirichlet prior adequately represents one's prior knowledge of the relevant categories. Readers unfamiliar with this material should try hard to think of convincing cases in which the number of times other categories occur does matter.

A severe limitation of the $\lambda-rules$ is that no finite sample can raise the probability of a generalization that affirms the nonemptiness of a specified subset of the categories above zero if the population is infinite. Hintikka and Niiniluoto (1976) discovered that such confirmation becomes possible if Johnson's postulate (equation 14) is relaxed to permit dependence on the number, c, of kinds or colors observed. Their "representative function," $f(n', n, c)$, like Johnson's $f(n', n)$, is linear in n':

$$f(n', n, c) = \mu(n, c) \frac{n' + \frac{\lambda}{K}}{n + K - c + \lambda} \quad (15)$$

where $\mu(n, c)$ does not depend on n' and $\lambda = \lambda(K)$ is given by

$$\lambda = \frac{Kf(1, K+1, K)}{1 - Kf(1, K+1, K)} - 1. \quad (15a)$$

Thus, the probability that a color whose sample proportion is n'/n will turn up next increases with both n' and the number c of colors in the sample.

Indeed, the $\lambda-rules$ enter the new family as a limiting case—the extreme of caution in generalizing. They satisfy

$$f(1, K-1) = \frac{1 + \frac{\lambda}{K}}{K - 1 + \lambda}$$

and Hintikka and Niiniluoto show that when

$$f(1, K-1, K-1) > \frac{1 + \frac{\lambda}{K}}{K - 1 + \lambda},$$

all predictions of the new system are more optimistic:

$$f(n', n, K-1) > \frac{n' + \frac{\lambda}{K}}{n + \lambda}.$$

More to the point, the posterior probability of the *constituent*, $C^{(K-1)}$, which affirms the nonemptiness of all but one of the possible kinds when all of these have occurred, is greater than 0 if and only if

$$f(1, n, K-1) > \frac{1 + \frac{\lambda}{K}}{n + \lambda} = f(1, n)$$

provided λ is a multiple of K. More generally, they show that when the parameters, $f(0, c, c)$, of the new system are chosen more "optimistically" than their Carnapian counterparts, the posterior probability of the simplest constituent, $C^{(c)}$, compatible with the sample approaches 1. The longer those "missing" kinds remain unsampled, the higher the probability that they do not exist.

This result, though qualitative, was an important clue to Rosenkrantz's Bayesian account of simplicity (Rosenkrantz 1977, Chap. 5), which says, roughly, that simpler theories are prized not because they are more probable a priori but because they are *more confirmable by conforming data*. A *simpler* theory is one that "fits" a smaller proportion of the possible outcomes of a relevant class of experiments or imposes stronger constraints on phenomena. If a theory T_2 *effectively* accommodates all of the possibilities that T_1 accommodates, along with many others besides, as with an ellipse of positive eccentricity (a *proper* ellipse) versus a circle, or a proper quadratic versus a linear polynomial, then T_1 is more strongly confirmed than a theory T_2 if what is observed is among the values or states of affairs allowed by T_1. Any account of simplicity and confirmation that failed to deliver this result would be a nonstarter. The main arguments Copernicus marshaled on behalf of heliocentrism nicely illustrate this Bayesian rationale (Rosenkrantz 1977, Chap. 7). But, of course, a simpler theory may be perceived as "too simple by half," as implausible—depending on the background knowledge. Few, if any, of Mendel's contemporaries would have credited (or did credit) his theory that inheritance is particulate and governed by simple combinatorial rules, much less Darwin's suggestion that all living species evolved from a single ancestral form.

There are other ways of modifying the Johnson-Carnap system. Exchangeable models often arise when the considered individuals are indiscernible, like electrons or copies of the same gene. By treating the categories as interchangeable, one arrives at a stronger concept, partition exchangeability, which requires the joint probability distribution of $X_1, \ldots, X_n$ to be invariant under both permutations of the indices (the trial numbers) and the category indices, $1, 2, \ldots, K$. For example, for a die ($K = 6$), the probability of an outcome sequence of n flips would then depend only on the "frequencies of the frequencies," that is, on the $a_r =$ the number of category counts, n_i, equal to r, so that, for example, the sequences 1,6,2,2,4,3,2,1,4,6 and 4,6,3,3,5,2,3,4,5,6 get the same probability, since they have the same *partition vector*:

$$(a_0, a_1, \ldots, a_K) = (1, 1, 3, 1, 0, \ldots, 0)$$

with $a_4 = a_5 = \ldots = 0$. Thus, one face is missing ($a_0 = 0$), one face turns up just once, three turn up twice, and one thrice, and *which* faces have these frequencies is immaterial. Notice, this definition would apply just as well if the number of possible categories were infinite or even indefinite (in which case a_0 is omitted).

Turing seems to have been the first to emphasize the relevance of these "abundancies," and after World War II, his statistical assistant, Good, published papers fleshing out this idea (see Good 1965, Chap. 8; Turing, Alan). If many kinds occur with roughly equal frequencies in a large sample with none predominant, then it seems likely that the relevant population contains many kinds (e.g., species of beetles), and one would expect to discover new ones. From this point of view, the relaxation of Johnson's postulate advanced by Hintikka and Niiniluoto was but a first step in limiting consideration to a_0 (through $c = n - a_0$), the first abundance (Zabell 1992, 218). "Ultimately," Zabell writes, "it is only partition exchangeability that captures the notion of complete ignorance about categories; any further restriction on a probability beyond that of category symmetry necessarily involves some assumption about the categories" (ibid.).

Zabell (1997) developed an imposing new system of inductive logic on this basis, in which the predictive probability of observing a species that has occurred n_i times in a sample of n depends only on n_i and n, as in Johnson and Carnap, but, in addition, the probability of observing a new species depends only on n and the number c of species instantiated in that sample, as in Hintikka and Niiniluoto. In addition, the trials are assumed to form an infinite exchangeable partition.

It would seem that these developments are far from Bayes and Price. However, in his appendix to Bayes's *Essay*, Price invited consideration of a die of an *indeterminate* number of sides; and de Morgan, an ardent Laplacian, also wrestled with the open-ended case where the possible species are not known in advance (Zabell 1992, 208–210). De Morgan constructed a simple urn model in which a ball is drawn at random from an urn containing a black "mutator" and t other balls of distinct colors and then replaced together with a ball of a new color if black is drawn and a ball of the same color otherwise. This led him to a rule of succession that may be considered an ancestral form of Zabell's rule. Finally, the results comprising Laplace's justification of induction were extended from Bernoulli sequences to exchangeable sequences by de Finetti ([1937] 1964, Chap. 3; [1938] 1980, 195–197).

Minimal Belief Change

Knowing nothing whatever about the horses in a three-way race, one's state of knowledge is unchanged by relabeling the entries. Hence, *mere consistency* demands that equal probabilities be assigned a given entry in these equivalent states, and the only distribution of probabilities invariant under all permutations of the horses' numbers or labels is, of course, the uniform distribution. In this manner, Jaynes (2003) has reinvented Laplace's hoary principle of indifference between events or "possibilities" as one of indifference (or equivalence) between problems. Namely, in two equivalent formulations of a problem, one must assign a given proposition the same probability. And two versions of a problem that fill in details left unspecified in the statement of the problem are *ipso facto* equivalent (Jaynes 1983, 144).

Suppose, next, that one of the entries is scratched. Provided no further information is supplied, it would be quite arbitrary to change one's relative odds on the remaining entries. If one's initial probabilities were p_1, p_2, and p_3, and horse 3 (H_3) drops out, the revised probabilities

$$\frac{P_1}{P_1 + P_2} \quad \frac{P_2}{P_1 + P_2} \quad 0$$

are merely the initial ones *renormalized*. One's partial beliefs undergo the *minimal* change dictated by the new information.

Suppose, instead, that one learns the relative frequency with which horse 2 (H_2) finished ahead of horse 1 (H_1) in a large sample of past races both entered. If nothing further is learned, one should be led to a constraint of the form

$$P_2 = rP_1 \qquad (*)$$

with $r > 0$. Call $r = o(H_2 : H_1)$ the *revealed* odds for H_2 versus H_1, where H_i is the proposition that horse i will win tonight's race. Frequentists and Bayesians would agree, presumably, up to this point.

The question now is this: How should one revise one's probabilities on H_1 given $(*)$? Notice, one cannot conditionalize on $(*)$, for one cannot assign $(*)$ probabilities conditional on H_1. Still, one can hew to the principle of moving the prior "just enough" to satisfy $(*)$. So the question becomes: What effect should $(*)$ have on H_3?

Intuitions about this are somewhat conflicted. One might think that merely shifting the relative probabilities of H_1 and H_2 should have no bearing on H_3. But what if $r = 10^{10}$, whose effect would be to virtually eliminate H_1? Eliminating H_1 would increase the probabilities of the other two horses by mere renormalization, and by the same factor. A continuity argument then applies to say that $P(H_3)$ should increase *whatever* the value of r, approaching its maximum increase as $r \to \infty$. At the same time, one feels that $P(H_3)$ should not increase by as much as $P(H_2)$ when $r > 1$. That is about as far as unaided intuition can take one; it cannot quantify these qualitative relations.

Given conflicting intuitions, the only rational recourse is to seek compelling general principles capable of at least narrowing the range of choices. Such a narrowing is achieved in an interesting paper by van Fraassen, Hughes, and Harman (1986).

Rewrite the initial probability vector, (P_1, P_2, P_3), as (1, S, T), where $S = P_2:P_1$ and $t = P_3:P_1$ are the initial odds on H_2 and H_3 against H_1. Then $P_1 = 1/(1 + S + T)$, $p_2 = S/(1 + S + T)$, and $P_3 = T/(1 + S + T)$. Clearly, the initial odds should not be altered if $R = S$, merely reinforcing the bettor's initial odds. Writing the updated odds vector as (1, R, $g(S, R, T)$), van Fraassen et al. (1986) developed an argument that entails $g(S, R, T) = T\gamma(S, R)$. Thus, the updating assumes the form

$$(1, S, T) \to (1, R, T\gamma(S, R)).$$

Next, van Fraassen et al. (1986) laid down a number of conditions on γ:

(i) $\gamma(S, R) = 1$ if $R = S$;
(ii) $\gamma(S, R)$ is continuous and $\lim_{r=\infty}\gamma(s, r) = r/s$;
(iii) γ should satisfy the functional equation, $\frac{S}{R}\gamma(S, R) = \gamma(\frac{1}{S}, \frac{1}{R})$.

To arrive at (iii), they ask what difference would it make if one's research had disclosed $P(H_1) = rP(H_2)$ instead of the other way around? "Really none," they answer. "It is the same problem as before," just as if the hypotheses have been relabeled (458). This is just Jaynes's principle. If the relabeling is carried out consistently, then the probabilities of H_3 in these two equivalent versions are:

$$\frac{T\gamma(S, R)}{1 + R + T\gamma(S, R)} \quad \text{and} \quad \frac{\frac{T}{S}\gamma(\frac{1}{S}, \frac{1}{R})}{1 + \frac{1}{R} + \frac{T}{S}\gamma(\frac{1}{S}, \frac{1}{R})}$$

since the initial and updated odds are now $1/s$ and $1/r$ for H_2 over H_1 and t/s for H_3 over H_1. After a little simple algebra, equating these two expressions for $P(H_3)$ then yields the functional equation of (iii).

Van Fraassen et al. (1986) introduce three rules that satisfy their conditions, whose representative functions are:

$$\text{MUD} : \gamma(S, R) = \max(1, r/s);$$

$$\text{MRE} : \gamma(S, R) = \left(\frac{R}{S}\right)^{\frac{R}{R+1}}; \text{ and}$$

$$\text{MTP} : \gamma(S, R) = \left(\frac{1+R}{1+\sqrt{RS}}\right)^2.$$

It is easy to verify that (i)–(iii) hold for all three rules. MRE makes the probability of H_3 grow at an intermediate rate: faster than MTP but slower than MUD. The burden of van Fraassen et al. (1986) is to argue that (i)–(iii) exhaust "the symmetries of the problem," hence the conditions one can reasonably impose; and since these three rules all meet their conditions, "the problem does not admit a unique solution" (453). They further support this conclusion with two sorts of empirical comparison and find that MRE "is not the best on either count" (453).

Their argument is quite persuasive. But is the discouraging conclusion they draw inescapable? The first thing one notices about MTP is that it is the special case, $m = 2$, of a continuum of rules:

$$\gamma_m(s,r) = \left(\frac{1+r^{2/m}}{1+(rs)^{1/m}}\right)^m$$

with $m > 0$ (real), all of which satisfy (i)–(iii). By parity of reasoning, van Fraassen et al. (1986) must concede that all of these rules are on an equal footing with MTP, MRE, and MUD, or any others satisfying (i)–(iii). Now it is easy to see that $\gamma_m(S,R) \to 1$ as $m \to \infty$. Thus, the "flat rule,"

$$(1, S, T) \to (1, R, T)$$

which leaves the probability of H_3 unchanged, is obtained as a limiting case of MTP. Recall, however, that the flat rule is the very one that van Fraassen et al. (1986) ruled out on grounds of continuity. Hence, one may exclude MTP on the very same grounds.

Consider, next, the case $R < S$ where the initial odds, $S = o(H_2 : H_1)$, overshoot the revealed odds. In this case, MUD produces

$$(1, S, T) \to (1, R, T)$$

leaving the odds, $o(H_3 : H_1)$, unchanged, even when $R < 1$. Again, when $S = T < R$, MUD yields:

$$(1, S, T) \to (1, R, R)$$

which raises the odds on H_3 against H_1 as much as H_2 does. This, too, represents extreme inductive behavior. MRE approaches this result as the limit of $r \to \infty$.

The constraint (*) is a very special case of a linear constraint, $\sum a_k P_k = 0$, or its continuous counterpart. Let an abstract rule of belief change operate on such a constraint, C, and a predistribution, P°, to produce a postdistribution, $P = P^\circ \circ C$.

As linear constraints are satisfied by mixtures, $\alpha P + (1-\alpha)R$, of distributions P and R that satisfy them, one can best view the constraint C as a *convex* set of distributions. Members of this set will be called "class-C" distributions.

Next, impose conditions on such a rule all of which can be interpreted as saying that *two ways of doing a calculation must agree*. The conditions are that the results should (1) be unique; (2) not depend on one's choice of a coordinate system; (3) preserve independence in the prior when the constraint implies no dependence; and (4) yield the same conditional distribution on a subset whether one applies the relevant constraint to the prior on that subset or condition the postdistribution of the entire system to that subset.

Shore and Johnson (1980) show that the one and only rule that satisfies these broader consistency conditions goes by minimizing the deviation from P° among all class-C distributions, with the deviation between distributions P and Q given by the *cross entropy*:

$$H(P, Q) = \sum p_i \ln(p_i/q_i). \quad (16)$$

Applied to the horse race problem, this rule of minimizing cross entropy (*MINXENT*) becomes MRE. Using the inequality, $\ln x \leq x - 1$, with equality if and only if $x = 1$, one shows that $H(P,Q) \geq 0$ with equality if and only if $P = Q$. Then using the convexity of $g(x) = x \ln x$, one shows that $H(P,Q)$ is convex in its first argument:

$$H(\alpha P + (1-\alpha)R, Q) \leq \alpha H(P, Q) + (1-\alpha)H(R, Q) \quad (17)$$

with strict inequality if $P \neq R$ and $0 < \alpha < 1$.

Now suppose there are distinct class-C distributions P, R, which both minimize the distance to Q. Then:

$$H(P, Q) = H(R, Q) = \alpha H(P, Q)$$
$$+ (1-\alpha)H(R, Q) > H(\alpha P + (1-\alpha)R, Q)$$

by convexity. Thus, the α – *mixture* of P and R has a smaller deviation from Q than either P or R. This contradiction establishes *uniqueness*. Moreover, a "nearest" distribution to P° among class-C distributions *exists*, provided C is closed under limits. Notice, too, that any belief rule that goes by minimizing a function, $I(x, y)$, satisfying $I(x, y) \geq 0$ with equality if and only if $x = y$ will have the *redundancy property*: $P \wedge C = P$ if $P \in C$ (i.e., if P already satisfies the constraint).

Minimizing cross entropy with respect to a uniform distribution,

$$H(P, U) = \sum P_i \ln(P_i/n^{-1}) = \sum P_i \ln P_i + \ln n$$

is equivalent to maximizing the (Shannon) entropy,

$$H(P) = -\sum P_i \ln P_i \quad (18)$$

a measure of the uncertainty embodied in the distribution P. The rule of maximizing the entropy (*MAXENT*), as mentioned in the introductory remarks, has also vastly expanded the arsenal of Bayesian statistics and modeling. To illustrate the third axiom governing independence, consider the alternative rule that goes by minimizing the repeat rate (RR), $\sum P_i^2$.

Like entropy, it is a continuous function of Pl that assumes its extreme values of $1/n$ and 1 at the extremes of uniformity and concentration. The RR is often considered a good approximation to (negative) entropy, and Table 2 shows how

Table 2 Maximum entropy and repeat rate

	1	2	3	4	5	6
Maxent	.103	.123	.146	.174	.207	.247
Repeat rate	.095	.124	.152	.181	.209	.238

closely the distribution of a die of mean 4 obtained by minimizing $\sum p_i^2$ approximates the maxent distribution obtained by maximizing the entropy.

A superficial examination might lead one to suppose that the RR rule performs about as well as MAXENT, just as van Fraassen et al. (1986) concluded that their rules performed as well as MINXENT. However, it can be shown that RR is inconsistent.

Consider, too, the more general family of rules that minimize a Csiszar divergence:

$$H_f(P, Q) = \sum q_i f(p_i/q_i)$$

with f being a convex function. This family includes MINXENT as the special case $f(x) = x \ln x$, as well as the chi-squared rule that minimizes $\sum \frac{(p_i - q_i)^2}{q_i}$, given by $f(x) = (x - 1)^2$.

There is much at stake here for Bayesian subjectivists, who wish to deny that prior information can ever single out one distribution of probabilities as uniquely reasonable. But if one grants that MINXENT is a uniquely reasonable way of modifying an initial distribution in the light of linear constraints, then its special case, MAXENT, singles out a prior satisfying given mean value constraints as uniquely reasonable. Alive to this threat, subjectivists have denied any special status to MINXENT or MAXENT, as in the paper by van Fraassen et al. (1986).

The upshot of Shore and Johnson's derivation is to validate Jaynes's 1957 conjecture that "deductions made from any other information measure will eventually lead to contradictions" (Jaynes 1983, 9). It places MINXENT on a par with Bayesian conditionalization, given Cox's demonstration that there is no other consistent way to update a discrete prior. In addition, MINXENT yields Bayes's rule as a special case (Williams 1980).

MAXENT also has a frequency connection (Jaynes 1983, 51–52). Of the k^N outcome sequences of N trials, the number that yields category counts $(n_1, \ldots, n_k)$ with $\sum n_i = N$ is given by the multinomial coefficient:

$$W = \frac{N!}{n_1! \ldots n_k!}.$$

Using Stirling's approximation to the factorial, one easily proves that

$$N^{-1} \ln W \to H(f_1, \ldots, f_k). \quad (19)$$

Hence, the MAXENT distribution is realized by the most outcomes. In fact, the peak is enormously sharp. Just how sharp is quantified by Jaynes's concentration theorem (1983, 322), which allows one to compute the fraction of possible outcome sequences whose category frequencies, f_i, have entropy in the range $H_{max} - \Delta H \leq H(f_1, \ldots, f_k) \leq H_{max}$, where H_{max} is the entropy of the maxent distribution.

Representing Prior Knowledge

A crucial part of the answer Efron (1986) offers to his question "Why isn't everyone a Bayesian?" is: "Frequentists have seized the high ground of objectivity." Orthodoxy has deemed the prior probability inputs needed for Bayesian inference as of no interest for science unless they are grounded in frequency data. Bayesian objectivists partially agree, as when Jaynes writes (1983, 117):

> Nevertheless, the author must agree with the conclusions of orthodox statisticians that the notion of personal probability belongs to the field of psychology and has no place in applied statistics. Or, to state the matter more constructively, objectivity requires that a statistical analysis should make use, not of anybody's personal opinions, but rather the specific factual data on which those opinions are based.

But that "factual data" need not be frequency data. It might comprise empirically given distributional constraints or the role a parameter plays in the data distribution. At any rate, it is clear that if the much-heralded "Bayesian revolution" is ever to reach fulfillment, the stain of subjectivism must be removed. For the attempts of Bayesian subjectivists to sweeten the pill are themselves rather hard to swallow.

The first coat of sugar is to draw a distinction between the "public" aspect of data analysis, the data distribution, $f(x|\theta)$, and the "personal" element, *viz.*, the investigator's beliefs about θ before sampling (Edwards, Lindman, and Savage 1965, 526). Subjectivism differs from orthodox statistics, in this view, only in its wish to *formally* incorporate prior beliefs into the final appraisal of the evidence. But, it is not always possible to cleanly separate the import of the data from one's prior beliefs. Inferences about a population mean are colored by one's beliefs about the population variance. Even Bayesians who sail under the flag of subjectivism routinely handle such so-called nuisance parameters by integrating them out of a joint posterior density, using either a prior chosen to represent sparse prior knowledge or one that is *minimally informative about the parameter(s) of interest*. (Priors based on these two principles usually turn out to be the same or nearly so.) Non-Bayesians, eschewing the formal inclusion of prior information, have no resources for dealing with the problem of nuisance parameters (Royall 1997, Chap. 7), and certainly not a single resource that has won anything like universal acceptance.

The second coat of sugar is the claim that application of Bayes's rule to the accumulating data will bring initially divergent opinions into near coincidence. "This approximate merging of initially divergent opinions is, we think, one reason why empirical research is called 'objective'" (Edwards et al. 1965, 523). Opinions will converge, however, only if the parties at variance, Abe and Babe, share the same prior information. For their posterior probabilities satisfy

$$P(H \mid D \wedge I) = P(H \mid I) \frac{P(D \mid H \wedge I)}{P(D \mid I)}$$

with $I = I_A$ for Abe and $I = I_B$ for Babe, and if, as often happens, $I_A \neq I_B$, it does not follow that

$$|P(H \mid D \wedge I_A) - P(H \mid D \wedge I_B)| < |P(H \mid I_A) - P(H \mid I_B)|.$$

Given the dependence of evidential import on prior opinion, one instantly senses trolls lurking under this placid surface.

Jaynes (2003, Sec. 5.3) reveals their presence. He notes that opinions may diverge when the parties distrust each other's sources of information (which he sees as a major cause of "polarization")—e.g., let H be the proposition that a drug is safe, and D that a well-known pundit has pronounced it unsafe. Abe considers the pundit reliable, Babe considers him a fraud. Both assign $P(D \mid \bar{H}) = 1$ but differ in their likelihoods: $P(D \mid HI_A) = 0.99$ and $P(D \mid HI_B) = 0.01$. Hence, if both assign H a fairly high prior probability, say, 0.9, their posterior probabilities of 0.899 and 0.083 are now far apart instead of identical.

More surprising is that Abe and Babe may diverge even when they are in total agreement about the pundit's reliability, assigning $P(D \mid H \wedge I_A) = P(D \mid H \wedge I_B) = a$ and $P(D \mid \bar{H} \wedge I_A) = P(D \mid \bar{H} \wedge I_B) = b$.

Given priors of $x = P(H \mid I_A)$ and $y = P(H \mid I_B)$, their posterior probabilities are:

$$P(H \mid D \wedge I_A) = \frac{ax}{ax + b(1-x)} \text{ and}$$

$$P(H \mid D \wedge I_B) = \frac{ay}{ay + b(1-y)}.$$

The necessary and sufficient condition for divergence works out to be

$$ab > [ax + b(1-x)][ay + b(1-y)],$$

which is easily satisfiable, by $a = 1/4$, $b = 3/4$, $x = 7/8$, and $y = 1/3$. Thus opinions may be driven further apart even when the parties place the same construction on the evidence.

Subjectivists view probabilities as partial beliefs to be elicited by introspection or betting behavior, but they require that they be "coherent" (i.e., consistent with the probability calculus). But if the subjectivist theory "is just a theory of consistency, plain and simple" (Zabell 1997, 365), how can subjectivists *consistently* disavow Jaynes's principle of equivalence or the consistency requirement that two calculations permitted by the rules must agree? The former underwrites the derivation of *uninformed* and the latter that of *informed* probability distributions, be they prior distributions or sampling distributions.

Thus, MAXENT, which derives from the latter requirement, leads to informed priors when the prior information takes the form of empirically given distributional constraints. On one hand, this can lead to *consensus priors* for experts who agree on those constraints, or, at worst, to a narrowing of the field and a clearer identification of the remaining areas of disagreement. On the other hand, the empirical success of many of the most ubiquitous probability models, like exponential decay or the normal (Gaussian) law of errors, is best explained by their derivation as maxent distributions satisfying commonly given constraints (see Jaynes 2003, Sec. 7.6). Arguably, this far better explains why empirical research is called objective.

Given scanty prior information, all but the most extreme subjectivists concede that some probability distributions are less faithful representations of that state than others. However, they insist that the "empty state" of knowledge is an illusory abstraction devoid of meaning (Lindley 1965, 18).

What lies behind this is the belief that earlier attempts to represent "total ignorance" all founder on the alleged "arbitrariness of parameterization" (see Hald 1998, sec. 15.6, for the tangled history of this objection, as well as Zabell 1988, esp. Sec. 6). For example, suppose one assigns a uniform distribution to the volume (V) of a liquid known to lie between only 1 and 2, but then, being equally ignorant of $D = V^{-1}$, one assigns it a uniform density on the corresponding interval $[\frac{1}{2}, 1]$. Since equal intervals of V, like $[1, \frac{3}{2}]$ and $[\frac{3}{2}, 1]$, correspond to unequal intervals of D, namely, $[\frac{1}{2}, \frac{2}{3}]$ and $[\frac{2}{3}, 1]$, there is a contradiction.

But as Jeffreys first pointed out, a log-uniform distribution of V with density

$$p(V)dV = V^{-1}$$

is the *same* distribution of V^k, for any power of V. For V is log-uniformly distributed in $[a,b]$, so that

$$P(c \leq V \leq d) = \frac{\ln d - \ln c}{\ln b - \ln a}$$

if and only if $\ln V$ is uniformly distributed in $[\ln a, \ln b]$, whence the term "log-uniform," and then $\ln V^k = k \ln V$ is uniformly distributed in $[k \ln a, k \ln b] = [\ln a^k, \ln b^k]$, so that V^k is log-uniformly distributed in $[a^k, b^k]$. Jaynes provided the justification Jeffreys only hinted at by noting that a log-uniform distribution is the only one invariant under changes of scale, while the uniform distribution is the only one that is translation invariant (Jaynes 1983, 126). Therefore, if all one knows about μ and θ is that μ is a *location parameter* and θ a *scale parameter* of the data distribution, $f(x \mid \mu, \theta)$, which means that the latter can be expressed in the form

$$f(x \mid \mu, \theta) = h\left(\frac{x - \mu}{\theta}\right).$$

Then *mere consistency* forces one to assign both μ and $\ln\theta$ a uniform density on $(-\infty, \infty)$. In practice, of course, one does know the (albeit vague) limits between which they lie, and so, in cases where the data are also scanty and it matters, one truncates these improper uniform densities to make them *proper*. Indeed, Jaynes recommends that Bayesians make a habit of such truncation as a kind of safety device (2003, 487).

For teaching purposes, however, nothing can match the mathematical simplicity of the improper Jeffreys priors that lead in a few lines of routine calculation to a joint posterior density, $p(\mu, \theta \mid DI)$, of the mean and variance of a normal population given a random sample (Lindley 1965, Sec. 5.4). Then by "marginalization," i.e., integrating with respect to θ, one is led to the posterior density of the mean (and, similarly, to one for the variance). The orthodox ("sampling theory") approach arrives, though much more laboriously, at interval estimates for μ that are numerically indistinguishable from their Bayesian counterparts owing to the mathematical quirk that the nuisance parameter can (in this special case of sampling a normal population) be eliminated (see de Groot 1986, Secs. 7.3–7.4). The same numerical agreement obtains for Bayesian "credence intervals" and orthodox "confidence intervals" of a binomial p, despite their radically different interpretation (Jaynes 1983, 170–171). Hence, orthodox acolytes of "performance characteristics" are hardly in a position to reject these Bayesian solutions on grounds of their inferior performance. Indeed, use of the Jeffreys priors realizes R. A. Fisher's ideal of "allowing the data to speak for themselves."

The Bayesian solution extends to the two-sample problem (Lindley 1965, Secs. 6.3–6.4), but the orthodox solution extends only if the two variances are known or are known to be equal. Behrens and Fisher gave a solution for the case where the two variances are known to be unequal that has never found widespread acceptance in the orthodox community but that follows rather easily from Bayes's rule. Moreover, in a definitive Bayesian treatment of this whole nexus of problems, Bretthorst *smoothly* extends the orthodox solutions of these two cases (variances known to be equal, known to be unequal) to a weighted average of their corresponding posterior densities, the weights being, of course, the respective posterior probabilities of being in each case (Bretthorst, 1993). Hence, a continuity argument comes into play here as well.

This "Bayes equivalence" of orthodox methods fails, however, whenever the latter are not based on *sufficient statistics,* functions of the data that yield the same posterior probability as the raw data (Lindley 1965, Sec. 5.5; de Groot 1986, Sec. 6.7). Such orthodox solutions become unavoidable when the given data distribution admits no nontrivial sufficient statistics, as in the famous example of the Cauchy distribution. In such cases, Bayes's rule (equation 1) will automatically pick the best interval estimate of a parameter for the sample actually observed, while the orthodox statistician must average over all samples that *might* be observed. The result is that the confidence coefficients attached to orthodox interval estimates are systematically misleading, and one will be able to define "good" and "bad" classes of samples in which the actual probability of including the true value of the parameter is better or worse than indicated (see Jaynes 1983, Chap. 9, examples 5, 6; Jaynes 2003, Sec. 17.1).

There is also pathology here. One may view orthodox methods—confidence intervals, unbiased estimators, chi-squared goodness-of-fit tests, etc.—as assorted ad hoc *approximations* to their Bayesian counterparts, joined by no unifying theoretical thread. When the approximation is satisfactory, one may expect the orthodox solution to perform about as well as the Bayesian, but where it is not satisfactory, orthodox solutions either fail to exist or yield absurd results. For examples, see Jaynes (1983, Chap. 9; 2003, Chap. 17). Necessary conditions under which orthodox solutions will closely approximate Bayes solutions are: (i) that they be based on sufficient statistics, (ii) that no nuisance parameters enter, and (iii) that prior information be sparse.

One could sum up the salient differences between the two approaches as follows. First, Bayesian methods solve the standard problems of statistics more simply, for they avoid the (often difficult) steps of choosing a suitable statistic (as test statistic or estimator) and finding its sampling distribution (or an approximation to it). Second, Bayesians are able to pose and solve problems involving nuisance parameters that lie wholly beyond the reach of orthodox theory. Above all, MINXENT and MAXENT have enormously expanded the powers of statistical inference and probability modeling. Third, Bayesianism offers a unified approach to all the problems of scientific method, inference, and decision. In particular, in place of an assortment of ad hocs, it offers a unified approach to the "modeler's dilemma" of deciding when the improved accuracy that normally accompanies a complication of theory is enough to compensate for the loss of simplicity.

All of these virtues are characteristic of any alleged new paradigm in the process of supplanting an old and entrenched theory in those upheavals termed "scientific revolutions" (see Kuhn, Thomas; Scientific Revolutions). By viewing orthodox solutions as approximate Bayes solutions, the Bayesian approach is also able to delineate the conditions of their validity—another very characteristic feature of a new paradigm.

There is yet another arena in which the mettle of Bayesianism can be tested, for, like MAXENT, symmetry has empirical implications when it is made to yield a data distribution. By utilizing this connection to the real world, two more stock objections to Laplace's principle of indifference are transformed into further triumphs of Jaynes's principle.

Naive application of Laplace's principle leads one to assign equal probabilities to the hypotheses, h_j, that j is the first significant digit in a table of numerical data, like the areas of the world's largest islands or lakes. But empirical investigation reveals that the probabilities decrease from $j = 1$ to $j = 9$. Nothing was said about the scale units, however, and the implied scale invariance leads to a log-uniform distribution:

$$p_j = P(h_j) = \log_{10}(j+1) - \log_{10} j = \log_{10}(1 + j^{-1}) \quad (20)$$

where $j = 1, 2, \ldots, 9$. Thus, $P_1 = \log_{10} 2 = 0.301$, $P_2 = \log_{10} 3 - 0.301 = 0.176$, $P_9 = 1 - \log_{10} 9 = 0.046$.

Benford discovered this now-famous "law of first digits" in 1938 but failed to explain it. Its derivation as the unique scale-invariant distribution explains why it works for ratio-scaled data, but Benford found that it also works for populations of towns or for street addresses. The explanation lies in Hill's recent discovery that "base invariance implies Benford's law" (Hill 1995). That is, equation 20 is the only distribution invariant under change of the base $b > 1$ of the number system employed. Any other distribution would yield different frequencies when the scale or base is changed. Hill even derives a far-reaching generalization of equation 20 that applies to blocks of $d > 1$ digits, hence, by marginalization, to 2nd, 3rd,..., as well as to first digits. About this example one may ask: What is the relevant "chance mechanism" that produces equation 20? What frequency data have led to it?

Bertrand's chord paradox asks for the probability that a "random chord" of a circle of radius R will exceed the side, $s = \sqrt{3}R$, of the inscribed equilateral triangle. Depending on how one defines a random chord, different answers result, and Bertrand himself seems to have attached no greater significance to the example than that "la question est mal posée." Jaynes, however, has given the problem a physical embodiment in which broomstraws are dropped onto a circular target from a height great enough to preclude skill. Nothing having been said about the exact size and location of the circle, the implied translation and scale invariance uniquely determine a density:

$$f(r, \theta) = \frac{1}{2\pi r R} \quad (21)$$

for the center (r, θ) of the chord in polar coordinates. And with $L = 2\sqrt{R^2 - r^2}$ the length of a chord whose center is at (r, θ), the relative length $x = L/2R$ of a chord has the induced density

$$p(x)dx = \frac{x}{\sqrt{1-x^2}}. \quad (21a)$$

Finally, since $L = \sqrt{3}R$ is the side length of the inscribed triangle, the probability sought is:

$$\int_{\sqrt{3}/2}^{1} p(x)dx = \frac{1}{2}\int_{0}^{1/4} u^{-1/2}du = \frac{1}{2}$$

with $u = 1 - x^2$.

All of these predictions of equation 21 can be put to the test (see Jaynes 1983, 143, for one such test and its outcome). In particular, equation 21 shows to which "hypothesis space" a uniform distribution should be assigned in order to get an empirically correct result, namely, to the linear distance between the centers of chord and circle. There is no claim, however, to be able to derive empirically correct distributions a priori, much less to conjure them out of "ignorance." All that has been shown is that any other distribution must violate one of the posited invariances. If, for example, the target circle is slightly displaced in the grid of lines determined by a rain of broomstraws, then the proportion of "hits" predicted by that distribution will be different for the two circles. But if, as Jaynes argues (1983, 142), the straws are tossed in a manner that precludes even the skill needed to make them fall across the circle, then, surely, the thrower will lack the microscopic control needed to produce different distributions on two circles that differ slightly in size or location. Hence, one is tempted to view equation 21 as the "objective" distribution for this experiment.

Have these arguments then answered all of the arguments mounted against the possibility of objectively representing information or states of knowledge in the language of probability? The method indicated, group invariance, applies as readily to data distributions, as in the last two examples, as to prior distributions. (And, as Jaynes remarks, "One man's data distribution is another man's prior.")

Attention has been focused on invariance under a suitable group of transformations to the exclusion of all the many other approaches because this method is, in Jaynes's formulation of it, so closely tied to the consistency principle that "in two situations where we have the same state of knowledge, we must assign the same probabilities." This requirement may be said to underwrite all sound applications of symmetry to probability, answering, in effect, the main question addressed in Zabell (1988). *Exactly those symmetry arguments are sound that rest on Jaynes's reinvented principle of indifference.*

When the empirical probability distributions such symmetry arguments yield prove inaccurate, that may be taken as indication that some symmetry-breaking element is at work, just as the failure of a MAXENT distribution indicates the presence of some additional constraint forcing the data into a proper subset of the otherwise allowed possibility space. Hence, Jaynes concludes (2003, 326), "We learn most when our predictions fail," a theme also emphasized by Jeffreys. But to learn from our mistakes, we need to be sure that those failures are not mere artifacts of poor inductive reasoning; hence, the relevant inferences "must be our *best* inferences, which make full use of all the knowledge we have."

In conclusion, mention must be made of a nest of paradoxes of continuous probability, which, according to Jaynes, are mass-produced in accordance with the following simple prescription:

1. Start with a mathematically well defined problem involving a finite set or a discrete or normalizable distribution, where the correct solution is evident.
2. Pass to a limit without specifying how that limit is approached.
3. Ask a question whose answer depends on how that limit is approached.

He adds that "as long as we look only at the limit, and not the limiting process, the source of the error is concealed from view" (485). Under this head, Jaynes defuses the nonconglomerability paradoxes, the Borel-Kolmogorov paradox (for which see de Groot 1986, Sec. 3.10), and the marginalization paradoxes of Dawid, Stone, and Zidek aimed at discrediting improper priors. Jaynes proposes to block all such paradoxes, which have more to do with the ambiguities surrounding continuous probability than with prior distributions per se, by adopting a "finite sets policy" in which probabilities on infinite sets are introduced only as well-defined limits of probabilities on finite sets.

There has been discussion of alternative approaches to representing prior states of partial knowledge—what Jaynes has called "that great neglected half of probability theory." It might also be called the new epistemology. Notable contributors, apart from Jeffreys and Jaynes, include Box and Tiao, Bernardo, Novick, and Hall, and Lindley, Hartigan, and Zellner (see Zellner and Min 1993). Apart from the satisfaction of seeing that various approaches all lead, in many cases, to the same prior, like the Jeffreys log-uniform prior, one may expect the different methods to generalize in different ways when applied to harder, multi-parameter problems. In any case, further work along these lines will undoubtedly contribute importantly to attempts to model expert opinion and heuristic reasoning in artificial intelligence or to arrive at consensus priors for policy decisions.

ROGER ROSENKRANTZ

References

Bayes, Thomas ([1763] 1958), "An Essay Towards Solving a Problem in the Doctrine of Chance." Originally published in the *Philosophical Transactions of the Royal Society* 53: 370–418. *Reprinted in Biometrika* 45: 293–315.

Bretthorst, G. L. (1993), "On the Difference in Means," in W. T. Grandy and P. W. Milonni (eds.), *Physics and Probabiity: Essays in Honor of Edwin T. Jaynes*. Cambridge: Cambridge University Press, 177–194.

BAYESIANISM

Brush, Stephen (1994), "Dynamics of Theory Change: The Role of Predictions," in *Philosophy of Science* 2 (Proceedings): 133–145.

Carnap, Rudolf (1952), *The Continuum of Inductive Methods*. Chicago: University of Chicago Press.

Cox, R. T. (1946), "Probability, Frequency, and Reasonable Expectation," *American Journal of Physics* 17: 1–13. Expanded in Cox (1961), *The Algebra of Probable Inference*, Baltimore: Johns Hopkins University Press.

Dale, A. I. (1999), *A History of Inverse Probability*. New York: Springer-Verlag.

Diaconis, Persi, and David Freedman (1980), "De Finetti's Generalizations of Exchangeability," in Richard C. Jeffrey (ed.), *Studies in Inductive Logic and Probability*, vol. 2. Berkeley and Los Angeles: University of California Press.

de Finetti, B. ([1937] 1964), "La prevision: Ses lois logiques, ses sources subjectives." Translated into English in Henry Kyburg and Howard Smokler (eds.), *Studies in Subjective Probability*. New York: Wiley. Originally published in *Annales de l'Institut Henri Poincare* 7: 1–38.

——— ([1938] 1980), "Sur la condition d'equivalence partielle." Translated by Paul Benacceraf and Richard Jeffrey in R. Jeffrey (ed.), *Studies in Inductive Logic and Probability*. Berkeley and Los Angeles: University of California Press, 193–205. Originally published in *Actualites scientifiques et industrielles* 739 (Colloque Geneve d'Octobre 1937 sur la Theorie des Probabilites, 6tieme partie).

——— (1972), *Probability, Induction and Statistics*. New York: Wiley.

de Groot, Morris (1986), *Probability and Statistics*, 2nd ed. Reading, MA: Addison-Wesley.

Edwards, W., H. Lindman, and L. J. Savage (1965), "Bayesian Statistical Inference for Psychological Research," in R. Duncan Luce, R. Bush, and Eugene Galanter (eds.), *Readings in Mathematical Psychology*, vol. 2. New York: Wiley.

Efron, Bradley (1986), "Why Isn't Everyone a Bayesian?" *American Statistician* 40: 1–4.

Fisher, R. A. (1956), *Statistical Methods and Scientific Inference*. Edinburgh: Oliver and Boyd.

Giere, R. N. (1983), "Testing Theoretical Hypotheses," in John Earman (ed.), *Testing Scientific Theories*. Minneapolis: University of Minnesota Press.

Good, I. J. (1965), *The Estimation of Probabilities: An Essay on Modern Bayesian Methods*. Cambridge, MA: MIT Press.

——— (1983), *Good Thinking: The Foundations of Probability and its Applications*. Minneapolis: University of Minnesota Press.

Hald, A. (1998), *A History of Mathematical Statistics from 1750 to 1930*. New York: Wiley.

Hill, T. R. (1995), "Base Invariance Implies Benford's Law," *Proceedings of the American Mathematical Society* 123: 887–895.

Hintikka, Jaakko, and Ilka Niiniluoto ([1976] 1980), "An Axiomatic Foundation for the Logic of Inductive Generalization," in R. Jeffrey (ed.), *Studies in Inductive Logic and Probability*. Berkeley and Los Angeles: University of California Press, 157–182. Originally published in M. Przelecki, K. Szaniawski, and R. Wojcicki (eds.) (1976), *Formal Methods in the Methodology of Empirical Sciences*. Dordrecht, Netherlands: D. Reidel.

Hodges, J. L., and E. L. Lehmann (1970), *Basic Concepts of Probability and Statistics*, 2nd ed. San Francisco: Holden-Day.

Jaynes, E. T. (1983), *Papers on Probability, Statistics and Statistical Physics*. Edited by R. D. Rosenkrantz. Dordrecht, Netherlands: D. Reidel.

——— (2003), *Probability Theory: The Logic of Science*. Cambridge: Cambridge University Press.

Jeffreys, H. ([1939] 1961), *Theory of Probability*, 3rd ed. Oxford: Clarendon Press. Jeffrey, Richard (ed.) (1980), *Studies in Inductive Logic and Probability*, vol. 2. Berkeley and Los Angeles: University of California Press.

Kullback, S., *Information Theory and Statistics*, New York: John Wiley.

Lehmann, E. L. (1959), *Testing Statistical Hypotheses*. New York: John Wiley.

Lindley, Dennis (1965), *Introduction to Probability and Statistics*, pt. 2: *Inference*. Cambridge: Cambridge University Press.

Loredo, T. J. (1990), "From Laplace to Supernova SN 1987A: Bayesian Inference in Astrophysics," in Paul Fougere (ed.), *Maximum Entropy and Bayesian Methods*, 81–142. Dordrecht, Netherlands: Kluwer Academic Publishers.

Mayo, Deborah G. (1996), *Error and the Growth of Experimental Knowledge*. Chicago: University of Chicago Press.

Popper, Karl (1995), *Conjectures and Refutations: The Growth of Scientific Knowledge*. New York: Harper Torchbooks.

Rosenkrantz, R. D. (1977), *Inference Method and Decision*. Dordrecht, Netherlands: D. Reidel.

——— (1992), "The Justification of Induction," *Philosophy of Science* 59: 527–539.

Royall, Richard, M. (1997), *Statistical Evidence: A Likelihood Paradigm*. London: Chapman-Hall.

Shore, J. E., and R. W. Johnson (1980), "Axiomatic Derivation of the Principle of Maximum Entropy and the Principle of Minimum Cross Entropy," *IEEE Transactions on Information Theory* IT-26: 26–37.

Smith, C. A. B. (1959), "Some Comments on the Statistical Methods Used in Linkage Investigations," *American Journal of Genetics* 11: 289–304.

Stigler, Stephen (1986), *The History of Statistics: The Measurement of Uncertainty Before 1900*. Cambridge, MA: Belknap Press of Harvard University, 1986.

van Fraassen, B., R. I. G. Hughes, and G. Harman (1986), "Discussion: A Problem for Relative Information Minimizers," *British Journal for the Philosophy of Science* 34: 453–475.

Zabell, S. L. (1988), "Symmetry and its Discontents," in Brian Skyrms and W. L. Harper (eds.), *Causation, Chance and Credence*. Dordrecht, Netherlands: Kluwer Academic Publishers, 155–190.

——— (1992), "Predicting the Unpredictable," *Synthese* 90: 205–232.

——— (1997), "The Continuum of Inductive Methods Revisited," in John Earman and John D. Norton (eds.), *The Cosmos of Science*. Pittsburgh: University of Pittsburgh Press, 351–385.

Zellner, Arnold, and Chung-ki Min (1993), "Bayesian Analysis, Model Selection and Prediction," in W. T. Grandy and P. W. Milonni (eds.), *Physics and Probability: Essays in Honor of Edwin T. Jaynes*. Cambridge: Cambridge University Press.

See also **Carnap, Rudolf; Confirmation Theory; Decision Theory; Epistemology; Induction, Problem of; Inductive Logic; Probability**

BEHAVIORISM

Behaviorism is regarded properly as a formal approach to psychology. The first to articulate systematically the tenets of the behaviorist approach was John B. Watson. In his essay "Psychology as a Behaviorist Views It," Watson (1913) attacked the predominant tendency to define psychology as the study of consciousness. His major target was Edward Titchener's structural psychology. Titchener (1898) had advanced psychology as the study of constituent elements of conscious states. This form of psychology favored introspection as the primary means to study the mind. To a lesser extent, Watson was also critical of functional psychology. Adherents of this view, such as James Angell (1907), placed emphasis on the biological significance of conscious processes. Critical of introspection and of the more general tendency to link the validity of psychological data to consciousness, Watson (1913) argued that psychology should be regarded as a purely objective experimental branch of natural science. He defined psychology as the prediction and control of behavior, and explicitly aligned psychology with the methods of physics and chemistry. He discarded conscious states and processes as the objects of observation and replaced them with stimulus–response connections.

Watson's views did not develop in isolation. (For a complete discussion of precursors to Watson, see O'Donnell 1985.) In the fist decade of the twentieth century, the Russian physiologist Ivan Pavlov advanced objectivity in psychology through research on the conditioned reflex. Pavlov demonstrated that a response that normally follows one particular stimulus could be produced by a different stimulus if the two stimuli were to occur together over a period of time. This technique was known as classical conditioning. Pavlov demonstrated the technique by inducing salivation in dogs through ringing a bell that had previously accompanied food. (A collection of English translations of Pavlov's major papers can be found in Pavlov 1955.) Pavlov's notion of classical conditioning became central to Watson's work.

After the publication of Watson's 1913 essay, behaviorism quickly became the mainstream approach in American psychology. The movement enjoyed immense popularity well into the 1950s through the work of such figures as Clark Hull (1943) and, more importantly, B. F. Skinner (1953). Skinner developed the notion of operant conditioning, which holds that behavior is shaped by its results, either positive or negative. Operant conditioning differed from the classical conditioning of Pavlov and Watson and was closely connected to the work of Edward Thorndike. In the early 1900s, Thorndike had formulated an approach known as connectionism. This approach held that learning consisted of connecting situations and responses, as opposed to connecting ideas. A central element in Thorndike's (1905) psychology was his *law of effect*. In essence, the law of effect held that the frequency with which a behavior occurred was related to the tendency the behavior had to produce positive or negative results. For instance, the likelihood that an animal will push on a lever is increased if doing so produces food, and is decreased if doing so produces pain. When incorporated into Skinner's concept of operant conditioning, the law of effect explained how new patterns of behavior emerge.

Skinner (1957) argued that higher cognitive processes, such as language, also could be treated as complex forms of operant behavior. In Skinner's view, verbal forms of expression differ from nonverbal forms only with respect to the contingencies that affect them. Even consciousness could be explained with reference to operant conditioning. According to Skinner, consciousness emerges as "stimulus control" designed to permit discriminations regarding one's own responding. Consciousness had often been considered a "private" or "first-person" event, but when treated as stimulus control, it was open to an operant analysis because it was placed in a functional relationship with the entire context of antecedent and consequent stimulation.

Behaviorism had profound philosophical implications. Prior to the rise of behaviorism, dualistic theories of mind enjoyed considerable popularity. In the philosophy of Descartes, for example, mind was interpreted as being of an essence that was distinct from matter. While philosophers such as Kant challenged such dualism, the views of behaviorists ultimately dealt dualism the most severe blow. Watson's attack on structural psychology, and on consciousness in particular, magnified the fact that consciousness was unobservable and, hence, outside the confines of science. Watson (1913) believed that mentality should be studied by focusing only on observable manifestations of the mind; namely, behavior. Skinner (1953) also argued against defining the mind in a way that located it in an unobservable realm. According to him, the goal of psychology was to describe the laws by which stimuli and behavior were connected, and the ways in which such connections were affected by changes in the physical environment. The laws governing these connections and their modifications were regarded by Skinner as being on a par with the laws of motion.

Skinner (1953) specifically argued that psychology could treat behavior as a function of the immediate physical environment and the environmental history. To say that behavior is a function of environmental history means that the way someone will behave in a given situation is determined by the physical stimuli to which that person has been exposed in the past. The implication of Skinner's view is that psychologists could eliminate entirely any reference to hidden entities and internal causes, leaving the concept of 'mind' to nonscientific forms of investigation. According to Skinner, this implication applied even to neural explanations. Neural factors could be eliminated, since they perform no function other than to describe behavior itself. To emphasize this point, Skinner presented what has become known as the theoretician's dilemma. If observable events are connected, no theory about internal mental states is needed because psychologists can predict one event from another without any reference to the theory. If the events are not connected, then the theory is not needed because it does not help make a prediction. (See Hempel 1958 for an early philosophical analysis of this problem.)

Behaviorist psychology became aligned closely with two interrelated philosophical movements. The first movement was operationalism. The goal of operationalism was to render the language and terminology of science more objective and precise and to rid science of those problems that were not physically demonstrable. The chief proponent of operationalism, Percy Bridgman (1927), held that a theoretical construct should be defined in terms of the physical procedures and operations used to study it. The implication for psychology is that a psychological construct is the same as the set of operations or procedures by which it is measured and used in experimentation. Since behaviorism eschewed consciousness and embraced instead observable activity, it corresponded neatly to operationalism (see Bridgman, Percy).

The second philosophical movement with which behaviorism was aligned was logical positivism. The positivist doctrine rested on the *verifiability theory of meaning*. The verifiability theory held that the meaningfulness of any scientific question was determined by asking whether there was observational evidence that could be collected to answer the question. Through the application of the verifiability theory, positivists held that the task of philosophy was to analyze the meaning of scientific statements in terms of the evidence that would confirm or disconfirm them. (A collection of essays on logical positivism can be found in Ayer 1959. See Logical Empiricism, Verifiability).

Behaviorism provided the positivists with the means for applying their project to psychology. Positivists argued that all descriptions of the mind were confirmed or disconfirmed solely by facts about a person's behavior in a given environment. According to the verifiability theory of meaning, these facts constitute the *meaning* of any psychological statement. This theory about the meanings of psychological statements became known as logical behaviorism. Carl Hempel (1949) defended logical behaviorism in his article "The Logical Analysis of Psychology." He argued that psychological statements that were meaningful, that is, verifiable, could be translated into statements that did not involve psychological concepts, but rather physical concepts. For example, statements about feelings of depression are meaningful because they can be translated into descriptions about the person's behavior and physical body. Hempel's position implied that meaningful statements of psychology were physicalistic statements and that proper psychology was an integral part of physics (see Hempel, Carl Gustav; Physicalism).

By the late 1950s, behaviorism faced serious criticisms. One major criticism corresponded to verbal behavior. Contrary to what Skinner (1957) had argued, it was not obvious that verbal behavior could be treated simply in terms of operant conditioning. Noam Chomsky (1959) argued that linguistic abilities could be explained only on the assumption that language was the result of complex mental processes that analyzed sentences into their grammatical and semantic components. In supporting this position, Chomsky pointed out that behaviorists wanted to claim that *all* behavior was a product of laws formulated in terms of responses to environmental stimuli. Yet, the only laws behaviorists were able to demonstrate arose in controlled experiments, usually with animals. In real situations in which natural language was used, Chomsky noted that psychologists could not know what the stimuli and environmental history were until a subject responded (see Chomsky, Noam).

Chomsky's (1959) own position assumed that stimuli must be described in terms of how the subject perceived them, rather than by objective physical characteristics. He was concerned that Skinner had failed to offer a characterization of the stimuli that permitted the connection of verbal behavior to objective physical properties of the environment. Chomsky argued that how a person behaved depended not only on the physical character of the stimulus but also on what was going on in the person's mind at the time the stimulus was presented. Chomsky consequently believed that psychologists could correctly predict what people would say only by considering their utterances to be the result of complex internal states of mind. This was particularly true of cases in which humans produced novel sentences. Many of the sentences that humans utter have never been produced before; hence there is no way that prior reinforcement would explain how they are learned.

Though some philosophers, such as Quine (1960), continued to defend it, eventually behaviorism fell out of favor (see Quine, Willard Van). It was replaced by an approach known as cognitivism. This approach placed greater emphasis on internal mental processing. Even though behaviorism is no longer a mainstream approach in psychology, recent scholarship (e.g., Thyer 1999) has attempted to highlight some potentially significant aspects of Skinner's views that may have been overlooked. This attempt rests on a distinction between Skinner's work and earlier behaviorism. In the earlier period, behaviorism was regarded as "methodological." Figures such as Watson and Hull considered behavior as important because it offered psychology epistemologically valid grounds for speaking about causal entities from a nonphysical dimension. Observable behaviors were means through which purely mental, or conscious, events could be studied scientifically. Skinner's behaviorism, on the other hand, was "radical." His approach treated behavior as a subject matter in its own right. Behavior was regarded as the interaction of organism and environment. Radical behaviorism held that mental events were appropriately regarded as aspects of the overall context, not as causes in a nonphysical dimension. Radical behaviorism consequently did not reduce behavior to physiological mechanisms. In comparison with methodological behaviorism, radical behaviorism may very well have unique consequences for major philosophical topics, including mind–body dualism, free will, and determinism.

ANDREW BACKE

References

Angell, James R. (1907), "The Province of Functional Psychology," *Psychological Review* 14: 61–91.
Ayer, A. J. (ed.) (1959), *Logical Positivism*. New York: Free Press.
Bridgman, Percy W. (1927), *The Logic of Modern Physics*. New York: Macmillan.
Chomsky, Noam (1959), "Review of *Verbal Behavior*, by B. F. Skinner," *Language* 35: 26–58.
Hempel, Carl G. (1949), "The Logical Analysis of Psychology," in H. Feigl and W. Sellars (eds.), *Readings in Philosophical Analysis*. New York: Appleton-Century-Crofts, 373–384.
——— (1958), "The Theoretician's Dilemma: A Study in the Logic of Theory Construction," *Minnesota Studies in the Philosophy of Science* 2: 37–98.
Hull, Clark L. (1943), *Principles of Behavior*. New York: Appleton.
O'Donnell, John M. (1985), *The Origins of Behaviorism: American Psychology, 1870–1920*. New York: NYU Press.
Pavlov, Ivan P. (1955), *Selected Works*. Moscow: Foreign Languages Publishing House.
Quine, Willard Van (1960), *Word and Object*. Cambridge, MA: MIT Press.
Skinner, Burrhus Frederic (1953), *Science and Human Behavior*. New York: Macmillan & Co.
——— (1957), *Verbal Behavior*. New York: Appleton.
Thorndike, Edward L. (1905), *The Elements of Psychology*. New York: A. G. Seiler.
Thyer, Bruce A. (ed.) (1999), *The Philosophical Legacy of Behaviorism*. London: Kluwer Academic Publishers.
Titchener, Edward B. (1898), "The Postulates of a Structural Psychology," *Philosophical Review* 7: 449–465.
Watson, John B. (1913), "Psychology as a Behaviorist Views It," *Psychological Review* 20: 158–177.

See also **Cognitive Science; Logical Empiricism; Mind–Body Problem; Physicalism; Psychology, Philosophy of; Verifiability**

BIOLOGICAL INFORMATION

Information is invoked by biologists in numerous contexts. Animal behaviorists examine the signaling between two organisms or attempt to delimit the structure of the internal map that guides an organism's behavior. Neurobiologists refer to the information passed along neurons and across synapses in brains and nervous systems. The way in which information terminology is used in these contexts has not so far been the main critical focus of philosophers of science. Philosophers of mind discuss animals' representation systems, such as bees' internal maps, and also focus on the way in which brains operate, with a view to shedding light on traditional problems in the philosophy of mind. In contrast, the focus of much discussion in philosophy of biology is the notion of information invoked to explain heredity and development: genetic information. The focus of this article will be on this latter form of biological information.

The ideas that genes are bearers of information and that they contain programs that guide organisms' development are pervasive ones, so much so in biology that they may seem hardly worth examining or questioning. Consulting any biology textbook will reveal that genes contain information in the form of DNA sequences and that this information provides instructions for the production of phenotypes. In contrast, an examination of the philosophical literature on biological information reveals that there are very few philosophers of biology who promote unqualified versions of either of these ideas. To understand how this situation has arisen requires first looking at the role that informational concepts play in biology.

Preliminaries

The two processes that are most relevant to the present context are evolution and development. There was much progress in conceptualizing evolutionary change when it was characterized in terms of changing gene frequencies in the 1930s and 1940s. Many evolutionary biologists discuss evolution entirely from a genetic perspective (see Evolution). After genes were established as the relevant heritable material, the next step was to conceptualize it in terms of molecular structure (see Genetics). In 1953 the structure of DNA was discovered and with this discovery came a mechanism for accounting for the duplication of heritable material and its transmission from one generation to the next. What the discovery of the structure of DNA also ushered in was a research focus for the developing field of molecular biology. An important part of this field is directed at uncovering aspects of organisms' development (see Developmental Biology).

Theory in developmental biology has often diverged from theory in evolutionary biology. Developmental biologists have periodically challenged views and approaches in evolutionary biology, including evolutionary biologists' focus on the gene. With the new techniques in molecular biology came the hope for a unified approach to evolution and development. In this approach, molecular evolutionary biology and molecular developmental biology would work consistently side by side (see Molecular Biology). The processes of development and evolution could be understood from a unified molecular perspective if the component of heredity in evolution were understood to be the passing on of DNA from one generation to the next and the component in development to be the production of proteins from DNA. In this picture, genes were discrete strands of DNA and each was responsible for the production of a particular polypeptide.

The linear structure of DNA and RNA reveals a role that a concept of information can play in understanding heredity and development. The bases in DNA and RNA can be helpfully construed as letters in an alphabet, and the relation between the triplets of letters in the DNA and the resulting polypeptide chain can be construed as a coding relation. So, the DNA contains the code for the polypeptide. Rather than *causing* the production of the relevant protein, the DNA sequence contains the *code* for it.

So, rather than genes being discrete strands of DNA passed on from one generation to the next, they can now be characterized as containing information that is transmitted across generations, and this information is the code for a particular polypeptide. What is relevantly transmitted across generations is the *information* in the DNA, encoded in the unique sequence of bases. Development can now be conceptualized as the faithful transmission of information from DNA to RNA, via complementary base pairs, and then the passing on of that information into the linear structure of the protein, via the coding relation between triplets of base pairs and specific amino acids. Molecular biologists have introduced terminology that is consistent with this approach: The information in DNA is "replicated" in cell division, "transcribed" from DNA to RNA, and "translated" from RNA into proteins.

Although the process of development includes every part of the life cycle of any particular organism, leading to the whole collection of the organism's phenotypic traits, the discussion that follows focuses on the part of the developmental process operating within cells that starts with the separation of DNA strands and concludes with the production of a protein. In some discussions, genetic information is presented as containing instructions for the production for phenotypic traits such as eyes, but these extensions of the concept present many additional problems to those reviewed below (Godfrey-Smith 2000).

The Pervasive Informational Gene Concept: History and Current Practice

In his provocative *What Is Life?* of 1944, the physicist Erwin Shrödinger said "these chromosomes . . . contain in some kind of code-script the entire pattern of the individual's future development and of its functioning in the mature state" (Shrödinger 1944, 20). He went on to explain his terminology:

> In calling the structure of the chromosome fibers a code-script we mean that the all-penetrating mind, once conceived by Laplace, to which every causal connection lay immediately open, could tell from their structure whether the egg would develop, under suitable conditions, into a black cock or into a speckled hen, into a fly or a maize plant, a rhododendron, a beetle, a mouse or a woman. (20–21)

As Morange (1998) put it, Shrödinger saw "genes merely as containers of information, as a code that determines the formation of the individual" (75). Shrödinger's proposals were made before the discovery of the structure of DNA. What is important is that his words were read by many of those who were instrumental in the development of molecular biology.

As Sarkar (1996) points out, Watson and Crick were the first to use the term "information" in the context of discussions of the genetic code:

> The phosphate-sugar backbone of our model is completely regular, but any sequence of the pairs of bases can fit into the structure. It follows that in a long molecule many different permutations are possible, and it therefore seems likely that the precise sequence of the bases is the code which carries the genetical information. (Watson and Crick 1953, 964)

Subsequently, Jacob and Monod also played roles in sustaining Shrödinger's language of the code, helping to reinforce the use of information language in the new field of molecular biology (Keller 2000). By the early 1960s this terminology was established there.

The informational gene concept also became pervasive in the work of theoretical evolutionary biologists. Perhaps the most influential formulation of the concept of heredity in terms of information was that of the evolutionary theorist George Williams. In his influential *Adaptation and Natural Selection*, Williams claims:

> In evolutionary theory, a gene could be defined as any hereditary information for which there is a favorable or unfavorable selection bias equal to several or many times the rate of endogenous change. (Williams 1966, 25)

And, later:

> A gene is not a DNA molecule; it is the transcribable information coded by the molecule. (Williams 1992, 11)

It should now be clear that information terminology is pervasive in disciplines of biology, and also at least somewhat clear why this is the case. There were some historical reasons for adopting the terminology, and there is some utility to the informational concepts. There are, however, some problems associated with construing genes informationally. Many of these problems have been introduced by philosophers of biology, but there has also been much discussion of the informational gene concept within biology.

Problems of the Informational Gene Concept

In several of his recent writings, the evolutionary biologist John Maynard Smith has invited philosophers to join the discussion about the informational gene concept. For example, he says that "given the role that ideas drawn from a study of human communication have played, and continue to play, in biology it is strange that so little attention has been paid to them by philosophers of biology. I think that it is a topic that would reward serious study" (Maynard Smith 2000, 192). While not addressing the concept of genetic information directly, philosophers of biology have been attending to these issues indirectly for some time in working on central problems in the philosophy of biology. For example, the notion of genes as information has played an important role in discussions of reductionism, units of selection, the replicator/interactor distinction, gene/environment interactions, and nativism (see Innate/Acquired Distinction, Population Genetics, Reductionism). Recently, philosophers' focus has turned more explicitly to the informational gene concept. Several philosophers are now engaged in the project of developing a general notion of information that fits best with biologists' aims when they invoke genetic information.

The informational definition of the gene introduced above says that genes contain information that is passed on from one generation to the next, information that codes for particular proteins and polypeptides. As Sterelny and Griffiths (1999) put it: "The classical molecular gene concept is a stretch of DNA that codes for a single polypeptide chain" (132). Genes, in this view, contain information about the phenotype, the protein that is expressed. While most biologists believe that genes contain information about the relevant phenotype, probably no one believes that the information in the genes is sufficient to produce the relevant phenotypes. Even those most routinely chastised for being genetic determinists understand that the information in the gene is expressed only with the aid of a whole host of cellular machinery. As a result, the standard view is that genes contain the relevant or important information guiding the development of the organism. All other cellular machinery merely assists in the expression of the information. One way to put this idea is that genes introduce information to the developmental process, while all other mechanisms make merely a causal contribution to development.

One move that philosophers (and some biologists) have made is to characterize the process of passing on the information in the gene by using terms from information theory. Information theory holds that

> an event carries information about another event to the extent that it is causally related to it in a systematic fashion. Information is thus said to be conveyed over a "channel" connecting the "sender" [or "signal"] with the "receiver" when a change in the receiver is causally related to a change in the sender. (Gray 2001, 190)

In this view information is reduced to causal covariance or systematic causal dependence. Philosophers of biology refer to this characterization of genetic information as the "causal" view. Sterelny and Griffiths (1999) illustrate how the causal information concept could work in the context of molecular biology:

> The idea of information as systematic causal dependence can be used to explain how genes convey developmental information. The genome is the signal and the rest of the developmental matrix provides channel conditions under which the life cycle of the organism contains (receives) information about the genome. (102)

It has been argued that the causal view suffers from serious problems. Sterelny and Griffiths (1999) point out that "it is a fundamental fact of information theory that the role of signal source and channel condition can be reversed" (102) as the signal/channel distinction is simply a matter of causal covariance. Further, the signal/channel distinction is a function of observers' interests. For example, one could choose to hold the developmental history of an organism constant, and from this perspective the organism's phenotype would carry information about its genotype. But if it is instead chosen to "hold all developmental factors other than (say) nutrient quantity constant, the amount of nutrition available to the organism will covary with, and hence also carry information about its phenotype" (102). The causal information concept is lacking, because it cannot distinguish the genes as the singular bearers of important or relevant information. Rather, in this view, genes are just one source of information; aspects of the organism's environment and cellular material also contain information. This position is called the parity thesis (Griffiths and Gray 1994). The parity thesis exposes the need for another information concept that elevates genes alone to the status of information bearers.

Alternative concepts of information have been examined in attempts to respond to this situation; one is referred to variously as intentional, semantic, or teleosemantic information. This notion of information has been defended most forcefully recently by Maynard Smith, and also by philosophers Daniel Dennett (1995) and Kim Sterelny (2000). The term *teleosemantics* is borrowed from "the philosophical program of reducing meaning to biological function (teleology) and then reducing biological function to" natural selection. (A good survey of relations between the philosophy of mind and genetic information concepts is provided in Godfrey-Smith 1999.) This view is articulated in the philosophy of mind as the thesis that a mental state token, such as a sentence, has the biological function of representing a particular state of the world and that this function arose as a result of selection.

Applying this view to the current problem results in the following: "A gene contains information about the developmental outcomes that it was selected to produce" (Sterelny and Griffiths 1999, 105). Maynard Smith puts the view as: "DNA contains information that has been programmed by natural selection" (Maynard Smith 2000, 190). Here the information in the gene is analogous to a sentence in the head. The gene contains information as a result not just of relevantly causal covariance with the phenotype, but of having the function of producing the relevant phenotype. Defenders of this view claim that this function allows for the information to stay the same even if the channel conditions change, in which case the information in the gene has simply been misinterpreted. This concept could solve the problem of rendering the genes as the sole information bearers, as "if other developmental causes do not contain [teleosemantic] information and genes do, then genes do indeed play a unique role in development" (Sterelny and Griffiths 1999, 104).

Although the teleosemantic view shows promise, the debate has not ended here. The teleosemantic view opens up a possibility: If a developmental cause—part of the cellular machinery, for example—is found to be heritable and performs the function of producing a particular developmental outcome, then, by definition, it also contains teleosemantic information. Many, including Sarkar (1996, 2000), Griffiths and Gray (1994), Gray (2001), Keller (2000), Sterelny (2000), have argued that indeed there are such mechanisms. These authors draw various conclusions from the demonstrated presence of mechanisms that are not genes, are heritable, and perform the function of producing a specific developmental outcome. Developmental systems theorists such as Griffiths and Gray take these findings to show that teleosemantic information succumbs to the parity thesis also. They go on to argue that no concept of information will distinguish genes as a special contributor to development. Genes are just fellow travelers alongside cellular machinery and the environment in shaping developmental outcomes. Others such as Sarkar and Keller are more cautious and hold out for a concept of information that can distinguish genes as a distinct kind of information bearer. On the other side, Maynard Smith and others have attempted to refine the notion of teleosemantic information to preserve a biological distinction that seems to be important: "The most fundamental distinction in biology is between nucleic acids, with their role as carriers of information, and proteins, which generate the phenotype" (Maynard Smith and Szathmary 1995, 61).

Three coherent options present themselves to answer the question, Where is biological information found?

1. Information is present in DNA and other nucleotide sequences. Other cellular mechanisms contain no information.
2. Information is present in DNA, other nucleotide sequences, and other cellular mechanisms (for example, cytoplasmic or extracellular proteins), and in many other media—for example, the embryonic environment or components of an organism's wider environment.
3. DNA and other nucleotide sequences do not contain information, nor do any other cellular mechanisms.

These options can be read either ontologically or heuristically. The ontological reading of (1) is that there is a certain kind of information that is present only in DNA and other nucleotide sequences. As a result, any workable concept of information is constrained. The concept adopted cannot be consistent with information of the relevant sort existing in any other media that are causally responsible for an organism's development. The heuristic reading of (1) is that viewing information as present in DNA and other nucleotides is the most reliable guide to good answers in research in developmental molecular biology. The philosophical discussion presented above focuses on developing or challenging accounts of information that are consistent with an ontological reading of (1). For example, Maynard Smith and others, such as Dennett, are defenders of an ontological version of (1).

Many assume that (2) makes sense ontologically only if one adopts a causal information concept, but some of the discussion already referred to indicates that other developmentally relevant media can be construed as containing teleosemantic information. Defenders of the developmental systems theory approach hold a version of (2), as does Sarkar (1996).

Only Waters (2000) seems to have provided a sustained defense of (3). Maynard Smith argues that to construe all processes of development in causal terms without recourse to the concept of genetic information is to relegate them to the hopelessly complex and to implicitly argue that no systematic explanations will be forthcoming (see e.g., Maynard Smith 1998, 5–6). Waters differs, arguing that informational talk in biology is misleading and can entirely be coherently substituted for by causal talk. Waters also argues that it is the intent of most practicing biologists to provide a causal account of development rather than one that invokes information.

In conclusion, philosophers are actively cooperating with theoretical biologists to develop fruitful concepts of information that help make sense of the information terminology widely used in biology. These discussions are as yet inconclusive, and as a result this is a potentially fertile area for future work.

Stephen M. Downes

The author acknowledges the helpful input of Sahotra Sarkar, University of Texas and Lindley Darden, University of Maryland.

References

Dennett, D. C. (1995), *Darwin's Dangerous Idea*. New York: Simon and Schuster.

Godfrey-Smith, P. (1999), "Genes and Codes: Lessons from the Philosophy of Mind?" in V. G. Hardcastle (ed.), *Where Biology Meets Psychology: Philosophical Essays*. Cambridge, MA: MIT Press: 305–332.

――― (2000), "On the Theoretical Role of 'Genetic Coding,'" *Philosophy of Science* 67: 26–44.

Gray, R. D. (2001), "Selfish Genes or Developmental Systems?" in R. S. Singh, C. B. Krimbas D. B. Paul, and J. Beatty (eds.), *Thinking About Evolution: Historical, Philosophical, and Political Perspectives*. Cambridge: Cambridge University Press, 184–207.

Griffiths, P. E., and R. D. Gray (1994), "Developmental Systems and Evolutionary Explanation," *Journal of Philosophy* 91: 277–304.

Keller, E. F. (2000), "Decoding the Genetic Program: Or, Some Circular Logic in the Logic of Circularity," in P. J. Beurton, R. Falk, and H. Rheinberger (eds.), *The Concept of the Gene in Development and Evolution*. Cambridge: Cambridge University Press, 159–177.

Maynard Smith, J. (1998), *Shaping Life*. New Haven, CT: Yale University Press.

――― (2000), "The Concept of Information in Biology," *Philosophy of Science* 67: 177–194.

Maynard Smith, J., and E. Szathmary (1995), *The Major Transitions in Evolution*. Oxford: Oxford University Press.

Morange, M. (1998), *A History of Molecular Biology*. Cambridge, MA: Harvard University Press.

Sarkar, S. (1996), "Biological Information: A Skeptical Look at Some Central Dogmas of Molecular Biology," in S. Sarkar (ed.), *The Philosophy and History of Molecular Biology: New Perspectives*. Dordrecht, Netherlands: Kluwer, 187–231.

――― (2000). "Information in Genetics and Developmental Biology," *Philosophy of Science* 67: 208–213.

Schrödinger, E. (1944), *What Is Life? The Physical Aspects of the Living Cell*. Cambridge: Cambridge University Press.

Sterelny, K. (2000), "The 'Genetic Program' Program: A Commentary on Maynard Smith on Information in Biology," *Philosophy of Science* 67: 195–201.

Sterelny, K., and P. E. Griffiths (1999), *Sex and Death*. Chicago: University of Chicago Press.

Waters, K. (2000), "Molecules Made Biological," *Revue Internationale de Philosophie* 4: 539–564.

Watson, J. D., and F. H. C. Crick (1953), "Genetical Implications of the Structure of Deoxyribonucleic Acid," *Nature* 171: 964–967.

Williams, G. C. (1966), *Adaptation and Natural Selection*. Princeton, NJ: Princeton University Press.

――― (1992), *Natural Selection: Domains, Levels and Challenges*. New York: Oxford University Press.

See also **Evolution; Genetics; Molecular Biology; Population Genetics**

PHILOSOPHY OF BIOLOGY

The philosophy of biology has existed as a distinct subdiscipline within the philosophy of science for about 30 years. The rapid growth of the field has mirrored that of the biological sciences in the same period. Today the discipline is well represented in the leading journals in philosophy of science, as well as in several specialist journals. There have been two generations of textbooks (see Conclusion), and the subject is regularly taught at the undergraduate as well as the graduate level. The current high profile of the biological sciences and the obvious philosophical issues that arise in fields as diverse as molecular genetics and conservation biology suggest that the philosophy of biology will remain an exciting field of enquiry for the foreseeable future.

Three Kinds of Philosophy of Biology

Philosophers have engaged with biological science in three quite distinct ways. Some have looked to biology to test general theses in philosophy of science (see Laws of Nature). Others have engaged with conceptual puzzles that arise within biology itself. Finally, philosophers have looked to biological science for answers to distinctively philosophical questions in such fields as ethics, the philosophy of mind, and epistemology (see Evolutionary Epistemology).

The debate that marked the beginning of contemporary philosophy of biology exemplified the first of these three approaches, the use of biological science as a testing ground for claims in general philosophy of science. In the late 1960s, Kenneth C. Schaffner applied the logical empiricist model of theory reduction to the relationship between classical, Mendelian genetics and the new molecular genetics (Schaffner 1967, 1969; Hull 1974). While the failure of this attempt in its initial form reinforced the near-consensus in the 1970s and 1980s that the special sciences are autonomous from the more fundamental sciences, it also led the formulation of increasingly more adequate models of theory reduction (Schaffner 1993; Sarkar 1998) (see Reductionism).

Another important early debate showed philosophy engaging biology in the second way, by confronting a conceptual puzzle within biology itself. The concept of reproductive fitness is at the heart of evolutionary theory, but its status has always been problematic (see Fitness). It has proved surprisingly hard for biologists to avoid the criticism that natural selection explains the reproductive success of organisms by citing their fitness, while defining their fitness in terms of their reproductive success (the so-called tautology problem). Philosophical analysis of this problem begins by noting that fitness is a supervenient property of organisms: The fitness of each particular organism is a consequence of some specific set of physical characteristics of the organism and its particular environment, but two organisms may have identical levels of fitness in virtue of very different sets of physical characteristics (Rosenberg 1978) (see Supervenience). The most common solution to the tautology problem is to argue that this supervenient property is a propensity—a probability distribution over possible numbers of offspring (Mills and Beatty 1979). Thus, although fitness is defined in terms of reproductive success, it is not a tautology that the fittest organisms have the most offspring. Fitness merely allows us to make fallible predictions about numbers of offspring that become more reliable as the size of the population tends to infinity. It remains unclear, however, whether it is possible to specify a probability distribution or set of distributions that can play all the roles actually played by fitnesses in population biology (Rosenberg and Bouchard 2002).

The third way in which philosophy has engaged with biology is by tracing out the wider ethical, epistemological, and metaphysical implications of biological findings. This has sometimes occurred in response to philosophical claims issuing from within biology itself. For example, some proponents of sociobiology—the application to humans of the models developed in behavioral ecology in the 1960s—suggested that the conventional social sciences could be reduced to or replaced by behavioral biology (see Evolutionary Psychology). Others claimed that certain aspects of human behavior result from strongly entrenched aspects of human biology and thus that public policy must be designed to work with and around such behavior rather than seeking to eradicate it. These claims were evaluated by leading philosophers of biology like Michael Ruse (1979), Alexander Rosenberg (1980), and Philip Kitcher (1985).

On other occasions, rather than responding to philosophical claims issuing from within biology, philosophers have actively sought from biology answers to questions arising in their own discipline that may not be of particular interest to working biologists. The extensive literature on biological teleology is a case in point (see Function). After a brief flurry of interest around the time of the modern synthesis (see Evolution), during which the term "teleonomy" was introduced to denote the specifically evolutionary interpretation of teleological language (Pittendrigh 1958), the ideas of function and goal directedness were regarded as relatively unproblematic by evolutionary biologists, and there was little felt need for any further theoretical elaboration of these notions. In the 1970s, however, philosophers started to look to biology to provide a solid, scientific basis for normative concepts, such as illness or malfunction (Wimsatt 1972; Wright 1973). These discussions eventually converged on an analysis of teleological language fundamentally similar to the view associated with the modern synthesis, although elaborated in far greater detail. According to the *etiological theory of function,* the functions of a trait are those activities in virtue of which the trait was selected (Brandon 1981; Millikan 1984; Neander 1991, 1995). Despite continued disputes over the scope and power of the etiological theory amongst philosophers of biology (Ariew, Cummins, and Perlman 2002), the idea of "etiological" or "proper" function has become part of the conceptual toolkit of philosophy in general and of the philosophy of language and of mind in particular.

These three approaches to doing philosophy of biology are exemplified in different combinations in philosophical discussion of the several biological disciplines.

The Philosophy of Evolutionary Biology

Evolutionary theory has been used as a case study in support of views of the structure of scientific theories in general, an approach that conforms to the "testing ground" conception of philosophy of biology described above (see Evolution). The example is most often thought to favor the "semantic view" of theories (Lloyd 1988) (see Theories). Most philosophical writing about evolutionary theory, however, is concerned with conceptual puzzles that arise inside the theory itself, and the work often resembles theoretical biology as much as pure philosophy of science. Elliott Sober's classic study *The Nature of Selection: Evolutionary Theory in Philosophical Focus* (Sober 1984) marks the point at which most nonspecialists became aware of the philosophy of biology as a major new field. In this work Sober analyzed the structure of selective explanations via an analogy with the composition of forces in dynamics, treating the actual change in gene frequencies over time as the result of several different "forces," such as selection, drift, and mutation (see Natural Selection). Sober's book also introduced the widely used distinction between "selection for" and "selection of." Traits that are causally connected to reproductive success and can therefore be used to *explain* reproductive success are said to be selected *for* (or to be "targets" of selection). In contrast, there is selection *of* traits that do not have this property but nevertheless are statistically associated with reproductive success, usually because they are linked in some way to traits that *do* have the property. For example, when two DNA segments are "linked" in the classical sense of being close to one another on the same chromosome, they have a high probability of being inherited together (see Genetics). If only one of the two segments has any effect on the phenotype, it is the presence of this segment alone that explains the success of both. There is selection *for* the causally active segment but only selection *of* its passive companion.

Robert Brandon's (1990) analysis of the concept of the environment is, similarly, of as much interest to biologists as to philosophers. Several biological authors have criticized the idea that the "environment," in the sense in which organisms are adapted to it, can be described independently of the organisms themselves. Brandon defines three different notions of 'environment,' all of which are needed to make sense of the role of environment in natural selection. All organisms in a particular region of space and time share the "external environment," but to understand the particular selective forces acting on one lineage of organisms it is necessary to pick out a specific "ecological environment" consisting of those environmental parameters whose value affects the reproductive output of members of the lineage. The ecological environment of a fly will be quite different from that of a tree, even if they occupy the same external environment. Finally, the "selective environment" is that part of the ecological environment that *differentially* affects the reproductive output of variant forms in the evolving lineage. It is this last that contains the sources of adaptive evolutionary pressures on the lineage.

Part of the early philosophical interest in selective explanation arose due to philosophical interest in sociobiology. Sociobiology was widely criticized for its "adaptationism," or an exclusive focus on selection to the exclusion of other evolutionary factors (see Adaptation and Adaptationism). This gave rise to several important papers on the concept of "optimality" in evolutionary modeling (Dupré 1987). Philosophers have now distinguished several distinct strands of the adaptationism debate, and many of the remaining issues are clearly empirical rather than conceptual, as is made clear in the latest collection of papers on this issue (Orzack and Sober 2001).

The sociobiology debate, and related discussion of the idea that the fundamental unit of evolution is the individual Mendelian allele (Dawkins 1976), also drove the explosion of philosophical work on the "units of selection" question in the 1980s (Brandon and Burian 1984). Philosophical work on the units-of-selection question has tended to favor some form of pluralism, according to which there may be units of selection at several levels within the hierarchy of biological organization— DNA segments, chromosomes, cells, organisms, and groups of organisms. Arguably, philosophers made a significant contribution to the rehabilitation of some forms of "group selection" in evolutionary biology itself, following two decades of neglect (Sober and Wilson 1998).

More recently, a heated debate has developed over the ontological status of the probabilities used in population biology (see Evolution). On the one hand, the best models of the evolutionary process assign organisms a certain probability of reproducing (fitness) and make probabilistic predictions about the evolutionary trajectory of populations. On the other hand, the actual process of evolution is the aggregation of the lives of many individual organisms, and those organisms lived, died, and reproduced in accordance with deterministic, macro-level physical laws. Hence, it has been argued, the evolutionary process itself is deterministic, a vast soap opera in which each member of the cast has an eventful history determined by particular causes; and the probabilities in evolutionary models are introduced because one cannot follow the process in all its detail (Rosenberg 1994; Walsh 2000). If correct, this argument has some interesting implications. It would seem to follow, for example, that there is no real distinction in nature between the process of drift and the process of natural selection. Robert Brandon and Scott Carson (1996) have strongly rejected this view, insisting that evolution is a genuinely indeterministic process and that the probabilistic properties ascribed to organisms by evolutionary models should be accepted in the same light as the ineliminable explanatory posits of other highly successful theories.

The Philosophy of Systematic Biology

Philosophical discussion of systematics was a response to a "scientific revolution" in that discipline in the 1960s and 1970s, which saw the discipline first transformed by the application of quantitative methods and then increasingly dominated by the "cladistic" approach, which rejects the view that systematics should sort organisms into a hierarchy of groups representing a roughly similar amount of diversity, and argues that its sole aim should be to represent evolutionary relationships between groups of organisms (phylogeny). Ideas from the philosophy of science were used to argue for both transformations, and the philosopher David L. Hull (1988) was an active participant throughout this whole period. Another major treatment of cladism was by Sober (1988).

The best-known topic in the philosophy of systematics was introduced by the biologist Michael Ghiselin (1974), when he suggested that traditional systematics was fundamentally mistaken about the ontological status of biological species (see also Hull 1976). Species, it was argued, are not natural kinds of organisms in the way that chemical elements are natural kinds of matter. Instead, they are historical particulars like families or nations (see Individuality). However, the view that species are historical particulars leaves other important questions about species unsolved and raises new problems of its own. As many as 20 different so-called species concepts are represented in the current biological literature, and their merits, interrelations, and mutual consistency or inconsistency have been a major topic of philosophical discussion (the papers collected in Ereshefsky [1992] provide a good introduction to these debates) (see also Species).

The philosophy of systematics has influenced general philosophy of science and, indeed, metaphysics, through its challenge to one of the two classical examples of a "natural kind," viz., biological species. The result has been a substantial reevaluation of what is meant by a natural kind, whether there are natural kinds, and whether traditional views about the nature of science that rely on the idea of natural kinds must be rejected (Wilkerson 1993; Dupré 1993; Wilson 1999).

The Philosophy of Molecular Biology

As mentioned above, one of the first topics to be discussed in the philosophy of biology was the reduction of Mendelian to molecular genetics. The initial debate between Schaffner and Hull was followed by the "anti-reductionist consensus," embodied in Philip Kitcher's (1984) classic paper *1953 and All That: A Tale of Two Sciences*. The reductionist position was revived in a series of important papers by Kenneth Waters (1990, 1994) and debate over the cognitive relationship between these two theories continues today, although the question is not now framed as a simple choice between reduction and irreducibility. For example, William Wimsatt has tried to understand 'reduction' not as a judgment on the fate of a theory, but as one amongst several strategies that scientists can deploy when trying to unravel complex systems (see Reductionism). The philosophical interest lies in understanding the strengths and weaknesses of this strategy (Wimsatt 1976, 1980). Lindley Darden, Schaffner, and others have argued that explanations in molecular biology are not neatly confined to one ontological level, and hence that ideas of "reduction" derived from classical examples like the reduction of the phenomenological gas laws to molecular kinematics in nineteenth-century physics are simply inapplicable (Darden and Maull 1977; Schaffner 1993). Moreover, molecular biology does not have the kind of grand theory based around a set of laws or a set of mathematical models that is familiar from the physical sciences. Instead, highly specific mechanisms that have been uncovered in detail in one model organism seem to act as "exemplars" allowing the investigation of similar, although not necessarily identical, mechanisms in other organisms that employ the same, or related, molecular interactants. Darden and collaborators have argued that these "mechanisms"—specific collections of entities and their distinctive activities—are the fundamental unit of scientific discovery and scientific explanation, not only in molecular biology, but in a wide range of special sciences (Machamer, Darden, and Craver 2000) (see Mechanisms).

An important strand in the early debate over reduction concerned the different ways in which the gene itself is understood in Mendelian and molecular genetics. The gene of classical Mendelian genetics has been replaced by a variety of structural and functional units in contemporary molecular genetics (see Genetics). One response to this is pluralism about the gene (Falk 2000). Another is to identify a central tendency that unifies the various different ways in which the term 'gene' is used (Waters 1994, 2000). Identifying the different ways in which genes are conceived in different areas of molecular biology and their relations to one another is a major focus of current research (Beurton, Falk, and Rhineberger 2000; Moss 2002; Stotz, Griffiths, and Knight 2004). Another very active topic is the concept of genetic information, or developmental information more generally (Sarkar 1996a, 2004; Maynard Smith 2000; Griffiths 2001; Jablonka 2002) (see Biological Information; Molecular Biology).

The Philosophy of Developmental Biology

Developmental biology has received growing attention from philosophers in recent years. The debate over "adaptationism" introduced philosophers to the idea that explanations of traits in terms of natural selection have time and time again in the history of Darwinism found themselves in competition with explanations of the same traits from developmental biology.

Developmental biology throws light on the kinds of variation that are likely to be available for selection, posing the question of how far the results of evolution can be understood in terms of the options that were available ("developmental constraints") rather that the natural selection of those options (Maynard Smith et al. 1985). The question of when these explanations compete and when they complement one another is of obvious philosophical interest. The debate over developmental constraints looked solely at whether developmental biology could provide answers to evolutionary questions. However, as Ron Amundson pointed out, developmental biologists are addressing questions of their own, and, he argued, a different concept of 'constraint' is needed to address those questions (Amundson 1994). In the last decade several other debates in the philosophy of biology have taken on a novel aspect by being viewed from the standpoint of developmental biology. These include the analysis of biological teleology (Amundson and Lauder 1994), the units-of-selection debate (Griffiths and Gray 1994), and the nature of biological classification, which from the perspective of development is as much a debate about classifying the parts of organisms as about classifying the organisms themselves (Wagner 2001). The vibrant new field of evolutionary developmental biology is transforming many evolutionary questions within biology itself and hence causing philosophers to revisit existing positions in the philosophy of evolutionary biology (Brandon and Sansom 2005).

Increasing philosophical attention to developmental biology has also led philosophers of biology to become involved in debates over the concept of innateness, the long tradition of philosophical literature on this topic having previously treated innateness primarily as a psychological concept (Ariew 1996; Griffiths 2002).

The Philosophy of Ecology and Conservation Biology

Until recently this was a severely underdeveloped field in the philosophy of biology, which was surprising, because there is obvious potential for all three of the approaches to philosophy of biology discussed above. First, ecology is a demanding testing ground for more general ideas about science, for reasons explained below (see Ecology). Second, there is a substantial quantity of philosophical work in environmental ethics, and it seems reasonable to suppose that answering the questions that arise there would require a critical methodological examination of ecology and conservation biology. Finally, ecology contains a number of deep conceptual puzzles, which ecologists themselves have recognized and discussed extensively.

The most substantial contributions to the field to date include works by Kristin Shrader-Frechette and Earl McCoy (1993), Gregory Cooper (2003), and Lev Ginzburg and Mark Colyvan (2004). Cooper focuses on the particular methodological problems that confront ecology as a result of its subject matter—massively complex, and often unique, systems operating on scales that frequently make controlled experiment impractical—and on the consequent lack of connection between the sophisticated mathematical modeling tradition in ecology and ecological field work. Shrader-Frechette and McCoy's book is concerned primarily with how practical conservation activity can be informed by ecological theory despite the problems addressed by Cooper (for a related discussion, see Sarkar 1996b). Ginzburg and Colyvan, in contrast, argue forcefully that ecology may still produce simple, general theories that will account for the data generated by ecological field work in as satisfactory a manner as Newtonian dynamics accounted for the motion of the planets.

The concept of the niche stands in marked contrast to other ecological concepts in that 'niche' has been widely discussed by philosophers of biology (summarized in Sterelny and Griffiths 1999, 268–279). This, however, reflects the importance of the niche concept in *evolutionary* biology. Topics that merit much more attention than the little they have received to date include the concept of biodiversity and of stability (or, in its popular guise, the "balance of nature"). A recent extended philosophical discussion of these concepts, integrating themes from the philosophy of ecology and conservation biology with more traditional environmental philosophy, is found in Sarkar (2005) (see also Conservation Biology).

Conclusion

The philosophy of biology is a flourishing field, partly because it encompasses all three of the very different ways in which philosophy makes intellectual contact with the biological sciences, as discussed above. The scope of philosophical discussion has extended from its starting points in evolutionary biology to encompass systematics, molecular biology, developmental biology, and, increasingly, ecology and conservation biology. For those who wish to explore the field beyond this article and the related articles in this volume, recent textbooks include Sober (1993) and Sterelny and Griffiths (1999). Two valuable edited collections designed to supplement such a text are Sober (1994), which collects the classic papers on core debates, and Hull and Ruse (1998), which aims at a comprehensive survey using recent papers. Keller and Lloyd (1992) have edited an excellent collection on evolutionary biology aimed primarily at philosophers of biology.

PAUL E. GRIFFITHS

References

Amundson, Ron (1994), "Two Concepts of Constraint: Adaptationism and the Challenge from Developmental Biology," *Philosophy of Science* 61: 556–578.
Amundson, Ron, and George V. Lauder (1994), "Function Without Purpose: The Uses of Causal Role Function in Evolutionary Biology," *Biology and Philosophy* 9: 443–470.
Ariew, Andre (1996), "Innateness and Canalization," *Philosophy of Science* 63(suppl): S19–S27.
Ariew, Andre, Robert Cummins, and Mark Perlman (eds.) (2002), *Functions: New Essays in the Philosophy of Psychology and Biology*. New York and Oxford: Oxford University Press.
Beurton, Peter, Raphael Falk, and Hans-Joerg Rhineberger (2000), *The Concept of the Gene in Development and Evolution*. Cambridge: Cambridge University Press.
Brandon, Robert N. (1981), "Biological Teleology: Questions and Explanations," *Studies in the History and Philosophy of Science* 12: 91–105.
——— (1990), *Adaptation and Environment*. Princeton, NJ: Princeton University Press.
Brandon, Robert N., and Richard M. Burian (eds.) (1984), *Genes, Organisms, Populations: Controversies Over the Units of Selection*. Cambridge, MA: MIT Press.
Brandon, Robert N., and Scott Carson (1996), The Indeterministic Character of Evolutionary Theory: No "Hidden Variables Proof" but No Room for Determinism Either," *Philosophy of Science* 63: 315–337.
Brandon, Robert N., and Roger Sansom (eds.) (2005), *Integrating Evolution and Development*. Cambridge: Cambridge University Press.
Cooper, Gregory (2003), *The Science of the Struggle for Existence: On the Foundations of Ecology*. Edited by M. Ruse. Cambridge Studies in Biology and Philosophy. Cambridge: Cambridge University Press.
Darden, Lindley, and Nancy Maull (1977), "Interfield Theories," *Philosophy of Science* 44: 43–64.
Dawkins, Richard (1976), *The Selfish Gene*. Oxford: Oxford University Press.
Dupré, John (1993), *The Disorder of Things: Metaphysical Foundations of the Disunity of Science*. Cambridge, MA: Harvard University Press.
——— (ed.) (1987), *The Latest on the Best: Essays on Optimality and Evolution*. Cambridge, MA: MIT Press.
Ereshefsky, Marc (ed.) (1992), *The Units of Evolution*. Cambridge, MA: MIT Press.
Falk, Raphael (2000), "The Gene: A Concept in Tension," in P. Beurton, R. Falk, and H.-J. Rheinberger (eds.), *The Concept of the Gene in Development and Evolution*. Cambridge: Cambridge University Press.
Ghiselin, Michael T. (1974), "A Radical Solution to the Species Problem," *Systematic Zoology* 23: 536–544.
Ginzburg, Lev, and Mark Colyvan (2004), *Ecological Orbits: How Planets Move and Populations Grow*. Oxford and New York: Oxford University Press.
Griffiths, Paul E. (2001), "Genetic Information: A Metaphor in Search of a Theory," *Philosophy of Science* 68: 394–412.
——— (2002), "What Is Innateness?" *The Monist* 85: 70–85.
Griffiths, P. E, and R. D Gray (1994), "Developmental Systems and Evolutionary Explanation," *Journal of Philosophy* XCI: 277–304.
Hull, David L. (1974), *Philosophy of Biological Science*. Englewood Cliffs, NJ: Prentice-Hall Inc.
——— (1976), "Are Species Really Individuals?" *Systematic Zoology* 25: 174–191.
——— (1988), *Science as a Process: An Evolutionary Account of the Social and Conceptual Development of Science*. Chicago: University of Chicago Press.

Hull, David L., and Michael Ruse (eds.) (1998), *The Philosophy of Biology*. Oxford: Oxford University Press.

Jablonka, Eva (2002), "Information Interpretation, Inheritance, and Sharing," *Philosophy of Science* 69: 578–605.

Keller, Evelyn Fox, and Elisabeth A. Lloyd (eds.) (1992), *Keywords in Evolutionary Biology*. Cambridge, MA: Harvard University Press.

Kitcher, Philip (1984), "1953 and All That: A Tale of Two Sciences," *Philosophical Review* 93: 335–373.

——— (1985), *Vaulting Ambition: Sociobiology and the Quest for Human Nature*. Cambridge, MA: MIT Press.

Lloyd, Elizabeth A. (1988), *The Structure and Confirmation of Evolutionary Theory*. Westport, CT: Greenwood Press.

Machamer, Peter, Lindley Darden, and Carl Craver (2000), "Thinking about Mechanisms," *Philosophy of Science* 67: 1–25.

Maynard Smith, John (2000), "The Concept of Information in Biology," *Philosophy of Science* 67: 177–194.

Maynard Smith, J., Richard M. Burian, Stuart Kauffman, Pere Alberch, J. Campbell, B. Goodwin, et al. (1985), "Developmental Constraints and Evolution," *Quarterly Review of Biology* 60: 265–287.

Millikan, Ruth G. (1984), *Language, Thought and Other Biological Categories*. Cambridge, MA: MIT Press.

Mills, Susan, and John Beatty (1979), "The Propensity Interpretation of Fitness," *Philosophy of Science* 46: 263–286.

Moss, L. (2002), *What Genes Can't Do*. Cambridge, MA: MIT Press.

Neander, Karen (1991), "Functions as Selected Effects: The Conceptual Analyst's Defense," *Philosophy of Science* 58: 168–184.

——— (1995), "Misrepresenting and Malfunctioning," *Philosophical Studies* 79: 109–141.

Orzack, Steve, and Elliott Sober (eds.) (2001), *Optimality and Adaptation*. Cambridge: Cambridge University Press.

Pittendrigh, C. S. (1958), "Adaptation, Natural Selection and Behavior," in A. Roe and G. Simpson (eds.), *Behavior and Evolution*. New York: Academic Press.

Rosenberg, Alexander (1978), "The Supervenience of Biological Concepts," *Philosophy of Science* 45: 368–386.

——— (1980), *Sociobiology and the Preemption of Social Science*. Baltimore: Johns Hopkins University Press.

——— (1994), *Instrumental Biology or The Disunity of Science*. Chicago: Chicago University Press.

Rosenberg, Alexander, and Frederic Bouchard (2002), "Fitness," *Stanford Encyclopedia of Philosophy*. Available at http://plato.stanford.edu/archives/win2002/entries/fitness. Accessed August 2005.

Ruse, Michael (1979), *Sociobiology: Sense or Nonsense*. Dordrecht, Holland: Reidel.

Sarkar, Sahotra (1996a), "Biological Information: A Sceptical Look at Some Central Dogmas of Molecular Biology," in *The Philosophy and History of Molecular Biology: New Perspectives*. Dordrecht, Holland: Kluwer Academic Publishers.

——— (1996b), "Ecological Theory and Anuran Declines," *BioScience* 46: 199–207.

——— (1998), *Genetics and Reductionism*. Cambridge: Cambridge University Press.

——— (2004), "Molecular Models of Life: Philosophical Papers on Molecular Biology," Cambridge, MA: MIT Press.

——— (2005), *Biodiversity and Environmental Philosophy: An Introduction*. Cambridge: Cambridge University Press.

Schaffner, Kenneth F. (1967), "Approaches to Reduction," *Philosophy of Science* 34: 137–47.

——— (1969), "The Watson-Crick Model and Reductionism," *British Journal for the Philosophy of Science* 20: 325–348.

——— (1993), *Discovery and Explanation in Biology and Medicine*. Chicago and London: University of Chicago Press.

Shrader-Frechette, Kristin S., and Earl D. McCoy (1993), *Method in Ecology: Strategies for Conservation*. Cambridge and New York: Cambridge University Press.

Sober, Elliott (1984), *The Nature of Selection: Evolutionary Theory in Philosophical Focus*. Cambridge, MA: MIT Press.

——— (1988), *Reconstructing the Past: Parsimony, Evolution and Inference*. Cambridge, MA: MIT Press.

——— (1993), *Philosophy of Biology*. Boulder, CO: Westview Press.

——— (ed.) (1994), *Conceptual Issues in Evolutionary Biology: An Anthology* (2nd ed.). Cambridge, MA: MIT Press.

Sober, Elliott, and David S. Wilson (1998), *Unto Others: The Evolution and Psychology of Unselfish Behavior*. Cambridge, MA: Harvard University Press.

Sterelny, Kim, and Paul E. Griffiths (1999), *Sex and Death: An Introduction to the Philosophy of Biology*. Chicago: University of Chicago Press.

Stotz, Karola, Paul E. Griffiths, and Rob D. Knight (2004), "How Scientists Conceptualise Genes: An Empirical Study," *Studies in History and Philosophy of Biological and Biomedical Sciences* 35: 647–673.

Wagner, Günther P. (ed.) (2001), *The Character Concept in Evolutionary Biology*. San Diego: Academic Press.

Walsh, Dennis M. (2000), "Chasing Shadows: Natural Selection and Adaptation," *Studies in the History and Philosophy of the Biological and Biomedical Sciences* 31C: 135–154.

Waters, C. Kenneth (1990), "Why the Antireductionist Consensus Won't Survive the Case of Classical Mendelian Genetics," in A. Fine, M. Forbes, and L. Wessells (eds.), *Proceedings of the Biennial Meeting of the Philosophy of Science Association*. East Lansing, MI: Philosophy of Science Association.

——— (1994), "Genes Made Molecular," *Philosophy of Science* 61: 163–185.

——— (2000), "Molecules Made Biological," *Revue Internationale de Philosophie* 4: 539–564.

Wilkerson, Timothy E. (1993), "Species, Essences and the Names of Natural Kinds," *Philosophical Quarterly* 43: 1–9.

Wilson, Robert A. (ed.) (1999), *Species: New Interdisciplinary Essays*. Cambridge, MA: MIT Press.

Wimsatt, William C. (1972), "Teleology and the Logical Structure of Function Statements," *Studies in the History and Philosophy of Science* 3: 1–80.

——— (1976), "Reductive Explanation: A Functional Account," in R. S. Cohen (ed.), *Proceedings of the Philosophy of Science Association, 1974*. East Lansing, MI: Philosophy of Science Association.

——— (1980), "Reductionistic Research Strategies and Their Biases in the Units of Selection Controversy," in

T. Nickles (ed.), *Scientific Discovery: Case Studies*. Dordrecht, Holland: D. Reidel Publishing Company.

Wright, Larry (1973), "Functions," *Philosophical Review* 82: 139–168.

See also **Altruism; Biological Individual; Biological Information; Conservation Biology; Ecology; Emergence; Evolution; Evolutionary Epistemology; Evolutionary Psychology; Fitness; Function; Genetics; Immunology; Innate/Acquired Distinction; Laws of Nature; Mechanisms; Molecular Biology; Natural Selection; Neurobiology; Prediction; Probability; Reductionism; Scientific Models; Species; Supervenience**

PERCY WILLIAMS BRIDGMAN

(21 April 1882–20 August 1961)

Percy Williams Bridgman won a Nobel Prize for his experimental work in high-pressure physics in 1946. In addition, his constant interest in understanding and improving the scientific method led him to make significant and lasting contributions to the philosophy of science. Specifically, Bridgman is responsible for identifying and carefully explicating a method for defining scientific concepts called *operationalism*. As a philosopher of science, Bridgman recognized that the importance of Einstein's theory of special relativity was not limited to mechanics or even just to the physical sciences. He saw Einstein as looking for the meaning of simultaneity and finding that meaning by analyzing the physical operations necessary in order to use the concept in any concrete situation. Bridgman revealed his unwavering empiricist roots by claiming that scientific theories are not valuable for their so-called metaphysical consequences, but for what they actually do. Likewise, Bridgman thought that concepts in theories should not be abstract, metaphysical ideas, but rather concrete operations.

Life

Bridgman was born April 21, 1882, in Cambridge, Massachusetts, and attended public schools in Newton. He matriculated at Harvard College in 1900 and graduated *summa cum laude* in 1904. He immediately began his graduate studies in physics at Harvard, where he received his M.A. in 1905 and his Ph.D. in 1908. Bridgman remained at Harvard for his entire career, becoming an instructor of physics in 1910, assistant professor in 1913, and professor of physics in 1919. In 1926, he was named Hollis Professor of Mathematics and Natural Philosophy, and, in 1950, the Higgins University Professor. He retired in 1954 and became professor emeritus. Believing that people should not outlive their usefulness, Bridgman took his own life on August 20, 1961 (Walter 1990).

Physics

Bridgman was awarded the 1946 Nobel Prize for physics for his invention of an apparatus designed to obtain extremely high pressures and for the discoveries he made using it. Prior to Bridgman, the greatest pressures achieved in the laboratory were around 3,000 kg/cm^2 (Lindh 1946). At this time there were two limitations to reaching greater pressures. The first, which continues to be a constraint, is the strength of the containing material. However, this limitation diminishes as stronger materials are developed. The second limitation on attaining higher pressures was the problem of leakage that occurs even before the materials fail. This limitation was completely eliminated by Bridgman's apparatus. Bridgman's apparatus consists of a vessel containing the liquid to be compressed, surrounded by a soft packing material. It is designed so that the pressure in the packing material is automatically maintained at a fixed higher percentage than the pressure of the liquid, making leaks impossible (Bridgman 1946). Consequently, Bridgman was able to reach unprecedented pressures as high as 500,000 kg/cm^2. Lindh (1946) points out that the tremendous pressures made possible by his apparatus led Bridgman to discover many new polymorphous substances and new modifications of substances, and enabled him to amass a wealth of data about the properties of matter at high pressures. For example, Bridgman discovered two new modifications of both ordinary and heavy water in solid form, as well as two new modifications of phosphorous. He worked extensively on the effect of high pressures on electric resistance. This research led to the discovery of the existence of a resistance minimum for certain metals at high pressures. He also investigated the effect of high pressures on thermoelectric phenomena, heat conduction in gases, fluid viscosity, and the elastic properties of solid bodies. In addition, he made significant advances with his investigations of materials for containing substances under high pressures.

Philosophy

Long after his high-pressure results are made obsolete by new technologies and materials, Bridgman's contributions as a philosopher of science will remain influential. His most important contribution in this area is his treatment of operationalism. Bridgman should not be named the inventor of operationalism; he explicitly gives credit to Einstein for using it in developing the theory of special relativity (Bridgman 1927). In fact, Bridgman suggests that perhaps even Einstein was not the first to make progress by operationalizing concepts in scientific theories (Bridgman 1955). However, Bridgman deserves proper credit for explicitly identifying, analyzing, and explaining this important principle. The defining feature of operationalism lies in the idea that concepts are given meaning not by abstract, metaphysical musings, but by the processes used to measure them. This is contrasted with the former method exemplified by Newton's definition of absolute time as that which flows uniformly, independent of material happenings. For example, instead of an abstract definition of 'length' such as "extent in space," the operational definition of the 'length of x' would be the act of comparing a standard unit to x.

In his 1936 book *The Nature of Physical Theory*, Bridgman (1936) explains that by operationalizing concepts, their meanings are determined by physical processes—the great advantage of which is never having to retract theories:

The more particular and important aspect of the operational significance of meaning is suggested by the fact that Einstein recognized that in dealing with physical situations the operations which give meaning to our physical concepts should properly be physical operations, actually carried out. For in so restricting the permissible operations, our theories reduce in the last analysis to descriptions of operations actually carried out in actual situations, and so cannot involve us in inconsistency or contradiction, since these do not occur in actual physical situations. Thus is solved at one stroke the problem of so constructing our fundamental physical concepts that we shall never have to revise them in the light of new experience. (9)

Bridgman did not believe that operationalism was restricted to scientific investigation. He was clear that, for instance, any question for which he could not conceive of a process by which to check the correctness of the answer must be regarded as a meaningless question. Bridgman thought that traditional metaphysical, philosophical questions, such as whether or not we have free will, should be considered meaningless because they cannot be operationalized. In this way Bridgman's philosophy is perfectly consistent with early logical empiricism.

Criticisms of Operationalism

In *The Logic of Modern Physics* Bridgman emphasizes that "the concept is synonymous with the corresponding set of operations" (1927, 5). This claim of synonymy raises problems for operationalism. For instance, it undermines the use of qualitative and dispositional properties such as hardness, which are very difficult to operationalize. Even in the realm of quantitative properties, Hempel (1966) argues that problems arise where more than one operation is possible. Consider the example of measuring temperature with a mercury thermometer and with an alcohol thermometer. According to operationalism, these have to be considered two *different* concepts—mercury-temperature and alcohol-temperature. The operations for each concept become more and more specific to particular conditions, to the point where theories adhering strictly to Bridgman's doctrine would become overburdened; and worse, they would lose generality. Hempel (1966) claims that "this would defeat one of the principal purposes of science; namely the attainment of a simple, systematically unified account of empirical phenomena" (94). Hempel agrees that operationalism is useful in certain contexts; however, he argues that it gives only "partial interpretations" of concepts. He claims that a scientific concept cannot be understood without knowing its systematic role. Concept formation and theory formation are interdependent; new theories are often generated out of discoveries about shared characteristics between concepts in different theories. This requires a certain flexibility of concept and theory formation not consistent with operationalism. Hempel gives the example of using the Sun to measure time. If a unit is marked as the Sun's return to a point in the sky each day, it cannot be questioned that the length of days are equal—it is true by definitional convention. However, when new operations are discovered for measuring the same phenomena, such as the invention of the pendulum clock, then it becomes possible to revise previous operations that turn out to be approximations. Hempel points out that this kind of concept revision can lead to scientific progress.

IAN NYBERG

References

Bridgman, Percy Williams (1927), *The Logic of Modern Physics*. New York: The Macmillan Company.

——— (1936), *The Nature of Physical Theory*. New York: Dover Publications.

——— (1946), "General Survey of Certain Results in the Field of High-Pressure Physics," in *Nobel Lectures, Physics 1942–1962*. Amsterdam: Elsevier Publishing.

——— (1955), *Reflections of A Physicist*. New York: Philosophical Library.

Hempel, Carl G. (1966), *Philosophy of Natural Science*. Englewood Cliffs, NJ: Prentice-Hall.

Lindh, A. E. (1946), "Presentation Speech for the Nobel Prize in Physics 1946," in *Nobel Lectures, Physics 1942–1962*. Amsterdam: Elsevier Publishing.

Walter, Maila L. (1990), *Science and Cultural Crisis: An Intellectual Biography of Percy Williams Bridgman (1882–1961)*. Stanford, CA: Stanford University Press.

C

RUDOLF CARNAP

(18 May 1891–14 September 1970)

Rudolf Carnap (1891–1970), preeminent member of the Vienna Circle, was one of the most influential figures of twentieth-century philosophy of science and analytic philosophy (including the philosophies of language, logic, and mathematics). The Vienna Circle was responsible for promulgating a set of doctrines (initially in the 1920s) that came to be known as logical positivism or logical empiricism (see Logical Empiricism; Vienna Circle). This set of doctrines has provided the point of departure for most subsequent developments in the philosophy of science. Consequently Carnap must be regarded as one of the most important philosophers of science of the twentieth century. Nevertheless, his most lasting positive contributions were in the philosophy of logic and mathematics and the philosophy of language. Meanwhile, his systematic but ultimately unsuccessful attempt to construct an inductive logic has been equally influential, since its failure has convinced most philosophers that such a project must fail (see Inductive Logic).

Carnap was born in 1891 in Ronsdorf, near Barmen, now incorporated into the city of Wuppertal, in Germany (Carnap 1963a). In early childhood he was educated at home by his mother, Anna Carnap (neé Dörpfeld), who had been a schoolteacher. From 1898, he attended the Gymnasium at Barmen, where the family moved after his father's death that year. In school, Carnap's chief interests were in mathematics and Latin. From 1910 to 1914 Carnap studied at the Universities of Jena and Freiburg, concentrating first on philosophy and mathematics and, later, on philosophy and physics. Among his teachers in Jena were Bruno Bauch, a prominent neo-Kantian, and Gottlob Frege, a founder of the modern theory of quantification in logic. Bauch impressed upon him the power of Kant's conception that the geometrical structure of space was determined by the form of pure intuition. Though Carnap was impressed by Frege's ongoing philosophical projects, Frege's real (and lasting) influence came only later through a study of his writings. Carnap's formal intellectual work was interrupted between 1914 and 1918 while he did military service during World War I. His political views had already been of a mildly socialist/pacifist nature. The horrors of the war served to make them more explicit and more conscious, and to codify them somewhat more rigorously.

Space

After the war, Carnap returned to Jena to begin research. His contacts with Hans Reichenbach and others pursuing philosophy informed by current science began during this period (see Reichenbach, Hans). In 1919 he read Whitehead and Russell's *Principia Mathematica* and was deeply influenced by the clarity of thought that could apparently be achieved through symbolization (see Russell, Bertrand). He began the construction of a putative axiom system for a physical theory of space-time. The physicists—represented by Max Wien, head of the Institute of Physics at the University of Jena—were convinced that the project did not belong in physics. Meanwhile, Bauch was equally certain that it did not belong in philosophy. This incident was instrumental in convincing Carnap of the institutional difficulties faced in Germany of doing interdisciplinary work that bridged the chasm between philosophy and the natural sciences. It also probably helped generate the attitude that later led the logical empiricists to dismiss much of traditional philosophy, especially metaphysics. By this point in his intellectual development (the early 1920s) Carnap was already a committed empiricist who, nevertheless, accepted both the analyticity of logic and mathematics and the Frege-Russell thesis of logicism, which required that mathematics be formally constructed and derived from logic (see Analyticity).

Faced with this lack of enthusiasm for his original project in Jena, Carnap (1922) abandoned it to write a dissertation on the philosophical foundations of geometry, which was subsequently published as *Der Raum*. Most traditional commentators have regarded the dissertation as a fundamentally neo-Kantian work because it included a discussion of "intuitive space," determined by pure intuition, independent of all contingent experience, and distinct from both mathematical (or formal) space and physical space (see Friedman 1999). However, recent reinterpretations argue for a decisive influence of Husserl (Sarkar 2003). In contrast to Kant, Carnap restricted what could be grasped by pure intuition to some topological and metric properties of finite local regions of space. He identifies this intuitive space with an infinitesimal space and goes on to postulate that a global space may be constructed from it by iterative extension. In agreement with Helmholtz and Moritz Schlick (a physicist-turned-philosopher, and founder of the Vienna Circle) (see Schlick, Moritz), the geometry of physical space was regarded as an empirical matter. Carnap included a discussion of the role of non-Euclidean geometry in Einstein's theory of general relativity. By distinguishing among intuitive, mathematical, and physical spaces, Carnap attempted to resolve the apparent differences among philosophers, mathematicians, and physicists by assigning the disputing camps to different discursive domains. In retrospect, this move heralded what later became the most salient features of Carnap's philosophical work: tolerance for diverse points of view (so long as they met stringent criteria of clarity and rigor) and an assignment of these viewpoints to different realms, the choice between which is to be resolved not by philosophically substantive (e.g., epistemological) criteria but by pragmatic ones (see Conventionalism).

The Constructionist Phase

During the winter of 1921, Carnap read Russell's *Our Knowledge of the External World*. Between 1922 and 1925, this work led him (Carnap 1963a) to begin the analysis that culminated in *Der logische Aufbau der Welt* ([1928] 1967), which is usually regarded as Carnap's first major work. The purpose of the *Aufbau* was to construct the everyday world from a phenomenalist basis (see Phenomenalism). The phenomenalist basis is an epistemological choice (§§54, 58). Carnap distinguished between four domains of objects: autopsychological, physical, heteropsychological, and cultural (§58). The first of these consists of objects of an individual's own psychology; the second, of physical entities (Carnap does not distinguish between everyday material objects and the abstract entities of theoretical physics); the third consists of the objects of some other individual's psychology; and the fourth, of cultural entities (*geistige Gegenstände*), which include historical and sociological phenomena.

From Carnap's ([1928] 1967) point of view, "[a]n object ... is called *epistemically primary* relative to another one ... if the second one is recognized through the mediation of the first and thus presupposes, for its recognition, the recognition of the first" (§54). Autopsychological objects are epistemically primary relative to the others in this sense. Moreover, physical objects are epistemically primary to heteropsychological ones because the latter can be recognized only through the mediation of the former—an expression on a face, a reading in an instrument, etc. Finally, heteropsychological objects are epistemically primary relative to cultural ones for the same reason.

The main task of the *Aufbau* is construction, which Carnap ([1928] 1967) conceives of as the converse of what he regarded as reduction (which is far from what was then—or is now—conceived of as "reduction" in Anglophone philosophy) (see Reductionism):

> [A]n object is 'reducible' to others ... if all statements about it can be translated into statements which speak only about these other objects .... By *constructing* a concept from other concepts, we shall mean the indication of its "constructional definition" on the basis of other concepts. By a *constructional definition* of the concept *a* on the basis of the concepts *b* and *c*, we mean a rule of translation which gives a general indication how any propositional function in which *a* occurs may be transformed into a coextensive propositional function in which *a* no longer occurs, but only *b* and *c*. If a concept is reducible to others, then it must indeed be possible to construct it from them (§35).

However, construction and reduction present different formal problems because, except in some degenerate cases (such as explicit definition), the transformations in the two directions may not have any simple explicit relation to each other. The question of reducibility/constructibility is distinct from that of epistemic primacy. In an important innovation in an empiricist context, Carnap argues that both the autopsychological and physical domains can be reduced to each other (in his sense). Thus, at the formal level, either could serve as the basis of the construction. It is epistemic primacy that dictates the choice of the former.

Carnap's task, ultimately, is to set up a constructional system that will allow the construction of the cultural domain from the autopsychological through the two intermediate domains. In the *Aufbau*, there are only informal discussions of how the last two stages of such a construction are to be executed; only the construction of the physical from the autopsychological is fully treated formally. As the basic units of the constructional system, Carnap chose what he calls "elementary experiences" (*Elementarerlebnisse*, or *elex*) (an extended discussion of Carnap's construction is to be found in Goodman 1951, Ch. 5). These are supposed to be instantaneous cross-sections of the stream of experience—or at least bits of that stream in the smallest perceivable unit of time—that are incapable of further analysis. The only primitive relation that Carnap introduces is "recollection of similarity" (*Rs*). (In the formal development of the system, *Rs* is introduced first and the *elex* are defined as the field of *Rs*.) The asymmetry of *Rs* is eventually exploited by Carnap to introduce temporal ordering.

Since the *elex* are elementary, they cannot be further analyzed to define what would be regarded as constituent qualities of them such as partial sensations or intensity components of a sensation. Had the *elex* not been elementary, Carnap could have used "proper analysis" to define such qualities by isolating the individuals into classes on the basis of having a certain (symmetric) relationship with each other. Carnap defines the process of "quasi-analysis" to be formally analogous to proper analysis but only defining "quasi-characteristics" or "quasi-constituents" because the *elex* are unanalyzable. Thus, if an *elex* is both *c* in color and *t* in temperature, *c* or *t* can be defined as classes of every *elex* having *c* or *t*, respectively. However, to say that *c* or *t* is a quality would imply that an *elex* is analyzable into simpler constituents. Quasi-analysis proceeds formally in this way (as if it is proper analysis) but defines only quasi-characteristics, thus allowing each *elex* to remain technically unanalyzable Quasi-analysis based on the relation "part similarity" (*Ps*), itself defined from *Rs*, is the central technique of the *Aufbau*. It is used eventually to define sense classes and, then, the visual sense, visual field places, the spatial order of the visual field, the order of colors and, eventually, sensations. Thus the physical domain is constructed out of the autopsychological. Carnap's accounts of the construction between the other two domains remain promissory sketches.

Carnap was aware that there were unresolved technical problems with his construction of the physical from the autopsychological, though he probably underestimated the seriousness of these problems. The systematic problems are that when a quality is defined as a class selected by quasi-analysis on the basis of a relation: (i) two (different) qualities that happen always to occur together (say, red and hot) will never be separated, and (ii) quality classes may emerge in which any two members bear some required relation to each other, but there may yet be no relation that holds between all members of the class. Carnap's response to these problems was extrasystematic: In the complicated construction of the world from the *elex*, he hoped that such examples would never or only very rarely arise. Nevertheless, because of these problems, and because the other constructions are not carried out, the attitude of the *Aufbau* is tentative and exploratory: The constructional system is presented as essentially unfinished. (Goodman 1951 also provides a lucid discussion of these problems.)

Some recent scholarship has questioned whether Carnap had any traditional epistemological concerns in the *Aufbau*. In particular, Friedman (e.g., 1992) has championed the view that Carnap's concerns in that work are purely ontological: The *Aufbau* is not concerned with the question of the source or status of knowledge of the external world (see Empiricism); rather, it investigates the bases on which such a world may be constructed (see Richardson 1998). Both Friedman and Richardson—as well as Sauer (1985) and Haack (1977) long before them—emphasize the Kantian roots of the *Aufbau*. If this reinterpretation is correct, then what exactly the *Aufbau* owes to Russell (and traditional empiricism) becomes uncertain. However, as Putnam (1994, 281) also points out, this reinterpretation goes too far: Though the project of the *Aufbau* is not identical to that of Russell's external world program, there is sufficient congruence between the two projects for Carnap to have correctly believed that he was carrying out Russell's program. In particular, the formal constructions of the *Aufbau* are a necessary prerequisite for the development of the epistemology that Russell had in mind: One must be able to construct the world formally from a phenomenalist basis before one can suggest that this construction shows that the phenomena are the source of knowledge of the world. Moreover, this reinterpretation ignores the epistemological remarks scattered throughout the *Aufbau* itself, including Carnap's concern for the epistemic primacy of the basis he begins with. Savage (2003) has recently pointed out that the salient difference between Russell's and Carnap's project is that whereas the former chose sense data as his point of departure, the latter chose elementary experiences. But this difference is simply a result of Carnap's having accepted the results of Gestalt psychology as having definitively shown what may be taken as individual experiential bases; other than that, that is, with respect to the issue of empiricism, it has no philosophical significance.

In any case, by this time of his intellectual development, Carnap had fully endorsed not only the logicism of the *Principia*, but also the form that Whitehead and Russell had given to logic (that is, the ramified theory of types including the axioms of infinity and reducibility) in that work. However, Henri Poincaré also emerges as a major influence during this period. Carnap did considerable work on the conceptual foundations of physics in the 1920s, and some of this work—in particular, his analysis of the relationship between causal determination and the structure of space—shows strong conventionalist attitudes (Carnap 1924; see also 1923 and 1926) (see Conventionalism; Poincaré, Henri).

Viennese Positivism

In 1926, at Schlick's invitation, Carnap moved to Vienna to become a *Privatdozent* (instructor) in philosophy at the University of Vienna for the next five years (see Vienna Circle). An early version of the *Aufbau* served as his *Habilitationsschrift*. He was welcomed into the Vienna Circle, a scientific philosophy discussion group organized by (and centered around) Schlick, who had occupied the chair for philosophy of the inductive sciences since 1922. In the meetings of the Vienna Circle, the typescript of the *Aufbau* was read and discussed. What Carnap seems to have found most congenial in the Circle—besides its members' concern for science and competence in modern logic—was their rejection of traditional metaphysics. Over the years, besides Carnap and Schlick, the Circle included Herbert Feigl, Kurt Gödel, Hans Hahn, Karl Menger, Otto Neurath, and Friedrich Waismann, though Gödel would later claim that he had little sympathy for the antimetaphysical position of the other members. The meetings of the Circle were characterized by open, intensely critical, discussion with no tolerance for ambiguity of formulation or lack of rigor in demonstration. The members of the Circle believed that philosophy was a collective enterprise in which progress could be made. These attitudes, even more than any canonical set of positions, characterized the philosophical movement—initially known as logical positivism and later as logical empiricism—that emerged from the work of the members of the Circle and a few others, especially Hans Reichenbach (see Reichenbach, Hans). However, besides rejecting traditional metaphysics, most members of the Circle accepted logicism and a sharp distinction between analytic and synthetic truths. The analytic was identified with the a priori; the synthetic with the *a posteriori* (see Analyticity). A. J. Ayer, who attended some meetings of the Circle in 1933 (after Carnap had left—see below), returned to Britain and published *Language, Truth and Logic* (Ayer 1936) (see Ayer, Alfred Jules). This short book did much to popularize the views of the Vienna Circle among Anglophone philosophers, though it lacks the sophistication that is found in the writings of the members of the Circle, particularly Carnap.

Under Neurath's influence, during his Vienna years, Carnap abandoned the phenomenalist language he had preferred in the *Aufbau* and came to accept physicalism (see Neurath, Otto; Physicalism). The epistemically privileged language is one in which sentences reporting empirical knowledge of the world ("protocol sentences") employ terms referring to material bodies and their observable properties (see Phenomenalism, Protocol Sentences). From Carnap's point of view, the chief advantage of a physicalist language is its intersubjectivity. Physicalism, moreover, came hand-in-hand with the thesis of the "unity of science," that is, that the different empirical sciences (including the social sciences) were merely different branches of a single unified science (see Unity of Science Movement). To defend this thesis, it had to be demonstrated that psychology could be based on a physicalist language In an important paper only published somewhat later, Carnap ([1932] 1934) attempted that demonstration (see Unity and Disunity of Science). Carnap's adoption of physicalism was final; he never went back to a phenomenalist language. However, what he meant by "physicalism" underwent radical transformations over the years. By the end of his life, it meant no more than the adoption of a nonsolipsistic language, that is, one in which intersubjective is possible (Carnap 1963b).

In the Vienna Circle, Wittgenstein's *Tractatus* was discussed in detail. Carnap found Wittgenstein's rejection of metaphysics concordant with the views he had developed independently. Partly because of Wittgenstein's influence on some members of the Circle (though not Carnap), the rejection of metaphysics took the form of an assertion that the sentences of metaphysics are meaningless in the sense of being devoid of cognitive content. Moreover, the decision whether a sentence is meaningful was to be made on the basis of the principle of verifiability, which claims that the meaning of a sentence is given by the conditions of its (potential) verification (see Verifiability). Observation terms are directly meaningful on this account (see Observation). Theoretical terms acquire meaning only through explicit definition from observation terms. Carnap's major innovation in these discussions within the Circle was to suggest that even the thesis of realism—asserting the "reality" of the external world—is meaningless, a position not shared by Schlick, Neurath, or Reichenbach. Problems generated by meaningless questions became the celebrated "pseudo-problems" of philosophy (Carnap [1928] 1967).

Wittgenstein's principle of verifiability posed fairly obvious problems in any scientific context. No universal generalization can ever be verified. Perhaps independently, Karl Popper perceived the same problem (see Popper, Karl). This led him to replace the requirement of verifiability with that of falsifiability, though only as a criterion to demarcate science from metaphysics, and not as one to be also used to demarcate meaningful from meaningless claims. It is also unclear what the status of the principle itself is, that is, whether it is meaningful by its own criterion of meaningfulness. Carnap, as well as other members of the Vienna Circle including Hahn and Neurath, realized that a weaker criterion of meaningfulness was necessary. Thus began the program of the "liberalization of empiricism." There was no unanimity within the Vienna Circle on this point. The differences between the members are sometimes described as those between a conservative "right" wing, led by Schlick and Waismann, which rejected both the liberalization of empiricism and the epistemological antifoundationalism of the move to physicalism, and a radical "left" wing, led by Neurath and Carnap, which endorsed the opposite views. The "left" wing also emphasized fallibilism and pragmatics; Carnap went far enough along this line to suggest that empiricism itself was a proposal to be accepted on pragmatic grounds. This difference also reflected political attitudes insofar as Neurath and, to a lesser extent, Carnap viewed science as a tool for social reform.

The precise formulation of what came to be called the criterion of cognitive significance took three decades (see Hempel 1950; Carnap 1956 and 1961) (see Cognitive Significance). In an important pair of papers, "Testability and Meaning," Carnap (1936-1937) replaced the requirement of verification with that of confirmation; at this stage, he made no attempt to quantify the latter. Individual terms replace sentences as the units of meaning. Universal generalizations are no longer problematic; though they cannot be conclusively verified, they can yet be confirmed. Moreover, in "Testability and Meaning," theoretical terms no longer require explicit definition from observational ones in order to acquire meaning; the connection between the two may be indirect through a system of implicit definitions. Carnap also provides an important pioneering discussion of disposition predicates.

The Syntactic Phase

Meanwhile, in 1931, Carnap had moved to Prague, where he held the chair for natural philosophy at the German University until 1935, when, under the shadow of Hitler, he emigrated to the United States. Toward the end of his Vienna years, a subtle but important shift in Carnap's philosophical interests had taken place. This shift was from a predominant concern for the foundations of physics to that for the foundations of mathematics and logic, even though he remained emphatic that the latter were important only insofar as they were used in the empirical sciences, especially physics.

In Vienna and before, following Frege and Russell, Carnap espoused logicism in its conventional sense, that is, as the doctrine that held that the concepts of mathematics were definable from those of logic, and the theorems of mathematics were derivable from the principles of logic. In the aftermath of Gödel's (1931) incompleteness theorems, however, Carnap abandoned this type of logicism and opted instead for the requirement that the concepts of mathematics and logic always have their customary (that is, everyday) interpretation in all contexts. He also began to advocate a radical conventionalism regarding what constituted "logic."

Besides the philosophical significance of Gödel's results, what impressed Carnap most about that work was Gödel's arithmetization of syntax. Downplaying the distinction between an object language and its metalanguage, Carnap interpreted this procedure as enabling the representation of the syntax of a language within the language itself. At this point Carnap had not yet accepted the possibility of semantics, even though he was aware of some of Tarski's work and had had some contact with the Polish school of logic. In this context, the representation of the syntax of a language within itself suggested to Carnap that all properties of a language could be studied within itself through a study of syntax.

These positions were codified in Carnap's major work from this period, *The Logical Syntax of Language* (Carnap 1934b and 1937). The English translation includes material that had to be omitted from the German original due to a shortage of paper; the omitted material was separately published in German as papers (Carnap 1934a and 1935). Conventionalism about logic was incorporated into the well-known principle of tolerance:

> *It is not our business to set up prohibitions but to arrive at conventions* [about what constitutes a logic] .... *In logic, there are no morals.* Every one is at liberty to build up his own logic, i.e., his own form of language, as he wishes. All that is required is that, if he wishes to discuss it, he must state his method clearly, and give syntactic rules instead of philosophical arguments. (Carnap 1937, 51–52; emphasis in the original)

Logic, therefore, is nothing but the syntax of language.

In *Syntax*, the principle of tolerance allows Carnap to navigate the ongoing disputes between logicism, formalism, and intuitionism/constructivism in the foundations of mathematics without abandoning any insight of interest from these schools. Carnap begins with a detailed study of the construction of two languages, I and II. The last few sections of *Syntax* also present a few results regarding the syntax of any language and also discuss the philosophical ramifications of the syntactic point of view. (Sarkar 1992 attempts a comprehensible reconstruction of the notoriously difficult formalism of *Syntax*.)

Language I, which Carnap calls "definite," is intended as a neutral core of all logically interesting languages, neutral enough to satisfy the strictures of almost any intuitionist or constructivist. It permits the definition of primitive recursive arithmetic and has bounded quantification (for all x up to some upper bound) but not much more. Its syntax is fully constructed formally. Language II, which is "indefinite" for Carnap, is richer. It includes Language I and has sufficient resources for the formulation of all of classical mathematics, and is therefore nonconstructive. Moreover, Carnap permits descriptive predicates in each language. Thus, the resources of Language II are strong enough to permit, in principle, the formulation of classical physics. The important point is that because of the principle of tolerance, the choice between Languages I and II or, for that matter, any other syntactically specified language, is not based on factual considerations. If one wants to use mathematics to study physics in the customary way, Language II is preferable, since as yet, nonconstructive mathematics remains necessary for physics. But the adoption of Language II, dictated by the pragmatic concern for doing physics, does not make Language I incorrect. This was Carnap's response to the foundational disputes of mathematics: By tolerance they are defined out of existence.

The price paid if one adopts the principle of tolerance is a radical conventionalism about what constitutes logic. Conventionalism, already apparent in Carnap's admission of both a phenomenalist and a physicalist possible basis for construction in the *Aufbau*, and strongly present in the works on the foundations of physics in the 1920s, had now been extended in *Syntax* to logic. As a consequence, what might be considered to be the most important question in any mathematical or empirical context—the choice of language—became pragmatic. This trend of relegating troublesome questions to the realm of pragmatics almost by fiat, thereby excusing them from systematic philosophical exploration, became increasingly prevalent in Carnap's views as the years went on.

Syntax contained four technical innovations in logic that are of significance: (i) a definition of analyticity that, as was later shown by S. C. Kleene, mimicked Tarski's definition of truth for a formalized language; (ii) a proof, constructed by Carnap independently of Tarski, that truth cannot be defined as a syntactic predicate in any consistent formalized language; (iii) a rule for infinite induction (in Language I) that later came to be called the omega rule; and (iv), most importantly, a generalization of Gödel's first incompleteness theorem that has come to be called the fixed-point lemma. With respect to (iv), what Carnap proved is that in a language strong enough to permit arithmetization, for any syntactic predicate, one can construct a sentence that would be interpreted as saying that it satisfies that predicate. If the chosen predicate is unprovability, one gets Gödel's result.

Besides the principle of tolerance, the main philosophical contribution of *Syntax* was the thesis that philosophy consisted of the study of logical syntax. Giving a new twist to the Vienna Circle's claim that metaphysical claims were meaningless, Carnap argues and tries to show by example that sentences making metaphysical claims are all syntactically ill-formed. Moreover, since the arithmetization procedure shows that all the syntactic rules of a language can be formulated within the language, even the rules that determine what sentences are meaningless can be constructed within the language. All that is left for philosophy is a study of the logic of science. But, as Carnap (1937) puts it: "The *logic of science* (logical methodology) is nothing else than the *syntax of the language of science*.... To share this view is to *substitute logical syntax for philosophy*" (7–8; emphasis in original). The claims of *Syntax* are far more grandiose—and more flamboyant—than anything in the *Aufbau*.

Semantics

In the late 1930s, Carnap abandoned the narrow syntacticism of *Syntax* and, under the influence of Tarski and the Polish school of logic, came to accept semantics. With this move, Carnap's work enters its final mature phase. For the first time, he accepted that the concept of truth can be given more than pragmatic content. Thereupon, he turned to the systematization of semantics with characteristic vigor, especially after his immigration to the United States, where he taught at the University of Chicago from 1936 to 1952. In his contribution to the *International Encyclopedia of Unified Science* (Carnap 1939), on the foundations of logic and mathematics, the distinctions among syntactic, semantic, and pragmatic considerations regarding any language are first presented in their mature form.

Introduction to Semantics, which followed in 1942, develops semantics systematically. In *Syntax* Carnap had distinguished between two types of transformations on sentences: those involving "the method of derivation" or "*d*-method," and those involving the "method of consequence" or "*c*-method." Both of these were supposed to be syntactic, but there is a critical distinction between them. The former allows only a finite number of elementary steps. The latter places no such restriction and is, therefore, more "indefinite." Terms defined using the *d*-method ("*d*-terms") include "derivable," "demonstrable," "refutable," "resoluble," and "irresoluble"; the corresponding *c*-terms are "consequence," "analytic," "contradictory," "L-determinate," and "synthetic." After the conversion to semantics, Carnap proposed that the *c*-method essentially captured what semantics allowed; the *c*-terms referred to semantic concepts.

Thus semantics involves a kind of formalization, though one that is dependent on stronger inference rules than the syntactical ones. In this sense, as Church (1956, 65) has perceptively pointed out, Carnap—and Tarski—reduce semantics to formal rules, that is, syntax. Thus emerges the interpretation of deductive logic that has since become the textbook version, so commonly accepted that is has become unnecessary to refer to Carnap when one uses it. For Carnap, the semantic move has an important philosophical consequence: Philosophy is no longer to be replaced just by the syntax of the language of science; rather, it is to be replaced by the syntax and the semantics of the language of science.

Carnap's (1947) most original—and influential—work in semantics is *Meaning and Necessity*, where the basis for an intensional semantics was laid down. Largely following Frege, intensional concepts are distinguished from extensional ones. Semantical rules are introduced and the analytic/synthetic distinction is clarified by requiring that any definition of analyticity must satisfy the (meta-) criterion that analytic sentences follow from the semantical rules alone. By now Carnap had fully accepted that semantic concepts and methods are more fundamental than syntactic ones: The retreat from the flamboyance of *Syntax* was complete. The most important contribution of *Meaning and Necessity* was the reintroduction into logic, in the new intensional framework, of modal concepts that had been ignored since the pioneering work of Lewis (1918). In the concluding chapter of his book, Carnap introduced an operator for necessity, gave semantic rules for its use, and showed how other modal concepts such as possibility, impossibility, necessary implication, and necessary equivalence can be defined from this basis.

By this point, Carnap had begun to restrict his analyses to exactly constructed languages, implicitly abandoning even a distant hope that they would have any direct bearing on natural languages. The problem with the latter is that their ambiguities made them unsuited for the analysis of science, which, ultimately, remained the motivation of all of Carnap's work. Nevertheless, Carnap's distinction between the analytic and the synthetic came under considerable criticism from many, including Quine (1951), primarily on the basis of considerations about natural languages (see Analyticity; Quine, Willard Van). Though philosophical fashion has largely followed Quine on this point, at least until recently, Carnap was never overly impressed by this criticism (Stein 1992). The analytic/synthetic distinction continued to be fundamental to his views, and, in a rejoinder to Quine, Carnap argued that nothing prevented empirical linguistics from exploring intensions and thereby discovering cases of synonymy and analyticity (Carnap 1955).

Carnap's (1950a) most systematic exposition of his final views on ontology is also from this period. A clear distinction is maintained between questions that are internal to a linguistic framework and questions that are external to it. The choice of a linguistic framework is to be based not on cognitive but on pragmatic considerations. The external question of "realism," which ostensibly refers to the "reality" of entities of a framework in some sense independent of it, rather than to their "reality" within it after the framework has been accepted, is rejected as noncognitive (see Scientific Realism). This appears to be an anti-"realist" position, but it is not in the sense that within a framework, Carnap is tolerant of the abstract entities that bother nominalists. The interesting question becomes the pragmatic one, that is, what frameworks are fruitful in which contexts, and Carnap's attitude toward the investigation of various alternative frameworks remains characteristically and consistently tolerant.

Carnap continued to explore questions about the nature of theoretical concepts and to search for a criterion of cognitive significance, preoccupations of the logical empiricists that date back to the Vienna Circle. Carnap (1956) published a detailed exposition of his final views regarding the relation between the theoretical and observational parts of a scientific language. This paper emphasizes the methodological and pragmatic aspects of theoretical concepts. It also contains his most subtle, though not his last, attempt to explicate the notion of the cognitive significance of a term and thus establish clearly the boundary between scientific and nonscientific discourse. However, the criterion he formulates makes theoretical terms significant only with respect to a class of terms, a theoretical language, an observation language, correspondence rules between them, and a theory. Relativization to a theory is critical to avoiding the problems that beset earlier attempts to find such a criterion. Carnap proves several theorems that are designed to show that the criterion does capture the distinction between scientific and nonscientific discourse. This criterion was criticized by Roozeboom (1960) and Kaplan (1975), but these criticisms depend on modifying Carnap's original proposal in important ways. According to Kaplan, Carnap accepted his criticism, though there is apparently no independent confirmation of that fact. However, Carnap (1961) did turn to a different formalism (Hilbert's ϵ-operator) in what has been interpreted as his last attempt to formulate such a criterion (Kaplan 1975), and this may indicate dissatisfaction with the 1956 attempt. If so, it remains unclear why: That attempt did manage to avoid the technical problems associated with the earlier attempts of the logical empiricists (see Cognitive Significance).

Probability and Inductive Logic

From 1941 onward, Carnap also began a systematic attempt to analyze the concepts of probability and to formulate an adequate inductive logic (a logic of confirmation), a project that would occupy him for the rest of his life. Carnap viewed this work as an extension of the semantical methods that he had been developing for the last decade. This underscores an interesting pattern in Carnap's intellectual development. Until the late 1930s Carnap viewed syntactic categories only as nonpragmatically specifiable; questions of truth and confirmation were viewed as pragmatic. His conversion to semantics saw the recovery of truth from the pragmatic to the semantic realm. Now, confirmation followed truth down the same pathway.

In *Logical Foundations of Probability* (1950b), his first systematic analysis of probability, Carnap distinguished between two concepts of probability: "statistical probability," which was the relevant concept to be used in empirical contexts and generally estimated from the relative frequencies of events, and "logical probability," which was to be used in contexts such as the confirmation of scientific hypotheses by empirical data. Though the latter concept, usually called the "logical interpretation" of probability, went back to Keynes (1921), Carnap provides its first systematic explication (see Probability).

Logical probability is explicated from three different points of view (1950b, 164–8): (i) as a conditional probability $c(h,e)$, which measures the degree of confirmation of a hypothesis h on the basis of evidence e (if $c(h,e) = r$, then r is determined by logical relations between h and e); (ii) as a rational degree of belief or fair betting quotient (if $c(h,e) = r$, then r represents a fair bet on h if e correctly describes the total knowledge available to a bettor); and (iii) as the limit of relative frequencies in some cases. According to Carnap, the first of these, which specifies a confirmation function ("c-function"), is the concept that is most relevant to the problem of induction. In the formal development of the theory, probabilities are associated with sentences of a formalized language.

In *Foundations*, Carnap (1950b) believed that a unique measure $c(h,e)$ of the degree of confirmation can be found, and he even proposed one (*viz.*, Laplace's rule of succession), though he could not prove its uniqueness satisfactorily. His general strategy was to augment the standard axioms of the probability calculus by a set of "conventions on adequacy" (285), which turned out to be equivalent to assumptions about the rationality of degrees of belief that had independently been proposed by both Ramsey and de Finetti (Shimony 1992). In a later work, *The Continuum of Inductive Methods*, using the conventions on adequacy and some plausible symmetry principles, Carnap (1952) managed to show that all acceptable c-functions could be parameterized by a single parameter, a real number, $\lambda \in [0,\infty]$. The trouble remained that there is no intuitively appealing a priori strategy to restrict λ to some preferably very small subset of $[0,\infty]$. At one point, Carnap even speculated that it would have to be fixed empirically. Unfortunately, some higher-order induction would then be required to justify the procedure for its estimation, and potentially, this leads to infinite regress (see Confirmation Theory; Inductive Logic).

Carnap spent 1952–1954 at the Institute for Advanced Study at Princeton, New Jersey, where he continued to work on inductive logic, often in collaboration with John Kemeny. He also returned to the foundations of physics, apparently motivated by a desire to trace and explicate the relations between the physical concept of entropy and an abstract concept of entropy appropriate for inductive logic. His discussion with physicists proved to be disappointing and he did not publish his results. (These were edited and published by Abner Shimony [Carnap 1977] after Carnap's death.)

In 1954 Carnap moved to the University of California at Los Angeles to assume the chair that had become vacant with Reichenbach's death in 1953. There he continued to work primarily on inductive logic, often with several collaborators, over the next decade. There were significant modifications of his earlier attempts to formulate a systematic inductive logic (see Carnap and Jeffrey 1971 and Jeffrey 1980. An excellent introduction to this part of Carnap's work on inductive logic is Hilpinen 1975). Obviously impressed by the earlier work of Ramsey and de Finetti, Carnap (1971a) returned to the second of his three 1950 explications of logical probability and emphasized the use of inductive logic in decision problems.

More importantly, Carnap, in "A Basic System of Inductive Logic" (1971b and 1980), finally recognized that attributing probabilities to sentences was too restrictive. If a conceptual system uses real numbers and real-valued functions, no language can express all possible cases using only sentences or classes of sentences. Because of this, he now began to attribute probabilities to events or propositions (which are taken to be synonymous). This finally brought some concordance between his formal methods and those of mathematical statisticians interested in epistemological questions. Propositions are identified with sets of models; however, the fields of the sets are defined using the atomic propositions of a formalized language. Thus, though probabilities are defined as measures of sets, they still remain relativized to a particular formalized language. Because of this, and because the languages considered remain relatively simple (mostly monadic predicate languages), much of this work remains similar to the earlier attempts.

By this point Carnap had abandoned the hope of finding a unique c-function. Instead, he distinguished between subjective and objective approaches in inductive logic. The former emphasizes individual freedom in the choice of necessary conventions; the latter emphasizes the existence of limitations. Though Carnap characteristically claimed to keep an open mind about these two approaches, his emphasis was on finding rational a priori principles that would systematically limit the choice of c-functions. Carnap was still working on this project when he died on September 14, 1970. He had not finished revising the last sections of the second part of the "Basic System," both parts of which were published only posthumously.

Toward the end of his life, Carnap's concern for political and social justice had led him to become an active supporter of an African-American civil rights organization in Los Angeles. According to Stegmüller (1972, lxvi), the "last photograph we have of Carnap shows him in the office of this organization, in conversation with various members. He was the only white in the discussion group."

The Legacy

Thirty-five years after Carnap's death it is easier to assess Carnap's legacy, and that of logical empiricism, than it was in the 1960s and 1970s, when a new generation of analytic philosophers and philosophers of science apparently felt that they had to reject that work altogether in order to be able to define their own philosophical agendas. This reaction can itself be taken as evidence of Carnap's seminal influence, but, nevertheless, it is fair to say that Carnap and logical empiricism fell into a period of neglect in the 1970s from which it began to emerge only in the late 1980s and early 1990s. Meanwhile it became commonplace among philosophers to assume that Carnap's projects had failed.

Diagnoses of this failure have varied. For some it was a result of the logical empiricists' alleged inability to produce a technically acceptable criterion for cognitive significance (see Cognitive Significance). For others, it was because of Quine's dicta against the concept of analyticity and the analytic/synthetic distinction (see Analyticity; Quine, Willard Van). Some took Popper's work to have superseded that of Carnap and the logical empiricists (see Popper, Karl Raimund). Many viewed Thomas Kuhn's seminal work on scientific change to have shown that the project of inductive logic was misplaced; they, and others, generally regarded Carnap's attempt to explicate inductive logic to have been a failure (see Kuhn, Thomas; Scientific Change). Finally, a new school of "scientific realists" attempted to escape Carnap's arguments against external realism (see Realism).

There can be little doubt that Carnap's project of founding inductive logic has faltered. He never claimed that he had gone beyond preliminary explorations of possibilities, and, though there has been some work since, by and large, epistemologists of science have abandoned that project in favor of less restrictive formalisms, for instance, those associated with Bayesian or Neyman-Pearson statistics (see Bayesianism; Statistics, Philosophy of). But, with respect to every other case mentioned in the last paragraph, the situation is far less clear. It has already been noted that Carnap's final criterion for cognitive significance does not suffer from any technical difficulty no matter what its other demerits may be. Quine's dicta against analyticity no longer appear as persuasive as they once did (Stein 1992); Quine's preference for using natural—rather than formalized—language in the analysis of science has proved to be counterproductive; and his program of naturalizing epistemology has yet to live up to its initial promise. Putnam's "internal realism" is based on and revives Carnap's views on ontology, and Kuhn is perhaps now better regarded as having contributed significantly to the sociology rather than to the epistemology of science.

However, to note that some of the traditionally fashionable objections to Carnap and logical empiricism cannot be sustained does not show that that work deserves a positive assessment on its own. There still remains the question: What, exactly, did Carnap contribute? The answer turns out to be straightforward: The textbook picture of deductive logic that is in use today is the one that Carnap produced in the early 1940s after he came to acknowledge the possibility of semantics. The fixed-point lemma has turned out to be an important minor contribution to logic. The reintroduction of modal logic into philosophy opened up new vistas for Kripke and others in the 1950s and 1960s. Carnap's views on ontology continue to influence philosophers today. Moreover, even though the project of inductive logic seems unsalvageable to most philosophers, it is hard to deny that Carnap managed to clarify significantly the ways in which concepts of probability must be deployed in the empirical sciences and why the problem of inductive logic is so difficult. But, most of all, Carnap took philosophy to a new level of rigor and clarity, accompanied by an open-mindedness (codified in the principle of tolerance) that, unfortunately, is not widely shared in contemporary analytic philosophy.

SAHOTRA SARKAR

References

Ayer, A. J. (1936), *Language, Truth and Logic*. London: Gollancz.

Carnap, R. (1922), *Der Raum*. Berlin: von Reuther and Reichard.

——— (1923), "Über die Aufgabe der Physik und die Anwendung des Grundsatzes der Einfachsteit," *Kant-Studien* 28: 90–107.

——— (1924), "Dreidimensionalität des Raumes und Kausalität: Eine Untersuchung über den logischen Zusammenhang zweier Fiktionen," *Annalen der Philosophie und philosophischen Kritik* 4: 105–130.

——— (1926), *Physikalische Begriffsbildung*. Karlsruhe, Germany: Braun.

——— ([1928] 1967), *The Logical Structure of the World and Pseudoproblems in Philosophy*. Berkeley and Los Angeles: University of California Press.

——— ([1932] 1934. *The Unity of Science*. London: Kegan Paul, Trench, Trubner & Co.

—— (1934a), "Die Antinomien und die Unvollständigkeit der Mathematik," *Monatshefte für Mathematik und Physik* 41: 42–48.

—— (1934b), *Logische Syntax der Sprache*. Vienna: Springer.

—— (1935), "Ein Gültigskriterium für die Sätze der klassischen Mathematik," *Monatshefte für Mathematik und Physik* 42: 163–190.

—— (1936–1937), "Testability and Meaning," *Philosophy of Science* 3 and 4: 419–471 and 1–40.

—— (1937), *The Logical Syntax of Language*. London: Kegan Paul, Trench, Trubner & Co.

—— (1939), *Foundations of Logic and Mathematics*. Chicago: University of Chicago Press.

—— (1947), *Meaning and Necessity: A Study in Semantics and Modal Logic*. Chicago: University of Chicago Press.

—— (1950a), "Empiricism, Semantics, and Ontology," *Revue Internationale de Philosophie* 4: 20–40.

—— (1950b), *Logical Foundations of Probability*. Chicago: University of Chicago Press.

—— (1952), *The Continuum of Inductive Methods*. Chicago: University of Chicago Press.

—— (1955), "Meaning and Synonymy in Natural Languages," *Philosophical Studies* 7: 33–47.

—— (1956), "The Methodological Character of Theoretical Concepts," in H. Feigl and M. Scriven (eds.), *The Foundations of Science and the Concepts of Psychology and Psychoanalysis*. Minneapolis: University of Minnesota Press, 38–76.

—— (1961), "On the Use of Hilbert's ϵ-Operator in Scientific Theories," in Y. Bar-Hillel, E. I. J. Poznanski, M. O. Rabin, and A. Robinson (eds.), *Essays on the Foundations of Mathematics*. Jerusalem: Magnes Press, 156–164.

—— (1963a), "Intellectual Autobiography," in P. A. Schilpp (ed.), *The Philosophy of Rudolf Carnap*. La Salle, IL: Open Court, 3–84.

—— (1963b), "Replies and Systematic Expositions," in P. A. Schilpp (ed.), *The Philosophy of Rudolf Carnap*. La Salle, IL: Open Court, 859–1013.

—— (1971a), "Inductive Logic and Rational Decisions," in R. Carnap and R. C. Jeffrey (eds.), *Studies in Inductive Logic and Probability* (Vol. 1). Berkeley and Los Angeles: University of California Press, 5–31.

—— (1971b), "A Basic System of Inductive Logic, Part I," in R. Carnap and R. C. Jeffrey (eds.), *Studies in Inductive Logic and Probability* (Vol. 1). Berkeley and Los Angeles: University of California Press, 33–165.

—— (1977), *Two Essays on Entropy*. Berkeley and Los Angeles: University of California Press.

—— (1980), "A Basic System of Inductive Logic, Part II," in R. C. Jeffrey (ed.), *Studies in Inductive Logic and Probability* (Vol. 2). Berkeley and Los Angeles: University of California Press, 8–155.

Carnap, R., and R. C. Jeffrey (eds.) (1971), *Studies in Inductive Logic and Probability* (Vol. 1). Berkeley and Los Angeles: University of California Press.

Church, A. (1956), *Introduction to Mathematical Logic*. Princeton: Princeton University Press.

Friedman, M. (1992), "Epistemology in the *Aufbau*," *Synthese* 93: 191–237.

—— (1999), *Reconsidering Logical Empiricism*. New York: Cambridge University Press.

Gödel, K. (1931), "Über formal unentscheidbare Sätze der Principia Mathematica und verwandter Systeme 1," *Monatshefte für Mathematik und Physik* 38: 173–198.

Goodman, N. (1951), *The Structure of Experience*. Cambridge, MA: Harvard University Press.

Haack, S. (1977), "Carnap's *Aufbau*: Some Kantian Reflections," *Ratio* 19: 170–175.

Hempel, C. G. (1950), "Problems and Changes in the Empiricist Criterion of Meaning," *Revue Internationale de Philosophie* 11: 41–63.

Hilpinen, R. (1975), "Carnap's New System of Inductive Logic," in J. Hintikka (ed.), *Rudolf Carnap, Logical Empiricist: Materials and Perspectives*. Dordrecht, Netherlands: Reidel, 333–359.

Jeffrey, R. C. (ed.) (1980), *Studies in Inductive Logic and Probability* (Vol. 2). Berkeley and Los Angeles: University of California Press.

Kaplan, D. (1975), "Significance and Analyticity: A Comment on Some Recent Proposals of Carnap," in J. Hintikka (ed.), *Rudolf Carnap, Logical Empiricist: Materials and Perspectives*. Dordrecht, Netherlands: Reidel, 87–94.

Keynes, J. M. (1921), *A Treatise on Probability*. London: Macmillan & Co.

Lewis, C. I. (1918), *A Survey of Symbolic Logic*. Berkeley and Los Angeles: University of California Press.

Putnam, H. (1994), "Comments and Replies," in P. Clark and B. Hale (eds.), *Reading Putnam*. Oxford: Blackwell, 242–295.

Quine, Willard Van (1951), "Two Dogmas of Empiricism," *Philosophical Review* 60: 20–43.

Richardson, A. (1998), *Carnap's Construction of the World*. Cambridge: Cambridge University Press.

Roozeboom, W. (1960), "A Note on Carnap's Meaning Criterion," *Philosophical Studies* 11: 33–38.

Sarkar, S. (1992), "'The Boundless Ocean of Unlimited Possibilities': Logic in Carnap's *Logical Syntax of Language*," *Synthese* 93: 191–237.

—— (2003), "Husserl's Role in Carnap's *Der Raum*," in T. Bonk (ed.), *Language, Truth and Knowledge: Contributions to the Philosophy of Rudolf Carnap*. Dordrecht, Netherlands: Kluwer, 179–190.

Sauer, W. (1985), "Carnap's '*Aufbau*' in Kantianishcer Sicht," *Grazer Philosophische Studien* 23: 19–35.

Savage, C. W. (2003), "Carnap's *Aufbau* Rehabilitated," in T. Bonk (ed.), *Language, Truth and Knowledge: Contributions to the Philosophy of Rudolf Carnap*. Dordrecht, Netherlands: Kluwer, 79–86.

Shimony, A. (1992), "On Carnap: Reflections of a Metaphysical Student," *Synthese* 93: 261–274.

Stegmüller, W. (1972), "Homage to Rudolf Carnap," in R. C. Buck and R. S. Cohen (eds.), *PSA 1970*. Dordrecht, Netherlands: Reidel, lii–lxvi.

Stein, H. (1992), "Was Carnap Entirely Wrong, After All?" *Synthese* 93: 275–295.

See also **Analyticity; Confirmation Theory; Cognitive Significance; Conventionalism; Hahn, Hans; Hempel, Carl Gustav; Induction, Problem of; Inductive Logic; Instrumentalism; Logical Empiricism; Neurath, Otto; Phenomenalism; Protocol Sentences; Quine, Willard Van; Reichenbach, Hans; Scientific Realism; Schlick, Moritz; Space-Time; Verifiability; Vienna Circle**

CAUSALITY

Arguably no concept is more fundamental to science than that of causality, for investigations into cases of existence, persistence, and change in the natural world are largely investigations into the causes of these phenomena. Yet the metaphysics and epistemology of causality remain unclear. For example, the ontological categories of the causal relata have been taken to be objects (Hume [1739] 1978), events (Davidson 1967), properties (Armstrong 1978), processes (Salmon 1984), variables (Hitchcock 1993), and facts (Mellor 1995). (For convenience, causes and effects will usually be understood as events in what follows.) Complicating matters, causal relations may be singular (*Socrates' drinking hemlock* caused *Socrates' death*) or general (*Drinking hemlock* causes *death*); hence the relata might be tokens (e.g., instances of properties) or types (e.g., types of events) of the category in question. Other questions up for grabs are: Are singular causes metaphysically and/or epistemologically prior to general causes or vice versa (or neither)? What grounds the intuitive asymmetry of the causal relation? Are macrocausal relations reducible to microcausal relations? And perhaps most importantly: Are causal facts (e.g., the holding of causal relations) reducible to noncausal facts (e.g., the holding of certain spatiotemporal relations)?

Some Issues in Philosophy of Causality

The Varieties of Causation

Causes can apparently contribute to effects in a variety of ways: by being background or standing conditions, "triggering events," omissions, factors that enhance or inhibit effects, factors that remove a common preventative of an effect, etc. Traditionally accounts of causation have focussed on triggering events, but contemporary accounts are increasingly expected to address a greater range of this diversity.

There may also be different notions of cause characteristic of the domains of different sciences (see Humphreys 1986; Suppes 1986): The seemingly indeterministic phenomena of quantum physics may require treatment different from either the seemingly deterministic processes of certain natural sciences or the "quasi-deterministic" processes characteristic of the social sciences, which are often presumed to be objectively deterministic but subjectively uncertain. Also relevant here is the distinction between teleological (intentional, goal-oriented) and nonteleological causality: While the broadly physical sciences tend not to cite motives and purposes, the plant, animal, human, and social sciences often explicitly do so. Contemporary treatments of teleological causality generally aim to avoid the positing of anything like entelechies or "vital forces" (of the sort associated with nineteenth-century accounts of biology), and also to avoid taking teleological goals to be causes that occur after their effects (see Salmon 1989, Sec. 3.8, for a discussion). On Wright's (1976) account, consequence etiology teleological behaviors (e.g., stalking a prey) are not caused by future catchings (which, after all, might not occur), but rather by the fact that the behavior in question has been often enough successful in the past that it has been evolutionarily selected for and for creatures capable of intentional representation, alternative explanations may be available. While teleological causes raise interesting questions for the causal underpinnings of behavior (especially concerning whether a naturalistically acceptable account of intentionality can be given), the focus in what follows will be on nonteleological causality, reflecting the primary concern of most contemporary philosophers of causation.

Singular vs. General Causation

Is all singular causation ultimately general? Different answers reflect different understandings of the notion of 'production' at issue in the platitude "Causes produce their effects." On generalist (or covering-law) accounts (see the section on "Hume and Pearson: Correlation, Not Causation" below), causal production is a matter of law: Roughly, event c causes event e just in case c and e are instances of terms in a law connecting events of c's type with events of e's type. The generalist interpretation is in part motivated by the need to ground inductive reasoning: Unless causal relations are subsumed by causal laws, one will be unjustified in inferring that events of c's type will, in the future, cause events of e's type. Another motivation stems from thinking that identifying a sequence of events as causal requires identifying the sequence as falling under a (possibly unknown) law.

Alternatively, singularists (see "Singularist Accounts" below) interpret causal production as involving a singular causal process (variously construed) that is metaphysically prior to laws. Singularists also argue for the epistemological priority of singular causes, maintaining that one can identify a sequence as causal without assuming that the sequence falls under a law, even when the sequence violates modal presuppositions (as in Fair's [1979] case: Intuitively, one could recognize a glass's breaking as causal, even if one antecedently thought glasses of that type were unbreakable).

Counterfactual accounts (see "Counterfactual Accounts" below) analyze singular causes in terms of counterfactual conditionals (as a first pass, event c causes event e just in case if c had not occurred, then e would not have occurred). Whether a counterfactual account should be considered singularist, however, depends on whether the truth of the counterfactuals is grounded in laws connecting types of events or in, for example, propensities (objective single-case chances) understood as irreducible to laws. Yet another approach to the issue of singular versus general causes is to deny that either is reducible to the other, and rather to give independent treatments of each type (as in Sober 1984).

Reduction vs. Nonreduction

There are at least three questions of reducibility at issue in philosophical accounts of causation, which largely cut across the generalist/singularist distinction. The first concerns whether causal facts (e.g., the holding of causal relations) are reducible to noncausal facts (e.g., the holding of certain spatiotemporal relations). Hume's generalist reduction of causality (see "Hume and Pearson: Correlation, Not Causation" below) has a projectivist or antirealist flavor: According to Hume, the seeming "necessary connexion" between cause and effect is a projection of a psychological habit of association between ideas, which habit is formed by regular experience of events of the cause type being spatially contiguous and temporally prior to events of the effect type. Contemporary neo-Humeans (see "Hempel: Explanation, Not Causation" and "Probabilistic Relevance Accounts" below) dispense with Hume's psychologism, focusing instead on the possibility of reducing causal relations and laws to objectively and noncausally characterized associations between events. (Whether such accounts are appropriately deemed antirealist is a matter of dispute, one philosopher's reductive elimination being another's reductive introduction.) By way of contrast, nonreductive generalists (often called realists—see "Causal Powers, Capacities, Universals, Forces" below) take the modally robust causal connection between event types to be an irreducible feature of reality (see Realism). Singularists also come in reductive or realist varieties (see "Singularist Accounts" below).

A second question of reducibility concerns whether a given account of causation aims to provide a conceptual analysis of the concept (hence to account for causation in bizarre worlds, containing magic, causal action at a distance, etc.) or instead to account for the causal relation in the actual world, in terms of physically or metaphysically more fundamental entities or processes. These different aims make a difference in what sort of cases and counterexamples philosophers of causation take to heart when developing or assessing theories. A common intermediate methodology focuses on central cases, leaving the verdict on far-fetched cases as "spoils for the victor."

A third question of reducibility concerns whether macro-causal relations (holding between entities, or expressed by laws, in the special sciences) are reducible to micro-causal relations (holding between entities, or expressed by laws, in fundamental physics). This question arises from a general desire to understand the ontological and causal underpinnings of the structured hierarchy of the sciences, and from a need to address, as a special case, the "problem of mental causation," of whether and how mental events (e.g., a feeling of pain) can be causally efficacious vis-à-vis certain effects (e.g., grimacing) that appear also to be caused by the brain events (and ultimately, fundamental physical events) upon which the mental events depend.

Causal reductionists (Davidson 1970; Kim 1984) suggest that mental events (more generally, macro-level events) are efficacious in virtue of supervening on (or being identical with) efficacious physical events. Many worry, however, that these approaches render macro-level events causally irrelevant (or "epiphenomenal"). Nonreductive approaches to macro-level causation come in both physicalist and nonphysicalist varieties (Wilson 1999 provides an overview; see Physicalism). Some physicalists posit a relation (e.g., the determinable/determinate relation or proper parthood) between macro- and micro-level events that entails that the set of causal powers of a given macro-level event m (roughly, the set of causal interactions that the event, in appropriate circumstances, could enter into) is a proper subset of those of the micro-level event p upon which m depends. On this "proper subset" strategy, the fact that the sets of causal powers are different provides some grounds for claiming that m is efficacious in its own right, but since each individual causal power of m is identical with a causal power of p, the two events are not in causal competition. On another nonreductive strategy—emergentism—the causal efficacy of at least some macro-level events (notably, mental events) is due to their having genuinely new causal powers not possessed by the physical events on which the mental events depend (see Emergence). When the effect in question is physical, such powers violate the causal closure of the physical (the claim that every physical effect has a fully sufficient physical cause); but such a violation arguably is not at odds with any cherished scientific principles, such as conservation laws (see McLaughlin 1992).

Features of Causality: Asymmetry, Temporal Direction, Transitivity

Intuitively, causality is asymmetric: if event c causes event e, then e does not cause c. Causality also generally proceeds from the past to the future. How to account for these data remains unclear. The problem of accounting for asymmetry is particularly pressing for accounts that reductively analyze causality in terms of associations, for it is easy to construct cases in which the associations are reversible, but the causation is not (as in Sylvain Bromberger's case in which the height h of a flagpole is correlated with the length l of the shadow it casts, and *vice versa*, and intuitively h causes l, but l does not cause h). Both the asymmetry and temporal direction of causality can be accommodated (as in Hume) by stipulatively identifying causal with temporal asymmetry: causes differ from their effects in being prior to their effects. This approach correctly rules out l's causing h in the case above. But it also rules out simultaneous and backwards causation, which are generally taken to be live (or at least not too distant) possibilities.

Accounts on which the general temporal direction of causation is determined by physical or psychological processes may avoid the latter difficulties. On Reichenbach's (1956) account, the temporal direction of causation reflects the direction of "conjunctive forks": processes where a common cause produces joint effects, and where, in accordance with what Reichenbach called "the principle of the common cause", the probabilistic dependence of the effects on each other is "screened off"—goes away—when the common cause is taken into account. Such forks are, he claimed, always (or nearly always) open to the future and closed to the past. Some have suggested that the direction of causation is fixed by the direction of increasing entropy, or (more speculatively) by the direction of collapse of the quantum wave packet. Alternatively, Price (1991) suggests that human experience of manipulating causes provides a basis for the (projected) belief that causality is forward-directed in time (see "Counterfactuals and Manipulability" below). These accounts explain the usual temporal direction of causal processes, while allowing the occasional exception.

It remains the case, however, that accommodating the asymmetry of causation by appeal to the direction of causation rules out reducing the direction of time to the (general) direction of causation, which some (e.g., Reichenbach) have wanted to do. More importantly, neither stipulative nor nonstipulative appeals to temporal direction seem to adequately explain the asymmetry of causation, which intuitively has more to do with causes producing their effects (in some robust sense of 'production') than with causes being prior to their effects. Nonreductive accounts on which causality involves manifestations of powers or transfers of energy (or other conserved quantities) may be better situated to provide the required explanation, if such manifestations or transfers can be understood as directed (which remains controversial).

Another feature commonly associated with causality is transitivity: if c causes d, and d causes e, then c causes e. This assumption has come in for question of late, largely due to the following sort of case (see Kvart 1991): A man's finger is severed in a factory accident; a surgeon reattaches the finger, which afterwards becomes perfectly functional. The accident caused the surgery, and surgery caused the finger's functionality; but it seems odd to say that the accident caused the finger's functionality (see Hall 2000 for further discussion).

Challenges to Causality

Galileo, Newton, and Maxwell—How, Not Why

From the ancient through modern periods, accounts of natural phenomena proceeded by citing the powers and capacities of agents, bodies, and mechanisms to bring about effects (see Hankinson 1998; Clatterbaugh 1999). Galileo's account of the physics of falling bodies initiated a different approach to scientific understanding, on which this was a matter of determining how certain measurable quantities were functionally correlated (the "how" of things, or the kinematics), as opposed to determining the causal mechanisms responsible for these correlations (the "why" of things, or the dynamics). This descriptive approach enabled scientific theories to be formulated with comparatively high precision, which in turn facilitated predictive and retrodictive success; by way of contrast, explanations in terms of (often unobservable) causal mechanisms came to be seen as explanatorily otiose at best and unscientific at worst.

Newton's famous claim in the *Principia* ([1687] 1999) that "*hypotheses non fingo*" ("I frame no hypotheses"), regarding gravitation's "physical causes and seats" is often taken as evidence that he advocated a descriptivist approach (though he speculated at length on the causes of gravitational forces in the *Optics*). And while Maxwell drew heavily upon Faraday's qualitative account of causally efficacious electromagnetic fields (and associated lines of force) in the course of developing his theories of electricity and magnetism, he later saw such appeals to underlying causes as heuristic aids that could be dropped from the final quantitative theory.

Such descriptivist tendencies have been encouraged by perennial worries about the metaphysical and epistemological presuppositions of explicitly causal explanations (see "Causal Powers, Capacities, Universals, Forces" below) and the concomitant seeming availability of eliminativist or reductivist treatments of causal notions in scientific laws. For example, Russell (1912) influentially argued that since the equations of physics do not contain any terms explicitly referring to causes or causal relations and moreover (in conflict with the presumed asymmetry of causality) appear to be functionally symmetric (one can write $a = F/m$ as well as $F = ma$), causality should be eliminated as "a relic of a bygone age." Jammer (1957) endorsed a view in which force-based dynamics is a sophisticated form of kinematics, with force terms being mere "methodological intermediaries" enabling the convenient calculation of quantities (e.g., accelerations) entering into descriptions. And more recently, van Fraassen (1980) has suggested that while explanations going beyond descriptions may serve various pragmatic purposes, these have no ontological or causal weight beyond their ability to "save the phenomena."

Whether science really does, or should, focus on the (noncausally) descriptive is, however, deeply controversial. Galileo himself sought for explanatory principles going beyond description (see Jammer 1957 for a discussion), and, notwithstanding Newton's professed neutrality about their physical seats, he took forces to be the "causal principle[s] of motion and rest." More generally, notwithstanding the availability of interpretations of scientific theories as purely descriptive, there are compelling reasons (say, the need to avoid a suspect action at a distance) for taking the causally explanatory posits of scientific theories (e.g., fields and forces) ontologically seriously. The deeper questions here, of course, concern how to assess the ontological and causal commitments of scientific theories; and at present there is no philosophical consensus on these important matters. In any case it is not enough, in assessing whether causes are implicated by physical theories, to note that terms like 'cause' do not explicitly appear in the equations of the theory, insofar as the commitments of a given theory may transcend the referents of the terms appearing in the theory, and given that many terms—force, charge, valence—that do appear are most naturally defined in causal terms ('force,' for example, is usually defined as that which causes acceleration). It is also worth noting that the apparent symmetry of many equations, as well as the fact that cause terms do not explicitly appear in scientific equations, may be artifacts of scientists' using the identity symbol as an all-purpose connective between functional quantities, which enables the quantities to be manipulated using mathematical techniques but is nonetheless implicitly understood as causally directed, as in $F = ma$.

Nor does scientific practice offer decisive illumination of whether scientific theorizing is or is not committed to causal notions: As with Maxwell, it remains common for scientists to draw upon apparently robustly causal notions when formulating or explaining a theory, even while maintaining that the theory expresses nothing beyond descriptive functional correlations of measurable quantities. Perhaps it is better to attend to what scientists do rather than what they say. That they rarely leave matters at the level of descriptive laws linking observables is some indication that they are not concerned with just the "how" question—though, to be sure, the tension between descriptive and causal/explanatory questions may recur at levels below the surface of observation.

Hume and Pearson: Correlation, Not Causation

The Galilean view of scientific understanding as involving correlations among measurable quantities was philosophically mirrored in the empiricist view that all knowledge (and meaning) is ultimately grounded in sensory experience (see Empiricism). The greatest philosophical challenge to causality came from Hume, who argued that there is no experience of causes being efficacious, productive, or powerful vis-à-vis their effects; hence "we only learn by experience the frequent *conjunction* of objects, without being ever able to comprehend any thing like *connexion* between them" ([1748] 1993, 46). In place of realistically interpreted "producing" theories of causation (see Strawson 1987 for a taxonomy), Hume offered the first regularity theory of causation, according to which event c causes event e just in case c and e occur, and events of c's type have (in one's experience) been universally followed by, and spatially contiguous to, events of e's type. (As discussed, such constant conjunctions were the source of the psychological imprinting that was, for Hume, the ultimate locus of causal connection.) Hume's requirement of contiguity may be straightforwardly extended to allow for causes to produce distant effects, via chains of spatially contiguous causes and effects.

Hume's requirement of universal association is not sufficient for causation (night always follows day but day does not cause night). Nor is Hume's requirement necessary, for even putting aside the requirement that one experience the association in question, there are many causal events that happen only once (e.g., the big bang). The immediate move of neo-Humeans (e.g., Hempel and Oppenheim 1948; Mackie 1965) is to understand the generalist component of the account in terms of laws of nature, which express the lawful sufficiency of the cause for the effect, and where (reflecting Hume's reductive approach) the sufficiency is not to be understood as grounded in robust causal production. One wonders, though, what is grounding the laws in question if associations are neither necessary nor sufficient for their holding and robust production is not allowed to play a role (see Laws of Nature). If the laws are grounded in brute fact, it is not clear that the reductive aim has been served (but see the discussion of Lewis's account of laws, below).

In any case, neo-Humean accounts face several problems concerning events that are inappropriately deemed causes ("spurious causes"). One is the problem of joint effects, as when a virus causes first a fever, and then independently causes a rash: Here the fever is lawfully sufficient for, hence incorrectly deemed a cause of, the rash. Another involves violations of causal asymmetry: Where events of the cause's type are lawfully necessary for events of the effect's type, the effect will be lawfully sufficient for (hence inappropriately deemed a cause of) the cause. Cases of preemption also give rise to spurious causes: Suzy's and Billy's rockthrowings are each lawfully sufficient for breaking the bottle; but given that Suzy's rock broke the bottle (thereby preempting Billy's rock from doing so), how is one to rule out Billy's rockthrowing as a cause?

The above cases indicate that lawful sufficiency alone is not sufficient for causality. One response is to adopt an account of events in which these are finely individuated, so that, for example, the bottle-breaking resulting from Suzy's rock-throwing turns out to be of a different event type than a bottle-breaking resulting from Billy's rock-throwing (in which case Billy's rock-throwing does not instantiate a rock-throwing–bottle-breaking law, and so does not count as a cause). Another response incorporates a proviso that the lawful sufficiency at issue is sufficiency in the circumstances (as in Mackie's "INUS" condition account, in which a cause is an *i*nsufficient but *n*ecessary part of a condition that is, in the circumstances, *u*nnecessary but *s*ufficient for the effect).

Nor is lawful sufficiency alone necessary for causality, as the live possibility of irreducibly probabilistic causality indicates. This worry is usually sidestepped by a reconception of laws according to which these need express only some lawlike pattern of dependence; but this reconception makes it yet more difficult for regularity theorists to distinguish spurious from genuine causes and laws (see "Probabilistic Relevance Accounts" below for developments). One neo-Humean response is to allow certain a priori constraints to enter into determining what laws there are in a world (as in the "best system" theory of laws of Lewis 1994) in which the laws are those that systematize the phenomena with the best combination of predictive strength and formal simplicity, so as to accommodate probabilistic (and even uninstantiated) laws (see also "Hempel: Explanation, Not Causation" below).

The view that causation is nothing above (appropriately complex) correlations was widespread following the emergence of social statistics in the nineteenth century and was advanced by Karl Pearson, one of the founders of modern statistics, in 1890 in *The Grammar of Science*. Pearson's endorsement of this view was, like Hume's, inspired by a rejection of causality as involving mysterious productive powers, and contributed to causes (as opposed to associations) being to a large extent expunged from statistics and from the many sciences relying upon statistics. An intermediate position between these extremes, according to which causes are understood to go beyond correlations but are not given any particular metaphysical interpretation (in particular, as involving productive powers), was advanced by the evolutionary biologist Sewall Wright, the inventor of path analysis. Wright (1921) claimed that path analysis not only enabled previously known causal relations to be appropriately weighted, but moreover enabled the testing of causal hypotheses in cases where the causal relations were (as yet) unknown. Developed descendants and variants of Wright's approach have found increasing favor of late (see "Bayesian Networks and Causal Models" below), contributing to some rehabilitation of the notion of causality in the statistical sciences.

Hempel: Explanation, Not Causation

Logical empiricists were suspicious of causation understood as a metaphysical connection in nature; instead they located causality in language, interpreting causal talk as talk of explanation (see Salmon 1989 for a discussion). On Hempel and Oppenheim's (1948) influential D-N (deductive-nomological) model of scientific explanation, event c explains event e just in case a statement expressing the occurrence of e is the conclusion of an argument with premises, one of which expresses the holding of a universal generalization to the effect that events of c's type are associated with events of e's type and another of which expresses the fact that c occurred (see Explanation). Imposing certain requirements on universal generalizations (e.g., projectibility) enabled the D-N account to avoid cases of spurious causation due to accidental regularities (s's being a screw in Smith's car does not explain why s is rusty, even if all the screws in Smith's car rusty). And requiring that explanations track temporal dependency relations (as a variation on identifying causal with temporal asymmetry) can prevent, for example, the length of the shadow from explaining the height of the flagpole. It is less clear, however, how to deal with explanatory preemption, as when Jones, immediately after ingesting a pound of arsenic, is run over by a bus and dies. Hempel's account allows us to explain Jones' death by citing the law that anyone who ingests a pound of arsenic dies within 24 hours, along with the fact that Jones ate a pound of arsenic; but such an explanatory arguments only cite laws and facts that are causally relevant to the event being explained; but this is in obvious tension with the empiricist's goal of characterizing causation in terms of explanation.

To accommodate the possibility of irreducibly probabilistic association, as well as explanations (characteristic of the social sciences) proceeding under conditions of partial uncertainty, Hempel (1965) proposed an inductive-statistical (I-S) model, in which event c explains event e if c occurs and it is an inductively grounded law that the probability of an event of type e given an event of type c is high (see Explanation; Inductive Logic). This account is subject of counterexamples in which a cause produces an effect but with a low probability, as in Scriven's case (discussed by him prior to Hempel's extension, and developed in Scriven 1975), where the probability of paresis given syphilis is low, but when paresis occurs, syphilis is the reason. A similar point applies to many quantum processes. Such cases gave rise to two different approaches to handling probabilistic explanation (or causation, by those inclined to accept this notion). One approach (see Railton 1978) locates probabilistic causality in propensities; the other (see "Probabilistic Relevance Accounts" below) in more sophisticated probabilistic relations.

At this point the line between accounts that are reductive (in the sense of reducing causal to noncausal goings-on) and nonreductive, as well as the line between singularist and generalist accounts, begins to blur. For while a propensity-based account of causality initially looks nonreductive and singularist, some think that propensities can be accommodated on a sophisticated associationist account of laws; and while an account based on relations of probabilistic relevance initially looks reductive and generalist, whether it is so depends on how the probabilities are interpreted (as given by frequencies, irreducible propensities, etc.).

Contemporary Generalist Accounts

Probabilistic Relevance Accounts

A natural response to Scriven-type cases is to understand positive causal relevance in terms of probability raising (Suppes 1970): Event c causes event e just in case the probability of events of e's type is higher given events of c's type than without. (Other relevance relations, such as being a negative causal factor, can be defined accordingly.) A common objection to such accounts (see Rosen 1978) proceeds by constructing "doing the hard way" cases in which it seems that causes lower the probability of their effects (e.g., where a mishit golf ball ricochets off a tree, resulting in a hole-in-one; or where a box contains a radioactive substance s that produces decay particles but the presence of s excludes the more effective radioactive substance s'). Such cases can often be handled, however, by locating a neutral context (where the golf ball is not hit at all, or where no radioactive substance is in the box) relative to which events of the given type do raise the probability of events of the effect type.

A more serious problem for probability-raising accounts is indicated by Simpson's paradox, according to which any statistical relationship between two variables may be reversed by including additional factors in the analysis. The "paradox" reflects the possibility that a variable C can be positively correlated with a variable E in a population and yet C be negatively correlated with E in every partition of the population induced by a third variable X. Just this occurred in the Berkeley sex discrimination case. Relative to the population of all men and women applying to graduate school at the University of California, Berkeley, being male (C) was positively correlated with being admitted (E). But relative to every partition of the population containing the men and women applying to a particular department (X), this correlation was reversed. In this case the difference between the general and specific population statistics reflected the fact that while in every department it was easier for women to be admitted than men, women were more likely to apply to departments that were harder (for *everyone*) to get into. The general population statistic was thus confounded: Being a male, *simpliciter*, was not in fact causally relevant to getting into graduate school at UC Berkeley; rather (assuming no other confounding was at issue), applying to certain departments rather than others was what was relevant.

To use probabilistic accounts as a basis for testing hypotheses and making predictions—and especially in order to identify effective strategies (courses of action) in the social sciences and, indeed, in everyday life—statistical confounding needs to be avoided. Nancy Cartwright (1979) suggested that avoiding confounding requires that the relevant probabilities be assessed relative to background contexts within which all other causal factors (besides the variable C, whose causal relevance is at issue) are held fixed. Opinions differ regarding whether events of type C must raise the probability of events of type E in at least one such context, in a majority of contexts, or in every such context. So one might take C to be a positive causal factor for E just in case $P(E \mid C \wedge X_i) \geq P(E \mid \neg C \wedge X_i)$ for all background contexts X_i, with strict inequality for at least one X_i. On this approach, smoking would be a positive causal factor for having lung cancer just in case smoking increases the chance of lung cancer in at least one background context and does not lower it in any background context.

Practically, Cartwright's suggestion has the disadvantage that one is frequently not in a position to control for all alternative causal factors (though in some circumstances one can avoid having to do this; see "Bayesian Networks and Causal Models" below). Philosophically, the requirement threatens reductive versions of probabilistic accounts with circularity. Attempts have been made to provide a noncircular means of specifying the relevant background contexts (e.g., Salmon 1984), but it is questionable whether these attempts succeed, and many are presently prepared to agree with Cartwright: "No causes in, no causes out."

As mentioned, probabilistic accounts may or may not be reductive, depending on whether the probabilities at issue are understood as grounded in associations (as in Suppes 1970) or else in powers, capacities, or propensities (as in Humphreys 1989 and Cartwright 1989). In the latter interpretation, further divisions are introduced: If the propensities are taken to be irreducible to laws (as in Cartwright's account), then the associated probabilistic relevance account is more appropriately deemed singularist. Complicating the taxonomy here is the fact that most proponents of probabilistic accounts are not explicit as regards what analysis should be given of the probabilities at issue.

Bayesian Networks and Causal Models

Philosophical worries concerning whether statistical information adequately tracks causal influence are echoed in current debates over the interpretation of the statistical techniques used in the social sciences. As noted above ("Hume and Pearson: Correlation, Not Causation"), these techniques have frequently been interpreted as relating exclusively to correlations, but recently researchers in computer science, artificial intelligence, and statistics have developed interpretations of these approaches as encoding explicitly causal information (see Spirtes, Glymour, and Scheines 1993; Pearl 2000).

In the causal modeling approach, one starts with a set of variables (representing properties) and a probability distribution over the variables. This probability distribution, partially interpreted with prior causal knowledge, is assumed to reflect a causal structure (a set laws expressing the causal relations between the variables, which laws may be expressed either graphically or as a set of structurally related functional equations). Given that certain conditions (to be discussed shortly) hold between the probabilities and the causal structure, algorithmic techniques are used to generate the set of all causal structures consistent with the probabilities and the prior causal knowledge. Techniques also exist for extracting information regarding the results of interventions (corresponding to manipulations of variables). Such strategies appear to lead to improved hypothesis testing and prediction of effects under observation and intervention.

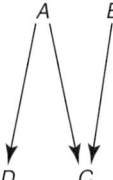

Fig. 1. *A* is a common causal factor of *D* and *C*.

Another advantage claimed for such accounts is that they provide a means of avoiding confounding without imposing the often impracticable requirement that the relevant probabilities be assessed against background contexts taking into account all causal factors, it rather being sufficient to take into account all common causal factors. As a simple illustration, suppose that *A* and *B* are known to causally influence *C*, as in Figure 1.

To judge whether *D* influence *C*, Cartwright (1979) generally recommends holding fixed both *A* and *B*, while Spirtes *et al.* (1993) instead recommend holding fixed only *A*. Cartwright allows, however, that attention to just common causal factors is possible when the causal Markov condition holds. Since this is one of the conditions that must hold in order to implement the causal modeling approach, the restriction to common causal factors is not really an advantage over Cartwright's account.

While causal modeling approaches may lead to improved causal inference concerning complex systems (in which case they are of some epistemological interest), it is unclear what bearing they have on the metaphysics of causality. Spirtes *et al.* (1993) present their account not so much as an analysis of causality as a guide to causal inference. Pearl (2000), however, takes the appeal to prior causal intuitions to indicate that causal modeling approaches are nonreductive (and moreover based on facts about humans' cognitive capacities to make effective causal inferences in simple cases). In any case, the potential of causal models to provide a basis for a general theory of causality is limited by the need for certain strong conditions to be in place in order for the algorithms to be correctly applied.

One of these is the aforementioned causal Markov condition (of which Reichenbach's [1956] "principle of the common cause" was a special case), which says that once one conditions on the complete set P_v of "causal parents" (direct causes) of a variable *V*, *V* will be probabilistically independent of all other variables except *V*'s descendants; that is for all variables *X*, where *X* is not one of *V*'s descendants, $P(V|P_v \land X) = P(V|P_v)$. (In particular, where *V* is a joint effect, conditioning on the causal parents of *V* screens off the probabilistic influence of the other joint effects on *V*.) Here again there is the practical problem that in the social sciences, where the approaches are supposed to be applicable, one is often not in a position to specify states with sufficient precision to guarantee that the condition is met. A metaphysical problem is that (contrary to Reichenbach's apparent assumption that the condition holds in all cases involving a common cause of joint effects) the causal Markov condition need not hold in cases of probabilistic causation: When a particle may probabilistically decay either by emitting a high-energy electron and falling into a low-energy state or by emitting a low-energy electron and falling into a different energy state, the joint effects in either case will not be probabilistically independent of each other, even conditioning on the cause; and certain cases of macrocausation appear also to violate the condition.

A second assumption of the causal modeling technique is what Spirtes *et al.* (1993) call "faithfulness" (also known as "stability" in Pearl 2000), according to which probabilistic dependencies faithfully reveal causal connections. In particular, if *Y* is probabilistically independent of *X*, given *X*'s parents, then *X* is assumed not to cause *Y*. Again, this condition cannot be assumed to hold in all cases, since some variables (properties) may sometimes prevent and sometimes produce and effect (as when birth control pills are a cause of thrombosis yet also prevent thrombosis, insofar as pregnancy causes thrombosis and the pills prevent pregnancy). In circumstances where the positive and negative contributions of *X* to *Y* are equally effective, the probabilistic dependence of effect on cause may cancel out, and thus *X* may inappropriately be taken not to be causally relevant to *Y*.

Causal Powers, Capacities, Universals, Forces

As mentioned, some proponents of probabilistic relevance accounts endorse metaphysical interpretations of the probabilities at issue. Such positions fall under the broader category of nonreductive ("realist") covering-law theories, in which laws express (or are grounded in) more than mere associations. The job such accounts face is to provide an alternative basis for causal laws. Among other possibilities, these bases are taken to be relations of necessitation or "probabilification" among universals (Dretske 1977; Tooley 1977; Armstrong 1978), (law-based) capacities or powers associated with objects or properties (Shoemaker 1980; Martin

1993), or fundamental forces or interactions (Bohm 1957; Strawson 1987).

Such accounts sidestep many of the problems associated with reductive covering law accounts. Since laws are not just a matter of association, a realist has the means to deny, in the virus–fever–rash case, that there is a law connecting fevers with rashes; similarly, in cases of preemption a realist may claim that, for example, Billy's rock throwing and the bottle's breaking did not instance the law in question (even without endorsing a fine-grained account of event individuation). Of course, much depends here on the details of the proposed account of laws. In the Dretske-Tooley-Armstrong account, causal laws are contingent, brute relations between universals. Some find this less than satisfying from a realist point of view, insofar as it is compatible with, for example, the property of having spin 1 bestowing completely different causal powers (say, all those actually bestowed by having spin $\frac{1}{2}$) on its possessing particulars. Other realists are more inclined to see the nature of properties and particulars as essentially dependent on the causal laws that actually govern them, a view that is not implausible for scientific entities: "[C]ausal laws are not like externally imposed legal restrictions that, so to speak, merely limit the course of events to certain prescribed paths .... [T]he causal laws satisfied by a thing ... are inextricably bound up with the basic properties of the thing which helps to define what it is" (Bohm 1957, 14).

The primary problem facing realist accounts is that they require accepting entities and relations (universals, causal powers, forces) that many philosophers and scientists find metaphysically obscure and/or epistemologically inaccessible. How one evaluates these assessments often depends on one's other commitments. For example, many traditional arguments against realist accounts (e.g., Hume's arguments) are aimed at showing that these do not satisfy a strict epistemological standard, according to which the warranted posit of a contingent entity requires that the entity be directly accessible to experience (or a construction from entities that are so accessible). But if inference to the existence of an unexperienced entity (as the best explanation of some phenomenon) is at least sometimes an acceptable mode of inference, such a strict epistemological standard (and associated arguments) will be rejected; and indeed, positive arguments for contemporary realist accounts of causality generally proceed via such inferences to the best explanation—often of the patterns of association appealed to by reductivist accounts.

Contemporary Singularist Accounts

Singularists reject the claim that causes follow laws in the order of explanation, but beyond this there is considerable variety in their accounts. *Contra* Hume, Anscombe (1971) takes causation to be a (primitive) relation that may be observed in cuttings, pushings, fallings, etc. It is worth noting that a primitivist approach to causality is compatible with even a strict empiricism (compare Hume's primitivist account of the resemblance relation). The empiricist Ducasse (1926) also locates causation in singular observation, but nonprimitively: A cause is the change event observed to be immediately prior and spatiotemporally contiguous to an effect event. While interesting in allowing for a non-associative, non-primitivist, empiricist causality, Ducasse's account is unsatisfactory in allowing only the coarse-grained identification of causes (as some backward temporal segment of the entire observed change); hence it fails to account for most ordinary causal judgments. Note that singularists basing causation on observation need not assert that one's knowledge of causality proceeds only via observations of the preferred sort; they rather generally maintain that such experiences are sufficient to account for one's acquiring the concept of causation, then both allow that causation need not be observed and that confirming singular causal claims may require attention to associations.

Another singularist approach takes causation to be theoretically inferred, as that relation satisfying (something like) the Ramsey sentence consisting of the platitudes about causality involving asymmetry, transitivity, and so on (see Tooley 1987). One problem here is that, as may be clear by now, such platitudes do not seem to uniformly apply to all cases. Relatedly, one may wonder whether the platitudes are consistent; given the competing causal intuitions driving various accounts of causality, it would be surprising if they were.

Finally, a wide variety of singularist accounts analyze causality in terms of singular processes. Such accounts are strongly motivated by the intuition that in a case of preemption such as that of Suzy and Billy, what distinguishes Suzy's throw as a cause is that it initiates a process ending in the bottle breaking, while the process initiated by Billy's throw never reaches completion (see Menzies 1996 for a discussion). Commonly, process for singularists attempt (like Ducasse) to provide a non-primitivist causality that is both broadly empiricist, in not appealing to any properly metaphysical elements, and non-associationist, in recognition of the difficulties that associationist accounts have (both with preemption and with distinguishing genuine causality from accidental regularity). Hence they typically fill in the "process" intuition by identifying causality with fundamental physical processes, including transfers or interactions, as in Fair's (1979) account of causation as identical with the transfer of energy momentum, Salmon's (1984) "mark transmission" account, and Dowe's (1992) account in which the transfer of any conserved quantity will suffice.

An objection to the claim that physical processes are sufficient for causality is illustrated by Cartwright's (1979) case of a plant sprayed with herbicide that improbably survives and goes on to flourish (compare also Kvart's 1991 finger-severing case, discussed previously). While transfers and interactions of the requisite sort can be traced from spraying to flourishing, intuitively the former did not cause the latter; however, accepting that the spraying did cause the flourishing may not be an overly high price to pay. A deeper worry concerns the epistemological question of how accounts of physical process link causation, understood as involving theoretical relations or processes of fundamental physics, with causation as ordinarily experienced. Fair suggests that ordinary experience involves macroprocesses, which are in turn reducible to the relevant physical processes; but even supposing that such reductions are in place, ordinary causal judgments do not seem to presuppose them.

Contemporary Counterfactual Accounts

Counterfactual accounts of causality, which may also be traced back to Hume, take as their starting point the intuition that a singular cause makes an important difference in what happens. As a first pass, c causes e (where c and e are actually occurring events) only if, were c not to occur, then e would not occur. As a second pass, c causes e only if c and e are connected by a chain of such dependencies (see Lewis 1973), so as to ensure that causation is transitive (causation is thus the "ancestral," or transitive closure, of counterfactual dependence). In addition to the requirement of counterfactual necessity of causes for effects, counterfactual accounts also commonly impose a requirement of counterfactual sufficiency of causes for effects: If c were to occur, then e would occur. Insofar as counterfactual accounts are standardly aimed at reducing causal to noncausal relations, and given plausible assumptions concerning evaluation of counterfactuals, the latter requirement is satisfied just by c and e's actually occurring (which occurrences, as above, are assumed); hence standard counterfactual accounts do not have a nontrivial notion of counterfactual sufficiency. A nontrivial notion of counterfactual sufficiency can be obtained by appeal to nested counterfactuals (see Vihvelin 1995): c causes e only if, if neither c nor e had occurred, then if c had occurred, e would have occurred.

Problems, Events, and Backtrackers

While counterfactual accounts are often motivated by a desire to give a reductive account of causality that avoids problems with reductive covering-law accounts (especially those of joint effects and of preemption), it is unclear whether counterfactual accounts do any better by these problems. First, consider the problem of joint effects. Suppose a virus causes first a fever, then a rash, and that the fever and rash could only have been caused by the virus. It seems correct to reason in the following "backtracking" fashion: If the fever had not occurred, then the viral infection would not have occurred, in which case the rash would not have occurred. But then the counterfactual "If the fever had not occurred, the rash would not have occurred" turns out true, which here means that the fever causes the rash, which is incorrect. Proponents of counterfactual accounts have responses to these objections, which require accepting controversial accounts of the truth conditions for counterfactuals (see Lewis 1979). Even so, the responses appear not to succeed (see Bennett 1984 for a discussion)

Second, consider the problem of preemption. In the Suzy-Billy case, it seems correct to reason that if Suzy had not thrown her rock, then Billy's rock would have gotten through and broken the bottle. Hence the counterfactual "If Suzy's throw had not occurred, the bottlebreaking would not have occurred" turns out false; so Suzy's rockthrowing turns out not to be a cause, which is incorrect. In cases (as here) of so-called "early preemption," where it makes sense to suppose that there was an intermediate event d between the effect and the cause on which the effect depended, this result can be avoided: Although the breaking does not counterfactually depend on Suzy's rockthrowing, there is a chain of counterfactual dependence linking the breaking to Suzy's rockthrowing (and no such chain linking the breaking to Billy's), and so her throw does end up being a cause (and Billy's does not). But the appeal to an intermediate event seems *ad hoc*, and in any case cannot resolve cases of "late preemption." Lewis (developing an idea broached in Paul 1998) eventually responded to such cases by allowing that an event may be counted as a cause if it counterfactually influences the mode of occurrence of the effect (e.g., how or when it occurs), as well as if it counterfactually influences the occurrence of the effect, *simpliciter.*

Counterfactuals and Manipulability

Where counterfactual accounts may be most useful is in providing a basis for understanding or formalizing the role that manipulability plays in the concept of causation. One such approach sees counterfactuals as providing the basis for an epistemological, rather than a metaphysical, account of causation (see Pearl 2000 for discussion). The idea here is that counterfactuals nicely model the role manipulability (actual or imagined) plays in causal inference, for a natural way to determine whether a counterfactual is true is to manipulate conditions so as to actualize the antecedent. Another approach takes counterfactuals to provide a basis for a generalist account of causal explanation (see Woodward 1997), according to which such explanations track stable or invariant connections and the notion of invariance is understood nonepistemologically in terms of a connection's continuing to hold through certain counterfactual (not necessarily human) "interventions." Whether the notion of manipulability is itself a causal notion, and so bars the reduction of causal to noncausal facts, is still an open question.

JESSICA WILSON

References

Anscombe, G. E. M. (1971), *Causality and Determination.* Cambridge: Cambridge University Press.
Armstrong, David M. (1978), *Universals and Scientific Realism*, vol. 2: *A Theory of Universals.* Cambridge: Cambridge University Press.
Bennett, Jonathan (1984), "Counterfactuals and Temporal Direction," *Philosophical Review* 93: 57–91.
Bohm, David (1957), *Causality and Chance in Modern Physics.* London: Kegan Paul.
Cartwright, Nancy (1979), "Causal Laws and Effective Strategies," *Noûs* 13: 419–438.
——— (1989), *Nature's Capacities and Their Measurement.* Oxford: Clarendon Press.
Clatterbaugh, Kenneth (1999), *The Causation Debate in Modern Philosophy, 1637–1739.* New York: Routledge.
Davidson, Donald (1967), "Causal Relations," *Journal of Philosophy* 64: 691–703. Reprinted in Davidson (2001), *Essays on Actions and Events.* Oxford: Oxford University Press, 149–162.
——— (1970), "Mental Events," in L. Foster and J. Swanson (eds.), *Experience and Theory.* Amherst: Massachusetts University Press. Reprinted in Davidson (2001), *Essays on Actions and Events.* Oxford: Oxford University Press, 207–224.

Dowe, Phil (1992), "Wesley Salmon's Process Theory of Causality and the Conserved Quantity Theory," *Philosophy of Science* 59: 195–216.
Dretske, Fred (1977), "Laws of Nature," *Philosophy of Science* 44: 248–268.
Ducasse, C. J. (1926), "On the Nature and Observability of the Causal Relation," *Journal of Philosophy* 23: 57–68.
Fair, David (1979), "Causation and the Flow of Energy," *Erkenntnis* 14: 219–250.
Hall, Ned (2000), "Causation and the Price of Transitivity," *Journal of Philosophy* 97: 198–222.
Hankinson, R. J. (1998), *Cause and Explanation in Ancient Greek Thought.* Oxford: Oxford University Press.
Hempel, Carl (1965), *In Aspects of Scientific Explanation and Other Essays in the Philosophy of Science.* New York: Free Press.
Hempel, Carl, and Paul Oppenheim (1948), "Studies in the Logic of Explanation," *Philosophy of Science* 15: 135–175.
Hitchcock, Christopher (1993), "A Generalized Probabilistic Theory of Causal Relevance," *Synthese* 97: 335–364.
Hume, David ([1739] 1978), *A Treatise of Human Nature.* Edited by L. A. Selby-Bigge. Oxford: Oxford University Press.
——— ([1748] 1993), *An Enquiry Concerning Human Understanding.* Edited by Eric Steinberg. Indianapolis: Hackett Publishing.
Humphreys, Paul (1986), "Causation in the Social Sciences: An Overview," *Synthese* 68: 1–12.
——— (1989), *The Chances of Explanation.* Princeton, NJ: Princeton University Press.
Jammer, Max (1957), *Concepts of Force.* Cambridge, MA: Harvard University Press.
Kim, Jaegwon (1984), "Epiphenomenal and Supervenient Causation," *Midwest Studies in Philosophy IX: Causation and Causal Theories*, 257–270.
——— (1993), *Supervenience and Mind: Selected Philosophical Essays.* Cambridge: Cambridge University Press.
Kvart, Igal (1991), "Transitivity and Preemption of Causal Relevance," *Philosophical Studies* LXIV: 125–160.
Lewis, David (1973), "Causation," *Journal of Philosophy* 70: 556–567.
——— (1979), "Counterfactual Dependence and Time's Arrow," *Noûs* 13:455–476.
——— (1983), *Philosophical Papers*, vol. 1. Oxford: Oxford University Press.
——— (1994), "Humean Supervenience Debugged," *Mind* 1: 473–490.
Mackie, John L. (1965), "Causes and Conditions," *American Philosophical Quarterly* 2: 245–264.
Martin, C. B. (1993), "Power for Realists," in Keith Cambell, John Bacon, and Lloyd Reinhardt (eds.), *Ontology, Causality, and Mind: Essays on the Philosophy of D. M. Armstrong.* Cambridge: Cambridge University Press.
McLaughlin, Brian (1992), "The Rise and Fall of British Emergentism," in Ansgar Beckerman, Hans Flohr, and Jaegwon Kim (eds.), *Emergence or Reduction? Essays on the Prospects of Nonreductive Physicalism.* Berlin: De Gruyter, 49–93.
Mellor, D. H. (1995), *The Facts of Causation.* Oxford: Oxford University Press.

Menzies, Peter (1996), "Probabilistic Causation and the Pre-Emption Problem," *Mind* 105: 85–117.
Newton, Isaac ([1687] 1999), *The Principia: Mathematical Principles of Natural Philosophy*. Translated by I. Bernard Cohen and Anne Whitman. Berkeley and Los Angeles: University of California Press.
Paul, Laurie (1998), "Keeping Track of the Time: Emending the Counterfactual Analysis of Causation," *Analysis* LVIII: 191–198.
Pearl, Judea (2000), *Causality*. Cambridge: Cambridge University Press.
Price, Huw (1992), "Agency and Causal Asymmetry," *Mind* 101: 501–520.
Railton, Peter (1978), "A Deductive-Nomological Model of Probabilistic Explanation," *Philosophy of Science* 45: 206–226.
Reichenbach, Hans (1956), *The Direction of Time*. Berkeley and Los Angeles: University of California Press.
Rosen, Deborah (1978), "In Defense of a Probabilistic Theory of Causality," *Philosophy of Science* 45: 604–613.
Russell, Bertrand (1912), "On the Notion of Cause," *Proceedings of the Aristotelian Society* 13: 1–26.
Salmon, Wesley (1984), *Scientific Explanation and the Causal Structure of the World*. Princeton, NJ: Princeton University Press.
——— (1989), "Four Decades of Scientific Explanation," in Philip Kitcher and Wesley Salmon (eds.), *Scientific Explanation*, 3–219.
Scriven, Michael (1975), "Causation as Explanation," *Noûs* 9: 238–264.
Shoemaker, Sydney (1980), "Causality and Properties," in Peter van Inwagen (ed.), *Time and Cause*. Dordrecht, Netherlands: D. Reidel, 109–135.
Sober, Elliott (1984), "Two Concepts of Cause," in Peter D. Asquith and Philip Kitcher (eds.), *PSA 1984*. East Lansing, MI: Philosophy of Science Association, 405–424.
Sosa, Ernest, and Michael Tooley (eds.) (1993), *Causation*. Oxford: Oxford University Press..
Spirtes, P., C. Glymour, and R. Scheines (1993), *Causation, Prediction, and Search*. New York: Springer-Verlag.
Strawson, Galen (1987), "Realism and Causation," *Philosophical Quarterly* 37: 253–277.
Suppes, Patrick (1970), *A Probabilistic Theory of Causality*. Amsterdam: North-Holland.
——— (1986), "Non-Markovian Causality in the Social Sciences with Some Theorems on Transitivity," *Synthese* 68: 129–140.
Tooley, Michael (1977), "The Nature of Laws," *Canadian Journal of Philosophy* 7: 667–698.
——— (1987), *Causation: A Realist Approach*. Oxford: Oxford University Press.
van Fraassen, Bas (1980), *The Scientific Image*. Oxford: Oxford University Press.
Vihvelin, Kadhri (1995), "Causes, Effects, and Counterfactual Dependence," *Australasian Journal of Philosophy* 73: 560–583.
Wilson, Jessica (1999), "How Superduper Does a Physicalist Supervenience Need to Be?" *Philosophical Quarterly* 49: 33–52.
Woodward, James (1997), "Explanation, Invariance, and Intervention," *Philosophy of Science* 64 (Proceedings): S26–S41.
Wright, Larry (1976), *Teleological Explanation*. Berkeley and Los Angeles: University of California Press.
Wright, Sewall (1921), "Correlation and Causation," *Journal of Agricultural Research* 20: 557–585.

CHAOS THEORY

See **Prediction**

PHILOSOPHY OF CHEMISTRY

Although many influential late-nineteenth- and early-twentieth-century philosophers of science were educated wholly or in part as chemists (Gaston Bachelard, Pierre Duhem, Emile Myerson, Wilhelm Ostwald, Michael Polanyi), they seldom reflected directly on the epistemological, methodological, or metaphysical commitments of their science. Subsequent philosophers of science followed suit, directing little attention to chemistry in comparison with physics and biology despite the industrial, economic, and academic success of the chemical sciences. Scattered examples of philosophical reflection on chemistry by chemists do exist; however, philosophically sensitive historical analysis and sustained conceptual analysis are relatively recent phenomena. Taken together, these two developments demonstrate that chemistry addresses general issues in the philosophy of science and, in addition, raises important questions in the interpretation of chemical theories, concepts, and experiments.

Historians of chemistry have also raised a number of general philosophical questions about chemistry. These include issues of explanation, ontology, reduction, and the relative roles of theories, experiments, and instruments in the advancement of the science.

Worries about the explanatory nature of the alchemical, corpuscularian, and phlogiston theories are well documented (cf. Bensaude-Vincent and Stengers 1996; Brock 1993). Lavoisier and—to a lesser extent historically—Dalton initiated shifts in the explanatory tasks and presuppositions of the science. For instance, prior to Lavoisier many chemists "explained" a chemical by assigning it to a type associated with its experimental dispositions (e.g., flammability, acidity, etc.). After Lavoisier and Dalton, "explanation" most often meant the isolation and identification of a chemical's constituents. Eventually, the goal of explanation changed to the identification of the transformation processes in the reactants that gave rise to the observed properties of the intermediaries and the products. It is at that time that chemists began to write the now familiar reaction equations, which encapsulate this change. These transformations cannot be described simply as the coming of new theories; they also involved changes in explanatory presuppositions and languages, as well as new experimental techniques (Bensaude-Vincent and Stengers 1996; Nye 1993).

Thinking in terms of transformation processes led chemists to postulate atoms as the agents of the transformation process. But atoms were not observable in the nineteenth century. How is the explanatory power of these unobservable entities accounted for? Also, do chemical elements retain their identity in compounds (Paneth 1962)? Something remains the same, yet the properties that identify elements (e.g., the green color of chlorine gas) do not exist in compounds (e.g., sodium chloride or common table salt).

The history of chemistry also raises a number of interesting questions about the character of knowledge and understanding in the science. Whereas philosophers have historically identified theoretical knowledge with laws or sets of propositions, the history of chemistry shows that there are different kinds of knowledge that function as a base for understanding. Much of chemistry is experimental, and much of what is known to be true arises in experimental practice independently of or only indirectly informed by theoretical knowledge. When chemists have theorized, they have done so freely, using and combining phenomenological, constructive (in which the values of certain variables are given by experiment or other theory), and deductive methods. Chemists have rarely been able to achieve anything like a strict set of axioms or first principles that order the phenomena and serve as their explanatory base (Bensaude-Vincent and Stengers 1996; Gavroglu 1997; Nye 1993).

Historical research has also raised the issue of whether theory has contributed most to the progress of chemistry. While philosophers often point to the conceptual "revolution" wrought by Lavoisier as an example of progress, historians more often point to the ways in which laboratory techniques have been an important motor of change in chemistry, by themselves or in tandem with theoretical shifts (Bensaude-Vincent and Stengers 1996; Nye 1993). For example, during the nineteenth century, substitution studies, in which one element is replaced by another in a compound, were driven largely by experimental practices. Theoretical concepts did not, in the first instance, organize the investigation (Klein 1999). Similar remarks can be made regarding the coming of modern experimental techniques such as various types of chromatography and spectroscopy (Baird 2000; Slater 2002). Chemists often characterize a molecule using such techniques, and while the techniques are grounded in physical theory, the results must often be interpreted in chemical language. These examples lead back to questions about the nature of chemical knowledge. Arguably, the knowledge appears to be a mix of "knowing how" and "knowing that" which is not based solely in the theory (chemical or physical) available at the time.

A number of the philosophical themes raised in the history of chemistry continue to reverberate in current chemistry. For instance, what is the proper ontological base for chemical theory, explanation, and practice? While many chemists would unfailingly resort to molecular structure as the explanation of what is seen while a reaction is taking place, this conception can be challenged from two directions. From one side, echoing the eighteenth-century conception of the science, chemistry begins in the first instance with conceptions and analyses of the qualitative properties of material stuff (Schummer 1996; van Brakel 2001). One might call the ontology associated with this conception a metaphysically nonreductive dispositional realism, since it focuses on properties and how they appear under certain conditions and does not attempt to interpret them in any simpler terms. In this conception, reference to the underlying molecular structure is subsidiary or even otiose, since the focus is on the observable properties of the materials. Given that molecular structure is difficult to justify within quantum mechanics (see below), one can argue that it is justifiable to remain with the observable properties. If this view is adopted, however, the justification of the ontology of material stuff becomes a pressing matter. Quantum mechanics will not supply the justification, since it does not deliver the qualitative properties of materials. Further, it still seems necessary to account for the phenomenal success that molecular explanations afford in planning and interpreting chemical structures and reactions.

From the other side, pure quantum mechanics makes it difficult to speak of the traditional atoms-within-a-molecule approach referred to in the reaction equations and structural diagrams (Primas 1983; Weininger 1984). Quantum mechanics tells us that the interior parts of molecules should not be distinguishable; there exists only a distribution of nuclear and electronic charges. Yet chemists rely on the existence and persistence of atoms and molecules in a number of ways. To justify these practices, some have argued that one can forgo the notion of an atom based on the orbital model and instead identify spatial regions within a molecule bounded by surfaces that have a zero flux of energy across the surfaces (Bader 1990). Currently, it is an open question whether this representation falls naturally out of quantum mechanics, and so allows one to recover atoms as naturally occurring substituents of molecules, or whether the notion of 'atom' must be presupposed in order for the identification to be made. Here again there is a question of the character of the theory that will give the desired explanation.

A number of examples supporting the claim that quantum mechanics and chemistry are uneasy bedfellows will be discussed below as they relate to the issue of reductionism, but their relationship also raises forcefully the long-standing issue of how theory guides chemical practice. Even in this era of supercomputers, only the energy states of systems with relatively few electrons or with high degrees of symmetry can be calculated with a high degree of faithfulness to the complete theoretical description. For most chemical systems, various semi-empirical methods must be used to get theoretically guided results. More often than not, it is the experimental practice independent of any theoretical calculation that gets the result. Strictly theoretical predictions of novel properties are rather rare, and whether they are strictly theoretical is a matter that can be disputed. A case in point is the structure of the CH_2 molecule. Theoretical chemists claimed to have predicted novel properties of the molecule, *viz.*, its nonlinear geometry, prior to any spectroscopic evidence (Foster and Boys 1960). While it is true that the spectroscopic evidence was not yet available, it may have been the case that reference to analogous molecules allowed the researchers to set the values for some of the parameters in the equations. So the derivation may not have been as a priori as it seemed.

Like the other special sciences, chemistry raises the issue of reductionism quite forcefully. However, perhaps because most philosophers have accepted at face value Dirac's famous dictum that chemistry has become nothing more than the application of quantum mechanics to chemical problems (cf. Nye 1993, 248), few seem to be aware of the difficulties of making good on that claim using the tools and concepts available in traditional philosophical analyses of the sciences. No one doubts that chemical forces are physical in nature, but connecting the chemical and physical mathematical structures and/or concepts proves to be quite a challenge. Although problems involving the relation between the physical and the chemical surround a wide variety of chemical concepts, such as aromaticity, acidity (and basicity), functional groups, and substituent effects (Hoffmann 1995), three examples will be discussed here to illustrate the difficulties: the periodic table, the use of orbitals to explain bonding, and the concept of molecular shape. Each also raises issues of explanation, representation, and realism.

Philosophers and scientists commonly believe that the periodic table has been explained by—and thus reduced to—quantum mechanics. This is taken to be an explanation of the configuration of the electrons in the atom, and, as a result of this, an explanation of the periodicity of the table. In the first case, however, configurations of electrons in atoms and molecules are the result of a particular approximation, in which the many-electron quantum wavefunction is rewritten as a series of one-electron functions. In practice, these one-electron functions are derived from the hydrogen wavefunction and, upon integration, lead to the familiar spherical *s* orbital, the dumbbell-shaped *p* orbitals, and the more complicated *d* and *f* orbitals. Via the Pauli exclusion principle, which states that the spins are to be paired if two electrons are to occupy one orbital and no more than two electrons may occupy an orbital, electrons are assigned to these orbitals. If the approximation is not made (and quantum mechanics tells us it should not be, since the approximation relies on the distinguishability of electrons), the notion of individual quantum numbers—and thus configurations—is no longer meaningful. In addition, configurations themselves are not observable; absorption and emission spectra are observed and interpreted as energy transitions between orbitals of different energies.

In the second case, quantum mechanics explains only part of the periodic table, and often it does not explain the features of the table that are of most interest to chemists (Scerri 1998). Pauli's introduction of the fourth quantum number, "spin" or "spin angular momentum," leads directly to the *Aufbau* principle, which states that the periodic table is constructed by placing electrons in lower energy levels first and then demonstrating that atoms with similar configurations have similar chemical properties. In this way, one can say that because chlorine and fluorine need one more electron to achieve a closed shell, they will behave similarly. However, this simple, unqualified explanation suffers from a number of anomalies. First, the filling sequence is not always strictly obeyed. Cobalt, nickel, and copper fill their shells in the sequence $3d^74s^2$, $3d^84s^2$, $3d^{10}4s^1$. The superscripts denote the number of electrons in the subshell; the *s* shell can hold a maximum of two electrons and the five *d* orbitals ten. The observed order of filling is curious from the perspective of the unmodified *Aufbau* principle for a number of reasons. The 4*s* shell, which is supposed to be higher in energy than the 3*d* shell, has been occupied and closed first. Then, there is the "demotion" of one 4*s* electron in nickel to a 3*d* electron in copper. Second, configurations are supposed to explain why elements falling into the same group behave similarly, as in the example of chlorine and fluorine. Yet nickel, palladium, and platinum are grouped together because of their marked chemical similarities despite the fact that their outer shells have different configurations ($4s^2$, $5s^0$ and $6s^1$, respectively). These and other anomalies can be resolved using alternative derivations more closely tied to fundamental quantum mechanics, but the derivations require that the orbital approximation be dropped, and it was that approximation that was the basis for the assignment into the *s*, *p*, and *d* orbitals in the first place. It thus becomes an open question whether quantum mechanics, via the *Aufbau* principle, has explained the chemical periodicities encapsulated in the table.

Similar questions about the tenuous relation between physics and chemistry surround the concept of bonding. At a broad level, chemists employ two seemingly inconsistent representations, the valence bond (VB) and molecular orbital (MO) theories, to explain why atoms and molecules react. Both are calculational approximations inherited from atomic physics. The relations between the two theories and respective relations to the underlying quantum mechanics raise many issues of theory interpretation and realism about the chemical concepts (see below). Subsidiary concepts such as resonance are also invoked to explain the finer points of bonding. Chemists have offered competing realist interpretations of this concept, and philosophers have offered various realist and instrumentalist interpretations of it as well (Mosini 2000).

More specifically, a host of philosophical issues are raised within the molecular orbital theory. Here, bonding is pictured as due to the interaction of electrons in various orbitals. As noted earlier, the familiar spherical and dumbbell shapes arise only because of the orbital approximation. However, there is no reason to expect that the hydrogenic wavefunctions will look anything like the molecular ones (Bader 1990; Woody 2000). After the hydrogenic wavefunctions have been chosen as the basis for the calculation, they must be processed mathematically to arrive at a value for the energy of the orbital that is at all close to the experimentally observed value. The molecular wave equation is solved by taking linear combinations of the hydrogenic wave functions, forming the product of these combinations (the "configuration interaction" approach), and using the variational method to produce a minimal energy solution to the equation. The familiar orbitals appear only when these three steps in the complete solution have been omitted (Woody 2000). Thus, the idea that the familiar orbitals are responsible for the bonding is thrown into question. Yet the orbitals classify and explain how atoms and molecules bond extremely well. That they do provide deep, unified, and fertile representations and explanations seems curious from the perspective of fundamental quantum mechanics. Clearly, more analysis is required to understand the relation clearly. If one insists on a philosophical account of reduction that requires the mathematical or logical derivability of one theory from another, how orbitals achieve their power remains obscure. Even when one abandons that philosophical account, it is not clear how the representations have the organizing and explanatory power they do (see Reductionism).

As a final illustration of the difficulty of connecting physics and chemistry in any sort of strict fashion, consider the concept of molecular shape. Partly through the tradition of orbitals described above and partly through a historical tradition of oriented bonding that arose well before the concept of orbitals was introduced, chemists commonly explain many behaviors of molecules as due to their three-dimensional orientiation in space. Molecules clearly react as if they are oriented in three-dimensional space. For example, the reaction $I^- + CH_3Br \rightarrow ICH_3 + Br^-$ is readily explained by invoking the notion that the iodine ion (I^-) attacks the carbon (C) on the side away from the bromine (Br) atom. (This can be detected by substituting deuterium atoms for one of the hydrogens [H] and measuring subsequent changes in spectroscopic properties.) As noted before, however, such explanations are suspect within quantum mechanics, since talk of oriented bonds and quasi-independent substituents in the reaction is questionable. Orientations must be "built into" the theory by parameterizing some of the theoretical variables. Unfortunately, there is no strict quantum mechanical justification for the method by which orientation in space is derived. Orientation relies on a notion of a nuclear frame surrounded by electrons. This notion is constructed via the Born-Oppenheimer approximation, which provides a physical rationale for why the nuclear positions should be slowly varying with respect to the electronic motions. In some measurement regimes, the approximation is invalid, and correct predictions are achieved only by resorting to a more general molecular Hamiltonian. In more common measurement regimes, the approximation is clearly valid. But, as with the issues involved in the case of the periodic table and orbitals, the physics alone do not tell us why it is valid. There is a physical justification for the procedure, but this justification has no natural representation with the available physical theory (Weininger 1984). Should justification based on past experience be trusted, or should the theory correct the interpretive practice? In any case, the chemistry is consistent with, but not yet derivable from, the physics.

All three examples are connected with a methodological issue mentioned earlier, *viz.*, the type of theory that chemists find useful. As previously noted, chemists often must parameterize the physical theories at their disposal to make them useful. All three of the cases described above involve such parameterization, albeit in different ways. How is it that such parameterizations uncover useful patterns in the data? Are they explanatory? When are they acceptable and when not (Ramsey 1997)? These and a host of similar questions remain to be answered.

Other epistemological and ontological issues raised in the practice of chemistry remain virtually unexplored. For instance, the question of whether one molecule is identical to another is answered by referring to some set of properties shared by the two samples. Yet the classification of two molecules as of the "same" type will vary, since different theoretical representations and experimental techniques detect quite different properties (Hoffmann 1995). For instance, reference can be made to the space-filling property of molecules, their three-dimensional structure, or the way they respond to an electric field. Additionally, the determination of sameness must be made in light of the question, For what function or purpose? For instance, two molecules of hemoglobin, which are large biological molecules, might have different isotopes of oxygen at one position. While this difference might be useful in order to discover the detailed structure of the hemoglobin molecule, it is usually irrelevant when talking about the molecule's biological function.

How the explanatory practices of chemistry stand in relation to the available philosophical accounts and to the practices of other sciences remains an important question. Chemical explanations are very specific, often lacking the generality invoked in philosophical accounts of explanation. Moreover, chemists invoke a wide variety of models, laws, theories, and mechanisms to explain the behavior and structure of molecules. Finally, most explanations require analogically based and/or experimentally derived adjustments to the theoretical laws and regularities in order for the account to be explanatory.

As mentioned earlier, chemistry is an extremely experimental science. In addition to unifying and fragmenting research programs in chemistry, new laboratory techniques have dramatically changed the epistemology of detection and observation in chemistry (e.g., from tapping manometers to reading NMR [nuclear magnetic resonance] outputs). As yet, however, there is no overarching, complete study of the changes in the epistemology of experimention in chemistry: for example, what chemists count as observable (and how this is connected to what they consider to be real), what they assume counts as a complete explanation, what they assume counts as a successful end to an experiment, etc.

Additionally, what are the relations between academic and industrial chemistry? What are the relative roles of skill, theory, and experiment in these two arenas of inquiry? Last but not least, there are pressing ethical questions. The world has been transformed by chemical products. Chemistry has blurred the distinction between the natural and the artificial in confusing ways (Hoffmann 1995). For instance, catalytically produced ethanol is chemically identical to the ethanol produced in fermentation. So is the carbon dioxide produced in a forest fire and in a car's exhaust. Why are there worries about the exhaust fumes but not the industrially produced ethanol? Is this the appropriate attitude? Last but not least, chemicals have often replaced earlier dangerous substances and practices; witness the great number of herbicides and insecticides available at the local garden center and the prescription medicines available at the pharmacy. Yet these replacements are often associated with a cost. One need think only of DDT or thalidomide to be flung headlong into ethical questions regarding the harmfulness and use of human-made products.

Many of the above topics have not been analyzed in any great depth. Much remains to be done to explore the methodology and philosophy of the chemical sciences.

JEFFRY L. RAMSEY

References

Bader, R. (1990), Atoms in Molecules: a Quantum Theory. New York: Oxford University Press.
Baird, D. (2000), "Encapsulating Knowledge: The Direct Reading Spectrometer," *Foundations of Chemistry* 2: 5–46.
Bensaude-Vincent, B., and I. Stengers (1996), *A History of Chemistry*. Translated by D. van Dam. Cambridge, MA: Harvard University Press.
Brock, W. (1993), *The Norton History of Chemistry*. New York: W. W. Norton.
Foster, J. M., and S. F. Boys (1960), "Quantum Variational Calculations for a Range of CH_2 Configurations," *Reviews of Modern Physics* 32: 305–307.
Gavroglu, K. (1997), "Philosophical Issues in the History of Chemistry," *Synthese* 111: 283–304.
Hoffmann, R. (1995), *The Same and Not the Same*. New York: Columbia University Press.
Klein, U. (1999), "Techniques of Modeling and Paper-Tools in Classical Chemistry," in M. Morgan and M. Morrison (eds.), *Models as Mediators*. New York: Cambridge University Press, 146–167.
Mosini, V. (2000), "A Brief History of the Theory of Resonance and of Its Interpretation," *Studies in History and Philosophy of Modern Physics* 31B: 569–581.
Nye, M. J. (1993), *From Chemical Philosophy to Theoretical Chemistry*. Berkeley and Los Angeles: University of California Press.
Paneth, F. (1962), "The Epistemological Status of the Concept of Element," *British Journal for the Philosophy of Science* 13: 1–14, 144–160.
Primas, H. (1983), *Chemistry, Quantum Mechanics and Reductionism*. New York: Springer Verlag.
Ramsey, J. (1997), "Between the Fundamental and the Phenomenological: The Challenge of 'Semi-Empirical' Methods," *Philosophy of Science* 64: 627–653.
Scerri, E. (1998), "How Good Is the Quantum Mechanical Explanation of the Periodic System?" *Journal of Chemical Education* 75: 1384–1385.
Schummer, J. (1996), *Realismus und Chemie*. Wuerzburg: Koenigshausen and Neumann.
Slater, L. (2002), "Instruments and Rules: R. B. Woodward and the Tools of Twentieth-century Organic Chemistry," *Studies in History and Philosophy of Science* 33: 1–32.
van Brakel, J. (2001), *Philosophy of Chemistry: Between the Manifest and the Scientific Image*. Leuven, Belgium: Leuven University Press.
Weininger, S. (1984), "The Molecular Structure Conundrum: Can Classical Chemistry Be Reduced to Quantum Chemistry?" *Journal of Chemical Education* 61: 939–944.
Woody, A. (2000), "Putting Quantum Mechanics to Work in Chemistry: The Power of Diagrammatic Representation," *Philosophy of Science* 67(suppl): S612–S627.

See also **Explanation; Laws of Nature; Quantum Mechanics; Reductionism**

NOAM CHOMSKY

(7 December 1928–)

Avram Noam Chomsky received his Ph.D in linguistics from the University of Pennsylvania and has been teaching at Massachusetts Institute of Technology since 1955, where he is currently Institute Professor. Philosophers are often familiar with the early work of Chomsky (1956, 1957, 1959a, and 1965), which applied the methods of formal language theory to empirical linguistics, but his work has also incorporated a number of philosophical assumptions about the nature of scientific practice—many of which are defended in his writings.

This entry will first describe the development and evolution of Chomsky's theory of generative linguistics, highlighting some of the philosophical assumptions that have been in play. It will then turn to some of the methodological debates in generative linguistics (and scientific practice more generally), focusing on Chomsky's role in these debates.

The Development and Evolution of Generative Grammar

A number of commentators have suggested that Chomsky's early work in generative linguistics initiated a kind of Kuhnian paradigm shift in linguistic theory. While Chomsky himself would reject this characterization (at least for his initial work in generative grammar), it is instructive to examine the development of generative linguistics, for it provides an excellent laboratory for the study of the development of a young science, and in particular it illuminates some of the philosophical prejudice that a young science is bound to encounter.

Chomsky's role in the development of linguistic theory and cognitive science generally can best be appreciated if his work is placed in the context of the prevailing intellectual climate in the 1950s—one in which behaviorism held sway in psychology departments and a doctrine known as American Structuralism was prevalent in linguistics departments.

American Structuralism, in particular as articulated by Bloomfield (1933 and 1939), adopted a number of key assumptions that were in turn adopted from logical empiricism (see Logical Empiricism). Newmeyer (1986, Ch. 1) notes that the following assumptions were in play:

- All useful generalizations are inductive generalizations.
- Meanings are to be eschewed because they are occult entities—that is, because they are not directly empirically observable.
- Discovery procedures like those advocated in logical empiricism should be developed for the proper conduct of linguistic inquiry.
- There should be no unobserved processes.

One of the ways in which these assumptions translated into theory was in the order that various levels of linguistic description were to be tackled. The American Structuralists identified four levels: phonemics (intuitively the study of sound patterns), morphemics (the study of words, their prefixes and suffixes), syntax (the study of sentence-level structure), and discourse (the study of cross-sentential phenomena). The idea was that proper methodology would dictate that one begin at the level of phonemics, presumably because it is closer to the data; then proceed to construct a theory of morphemics on the foundations of phonemics; and then proceed to construct a theory of syntax, etc.

Notice the role that the concepts of logical empiricism played in this proposed methodology. One finds radical reductionism in the idea that every level must be reducible to the more basic phonemic level; verificationism in the contention that the phonemic level is closely tied to sense experience; and discovery procedures in the suggestion that this overall order of inquiry should be adopted (see Reductionism; Verifiability).

Chomsky rejected most if not all of these assumptions early on (see Chomsky [1955] 1975, introduction, for a detailed discussion). As regards discovery procedures, for example, he rejected them while still a matriculating graduate student, then holding a position in the Harvard Society of Fellows:

> By 1953, I came to the same conclusion [as Morris Halle] if the discovery procedures did not work, it was not because I had failed to formulate them correctly, but because the entire approach was wrong.... [S]everal years of intense effort devoted to improving discovery procedures had come to naught, while work I had been doing during the same period on generative grammars and explanatory theory, in almost complete isolation, seemed to be consistently yielding interesting results. (1979: 131)

Chomsky also rejected the assumption that all processes should be "observable"—early theories of transformational grammar offered key examples of unobservable processes. For example, in his "aspects theory" of generative grammar (Chomsky 1965), the grammar is divided into two different "levels of representation," termed initially *deep structure* and *surface structure*. The deep-structure representations were generated by a context-free phase structure grammar—that is, by rules (of decomposition, "→") of the following form, where S stands for sentence, NP for noun phrase, VP for verb phrase, etc.

S → NP VP
VP → V NP
NP → John
NP → Bill
V → saw

These rewriting rules then generated linguistic representations of the following form:

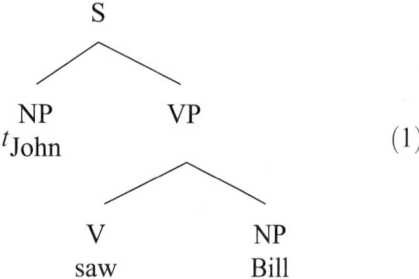

(1)

Crucially for Chomsky, the objects of analysis in linguistic theory were not the terminal strings of words, but rather *phrase markers*—structured objects like (1). Transformational rules then operated on these deep-structure representations to yield surface-structure representations. So, for example, the operation of passivization would take a deep-structure representation like (1) and yield the surface-structure representation (abstracting from detail) in (2):

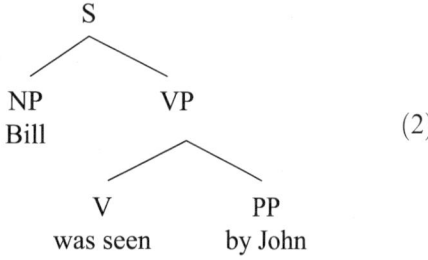

(2)

The sentence in (3) is therefore a complex object consisting of (at a minimum) an ordered pair of the two representations corresponding to (1) and (2):

    Bill was seen by John.   (3)

Clearly, Chomsky was committed not only to "unobserved" processes in the guise of transformations, but also to unobserved levels of representation. No less significant was the nature of the data that Chomsky admitted—not utterances or written strings, but rather speakers' judgments of acceptability and meaning. Thus, (3) is not a datum because it has been written or spoken, but rather because speakers have intuitions that it is (would be) an acceptable utterance. Here again, Chomsky broke with prevailing methodology in structuralist linguistics and, indeed, behaviorist psychology, by allowing intuitions rather than publicly available behaviors as data.

Generative grammar subsequently evolved in response to a number of internal pressures. Crucially, the number of transformations began to proliferate in a way that Chomsky found unacceptable. Why was this proliferation unacceptable? Early on in the development of generative grammar, Chomsky had made a distinction between the *descriptive adequacy* and the *explanatory adaquacy* of an empirical linguistic theory (Chomsky 1965 and 1986b). In particular, if a linguistic theory is to be explanatorially adequate, it must not merely describe the facts, but must do so in a way that explains how humans are able to learn languages. Thus, linguistics was supposed to be embeddable into cognitive science more broadly. But if this is the case, then there is a concern about the unchecked proliferation of rules—such rule systems might be descriptively adequate, but they would fail to account for how we learn languages (perhaps due to the burden of having to learn all those language-specific rules).

Chomsky's initial (1964 and 1965) solution to this problem involved the introduction of conditions on transformations (or constraints on movement), with the goal of reducing the complexity of the descriptive grammar. In Chomsky (1965), for example, the recursive power of the grammar is shifted from the transformations to the phrase structure rules alone. In the "extended standard theory" of the 1970s, there was a reduction of the phrase structure component with the introduction of "X-bar theory," and a simplification of the constraints on movement. This was followed by a number of proposals to reduce the number and types of movement rules themselves. This came to a head (Chomsky 1977 and 1981a) with the abandonment of specific transformations altogether for a single rule ("move-α"), which stated, in effect, that one could move anything anywhere. This one rule was then supplemented with a number of constraints on movement. As Chomsky and a number of other generative linguists were able to show, it was possible to reduce a great number of transformations to a single-rule move-α and to a handful of constraints on movement.

Chomsky (1981b, 1982, and 1986a) synthesized subsequent work undertaken by linguists working in a number of languages, ranging from the romance languages to Chinese and Japanese, showing that other natural languages had similar but not identical constraints, and it was hypothesized that the variation was due to some limited parametric variation among human languages. This work established the "principles and parameters" framework of generative grammar. To get an idea of this framework, consider the following analogy: Think of the language faculty as a prewired box containing a number of switches. When exposed to environmental data, new switch settings are established. Applying this metaphor, the task of the linguist is to study the initial state of the language faculty, determine the possible parametric variations (switch settings), and account for language variation in terms of a limited range of variation in parameter settings. In Chomsky's view (2000, 8), the principles-and-parameters framework "gives at least an outline of a genuine theory of language, really for the first time." Commentators (e.g., Smith 2000, p. xi) have gone so far as to say that it is "the first really novel approach to language of the last two and a half thousand years." In what sense is it a radical departure? For the first time it allowed linguists to get away from simply constructing rule systems for individual languages and to begin exploring in a deep way the underlying similarities of human languages (even across different language families), to illuminate the principles that account for those similarities, and ultimately to show how those principles are grounded in the mind/brain. In Chomsky's view, the principles-and-parameters framework has yielded a number of promising results, ranging from the discovery of important similarities between *prima facie* radically different languages like Chinese and English to insights into the related studies of language acquisition, language processing, and acquired linguistic deficits (e.g., aphasia). Perhaps most importantly, the principles-and-parameters framework offered a way to resolve the tension between the two goals of descriptive adequacy and explanatory adequacy.

Still working within the general principles-and-parameters framework, Chomsky (1995 and 2000, Ch. 1) has recently articulated a research program that has come to be known as the "minimalist program," the main idea behind which is the working hypothesis that the language faculty is not the product of messy evolutionary tinkering—for example, there is no redundancy, and the only resources at work are those that are driven by "conceptual necessity." Chomsky (1995, ch. 1) initially seemed to hold that in this respect the language faculty would be unlike other biological functions, but more recently (2001) he seems to be drawn to D'Arcy Thompson's theory that the core of evolutionary theory consists of physical/mathematical/chemical principles that sharply constrain the possible range of organisms. In this case, the idea would be not only that those principles constrain low-level biological processes (like sphere packing in cell division) but also that such factors might be involved across the board—even including the human brain and its language faculty.

In broadest outline, the minimalist program works as follows: There are two levels of linguistic representation, phonetic form (PF) and logical form (LF), and a well-formed sentence (or linguistic *structure*) must be an ordered pair $<\pi, \lambda>$ of these representations (where π is a phonetic form and λ is the logical form). PF is taken to be the level of representation that is the input to the performance system (e.g., speech generation), and LF is, in Chomsky's terminology, the input to the conceptual/intensional system. Since language is, if nothing else, involved with the pairing of sounds and meanings, these two levels of representation are conceptually necessary. A minimal theory would posit no other levels of representation.

It is assumed that each sentence (or better, *structure*, Σ) is constructed out of an *array* or *numeration*, N, of lexical items. Some of the items in the numeration will be part of the pronounced (written) sentence, and others will be part of a universal inventory of lexical items freely inserted into all numerations. Given the numeration N, the computational system (C_{HL}) attempts to derive (compute) well-formed PF and LF representations, converging on the pair $<\pi, \lambda>$. The derivation is said to *converge* at a certain level if it yields a representation that is interpretable at that level. If it fails to yield an interpretable representation, the derivation *crashes*. Not all converging derivations yield structures that belong to a given language L. Derivations must also meet certain economy conditions.

Chomsky (2000, 9) notes that the import of the minimalist program is not yet clear. As matters currently stand, it is a subresearch program within the principles-and-parameters framework that is showing some signs of progress—at least enough to encourage those working within the program. As always, the concerns are to keep the number of principles constrained, not just to satisfy economy constraints, but to better facilitate the embedding of linguistics into theories of language acquisition, cognitive psychology, and, perhaps most importantly, general biology.

Some Conceptual Issues in Generative Grammar

While Chomsky would argue that he does not have a philosophy of science per se and that his philosophical observations largely amount to common sense, a number of interesting debates have arisen in the wake of his work. The remainder of this entry will review some of those debates.

On the Object of Study

Chomsky (1986b) draws the distinction between the notions of *I*-language and *E*-language, where *I*-language is the language faculty discussed above, construed as a chapter in cognitive psychology and ultimately human biology. *E*-language, on the other hand, comprises a loose collection of theories that take language to be a shared social object, established by convention and developed for purposes of communication, or an abstract mathematical object of some sort.

In Chomsky's view (widely shared by linguists), the notion of a 'language' as it is ordinarily construed by philosophers of language is fundamentally incoherent. One may talk about "the English language" or "the French language" but these are loose ways of talking. Typically, the question of who counts as speaking a particular language is determined more by political boundaries than actual linguistic variation. For example, there are dialects of German that, from a linguistic point of view, are closer to Dutch than to standard German. Likewise, in the Italian linguistic situation, there are a number of so-called dialects only some of which are recognized as "official" languages by the Italian government. Are the official languages intrinsically different from the "mere" dialects? Not in any linguistic sense. The decision to recognize the former as official is entirely a political decision. In the words attributed to Max Weinreich: A language is a dialect with an army and a navy. In this case, a language is a dialect with substantial political clout and maybe a threat of separatism.

Chomsky (1994 and 2000, Chap. 2) compares talk of languages (i.e., *E*-languages) to saying that two cities are "near" each other; whether two cities are near depends on one's interests and one's mode of transportation and very little on brute facts of geography. In the study of language, the notion of 'sameness' is no more respectable than that of 'nearness' in geography. Informally we might group together ways of speaking that seem to be similar (relative to one's interests), but such groupings have no real scientific merit. As a subject of natural inquiry, the key object of study has to be the language faculty and its set of possible parametric variations.

Not only is the notion of an *E*-language problematic, but it will not help to retreat to talk about *E*-dialects. The problem is that what counts as a separate *E*-dialect is also incoherent from a scientific point of view. For example, Chomsky (2000, 27) reports that in his idiolect, the word "ladder" rhymes with "matter" but not with "madder." For others, the facts do not cut in this way. Do they speak the same dialect as Chomsky or not? There is no empirical fact of the matter here; it all depends on individuals' desires to identify linguistically with each other. Even appeals to mutual intelligibility will not do, since what one counts as intelligible will depend much more on one's patience, one's ambition, and one's familiarity with the practices of one's interlocutors than it will on brute linguistic facts.

If the notion of *E*-language and *E*-dialect are incoherent, is it possible to construct a notion of *E-idiolect*?—that is, to identify idiolects by external criteria like an individual's spoken or written language? Apparently it is not. Included in what a person says or writes are numerous slips of the tongue, performance errors, etc. How is one we to rule those out of the individual's *E*-idiolect? Appeal to the agent's linguistic community will not do, since that would in turn require appeal to an *E*-language, and, for the reasons outlined above, there is no meaningful way to individuate *E*-languages. In the *I*-language approach, however, the problem of individuating *I*-idiolects takes the form of a coherent empirical research project. The idiolect (*I*-idiolect) is determined by the parametric state of *A*'s language faculty, and the language faculty thus determines *A*'s linguistic *competence*. Speech production that diverges from this competence can be attributed to *performance* errors. Thus, the competence/performance distinction is introduced to illuminate the distinction between sounds and interpretations that are part of *A*'s grammar and those that are simply mistakes. The *E*-language perspective has no similar recourse.

The Underdetermination of Theory by Evidence

One of the first philosophical issues to fall out from the development of generative grammar has been the dispute between Quine (1970) and Chomsky (1969 and 2000, Chap. 3) on the indeterminacy of grammar. Similar to his argument for the indeterminacy of meaning, Quine held that there is no way to adjudicate between two descriptively adequate sets of grammatical rules. So, for example, imagine two rule sets, one envisioning structures like (1) above, and another positing the following:

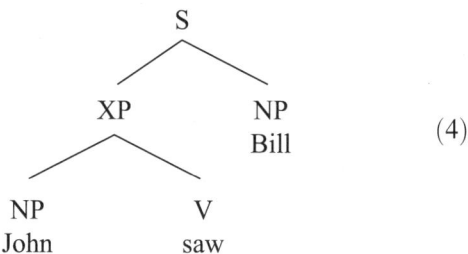

(4)

Chomsky maintained that Quine's argument is simply a recapitulation of the standard scientific problem of the underdetermination of theory by evidence. And in any case it is not clear that there are no linguistic tests that would allow us to choose between (1) and (4)—there are, after all, "constituency tests" (e.g., involving movement possibilities) that would allow us to determine whether the XP or the VP is the more plausible constituent.

Even if there are several grammars that are consistent with the available linguistic facts (the facts are not linguistic *behavior*, for Chomsky, but intuitions about acceptability and possible interpretation), one still has the additional constraint of which theory best accounts for the problem of language acquisition, acquired linguistic deficits (e.g., from brain damage), linguistic processing, etc. In other words, since grammatical theory is embedded within cognitive psychology, the choice between candidate theories can, in principle, be radically constrained. But further, even if there were two descriptively adequate grammars, each of which could be naturally embedded within cognitive psychology, there remain standard best-theory criteria (simplicity, etc.) that can help us to adjudicate between the theories.

Intrinsic Versus Relational Properties in Linguistics

In the philosophy of science, it is routine to make the distinction between "intrinsic" and "relational" properties. So, for example, the rest mass of an object would be an intrinsic property, and its weight would be a relational property (since it depends upon the mass of the body that the object is standing on). Similarly, in the philosophy of psychology there is a common distinction between "individualistic" and "externalist" properties. So, for example, there is the question of whether psychological states supervene on individualistic properties (intrinsic properties that hold of the individual in isolation) or whether they supervene on "externalist" properties (in effect, on relations between the agent and its environment) (see Supervenience). Chomsky has argued that all scientifically interesting psychological and linguistic properties supervene upon individualist (intrinsic) properties. In particular, there is a brute fact about the state of an individual's language faculty, and that fact is determined in turn by facts about the individual in isolation—not by the environment in which the individual is embedded. Because the language faculty is part of one's biological endowment, the nature of the representations utilized by the language faculty are fixed by biology and are not sensitive to environmental issues such as whether one is moving about on Earth or a phenomenologically identical planet with a different microstructure (e.g., Putnam's "Twin Earth").

Thus Chomsky takes issue with philosophers like Burge (1986), who argues in *Individualism and Psychology* that the content of the representations posited in psychology are determined at least in part by environmental factors. Chomsky holds that if the notion of content involves externalist or environmental notions, then it is not clear that it can play an interesting role in naturalistic inquiry in cognitive psychology (see Psychology, Philosophy of).

If environmentalism is to be rejected in psychology, then it naturally must be rejected in the semantics of natural language as well. That is: if the task of the linguist is to investigate the nature of *I*-language; if the nature of *I*-language is a chapter of cognitive psychology; and if cognitive psychology is an individualistic rather than a relational science, semantics will want to eschew relational properties like reference (where 'reference' is construed as a relation between a linguistic form and some object in the external environment). Thus Chomsky (1975 and 2000) rejects the notion of reference that has been central to the philosophy of language for the past three decades, characterizing it as an ill-defined technical notion (certainly one with no empirical applications), and, following Strawson, suggests that in the informal usage individuals 'refer' but linguistic objects do not.

This conclusion has immediate results for the notion of theory change in science. If the notion of reference is suspect or incoherent, then it can hardly be employed in an account of theory change (as in Putnam 1988). How then can one make sense of theory change? Chomsky (2000) suggests the following:

> Some of the motivation for externalist approaches derives from the concern to make sense of the history

of science. Thus, Putnam argues that we should take the early Niels Bohr to have been referring to electrons in the quantum-theoretic sense, or we would have to "dismiss all of his 1900 beliefs as totally wrong" (Putnam 1988), perhaps on a par with someone's beliefs about angels, a conclusion that is plainly absurd....

Agreeing... that an interest in intelligibility in scientific discourse across time is a fair enough concern, still it cannot serve as the basis for a general theory of meaning; it is, after all, only one concern among many, and not a central one for the study of human psychology. Furthermore, there are internalist paraphrases. Thus we might say that in Bohr's earlier usage, he expressed beliefs that were literally false, because there was nothing of the sort he had in mind in referring to electrons; but his picture of the world and articulation of it was structurally similar enough to later conceptions so that we can distinguish his beliefs about electrons from beliefs about angels. (43)

Against Teleological Explanation in Linguistics

A number of philosophers and linguists have thought that some progress can be made in the understanding of language by thinking of it as principally being a medium of communication. Chomsky rejects this conception of the nature of language, arguing that the language faculty is not *for* communication in any interesting sense. Of course, by rejecting the contention that language is a social object established for purposes of communication, Chomsky has not left much room for thinking of language in this way. But he also rejects standard claims that *I*-language must have evolved for the selectional value of communication; he regards such claims as without basis in fact. On this score, Chomsky sides with Gould (1980), Lewontin (1990), and many evolutionary biologists in supposing that many of our features (cognitive or anatomical) did not necessarily evolve for selectional reasons, but may have been the result of arbitrary hereditary changes that have perhaps been co-opted (see Evolution). Thus the language faculty may not have evolved for purposes of communication but may have been co-opted for that purpose, despite its nonoptimal design for communicative purposes.

In any case, Chomsky cautions that even if there was selectional pressure for the language faculty to serve as a means of communication, selection is but one of many factors in the emerging system. Crucially (2000, 163), "physical law provides narrow channels within which complex organisms may vary," and natural selection is only one factor that determines how creatures may vary within these constraints. Other factors (as Darwin himself noted) will include nonadaptive modifications and unselected functions that are determined from structure (see Function).

Inductive Versus Abductive Learning

A number of debates have turned on whether language acquisition requires a dedicated language faculty or whether "general intelligence" is enough to account for linguistic competence. Chomsky considers the general-intelligence thesis hopelessly vague, and argues that generalized inductive learning mechanisms make the wrong predictions about which hypotheses children would select in a number of cases. Consider the following two examples from Chomsky (1975 and [1980] 1992a):

The man is tall (5)

Is the man tall? (6)

Chomsky observes that confronted with evidence of question formation like that in (5)–(6) and given a choice between hypothesis (H1) and (H2), the generalized inductive learning mechanism will select (H1):

Move the first "is" to the front of the sentence. (H1)

Move the first "is" following the first NP to the front of the sentence. (H2)

But children apparently select (H2), since in forming a question from (7), they never make the error of producing (8), but always opt for (9):

The man who is here is tall. (7)

*Is the man who here is tall? (8)

Is the man who is here tall? (9)

Note that this is true despite the fact that the only data they have been confronted with previous to encountering (7) is simple data like (5)–(6). Chomsky's conclusion is that whatever accounts for children's acquisition of language, it cannot be generalized inductive learning mechanisms, but rather must be a system with structure-dependent principles/rules. Chomsky (1975, ch.1; 2000, 80) compares such learning to Peircian abduction. In other contexts this has been cast as a thesis about the "modularity" of language—that is, that there is a dedicated language acquisition device, and it, rather than some vague notion of general intelligence, accounts for the acquisition of language (see Evolutionary Psychology).

Putnam (1992) once characterized Chomsky's notion of a mental organ like the language faculty as being "biologically inexplicable," but Chomsky ([1980] 1992b) has held that it is merely "unexplained" and not inexplicable; in his view the language faculty is thus on the same footing as many other features of our biology (see Abduction; Inductive Logic).

The Science-Forming Faculty

On a more general and perhaps more abstract level, Chomsky has often spoken of a "science-forming faculty," parallel to our language faculty. The idea is that science could not have formed in response to mere inductive generalizations, but that human beings have an innate capacity to develop scientific theories (Chomsky credits C. S. Peirce with this basic idea). Despite Star Trekkian assumptions to the contrary, extraterrestrials presumably do not have a science-forming faculty like humans do, and may go about theorizing in entirely different ways, which would not be recognized as "scientific" by humans.

While the science-forming faculty may be limited, Chomsky would not concede that its limits are necessarily imposed by the selectional pressures of the prehistoric past. Just as the language faculty may not have evolved principally in response to selectional pressures, so too the science-forming faculty may have emerged quite independently of selectional considerations. As Chomsky (2000, Chap. 6) notes (citing Lewontin 1990), insects may seem marvelously well adapted to flowering plants, but in fact insects evolved to almost their current diversity and structure millions of years before flowering plants existed. Perhaps it is a similar situation with the science-forming faculty. That is, perhaps it is simply a matter of good fortune for humans that the science-forming faculty is reliable and useful, since it could not have evolved to help us with quantum physics, for example.

The Limits of Science

The notion of a science-forming faculty also raises some interesting questions about the limits of human ability to understand the world. Since what one can know through naturalistic endeavors is bounded by the human science-forming faculty, which in turn is part of the human biological endowment, it stands to reason that there are questions that will remain mysteries—or at least outside the scope of naturalistic inquiry:

> Like other biological systems, SFF [the science-forming faculty] has its potential scope and limits; we may distinguish between *problems* that in principle fall within its range, and *mysteries* that do not. The distinction is relative to humans; rats and Martians have different problems and mysteries and, in the case of rats, we even know a fair amount about them. The distinction also need not be sharp, though we certainly expect it to exist, for any organism and any cognitive faculty. The successful natural sciences, then, fall within the intersection of the scope of SFF and the nature of the world; they treat the (scattered and limited) aspects of the world that we can grasp and comprehend by naturalistic inquiry, in principle. The intersection is a chance product of human nature. Contrary to speculations since Peirce, there is nothing in the theory of evolution, or any other intelligible source, that suggests that it should include answers to serious questions we raise, or even that we should be able to formulate questions properly in areas of puzzlement. (2000, Ch. 4)

The question of what the "natural" sciences are, then, might be answered, narrowly, by asking what they have achieved; or more generally, by inquiry into a particular faculty of (the human) mind, with its specific properties.

The Mind/Body Problem and the Question of Physicalism

Chomsky has consistently defended a form of methodological monism (he is certainly no dualist); but, for all that, he is likewise no materialist. In Chomsky's view the entire mind/body question is ill-formed, since there is no coherent notion of physical body. This latter claim is not in itself unique; Crane and Mellor (1990) have made a similar point. There is a difference, however; for Crane and Mellor, developments in twentieth-century science have undermined physicalism, but for Chomsky the notion of physical body was already undermined by the time of Newton:

> Just as the mechanical philosophy appeared to be triumphant, it was demolished by Newton, who reintroduced a kind of "occult" cause and quality, much to the dismay of leading scientists of the day, and of Newton himself. The Cartesian theory of mind (such as it was) was unaffected by his discoveries, but the theory of body was demonstrated to be untenable. To put it differently, Newton eliminated the problem of "the ghost in the machine" by exorcising the machine; the ghost was unaffected. (2000, 84)

In Chomsky's view, then, investigations into the mind (in the guise of cognitive science generally or linguistics in particular) can currently proceed without worrying about whether they hook up with what is known about the brain, or even fundamental particles. The unification of science remains a goal, but in Chomsky's view it is not the study of mind that must be revised so as to conform to physical theory, but rather physical theory may eventually have to incorporate what is learnt in the study of the mind. According to Chomsky this is parallel to the situation that held prior to the unification of chemistry and physics; it was not chemistry that needed to be modified to account for what was known about physics, but in fact just the opposite:

> Large-scale reduction is not the usual pattern; one should not be misled by such dramatic examples as the reduction of much of biology to biochemistry in the middle of the twentieth century. Repeatedly, the more "fundamental" science has had to be revised, sometimes radically, for unification to proceed. (2000, 82)

Conclusion

The influence of Chomsky's work has been felt in a number of sciences, but perhaps the greatest influence has been within the various branches of cognitive science. Indeed, Gardner (1987) has remarked that Chomsky has been the single most important figure in the development of cognitive science. Some of Chomsky's impact is due to his role in arguing against behaviorist philosophers such as Quine; some of it is due to work that led to the integration of linguistic theory with other sciences; some of it is due to the development of formal tools that were later employed in disciplines ranging from formal language theory (cf. the "Chomsky Hierarchy") to natural language processing; and some of it is due to his directly engaging psychologists on their own turf. (One classic example of this was Chomsky's [1959b] devastating review of Skinner's [1957] *Verbal Behavior*. See also his contributions to Piatelli-Palmerini 1980) (See Behaviorism).

With respect to his debates with various philosophers, Chomsky has sought to expose what he has taken to be double standards in the philosophical literature. In particular he has held that while other sciences are allowed to proceed where inquiry takes them without criticism by armchair philosophers, matters change when the domain of inquiry shifts to mind and language:

> The idea is by now a commonplace with regard to physics; it is a rare philosopher who would scoff at its weird and counterintuitive principles as contrary to right thinking and therefore untenable. But this standpoint is commonly regarded as inapplicable to cognitive science, linguistics in particular. Somewhere in-between, there is a boundary. Within that boundary, science is self-justifying; the critical analyst seeks to learn about the criteria for rationality and justification from the study of scientific success. Beyond that boundary, everything changes; the critic applies independent criteria to sit in judgment over the theories advanced and the entitities they postulate. This seems to be nothing more than a kind of "methodological dualism," far more pernicious than the tradtional metaphysical dualism, which was a scientific hypothesis, naturalistic in spirit. Abandoning this dualist stance, we pursue inquiry where it leads. (2000, 112)

PETER LUDLOW

References

Bloomfield, L. (1933), *Language*. New York: Holt, Rinehart and Winston.

―――― (1939), *Linguistic Aspects of Science: International Encyclopedia of Unified Science* (Vol. 1, No. 4). Chicago: University of Chicago Press.

Burge, T. (1986), "Individualism and Psychology," *Philosophical Review* 95: 3–45.

Chomsky, N. ([1955] 1975), *The Logical Structure of Linguistic Theory*. Chicago: University of Chicago Press.

―――― (1956), "Three Models for the Description of Language," *I.R.E. Transactions of Information Theory* IT-2: 113–124.

―――― (1957), *Syntactic Structures*. The Hague: Mouton.

―――― (1959a), "On Certain Formal Properties of Grammars," *Information and Control* 2: 137–167.

―――― (1959b), "Review of B. F. Skinner, *Verbal Behavior*," *Language* 35, 26–57.

―――― (1964), *Current Issues in Linguistic Theory*. The Hague: Mouton & Co.

―――― (1965), *Aspects of the Theory of Syntax*. Cambridge, MA: MIT Press.

―――― (1969), "Quine's Empirical Assumptions," in D. Davidson and J. Hintikka (eds.), *Words and Objections: Essays on the Work of Willard Van Quine*. Dordrecht, Netherlands: D. Reidel.

―――― (1975), *Reflections on Language*. New York: Pantheon.

―――― (1977), "Conditions on Rules of Grammar," in *Essays on Form and Interpretation*. Amsterdam: Elsevier North-Holland, 163–210.

―――― (1979), *Language and Responsibility*. New York: Pantheon.

―――― (1981a), *Lectures on Government and Binding*. Dordrecht, Netherlands: Foris Publications.

―――― (1981b), "Principles and Parameters in Syntactic Theory," in N. Horstein and D. Lightfoot (eds.), *Explanation in Linguistics: The Logical Problem of Language Acquisition*. London: Longman.

―――― (1982), *Some Concepts and Consequences of the Theory of Government and Binding*. Cambridge, MA: MIT Press.

―――― (1986a), *Barriers*. Cambridge, MA: MIT Press.

―――― (1986b), *Knowledge of Language*. New York: Praeger.

―――― ([1980] 1992a), "On Cognitive Structures and Their Development," in B. Beakley and P. Ludlow (eds.), *The*

Philosophy of Mind: Classical Problems/Contemporary Issues. Cambridge, MA: MIT Press, 393–396.
——— ([1980] 1992b), "Discussion of Putnam's Comments," in B. Beakley and P. Ludlow (eds.), *The Philosophy of Mind: Classical Problems/Contemporary Issues*. Cambridge, MA: MIT Press, 411–422.
——— ([1988] 1992c), "From *Language and Problems of Knowledge*," in B. Beakley and P. Ludlow (eds.), *The Philosophy of Mind: Classical Problems/Contemporary Issues*. Cambridge, MA: MIT Press, 47–50.
——— (1994), "Noam Chomsky," in S. Guttenplan (ed.), *A Companion to the Philosophy of Mind*. Oxford: Blackwell, 153–167.
——— (1995), *The Minimalist Program*. Cambridge, MA: MIT Press.
——— (2000), *New Horizons in the Study of Language and Mind*. Cambridge: Cambridge University Press.
——— (2001), "Beyond Explanatory Adequacy," *MIT Occasional Papers in Linguistics*, no. 20. MIT Department of Linguistics.
Crane, T., and D. H. Mellor (1990), "There Is No Question of Physicalism," *Mind* 99: 185–206.
Gardner, H. (1987), *The Mind's New Science: A History of the Cognitive Revolution*. New York: Basic Books.
Gould, S. J. (1980), *The Panda's Thumb: More Reflections in Natural History*. New York: W. W. Norton.
Lewontin, R. (1990), "The Evolution of Cognition," in D. N. Osherson and E. E. Smith (eds.), *An Invitation to Cognitive Science* (Vol. 3). Cambridge, MA: MIT Press, 229–246.
Newmeyer, Frederick (1986), *Linguistic Theory in America* (2nd ed.). San Diego: Academic Press.
Piatelli-Palmerini, M. (ed.) (1980), *Language and Learning: The Debate between Jean Piaget and Noam Chomsky*. Cambridge, MA: Harvard University Press.
Putnam, H. (1988), *Representation and Reality*. Cambridge, MA: MIT Press.
Quine, Willard Van (1970), "Methodological Reflections on Current Linguistic Theory," *Synthese*, 21: 386–398.
Skinner, B. F. (1957), *Verbal Behavior*. New York: Appleton-Century-Crofts.
Smith, N. (2000), Foreword, in Chomsky, *New Horizons in the Study of Language and Mind*. Cambridge: Cambridge University Press.

See also **Behaviorism; Cognitive Science; Innate/ Acquired Distinction; Linguistics, Philosophy of; Psychology, Philosophy of; Physicalism**

CLASSICAL MECHANICS

Over the centuries classical mechanics has been a steady companion of the philosophy of science. It has played different parts, ranging (i) from positing its principles as a priori truths to the insight—pivotal for the formation of a modern philosophy of science—that modern physics requires a farewell to the explanatory ideal erected upon mechanics; (ii) from the physiological analyses of mechanical experiences to axiomatizations according to the strictest logical standards; and (iii) from the mechanistic philosophy to conventionalism (see Conventionalism; Determinism; Space-Time). Core structures of modern science, among them differential equations and conservation laws, as well as core themes of philosophy, among them determinism and the ontological status of theoretical terms, have emerged from this context. Two of the founders of classical mechanics, Galileo and Sir Isaac Newton, have often been identified with the idea of modern science as a whole. Owing to its increasing conceptual and mathematical refinement during the nineteenth century, classical mechanics gave birth to the combination of formal analysis and philosophical interpretation that distinguished the modern philosophy of science from the earlier *Naturphilosophie*. The works of Helmholtz, Mach, and Poincaré molded the historical character of the scientist-philosopher that would fully bloom during the emergence of relativity theory and quantum mechanics (see Mach, Ernest; Quantum Mechanics; Poincaré, Henri; Space-Time).

Mechanics became "classical" at the latest with the advent of quantum mechanics. Already relativistic field theory had challenged mechanics as the leading scientific paradigm. Consequently, present-day philosophers of science usually treat classical mechanics within historical case studies or as the first touchstone for new proposals of a general kind. Nonetheless, there are at least two lines on which classical mechanics in itself remains a worthy topic for philosophers. On the one hand, ensuing from substantial mathematical progress during the

CLASSICAL MECHANICS

twentieth century, classical mechanics has developed into the theory of classical dynamical systems. Among the problems of interest to formally minded philosophers of science are the relationships between the different conceptualizations, issues of stability and chaotic behavior, and whether classical mechanics is a special case of the conceptual structures characteristic of other physical theories. On the other hand, classical mechanics or semiclassical approaches continue to be applied widely by working scientists and engineers. Those real-world applications involve a variety of features that substantially differ from the highly idealized textbook models to which physicists and philosophers are typically accustomed, and they require solution strategies whose epistemological status is far from obvious.

Often classical mechanics is used synonymously with Newtonian mechanics, intending that its content is circumscribed by Newton's famous three laws. The terms 'analytical mechanics' and 'rational mechanics' stress the mathematical basis and theoretical side of mechanics as opposed to 'practical mechanics,' which originally centered on the traditional simple machines: lever, wedge, wheel and axle, tackle block, and screw. But the domain of mechanics was being constantly enlarged by the invention of new mechanical machines and technologies. Most expositions of mechanics also include the theory of elasticity and the mechanics of continua. The traditional distinction between mechanics and physics was surprisingly long-lived. One reason was that well into the nineteenth century, no part of physics could live up to the level of formal sophistication that mechanics had achieved. Although of little importance from a theoretical point of view, it is still common to distinguish statics (mechanical systems in equilibrium) and dynamics that are subdivided into a merely geometrical part, kinematics, and kinetics.

Ancient Greek and Early Modern Mechanics

In Greek antiquity, mechanics originally denoted the art of voluntarily causing motions against the nature of the objects moved. Other than physics, it was tractable by the methods of Euclidean geometry. Archimedes used mechanical methods to determine the center of mass of complex geometrical shapes but did not recognize them as valid geometrical proofs. His Euclidean derivation of the law of the lever, on the other hand, sparked severe criticism from Mach ([1883] 1960), who held that no mathematical derivation whatsoever could replace experience.

Pivotal for bringing about modern experimental science was Galileo's successful criticism of Aristotelian mechanics. Aristotle had divided all natural motions into celestial motions, which were circular and eternal, and terrestrial motions, which were rectilinear and finite. Each of the four elements (earth, water, fire, air) moved toward its natural place. Aristotle held that heavy bodies fell faster than light ones because velocity was determined by the relation of motive force (weight) and resistance. All forces were contact forces, such that a body moving with constant velocity was continuously acted upon by a force from the medium. Resistance guaranteed that motion remained finite in extent. Thus there was no void, nor motion in the void. Galileo's main achievement was the idealization of a constantly accelerated motion in the void that is slowed down by the resistance of the medium. This made free fall amenable to geometry. For today's reader, the geometrical derivation of the law is clumsy; the ratio of different physical quantities $v = s/t$ (where s stands for a distance, t for a time interval, and v for a velocity) was not yet meaningful. Galileo expressly put aside what caused bodies to fall and referred to experimentation. Historians and philosophers have broadly discussed what and how Galileo reasoned from empirical evidence. Interpreters wondered, in particular, whether the thought experiment establishing the absurdity of the Aristotelian position was conclusive by itself or whether Galileo had simply repackaged empirical induction in the deductive fashion of geometry.

Huygens' most important contribution to mechanics was the derivation of the laws of impact by invoking the principle of energy conservation. His solution stood at the crossroads of two traditions. On the one hand, it solved the foundational problem of Cartesian physics, the program of which was to reduce all mechanical phenomena to contact forces exchanged in collision processes. On the other hand, it gave birth to an approach based upon the concept of energy, which became an alternative to the Newtonian framework narrowly understood.

The work of Kepler has repeatedly intrigued philosophers of diverging orientations. Utilizing the mass of observational data collected by Brahe, Kepler showed that the planetary orbits were ellipses. Kepler's second law states that the line joining the sun to the planet sweeps through equal areas in equal times, and the third law states that the square of the periods of revolution of any two planets are in the same proportion as the cubes of their semi-major axes. All three laws were merely kinematical. In his *Mysterium cosmographicum*, Kepler

([1596] 1981) identified the spacings of the then known six planets with the five platonic solids. It will be shown below that the peculiar shape of the solar system, given Newton's laws of universal gravitation, can be explained without reference to Kepler's metaphysical belief in numerical harmony.

On the Status of Newton's Laws

The most important personality in the history of classical mechanics was Newton. Owing to the activities of his popularizers, he became regarded as the model scientist; this admiration included a devotion to the methodology of his *Philosophiae naturalis principia mathematica* (Newton [1687, 1713] 1969), outlined in a set of rules preceding book III. But interpreters disagree whether Newton really pursued the Baconian ideal of science and licensed induction from phenomena or must be subsumed under the later descriptivist tradition that emerged with Mach ([1883] 1960) and Kirchhoff (1874). At any rate, the famous declaration not to feign hypothesis targeted the Cartesian and Leibnizian quest for a metaphysical basis of the principles of mechanics.

After the model of Euclid, the *Principia* began with eight definitions and three axioms or laws. Given Newton's empiricist methodology, interpreters have wondered about their epistemological status and the logical relations among them. Certainly, the axioms were neither self-evident truths nor mutually independent:

1. Every body continues in its state of rest or uniform motion in a right (i.e., straight) line, unless it is compelled to change that state by forces impressed upon it.
2. The change of motion is proportional to the motive force impressed and is made in the direction of the right line in which that force is impressed.
3. To every action there is always opposed an equal reaction; or the mutual actions of two bodies upon each other are always equal, and directed to contrary parts.

The proper philosophical interpretation of these three laws remained contentious until the end of the nineteenth century. Kant claimed to have deduced the first and third laws from the synthetic a priori categories of causality and reciprocity, respectively. The absolute distinction between rest and motion in the first law was based on absolute space and time. Within Kant's transcendental philosophy, Euclidean space-time emerged from pure intuition a priori. This was the stand against which twentieth-century philosophy of science rebelled, having relativity theory in its support (see Space-Time).

Admitting that the laws were suggested by previous experiences, some interpreters, including Poincaré, considered them as mere conventions (see Conventionalism). Positing absolute Euclidean space, the first law states how inertial matter moves in it. But it is impossible to obtain knowledge about space independent of everything else. Thus, the first law represents merely a criterion for choosing a suitable geometry. Mach ([1883] 1960) rejected absolute space and time as metaphysical. All observable motion was relative, and thus, at least in principle, all material objects in the universe were mutually linked. Mach's principle, as it became called, influenced the early development of general relativity but is still controversial among philosophers (see Barbour and Pfister 1995; Pooley and Brown 2002). According to Mach, the three laws were highly redundant and followed from a proper empiricist explication of Newton's definitions of mass and force. Defining force, with Newton, as an action exerted upon a body to change its state, that is, as an acceleration, the first and second laws can be straightforwardly derived. Mach rejected Newton's definition of mass as quantity of matter as a pseudo-definition and replaced it with the empirical insight that mass is the property of bodies determining acceleration. This was equivalent to the third law. Mach's criticism became an important motivation for Albert Einstein's special theory of relativity. Yet, even the members of the Vienna Circle disagreed whether this influence was formative or Mach merely revealed the internal contradictions of the Newtonian framework (see Vienna Circle).

As to the second law, one may wonder whether the forces are inferred from the observed phenomena of motion or the motions are calculated from given specific forces. Textbooks often interpret the second law as providing a connection from forces to motion, but this is nontrivial and requires a proper superposition of the different forces and a due account of the constraints. The opposite interpretation has the advantage of not facing the notorious problem of characterizing force as an entity in its own right, which requires a distinction between fundamental forces from fictitious or inertial forces.

Book III of Newton's *Principia* introduced the law of universal gravitation, which finally unified the dynamics of the celestial and terrestrial spheres. But, as it stood, it involved an action at a distance through the vacuum, which Newton regarded as the greatest absurdity. To Bentley he wrote: "Gravity must be caused by an Agent acting according to certain Laws; but whether this Agent be material or immaterial, I have left to the Consideration of my readers" (Cohen 1958, 303). Given the law of universal gravity, the peculiar shape of the solar system was a matter of initial conditions, a fact that Newton ascribed to the contrivance of a voluntary agent.

The invention of the calculus was no less important for the development of mechanics than was the law of gravitation. But its use in the *Principia* was not consistent and intertwined with geometrical arguments, and it quickly turned out that Leibniz's version of the calculus was far more elegant. That British scientists remained loyal to Newton's fluxions until the days of Hamilton proved to be a substantial impediment to the use of calculus in science.

Celestial Mechanics and the Apparent Triumph of Determinism

No other field of mechanics witnessed greater triumphs of prediction than celestial mechanics: the return of Halley's comet in 1758; the oblateness of the Earth in the 1740s; and finally the discovery of Neptune in 1846 (cf. Grosser 1962). However, while Neptune was found at the location that the anomalies in the motion of Uranus had suggested, Le Verrier's prediction of a planet Vulcan to account for the anomalous perihelion motion of Mercury failed. Only general relativity would provide a satisfactory explanation of this anomaly (see Space-Time).

But in actual fact little follows from Newton's axioms and the inverse square law of gravitation alone. Only the two-body problem can be solved analytically by reducing it to a one-body problem for relative distance. The three-body problem requires approximation techniques, even if the mass of one body can be neglected. To ensure the convergence of the respective perturbation series became a major mathematical task. In his lunar theory, Clairaut could derive most of the motion of the lunar apsides from the inverse-square law, provided the approximation was carried far enough. But d'Alembert warned that further iterations might fail to converge; hence subsequent analysts calculated to higher and higher orders. D'Alembert's derivation of the precession and nutation of the Earth completed a series of breakthroughs around 1750 that won Newton's law a wide acceptance. In 1785, Laplace explained the remaining chief anomalies in the solar system and provided a (flawed) indirect proof for the stability of the solar system from the conservation of angular momentum (cf. Wilson 1995).

One might draw an inductivist lesson from this history and conceive the increased precision in the core parameters of the solar system, above all the planetary masses, as a measure of explanatory success for the theory containing them (cf. Harper and Smith 1995). Yet, the historically more influential lesson for philosophers consisted in the ideal of Laplace's Demon, the intellect that became the executive officer of strict determinism:

> Given for one instant an intelligence which could comprehend all the forces by which nature is animated and the respective situation of the beings who compose it—an intelligence sufficiently vast to submit these data to analysis—it would embrace in the same formula the greatest bodies of the universe and those of the lightest atom; for it, nothing would be uncertain, and the future, as the past, would be present to its eyes. (Laplace [1795] 1952, 4)

But the Demon is threatened by idleness in many ways. If the force laws are too complex, determinism becomes tautologous; if Newton's equations cannot be integrated, perturbative and statistical strategies are mandatory; calculations could still be too complex; exact knowledge of the initial state of a system presupposes that the precision of measurement could be increased at will.

A further idealization necessary to make the Laplacian ideal thrive was the point mass approach of Boscovich ([1763] 1966). But some problems cannot be solved in this way—for instance, whether a point particle moving in a head-on orbit directed at the origin of a central force is reflected by the singularity or goes right through it. In many cases, celestial mechanics treats planets as extended bodies or gyroscopes rather than point masses. Non-rigid bodies or motion in resistant media require even further departures from the Laplacian ideal. As Wilson put it, "applied mathematicians are often forced to pursue roundabout and shakily rationalized expedients if any progress is to be made" (2000, 296).

Only some of these expedients can be mathematically justified. This poses problems for the relationship of mathematical and physical ontology. Often unsolvable equations are divided into tractable satellite equations. Continuum mechanics is a case in point. The Navier-Stokes equations are virtually intractable by analytic methods; Prandtl's boundary layer theory splits a flow in a pipe into one equation for the fluid's boundary and one for the middle of the flow. Prandtl's theory failed in the case of turbulence, while statistical investigations brought useful results. Accordingly, in this case predictability required abandoning determinism altogether years before the advent of quantum mechanics (cf. von Mises 1922).

From Conserved Quantities to Invariances: Force and Energy

The framework of Newton's laws was not the only conception of mechanics used by 1800 (cf. Grattan-Guinness 1990). There were purely algebraic versions of the calculus based on variational problems developed by Euler and perfected in Lagrange's *Analytical Mechanics* ([1788] 1997). Engineers developed a kind of energy mechanics that emerged from Coulomb's friction studies. Lazare Carnot developed it into an alternative to Lagrange's reduction of dynamics to equilibrium, that is, to statics. Dynamics should come first, and engineers had to deal with many forces that did not admit a potential function.

The nature of energy—primary entity or just inferred quantity—was no less in dispute than that of force. Both were intermingled in content and terminology. The eighteenth century was strongly influenced by the *vis viva* controversy launched by Leibniz. What was the proper quantity in the description of mechanical processes: mv or mv^2 (where m = mass and v = velocity)? After the Leibniz-Clarke correspondence, the issue became associated with atomism and the metaphysical character of conserved quantities. In the nineteenth century, 'energy,' or work, became a universal principle after the discovery of the mechanical equivalent of heat. Helmholtz gave the principle a more general form based upon the mechanical conception of nature. For many scientists this suggested a reduction of the other domains of physical science to mechanics. But this program failed in electrodynamics.

Through the works of Ostwald and Helm, the concept of energy became the center of a movement that spellbound German-speaking academia from the 1880s until the 1900s. But, as Boltzmann untiringly stressed, energeticists obtained the equations of motion only by assuming energy conservation in each spatial direction. But why should Cartesian coordinates have a special meaning?

Two historico-critical studies of the energy concept set the stage for the influential controversy between Planck and Mach (1908–1910). While Mach ([1872] 1911) considered the conservation of work as an empirical specification of the instinctive experience that perpetual motion is impossible, Planck (1887) stressed the independence and universality of the principle of energy conservation. Planck later lauded Mach's positivism as a useful antidote against exaggerated mechanical reductionism but called for a stable world view erected upon unifying variational principles and invariant quantities.

Group theory permitted mathematical physicists to ultimately drop any substantivalist connotation of conserved quantities, such as energy. The main achievement was a theorem of Emmy Noether that identified conserved quantities, or constants of motion, with invariances under one-parameter group transformations (cf. Arnold 1989, 88–90).

Variational Principles and Hamiltonian Mechanics

In 1696–1697 Johann Bernoulli posed the problem of finding the curve of quickest descent between two points in a homogeneous gravitational field. While his own solution used an analogy between geometrical optics and mechanics, the solutions of Leibniz and Jacob Bernoulli assumed that the property of minimality was present in the small as in the large, arriving thus at a differential equation. The title of Euler's classic treatise describes the scope of the variational calculus: *The Method of Finding Curved Lines That Show Some Property of Maximum or Minimum* (1744). The issue of minimality sparked philosophical confusion. In 1746, Pierre Moreau de Maupertuis announced that in all natural processes the quantity $\int vds = \int v^2 dt$ attains its minimum; he interpreted this as a formal teleological principle that avoided the outgrowths of the earlier physicotheology of Boyle and Bentley. Due to his defective examples and a priority struggle, the principle lost much of its credit. Lagrange conceived of it simply as an effective problem-solving machinery. There was considerable mathematical progress during the nineteenth century, most notably by Hamilton, Gauss, and Jacobi. Only Weierstrass found the first sufficient condition for the variational integral to actually attain its minimum value. In 1900, Hilbert urged mathematicians to systematically develop the variational calculus (see Hilbert, David). He made action principles a core element of his axiomatizations of physics. Hilbert and Planck cherished the principles' applicability, after appropriate specification, to all fields of reversible physics; thus they promised a formal unification instead of the discredited mechanical reductionism.

CLASSICAL MECHANICS

There exist differential principles, among them d'Alembert's principle and Lagrange's principle of virtual work, that reduce dynamics to statics, and integral action principles that characterize the actual dynamical evolution over a finite time by the stationarity of an integral as compared with other possible evolutions. Taking M_q as the space of all possible motions $q(t)$ between two points in configuration space, the action principle states that the actual motion q extremizes the value of the integral $W[q] = \int_a^b F(t, q(t), \dot{q}(t))dt$ in comparison with all possible motions, the varied curves $(q + \delta q) \in M_q$ (the endpoints of the interval $[a, b]$ remain fixed).

If one views the philosophical core of action principles in a temporal teleology, insofar as a particle's motion between a and b is also determined by the fixed state, this contradicts the fact that almost all motions that can be treated by way of action principles are reversible. Yet, already the mathematical conditions for an action principle to be well-defined suggest that the gist of the matter lies in its global and modal features. Among the necessary conditions is the absence of conjugate points between a and b through which all varied curves have to pass. Sufficient conditions typically involve a specific field embedding of the varied curves; also the continuity properties of the $q \in M_q$ play a role. Thus initial and final conditions have to be understood in the same vein as boundary conditions for partial differential equations that have to be specified beforehand so that the solution cannot be grown stepwise from initial data.

There are also different ways of associating the possible motions. In Hamilton's principle, F is the Lagrangian $L = T - V$, where T is the kinetic and V the potential energy; time is not varied such that all possible motions take equal time but correspond to higher energies than the actual motion. For the original principle of least action, F equals T, and one obtains the equations of motion only by assuming energy conservation, such that at equal total energy the varied motions take a longer time than the actual ones.

Butterfield (2004) spots three types of modality in the sense of Lewis (1986) here. While all action principles involve a modality of changed initial conditions and changed problems, Hamilton's principle also involves changed laws because the varied curves violate energy conservation. Energeticists and Mach ([1883] 1960) held that u is uniquely determined within M_u as compared with the other motions that appear pairwise or with higher degeneracy, but they disagreed whether one could draw conclusions about modal characteristics. Albeit well versed in their mathematical intricacies, logical empiricists treated action principles with neglect in order to prevent an intrusion of metaphysics (Stöltzner 2003).

The second-order Euler-Lagrange equations, which typically correspond to the equations established by way of Newton's laws, can be reformulated as a pair of first-order equations, Hamilton's equations, or a single partial differential equation, viz., the Hamilton-Jacobi equation. Hamilton's equations emerge from the so-called Legendre transformation, which maps configuration space $(q, \dot{q})$ into phase space (q, p), thus transforming the derivative of position $\dot{q}$ into momentum p. If this transformation fails, constraints may be present. The core property of Hamilton's equations is their invariance under the so-called canonical transformations. The canonical transformation $(q, p) \to (Q, P)$ that renders the Hamiltonian $H(Q, P) = T + V$ equal to zero leads to the Hamilton-Jacobi equation. Its generator $W = S - E_t$ (where E_t is the total energy) can be interpreted as an action functional corresponding to moving wave fronts in ordinary space that are orthogonal to the extremals of the variational problem (cf. Lanczos 1986).

The analogy between mechanics and geometric optics is generic and played an important role in Schrödinger's justification of his wave equation. Hamilton-Jacobi theory was also a motivation for Bohm's reformulation of the de Broglie pilot wave theory (see Quantum Mechanics). For periodical motions one can use S to generate a canonical transformation that arrives at action and angle variables (J, ω), where $J = \oint p dq$. This integral was quantized in the older Bohr-Sommerfeld quantum theory. The space of all possible motions associated with a variational problem M_q can be considered as an ensemble of trajectories. Arguing that in quantum theory all possible motions are realized with a certain possibility provides some intuitive motivation for the Feynman path integral approach (see Quantum Field Theory).

Variational principles played a central role in several philosophically influential treatises of mechanics in the late nineteenth century. Most radical was that of Hertz ([1894] 1956), who used Gauss's (differential) principle of least constraints to dispense with the notion of force altogether. There were no single mass points but only a system of mass points connected by constraints. Hertz's problem was to obtain a geometry of straight line for this system of mass points. This task was complicated by Hertz's Kantian preference for Euclidean geometry as a basis for the non-Euclidean geometry of mass points. Contemporaries deemed Hertz's geometrization of mechanics a "God's eye view." Boltzmann (1897, 1904) praised its coherence but judged Hertz's picture as inapplicable to problems easily tractable by means of forces.

Hertz held that theoretical pictures had to be logically permissible, empirically correct, and appropriate (that is, sufficiently complete and simple). Boltzmann criticized permissibility as an unwarranted reliance upon a priori laws of thought and held instead that pictures were historically acquired and corroborated by success. Around 1900, treating theories (e.g., atomism) as pictures represented an alternative to mechanical reductionism and positivist descriptivism, until the idea of coordinative definitions between symbolic theory and empirical observations won favor (see Vienna Circle).

Mathematical Mechanics and Classical Dynamical Systems

After classical mechanics was finally dethroned as the governing paradigm of physics, its course became largely mathematical in kind and was strongly influenced by concepts and techniques developed by the new-frontier theories of physics: relativity theory and quantum physics. There was substantial progress in the variational calculus, but the main inspirations came from differential geometry, group theory, and topology, on the one hand, and probability theory and measure theory, on the other. This has led to some rigorous results on the n-body problem. The advent of the modern computer not only opened a new era of celestial mechanics, it also revealed that chaotic behavior, ignored by physicists in spite of its early discovery by Poincaré, occurred in a variety of simple mechanical systems.

Geometrization has drastically changed the appearance of classical mechanics. Configuration space and phase space have become the tangent and cotangent bundles on which tensors, vector fields, and differential forms are defined. Coordinates have turned into charts, and the theory of differentiable manifolds studies the relationships between the local level, where everything looks Euclidean, and the global level, where topological obstructions may arise. The dynamics acting on these bundles are expressed in terms of flows defined by vector fields, which gives a precise meaning to variations. The picture of flows continues Hamilton-Jacobi theory. This elucidates that the intricacies of variational calculus do not evaporate; they transform into obstructions of and conditions for the application of the whole geometrical machinery, e.g., how far a local flow can be extended. The constants of the motion define invariant submanifolds that restrict the flow to a manifold of lower dimension; they act like constraints. If a Hamiltonian system has, apart from total energy, enough linearly independent constants of motion and if their Poisson brackets $\left(\frac{\partial F}{\partial q_i}\frac{\partial G}{\partial p_i} - \frac{\partial G}{\partial q_i}\frac{\partial F}{\partial p_i}\right)$, where F and G are such constants, mutually vanish, the system is integrable (the equations of motion can be solved) and the invariant submanifold can be identified with a higher-dimensional torus.

The geometrical structure of Hamiltonian mechanics is kept together by the symplectic form $\omega = dq^i \wedge dp_i$ that is left invariant by canonical transformations. It plays a role analogous to the metric in relativity theory. The skew-symmetric product $\wedge$ of forms and the exterior derivative are the main tools of the Cartan calculus and permit an entirely coordinate-free formulation of dynamics. Many equations thus drastically simplify, but quantitative results still require the choice of a specific chart.

Mathematical progress went hand in hand with a shift of emphasis from quantitative to qualitative results that started with Poincaré's work on the n-body problem. Some deep theorems of Hamiltonian mechanics become trivialities in this new language, among them the invariance of phase space volume (Liouville's theorem) and the fact that almost all points in a phase space volume eventually—in fact, infinitely often—return arbitrarily close to their original position (Poincaré recurrence) (see also Statistical Mechanics).

For the small number of mass points characteristic of celestial mechanics, there has been major progress along the lines of the questions asked by Thirring (1992, 6): "Which configurations are stable? Will collisions ever occur? Will particles ever escape to infinity? Will the trajectory always remain in a bounded region of phase space?" In the two-body problem, periodic elliptic motions, head-on collisions, and the hyperbolic and parabolic trajectories leading to infinity are neatly separated by initial data.

For the three-body problem the situation is already complex; there do not exist sufficient constants of motion. Two types of exact solutions were quickly found: The particles remain collinear (Euler) or they remain on the vertices of an equilateral triangle (Lagrange). The equilateral configuration is realized by the Trojan asteroids, Jupiter and the sun. In general, however, even for a negative total energy, two bodies can come close enough to expel the third to infinity. In the four-body problem, there exist even collinear trajectories on which a particle might reach spatial infinity in a finite time (Mather and McGehee 1975). Five bodies, one of them lighter than the others, can even be arranged in such a manner that this happens without any previous collisions because the fifth particle oscillates faster and faster between the two escaping planes, in each of which two particles rotate around one another (Xia 1992; Saari 2005). Both examples blatantly violate special relativity. But rather than hastily resort to arguments of what is 'physical,' the philosopher should follow the mathematical physicist in analyzing the logical structure of classical mechanics, including collisions and other singularities.

Can an integrable system remain stable under perturbation? In the 1960s Kolmogorov, Arnold, and Moser (KAM) gave a very general answer that avails itself of the identification of integrable systems and invariant tori. The KAM theorem shows that sufficiently small perturbations deform only the invariant tori. If the perturbation increases, the tori in resonance with it break first; or more precisely, the more irrational the ratio of the frequency of an invariant torus (in action and angle variables) and the frequency of the perturbation, the more stable is the respective torus. If all invariant tori are broken, the system becomes chaotic (cf. Arnold 1989; Thirring 1992). The KAM theorem also provides the rigorous basis of Kepler's association of planetary orbits and platonic solids. The ratios of the radii of platonic solids to the radii of inscribed platonic solids are irrational numbers of a kind that is badly approximated by rational numbers. And, indeed, among the asteroids between Mars and Jupiter, one finds significant gaps for small ratios of an asteroid's revolution time to that of the perturbing Jupiter.

If all invariant tori have broken up, the only remaining constant of motion is energy, such that the system begins to densely cover the energy shell and becomes ergodic. No wonder that concepts of statistical physics, among them ergodicity, entropy, and mixing, have been used to classify chaotic behavior, even though chaotic behavior does not succumb to statistical laws. They are supplemented by topological concepts, such as the Hausdorff dimension, and concepts of dynamical systems theory, such as bifurcations (nonuniqueness of the time evolution) and attractors (in the case of dissipative systems where energy is no longer conserved).

MICHAEL STÖLTZNER

References

Arnold, Vladimir I (1989), *Mathematical Methods of Classical Mechanics*, 2nd ed. New York and Berlin: Springer.

Barbour, Julian B., and Herbert Pfister (eds.) (1995), *Mach's Principle: From Newton's Bucket to Quantum Gravity*. Boston and Basel: Birkhäuser.

Boltzmann, Ludwig (1897 and 1904), *Vorlesungen über die Principe der Mechanik*. 2 vols. Leipzig: Barth.

Boscovich, Ruder J. ([1763] 1966), *A Theory of Natural Philosophy*. Cambridge, MA: MIT Press.

Butterfield, Jeremy (2004), "Some Aspects of Modality in Analytical Mechanics," in Michael Stöltzner and Paul Weingartner (eds.), *Formale Teleologie und Kausalität*. Paderborn, Germany: Mentis.

Cohen, I. Bernhard (ed.) (1958), *Isaac Newton's Papers and Letters on Natural Philosophy*. Cambridge, MA: Harvard University Press.

Euler, Leonhard (1744), *Methodus Inveniendi Lineas Curvas Maximi Minimive Proprietate Gaudentes sive Solutio Problematis Isoperimetrici Latissimo Sensu Accepti*. Lausanne: Bousquet.

Grattan-Guinness, Ivor (1990), "The Varieties of Mechanics by 1800," *Historia Mathematica* 17: 313–338.

Grosser, Morton (1962), *The Discovery of Neptune*. Cambridge, MA: Harvard University Press.

Harper, William, and George E. Smith (1995), "Newton's New Way of Inquiry," in J. Leplin (ed.), *The Creation of Ideas in Physics: Studies for a Methodology of Theory Construction*. Dordrecht, Netherlands: Kluwer, 113–166.

Hertz, Heinrich ([1894] 1956), *The Principles of Mechanics: Presented in a New Form*. New York: Dover Publications.

Kepler, Johannes ([1596] 1981), *Mysterium cosmographicum*. New York: Abaris.

Kirchhoff, Gustav Robert (1874), *Vorlesungen über analytische Mechanik*. Leipzig: Teubner.

Lagrange, Joseph Louis ([1788] 1997), *Analytical Mechanics*. Dordrecht, Netherlands: Kluwer.

Lanczos, Cornelius (1986), *The Variational Principles of Mechanics*. New York: Dover.

Laplace, Pierre Simon de ([1795] 1952), *A Philosophical Essay on Probabilities*. New York: Dover.

Lewis, David (1986), *On the Plurality of Worlds*. Oxford: Blackwell.

Mach, Ernest ([1872] 1911), *History and Root of the Principle of the Conservation of Energy*. Chicago: Open Court.

——— ([1883] 1960), *The Science of Mechanics: A Critical and Historical Account of Its Development*. La Salle, IL: Open Court. German first edition published by Brockhaus, Leipzig, 1883.

Mather, John, and Richard McGehee, "Solutions of the Collinear Four-Body Problem Which Become Unbounded in Finite Time," in Jürgen Moser (ed.), *Dynamical Systems Theory and Applications*. New York: Springer, 573–587.

Newton, Isaac ([1687, 1713] 1969), *Mathematical Principles of Natural Philosophy*. Translated by Andrew Motte (1729) and revised by Florian Cajori. 2 vols. New York: Greenwood.

Planck, Max (1887), *Das Princip der Erhaltung der Energie*, 2nd ed. Leipzig: B. G. Teubner.

Pooley, Oliver, and Harvey Brown (2002), "Relationalism Rehabilitated? I: Classical Mechanics," *British Journal for the Philosophy of Science* 53: 183–204.

Saari, Donald G. (2005), *Collisions, Rings, and Other Newtonian N-Body Problems*. Providence, RI: American Mathematical Society.

Stöltzner, Michael (2003), "The Principle of Least Action as the Logical Empiricists' Shibboleth," *Studies in History and Philosophy of Modern Physics* 34: 285–318.

Thirring, Walter (1992), *A Course in Mathematical Physics 1: Classical Dynamical Systems*, 2nd ed. Translated by Evans M. Harrell. New York and Vienna: Springer.

von Mises, Richard (1922), "Über die gegenwärtige Krise der Mechanik," *Die Naturwissenschaften* 10: 25–29.

Wilson, Curtis (1995), "The Dynamics of the Solar System," in Ivor Grattan-Guinness (ed.), *Companion Encyclopedia of the History and Philosophy of the Mathematical Sciences*. London: Routledge.

Wilson, Mark (2000), "The Unreasonable Uncooperativeness of Mathematics in the Natural Sciences," *The Monist* 83 (2000): 296–314. See also his very instructive entry "Classical Mechanics" in the *Routledge Encyclopedia of Philosophy*.

Xia, Zhihong (1992), "The Existence of Noncollision Singularities in Newtonian Systems," *Annals of Mathematics* 135: 411–468.

COGNITIVE SCIENCE

Cognitive science is a multidisciplinary approach to the study of cognition and intelligence that emerged in the late 1950s and 1960s. "Core" cognitive science holds that the mind/brain is a kind of computer that processes information in the form of mental representations. The major disciplinary participants in the cognitive science enterprise are psychology, linguistics, neuroscience, computer science, and philosophy. Other fields that are sometimes included are anthropology, education, mathematics, biology, and sociology.

There has been considerable philosophical discussion in recent years that relates, in one way or another, to cognitive science. Arguably, not all of this discussion falls within the tradition of philosophy of science. There are two fundamentally different kinds of question that philosophers of science typically raise about a specific scientific field: "external" questions and "internal" questions. If one assumes that a scientific field provides a framework of inquiry, external questions will be about that framework as a whole, from some external point of view, or about its relation to other scientific frameworks. In contrast, an internal question will be one that is asked *within* the framework, either with respect to some entity or process that constitutes part of the framework's foundations or with respect to some specific theoretical/empirical issue that scientists committed to that framework are addressing. Philosophers have dealt extensively with both external and internal questions associated with cognitive science.

Key External Questions

There are three basic groups of external questions: those concerning the nature of a scientific field X, those concerning the relations of X to other scientific fields, and those concerning the scientific merits of X. An interesting feature of such discussions is that they often draw on, and sometimes even contribute to, the literature on a relevant prior meta-question. For example, discussions concerning the scientific nature of a particular X (e.g., cognitive science) may require prior consideration of the question, What is the best way to characterize X (in general) scientifically (e.g., in terms of its theories, explanations, paradigm, research program)?

What Is Cognitive Science?

There are many *informal* descriptions of cognitive science in the literature, not based on any serious consideration of the relevant meta-question (e.g., Simon 1981; Gardner 1985). The only systematic formal treatment of cognitive science is that of von Eckardt (1993), although there have been attempts to describe aspects of *cognitive psychology* in terms of Kuhn's notion of a paradigm. Rejecting Kuhn's notion of a paradigm as unsuitable for the characterization of an *immature* field, von Eckardt (1993) proposes, as an alternative, the notion of a *research framework*. A research framework consists of four sets of elements D, Q, SA, MA, where D is a set of assumptions that provide a pretheoretical specification of the domain under study; Q is a set of basic empirical research questions, formulated pretheoretically; SA is a set of substantive assumptions that embody the approach being taken in answering the basic questions and that constrain possible answers to those questions; and MA is a set of methodological assumptions.

One can think of what Kuhn (1970) calls "normal science" as problem-solving activity. The fundamental problem facing the community of researchers committed to a research framework is to answer each of the basic empirical questions of that framework, subject to the following constraints:

1. Each answer is scientifically acceptable in light of the scientific standards set down by the shared and specific methodological assumptions of the research framework.
2. Each answer is consistent with the substantive assumptions of the research framework.
3. Each answer to a theoretical question makes significant reference to the entities and processes posited by the substantive assumptions of the research framework.

According to von Eckardt (1993), cognitive science consists of a *set* of overlapping research frameworks, each concerned with one or another aspect of human cognitive capacities. Arguably, the *central* research framework concerns the study of adult, normal, typical cognition, with subsidiary frameworks focused on individual differences, group differences (e.g., expert vs. novice, male vs. female), cultural variation, development, pathology, and neural realization. The description offered in von Eckardt (1993), summarized in Table 1, is intended to be of only the central research framework. In addition, it is claimed to be a rational reconstruction as well as transdisciplinary, that is, common to cognitive scientists of all disciplines. Von Eckardt claims that research frameworks can evolve and that the description in question reflects commitments of the cognitive science community at only one period of its history (specifically, the late 1980s/early 1990s).

Cognitive science is a very complex and rapidly changing field. In light of this complexity and change, von Eckardt's original reconstruction should, perhaps, be modified as follows. First, it needs to be acknowledged that there exists a fundamental disagreement within the field as to its domain. The narrow conception, embraced by most psychologists and linguists, is that the domain is *human cognition;* the broad conception, embraced by artificial intelligence researchers, is that the domain is *intelligence in general.* (Philosophers seem to be about evenly split.) A further area of disagreement concerns whether cognitive science encompasses only cognition/intelligence or includes all aspects of mind, including touch, taste, smell, emotion, mood, motivation, personality, and motor skills. One point on which there now seems to be unanimity is that cognitive science must address the phenomenon of consciousness.

Second, a modified characterization of cognitive science must describe it as evolving not only with respect to the computational assumption (from an exclusive focus on symbol systems to the inclusion of connectionist devices) but also with respect to the role of neuroscience. Because cognitive science originally emerged from cognitive psychology, artificial intelligence research, and generative linguistics, in the early years neuroscience was often relegated to a secondary role. Currently, most nonneural cognitive scientists believe that research on the mind/brain should proceed in an interactive way—simultaneously both top-down and bottom-up. Ironically, a dominant emerging view of cognitive neuroscience seems to be that it, rather than cognitive *science*, will be the locus of an interdisciplinary effort to develop, from the bottom up, a computational or information-processing theory of the mind/brain.

There are also research programs currently at the periphery of cognitive science—what might be called alternative cognitive science. These are research programs that investigate some aspect of cognition or intelligence but whose proponents reject one or more of the major guiding or methodological assumptions of mainstream cognitive science. Three such programs are research on situated cognition, artificial life, and the dynamical approach to cognition.

Interfield Relations Within Cognitive Science

The second group of external questions typically asked by philosophers of some particular scientific field concerns the relation of that field to other scientific fields. Because cognitive science is itself multidisciplinary, the most pressing interfield questions arise about the relationship of the various subdisciplines *within* cognitive science. Of the ten possible two-place relations among the five core disciplines of cognitive science, two have received the most attention from philosophers: relations between linguistics and psychology (particularly psycholinguistics) and relations between cognitive psychology and neuroscience.

Discussion of the relation between linguistics and psychology has addressed primarily Noam

Table 1. The central research framework of cognitive science

Domain-specifying assumptions

D1 (Identification assumption): The domain consists of the human cognitive capacities.

D2 (Property assumption): Pretheoretically conceived, the human cognitive capacities have a number of *basic general properties:*

- Each capacity is *intentional*; that is, it involves states that have content or are "about" something.
- Virtually all of the capacities are *pragmatically evaluable;* that is, they can be exercised with varying degrees of success.
- When successfully exercised, each of the evaluable capacities has a certain *coherence* or cogency.
- Most of the evaluable capacities are *reliable*; that is, typically, they are exercised successfully (at least to some degree) rather than unsuccessfully.
- Most of the capacities are *productive*; that is, once a person has the capacity in question, he or she is typically in a position to manifest it in a practically unlimited number of novel ways
- Each capacity involves one or more *conscious* states.[a]

D3 (Grouping assumption): The cognitive capacities of the normal, typical adult make up a theoretically coherent set of phenomena, or a *system*. This means that with sufficient research, it will be possible to arrive at a set of answers to the basic questions of the research framework that constitute a unified theory and are empirically and conceptually acceptable.

Basic questions

Q1: For the normal, typical adult, what precisely is the human capacity to _____?

Q2: In virtue of what does a normal, typical adult have the capacity to _____ (such that this capacity is intentional, pragmatically evaluable, coherent, reliable, and productive and involves consciousness?[a])

Q3: How does a normal, typical adult exercise his or her capacity to _____?

Q4: How does the capacity to _____ of the normal, typical adult interact with the rest of his or her cognitive capacities?

Substantive assumptions

SA1: The computational assumption

C1: (Linking assumption): The human, cognitive mind/brain is a computational device (computer); hence, the human cognitive capacities consist, to a large extent, of a system of computational capacities.

C2: (System assumption): A computer is a device capable of automatically inputting, storing, manipulating, and outputting information in virtue of inputting, storing, manipulating, and outputting representations of that information. These information processes occur in accordance with a finite set of rules that are effective and that are, in some sense, in the machine itself.

SA2: The representational assumption

R1 (Linking assumption): The human, cognitive mind/brain is a representational device; hence, the human cognitive capacities consist of a system of representational capacities.

R2 (System assumption): A representational device is a device that has states or that contains within it entities that are representations. Any representation will have four aspects essential to its being a representation: (1) It will be realized by a representation bearer, (2) it will represent one or more representational objects, (3) its representation relations will be grounded somehow, and (4) it will be interpretable by (will function as a representation *for*) some currently existing interpreter. In the mind/brain representational system, the nature of these four aspects is constrained as follows:
- R2.1 (The representation bearer): The representation bearer of a mental representation is a computational structure or process, considered purely formally.
  (a) This structure or process has *constituent structure*.[b]
- R2.2 (The representational content): Mental representations have a number of semantic properties.[c]
  (a) They are semantically *selective*.
  (b) They are semantically *diverse*.
  (c) They are semantically *complex*.
  (d) They are semantically *evaluable*.
  (e) They have *compositional* semantics.[b]

(Continued)

Table 1. (*Continued*)

- R2.3 (The ground): The ground of a mental representation is a property or relation that determines the fact that the representation has the object (or content) it has.
  a) This ground is *naturalistic* (that is, nonsemantic and nonintentional).
  b) This ground may consist of either internal or external factors. However, any such factor must satisfy the following restriction: If two token representations have different grounds, and this ground difference determines a difference in content, then they must also differ in their causal powers to produce relevant effects across nomologically possible contexts.[b]
- R2.4 (The interpretant): Mental representations are significant for the person in whose mind they "reside." The interpretant of a mental representation R for some subject S consists of the set of all possible determinate computational processes contingent upon entertaining R in S.

Methodological assumptions

M1: Human cognition can be successfully studied by focusing exclusively on the individual cognizer and his or her place in the natural environment. The influence of society or culture on individual cognition can always be explained by appealing to the fact that this influence is mediated through individual perception and representation.

M2: The human cognitive capacities are sufficiently autonomous from other aspects of mind such as affect and personality that, to a large extent, they can be successfully studied in isolation.

M3: There exists a partitioning of cognition in general into individual cognitive capacities such that each of these individual capacities can, to a large extent, be successfully studied in isolation from each of the others.

M4: Although there is considerable variation in how adult human beings exercise their cognitive capacities, it is meaningful to distinguish, at least roughly, between "normal" and "abnormal" cognition.

M5: Although there is considerable variation in how adult human beings exercise their cognitive capacities, adults are sufficiently alike when they cognize that it is meaningful to talk about a "typical" adult cognizer and it is possible to arrive at generalizations about cognition that hold (at least approximately) for all normal adults.

M6: The explanatory strategy of cognitive science is sound. In particular, answers to the original basic questions can, to a large extent, be obtained by answering their narrow information-processing counterparts (that is, those involving processes purely "in the head").

M7: In choosing among alternative hypothesized answers to the basic questions of the research framework, one should invoke the usual canons of scientific methodology. That is, ultimately, answers to the basic questions should be justified on empirical grounds.

M8: A complete theory of human cognition will not be possible without a substantial contribution from each of the subdisciplines of cognitive science.

M9: Information-processing answers to the basic questions of cognitive science are constrained by the findings of neuroscience.[c]

M10: The optimal research strategy for developing an adequate theory of the cognitive mind/brain is to adopt a coevolutionary approach—that is, to develop information-processing answers to the basic questions of cognitive science on the basis of empirical findings from both the nonneural cognitive sciences and the neurosciences.[c]

M11: Information-processing theories of the cognitive mind/brain can explain certain features of cognition that cannot be explained by means of lower-level neuroscientific accounts. Such theories are thus, in principle, explanatorily ineliminable.[c]

Key: [a]Not in von Eckardt (1993); [b]Controversial; [c]Included on normative grounds (given the commitment of cognitive science to X, cognitive scientists should be committed to Y)

Chomsky's claim that the aims of linguistics are, first, to develop hypotheses (in the form of generative grammars) about what the native speaker of a language *knows* (tacitly) or that capture the native speaker's linguistic *competence* and, second, to develop a theory of the innate constraints on language ("universal grammar") that the child brings to bear in learning any given language. Philosophers have focused on the relation of competence models to so-called performance models, that is, models developed by psychologists to describe the information processes involved in understanding and producing language. Earlier discussion focused on syntax; more recent debates have looked at semantics.

There are several views. One, favored by Chomsky himself, is that a representation of the grammar of a language L constitutes a *part* of the mental apparatus involved in the psychological processes underlying language performance and, hence, is causally implicated in the production of that performance (Chomsky and Katz 1974). A second, suggested by Marr (1982) and others, is that competence theories describe a native speaker's linguistic input-output *capacity* (Marr's "computational" level) without describing the processes underlying that capacity (Marr's "algorithmic" level). For example, the capacity to understand a sentence *S* can be viewed as the capacity that maps a phonological representation of *S* onto a semantic representation *S*. Both views have been criticized. Against the first, it has been argued that because of the linguist's concern with simplicity and generality, it is unlikely that the formal structures utilized by optimal linguistic theories will be isomorphic to the internal representations posited by psycholinguists (Soames 1984). A complementary point is that because processing models are sensitive to architectural and resource constraints, the ways in which they implement syntactic knowledge have turned out to be much less transparent than followers of Chomsky had hoped (Mathews 1991). At best, they permit a psycholinguistic explanation of why a particular set of syntactic rules and principles correctly characterize linguistic competence. Against the second capacity view, it has been argued that grammars constitute idealizations (for example, there are no limitations on the length of permissible sentences or on their degree of internal complexity) and so do not describe actual speakers' linguistic capacities (Franks 1995; see Philosophy of Psychology; Neurobiology).

Internal questions have been raised about both foundational assumptions and concepts and particular theories and findings within cognitive science. Foundational discussions have focused on the core substantive assumptions: (1) The cognitive mind/brain is a computational system and (2) it is a representational system.

The Computational Assumption

Cognitive science assumes that the mind/brain is a kind of computer. But what *is* a computer? To date, two *kinds* of computer have been important to cognitive science modeling—*classical* machines ("conventional," "von Neumann," or "symbol system") and *connectionist* machines ("parallel distributed processing"). Classical machines have an architecture similar to ordinary personal computers (PCs). There are separate components for processing and memory, and processing occurs by the manipulation of data structures. In contrast, connectionist computers are more like brains, consisting of interconnected networks of neuron-like elements. Philosophers interested in the computational assumption have focused on two questions: What is a computer *in general* (such that both classical and connectionist machines count as computers), and is there any reason to believe, at the current stage of research, that human mind/brains are one sort of computer rather than another? The view that the human mind/brain is, or is importantly like, a classical computer is called classicism; the view that it is, or is importantly like, a connectionist computer is called connectionism.

The theory of computation in mathematics defines a number of different kinds of *abstract* machines in terms of the sets of functions they can compute. Of these, the most relevant to cognitive science is the universal Turing machine, which, according to the Church-Turing thesis, can compute any function that can be computed by an *effective* method, that is, a method specified by a finite number of exact instructions, in a finite number of steps, carried out by a human unaided by any machinery except paper and pencil and demanding no insight or ingenuity. Many philosophers and cognitive scientists think that the notion of a computer relevant to cognitive science can be adequately captured by the notion of a Turing machine. In fact, that is not the case. To say that a computer (and hence the mind/brain) is simply a device *equivalent* to a universal Turing machine says nothing about the device's internal structure, and to say that a computer (and hence the mind/brain) is an *implementation* of a Turing machine seems flat-out false. There are many dissimilarities. A Turing machine is in only one state at a time, while humans are, typically, in many mental or brain states simultaneously. Further, the memory of a Turing machine is a "tape" with only a simple linear structure, while human memory appears to have a complex multidimensional structure.

An alternative approach is to provide an architectural characterization, but at a fairly high level of abstraction. For example, von Eckardt (1993, 114) claims that cognitive science's computational assumption takes a computer to be "a device capable of automatically inputting, storing, manipulating, and outputting information in virtue of inputting, storing, manipulating, and outputting representations of that information," where "[t]hese information processes occur in accordance with a finite set of rules that are effective and that are, in some sense, in the machine itself." Another proposal, presented by Copeland (1996) and specifically intended to include functions that a Turing machine cannot compute, is that a computer is a device capable of solving a class of problems, that can represent both the problems and their solution, contains some number of primitive operations (some of which may not be Turing computable), can sequence these operations in some predetermined way, and has a provision for feedback (see Turing, Alan).

To the question of whether the mind/brain is a connectionist versus a classical computer, discussion has centered around Fodor and Pylyshyn's (1988) argument in favor of classicism. In their view, the claim that the mind/brain is a connectionist computer should be rejected on the grounds of the premises that

1. cognitive capacities exhibit *systematicity* (where 'systematicity' is the fact that some capacities are intrinsically connected to others—e.g., if a native speaker of English knows how to say "John loves the girl," the speaker will know how to say "The girl loves John") and
2. this feature of systematicity cannot be explained by reference to connectionist models (unless they are implementations of classical models), whereas
3. it can be explained by reference to classical models.

Fodor and Pylyshyn's reasoning is that it is "characteristic" of classical systems but not of connectionist systems to be "symbol processors," that is, systems that posit mental representations with constituent structure and then process these representations in a way that is sensitive to that structure. Given such features, classical models can explain systematicity; but without such features, as in connectionist machines, it is a mystery. Fodor and Pylyshyn's challenge has generated a host of responses, from both philosophers and computer scientists. In addition, the argument itself has evolved in two significant ways. First, it has been emphasized that to be a counterexample to premise 2, a connectionist model must not only exhibit systematicity, it must also *explain* it. Second, it has been claimed that what needs explaining is not just that human cognitive capacities are systematic; it is that they are *necessarily* so (on the basis of scientific law).

The critical points regarding premise 3 are especially important. The force of Fodor and Pylyshyn's argument rests on classical systems, being able to do something that connectionist systems (at a cognitive, nonimplementational level) cannot. However, it has been pointed out that the mere fact that a system is a symbol processor (and hence classical) does not *by itself* explain systematicity, much less the lawful necessity of it; additional assumptions must be made about the system's computational resources. Thus, if the critics are right that connectionist systems of *specific* sorts can also explain systematicity, then the explanatory asymmetry between classicism and connectionism will no longer hold (that is, neither the fact that a system is classical nor the fact that it is connectionist can, taken by itself, explain systematicity, while either fact can explain systematicity when appropriately supplemented), and Fodor and Pylyshyn's argument would be unsound.

The Representational Assumption

Following Peirce (Hartshorne, Weiss, and Burks 1931–58), one can say that any representation has four aspects essential to its being a representation: (i) It is realized by a representation bearer; (ii) it has *semantic content* of some sort; (iii) its semantic content is *grounded* somehow; and (iv) it has *significance* for (that is, it can function as a representation for) the person who has it. (Peirce's terminology was somewhat different. He spoke of a representation's bearer as the "material qualities" of the representation and the content as the representation's "object.")

In Peirce, a representation has significance for a person insofar as it produces a certain effect or "interpretant." This conception of representation is extremely useful for exploring the view of cognitive science vis-à-vis mental representation, for it leads one to ask: What is the representation bearer, semantic content, and ground of *mental* representation, and how do mental representations have significance for the people in whose mind/brains they reside?

A representation bearer is an entity or state that has semantic properties considered with respect to its nonrepresentational properties. The representation bearers for mental representations, according to cognitive science, are computational structures or states. If the mind/brain is a classical computer, its representation bearers are data structures; if it is a connectionist computer, its *explicit* representation bearers are patterns of activation of nodes in a network. It is also claimed that connectionist computers *implicitly* represent by means of their connection weights.

Much of the theoretical work of cognitive psychologists consists of claims regarding the content of the representations used in exercising one or another cognitive capacity. For example, psycholinguists posit that in understanding a sentence, people unconsciously form representations of the sentence's phonological structure, words, syntactic structure, and meaning. Although psychologists do not know enough about mental representation as a system to theorize about its semantics in the sense in which linguistics provides the formal semantics of natural language, if one reflects on what it is that mental representations are hypothesized to explain—certain features of cognitive capacities—one can plausibly infer that the semantics of human mental representation systems must have certain characteristics. In particular, a case can be made that people must be able to represent (i) specific objects; (ii) many different kinds of objects including concrete objects, sets, properties, events and states of affairs in the world, in possible worlds, and in fictional worlds, as well as abstract objects such as universals and numbers; (iii) both objects (*tout court*) and aspects of those objects (or something like extension and intension); and (iv) both correctly and incorrectly.

Although cognitive psychologists have concerned themselves primarily with the representation bearers and semantic content of mental representations, the hope is that eventually there will be an account of how the computational states and structures that function as representation bearers come by their content. In virtue of what, for example, do lexical representations represent specific words? What makes an edge detector represent edges? Such theories are sometimes described as one or another form of "semantics" (e.g., informational semantics, functional role semantics), but this facilitates a confusion between theories of content and theories of what determines that content. A preferable term is *theory of content determination*. Philosophers have been concerned both (a) to delineate the basic relation between a representation's having a certain content and its having a certain ground and (b) to sketch alternative theories of content determination, that is, theories of what that ground might be. A common view is that the basic relation is strong supervenience, as defined by Kim (1984). Others, such as Poland (1994), believe that a stronger relation of *realization* is required.

How is the content of mental representations determined? Proposals appeal either exclusively or in some combination to the structure of the representation bearer (Palmer 1978); actual historical (Devitt 1981) or counterfactual causal relations (Fodor 1987) between the representation bearer and phenomena in the world; actual and counterfactual (causal, computational, inferential) relations between the representation-bearer state and other states in the mind/brain (Harman 1987; Block 1987); or the information-carrying or other *functions* of the representation-bearer state and associated components (based on what they were selected for in evolution or learning) (Millikan 1984; Papineau 1987).

Arguably, any adequate theory of content determination will be able to account for the full range of semantic properties of mental representational systems. On this criterion, all current theories of content determination are inadequate. For example, theories that ground representational content in an isomorphism between some aspect of a representation (usually, either its formal structure or its functional role) and what it represents do not seem to be able to explain how one can represent *specific* objects, such as a favorite coffee mug. In contrast, theories that rely on actual, historical causal relations can easily explain the representation of specific objects but do not seem to be able to explain the representation of sets or kinds of objects (e.g., all coffee mugs). It is precisely because single-factor theories of content determination do not seem to have the resources to explain all aspects of people's representational capacities that many philosophers have turned to two-factor theories, such as theories that combine causal relations with function (so-called teleofunctional theories) and theories that combine causal relations with functional role (so-called two-factor theories).

The fourth aspect of a mental representation, in the Peircian view, is that the content of a representation must have significance *for* the representer. In the information-processing paradigm, this amounts to the fact that for each representation, there will be a set of computational consequences of that representation being "entertained" or "activated," and in particular a set of computational consequences that are *appropriate* given the content of the representation (von Eckardt 1993, Ch. 8).

The Viability of Cognitive Science

Cognitive science has been criticized on several grounds. It has been claimed that there are phenomena within its domain that it does not have the conceptual resources to explain; that one or more of its foundational assumptions are problematic and, hence, that the research program grounded in those assumptions can never succeed; and that it can, in principle, be eliminated in favor of pure neuroscience.

The list of mental phenomena that, according to critics, cognitive science will never be able to explain include that of people "making sense" in their actions and speech, of their having sensations, emotions, and moods, a self and a sense of self, consciousness, and the capabilities of insight and creativity and of interacting closely and directly with their physical and social environments. Although no impossibility proofs have been offered, when cognitive science consisted simply of a top-down, "symbolic" computational approach, this explanatory challenge had a fair amount of intuitive plausibility. However, given the increasing importance of neuroscience within cognitive science and the continuing evolution of the field, it is much less clear today that no cognitive science explanation of such phenomena will be forthcoming. The biggest challenge seems to be the "explanatory gap" posed by phenomenal consciousness (Levine 1983) (see Consciousness).

The major challenge against the computational assumption is the claim that the notion of a computer employed within cognitive science is vacuous. Specifically, Putnam (1988) has offered a proof that every ordinary open system realizes every abstract finite automaton, and Searle (1990) has claimed that implementation is not an objective relation and, hence, that any given system can be seen to implement any computation if interpreted appropriately. Both claims have been disputed. For example, it has been pointed out that Putnam's result relies on employing both an inappropriate computational formalism (the formalism of finite state automata, which have only *monadic* states), an inappropriately weak notion of implementation (one that does not require the mapping from computational to physical states to satisfy *counterfactual* or lawful state–state transitions), and an inappropriately permissive notion of a physical state (one that allows for rigged disjunctions). When these parameters of the problem are made more restrictive, implementations are much harder to come by (Chalmers 1996). However, in defense of Putnam, it has been suggested that this response does not get to the heart of Putnam's challenge, which is to develop a theory that provides necessary and sufficient criteria to determine whether a class of computations is implemented by a class of physical systems (described at a given level). An alternative approach to implementation might be based on the notion of realization of a (mathematical) function by a "digital system" (Scheutz 1999).

The major challenge to the representational assumption is the claim that the project of finding an adequate theory of content determination is doomed to failure. One view is that the attempt by cognitive science to explain the intentionality of cognition by positing mental representations is fundamentally confused (Horst 1996). The argument is basically this: According to cognitive science, a mental state, as ordinarily construed, has some content property in virtue of the fact that it is identical to (supervenes on) a representational state that has some associated semantic property. There are four ways of making sense of this posited semantic property. The semantic property in question is identical to one of the following options:

1. The original content property,
2. An interpreter-dependent semiotic-semantic property,
3. A pure semiotic-semantic property of the sort posited in linguistics, or
4. Some new theoretical "naturalized" property.

However, according to Horst (1996), none of these options will work. Options 1 and 2 are circular and hence uninformative. Option 3 is ruled out on the grounds that there is no reason to believe that there are such properties. And option 4 is ruled out on the grounds that naturalized theories of content determination do not have the conceptual resources to deliver the kind of explanation required.

In response, it has been argued that Horst's case against option 3 is inadequate and that his argument against option 4 shows a misunderstanding of what theories of content determination are trying to achieve (von Eckardt 2001). Horst's reason for ruling out option 3 is that he thinks that people who believe in the existence of pure semantic properties or relations do so because there are formal semantic theories that posit such properties and such theories have met with a certain degree of success. But (he argues) when scientists develop models (theories) in science, they do so by a process of abstraction from the phenomena *in vivo* that they wish to characterize. Such abstractions can be viewed either as models *of* the real-world phenomena they were abstracted from or as descriptions of the mathematical properties of those phenomena. Neither view provides a license for new ontological claims. Thus, Horst's argument rests on a nonstandard conception of both theory construction (as nothing but abstraction) and models or theories (as purely mathematical). If, *contra* Horst, a more standard conception is substituted, then linguists have as much right to posit pure semantic properties as physicists have to posit strange subatomic particles. Furthermore, the existence of the posited entities will depend completely on how successful they are epistemically.

Horst's case against option 4, again, exhibits a misunderstanding of the cognitive science project. He assumes that the naturalistic ground N of the content of a mental representation C will be such that the truth of the statement "X is N" conjoined with the necessary truths of logic and mathematics will be *logically sufficient* for the truth of the statement "X has C" and that, further, this entailment will be epistemically transparent. He then argues against there being good prospects for a naturalistic theory of content on the grounds that naturalistic discourse does not have the conceptual resources to build a naturalistic theory that will entail, in an epistemically transparent way, the truths about intentionality. However, as von Eckardt (2001) points out, Horst's conception of naturalization is much stronger than what most current theory of content determination theorists have in mind, *viz.*, strong supervenience or realization (see Supervenience). As a consequence, his arguments that naturalization is implausible given the conceptual resources of naturalistic discourse are seriously misguided.

BARBARA VON ECKARDT

References

Block, N. (1987), "Functional Role and Truth Conditions," *Proceedings of the Aristotelian Society* 61: 157–181.
Chalmers, D. (1996), "Does a Rock Implement Every Finite-State Automaton?" *Synthese* 108: 309–333.
Chomsky, N., and J. Katz (1974), "What the Linguist Is Talking About," *Journal of Philosophy* 71: 347–367.
Copeland, B. (1996), "What Is Computation?" *Synthese* 108: 336–359.
Devitt, M. (1981), *Designation*. New York: Columbia University Press.
Fodor, J. (1987), *Psychosemantics*. Cambridge, MA: MIT Press.
Fodor, J., and Z. W. Pylyshyn (1988), "Connectionism and Cognitive Architecture: A Critical Analysis," in S. Pinker and J. Miller (eds.), *Connections and Symbols*. Cambridge, MA: MIT Press, 1–71.
Franks, B. (1995), "On Explanation in the Cognitive Sciences: Competence, Idealization and the Failure of the Classical Cascade," *British Journal for the Philosophy of Science* 46: 475–502.
Gardner, H. (1985), *The Mind's New Science: A History of the Cognitive Revolution*. New York: Basic Books.
Harman, G. (1987), "(Non-Solipsistic) Conceptual Role Semantics," in E. Lepore (ed.), *New Directions in Semantics*. London: Academic Press.
Hartshorne, C., P. Weiss, and A. Burks (eds.) (1931–58), *Collected Papers of Charles Sanders Peirce*. Cambridge, MA: Harvard University Press.
Horst, S. W. (1996), *Symbols, Computation, and Intentionality*. Berkeley and Los Angeles: University of California Press.
Kim, J. (1984), "Concepts of Supervenience," *Philosophical and Phenomenological Research* 45: 153–176.
Kuhn, T. (1970), *The Structure of Scientific Revolutions*. Chicago: University of Chicago Press.
Levine, J. (1983), "Materialism and Qualia: The Explanatory Gap," *Pacific Philosophical Quarterly* 64: 354–361.
Marr, D. (1982), *Vision: A Computational Investigation into the Human Representation and Processing of Visual Information*. San Francisco: W. H. Freeman.
Mathews, R. (1991), "Psychological Reality of Grammars," in A. Kasher (ed.), *The Chomskyan Turn*. Oxford and Cambridge, MA: Basil Blackwell.
Millikan, R. (1984), *Language, Thought, and Other Biological Categories*. Cambridge, MA: MIT Press.
Palmer, S. (1978), "Fundamental Aspects of Cognitive Representation," in E. Rosch and B. Lloyd (eds.), *Cognition and Categorization*. Hillsdale, NJ: Erlbaum.
Papineau, D. (1987), *Reality and Representation*. Oxford: Blackwell.
Poland, J. (1994), *Physicalism: The Philosophical Foundations*. Oxford: Oxford University Press.
Putnam, H. (1988), *Representation and Reality*. Cambridge, MA: MIT Press.
Scheutz, M. (1999), "When Physical Systems Realize Functions," *Minds and Machines* 9: 161–196.
Searle, J. (1990), "Is the Brain a Digital Computer?" *Proceedings and Addresses of the American Philosophical Association* 64: 21–37.
Simon, H. (1981), "Cognitive Science: The Newest Science of the Artificial," in D. A. Norman (ed.), *Perspectives on Cognitive Science*. Norwood, NJ: Ablex Publishing, 13–26.
Soames, S. (1984), "Linguistics and Psychology," *Linguistics and Philosophy* 7: 155–179.
von Eckardt, B. (2001), "In Defense of Mental Representation," in P. Gardenfors, K. Kijania-Placek, and J. Wolenski (eds.), *Proceedings of the 11th International Congress of Logic, Methodology and Philosophy of Science*. Dordrecht, Netherlands: Kluwer.
——— (1993), *What Is Cognitive Science?* Cambridge, MA: MIT Press.

See also **Consciousness; Physicalism; Psychology, Philosophy of; Supervenience**

COGNITIVE SIGNIFICANCE

One of the main objectives of logical empiricism was to develop a formal criterion by which cognitively significant statements, which are true or false, could be delineated from meaningless ones, which are neither. The desired criterion would specify, and in some way justify, the logical empiricists' conviction that scientific statements were exemplars of significance and metaphysical ones were decidedly not (see Logical Empiricism). Finding such a criterion was crucial to logical empiricism. Without it there seemed to be no defensible way to distinguish metaphysics from science and, consequently, no defensible way to exclude metaphysics from subjects that deserved serious philosophical attention (see Demarcation, Problem of). Accordingly, several logical empiricists devoted attention to developing a criterion of cognitive significance, including Carnap, Schlick, Ayer, Hempel, and, to a lesser degree, Reichenbach.

Scientific developments also motivated the project in two related ways. First, physics and biology were demonstrating that a priori metaphysical speculations about empirical matters were usually erroneous and methodologically misguided. Hans Driesch's idea of an essential entelechy was no longer considered scientifically respectable, and the intuitive appeal of the concept of absolute simultaneity was shown to be misleading by Albert Einstein (Feigl 1969). Scientific results demonstrated both the necessity and the fruitfulness of replacing intuitive convictions with precise, empirically testable hypotheses, and logical empiricists thought the same methodology should be applied to philosophy. Formulating a defensible criterion that ensured the privileged epistemological status of science, and revealed the vacuity of metaphysics, was thought crucial to the progress and respectability of philosophy.

Second, many scientific discoveries and emerging research programs, especially in theoretical physics, were considerably removed from everyday observable experience and involved abstract, mathematically sophisticated theories. The logical empiricists felt there was a need for a formal systematization of science that could clarify theoretical concepts, their interrelations, and their connection with observation. The emerging tools of modern mathematical logic made this task seem imminently attainable. With the desire for clarity came the pursuit of a criterion that could sharply distinguish these scientific developments, which provided insights about the world and constituted advances in knowledge, from the obfuscations of metaphysics.

Formulation of a cognitive significance criterion requires an empirical significance criterion to delineate empirical from nonempirical statements and a criterion of analyticity to delineate analytic from synthetic statements (see Analyticity). Most logical empiricists thought analytically true and false statements were meaningful, and most metaphysicians thought their claims were true but not analytically so. In their search for a cognitive significance criterion, as the principal weapon of their antimetaphysical agenda, the logical empiricists focused on empirical significance.

The Verifiability Requirement

The first attempts to develop the antimetaphysical ideas of the logical empiricists into a more rigorous criterion of meaningfulness were based on the verifiability theory of meaning (see Verifiability). Though of auxiliary importance to the rational reconstruction in the *Aufbau*, Carnap (1928a) claimed that a statement was verifiable and thereby meaningful if and only if it could be translated into a constructional system; for instance, by reducing it (at least in principle) to a system about basic physical objects or elementary experiences (§179) (see Carnap, Rudolf). Meaningful questions have verifiable answers; questions that fail this requirement are pseudo-questions devoid of cognitive content (§180).

The first explicit, semiformal criterion originated with Carnap in 1928. With the intention of demonstrating that the realism/idealism debate, and many other philosophical controversies, were devoid of cognitive significance, Carnap (1928b) presented a criterion of factual content:

> If a statement p expresses the content of an experience E, and if the statement q is either the same as p or can be derived from p and prior experiences, either through deductive or inductive arguments, then we say that q is 'supported by' the experience E.... A statement p is said to have 'factual content' if experiences which would support p or the contradictory of p are at least conceivable, and if their characteristics can be indicated. (Carnap 1967, 327)

Only statements with factual content are empirically meaningful. Notice that a fairly precise inferential method is specified and that a statement has factual content if there are conceivable experiments that could support it. Thus, the earliest formal significance criterion already emphasized that possible, not necessarily actual, connection to experience made statements meaningful.

Carnap ([1932] 1959) made three significant changes to his proposal. First, building on an earlier example (1928b, §7), he developed in more detail the role of syntax in determining the meaningfulness of words and statements in natural languages. The "elementary" sentence form for a word is the simplest in which it can occur. For Carnap, a word had to have a fixed mode of occurrence in its elementary sentence form to be significant. Besides failing to connect with experience in some way, statements could also be meaningless because they contained sequences of words that violated the language's syntactic rules, or its "logical syntax." According to Carnap, "Dog boat in of" is meaningless because it violates grammatical syntax, and "Our president is definitely a finite ordinal," is meaningless because it violates logical syntax, 'president' and 'finite ordinal' being members of different logical categories. The focus on syntax led Carnap to contextualize claims of significance to specific languages. Two languages that differ in syntax differ in whether words and word sequences are meaningful.

Second, Carnap ([1932] 1959) no longer required statements to be meaningful by expressing conceivable states of affairs. Rather, statements are meaningful because they exhibit appropriate deducibility relations with protocol statements whose significance was taken as primitive and incorrigible by Carnap at that time (see Protocol Sentences). Third, Carnap did not specify exactly how significant statements must connect to protocol statements, as he had earlier (1928b). In 1932, Carnap would ascertain a word's meaning by considering the elementary sentence in which it occurred and determining what statements entailed and were entailed by it, the truth conditions of the statement, or how it was verified—considerations Carnap then thought were equivalent. The relations were probably left unspecified because Carnap came to appreciate how difficult it was to formalize the significance criterion, and realized that his earlier criterion was seriously flawed, as was shown of Ayer's first formal criterion (see below).

In contrast to antimetaphysical positions that evaluated metaphysical statements as false, Carnap believed his criterion justified a radical elimination of metaphysics as a vacuous enterprise. The defensibility of this claim depended upon the status of the criterion—whether it was an empirical hypothesis that had to be supported by evidence or a definition that had to be justified on other grounds. Carnap ([1932] 1959) did not address this issue, though he labels the criterion as a stipulation. Whether this stipulation was defensible in relation to other possible criteria or whether the statement of the criterion satisfied the criterion itself were questions left unanswered.

In his popularization of the work of Carnap ([1932] 1959) and Schlick ([1932] 1979), Ayer (1934) addressed these questions and stated that a significance criterion should not be taken as an empirical claim about the linguistic habits of the class of people who use the word 'meaning' (see Ayer, Alfred Jules; Schlick, Moritz). Rather, it is a different kind of empirical proposition, which, though conventional, has to satisfy an adequacy condition. The criterion is empirical because, to be adequate, it must classify "propositions which by universal agreement are given as significant" as significant, and propositions that are universally agreed to be nonsignificant as nonsignificant (Ayer 1934, 345).

Ayer developed two formalizations of the criterion. The first edition of *Language, Truth, and Logic* contained the proposal that "a statement is verifiable ... if some observation-statement can be deduced from it in conjunction with certain other premises, without being deducible from those other premises alone," where an observation statement is one that records any actual or possible observation (Ayer 1946, 11).

Following criticisms (see the following section) of his earlier work, a decade later Ayer (1946) proposed a more sophisticated criterion by distinguishing between directly verifiable statements and indirectly verifiable ones. In conjunction with a set of observation statements, *directly* verifiable statements entail at least one observation statement that does not follow from the set alone. *Indirectly* verifiable statements satisfy two requirements: (1) In conjunction with a set of premises, they entail at least one directly verifiable statement that does not follow from the set alone; and (2) the premises can include only statements that are either analytic or directly verifiable or can be indirectly verified on independent grounds. Nonanalytic statements that are directly or indirectly verifiable are meaningful, whereas analytic statements are meaningful but do not assert anything about the world.

Early Criticisms of the Verifiability Criterion

The verifiability criterion faced several criticisms, which took two general forms. The first, already mentioned in the last section, questioned its status—specifically, whether the statement of the criterion satisfies the criterion. The second questioned its adequacy: Does the criterion ensure that obviously meaningful statements, especially scientific ones, are labeled as meaningful and that obviously meaningless statements are labeled as meaningless?

Criticisms of the first form often mistook the point of the criterion, construing it as a simple empirical hypothesis about how the concept of meaning is understood or a dogmatic stipulation about how it should be understood (Stace 1935). As mentioned earlier, Ayer (1934, 1946) clearly recognized that it was not this type of empirical claim, nor was it an arbitrary definition. Rather, as Hempel (1950) later made clear, the criterion was intended to clarify and explicate the idea of a meaningful statement. As an explication it must accord with intuitions about the meaningfulness of common statements and suggest a framework for understanding how theoretical terms of science are significant (see Explication). The metaphysician can deny the adequacy of this explication but must then develop a more liberal criterion that classifies metaphysical claims as significant while evaluating clearly meaningless assertions as meaningless (Ayer 1934).

Criticisms of the second form often involved misinterpretations of the details of the criterion, due partially to the ambiguity of what was meant by 'verifiability.' For example, in a criticism of Ayer (1934), Stace (1935) argued that the verifiability criterion made all statements about the past meaningless, since it was in principle impossible to access the past and therefore verify them. His argument involved two misconceptions. First, Stace construed the criterion to require the possibility of conclusive verification, for instance a complete reduction of any statement to (possible) observations that could be directly verified. Ayer (1934) did not address this issue, but Schlick ([1932] 1979), from whose work Ayer drew substantially, emphasized that many meaningful propositions, such as those concerning physical objects, could never be verified conclusively. Accepting Neurath's criticisms in the early 1930s, Carnap accepted that no statement, including no protocol statement, was conclusively verified (see Neurath, Otto). Recall also that Carnap (1928b) classified statements that were "supported by" conceivable experiences—not conclusively verified—as meaningful.

Second, Stace's argument depended on the ambiguity of "possible verification," which made early formulations of the criterion misleadingly unclear (Lewis 1934). The possibility of verification can have three senses: practical possibility, empirical possibility, and logical possibility. Practical possibility was not the intended sense: "There are 10,000-foot mountains on the moon's far side" was meaningful in the 1930s, though its verification was practically impossible (Schlick [1932] 1979).

However, Carnap (1928a, 1928b, [1932] 1959), Schlick ([1932] 1979), and Ayer (1934) were silent on whether empirical or logical possibility divided the verifiable from the unverifiable. Stace thought time travel was empirically impossible. The question was therefore whether statements about past events were meaningful for which no present evidence was available, and no future evidence would be.

In the first detailed analysis of the verifiability criterion, Schlick ([1936] 1979) stated that the logical impossibility of verification renders a statement nonsignificant. Empirical impossibility, which Schlick understood as contradicting the "laws of nature," does not entail non-verifiability. If it did, Schlick argued, the meaningfulness of a putative statement could be established only by empirical inquiry about the laws of nature. For Schlick, this conflated a statement's meaning with its truth. The meaning of a statement is determined ("bestowed") by logical syntax, and only with meaning fixed a priori can its truth or falsity be assessed. Furthermore, since some lawlike generalizations are yet to be identified and lawlike generalizations are never established with absolute certainty, it seems that a sharp boundary between the empirically impossible and possible could never be determined. Hence, there would be no sharp distinction between the verifiable and unverifiable, which Schlick found unacceptable.

For Schlick ([1936] 1979), questions formulated according to the rules of logical grammar are meaningful if and only if it is logically possible to verify their answers. A state of affairs is logically possible for Schlick if the statement that describes it conforms to the logical grammar of language. Hence, meaningful questions may concern states of affairs that contradict well-supported lawlike generalizations. Schlick's position implies that the set of meaningful questions is an extension of the set of questions for which verifiable answers can be imagined. Questions about velocities greater than light are meaningful according to Schlick, but imagining how they could be verified surpasses our mental capabilities.

Schlick's emphasis on logical possibility was problematic because it was unclear that the verification conditions of most metaphysical statements are, or entail, logical impossibilities. In contrast, Carnap ([1936–1937] 1965) and Reichenbach (1938) claimed that metaphysical statements were nonsignificant because no *empirically* possible process of confirmation could be specified for them (see Reichenbach, Hans). Furthermore, if only the *logical* possibility of verification were required for significance, then the nonsignificance of metaphysical statements could no longer be demonstrated by demanding an elucidation of the circumstances in which they could be verified. Metaphysicians can legitimately respond that such circumstances may be difficult or impossible to conceive because they are not empirically possible. Nevertheless, the circumstances may be logically possible, and hence the metaphysical statements may be significant according to Schlick's position.

Faced with the problematic vagueness of the early criteria, a formal specification of the criterion was thought to be crucial. Berlin (1939) pointed out that the early verifiability criteria were open to objections from metaphysicians because the details of the experiential relevance required of meaningful statements were left unclear: "Relevance is not a precise logical category, and fantastic metaphysical systems may choose to claim that observation data are 'relevant' to their truth" (233).

With formalizations of the criterion, however, came more definitive criticisms. Ayer's (1946, 39) first proposal was seriously flawed because it seemed to make almost all statements verifiable. For any grammatical statement S—for instance "The Absolute is peevish"—any observation statement O, and the conditional $S \rightarrow O$, S and $S \rightarrow O$ jointly entail O, though neither of them alone usually does. According to Ayer's criterion, therefore, S and $S \rightarrow O$ are meaningful except in the rare case that $S \rightarrow O$ entails O (Berlin 1939).

Church (1949) presented a decisive criticism of Ayer's (1946, 13) second proposal. Consider three logically independent observation statements O_1, O_2, and O_3 and any statement S. The disjunction $(\neg O_1 \wedge O_2) \vee (\neg S \wedge O_3)$ is directly verifiable, since in conjunction with O_1 it entails O_3. Also, $(\neg O_1 \wedge O_2) \vee (\neg S \wedge O_3)$ and S together entail O_2. Hence, by Ayer's criterion, S is indirectly verifiable, unless $(\neg O_1 \wedge O_2) \vee (\neg S \wedge O_3)$ alone entails O_2, which implies $\neg S$ and O_3 entail O_2 so that $\neg S$ is directly verifiable. Thus, according to Ayer's criterion, any statement is indirectly verifiable, and therefore significant, or its negation is directly verifiable, and thereby meaningful.

Nidditch (1961) pointed out that Ayer's (1946) proposal could be amended to avoid Church's (1949) criticism by specifying that the premises could only be analytic, directly verifiable, or indirectly verifiable on independent grounds *and* could only be composed of such statements. Thus that $(\neg O_1 \wedge O_2) \vee (\neg S \wedge O_3)$ and S together entail O_2 does not show that S is indirectly verifiable because $(\neg O_1 \wedge O_2) \vee (\neg S \wedge O_3)$ contains a statement (S) that has not been shown to be analytic, directly verifiable, or independently verifiable on independent grounds. Unfortunately, Scheffler (1963) pointed out that according to Nidditch's (1961) revised criterion, an argument similar to Church's (1949) with the disjunction $\neg O_2 \vee (S \wedge O_1)$ shows that any statement S is significant, unless it is a logical consequence of an observation statement. Scheffler (1963) also pointed out that Ayer's second proposal makes any statement of the form $S \wedge (O_1 \rightarrow O_2)$ significant, where O_1, O_2 are logically independent observation sentences and S is any statement. $S \wedge (O_1 \rightarrow O_2)$ entails O_2 when conjoined with O_1 and neither the conjunction nor O_1 entails O_2 alone.

Beyond Verifiability: Carnap and Hempel

While Ayer first attempted to formalize the verifiability criterion, Carnap ([1936–7] 1965) recognized the obvious weaknesses of verifiability-based significance criteria. At roughly the same time, in the light of Tarski's rigorous semantic account of truth, Carnap was coming to accept that a systematic (that is, nonpragmatic) account might be possible for other concepts, such as 'confirmation.' He subsequently refocused the question of cognitive significance away from verifiability, which seemed to connote the possibility of definitive establishment of truth, to confirmability—the possibility of obtaining evidence, however partial, for a statement. In particular, Carnap thought a justifiable significance criterion could be formulated if an adequate account of the confirmation of theory by observation were available. A better understanding of the latter would provide a clearer grasp of how scientific terms are significant due to their connection to observation and prediction and how metaphysical concepts are not, because they lack this connection. Yet, insights into the nature of confirmation of theory by observation do not alone determine the form of an adequate significance criterion. Rather, these insights were important because Carnap ([1936–7] 1965) radically changed the nature of the debate over cognitive significance.

Carnap reemphasized (from his work in 1932) that what expressions are cognitively significant depends upon the structure of language, and hence a criterion could be proposed relative to only a specific language. He distinguished two kinds of questions about cognitive significance: those concerning "historically given language system[s]" and those concerning constructible ones (Carnap [1932] 1959, 237). Answers to the two kinds of questions are evaluated by different standards. To be meaningful in the first case, an expression E must be a sentence of L, which is determined by the language's syntax, and it must "fulfill the empiricist criterion of meaning" (167) for L. Carnap does not disclose the exact form of the criterion—verifiability, testability, or confirmability—for a particular language, such as English.

The reason for Carnap's silence, however, was his belief that the second type of question posed a more fruitful direction for the debate. The second type of question is practical, and the answers are proposals, not assertions. Carnap ([1936–7] 1965) remarked that he was no longer concerned with arguing directly that metaphysical statements are not cognitively significant (236). Rather, his strategy was to construct a language L in which every nonanalytic statement was confirmable by some experimental procedure. Given its designed structure, L will clearly indicate how theoretical statements can be confirmed by observational ones, and it will not permit the construction of metaphysical statements. If a language such as L can be constructed that accords with intuitions about the significance of common statements and is sufficient for the purposes of science, then the onus is on the metaphysician to show why metaphysical statements are significant in anything but an emotive or attitude-expressing way.

In a review paper more than a decade later, Hempel (1950) construed Carnap's ([1936–1937] 1965) position as proposing a translatability criterion—a sentence is cognitively significant if and only if it is translatable into an empiricist language (see Hempel, Carl). The vocabulary of an empiricist language L contains observational predicates, the customary logical constants, and any expression constructible from these; the sentence formation rules of L are those of *Principia Mathematica*. The problem Carnap ([1936–1937] 1965) attempted to rectify was that many theoretical terms of science cannot be defined in L.

Hempel's interpretation, however, slightly misconstrued Carnap's intention. Carnap ([1936–7] 1965) did not try to demonstrate how theoretical terms could be connected to observational ones in order to *assert* translatability as a criterion of cognitive significance. Rather, in accord with the *principle of tolerance* (Carnap 1934) Carnap's project in 1936–1937 was to construct an alternative to metaphysically infused language. The features of the language are then evaluated with respect to the purposes of the language user on pragmatic grounds. Although it seems to conflict with his position in 1932, Carnap (1963) clarified that a "neutral attitude toward various forms of language based on the principle that everyone is free to use the language most suited to his purposes, has remained the same throughout my life" (18–19). Carnap ([1936–1937] 1965) tried to formulate a replacement for metaphysics, rather than directly repudiate it on empiricist grounds.

Three definitions were important in this regard. The forms presented here are slightly modified from those given by Carnap ([1936–1937] 1965):

1. The confirmation of a sentence S is completely reducible to the confirmation of a class of sentences C if S is a consequence of a finite subclass of C.
2. The confirmation of S directly incompletely reduces to the confirmation of C if (a) the confirmation of S is not completely reducible to C and (b) there is an infinite subclass C' of mutually independent sentences of C such that S entails, by substitution alone, each member of C'.
3. The confirmation of a predicate P reduces to the confirmation of a class of predicates Q if the confirmation of every full sentence of P with a particular argument (e.g., P(a), in which a is a constant of the language) is reducible to the confirmation of a consistent set of predicates of Q with the same argument, together with their negations.

With these definitions Carnap ([1936–1937] 1965) showed how dispositional predicates (for instance, "is soluble in water") S could be introduced into an empiricist language by means of reduction postulates or finite chains of them. These postulates could take the simple form of a reduction pair:

$$(\forall x)(Wx \to (Dx \to Sx));$$
$$(\forall x)(Fx \to (Rx \to \neg Sx));$$

in which W, D, F, and R designate observational terms and S is a dispositional predicate. (In the solubility example, Sx = "x is soluble in water"; Dx = "x dissolves in water"; Wx = "x is placed in water"; and R and F are other observational terms.) If $(Vx)(Dx \leftrightarrow \neg Rx)$ and $(Vx)(Wx \leftrightarrow Fx)$, then the reduction pair is a bilateral reduction sentence:

$$(\forall x)(Wx \to (Dx \leftrightarrow Sx)).$$

The reduction postulates introduce, but do not explicitly define, terms by specifying their logical relations with observational terms. They also provide confirmation relations between the two types of terms. For instance, the above reduction pair entails that the confirmation of S reduces to that of the confirmation of the set $\{W, D, F, R\}$. Carnap ([1936–1937] 1965) defined a sentence or a predicate to be confirmable (following definitions 1–3 above) if its confirmation reduces to that of a class of observable predicates (156–157). Reduction postulates provide such a reduction for disposition terms such as S. The reduction pair does not define S in terms of observational terms. If $\neg Wx$ and $\neg Fx$, then Sx is undetermined. However, the conditions in which S or its negation hold can be extended by adding other reduction postulates to the language. Carnap thought that supplementing an empiricist language to include terms that could be introduced by means of reduction postulates or chains of them (for example, if Wx is introduced by a reduction pair) would adequately translate all theoretical terms of scientific theories.

Although it set a more rigorous standard for the debate, Carnap's ([1936–1937] 1965) proposal encountered difficulties. Carnap believed that bilateral reduction sentences were analytic, since all the consequences of individual reduction sentences that contained only observation terms were tautologies. Yet, Hempel (1951) pointed out that two bilateral reduction sentences together sometimes entailed synthetic statements that contained only observation terms. Since the idea that the conjunction of two analytic sentences could entail synthetic statements was counterintuitive, Hempel made the important suggestion that analyticity and cognitive significance must be relativized to a specific language *and* a particular theoretical context. A bilateral reduction sentence could be analytic in one context but synthetic in a different context that contained other reduction postulates.

Hempel (1950) also argued that many theoretical terms, for instance "gravitational potential" or "electric field," could not be translated into an empiricist language with reduction postulates or chains of them. Introducing a term with reduction postulates provides some sufficient and necessary observation conditions for the term, but Hempel claimed that this was possible only in simple cases, such as electric fields of a simple kind. Introducing a theoretical term with reduction sentences also unduly restricted theoretical concepts to observation conditions. The concept of length could not be constructed to describe unobservable intervals, for instance 1×10^{-100} m, and the principles of calculus would not be constructible in such a language (Hempel 1951). Carnap's ([1936–1937] 1965) proposal could not accommodate most of scientific theorizing.

Although ultimately untenable, adequacy conditions for a significance criterion were included in Carnap's ([1936–1937] 1965) papers, generalized by Hempel (1951) as: If N is a nonsignificant sentence, then all truth-functional compound sentences that nonvacuously contain N must be nonsignificant. It follows that the denial of a nonsignificant sentence is nonsignificant and that a disjunction, conjunction, or conditional containing a nonsignificant component sentence is also nonsignificant. Yet Hempel (1951) was pessimistic that any adequate criterion satisfying this condition and yielding a sharp dichotomy between significance and nonsignificance could be found. Instead, he thought that cognitive significance was a matter of degree:

> Significant systems range from those whose entire extralogical vocabulary consists of observational terms, through theories whose formulation relies heavily on theoretical constructs, on to systems with hardly any bearing on potential empirical findings. (74).

Hempel suggested that it may be more fruitful to compare theoretical systems according to other characteristics, such as clarity, predictive and explanatory power, and simplicity. On these bases, the failings of metaphysical systems would be more clearly manifested.

Of all the logical empiricists' criteria, Carnap's (1956) criterion was the most sophisticated. It attempted to rectify the deficiencies of his 1936–7 work and thereby avoid Hempel's pessimistic conclusions. Scientific languages were divided into two parts, a theoretical language L_T and an observation language L_O. Let V_O be the class of descriptive constants of L_O, and V_T be the class of *primitive* descriptive constants of L_T. Members of V_O designate observable properties and relations such as 'hard,' 'white,' and 'in physical contact with.' The logical structure of L_O contains only an elementary logic, such as a simple first-order predicate calculus.

The descriptive constants of L_T, called theoretical terms, designate unobservable properties and relations such as 'electron' or 'magnetic field.' L_T contains the mathematics required by science along with the "entities" referred to in scientific physical, psychological, and social theories, though Carnap stressed that this way of speaking does not entail any ontological theses. A theory was construed as a finite set of postulates within L_T and represented by the conjunction of its members T. A finite set of correspondence rules, represented by the conjunction of its members C, connects terms of V_T and V_O.

Within this framework Carnap (1956) presented three definitions, reformulated as:

D1. A theoretical term M is *significant relative* to a class K with respect to L_T, L_O, T, and $C =_{df}$ if (i) $K \subset V_T$, (ii) $M \notin K$, and (iii) there are three sentences S_M, $S_K \in L_T$, and $S_O \in L_O$ such that:
  (a) S_M contains M as the only descriptive term.
  (b) The descriptive terms in S_K belong to K.
  (c) $(S_M \wedge S_K \wedge T \wedge C)$ is consistent.
  (d) $(S_M \wedge S_K \wedge T \wedge C)$ logically implies S_O.
  (e) $\neg[(S_K \wedge T \wedge C)$ logically implies $S_O]$.

D2. A theoretical term M_n is *significant* with respect to L_T, L_O, T, and $C =_{df}$ if there is a sequence of theoretical constants $<M_1, ... M_n>$ ($M_i \in V_T$) such that every M_i is significant relative to $\{M_1, ... M_{i-1}\}$ with respect to L_T, L_O, T, and C.

D3. An expression A of L_T is a significant sentence of $L_T =_{df}$ if (i) A satisfies the rules of formation of L_T and (ii) every descriptive term in A is significant, as in D2.

These definitions, especially D1 (d) and (e), are intended to explicate the idea that a significant term must make a predictive difference. Carnap was aware that observation statements can often be deduced only from theoretical statements containing several theoretical terms. With D2 Carnap implicitly distinguishes between theoretical terms whose significance depends on other theoretical terms and those that acquire significance independently of others. In contrast to his work in 1936–7, and in accord with Hempel's (1951) relativization of analyticity and cognitive significance, Carnap (1956) specified that the significance of theoretical terms is relativized to a particular language *and* a particular theory T.

With the adequacy of his proposal in mind, Carnap (1956, 54–6) proved an interesting result. Consider a language in which V_T is divided into empirically meaningful terms V_1 and empirically meaningless terms V_2. Assume that C does not permit any implication relation between those sentences that contain only V_1 or V_O terms and those sentences that contain only V_2 terms. For a given theory T that can be resolved into a class of statements T_1 that contain only terms from V_1, and T_2 that contain only terms of V_2, then a simple but adequate significance criterion can be given. Any theoretical term that occurs only in isolated sentences, which can be omitted from T without affecting the class of sentences of L_O that it entails in conjunction with C, is meaningless.

The problem is that this criterion cannot be utilized for a theory T' equivalent to T that cannot be similarly divided. Carnap (1956), however, showed by indirect proof that his criterion led to the desired conclusion that the terms of V_2 were not significant relative to T' (L_O, L_T, and C) and that therefore the criterion was not too liberal.

The Supposed Failure of Carnap

Kaplan (1975) raised two objections to Carnap's (1956) criterion that were designed to show that it was too liberal and too restrictive. Kaplan's first objection utilized the "deoccamization" of $T \wedge C$. The label is appropriate, since the transformation of $T \wedge C$ into its deoccamization $T' \wedge C'$ involves replacing all instances of some theoretical terms with disjunctions or conjunctions of new terms of the same type: an Occam-unfriendly multiplication of theoretical terms. Kaplan proved that any deductive systematization of L_O by $T \wedge C$ is also established by any of its deoccamizations. This motivates his intuition that deoccamization should preserve the empirical content of a theory and, therefore, not change the significance of its theoretical terms.

The objection is as follows: If any members of V_T are significant with respect to T, C, L_T, and L_O, then there must be at least one M_1 that is significant relative to an empty K (D2). Yet, if $T \wedge C$ is deoccamized such that M_1 is resolved into two new terms M_{11} and M_{12} that are never found apart, then the original argument that satisfied D1 can no longer be used, since $T' \wedge C'$ do not provide similar logical relationships for M_{11} and M_{12} individually. Hence, the sequence of theoretical terms required by D2 will have no first member. Subsequently, no chain of implications that establishes the significance of successive theoretical terms exists. Although deoccamization preserves the deductive systematization of L_O, according to Carnap's criterion it may render every theoretical term of $T' \wedge C'$ meaningless and therefore render $T' \wedge C'$ devoid of empirical content.

Creath (1976) vindicated the core of Carnap's (1956) criterion by generalizing it to accommodate *sets* of terms, reformulated as:

D1′. A theoretical term M is *significant relative* to a class K with respect to L_T, L_O, T, and $C =_{df}$ if (i) $K \subset V_T$, (ii) $M \notin K$, (iii) there is a class J such that $J \subset V_T$, $M \in J$, but J and K do not share any members, and (iv) there are sentences S_J, $S_K \in L_T$, and $S_O \in L_O$ such that:
(a) S_J contains members of J as the only descriptive terms.
(b) The descriptive terms in S_K belong to K.
(c) $(S_J \wedge S_K \wedge T \wedge C)$ *is* consistent.
(d) $(S_J \wedge S_K \wedge T \wedge C)$ logically implies S_O.
(e) $\neg[(S_K \wedge T \wedge C)$ logically implies $S_O]$.
(f) It is not the case that $(\exists J')(J' \subset J)$ and sentences $S_{J'}$, $S_{K'} \in L_T$, and $S_{O'} \in L_O$ such that:
(f1) $S_{J'}$ contains only terms of J' as its descriptive terms.
(f2) The descriptive terms of $S_{K'}$ belong to K.
(f3) $(S_{J'} \wedge S_{K'} \wedge T \wedge C)$ is consistent.
(f4) $(S_{J'} \wedge S_{K'} \wedge T \wedge C)$ logically implies $S_{O'}$.
(f5) $\neg[(S_{K'} \wedge T \wedge C)$ logically implies $S_{O'}]$.

D2′. A theoretical term M_n is *significant* with respect to L_T, L_O, T and $C =_{df}$ if there is a sequence of sets $\langle J_1, \ldots J_n \rangle$ ($M_n \in J_n$ and $J_i \subset V_T$) such that every member of every set J_i is significant relative to the union of

J_1 through J_{i-1} with respect to L_T, L_O, T and C.

Condition (f) ensures that each member of J is required for the significance of the entire set. Creath (1976) points out that any term made significant by D1 and D2 of Carnap (1956) is made significant by D1' and D2' and that according to the generalized criterion, Kaplan's (1975) deoccamization criticism no longer holds.

Kaplan (1975) and Rozeboom (1960) revealed an apparent second flaw in Carnap's (1956) proposal: As postulates (definitions for Kaplan's criticism) are added to $T \wedge C$, the theoretical terms it contains may change from cognitively significant to nonsignificant or vice versa. Consider an example from Kaplan (1975) in which $V_O = \{J_O, P_O, R_O\}$; L_O is the class of all sentences of first-order logic with identity that contain no descriptive constants or only those from V_O; $V_T = \{B_T, F_T, G_T, H_T, M_T, N_T\}$; and L_T is the class of all sentences of first-order logic with identity that contain theoretical terms from V_T. Let T be:

$$(T)[(\forall x)(H_T x \to F_T x)] \wedge [(\forall x)(H_T x \to (B_T x \vee \neg G_T x))] \wedge [(\forall x)(M_T x \leftrightarrow N_T x)];$$

and let C be:

$$(C)[(\forall x)(R_O x \to H_T x)] \wedge [(\forall x)(F_T x \to J_O x)] \wedge [(\forall x)(G_T x \to P_O x)].$$

G_T, F_T, and H_T are significant with respect to $T \wedge C$ relative to the empty set (see Carnap [1956] D1) and, hence, significant with respect to L_O, L_T, T, and C (see Carnap [1956] D2). R_O is significant relative to $K = \{G_T\}$; M_T and N_T are not significant.

Consider a definitional extension T' of T in an extended vocabulary V_T' and language L_T'. After adding two definitions to T:

$$(DEF1)(\forall x)(D1_T x \leftrightarrow (M_T x \wedge (\exists x) F_T x))$$

and

$$(DEF2)(\forall x)(D2_T x \leftrightarrow (M_T x \to (\exists x) G_T x)),$$

$D1_T$ is significant relative to the empty set and therefore significant with respect to T', C, L_O, and L_T' (D2). $D2_T$ is significant relative to $K = \{D1_T\}$, and therefore significant with respect to T', C, L_O, and L_T' (D2). M_T, which failed to be significant with respect to T, C, L_O, and L_T, is now significant with respect to T', C, L_O, and L_T'. A similar procedure makes N_T significant. Kaplan thought this showed that Carnap's (1956) criterion was too liberal. The procedure seems able to make any theoretical term significant with respect to some extended language and definition-extended theory, but "definitional extensions are ordinarily thought of as having no more empirical content than the original theory" (Kaplan 1975, 90).

Using the same basic strategy, Rozeboom (1960) demonstrated that extending $T \wedge C$ can transform an empirically significant term into an insignificant one. Consider a term M that is significant with respect to T, C, L_T, and L_O. Rozeboom showed that if postulates (not necessarily definitions) are added to T or to C to form T' or C', in some cases D1(e) will no longer be satisfied, and no other sentences S_M', S_K', S_O' exist by which M could be independently shown to be significant. Furthermore, if $T \wedge C$ is maximally L_O consistent, no theoretical term of L_T is significant, since D1(e) is never satisfied; for any S_O, if $T \wedge C$ is maximally L_O consistent then it alone implies S_O. Rozeboom (1960) took the strength of his criticism to depend upon the claim that for a criterion to be "intuitively acceptable," theoretical terms must retain significance if T or C is extended.

Carnap (1956) can be defended in at least two ways. First, as Kaplan (1975) notes, the criterion was restricted to primitive, nondefined theoretical terms. It was explicitly formulated to avoid criticisms derived from definitional extensions. Defined terms often play an important role in scientific theories, and it could be objected that any adequate criterion should apply directly to theories that contain them. Yet the amendment that any theoretical term within the definiens of a significant defined term must be antecedently shown significant quells these worries (Creath 1976).

Second, Carnap (1956) insisted that terms are significant only *within a particular language* and *for a particular T and C*. He did not intend to formulate a criterion of cognitive significance that held under theory or language change. If Carnap's (1956) work on a significance criterion was an explication of the idea of meaningfulness (Hempel 1950), the explicandum was the idea of a meaningful statement of a particular language in a particular theoretical context, not meaningfulness per se. Hence, Kaplan and Rozeboom's objections, which rely on questionable intuitions about the invariance of significance as $T \wedge C$ changes, are not appropriately directed at Carnap (1956). The fact that Carnap did not attempt such an account is not merely the result of a realization that so many problems would thwart the project. Rather, it is a consequence of the external/internal framework that he believed was the most fruitful approach to the philosophical questions (Carnap 1947).

Furthermore, Rozeboom's acceptability condition is especially counterintuitive, since changes in

T or *C* designate changes in the connections between theoretical terms themselves or theoretical terms and observation terms. Additional postulates that specify new connections, or changes in the connections, between these terms can obviously change the significance of a theoretical term. Scientific advances are sometimes made when empirical or theoretical discoveries render a theoretical term nonsignificant.

JAMES JUSTUS

References

Ayer, A. J. (1934), "Demonstration of the Impossibility of Metaphysics," *Mind* 43: 335–345.
——— (1946), *Language, Truth, and Logic*, 2nd ed. New York: Dover Publications.
Berlin, I. (1939), "Verification," *Proceedings of the Aristotelian Society* 39: 225–248.
Carnap, R. (1928a), *Der Logische Aufbau der Welt*. Berlin-Schlachtensee: Weltkreis-Verlag.
——— (1928b), *Scheinprobleme in der Philosophie: Das Fremdpsychische und der Realismusstreit*. Berlin-Schlachtensee: Weltkreis-Verlag.
——— ([1932] 1959), "The Elimination of Metaphysics Through Logical Analysis of Language," in A. J. Ayer (ed.), *Logical Positivism*. Glencoe, IL: Free Press, 60–81.
——— (1934), *The Logical Syntax of Language*. London: Routledge Press.
——— ([1936–7] 1965), "Testability and Meaning," in R. R. Ammerman (ed.), *Classics of Analytic Philosophy*. New York: McGraw-Hill, 130–195.
——— (1947), "Empiricism, Semantics, and Ontology," in *Meaning and Necessity*. Chicago: University of Chicago Press, 205–221.
——— (1956), "The Methodological Character of Theoretical Concepts," in H. Feigl and M. Scriven (eds.), *The Foundations of Science and the Concepts of Psychology and Psychoanalysis*. Minneapolis: University of Minnesota Press, 38–76.
——— (1963), "Intellectual Autobiography," in P. Schlipp (ed.), *The Philosophy of Rudolf Carnap*. Peru, IL: Open Court Press, 3–86.
——— (1967), *Logical Structure of the World and Pseudoproblems in Philosophy*. Berkeley and Los Angeles: University of California Press.
Church, A. (1949), "Review of the Second Edition of *Language, Truth and Logic*," *Journal of Symbolic Logic* 14: 52–53.
Creath, R. (1976), "Kaplan on Carnap on Significance," *Philosophical Studies* 30: 393–400.
Feigl, H. (1969), "The Origin and Spirit of Logical Positivism," in P. Achinstein and S. F. Barker (eds.), *The Legacy of Logical Positivism*. Baltimore, MD: Johns Hopkins Press, 3–24.
Hempel, C. (1950), "Problems and Changes in the Empiricist Criterion of Meaning," *Revue Internationale de Philosophie* 11: 41–63.
——— (1951), "The Concept of Cognitive Significance: A Reconsideration," *Proceedings of the American Academy of Arts and Sciences* 80: 61–77.
Kaplan, D. (1975), "Significance and Analyticity," in J. Hintikka (ed.), *Rudolf Carnap, Logical Empiricist*. Dordrecht, Netherlands: D. Reidel Publishing Co., 87–94.
Lewis, C. I. (1934), "Experience and Meaning," *Philosophical Review* 43: 125–146.
Nidditch, P. (1961), "A Defense of Ayer's Verifiability Principle Against Church's Criticism," *Mind* 70: 88–89.
Reichenbach, H. (1938), *Experience and Prediction*. Chicago: University of Chicago Press.
Rozeboom, W. W. (1960), "A Note on Carnap's Meaning Criterion," *Philosophical Studies* 11: 33–38.
Scheffler, I. (1963), *Anatomy of Inquiry*. New York: Alfred A. Knopf.
Schlick, M. ([1932] 1979), "Positivism and Realism," in H. L. Mulder and B. F. B. van de Velde-Schlick (eds.), *Moritz Schlick: Philosophical Papers (1926–1936)*, vol. 2. Dordrecht, Netherlands: D. Reidel Publishing Co., 259–284.
———([1936] 1979), "Meaning and Verification," in H. L. Mulder and B. F. B. van de Velde-Schlick (eds.), *Moritz Schlick: Philosophical Papers (1926–1936)*, vol. 2. Dordrecht, Netherlands: D. Reidel Publishing Co., 456–481.
Stace, W. T. (1935), "Metaphysics and Meaning," *Mind* 44: 417–438.

See also **Analyticity; Ayer, Alfred Jules; Carnap, Rudolf; Corroboration; Demarcation, Problem of; Explication; Feigel, Herbert; Hempel, Carl; Logical Empiricism; Neurath, Otto; Popper, Karl; Rational Reconstruction; Reichenbach, Hans; Schlick, Moritz; Verifiability; Vienna Circle**

COMPLEMENTARITY

The existence of indivisible interaction quanta is a crucial point that implies the *impossibility of any sharp separation between the behavior of atomic objects and the interaction with the measuring instruments that serve to define the conditions under which the phenomena appear*. In fact, the individuality of the typical quantum effects finds its proper expression in the circumstance that any attempt at subdividing the phenomena will demand a change in the experimental arrangement, introducing new possibilities of interaction between objects and measuring instruments, which in principle cannot be controlled. Consequently, evidence obtained under different experimental conditions cannot be comprehended within a single picture but must be regarded as "complementary" in the sense that only the totality of the phenomena exhausts the possible information about the objects (Bohr 2000, 209–10).

Complementarity is distinctively associated with the Danish physicist Niels Bohr and his attempt to understand the new quantum mechanics (QM) during the heyday of its invention, 1920–1935 (see Quantum Mechanics). Physicists know in practice how to extract very precise and accurate predictions and explanations from QM. Yet, remarkably, even today no one is confident about how to interpret it metaphysically ("M-interpret" it). That is, there is no compellingly satisfactory account of what sort of objects and relations make up the QM realm, and so, ultimately, no one knows why the precise answers are as they are. In classical mechanics (CM), physicists thought they had a lucidly M-interpretable theory: There was a collection of clearly specified entities, particles, or waves that interacted continuously according to simple laws of force so that the system state was completely specified everywhere and at all times—indeed, specification of just instantaneous position and momenta sufficed (see Classical Mechanics). Here the dynamic process specified by the laws of force, expressed in terms of energy and momentum (Bohr's "causal picture"), generated a uniquely unfolding system configuration expressed in terms of position and time (Bohr's "space-time picture"). To repeat this for QM, what is needed is a collection of equivalent quantum objects whose interactions and movements in space-time generate the peculiar QM statistical results in ways that are as intuitively clear as they are for CM. (However, even this appealing concept of CM proves too simplistic; there is continuing metaphysical perplexity underlying physics; see e.g., Earman 1986; Hooker 1991, note 13.)

That M-interpreting QM is not easy is nicely illustrated by the status of the only agreed-on general "interpretation" to which physicists refer, Born's rule. It specifies how to extract probabilities from the QM wave function (ψ ["psi"]-function). These are normally associated with particle-like events. But without further M-interpretive support, Born's rule becomes merely part of the recipe for extracting numbers from the QM mathematics. That it does not M-interpret the QM mathematics is made painfully clear by the fact that the obvious conception of the QM state it suggests—a statistical ensemble of particles, each in a definite classical state—is provably not a possible M-interpretation of QM. (For example, no consistent sense can then be made of a superposition of ψ-states, since it is a mathematical theorem that the QM statistics of a superposed state cannot all be deduced from any single product of classical statistical states.)

Bohr does not offer an M-interpretation of QM. He came to think the very idea of such an interpretation incoherent. (In that sense, the term "Copenhagen interpretation" is a misnomer; Bohr does not use this label.) Equally, however, Bohr does not eschew all interpretive discussion of QM, as many others do on the (pragmatic or positivist-inspired) basis that confining the use of QM strictly to deriving statistics will avoid error while allowing science to continue. Bohr's position is that this too is profoundly wrong, and ultimately harmful to physics. Instead he offers the doctrine of complementarity as a "framework" for understanding the "epistemological lesson" of QM (not its ontological lesson) and for applying QM consistently and as meaningfully as possible. (see Folse 1985 for a general introduction. For extensive, more technical analyses, see Faye and Folse 1994; Honner 1987; Murdoch 1987. For one of many critiques, and the opposing Bohrian M-interpretation, see Cushing 1994.)

Although Bohr considered it necessary ("unavoidable") to continue using the key descriptive concepts of CM, the epistemological lesson of QM was that the basic conditions for their well-defined use were altered by the quantization of QM interactions into discrete unanalyzable units. This, he argued, divided the CM state in two, the conditions for the well-defined use of (1) causal (energy-momentum) concepts and (2) configurational (space-time) concepts being now mutually exclusive, so that only one kind of description could be coherently provided at a time. Both kinds of description were necessary to capture all the aspects of a QM system, but they could not be simply conjoined as in CM. They were now complementary.

Heisenberg's uncertainty relations of QM, such as $\Delta x \Delta p \geq h/2\pi$, where Δx is the uncertainty in the position, Δp is the uncertainty in the momentum, and h is Planck's constant (or more generally the commutation relations, such as $[x, p_x] = ih/2\pi$ where h is again Planck's constant, the magnitude of quantization), are not themselves statements of complementarity. Rather, they specify the corresponding quantitative relationships between complementary quantities.

These complementary exclusions are implicit in QM ontologies. For instance, single-frequency ("pure") waves must, mathematically, occupy all space, whereas restricting their extent involves superposing waves of different frequencies, with a point-size wave packet requiring use of all frequencies; thus, frequency and position are not uniquely cospecifiable. And since the wavelength λ is the wave velocity V (a constant) divided by the frequency v ($\lambda = V/v$), uniquely specifying wavelength and uniquely specifying position equally are mutually exclusive. QM associates wavelength with momentum ($p = h/\lambda$) and energy with frequency ($E = hv$) in all cases with discrete values and for both radiation (waves) and matter (particles), yielding the QM exclusions. (Note, however, that wave/particle complementarity is but one aspect of causal/configurational complementarity, the aspect concerned with physical conditions that frame coherent superposition versus those that frame localization.)

Precisely why these particular associations (and similar QM associations) should follow from quantization of interaction is not physically obvious, despite Bohr's confident assertion. Of course such associations follow from the QM mathematics, but that presupposes rather than explains complementarity, and Bohr intended complementarity to elucidate QM. The physical and mathematical roots of quantization are still only partially understood. However, it is clear that discontinuity leads to a constraint, in principle, on joint precise specification. Consider initially any quantity that varies with time (t), for example, energy (E), so that $E = f(t)$. Then across some time interval, $t_1 \rightarrow t_2$, E will change accordingly: $E_1(t_1) \rightarrow E_2(t_2)$.

If E varies continuously, then both E and t are everywhere jointly precisely specifiable because for every intermediate value of t between t_1 and t_2 (say, t_{1+n}) there will be a corresponding value for E: $E_{1+n} = f(t_{1+n})$. Suppose, however, that E (but not t) is quantized, with no allowed value between E_1 and E_2. Then no intermediate E value is available, and energy must remain undefined during at least some part of the transition period. This conclusion can be generalized to any two or more related quantities. The problem is resolved if both quantities are quantized, but there is as yet no satisfactory quantization of space and time (Hooker 1991).

Such inherent mutual exclusions should not be mistaken for merely practical epistemic exclusions (some of Heisenberg's pronouncements notwithstanding). Suppose that the position of an investigated particle i is determined by bouncing ("scattering") another probe particle p off of it, determining the position of i from the intersection of the initial and final momenta of p. However, i will have received an altered momentum in the interaction, and it is tempting to conclude that we are thus excluded from knowing both the position and the momentum of i immediately after the interaction. But in CM the interaction may be retrospectively analyzed to calculate the precise change in momentum introduced by p to i, using conservation of momentum, and so establish both the position and the momentum of i simultaneously. More generally, it is in this manner possible to correct for all measurement interactions and arrive at a complete classical state specified independently of its method of measurement. This cannot, in principle, be done in QM, because of quantization.

Faye (1991) provides a persuasive account of the origins of Bohr's ideas about the applicability of physical concepts in the thought of the Danish philosopher Harald Høffding (a family friend and early mentor of Bohr's) and sets out Bohr's consequent approach. According to Høffding, objective description in principle required a separation between describing a subject and describing a known object (Bohr's "cut" between them) in a way that always permitted the object to be ascribed a unique (Bohr's "unambiguous") spatiotemporal location, state, and causal interaction. These ideas in turn originated in the Kantian doctrine that an objective description of nature requires a well-defined distinction between the knower and the object of knowledge, permitting the unique construction of a well-defined object state, specified in applications of concepts from the synthetic a priori (essentially Newtonian) construction of the external world (see Friedman 1992). We have just noted how CM satisfies this requirement.

Contrarily, Bohr insisted, the quantum of action creates an "indissoluble bond" between the measurement apparatus (m-apparatus, including the sentient observer) and the measured (observed) system, preventing the construction of a well-defined system state separate from observing interactions. This vitiates any well-defined, global cut between m-apparatus and system. Creating a set of complementary partial cuts is the best that can now be done. In fact, these circumstances are generalized to all interactions between QM systems; the lack of a global separation is expressed in their superposition, which defies reduction to any combination of objectively separate states. Consequently, Bohr regarded CM as an idealized physics (achieved, imperfectly, only in the limit $h \to 0$) and QM as a "rational generalization" of it, in the sense of the principle of affinity, the Kantian methodological requirement of continuity.

Bohr's conception of what is required of a physically intelligible theory T can thus be summarized as follows (Hooker 1991, 1994):

BI1. Each descriptive concept A of T has a set of well-defined, epistemically accessible conditions C_A under which it is unambiguously applicable.

BI2. The set of such concepts collectively exhausts, in a complementary way, the epistemically accessible features of the phenomena in the domain of T.

BI3. There is a well-defined, unified, and essentially unique formal structure $S(T)$ that structures and coordinates descriptions of phenomena so that each description is well defined (the various conditions C_A are consistently combined), $S(T)$ is formally complete (Bub 1974), and BI2 is met.

BI4. Bohr objectivity (BO) satisfies BI1–3 in the most empirically precise and accurate way available across the widest domain of phenomena while accurately specifying the interactive conditions under which such phenomena are accessible to us.

An objective representation of nature thus reflects the interactive access ("point of view") of the knowing subject, which cannot be eliminated. In coming to know nature, we also come to know ourselves as knowers—not fundamentally by being modeled in the theory as *objects* (although this too happens, in part), but by the way the very form of rational generalization reflects our being as knowing *subjects*.

A very different ideal of scientific intelligibility operates in classical physics, and in many proposed M-interpretations of QM. Contrary to BI1, descriptive concepts are taken as straightforwardly characterizing external reality (describing an M, even if it is a strange one). Hence, contrary to BI2, these concepts apply conjointly to describe reality completely. Contrary to BI3 and BI4, an objective theory completely and accurately describes the physical state at each moment in time and provides a unique interactive dynamic history of states for all systems in its domain. Accordingly, measurements are analyzed similarly as the same kinds of dynamic interactions, and statistical descriptions reflect (only) limited information about states and are not fundamental (contrary to common readings of QM). Here the objective representation of nature through invariances eliminates any inherent reference to any subject's point of view. Rather, in coming to know nature we also come to know ourselves as knowers by being modeled in the theory as some *objects* among others so as to remove ourselves from the form of the theory, disappearing as *subjects*. This shift in ideals of intelligibility and objectivity locates the full depth of Bohr's doctrine of complementarity.

References

Bohr, N. (2000), "Discussion with Einstein on Epistemological Problems in Atomic Physics," in P. A. Schilpp (ed.), *Albert Einstein: Philosopher-Scientist*, 3rd ed. La Salle, IL: Open Court, 199–242.

——— (1961), *Atomic Theory and the Description of Nature*. Cambridge: Cambridge University Press.

Bub, J. (1974), *The Interpretation of Quantum Mechanics*. Dordrecht, Netherlands: Reidel.

Cushing, J. T. (1994), "A Bohmian Response to Bohr's Complementarity," in J. Faye and H. J. Folse (eds.), *Niels Bohr and Contemporary Philosophy*. Dordrecht, Netherlands: Kluwer.

Earman, J. (1986), *A Primer on Determinism*. Dordrecht, Netherlands: Reidel.

Faye, J. (1991), *Niels Bohr: His Heritage and Legacy*. Dordrecht, Netherlands: Kluwer.

Faye, J., and H. J. Folse (eds.) (1994), *Niels Bohr and Contemporary Philosophy*. Dordrecht, Netherlands: Kluwer.

Folse, H. J. (1985), *The Philosophy of Niels Bohr*. Amsterdam: North-Holland.

Friedman, M. (1992), *Kant and the Exact Sciences*. Cambridge, MA: Harvard University Press.

Honner, J. (1987), *The Description of Nature*. Oxford: Clarendon.

Hooker, C. A. (1972), "The Nature of Quantum Mechanical Reality: Einstein versus Bohr," in R. G. Colodny (ed.), *Pittsburgh Studies in the Philosophy of Science*, vol. 5. Pittsburgh, PA: University of Pittsburgh Press.

——— (1991), "Physical Intelligibility, Projection, and Objectivity: The Divergent Ideals of Einstein and Bohr," *British Journal for the Philosophy of Science* 42, 491–511.

——— (1994), "Bohr and the Crisis of Empirical Intelligibility: An Essay on the Depth of Bohr's Thought and Our Philosophical Ignorance," in J. Faye and H. J. Folse (eds.), *Niels Bohr and Contemporary Philosophy*. Dordrecht, Netherlands: Kluwer.

Murdoch, D. (1987), *Niels Bohr's Philosophy of Physics*. Cambridge: Cambridge University Press.

C. A. HOOKER

See also **Classical Mechanics; Quantum Measurement Problem; Quantum Mechanics**

COMPLEXITY

See **Unity and Disunity of Science**

CONFIRMATION THEORY

When evidence does not conclusively establish (or refute) a hypothesis or theory, it may nevertheless provide *some* support for (or against) the hypothesis or theory. Confirmation theory is concerned almost exclusively with the latter, where conclusive support (or "countersupport") are limiting cases of confirmation (or disconfirmation). (Included also, of course, is concern for the case in which the evidence is confirmationally irrelevant.) Typically, confirmation theory concerns *potential* support, the impact that evidence *would* have on a hypothesis or theory if learned, where whether the evidence is actually learned or not is not the point; for this reason, confirmation theory is sometimes called the *logic* of confirmation. (For simplicity of exposition for now, theories will be considered separately below and not explicitly mentioned until then.)

It is relevant here to point out the distinction between deductive logic (or deductive evaluation of arguments) and inductive logic (or inductive evaluation of arguments). In deductive logic, the question is just *whether or not* the supposed truth of all the premises of an argument gives an *absolute guarantee* of truth to the conclusion of the argument. In inductive logic, the question is whether the supposed truth of all the premises of an argument gives *significant support* for the truth of the conclusion, where, ideally, some measure of *to what degree* the premises support the conclusion (which is sometimes called the inductive probability of an argument) would be provided (see Inductive Logic; Induction, Problem of; and Verisimilitude. As in each of these topics also, the question is one of either qualitative or quantitative support that premises or evidence provides to a conclusion or that a hypothesis has. See Carnap, Rudolf, for an idea of degree of confirmation based on his proposed "logical" interpretation of probability and degree of support.) In the theory of the logic of support, confirmation theory is concerned primarily with inductive support, where the theory of deductive support is supposed to be more fully understood.

The concept of confirmation can be divided into a number of subconcepts, corresponding to three distinctions. First, absolute confirmation and incremental confirmation may be distinguished. In the absolute sense, a hypothesis is confirmed by evidence if the evidence makes (or would make) the hypothesis highly supported; absolute confirmation is about how the evidence "leaves" the hypothesis. In the incremental sense, evidence confirms a hypothesis if the evidence makes the hypothesis *more* highly confirmed (in the absolute sense) than it is (in the absolute sense) without the evidence; incremental confirmation involves a comparison. Second, confirmation can be thought of either *qualitatively* or *quantitatively*. So, in the absolute sense of confirmation, a hypothesis can be, qualitatively, left *more or less* confirmed by evidence, where quantitative confirmation theory attempts to make sense of assigning *numerical degrees* of confirmational support ("inductive probabilities") to hypotheses in light of the evidence. In the incremental sense of confirmation, evidence E may, qualitatively, either confirm, disconfirm, or be evidentially irrelevant to a hypothesis H, where in the quantitative sense, a numerical magnitude (which can be measured, "inductive probabilistically," in different ways; see below) is assigned to the "boost" (positive or negative, if any) that E gives H. Finally, confirmation can be considered to be either comparative or noncomparative. Noncomparative confirmation concerns just one hypothesis/evidence pair. In comparative confirmation, one can compare how well an E supports an H with how well an E' supports the same H; or one may compare how well an E supports an H with how well the same E supports an H'; or one may compare how well an E supports an H with how well an E' supports an H'.

The exposition below will be divided into two main parts. The first part, "Nonprobabilistic Approaches," will concern different aspects of qualitative confirmation; and the second part, "Probabilistic Approaches," will consider some major quantitative approaches. Almost exclusively, as in the literature, the issue will be incremental confirmation rather than absolute confirmation. Both noncomparative and various kinds of comparative approaches will be described.

Nonprobabilistic Approaches

A simple and natural idea about the confirmation of a general hypothesis of the form "All Fs are Gs" is that an object confirms the hypothesis if and only if it is both an F and a G (a "positive instance" of the hypothesis), disconfirms the hypothesis if and only if it is an F but not a G (a "negative instance"), and is evidentially irrelevant if and only if it is not even an F (no kind of instance). Hempel ([1945] 1965) calls this Nicod's criterion (Nicod 1930). Another natural idea about the confirmation of hypotheses is that if hypotheses H and H' are logically equivalent, then evidence E confirms, disconfirms, or is irrelevant to H if and only if E confirms, disconfirms, or is irrelevant to H', respectively. Hempel calls this the *equivalence condition*, and distinguishes between criteria (definitions or partial definitions) of confirmation and the conditions of adequacy that the criteria should satisfy. Hempel points out that Nicod's criterion does not satisfy the equivalence condition (as long as confirmation, disconfirmation, and evidential irrelevance are mutually exclusive). For example, a hypothesis "All Fs are Gs" is logically equivalent to "All non-Gs are non-Fs," but Nicod's criterion implies that an object that is an F and a G would confirm the former but be irrelevant to the latter, thus violating the equivalence condition. Also, "All Fs are Gs" is logically equivalent to "Anything that is both an F and a non-G is both an F and a non-F," which Nicod's criterion implies that nothing can confirm (a positive instance would have to be both an F and a non-F).

Thus, Hempel suggests weakening Nicod's criterion. The idea that negative instances disconfirm (i.e., are *sufficient* to disconfirm) is retained. Further, Hempel endorses the positive-instance criterion, according to which positive instances are *sufficient* for confirmation. Nicod's criterion can be thought of as containing six parts: necessary and sufficient conditions for all three of confirmation, disconfirmation, and irrelevance. The positive-instance criterion is said to be one-sixth of Nicod's criterion, and it does not lead to the kind of contradiction that Nicod's full criterion does when conjoined with the equivalence condition.

However, the combination of the positive-instance criterion and the equivalence condition (i.e., the proposition that the positive-instance criterion satisfies the equivalence condition) does lead to what Hempel called *paradoxes of confirmation*, also known as the Ravens paradox and Hempel's paradox. Hempel's famous example is the hypothesis H: "All ravens are black." Hypothesis H is logically equivalent to hypothesis H': "All nonblack things are nonravens." According to the equivalence condition, anything that confirms H' confirms H. According to the positive-instance criterion, nonblack nonravens (positive instances of H') confirm H'. Examples of nonblack nonravens (positive instances of H') include white shoes, yellow pencils, transparent tumblers, etc. So it follows from the positive-instance criterion plus the equivalence condition that objects of the kinds just listed confirm the hypothesis H that all ravens are black. These conclusions seem incorrect or counterintuitive, and the paradox is that the two seemingly plausible principles, the positive-instance criterion and the equivalence condition, lead, by valid reasoning, to these seemingly implausible conclusions. Further paradoxical consequences can be obtained by noting that the hypothesis H is logically equivalent also to H'', "All things that are either a raven or not a raven (i.e., all things) are either black or not a raven," which has as positive instances any objects that are black and any objects that are not ravens.

Since the equivalence condition is so plausible (if H and H' are logically equivalent, they can be thought of as simply different formulations of the same hypothesis), attention has focused on the positive-instance criterion. Hempel defended the criterion, arguing that the seeming paradoxicalness of the consequences of the criterion is more of a psychological illusion than a mark of a logical flaw in the criterion:

> In the seemingly paradoxical cases of confirmation, we are often not actually judging the relation of the given evidence E alone to the hypothesis H .... [I]nstead, we tacitly introduce a comparison of H with a body of evidence which consists of E in conjunction with additional information that we happen to have at our disposal. (Hempel [1945] 1965, 19)

So, for example, if one is just given the information that an object is nonblack and a nonraven (where it may happen to be a white shoe or a yellow pencil, but this is not included in the evidence), then the idea is that one should intuitively judge the evidence as confirmatory, "and the paradoxes vanish" (20). To assess properly the seemingly paradoxical cases for their significance for the logic of confirmation, one must observe the "*methodological* fiction" (as Hempel calls it) that one is in a position to judge the relation between the given evidence *alone* (e.g., that an object is a positive instance of the contrapositive of a universalized conditional) and the hypothesis in question and that there is no other information. This approach has been challenged by some who have argued that confirmation should be thought of as a relation among three things: evidence, hypothesis, and background knowledge (see the section below on Probabilistic Approaches).

Given the equivalence condition, the Ravens paradox considers the question of which of *several* kinds of evidence confirm(s) what one can consider to be a single hypothesis. There is another kind of paradox, or puzzle, that arises in a case of a single body of evidence and multiple hypotheses. In Nelson Goodman's (1965) well-known Grue paradox or puzzle, a 'new' predicate is defined as follows. Let's say that an object A is "grue" if and only if *either* (i) A has been observed before a certain time t (which could be now or some time in the future) and A is green *or* (ii) A has not been observed before that time t and A is blue. Consider the hypothesis H that all emeralds are green and the hypothesis H' that all emeralds are grue. And consider the evidence E to be the observation of a vast number of emeralds, all of which have been green. Given that t is now or some time in the future, E is equivalent to E', that the vast number of emeralds observed have all been grue. It is taken that E (the "same" as E') confirms H (this is natural enough) but not H'—for in order for H' to be true, exactly all of the unobserved (by t) emeralds would have to be blue, which would seem to be disconfirmed by the evidence.

Yet, the evidence E' (or E) consists of positive instances of H'.

Since the positive-instance criterion can be formulated purely syntactically—in terms of simply the logical forms of evidence sentences and hypotheses and the logical relation between their forms—a natural lesson of the Grue example is that confirmation cannot be characterized purely syntactically. (It should be noted that an important feature of Hempel's ([1945] 1965) project was the attempt to characterize confirmation purely syntactically, so that evidence E should, strictly speaking, be construed as evidence statements or sentences, or "observation reports," as he put it, rather than as observations or the objects of observation.) And a natural response to this has been to try to find nonsyntactical features of evidence and hypothesis that differentiate cases in which positive instances confirm and cases in which they do not. And a natural idea here is to distinguish between *predicates* that are "projectible" (in Goodman's terminology) and those that are not. Goodman suggested "entrenchment" of predicates as the mark of projectibility—where a predicate is entrenched to the extent to which it has been used in the past in hypotheses that have been successfully confirmed. Quine (1969) suggested drawing the distinction in terms of the idea of natural kinds. A completely different approach would be to point out that the reason why one thinks the observation of grue emeralds (E' or E) disconfirms the grue hypothesis (H') is because of background knowledge about constancy of color (in our usual concept of color) of many kinds of objects, and to argue that the evidence in this case should be taken as actually confirming the hypothesis H, given the Hempelian methodological fiction. It should be pointed out that the positive-instance criterion applies to a limited, though very important, kind of hypothesis and evidence: universalized conditionals for the hypothesis and positive instances for the evidence. And it supplies only a sufficient condition for confirmation. Hempel ([1945] 1965) generalized, in a natural way, this criterion to his satisfaction criterion, which applies to different and more complex logical structures for evidence and hypothesis and provides explicit definitions of confirmation, disconfirmation, and evidential irrelevance. Without going into any detail about this more general criterion, it is worth pointing out what Hempel took to be evidence for its adequacy. It is the satisfaction, by the satisfaction criterion, of what Hempel took to be some intuitively obvious conditions of adequacy for definitions, or *criteria*, of confirmation. Besides the equivalence condition, two others are the entailment condition and the special-consequence condition. The entailment condition says that evidence that logically entails a hypothesis should be deemed as confirming the hypothesis. The special-consequence condition says that if evidence E confirms hypothesis H and if H logically entails hypothesis H', then E confirms H'. This last condition will be considered further in the section below on probabilistic approaches.

The criteria for confirmation discussed above apply in cases in which evidence reports and hypotheses are stated in the "same language," which Hempel took to be an observational language. Statements of evidence, for example, are usually referred to as observational reports in Hempel ([1945] 1965). What about confirmation of *theories*, though, which are often thought of as containing two kinds of vocabulary, observational and theoretical? Hypothetico-deductivism (HD) is the idea that theories and hypotheses are confirmed by their observational deductive consequences. This is different from the positive-instance criterion and the satisfaction criterion. For example, "A is an F and A is a G," which is a report of a positive instance of the hypothesis that all Fs are Gs, is not a deductive consequence of the hypothesis. The positive-instance and satisfaction criteria are formulations of the idea, roughly, that observations that are *logically consistent with* a hypothesis confirm the hypothesis, while HD says that *deductive consequences of* a hypothesis or theory confirm the hypothesis or theory.

As an example, Edmund Halley in 1705 published his prediction that a comet, now known as Halley's comet, would be visible from Earth sometime in December of 1758; he deduced this using Newtonian theory. The prediction was successful, and the December 1758 observation of the comet was taken by scientists to provide (further) very significant confirmation of Newtonian theory. Of course, the prediction was not deduced from Newtonian theory alone. In general, other needed premises include statements of *initial conditions* (in the case of the example, reports of similar or related observations at approximately 75-year intervals) and *auxiliary assumptions* (that the comet would not explode before December 1758; that other bodies in the solar system would have only an insignificant effect on the path of the comet; and so on). In addition, when the theory and the observation report share no nonlogical vocabulary (say, the theory is highly theoretical, containing no observational terms), then so-called bridge principles are needed to establish a deductive connection between theory and observation. An example of such a principle would be, "If there is an excess of electrons [theoretical] on the surface of a balloon [observational], then a sheet of paper [observational] will cling [observational] to it, in normal circumstances [auxiliary assumption]." Of course, if the prediction fails (an observational deductive consequence of a theory turns out to be false), then this is supposed to provide disconfirmation of the theory.

Two of the main issues or difficulties that have been discussed in connection with the HD idea have to do with what might be called distribution of credit and distribution of blame. The first has also been called the *problem of irrelevant conjunction*. If a hypothesis H logically implies an observation report E, then so does the conjunction, $H \wedge G$, where G can be any sentence whatsoever. So the basic idea of HD has the consequence that whenever an E confirms an H, the E confirms also $H \wedge G$, where G can be any (even irrelevant) hypothesis whatsoever. This problem concerns the distribution of credit. A natural response would be to refine the basic HD idea in a way to make it sensitive to the possibility that logically weaker parts of a hypothesis may suffice to deductively imply the observation report. The second issue has to do with the possibility of the failure of the prediction, of the observational deductive consequence of the hypothesis turning out to be false. This is also known as Duhem's problem (Duhem 1914) (see Duhem Thesis). If a hypothesis H plus statements of initial conditions I plus auxiliary assumptions A plus bridge principles B logically imply an observation report E (($H \wedge I \wedge A \wedge B) \Rightarrow E$), and E turns out to be false, then what one can conclude is that the four-part conjunction $H \wedge I \wedge A \wedge B$ is false. And the problem is how in general to decide whether the evidence should be counted as telling against the hypothesis H, the statements of initial conditions I, the auxiliary assumptions A, or the bridge principles B.

Clark Glymour (1980) catalogues a number of issues relevant to the assessment of HD (and accounts of confirmation in general) and proposes an alternative deductivist approach to confirmation called "the bootstrap strategy," which attempts to clarify the idea of different parts of a theory and how evidence can bear differently on them. In Glymour's bootstrap account, confirmation is a relation among a theory T, a hypothesis H, and evidence expressed as a sentence E, and Glymour gives an intricate explication of the idea that "E confirms H with respect to T," an explication that is supposed to be sensitive especially to the idea that evidence can be differently confirmationally relevant to different hypotheses that are parts of a complex theory.

Probabilistic Approaches

One influential probabilistic approach to various issues in confirmation theory is called Bayesian confirmation theory (see Bayesianism). The basic idea, on which several refinements may be based, is that evidence E confirms hypothesis H if and only if the conditional probability $P(H|E)$ (defined as $P(H \wedge E)|P(E)$) is greater than the unconditional probability $P(H)$ and where $P(S)$ is the probability that the statement, or proposition, or claim, or assertion, or sentence S is true (the status of what kind of entity S might be is a concern in metaphysics or the philosophy of language, as well as in the philosophy of science). Disconfirmation is defined by reversing the inequality, and evidential irrelevance is defined by changing the inequality to an equality.

Typically in Bayesian confirmation theory, the function P is taken to be a measure of an agent's subjective probabilities—also called degrees of belief, partial beliefs, or degrees of confidence. Much philosophical work has been done in the area of foundations of subjective probability, the intent being to clarify or operationalize the idea that agents (e.g., scientists) have (or should have only) more or less strong or weak beliefs in propositions, rather than simply adopting the attitudes of acceptance or rejection of them. One approach, called the *Dutch book argument*, attempts to clarify the idea of subjective probability in terms of odds that one is willing to accept when betting for or against the truth or falsity of propositions (see Dutch Book Argument). Another approach, characterized as a decision theoretical approach, assumes various axioms regarding rational preference (transitivity, asymmetry, etc.) and some structural conditions (involving the richness of the set of propositions, or acts, states, and outcomes, considered by an agent) and derives, from preference data, via representation theorems, a probability assignment P and a desirability (or utility) assignment DES such that an agent prefers an item A to an item B if and only if some kind of expected utility of A is numerically greater than the expected utility of B, when the expected utilities are calculated in terms of the derived P and DES functions (see Decision Theory). Various formulas for expected utility have been proposed. (Important work in foundations of subjective probability include Ramsey 1931; de Finetti 1937; Savage [1954] 1972; Jeffrey [1965] 1983; and Joyce 1999.)

Where P measures an agent's subjective degrees of belief, $P(H|E)$ is supposed to be the agent's degree of belief in H on the assumption that E is true, or the degree of belief that the agent would have in H were the agent to learn that E is true. If a person's degree of belief in H would increase if E were learned, then it is natural to say that for this agent, E is positively evidentially relevant to H, even when E is not in fact learned. Of course, different people will have different subjective probabilities, or degrees of belief, even if the different people are equally rational, this being due to different bodies of background knowledge or beliefs possessed (albeit possibly equally justifiable or excusable, depending on one's experience), so that in this approach to confirmation theory, confirmation is a relation among three things: evidence, hypothesis, and background knowledge. The reason this approach is called Bayesian is because of the use that is sometimes made of a mathematical theorem discovered by Thomas Bayes (Bayes 1764), a simple version of which is $P(H|E) = P(E|H)P(H)/P(E)$. This is significant because it is sometimes easier to figure out the probability of an evidence statement conditional on a hypothesis than it is to figure out the probability of a hypothesis on the assumption that the evidence statement is true (for example, when the hypothesis is statistical and the evidence statement reports the outcome of an experiment to which the hypothesis applies). Bayes's theorem can be used to link these two converse conditional probabilities when the priors, $P(H)$ and $P(E)$, are known (see Bayesianism).

Bayesian confirmation theory not only provides a qualitative definition of confirmation, disconfirmation, and evidential irrelevance, but also suggests measures of degree of evidential support. The most common is the *difference measure*: $d(H,E) = P(H|E) - P(H)$, where confirmation, disconfirmation, and evidential irrelevance correspond to whether this measure is greater than, less than, or equal to 0, and the degree is measured by the magnitude of the difference. Another commonly used measure is the *ratio measure*: $r(H,E) = P(H|E)/P(H)$ where confirmation, disconfirmation, and evidential irrelevance correspond to whether this measure is greater than, less than, or equal to 1, and the degree is measured by the magnitude of the ratio. Other measures have been defined as functions of likelihoods, or converse conditional probabilities, $P(E|H)$. One application of the idea of degree of evidential support has been in the Ravens paradox, discussed above. Such definitions of degree of evidential support provide a framework within which one can clarify intuitions that under certain conditions (contrapositive instances of the hypothesis that all ravens are black [i.e., nonblack nonravens]) confirm the hypothesis that all ravens are black, but to a minuscule degree compared with positive instances (black ravens).

What follows is a little more detail about the application of Bayesian confirmation theory to the positive-instance criterion and the Ravens paradox. (See Eells 1982 for a discussion and references.) Let H be the hypothesis that all ravens are black; let R_A symbolize the statement that object A is a raven; and let B_A symbolize the statement that object A is black. It can be shown that if H is probabilistically independent of R_A, (i.e., $P(H|R_A) = P(H)$), then a positive instance (or report of one), $R_A \wedge B_A$, of H confirms H in the Bayesian sense (i.e., $P(H|R_A \wedge B_A) > P(H)$) if and only if $P(B_A|R_A) < 1$ (which latter inequality can naturally be interpreted as saying that it was not *already* certain that A would be black if a raven). Further, on the same independence assumption, it can be shown that a positive instance, $R_A \wedge B_A$, confirms H more than a contrapositive instance, $\neg B_A \wedge \neg R_A$ if and only if $P(B_A|R_A) < P(\neg R_A|\neg B_A)$.

What about the assumption of probabilistic independence of H from R_A? I. J. Good (1967) has proposed counterexamples to the positive-instance criterion like the following. Suppose it is believed that *either* (1) there are just a few ravens in the world and they are all black *or* (2) there are lots and lots of ravens in the world, a very, very few of which are nonblack. Observation of a raven, even a black one (hence a positive instance of H), would tend to support supposition (2) against supposition 1 and thus undermine H, so that a positive instance would disconfirm H. But in this case the independence assumption does not hold, so that Bayesian confirmation theory can help to isolate the kinds of situations in which the positive-instance criterion holds and the kinds in which it may not.

Bayesian confirmation theory can also be used to assess Hempel's proposed conditions of adequacy for criteria of confirmation. Recall, for example, his special-consequence condition, discussed above: If E confirms H and H logically entails H', then E must also confirm H'. It is a theorem of probability theory that if H logically entails H', then $P(H')$ is at least as great as $P(H)$. If the inequality is strict, then an E can increase the probability of H while decreasing the probability of H', consistent with H logically entailing H'. This fact can be used to construct intuitively compelling examples of an H entailing an H' and an E confirming the H while disconfirming the H', telling against the special-consequence condition and also in favor of Bayesian confirmation theory (e.g., Eells 1982).

Some standard objections to Bayesian confirmation theory are characterized as the *problem of the priors* and the *problem of old evidence*. As to the first, while it is sometimes admitted that it makes sense to assign probabilities to evidence statements E conditional on some hypothesis H (even in the absence of much background knowledge), it is objected that it often does not make sense to assign unconditional, or "prior," probabilities to hypothesis H or to evidence statements (reports of observation) E. If H is a newly formulated physical hypothesis, for example, it is hard to imagine what would *justify* an assignment of probability to it prior to evidence—but that is just what the suggested criterion of confirmation, Bayes's theorem, and the measures of confirmation described above require. Such issues make some favor a likelihood approach to the evaluation of evidence—Edwards (1972) and Royall (1997), for instance, who represent a different approach and tradition in the area of statistical inference. According to one formulation of the likelihood account, an E confirms an H more than the E confirms an H' if and only if $P(E|H)$ is greater than $P(E|H')$. This is a comparative principle, an approach that separates the question of which hypothesis it is more justified to believe given the evidence (or the comparative acceptability of hypotheses given the evidence) from the question of what the comparative significance is of evidence for one hypothesis compared with the evidence's significance for another hypothesis. It is the latter question that the likelihood approach actually addresses, and it is sometimes suggested that the degree to which an E supports an H compared with the support of E for an H' is measured by the likelihood ratio, $P(E|H)/P(E|H')$. Also, likelihood measures of *degree* of confirmation of *single* hypotheses have been proposed, such as $L(H, E) = P(E|H)/P(E|\neg H)$, or the log of this ratio. (see Fitelson 2001 and Forster and Sober 2002 for recent discussion and references.)

Another possible response to the problem of priors is to point to convergence theorems (as in de Finetti 1937) and argue that initial settings of priors does not matter in the long run. According to such theorems, if a number of agents set different priors, are exposed to the same series of evidence, and update their subjective probabilities (degrees of belief) in certain ways, then, almost certainly, their subjective probabilities will eventually converge on each other and, under certain circumstances, upon the truth.

The problem of old evidence (Glymour 1980; Good 1968, 1985) arises in cases in which $P(E) = 1$. It is a theorem of probability theory that in such cases $P(H|E) = P(H)$, for any hypothesis H, so that in the Bayesian conception of confirmation as formulated above, an E with probability 1 cannot confirm any hypothesis H. But this seems to run against intuition in some cases. An often-cited such case is the confirmation that Albert Einstein's general theory of relativity apparently was informed by already known facts about the behavior of the perihelion of the planet Mercury. One possible Bayesian solution to the problem, suggested by Glymour (1980), would be to say that it is not the already known E that confirms the H after all, but rather a newly discovered logical or explanatory relation between the H and the E. Other solutions have been proposed, various versions of the problem have been distinguished (see Earman 1992 for a discussion), and the problem remains one of lively debate.

ELLERY EELLS

References

Bayes, Thomas (1764), "An Essay Towards Solving a Problem in the Doctrine of Chance," *Philosophical Transactions of the Royal Society of London* 53: 370–418.

de Finetti, Bruno (1937), "La prévision: Ses lois logiques, ses sources subjectives," *Annales de l'Institute Henri Poincaré* 7: 1–68.

Duhem, Pierre (1914), *The Aim and Structure of Physical Theory*. Princeton, NJ: Princeton University Press.

Earman, John (1992), *Bayes or Bust: A Critical Examination of Bayesian Confirmation Theory*. Cambridge, MA, and London: MIT Press.

Edwards, A. W. F. (1972), *Likelihood*. Cambridge: Cambridge University Press.

Eells, Ellery (1982), *Rational Decision and Causality*. Cambridge and New York: Cambridge University Press.

Fitelson, Branden (2001), *Studies in Bayesian Confirmation Theory*. Ph.D. dissertation, University of Wisconsin–Madison.

Forster, Malcolm, and Elliott Sober (2002), "Why Likelihood?" in M. Taper and S. Lee (eds.), *The Nature of Scientific Evidence*. Chicago: University of Chicago Press.

Glymour, Clark (1980), *Theory and Evidence*. Princeton, NJ: Princeton University Press.

Good, I. J. (1967), "The White Shoe Is a Red Herring," *The British Journal for the Philosophy of Science* 17: 322.

——— (1968), "Corroboration, Explanation, Evolving Probability, Simplicity, and a Sharpened Razor," *British Journal for the Philosophy of Science* 19: 123–143.

——— (1985), "A Historical Comment Concerning Novel Confirmation," *British Journal for the Philosophy of Science* 36: 184–186.

Goodman, Nelson (1965), *Fact, Fiction, and Forecast*. New York: Bobbs-Merrill.

Hempel, Carl G. ([1945] 1965), "Studies in the Logic of Confirmation," in *Aspects of Scientific Explanation and Other Essays in the Philosophy of Science*. New York: Free Press, 3–51. Originally published in *Mind* 54: 1–26 and 97–121.

Jeffrey, Richard C. ([1965] 1983), *The Logic of Decision*. Chicago and London: University of Chicago Press.

Joyce, James M. (1999), *The Foundations of Causal Decision Theory*. Cambridge and New York: Cambridge University Press.

Nicod, Jean (1930), *Foundations of Geometry and Induction*. New York and London: P. P. Wiener.

Quine, Willard Van (1969), "Natural Kinds," in *Ontological Relativity and Other Essays*. New York: Columbia University Press.

Ramsey, Frank Plumpton (1931), "Truth and Probability," in R. B. Braithwaite (ed.), *The Foundations of Mathematics and Other Logical Essays*. London: Routledge and Kegan Paul, 156–198.

Royall, Richard (1997), *Statistical Evidence: A Likelihood Paradigm*. Boca Raton, FL: Chapman and Hall.

Savage, Leonard ([1954] 1972), *The Foundations of Statistics*. New York: Dover Publications.

See also Bayesianism; Carnap, Rudolf; Decision Theory; Dutch Book Argument; Epistemology; Hempel, Carl Gustav; Induction, Problem of; Inductive Logic; Probability

CONNECTIONISM

Connectionist models, also known as models of parallel distributed processing (PDP) and artificial neural networks (ANN), have merged into the mainstream of cognitive science since the mid-1980s. Connectionism currently represents one of two dominant approaches (symbolic modeling is the other) within artificial intelligence used to develop computational models of mental processes.

Unlike symbolic modeling, connectionist modeling also figures prominently in computational neuroscience (where the preferred term is *neural network modeling*).

Connectionist modeling first emerged in the period 1940–1960, as researchers explored possible approaches to using networks of simple neurons to perform psychological tasks, but it fell into decline when limitations of early network designs became apparent around 1970. With the publication of the PDP Research Group volumes (Rumelhart, McClelland, et al. 1986; McClelland, Rumelhart, et al. 1986), connectionism was rescued from over a decade of popular neglect, and the way was opened for a new generation to extend the approach to fresh explanatory domains. (For a collection of historically significant papers from these neglected years and before, see Anderson and Rosenfeld 1988; Anderson, Pellionisz, and Rosenfeld 1990.) Despite some early claims that connectionism constituted a new, perhaps revolutionary, way of understanding cognition, the veritable flood of network-based research has ultimately occurred side by side with other, more traditional, styles of modeling and theoretical frameworks.

The renaissance in connectionist modeling is a result of convergence from many different fields. Mathematicians and computer scientists attempt to describe the formal, mathematical properties of abstract network architectures. Psychologists and neuroscientists use networks to model behavioral, cognitive, and biological phenomena. Roboticists also make networks the control systems for many kinds of embodied artificial agents. Finally, engineers employ connectionist systems in many industrial and commercial applications. Research has thus been driven by a broad spectrum of concerns, ranging from the purely theoretical to problem-solving applications for problems in various scientific domains to application-based or engineering needs.

Given these heterogeneous motivations, and the recent proliferation of network models, analytic techniques, applications, and theories, it is appropriate to ask whether connectionism constitutes a coherent research program or is instead primarily a modeling tool. Following Lakatos, connectionism could be construed as a research program involving a set of core theoretical principles about the role of networks in explaining and understanding cognition, a set of positive and negative heuristics that guide research, an ordering of the important commitments of connectionist modeling, and a set of principles and strategies dictating how recalcitrant empirical results are to be accounted for (see Lakatos, Imre; Research Programs). The greater the disunity in these factors, the less connectionism resembles a research program, and the more it appears to be a convenient tool for modeling certain phenomena. If it is a modeling tool, connectionism need not commit modelers to having anything in common beyond their use of the particular mathematical and formal apparatus itself.

This article will briefly describe the features of prevalent connectionist architectures and discuss a number of challenges to the use of these models. One challenge comes from symbolic models of cognition, which present an alternative representational framework and set of processing assumptions. Another comes from a purportedly nonrepresentational framework, that of nonlinear dynamical systems theory. Finally, there is the neuroscientific challenge to the disciplinary boundaries drawn around "cognitive" modeling by some connectionist psychologists. The status of connectionism is assessed in light of these challenges.

The Properties of Connectionist Models

Connectionist networks are built up from basic computational elements called units or nodes, which are linked to each other via weighted connections, called simply weights. Units take on a variable numerical level of activation, and they pass activation to each other via the weights. Weights determine how great an effect one unit has on other units. This effect may be positive (excitatory) or negative (inhibitory). The net input a unit receives at a time is the weighted sum of the activations on all of the units that are active and connected to it. Given the net input, an activation function (often nonlinear or imposing a threshold) determines the activation of the unit. In this way, activation is passed in parallel throughout the network. Connectionist networks compute functions by mapping vectors of activation values onto other such vectors.

Multilayer feedforward networks are the most intensively studied and widely used class of contemporary models. Units are arranged into layers, beginning with an input layer and passing through a number of intermediate hidden layers, terminating with an output layer. There are no reciprocal connections, so activation flows unidirectionally through the network. The modeler assigns representational significance to the activation vectors at the input and output layers of a network, thereby forging the link between the model and the cognitive task to be explained.

For example, Sejnowski and Rosenberg's NETtalk architecture is designed to map graphemic inputs (letters) onto phonemic outputs. It consists of an input layer of seven groups of 29 units, a hidden layer of 80 units, and an output layer of 26 units. Each layer is completely connected to the next one. Vectors of activity in the network represent aspects of the letter-reading task. The input groups are used to represent the seven letters of text that the network is perceiving at a time, and the output layer represents the phoneme corresponding to the fourth letter in the input string. The task of the network is to pronounce the text correctly, given the relevant context. When the values of the weights are set correctly, the network can produce the appropriate phoneme-representing vectors in response to the text-representing vectors.

Simple networks can be wired by hand to compute some functions, but in networks containing hundreds of units, this is impossible. Connectionist systems are therefore usually not programmed in the traditional sense, but are trained by fixing a learning rule and repeatedly exposing the network to a subset of the input-output mappings it is intended to learn. The rule then systematically adjusts the network's weights until its outputs are near the target output values. One way to classify learning rules is according to whether they require an external trainer. Unsupervised learning (e.g., Hebbian learning) does not require an external trainer or source of error signals. Supervised learning, on the other hand, requires that something outside the network indicate when its performance is incorrect. The most popular supervised learning rule currently in use is the backpropagation rule.

In backpropagation learning, the network's weights are initially set to random values (within certain bounds). The network is then presented with patterns from the training environment. Given random weights, the network's response will likely be far from the intended mapping. The difference between the output and the target is computed by the external trainer and used to send an error signal backward through the network. As the signal propagates, the weights between each layer are adjusted by a slight amount. Over many training cycles, the network's performance gradually approaches the target. When the output is within some criterion distance of the target, training ceases. Since the error is being reduced gradually, backpropagation is an instance of a gradient-descent learning algorithm.

Backpropagation-trained networks have been successful at performing in many domains, including past-tense transformation of verbs, generation of prototypes from exemplars, single word reading, shape-from-shading extraction, visual object recognition, modeling deficits arising in deep dyslexia, and more. Their formal properties are well known. However, they suffer from a number of problems. Among these is the fact that learning via backpropagation is extremely slow, and increasing the learning-rate parameter typically results in overshooting the optimum weights for solving the task.

Another problem facing feedforward networks generally is that individual episodes of processing inputs are independent of each other except for changes in weights resulting from learning. But often a cognitive agent is sensitive not just to what it has learned over many episodes, but to what it processed recently (e.g., the words prior to the preceding one). The primary way sensitivity to context has been achieved in feedforward networks has been to present a constantly moving window of input. For example, in NETtalk, the input specified the three phonemes before and three phonemes after the one to be pronounced. But this solution is clearly a kludge and suffers from the fact that it imposes a fixed window. If sensitivity to the item four back is critical to correct performance, the network cannot perform correctly.

An alternative architecture that is increasingly being explored is the simple recurrent network (SRN) (Elman 1991). SRNs have both feedforward and recurrent connections. In the standard model, an input layer sends activity to a hidden layer, which has two sets of outgoing connections: one to other hidden layers and eventually on to the output layer, and another to a specialized context layer. The weights to the units in the context layer enable it to construct a copy of the activity in the hidden units. The activation over these units is then treated as an additional input to the same hidden units at the next temporal stage of processing. This allows for a limited form of short-term memory, since activity patterns that were present during the previous processing cycle have an effect on the next cycle. Since the activity on the previous cycle was itself influenced by that on a yet earlier cycle, this allows for memory extending over several previous processing epochs (although sensitivity to more than one cycle back will be diminished).

Once trained in a variation of backpropagation, many SRNs are able to discover patterns in temporally ordered sets of events. Elman (1991) has trained SRNs on serially presented words in an attempt to teach them to predict the grammatical category of the next word in a sentence. The networks can achieve fairly good performance at this task. Since the networks were never supplied information about grammatical categories, this suggests that they induced a representation of a more abstract similarity among words than was present in the raw training data. There are many other kinds of neural network architecture. (For further details on their properties and applications, see Anderson 1995; Bechtel and Abrahamsen 2002.)

Connectionism and Symbolic Models

Within cognitive science, symbolic models of cognition have constituted the traditional alternative to connectionism. In symbolic models, the basic representational units are symbols having both syntactic form and typically an intuitive semantics that corresponds to the elements picked out by words of natural language. The symbols are discrete and capable of combining to form complex symbols that have internal syntactic structure. Like the symbol strings used in formal logic, these complex symbols exhibit variable binding and scope, function-and-argument structure, cross-reference, and so on. The semantics for complex symbols is combinatorial: The meaning of a complex symbol is determined by the meanings of its parts plus its syntax. Finally, in symbolic models the dynamics of the system are governed by rules that transform symbols into other symbols by responding to their syntactic or formal properties. These rules are intended to preserve the truth of the structures manipulated. Symbolic models are essentially proof-theoretic engines.

Connectionist models typically contain units that do not individually represent lexicalized semantic contents. (What are called *localist* networks are an exception. In these, individual units are interpretable as expressing everyday properties or propositions. See Page 2000.) More commonly, representations with lexicalized content are *distributed* over a number of units in a network (Smolensky 1988). In a distributed scheme, individual units may stand for repeatable but nonlexicalized microfeatures of familiar objects, which are themselves represented by vectors of such features. In networks of significant complexity, it may be difficult, if not impossible, to discern what content a particular unit is carrying.

In networks, there is no clear analog to the symbolicist's syntactic structures. Units acquire and transmit activation values, resulting in larger patterns of coactivation, but these patterns of units do not themselves syntactically compose. Also, there is no clear program/data distinction in connectionist systems. Whether a network is hand-wired or trained using a learning rule, the modifications are changes to the weights between units. The new weight settings determine the future course of activation in the network and simultaneously constitute the data stored in the network. There are no explicitly represented rules that govern the system's dynamics.

Classical theorists (Fodor and Pylyshyn 1988) have claimed that there are properties of cognition that are captured naturally in symbolic models but that connectionist models can capture them only in an ad hoc manner, if at all. Among these properties are the productivity and the systematicity of thought. Like natural language, thought is productive, in that a person can think a potentially infinite number of thoughts. For example, one can think that Walt is an idiot, that Sandra believes that Walt is an idiot, that Max wonders whether Sandra believes that Walt is an idiot, and so on. Thought is also systematic, in that a person who can entertain a thought can also entertain many other thoughts that are semantically related to it (Cummins 1996). If a person can think that Rex admires the butler's courage, that person can also think that the butler admires Rex's courage. Anyone who can think that dogs fear cats can think that cats fear dogs, and so on. Unless each thought is to be learned anew, these capacities need some finite basis.

Symbolicists argue that this basis is compositionality: Thought possesses a combinatorial syntax and semantics, according to which complex thoughts are built up from their constituent concepts, and those concepts completely determine the meaning of a complex thought. The compositionality of thought would explain both productivity and systematicity. Grasping the meaning of a set of primitive concepts and grasping the recursive rules by which they can be combined is sufficient for grasping the infinite number of thoughts that the concepts and rules can generate. Similarly, grasping a syntactic schema and a set of constituent concepts explains why the ability to entertain one thought necessitates the ability to entertain others: Concepts may be substituted into any permissible role slot in the schema.

The challenge symbolicists have put to connectionists is to explain productivity and systematicity in a principled fashion, without merely implementing a symbolic architecture on top of the connectionist network (for one such implementation, see Franklin and Garzon 1990). One option that some connectionists pursue is to deny that thought is productive or systematic, as symbolicists characterize these properties. For example, one might deny that it is possible to think any systematically structured proposition. Although one can compose the symbol string "The blackberry ate the bear," that does not entail that one is able to think such a thought. But clearly much of thought exhibits some degree of productivity and systematicity, and it is incumbent on connectionists to offer some account of how it is achieved.

Connectionists have advanced a number of proposals for explaining productivity and systematicity. Two of the most widely discussed are Pollack's recursive auto-associative memories (RAAMs) and Smolensky's tensor product networks (Pollack 1990; Smolensky, 1991). RAAMs are easier to understand. An auto-associative network is one trained to produce the same pattern on the output layer as is present on the input layer. If the hidden layer is smaller than the input and output layers, then the pattern produced on the hidden layer is a compressed representation of the input pattern. If the input layer contains three times the number of units as the hidden layer, then one can treat the input activation as consisting of three parts constituting three patterns (e.g., for different words) and recursively compose the hidden pattern produced by three patterns with two new ones. In such an encoding, one is implicitly ignoring the output units, but one can also ignore the input units and supply patterns to the hidden units, allowing the RAAM to generate a pattern on the output units. What is interesting is that even after several cycles of compression, one can, by recursively copying the pattern on one-third of the output units back onto the hidden units, re-create with a fair degree of accuracy the original input patterns.

If RAAMs are required to perform many cycles of recursive encoding, the regeneration of the original pattern may exhibit errors. But up to this point, the RAAM has exhibited a degree of productivity by composing representations of complex structures from representations of their parts. One can also use the compressed patterns in other processing (e.g., to train another feedforward network to construct the compressed representation of a passive sentence from the corresponding active sentence). RAAMs thus exhibit a degree of systematicity. But the compressed representations are not composed according to syntactic principles. Van Gelder (1990), accordingly, construes them as manifesting functional, not explicit, compositionality.

Symbolicists have rejected such functional compositionality as inadequate for explaining cognition. In many respects, this debate has reached a standoff. In part its resolution will turn on the issue posed earlier: To what degree do humans exhibit productivity and systematicity? But independently of that issue, there are serious problems in scaling up from connectionist networks designed to handle toy problems to ones capable of handling the sort of problems humans deal with regularly (e.g., communicating in natural languages). Thus, it is not clear whether solutions similar to those employed in RAAMs (or in SRNs, which also exhibit a degree of productivity and systematicity) will account for human performance. (Symbolic models have their own problems in scaling, and so are not significantly better off in practice.)

Connectionism and Dynamical Systems Theory

Although the conflict between connectionists and symbolicists reached a stalemate in the 1990s, within the broader cognitive science community a kind of accord was achieved. Connectionist approaches were added to symbolic approaches as parts of the modeling toolkit. For some tasks, connectionist models proved to be more useful tools than symbolic models, while for others symbolic models continued to be preferred. For yet other tasks, connectionist models were integrated with symbolic models into hybrids.

As this was happening, a new competitor emerged on the scene, an approach to cognition that challenged both symbolic and connectionist modeling insofar as both took seriously that cognitive activity involved the use of some form of representation. Dynamical systems theory suggested that rather than construing cognition as involving syntactic manipulation of representations or processing them through layers of a network, one should reject the notion of representation altogether. The alternative that these critics advanced was to characterize cognitive activity in terms of a (typically small) set of variables and to formulate (typically nonlinear) equations that would relate the values of different variables in terms of how they changed over time. In physics, dynamical systems theory provides a set of tools for understanding the changes over time of such systems of variables. For example, each variable can be construed as defining a dimension in a multidimensional *state space*, and many systems, although starting at different points in this space, will settle onto a fixed point or into a cycle of points. These points are known as *attractors*, and the paths to them as the *transients*. The structure of the state space can often be represented geometrically—for example, by showing different basins, each of which represents starting states that will end up in a different attractor.

Connectionist networks, especially those employing recurrent connections, are dynamical systems, and some connectionist modelers have embraced the tools of dynamical systems theory for describing their networks. Elman (1991), for example, employs such tools to understand how networks manage to learn to respect syntactic categories in processing streams of words. Others have made use of some of the more exotic elements in dynamical systems theory, such as activation functions that exhibit deterministic chaos, to develop new classes of networks that exhibit more complex and interesting patterns of behavior than simpler networks (e.g., jumping intermittently between two competing interpretations of an ambiguous perceptual figure such as the duck-rabbit (see van Leeuwen, Verver, and Brinkers 2000)). Some particularly extreme dynamicists, however, contend that once one has characterized a cognitive system in this way, there is no further point to identifying internal states as representations and characterizing changes as operations on these representations. Accordingly, they construe dynamical systems theory as a truly radical paradigm shift for cognitive science (van Gelder 1995).

This challenge has been bolstered by some notable empirical successes in dynamical systems modeling of complex cognition and behavior. For example, Busemeyer and Townsend (1993) offer a model of decision under uncertainty, decision field theory, that captures much of the empirical data about the temporal course of decision making using difference equations containing only seven parameters for psychological quantities such as attention weight, valence, preference, and so on. Connectionist models, by contrast, have activation state spaces of as many dimensions as they have units. There are dynamical systems models of many other phenomena, including coordination of finger movement, olfactory perception, infants' stepping behavior, the control of autonomous robot agents, and simple language processing. The relative simplicity and comprehensibility of their models motivates the antirepresentational claims advanced by dynamical systems theorists.

There is reason, though, to question dynamical systems theory's more radical challenge to cognitive modeling. One can accept the utility of characterizing a system in terms of a set of equations and portraying its transitions through a multidimensional state space without rejecting the utility of construing states within the system as representational and the trajectories through state space in terms of transitions between representations. This is particularly true when the system is carrying out what one ordinarily thinks of as a complex reasoning task such as playing chess, where the task is defined in terms of goals and the cognizer can be construed as considering different possible moves and their consequences. To understand why a certain system is able to play chess successfully, rather than just recognizing that it does, the common strategy is to treat some of its internal states as representing goals or possible moves. Moreover, insofar as these internal states are causally connected in appropriate ways, they do in fact carry information (in what is fundamentally an informational-theoretic sense, in which the state covaries with referents external to the system), which is then utilized by other parts of the system. In systems designed to have them, these information-carrying states may arise without their normal cause (as when a frog is subjected to a laboratory with bullets on a string moving in front of its eyes). In these situations the system responds as it would if the state were generated by the cause for which the response was designed or selected. Such internal states satisfy a common understanding of what a representation is, and the ability to understand why the system works successfully appeals to these representations. If this construal is correct, then dynamical systems theory is not an alternative to connectionist modeling, but should be construed as an extension of the connectionist toolkit (Bechtel and Abrahamsen 2002).

Connectionism and Neuroscience

Connectionist models are often described as being *neurally inspired*, as the term "artificial neural networks" implies. More strongly, many connectionists have claimed that their models enjoy a special sort of *neural plausibility*. If there is a significant similarity between the processing in ANNs and the activity in real networks of neurons, this might support connectionism for two reasons. First, since the cognitive description of the system closely resembles the neurobiological description, it seems that connectionist models in psychology have a more obvious account about how the mind might supervene on the brain than do symbolic models (see Supervenience). Second, connectionist models might provide a fairly direct characterization of the functioning of the neural level itself (e.g., the particular causal interactions among neurons). This functioning is not easily revealed by many standard neuroscience methods. For example, localization of mental activity via functional magnetic resonance imaging can indicate which brain regions are preferentially activated during certain tasks, but this alone does not give information about the specific computations being carried out within those regions. Connectionist modeling of brain function thus might supplement other neurobiological techniques.

This strategy can be illustrated within the domain of learning and memory. Neuropsychological studies suggest that the destruction of structures in the medial temporal lobes of the brain, especially the hippocampus, results in a characteristic pattern of memory deficits. These include (i) profound anterograde amnesia for information presented in declarative or verbal form, such as arbitrary paired associates, as well as memory for particular experienced events more generally (episodic memory); (ii) preserved implicit memory for new information, such as gradually acquired perceptual-motor skills; and (iii) retrograde amnesia for recently acquired information, with relative preservation of memories farther back in time. This triad of deficits was famously manifested by H.M., a patient who underwent bilateral removal of sections of his medial temporal lobes in the early 1950s in order to cure intractable epilepsy. Since H.M., this pattern of deficits has been confirmed in other human and animal studies.

McClelland, McNaughton, and O'Reilly (1995) offer an explanation of these deficits based on connectionist principles. One feature of backpropagation-trained networks is that once they are trained on one mapping, they cannot learn another mapping without "unlearning" the previously stored knowledge. This phenomenon is known as catastrophic interference. However, catastrophic interference can be overcome if, rather than fully training a network on one mapping, then training it on another, the training sets are interleaved so that the network is exposed to both mappings in alternation. When the training environment is manipulated in this way, the network can learn both mappings without overwriting either one.

As a model of all learning and memory, this technique suffers from being slow and reliant on a fortunate arrangement of environmental contingencies. However, McClelland et al. (1995) suggest that learning in the neocortex may be characterized by just such a process, if there is a neural mechanism that stores, organizes, and presents appropriately interleaved stimuli to it. They conjecture that this is the computational function of the hippocampus. Anatomically, the hippocampus receives convergent inputs from many sensory centers and has wide-ranging efferent connections to the neocortex. The pattern of deficits resulting from hippocampal lesions could be explained on the assumption that the hippocampus has a method of temporarily storing associations among stimuli without catastrophic interference. Ablation of the hippocampus results in anterograde amnesia for arbitrary associations and declarative information because the neocortex alone is incapable of learning these without the appropriately interleaved presentation. Implicit learning is preserved because it does not require rapid integration of many disparate representations into a single remembered experience; further, it typically takes many trials for mastery, as is also the case with backpropagation learning. Finally, the temporally graded retrograde amnesia is explained by the elimination of memory traces that are temporarily stored in the hippocampus itself. Older memories have already been integrated into the neocortex, and hence are preserved.

McClelland et al. (1995) did not model the hippocampus directly; rather, they implemented it as a black box that trained the neocortical network according to the interleaving regimen. Others have since presented more elaborate models. Murre, Graham, and Hodges (2001) have implemented a system called TraceLink that features a network corresponding to a simplified single-layer hippocampus, the neocortex, and a network of neuromodulatory systems (intended to correspond to the basal forebrain nuclei). TraceLink accounts for the data reviewed by McClelland et al. (1995), and also predicts several phenomena associated with semantic dementia. Other models incorporating a more elaborate multilayer hippocampal network have been presented by Rolls and Treves (1998) and O'Reilly and Rudy (2001). These models collectively support the general framework set out by McClelland et al. (1995) concerning the computational division of labor between the hippocampus and the neocortex in learning and memory.

These studies of complementary learning systems suggest a useful role for network-based modeling in neuroscience. However, this role is presently limited in several crucial respects. The models currently being offered are highly impoverished compared with the actual complexity of the relevant neurobiological structures. Assuming that these networks are intended to capture neuron-level interactions, they are several orders of magnitude short of the number of neurons and connections in the brain. Further, the backpropagation rule itself is biologically implausible if interpreted at the neuronal level, since it allows individual weights to take on either positive or negative values, while actual axonal connections are either excitatory or inhibitory, but never both. There is no network model that captures all of the known causal properties of neurons, even when the presence of glial cells, endocrine regulators of neural function, and other factors are abstracted away.

A common response to these objections is to interpret networks as describing only select patterns of causal activity among large populations of neurons. In many cases, this interpretation is appropriate. However, there are many kinds of network models in neuroscience, and they can be interpreted as applying to many different levels of organization within the nervous system, including individual synaptic junctions on dendrites, particular neurons within cortical columns, and interactions at the level of whole neural systems. No single interpretation appears to have any special priority over the others. The specific details of the network architecture are dictated in most cases by the particular level of neural analysis being pursued, and theorists investigate multiple levels simultaneously. There are likely to be at least as many distinct kinds of possible connectionist models in neuroscience as there are distinct levels of generalization within the nervous system.

Conclusions

At the beginning of this article, the question was asked whether connectionism is best thought of as a research program, a modeling tool, or something in between. Considering the many uses to which networks have been put, and the many disciplines that have been involved in cataloguing their properties, it seems unlikely that there will be any common unity of methods, heuristics, principles, etc., among them. Ask whether a neuroscientist using networks to model the development of receptive fields in the somatosensory system would have anything in common with a programmer training a network to take customers' airline reservations. Each of these might be a neural network theorist, despite having nothing significant in common besides the formal apparatus they employ. Across disciplines, then, connectionism lacks the characteristic unity one would expect from a research program.

The question may be asked again at the level of each individual discipline. This article has not surveyed every field in which networks have played a significant role but has focused on their uses in artificial intelligence, psychology, and neuroscience. Even within these fields, the characteristic unity of a research program is also largely absent, if one takes into account the diverse uses that are made of networks.

In psychology, for instance, there are some connectionists who conceive of their models as providing a theory of how mental structures might functionally resemble, and therefore plausibly supervene on, the organization of large-scale neuronal structures (see Psychology). However, there are just as many theorists who see their work as being only, in some quite loose sense, neurally inspired. The organization of the NETtalk network is not particularly neurally plausible, since it posits a simple three-layered linear causal process leading from the perception of letters to the utterance of phonemes. Being a connectionist in psychology does not appear to require agreement on the purpose of the models used or the possible data (e.g., neurobiological) that might confirm or disconfirm them. This is what one might expect of a tool rather than a research program.

It is perhaps ironic that this state of affairs was predicted by Rumelhart, one of the theorists who revitalized connectionism during the 1980s. In a 1993 interview, Rumelhart claimed that as networks become more widely used in a number of disciplines, "there will be less and less of a core remaining for neural networks per se and more of, 'Here's a person doing good work in [her] field, and [she's] using neural networks as a tool'" (Anderson and Rosenfeld 1998, 290). Within the fields, in turn, network modeling will "[d]isappear as an identifiable separate thing" and become "part of doing science or doing engineering" (291). Such a disappearance, however, may not be harmful. Connectionist networks, like other tools for scientific inquiry, are to be evaluated by the quality of the results they produce. In this respect, they have clearly proven themselves a worthy, and sometimes indispensible, component of research in an impressive variety of disciplines.

DAN WIESKOPF
WILLIAM BECHTEL

References

Anderson, James A. (1995), *An Introduction to Neural Networks*. Cambridge, MA: MIT Press.
Anderson, James A., and Edward Rosenfeld (eds.) (1988), *Neurocomputing: Foundations of Research*, vol. 1. Cambridge, MA: MIT Press.
——— (1998), *Talking Nets: An Oral History of Neural Networks*. Cambridge, MA: MIT Press.
Anderson, James A., Andras Pellionisz, and Edward Rosenfeld (eds.) (1990), *Neurocomputing: Directions for Research*, vol. 2. Cambridge, MA: MIT Press.
Bechtel, William, and Adele Abrahamsen (2002), *Connectionism and the Mind: Parallel Processing, Dynamics, and Evolution in Networks*. Oxford: Basil Blackwell.
Busemeyer, Jerome R., and James T. Townsend (1993), "Decision Field Theory: A Dynamic-Cognitive

CONNECTIONISM

Approach to Decision Making in an Uncertain Environment," *Psychological Review* 100: 423–459.

Cummins, Robert (1996), "Systematicity," *Journal of Philosophy* 93: 591–614.

Elman, Jeffrey L. (1991), "Distributed Representations, Simple Recurrent Networks, and Grammatical Structure," *Machine Learning* 7: 195–225.

Fodor, Jerry A., and Zenon Pylyshyn (1988), "Connectionism and Cognitive Architecture: A Critical Analysis," *Cognition* 28: 3–71.

Franklin, Stan, and Max Garzon (1990), "Neural Computability," in O. M. Omidvar (ed.), *Progress in Neural Networks*, vol. 1. Norwood, NJ: Ablex, 127–145.

McClelland, James L., Bruce L. McNaughton, and Randall C. O'Reilly (1995), "Why There Are Complementary Learning Systems in the Hippocampus and Neocortex: Insights from the Successes and Failures of Connectionist Models of Learning and Memory," *Psychological Review* 102: 419–457.

McClelland, James L., David E. Rumelhart, and the PDP Research Group (1986), *Parallel Distributed Processing: Explorations in the Microstructure of Cognition*, vol. 2. Cambridge, MA: MIT Press.

Murre, Jacob, Kim S. Graham, and John R. Hodges (2001), "Semantic Dementia: Relevance to Connectionist Models of Long-Term Memory," *Brain* 124: 647–675.

O'Reilly, Randall C., and J. W. Rudy (2001), "Conjunctive Representations in Learning and Memory: Principles of Cortical and Hippocampal Function," *Psychological Review* 108: 311–345.

Page, M. (2000), "Connectionist Modeling in Psychology: A Localist Manifesto," *Behavioral and Brain Sciences* 23: 443–512.

Pollack, Jordan B (1990), "Recursive Distributed Representations," *Artificial Intelligence* 46: 77–105.

Rolls, Edmund T., and Alessandro Treves (1998), *Neural Networks and Brain Function*. Oxford: Oxford University Press.

Rumelhart, David E., James L. McClelland, and the PDP Research Group (1986), *Parallel Distributed Processing: Explorations in the Microstructure of Cognition*, vol. 1. Cambridge, MA: MIT Press.

Smolensky, Paul (1988), "On the Proper Treatment of Connectionism," *Behavioral and Brain Sciences* 11: 1–74.

——— (1991), "Connectionism, Constituency, and the Language of Thought," in Barry Loewer and Georges Rey (eds.), *Meaning in Mind: Fodor and His Critics*. Oxford: Basil Blackwell.

van Gelder, Timothy (1990), "Compositionality: A Connectionist Variation on a Classical Theme," *Cognitive Science* 14: 355–384.

——— (1995), "What Might Cognition Be, If Not Computation?" *Journal of Philosophy* 111: 345–381.

van Leeuwen, Cees, S. Verver, and M. Brinkers (2000), "Visual Illusions, Solid/Outline Invariance and Non-Stationary Activity Patterns," *Connection Science* 12: 279–297.

See also **Artificial Intelligence; Cognitive Science; Neurobiology; Psychology, Philosophy of; Supervenience**

CONSCIOUSNESS

Consciousness is extremely familiar, yet it is at the limits—beyond the limits, some would say—of what one can sensibly talk about or explain. Perhaps this is the reason its study has drawn contributions from many fields, including psychology, neuroscience, philosophy, anthropology, cultural and literary theory, artificial intelligence, physics, and others. The focus of this article is on the varieties of consciousness, different problems that have been raised about these varieties, and prospects for progress on these problems.

Varieties of Consciousness

Creature Versus State Consciousness

One attributes consciousness both to people and to their psychological states. An agent can be conscious (as opposed to unconscious), and that agent's desire for a certain emotional satisfaction might be unconscious (as opposed to conscious). Rosenthal (1992) calls the former creature consciousness and the latter state consciousness. Most, but not all, discussion of consciousness in the contemporary literature concerns state consciousness rather than creature consciousness. Rosenthal (1992) goes on to propose an explanation of state consciousness in terms of creature consciousness, according to which a state is conscious just in case an agent who is in the state is conscious of it—but this proposal has proved controversial (e.g., Dretske 1993).

Essential Versus Nonessential Consciousness

Focusing on conscious states, one may distinguish those that are essentially conscious from those that are (or might be) conscious but not essentially so. The distinction is no doubt vague, but, to a first approximation, a state is essentially conscious just in case being in the state entails that it is conscious, and is not essentially conscious just in case this is not so.

Sensations are good candidates for states that are essentially conscious. If an agent is in pain, this state would seem to be conscious. (This is not to deny that the agent might fail to attend to it.) Beliefs, knowledge, and other cognitive states are good candidates for states that might be conscious but not essentially so. One may truly say that an agent knows the rules of his language even though this knowledge is unconscious. Perception presents a hard case, as is demonstrated by the phenomenon of blindsight, in which subjects report that they do not see anything in portions of the visual field and yet their performance on forced-choice tasks suggests otherwise (Weiskrantz 1986). Clearly *some* information processing is going on in such cases, but it is not obvious that what is going on is properly described as perception, or at least as perceptual experience. It is plausible to suppose that indecision about how to describe matters here derives in part from indecision about whether perceptual states or experiences are essentially conscious.

Transitive Versus Intransitive Consciousness

In the case of creature consciousness, one may speak of someone's being conscious *simpliciter* and of someone's being conscious *of* something or other. Malcolm calls the first "intransitive" and the second "transitive" consciousness (e.g., Armstrong and Malcolm 1984). To say that a person is conscious *simpliciter* is a way of saying that the person is awake or alert. So the study of creature intransitive consciousness may be assimilated to the study of what it is to be alert. The denial of consciousness *simpliciter* does not entail a denial of psychological states altogether. If an agent is fast asleep on a couch, one may truly say both that the agent is unconscious *and* that the agent believes that snow is white. Humphrey (1992) speculates that the *notion* of intransitive consciousness is a recent one, perhaps about 200 years old, but presumably people *were* on occasion intransitively conscious (i.e., alert or awake) prior to that date.

To say that a person is conscious of something seems to be a way of saying that the person knows or has beliefs about that thing. To say that one is conscious of a noise overhead is to say that one knows there is a noise overhead, though perhaps with the accompanying implication that one knows this only vaguely. So the study of creature transitive consciousness may be assimilated to the study of knowledge or beliefs. It is sometimes suggested that "consciousness" and "awareness" are synonyms. This is true only on the assumption that what is intended is creature transitive consciousness, since awareness is always *by* someone *of* something.

Intentional Versus Nonintentional Consciousness

While the transitive/intransitive distinction has no obvious analogue in the case of state consciousness—a state is not *itself* awake or alert, nor is it aware of anything—a related distinction is that between intentional and nonintentional conscious states. An intentional conscious state is *of* something in the sense that it represents the world as being a certain way—such states exhibit "intentionality," to adopt the traditional word. A nonintentional conscious state is a state that does not represent the world as being in some way. It is sometimes suggested that bodily sensations (itches, pains) are states of this second kind, while perceptual experiences (seeing a blue square on a red background) are cases of the first. But the matter is controversial given that to have a pain in one's foot seems to involve among other things representing one's foot as being in some condition or other, a fact that suggests that even here there is an intentional element in consciousness (see Intentionality).

Phenomenal Versus Access Consciousness

Block (1995) distinguishes two kinds of state consciousness: phenomenal consciousness and access consciousness. The notion of a phenomenally conscious state is usually phrased in terms of "What is it like ... ?" (e.g., Nagel 1974, "What is it like to be a bat?"). For a state to be phenomenally conscious is for there to be something it is akin to being, in that state. In the philosophical literature, the terms "qualia," "phenomenal character," and "experience" are all used as rough synonyms for phenomenal consciousness in this sense, though unfortunately there is no terminological consensus here.

For a state to be access conscious is, roughly, for the state to control rationally, or be poised to control rationally, thought and behavior. For creatures who have language, access consciousness closely correlates with reportability, the ability to express the state at will in language. Block suggests, among other things, that a state can be phenomenally conscious without its being access conscious.

For example, suppose one is engaged in intense conversation and becomes aware only at noon that there is a jackhammer operating outside—at five to twelve, one's hearing the noise is phenomenally conscious, but it is not access conscious. Block argues that many discussions both in the sciences and in philosophy putatively about phenomenal consciousness are in fact about access consciousness. Block's distinction is related to one made by Armstrong (1980) between experience and consciousness. For Armstrong, consciousness is attentional: A state is conscious just in case one attends to it. However, since one can have an experience without attending to it, it is possible to divorce experience and consciousness in this sense.

Within the general concept of access consciousness, a number of different strands may be distinguished. For example, an (epistemologically) normative interpretation of the notion needs to be separated from a nonnormative one. In the former case, the mere fact that one is in an access-conscious state puts one in a position to *know* or *justifiably* believe that one is; in the latter, being in an access-conscious state prompts one to think or believe that one is—here there is no issue of epistemic appraisal (see Epistemology). Further, an actualist interpretation of the notion needs to be separated from a counterfactual one. In the former case, what is at issue is whether one *does* know or think that one is in the state; in the latter, what is at issue is whether one *would*, provided other cognitive conditions were met. Distinguishing these various notions leads naturally into other issues. For example, consider the claim that it is essentially true of all psychological states that if one is in them, one would know that one is, provided one reflects and has the relevant concepts. That is one way of spelling out the Cartesian idea (recently defended by Searle [1992]) that the mind is "transparent" to itself.

There are hints both in Block (1995) and in related discussion (e.g., Davies and Humphreys 1993) that the phenomenal/access distinction is in (perhaps rough) alignment with both the intentional/nonintentional and the essential/nonessential distinction. The general idea is that phenomenally conscious states are *both* essentially so *and* nonintentional, while access-conscious states are neither. In view of the different interpretations of access consciousness, however, it is not clear that this is so. Psychological states might well be both essentially access conscious and phenomenally conscious. And, as indicated earlier, perhaps all conscious states exhibit intentionality in some form or other.

Self-Consciousness

Turning back from state consciousness to creature consciousness, a notion of importance here is self-consciousness, i.e., one's being conscious of oneself as an agent or self or (in some cases) a person. If to speak of a creature's being "conscious" of something is to speak about knowledge or beliefs, to attribute self-consciousness is to attribute to a creature knowledge or beliefs that the creature is a self or an agent. This would presumably require significant psychological complexity and perhaps cultural specificity. Proposals like those of Jaynes (1976) and Dennett (1992)—that consciousness is a phenomenon that emerges only in various societies—are best interpreted as concerning self-consciousness in this sense, which becomes more natural the more one complicates the underlying notion of self or agent. (Parallel remarks apply to any notion of group consciousness, assuming such a notion could be made clear.)

Problems of Consciousness

If these are the varieties of consciousness, it is easy enough to say in general terms what the problems of consciousness are, i.e., to explain or understand consciousness in all its varieties. But demands for explanation mean different things to different people, so the matter requires further examination.

To start with, one might approach the issue from an unabashedly scientific point of view. Consciousness is a variegated phenomenon that is a pervasive feature of the mental lives of humans and other creatures. It is desirable to have an explanation of this phenomenon, just as it is desirable to have an explanation of the formation of the moon, or the origin of HIV/AIDS. Questions that might be raised in this connection, and indeed have been raised, concern the relation of consciousness to neural structures (e.g., Crick and Koch 1998), the evolution of consciousness (e.g., Humphrey 1992), the relation of consciousness to other psychological capacities (e.g., McDermott 2001), the relation of consciousness to the physical and social environment of conscious organisms (e.g., Barlow 1987), and relations of unity and difference among conscious states themselves (e.g., Bayne and Chalmers 2003).

The attitude implicit in this approach—that consciousness might be studied like other empirical phenomena—is attractive, but there are at least four facts that need to be confronted before it can be completely adopted:

1. No framework of ideas has as yet been worked out within which the study of consciousness can proceed. Of course, this does not exclude the possibility that such a framework might be developed in the future, but it does make the specific proposals (such as that of Crick and Koch 1998) difficult to evaluate.
2. In the past, one of the main research tasks in psychology and related fields was to study psychological processes that were not conscious. This approach yielded a number of fruitful lines of inquiry—for example, Noam Chomsky's (1966) idea that linguistic knowledge is to be explained by the fact that people have unconscious knowledge of the rules of their language—but will presumably have to be abandoned when it comes to consciousness itself.
3. As Block (1995) notes, the standard concept of consciousness seems to combine a number of different concepts, which in turn raises the threat that consciousness in one sense will be confused with consciousness in another.
4. The issue of consciousness is often thought to raise questions of a philosophical nature, and this prompts further questions about whether a purely scientific approach is appropriate.

What are the philosophical aspects of the issue of consciousness? In the history of the subject, the issue of consciousness is usually discussed in the context of another, the traditional mind–body problem. This problem assumes a distinction between two views of human beings: the materialist or physicalist view, according to which human beings are completely physical objects; and the dualist view, according to which human beings are a complex of both irreducibly physical and irreducibly mental features. Consciousness, then, emerges as a central test case that decides the issue. The reason is that there are a number of thought experiments that apparently make it plausible to suppose that consciousness is distinct from anything physical. An example is the inverted spectrum hypothesis, in which it is imagined that two people might be identical except for the fact that the sensation provoked in one when looking at blood is precisely the sensation provoked in the other when looking at grass (Shoemaker 1981). If this hypothesis represents a genuine possibility, it is a very short step to the falsity of physicalism, which, setting aside some complications, entails that if any two people are identical physically, they are identical psychologically. On the other hand, if the inverted spectrum is possible, then two people identical physically may yet differ in respect of certain aspects of their conscious experience, and so differ psychologically. In short, physicalists are required to argue that the inverted spectrum hypothesis does not represent a genuine possibility. And this places the issue of consciousness at the heart of the mind–body problem.

In contemporary philosophy of mind, the traditional mind–body problem has been severely criticized. First, most contemporary philosophers do not regard the falsity of physicalism as a live option (Chalmers [1996] is an exception), so it seems absurd to debate something one already assumes to be true. Second, some writers argue that the very notions within which the traditional mind–body problem is formed are misguided (Chomsky 2000). Third, there are serious questions about the legitimacy of supposing that reflection of possible cases such as the inverted spectrum could even in principle decide the question of dualism or materialism, which are apparently empirical, contingent claims about the nature of the world (Jackson 1998).

As a result of this critique, many philosophers reject the mind–body problem in its traditional guise. However, it is mistaken to infer from this that concern with the inverted spectrum and related ideas has likewise been rejected. Instead, the theoretical setting of these arguments has changed. For example, in contemporary philosophy, the inverted spectrum often plays a role not so much in the question of whether physicalism is true, but rather in questions about whether phenomenal consciousness is in principle irreducible or else lies beyond the limits of rational inquiry. The impact of philosophical issues of this kind on a possible science of consciousness is therefore straightforward.

In the philosophical debates just alluded to, the notion of consciousness at issue is phenomenal consciousness. In other areas of philosophy, other notions are more prominent. In epistemology, for example, an important question concerns the intuitive difference between knowledge of one's own mental states—which seems in a certain sense privileged or direct—and knowledge of the external world, including the minds of others. This question has been made more acute by the impact of externalism, the thesis that one's psychological states depend constitutively on matters external to the subject, factors for which direct knowledge is not plausible (Davies 1997). Presumably these issues will be informed by the study of access consciousness. Similarly, in discussions of the notion of personal identity and related questions about how and why persons are objects of special moral concern, the notion of self-consciousness plays an important role. One might also regard both access consciousness and self-consciousness as topics for straightforward scientific study.

Prospects for Progress

Due to the influence of positivist and postpositivist philosophy of science in the twentieth century, it was at one time common to assume that some or all questions of consciousness were pseudo-questions. Recently it has been more common to concede that the questions are real enough. But what are the chances of progress here?

In light of the multifariousness of the issues, a formulaic answer to this question would be inappropriate. Access consciousness seems to be a matter of information processing, and there is reason to suppose that such questions might be addressed using contemporary techniques. Hence, many writers (e.g., Block 1995) find grounds for cautious optimism here, though this might be tempered depending on whether access consciousness is construed as involving a normative element. In the case of self-consciousness, the issue of normativity is also present, and there is the added complication that self-consciousness is partly responsive to questions of social arrangements and their impact on individual subjects.

But it is widely acknowledged that the hardest part of the issue is phenomenal consciousness. Here the dominant strategy has been an indirect one of attempting to reduce the overall number of problems. One way to implement this strategy is to attempt to explain the notion of phenomenal consciousness in terms of another notion, say, access consciousness or something like it. Some philosophers suggest that puzzlement about phenomenal consciousness is a cognitive illusion, generated by a failure to understand the special nature of concepts of phenomenal consciousness, and that once this puzzlement is dispelled, the way will be clear for a straightforward identification of phenomenal and access consciousness (e.g., Tye 1999). Others argue that discussions of phenomenal consciousness neglect the extent to which conscious states involve intentionality, and that once this is fully appreciated there is no bar to adopting the view that phenomenal consciousness is just access consciousness (e.g., Carruthers 2000).

The attractive feature of these ideas is that, if successful, they represent both philosophical and scientific progress. But the persistent difficulty is that the proposed explanations are unpersuasive. It is difficult to rid oneself of the feeling that what is special about concepts of phenomenal consciousness derives from only what it is that they are concepts of, and this makes it unlikely that the puzzles of phenomenal consciousness are an illusion. And, while it is plausible that phenomenally conscious states are intentional, emphasizing this fact will not necessarily shed light on the issue, for the intentionality of phenomenal consciousness might be just as puzzling as phenomenal consciousness itself.

However, even if one agrees that phenomenal consciousness represents a phenomenon distinct from these other notions, and therefore requires a separate approach, there is still a way in which one might seek to implement the strategy of reducing the number of problems, for, as noted earlier, phenomenal consciousness is thought to present both a philosophical and a scientific challenge. But what is the relation between these two issues? It is common to assume that the philosophical problem needs to be removed before one can make progress on the science. But perhaps the reverse is true. If the philosophical problems can be seen to be a reflection partly of ignorance in the scientific domain, there is no reason to regard them as a further impediment to scientific study. This might not seem like much of an advance. But the study of consciousness has been hampered by the feeling that it presents a problem of a different order from more straightforward empirical problems. In this context, to combat that assumption is to move forward, though slowly.

DANIEL STOLJAR

References

Armstrong, D. (1980), "What Is Consciousness?" in *The Nature of Mind and Other Essays*. Ithaca, NY: Cornell University Press, 55–67.
Armstrong, D., and N. Malcolm (1984), *Consciousness and Causality*. Oxford: Blackwell.
Barlow, H. (1987), "The Biological Role of Consciousness," in C. Blakemore and S. Greenfield (eds.), *Mindwaves*. Oxford: Basil Blackwell.
Bayne, T., and D. Chalmers (2003), "What Is the Unity of Consciousness?" in A. Cleeremans (ed.), *The Unity of Consciousness: Binding, Integration and Dissociation*. Oxford: Oxford University Press.
Block, N. (1995), "On a Confusion About a Function of Consciousness," *Behavioral and Brain Sciences* 18: 227–247.
Chalmers, D. (1996), *The Conscious Mind*. New York: Oxford University Press.
Chomsky, N. (1966), *Aspects of the Theory of Syntax*. Cambridge, MA: MIT Press.
——— (2000), *New Horizons in the Study of Mind and Language*. Cambridge: Cambridge University Press.
Crick, F., and C. Koch (1990), "Towards a Neurobiological Theory of Consciousness," *Seminars in the Neurosciences* 2: 263–275.

Carruthers, P. (2000), *Phenomenal Consciousness: A Naturalistic Theory*. Cambridge: Cambridge University Press.

Davies, M. (1997), "Externalism and Experience," in N. Block, O. Flanagan, and G. Guzeldere (eds.), *The Nature of Consciousness: Philosophical Debates*. Cambridge, MA: MIT Press, 309–328.

Davies, M., and G. Humphreys (1993), "Introduction," in M. Davies and G. Humphreys (eds.), *Consciousness: Psychological and Philosophical Essays*. Oxford: Blackwell.

Dennett, D. (1992), *Consciousness Explained*. Boston: Little, Brown and Co.

Dretske, D. (1993), "Conscious Experience," *Mind* 102: 263–283.

Humphrey, N. (1992), *A History of the Mind*. New York: Simon and Schuster.

Jackson, J. (1998), *From Metaphysics to Ethics*. Oxford: Clarendon.

Jaynes, J. (1976), *The Origin of Consciousness in the Breakdown of the Bicameral Mind*. Boston: Houghton Mifflin Co.

McDermott, D. (2001), *Mind and Mechanism*. Cambridge, MA: MIT Press.

Nagel, T. (1974), "What Is It Like to Be a Bat?" *Philosophical Review* 83: 7–22.

Rosenthal, D. (1992), "A Theory of Consciousness," in N. Block, O. Flanagan, and G. Guzeldere (eds.), *The Nature of Consciousness: Philosophical Debates*. Cambridge, MA: MIT Press, 729–754.

Searle, R. (1992), *The Rediscovery of the Mind*. Cambridge, MA: MIT Press.

Shoemaker, S. (1981), "The Inverted Spectrum," *Journal of Philosophy* 74: 357–381.

Tye, M. (1999), "Phenomenal Consciousness: The Explanatory Gap as a Cognitive Illusion," *Mind* 108: 432.

Weiskrantz, L. (1986), *Blindsight*. Oxford: Oxford University Press.

See also **Cognitive Science; Connectionism; Intentionality; Psychology, Philosophy of**

CONSERVATION BIOLOGY

Conservation biology emerged in the mid-1980s as a science devoted to the conservation of biological diversity, or *biodiversity*. Its emergence was precipitated by a widespread concern that anthropogenic development, especially deforestation in the tropics (Gómez-Pompa, Vázquez-Yanes, and Guevera 1972), had created an extinction crisis: a significant increase in the rate of species extinction (Soulé 1985). From its beginning, the primary objective of conservation biology was the design of conservation area networks (CANs), such as national parks, nature reserves, and managed-use zones that protect areas from anthropogenic transformation.

Conservation biology, then, is a normative discipline, in that it is defined in terms of a practical goal in addition to the accumulation of knowledge about a domain of nature. In this respect, it is analogous to medicine (Soulé 1985). Like medicine, conservation biology performs its remedial function in two ways: through intervention (for example, when conservation plans must be designed for species at risk of imminent extinction) and through prevention (for example, when plans are designed to prevent decline in species numbers long before extinction is imminent).

The normative status of conservation biology distinguishes it from ecology, which is not defined in terms of a practical goal (see Ecology). Moreover, besides using the models and empirical results of ecology, conservation biology also draws upon such disparate disciplines as genetics, computer science, operations research, and economics in designing and implementing CANs. Each of these fields contributes to a comprehensive framework that has recently emerged about the structure of conservation biology (see "The Consensus Framework" below).

Different views about the appropriate target of conservation have generated distinct methodologies within conservation biology. How 'biodiversity' is defined and, correspondingly, what conservation plans are designed will partly reflect ethical views about what features of the natural world are valuable (Norton 1994; Takacs 1996). There exists, therefore, a close connection between conservation biology and environmental ethics.

The Concept of Biodiversity

'Biodiversity' is typically taken to refer to diversity at all levels of biological organization: molecules, cells, organisms, species, and communities (e.g., Meffe and Carroll 1994). This definition does not, however, provide insight into the fundamental goal of conservation biology, since it refers to all biological entities (Sarkar and Margules 2002). Even worse, this definition does not exhaust all items of biological interest that are worth preserving: Endangered biological phenomena, such as the migration of the monarch butterfly, are not included within this definition (Brower and Malcolm 1991). Finally, since even a liberally construed notion of biodiversity does not capture the ecosystem processes that sustain biological diversity, some have argued that a more general concept of biological *integrity*, incorporating both diversity of entities and the ecological processes that sustain them, should be recognized as the proper focus of conservation biology (Angermeier and Karr 1994).

In response to these problems, many conservation biologists have adopted a pluralistic approach to biodiversity concepts (Norton 1994; Sarkar and Margules 2002; see, however, Faith [2003], who argues that this ready acceptance of pluralism confuses the [unified] concept of biodiversity with the plurality of different conservation strategies). Norton (1994), for example, points out that any measure of biodiversity presupposes the validity of a specific model of the natural world. The existence of several equally accurate models ensures the absence of any uniquely correct measure. Thus, he argues, the selection of a biodiversity measure should be thought of as a normative political decision that reflects specific conservation values and goals. Sarkar and Margules (2002) argue that the concept of biodiversity is implicitly defined by the specific procedure employed to prioritize places for conservation action (see "Place Prioritization" below). Since different contexts warrant different procedures, biodiversity should similarly be understood pluralistically.

Two Perspectives

Throughout the 1980s and early 90s, the discipline was loosely characterized by two general approaches, which Caughley (1994) described as the "small-population" and "declining-population" paradigms of conservation. Motivated significantly by the legal framework of the Endangered Species Act of 1973 and the National Forest Management Act of 1976, the small-population paradigm originated and was widely adopted in the United States (Sarkar 2005). It focused primarily on individual species threatened by extinction. By analyzing their distributions and habitat requirements, conservation biologists thought that "minimum viable populations" could be demonstrated, that is, population sizes below which purely stochastic processes would significantly increase the probability of extinction (see "Viability Analysis" below). It quickly became clear, however, that this methodology was inadequate. It required much more data than could be feasibly collected for most species (Caughley 1994) and, more importantly, failed to consider the interspecies dynamics essential to most species' survival (Boyce 1992).

Widely followed in Australia, the declining-population paradigm focused on deterministic, rather than stochastic, causes of population decline (Sarkar 2005). Unlike the small-population paradigm, its objective was to identify and eradicate these causes before stochastic effects became significant. Since the primary cause of population decline was, and continues to be, habitat loss, conservation biologists, especially in Australia, became principally concerned with protecting the full complement of regional species diversity and required habitat within CANs. With meager monetary resources for protection, conservation biologists concentrated on developing methods that identified representative CANs in minimal areas (see "Place Prioritization" below).

The Consensus Framework

Recently, a growing consensus about the structure of conservation biology has emerged that combines aspects of the small-population and declining-population paradigms into a framework that emphasizes the crucial role computer-based place prioritization algorithms play in conservation planning (Margules and Pressey 2000; Sarkar 2005). The framework's purpose is to make conservation planning more systematic, thereby replacing the ad hoc reserve design strategies often employed in real-world planning in the past. It focuses on ensuring the adequate representation and persistence of regional biodiversity within the socioeconomic and political constraints inherent in such planning.

Place Prioritization

The first CAN design methods, especially within the United States, relied almost exclusively on island biogeography theory (MacArthur and Wilson 1967) (see Ecology). It was cited as the basis for geometric design principles intended to minimize extinction rates in CANs as their ambient regions were anthropogenically transformed (Diamond 1975). Island biogeography theory entails that the particular species composition of an area is constantly changing while, at an equilibrium between extinction and immigration, its species richness remains constant. The intention behind design principles inspired by the theory, therefore, was to ensure persistence of the maximum number of species, not the specific species the areas currently contained. However, incisive criticism of the theory (Gilbert 1980; Margules, Higgs, and Rafe 1982) convinced many conservation biologists, especially in Australia, that *representation* of the *specific species* that areas now contain, rather than the *persistence* of the *greatest number of species* at some future time, should be the first goal of CAN design. Computer-based place prioritization algorithms supplied a defensible methodology for achieving the first goal and an alternative to the problematic reliance on island biogeography theory in CAN design.

Place prioritization involves solving a resource allocation problem. Conservation funds are usually significantly limited and priority must be given to protecting some areas over others. The Expected Surrogate Set Covering Problem (ESSCP) and the Maximal Expected Surrogate Covering Problem (MESCP) are two prioritization problems typically encountered in biodiversity conservation planning (Sarkar 2004). Formally, consider a set of individual places called cells $\{c_j : j = 1, \ldots, n\}$; cell areas $\{a_j : j = 1, \ldots, n\}$; biodiversity surrogates $\Lambda = \{s_i : i = 1, \ldots, m\}$; representation targets, one for each surrogate $\{t_i : i = 1, \ldots, m\}$; probabilities of finding s_i at c_j $\{p_{ij} : i = 1, \ldots, m; j = 1, \ldots, n\}$; and two indicator variables $X_j (j = 1, \ldots, n)$ and $Y_i (i = 1, \ldots, m)$ defined as follows:

$$X_j = \begin{cases} 1, & \text{if } c_j \in \Gamma; \\ 0, & \text{otherwise}; \end{cases}$$

$$Y_i = \begin{cases} 1, & \text{if } \sum_{c_j \in \Gamma} p_{ij} > t_i; \\ 0, & \text{otherwise}. \end{cases}$$

ESSCP is the problem:

$$\text{Minimize } \sum_{j=1}^{n} a_j X_j \text{ such that } \sum_{j=1}^{n} X_j p_{ij} \geq t_i \text{ f}$$

or $\forall s_i \in \Lambda$.

Informally, find the set of cells Γ with the smallest area such that every representation target is satisfied. MESCP is the problem:

$$\text{Maximize } \sum_{i=1}^{m} Y_i \text{ such that } \sum_{j=1}^{n} X_j = \mathbf{M},$$

where $\mathbf{M}$ is the number of protectable cells. Informally, given the opportunity to protect $\mathbf{M}$ cells, find those cells that maximize the number of representation targets satisfied.

The formal precision of these problems allows them to be solved computationally with heuristic or exact algorithms (Margules and Pressey 2000; Sarkar 2005). Exact algorithms, using mixed integer linear programming, guarantee optimal solutions: the smallest number of areas satisfying the problem conditions. Since ESSCP and MESCP are NP hard, however, exact algorithms are computationally intractable for practical problems with large datasets. Consequently, most research focuses on heuristic algorithms that are computationally tractable for large datasets but do not guarantee minimal solutions.

The most commonly used heuristic algorithms are "transparent," so called because the exact criterion by which each solution cell is selected is known. These algorithms select areas for incorporation into CANs by iteratively applying a hierarchical set of conservation criteria, such as rarity, richness, and complementarity. Rarity, for example, requires selecting the cell containing the region's most infrequently present surrogates, which ensures that endemic taxa are represented. The criterion most responsible for the efficiency of transparent heuristic algorithms is complementarity: Select subsequent cells that complement those already selected by adding the most surrogates not yet represented (Justus and Sarkar 2002). In policymaking contexts, transparency facilitates more perspicuous negotiations about competing land uses, which constitutes an advantage over nontransparent heuristic prioritization procedures such as those based on simulated annealing.

Methodologically, the problem of place prioritization refocused theoretical research in conservation biology from general theories to algorithmic procedures that require geographically explicit data (Sarkar 2004). In contrast to general theories, which abstract from the particularities of individual areas, place prioritization algorithms demonstrated that adequate CAN design critically depends upon these particularities.

Surrogacy

Since place prioritization requires detailed information about the precise distribution of a region's biota, devising conservation plans for specific areas would ideally begin with an exhaustive series of field surveys. However, owing to limitations of time, money, and expertise, as well as geographical and sociopolitical boundaries to fieldwork, such surveys are usually not feasible in practice. These limitations give rise to the problem of discovering a (preferably small) set of biotic or abiotic land attributes (such as subsets of species, soil types, vegetation types, etc.) the precise distribution of which is realistically obtainable and that adequately represents biodiversity as such. This problem is referred to as that of finding surrogates or "indicators" for biodiversity (Sarkar and Margules 2002).

This challenge immediately gives rise to two important conceptual problems. Given the generality of the concept of biodiversity as such, the first problem concerns the selection of those entities that should be taken to represent concretely biodiversity in a particular planning context. These entities will be referred to as the *true surrogate*, or objective parameter, as it is the parameter that one is attempting to estimate in the field. Clearly, selection of the true surrogate is partly conventional and will depend largely on pragmatic and ethical concerns (Sarkar and Margules 2002); in most conservation contexts the true surrogate will typically be species diversity, or a species at risk. Once a set of true surrogates is chosen, the set of biotic and/or abiotic land attributes that will be tested for their capacity to represent the true surrogates adequately must be selected; these are *estimator surrogates*, or indicator parameters.

The second problem concerns the nature of the relation of representation that should obtain between the estimator and the true surrogate: What does 'representation' mean in this context and under what conditions can an estimator surrogate be said to represent adequately a true surrogate? Once the true and estimator surrogates are selected and a precise (operational) interpretation of representation is determined, the adequacy with which the estimator surrogate represents the true surrogate becomes an empirical question. (Landres, Verner, and Thomas [1988] discuss the import of subjecting one's choice of estimator surrogate to stringent empirical testing.)

Very generally, two different interpretations of representation have been proposed in the conservation literature. The more stringent interpretation is that the distribution of estimator surrogates should allow one to *predict* the distribution of true surrogates (Ferrier and Watson 1997). The satisfaction of this condition would consist in the construction of a well-confirmed model from which correlations between a given set of estimator surrogates and a given set of true surrogates can be derived. Currently such models have met with only limited predictive success; moreover, since such models can typically be used to predict the distribution of only one surrogate at a time, they are not computationally feasible for practical conservation planning, which must devise CANs to sample a wide range of regional biodiversity. Fortunately, the solution to the surrogacy problem does not in practice require predictions of true-surrogate distributions.

A less stringent interpretation of representation assumes only that the set of places prioritized on the basis of the estimator surrogates adequately captures true-surrogate diversity up to some specified target (Sarkar and Margules 2002). The question of the *adequacy* of a given estimator surrogate then becomes the following: If one were to construct a CAN that samples the full complement of estimator-surrogate diversity, to what extent does this CAN also sample the full complement of true-surrogate diversity? Several different quantitative measurements can be carried out to evaluate this question (Sarkar 2004). At present, whether adequate surrogate sets exist for conservation planning remains an open empirical question.

Viability Analysis

Place prioritization and surrogacy analysis help identify areas that *currently* represent biodiversity. This is usually not, however, sufficient for successful biodiversity conservation. The problem is that the biodiversity these areas contain may be in irreversible decline and unlikely to persist. Since principles of CAN design inspired by island biogeography theory were, by the late 1980s, no longer believed to ensure persistence adequately, attention subsequently turned, especially in the United States, to modeling the probability of population extinction. These principles became known as population viability analysis (PVA).

PVA models focus primarily on factors affecting small populations, such as inbreeding depression, environmental and demographic stochasticity, genetic drift, and spatial structure. Drift and inbreeding depression, for example, may reduce the genetic variation required for substantial evolvability, without which populations may be more susceptible to disease, predation, or future environmental change. PVA modeling has not, however, provided a clear understanding of the general import of drift and inbreeding depression to population persistence in nature (Boyce 1992). Unfortunately, this kind of problem pervades PVA. In a trenchant review, for instance, Caughley (1994) concluded that the predominantly theoretical work done thus far in PVA had not been adequately tested with field data and that many of the tests that had been done were seriously flawed.

Models used for PVA have a variety of different structures and assumptions (see Beisinger and McCullough 2002). As a biodiversity conservation methodology, however, PVA faces at least three general difficulties:

1. Precise estimation of parameters common to PVA models requires enormous amounts of quality field data, which are usually unavailable (Fieberg and Ellner 2000) and cannot be collected, given limited monetary resources and the imperative to take conservation action quickly. Thus, with prior knowledge being uncommon concerning what mechanisms are primarily responsible for a population's dynamics, PVA provides little guidance about how to minimize extinction probability.
2. In general, the results of PVA modeling are extremely sensitive to model structure and parameter values. Models with seemingly slightly different structure may make radically different predictions (Sarkar 2004), and different parameter values for one model may produce markedly different predictions, a problem exacerbated by the difficulties discussed under (1).
3. Currently, PVA models have been developed almost exclusively for single species (occasionally two) and rarely consider more than a few factors affecting population decline. Therefore, only in narrow conservation contexts focused on individual species would they potentially play an important role. Successful models that consider the numerous factors affecting the viability of *multiple*-species assemblages, which are and should be the primary target of actual conservation planning, are unlikely to be developed in the near future (Fieberg and Ellner 2000).

For these reasons and others, PVA has not thus far uncovered nontrivial generalities relevant to CAN design. Consequently, attention has turned to more pragmatic principles, such as designing CANs to minimize distance to anthropogenically transformed areas. This is not to abandon the important goals of PVA or the need for sound theory about biodiversity persistence, but to recognize the weaknesses of existing PVA models and the imperative to act now given significant threats to biodiversity.

Multiple-Criterion Synchronization

The implementation and maintenance of CANs inevitably take place within a context of competing demands upon the allocation and use of land. Consequently, successful implementation strategies should ideally be built upon a wide consensus among agents with different priorities with respect to that usage. Thus, practical CAN implementation involves the attempt to optimize the value of a CAN amongst several different criteria simultaneously. Because different criteria typically conflict, however, the term "synchronization" rather than "optimization" is more accurate. The multiple-constraint synchronization problem involves developing and evaluating procedures designed to support such decision-making processes.

One approach to this task is to "reduce" these various criteria to a single scale, such as monetary cost. For example, cost-benefit analysis has attempted to do this by assessing the amount an agent is willing to pay to improve the conservation value of an area. In practice, however, such estimates are difficult to carry out and are rarely attempted (Norton 1987). Another method, based on multiple-objective decision-making models, does not attempt to reduce the plurality of criteria to a single scale; rather it seeks merely to eliminate those feasible CANs that are suboptimal when evaluated according to all relevant criteria. This method, of course, will typically not result in the determination of a uniquely best solution, but it may be able to reduce the number of potential CANs to one that is small enough so that decision-making bodies can bring other implicit criteria to bear on their ultimate decision (see Rothley 1999 and Sarkar 2004 for applications to conservation planning).

JUSTIN GARSON
JAMES JUSTUS

References

Angermeier, P. L., and J. R. Karr (1994), "Biological Integrity versus Biological Diversity as Policy Directives," *BioScience* 44: 690–697.
Beisinger, S. R., and D. R. McCullough (eds.) (2002), *Population Viability Analysis*. Chicago: University of Chicago Press.
Boyce, M. (1992), "Population Viability Analysis," *Annual Review of Ecology and Systematics* 23: 481–506.
Brower, L. P., and S. B. Malcolm (1991), "Animal Migrations: Endangered Phenomena," *American Zoologist* 31: 265–276.
Caughley, G. (1994), "Directions in Conservation Biology," *Journal of Animal Ecology* 63: 215–244.
Diamond, J. (1975), "The Island Dilemma: Lessons of Modern Biogeographic Studies for the Design of Natural Reserves," *Biological Conservation* 7: 129–146.
Faith, D. P. (2003), "Biodiversity," in Edward N. Zalta (ed.), *The Stanford Encyclopedia of Philosophy*. http://plato.stanford.edu/archives/sum2003/entries/biodiversity
Ferrier, S., and G. Watson (1997), *An Evaluation of the Effectiveness of Environmental Surrogates and Modelling Techniques in Predicting the Distribution of Biological*

Diversity: Consultancy Report to the Biodiversity Convention and Strategy Section of the Biodiversity Group, Environment Australia*. Armidale, New South Wales: Environment Australia.
Fieberg, J., and S. P. Ellner (2000), "When Is It Meaningful to Estimate an Extinction Probability?" *Ecology* 8: 2040–2047.
Gilbert, F. S. (1980), "The Equilibrium Theory of Island Biogeography: Fact or Fiction?" *Journal of Biogeography* 7: 209–235.
Gómez-Pompa, A., C. Vázquez-Yanes, and S. Guevera (1972), "The Tropical Rain Forest: A Nonrenewable Resource," *Science* 177: 762–765.
Justus, J., and S. Sarkar (2002), "The Principle of Complementarity in the Design of Reserve Networks to Conserve Biodiversity: A Preliminary History," *Journal of Biosciences* 27: 421–435.
Landres, P. B., J. Verner, and J. W. Thomas (1988), "Ecological Uses of Vertebrate Indicator Species: A Critique," *Conservation Biology* 2: 316–328.
MacArthur, R., and E. O. Wilson (1967), *The Theory of Island Biogeography*. Princeton, NJ: Princeton University Press.
Margules, C. R., and R. L. Pressey (2000), "Systematic Conservation Planning," *Nature* 405: 242–253.
Margules, C., A. J. Higgs, and R. W. Rafe (1982), "Modern Biogeographic Theory: Are There Lessons for Nature Reserve Design?" *Biological Conservation* 24: 115–128.
Meffe, G. K., and C. R. Carroll (1994), *Principles of Conservation Biology*. Sunderland, MA: Sinauer Associates.
Norton, B. G. (1987), *Why Preserve Natural Variety?* Princeton, NJ: Princeton University Press.
Norton, B. G. (1994), "On What We Should Save: The Role of Cultures in Determining Conservation Targets," in P. L. Forey, C. J. Humphries, and R. I. Vane-Wright (eds.), *Systematics and Conservation Evaluation*. Oxford: Clarendon Press, 23–29.
Rothley, K. D. (1999), "Designing Bioreserve Networks to Satisfy Multiple, Conflicting Demands," *Ecological Applications* 9: 741–750.
Sarkar, S. (2004), "Conservation Biology," in E. N. Zalta (ed.), *The Stanford Encyclopedia of Philosophy*. http://stanford.edu/entries/conservation-biology
——— (2005), *Biodiversity and Environmental Philosophy*. New York: Cambridge University Press.
Sarkar, S., and C. R. Margules (2002), "Operationalizing Biodiversity for Conservation Planning," *Journal of Biosciences* 27: 299–308.
Soulé, M. E. (1985), "What Is Conservation Biology?" *BioScience* 35: 727–734.
Takacs, D. (1996), *The Idea of Biodiversity: Philosophies of Paradise*. Baltimore: Johns Hopkins University Press.

CONSTRUCTIVE EMPIRICISM

See **Instrumentalism; Realism**

CONVENTIONALISM

Conventionalism is a philosophical position according to which the truth of certain propositions, such as those of ethics, aesthetics, physics, mathematics, or logic, is in some sense best explained by appeal to intentional human actions, such as linguistic stipulations. In this article, the focus will be on conventionalism concerning physics, mathematics, and logic. Conventionalism concerning empirical science and mathematics appears to have emerged as a distinctive philosophical position only in the latter half of the nineteenth century.

The philosophical motivations for conventionalism are manifold. Early conventionalists such as Pierre Duhem and Henri Poincaré, and more recently Adolf Grünbaum, have seen conventionalism about physical or geometrical principles as justified in part by what they regard as the underdetermination of a physical or geometrical theory by empirical observation (see Duhem Thesis; Poincaré, Henri; Underdetermination of Theories). An (empirically) arbitrary choice of a system from among empirically equivalent theories seems required.

A second motivation, also suggested by Duhem and Poincaré, and explicitly developed later by Rudolf Carnap, stems from their view that any description of the empirical world presupposes a suitable descriptive apparatus, such as a geometry, a metric, or a mathematics. In some cases there appears to be a choice as to which descriptive apparatus to employ, and this choice involves an arbitrary convention (see Carnap, Rudolf).

A third motivation for conventionalism, emphasized by philosophers such as Moritz Schlick, Hans Hahn, and Alfred Ayer, is that appeal to conventions provides a straightforward explanation of a priori propositional knowledge, such as knowledge of mathematical truths (see Ayer, Alfred Jules; Hahn, Hans; Schlick, Moritz). Such truths, it was thought, are known a priori in virtue of the fact that they are stipulated rather than discovered.

It is important to note that conventionalists do not regard the selection of a set of conventions to be wholly arbitrary, with the exception of the radical French conventionalist Edouard LeRoy, who did (see Giedymin 1982, 118–128). Conventionalists acknowledge that pragmatic or instrumental considerations involving human capacities or purposes might be relevant to the adoption of a set of conventions (see Instrumentalism). However, they insist that the empirical evidence does not compel one choice rather than another.

This article focuses on several major figures and issues, but there are a number of other important philosophers with broadly conventionalist leanings, including Duhem, Kazimierz Ajdukiewicz, and Ludwig Wittgenstein, that for the sake of brevity will not here receive the attention they deserve.

Poincaré and Geometric Conventionalism

The development of non-Euclidean geometries and subsequent research into the foundations of geometry provided both the inspiration and the model for much of the conventionalism of the early twentieth century (see Space-Time). The central figure in early conventionalism was the French mathematician and philosopher Poincaré (see Poincaré, Henri). Although a Kantian in his philosophy of mathematics, Poincaré shared with many late-nineteenth-century mathematicians and philosophers the growing conviction that the discovery of non-Euclidean geometries, such as the geometries of Lobatschevsky and Riemann, conjoined with other developments in the foundations of geometry, rendered untenable the Kantian treatment of geometrical axioms as a form of synthetic a priori intuition.

The perceived failings of Kant's analysis of geometry did not, however, lead Poincaré to an empiricist treatment of geometry. If geometry were an experimental science, he reasoned, it would be open to continual revision or falsified outright (the perfectly invariable solids of geometry are never empirically discovered, for instance) (Poincaré [1905] 1952, 49–50). Rather, geometrical axioms are conventions. As such, the axioms (postulates) of a systematic geometry constitute *implicit definitions* of such primitive terms of the system as "point" and "line" (50). This notion of implicit definition originated with J. D. Gergonne (1818), who saw an analogy between a set of sentences with n undefined terms and a set of equations with n unknowns. The two are analogous in that the roots that satisfy the equations are akin to interpretations of the undefined terms in the set of sentences under which the sentences are true. Of course, not every set of equations determines a set of values, and so too not every set of sentences (system of axioms) constitutes an implicit definition of its primitive terms. Poincaré accommodated this fact by recognizing two constraints on the admissibility of a set of axioms. First, the set must be consistent, and second, it must uniquely determine the objects defined (1952, 150–153).

Although the axioms were in his view conventional, Poincaré thought that experience nonetheless plays a role within geometry. The genesis of geometrical systems is closely tied to experience, in that the systems of geometry that are constructed are selected on the basis both of prior idealized empirical generalizations and of the simplicity and convenience that particular systems (such as Euclidean geometry) may afford their users ([1905] 1952, 70–71; 1952, 114–115). But these empirical considerations do not provide a test of empirically applied geometries. Once elevated to the status of a convention, a geometrical system is not an empirical theory, but is rather akin to a language used, in part, to frame subsequent empirical assertions. As a system of implicit definitions, it is senseless to speak of one geometry being "more true" than another ([1905] 1952, 50).

Poincaré therefore rejected the supposition that an empirical experiment could compel the selection of one geometry over another, provided that both satisfied certain constraints. He was inspired by Sophus Lie's (1959) group-theoretic approach to geometrical transformations, according to which, given an *n*-dimensional space and the possible transformations of figures within it, only a finite group of geometries with differing coordinate systems are possible for that space. Among these geometries there is no principled justification for selecting one (such as Plücker line geometry) over another (such as sphere geometry) (Poincaré [1905] 1952, 46–7). Indeed, one geometry within the group may be transformed into another given a certain method of translation, which Lie derived from Gergonne's theory of reciprocity. Poincaré conjoined the intertransformability of certain geometrical systems with the recognition that alternative geometries yield different conventions governing the notions of "distance" or "congruence" to argue that any attempt to decide by experiment between alternative geometries would be futile, since interpretations of the data presupposed geometric conventions. In a much-discussed passage he wrote:

> If Lobatchevsky's geometry is true, the parallax of a very distant star will be finite. If Riemann's is true, it will be negative. These are the results which seem within the reach of experiment.... But what we call a straight line in astronomy is simply the path of a ray of light. If, therefore, we were to discover negative parallaxes, or to prove that all parallaxes are higher than a certain limit, we should have a choice between two conclusions: we could give up Euclidean geometry, or modify the laws of optics, and suppose that light is not rigorously propagated in a straight line. ([1905] 1952, 72–73)

On one reading, advanced by Grünbaum, this passage affirms that there is no fact of the matter as to which is the "correct" metric, and hence supports the *conventionality of spatial metrics*. This reading is further discussed below. Another reading of this passage sees in it an argument closely akin to Duhem's argument leading to the denial of the possibility of a "crucial experiment" that could force the acceptance or elimination of a geometrical theory (see Duhem [1906] 1954, 180–90). Interestingly, Poincaré uses the argument to support a distinction between "conventional" truths and empirical truths, whereas Quine later adduces similar Duhemian considerations on behalf of his claim that there is no such distinction to be drawn (see Duhem Thesis; Quine, Willard Van).

However he may have intended his parallax example, Poincaré would probably have accepted both the conventionality of the spatial metric and the absence of any experimental test capable of deciding between two alternative geometries. As a further illustration of these issues, he proposed a well-known thought experiment ([1905] 1952, 65–67). Imagine a world consisting of a large sphere subject to the following temperature law: At the center of the sphere the temperature is greatest, and at the periphery it is absolute zero. The temperature decreases uniformly as one moves toward the circumference in proportion to the formula $R^2 - r^2$, where R is the radius of the sphere and r the distance from the center. Assume further that all bodies in the sphere are in perfect thermal equilibrium with their environment, that all bodies share a coefficient of dilation proportional to their temperature, and that light in this sphere is transmitted through a medium whose index of refraction is $1/(R^2 - r^2)$. Finally, suppose that there are inhabitants of this sphere and that they adopt a convention that allows them to measure lengths with rigid rods. If the inhabitants assume that these rods are invariant in length under transport, and if they further triangulate the positions in their world with light rays, they might well come to the conclusion that their world has a Lobatchevskian geometry. But they could also infer that their world is Euclidean, by postulating the universal physical forces just described. Again, no empirical experiment seems adequate to establish one geometry over the other, since both are compatible with all known possible observations given appropriate auxiliary hypotheses. Rather, a conventional choice of a geometry seems called for. The choice will be motivated by pragmatic factors, perhaps, but not "imposed" by the experimental facts (Poincaré [1905] 1952, 70–71).

Grünbaum's Conventionalism: The Metric and Simultaneity

Poincaré's parallax and sphere-world examples motivated his view that the choice of a spatial metric is conventional. Adolph Grünbaum has defended this conventionalist conclusion at length (Grünbaum 1968, 1973). Grünbaum claims that there is no unique metric "intrinsic" to a spatial manifold, since between any two points on a real number line, there is an uncountably infinite continuum of points. Hence, if one wishes to specify a metric for a manifold, one must employ "extrinsic" devices such as measuring rods, and must further stipulate the rods' behavior under transport.

Grünbaum defends what he takes to be Poincaré's metric conventionalism against a variety of objections. Perhaps the most serious objection is the claim that Poincaré's result illustrates only a *trivial* point about the conventionality of referring expressions, an objection first directed against Poincaré by Arthur Eddington and subsequently developed by Hilary Putnam and Paul Feyerabend. Eddington objected that Poincaré's examples illustrated not that metrical *relations* (such as the relations of equality or of the congruence of two rods) were conventional, but rather—and only—that the meanings of *words* such as "equal" and "congruent" were conventional. This, Eddington objected, was a trivial point about *semantical* conventionality, and the fact that the selection of different metrics could yield different but apparently equipollent geometries illustrated only that "the meaning assigned to length and distance has to go along with the meaning assigned to space" (Eddington 1953, 9–10). Eddington suggested, and Feyerabend later developed, a simple illustration of this point within the theory of gases. Upon the discovery that Boyle's law:

$$pv = RT$$

holds only approximately of real gases, one could *either* revise Boyle's law in favor of van der Waal's law:

$$(p + a/v^2)(v - b) = RT$$

or one could preserve Boyle's law by redefining pressure as follows:

$$\text{Pressure} =_{\text{Def}} (p + a/v^2)(1 - b/v).$$
(Feyerabend, quoted in Grünbaum 1973, 34)

The possibility of such a redefinition, the objection proceeds, is physically and philosophically uninteresting. Provided that a similar move is available in Poincaré's own examples with expressions like 'congruent,' Poincaré's examples appear to illustrate only platitudes about semantic conventionality.

Grünbaum tries to show that Poincaré's conventionalism is not merely an example of what Grünbaum calls "trivial semantic conventionality" by showing that the conventionality of the metric is the result of a certain absence of "intrinsic" structure within the space-time manifold, which structure can (or must) then be imposed "extrinsically":

> And the metric amorphousness of these continua then serves to *explain* that even *after* the word "congruent" has been pre-empted semantically as a spatial or temporal *equality* predicate by the axioms of congruence, congruence remains ambiguous in the sense that these axioms still allow an infinitude of mutually exclusive congruence classes of intervals. (Grünbaum 1973, 27)

Grünbaum developed a related argument to the effect that the simultaneity relation in special relativity (SR) involves a conventional element. A sympathetic reconstruction of the argument will be attempted here, omitting a number of details in order to present the gist of the argument as briefly and intuitively as possible. Within Newtonian mechanics (NM), there is no finite limit to the speed at which causal signals can be sent. Further, one might take it to be a defining feature of causal relations that effects cannot precede their causes. This asymmetry allows for a distinction between events that are temporally before or after a given event on a world line. There is then only one simultaneity relation within the space-time of NM meeting the constraints mentioned. Consider events (or space-time points, if one prefers) a and b on two distinct parallel world lines. Event a is simultaneous with b just in case a is not causally connectible to any event on the future part of b's world line but is causally connectible to any event on the past of b's world line. Within SR, however, there is an upper limit to the speed of causal signals. Thus the constraint that effects cannot precede their causes allows any one of a continuum of points along a parallel world line to be a candidate for simultaneity with a given event on the world line of an inertial observer. Grünbaum calls such points "topologically simultaneous" with a point o on the observer's world line. In claiming that simultaneity is (to some extent) conventional within the SR picture, Grünbaum is in good company, as Albert Einstein makes a claim to this effect in his original 1905 paper. (It is important to distinguish the issue of the *conventionality* of simultaneity within SR, which has been controversial, from the (observer or frame) relativity of simultaneity within SR, which is unchallenged; see Space-Time.)

Grünbaum claims that the existence of a continuum of possible "planes of simultaneity" through a given point is explained by, or reflects, the fact that the simultaneity relation is not "intrinsic" to the manifold of events within special relativity. One objection that a number of Grünbaum's critics such as Putnam (in Putnam 1975b) have had to his treatment of conventionalism (concerning both the metric and the simultaneity relation) is that the notion of an intrinsic feature, which plays a crucial role for Grünbaum, is never adequately clarified. David Malament (1977) provides a natural reading of 'intrinsic': An intrinsic property or relation of a system is "definable" in terms of a set of basic features. The intuitive idea is that if a relation is definable from relations that are uncontroversially intrinsic, then these ought to count as intrinsic as well. Malament goes on to show that if one takes certain fairly minimal relations to be given as intrinsic, then a unique simultaneity relation, in fact the standard one, is definable from them in a fairly well defined sense of 'definable.'

Malament's result has been taken by many to have definitively settled the issue of whether simultaneity is conventional within SR (see e.g., Norton 1992). In this view, the standard simultaneity relation is nonconventional, since it is definable, or "logically constructible," from uncontroversially intrinsic features of Minkowski space-time, the space-time of SR. Furthermore, standard simultaneity is the only such nonconventional simultaneity relation.

However, there are dissenters, including Grünbaum. A number of authors have attacked one or another of Malament's premises, including for instance the claim that simultaneity must be a transitive relation, that is, the requirement that for arbitrary events x, y, and z, if x sim y and y sim z, then x sim z. Sarkar and Stachel have shown that even if it is allowed that simultaneity must be an equivalence relation, other simultaneity relations (the backward and the forward light cones of events on the given world line) are definable from relations that Malament takes as basic or intrinsic (see Sarkar and Stachel 1999). Peter Spirtes (1981) shows that Malament's result is highly sensitive to the choice of basic or intrinsic relations, which fact might be taken to undercut the significance of Malament's result as well.

A conventionalist might raise a different sort of objection to Malament's results, questioning their relevance to a reasonable, although perhaps not the only reasonable, construal of the question as to whether simultaneity is conventional within SR. Suppose for simplicity that any relation that one is willing to call a simultaneity relation is an equivalence relation, that causes must always precede their effects, and that there is an upper bound to the speed of causal influences within SR. One may now ask whether more than one relation can be defined (extrinsically or otherwise) on the manifold of events (or "space-time points"), meeting these criteria within SR. That is, one might ask whether, given any model M of T (roughly, Minkowski space-time, the "standard" space-time of SR), there is a unique expansion of the model to a model M' for T', where T' adds to T sentences containing a symbol 'sim', which sentences in effect require that 'sim' be an equivalence relation and that causes are not 'sim' with their effects. It is a straightforward matter to see, as Grünbaum shows (the details are omitted here), that there is no such unique expansion of M. The constraints mentioned leave room for infinitely many possible interpretations of 'sim'. In this sense, the simultaneity relation (interpretation of 'sim') might naturally be said to be conventional within SR. In contrast, given a model for the causal structure of NM, the meaning constraints on the concept of simultaneity that are assumed here yield a unique expansion (a unique extension for 'sim').

It should be noted that the conventionality of simultaneity within SR and its nonconventionality within NM in the sense just described reflect structural differences between the two theories (or their standard models). Thus this form of conventionalism appears to escape the charge that the only sense in which simultaneity is conventional is the trivial sense in which the word 'simultaneous' can be used to denote different relations. As Grünbaum frequently emphasizes, the interesting cases of conventionality arise only when one begins with an already meaningful term or concept, whose meaning is constrained in some nontrivial way. It also allows one to interpret Einstein as making a substantive, yet fairly obvious, point when he claims that the standard simultaneity relation within SR is a conventional, or stipulated, choice.

How does this version of conventionalism relate to Malament's arguments? It will be helpful to note a puzzle before proceeding. Wesley Salmon (1977) argues at length that the one-way speed of light is not empirically identifiable within SR. (The basic problem is that the one-way speed seems measurable only by using distant synchronized clocks, but the synchronization of distant clocks would appear to involve light signals and presuppositions about their one-way speeds.) This fact appears to yield as an immediate consequence the conventionality (in the sense of empirical admissibility described above) of simultaneity within SR, given the various apparently admissible ways of synchronizing distant clocks within a reference frame. The puzzle is this: On one hand, the standard view has it that Malament's result effectively proves the falsehood of conventionalism (about simultaneity within SR); on the other hand, Salmon's arguments appear to entail the truth of conventionalism. Yet, his arguments for the empirical inaccessibility of the one-way speed of light seem sound. No one has convincingly shown what is wrong with them. All that the nonconventionalist seems able to provide is the indirect argument that Malament is right, so Salmon (and Grünbaum) must be wrong.

In the present analysis, this puzzle is dissolved by distinguishing two proposals, each of which may justifiably be called versions of conventionalism.

One version claims that a *relation* (thought of as a set or extension) is conventional within a model for a theory if it is not intrinsic within M, where 'intrinsic' might then be interpreted as definability (in the sense of logical constructibility) from uncontroversially intrinsic relations within M. The other version claims that the extension of a *concept* is conventional within a model (or theory) if the meaning of the concept does not entail a unique extension for the concept within the model (or within all models of the theory). Call these versions of conventionalism C_1 and C_2, respectively. Malament shows that simultaneity is not C_1 conventional within SR (leaving aside other possible objections to his result). Grünbaum and Salmon show that simultaneity is C_2 conventional within SR. Malament further considers relations definable from the causal connectibility relation (together with the world line of an inertial observer) within SR, and notes that only one is a nontrivial candidate simultaneity relation (relative to the corresponding inertial frame). He shows there is only one, and that therefore there is a unique intrinsic (and hence non–C_1 conventional) simultaneity relation within SR, the standard one. The C_2 conventionalist may respond that although Malament has pointed out an interesting feature of a particular candidate interpretation of the already meaningful term 'simultaneous,' he has not thereby ruled out all other candidate interpretations. One should not, the C_2 defender will argue, confuse the metalinguistic notion 'definable within M (or T)' with a meaning constraint on the concept of simultaneity, i.e., a constraint on potential interpretations of the already meaningful term 'simultaneous.' It would be absurd to claim that what was meant all along by 'simultaneous' required not only that any candidate be an equivalence relation and that it must "fit with" causal relations in that effects may not precede their causes, but also that for any two events, they are simultaneous if and only if they are in a relation that is definable from the causal connectibility relation.

A possible interpretation of the debate concerning the conventionality of simultaneity within SR, on the present construal, is that Grünbaum, perhaps in order to avoid the charge of simply rediscovering a form of trivial semantic conventionality, appealed to his notion of an intrinsic relation. Failure to express or denote an intrinsic feature was then said to characterize the interesting (nontrivial) cases of conventionality. This notion led to a number of difficulties for Grünbaum, culminating in Malament's apparent refutation of the conventionality of simultaneity within SR. However, it has been argued above that the appeal to intrinsicness turns out to be unnecessary in order to preserve a nontrivial form of conventionalism, C_2.

Grünbaum's arguments concerning the conventionality of the space-time metric for continuous manifolds involve more complex issues than those in the simultaneity case. But some key elements are common to both disputes. Grünbaum claims that continuous manifolds do not have intrinsic metrics, and critics complain that the notion of an intrinsic feature is unclear (see e.g., Stein 1977). Some of the claims that Grünbaum makes lend support to the idea that he is concerned with C_1 conventionalism. He claims, for example, that for a discrete manifold (such that between any two points there are only finitely many points), there is an intrinsic and hence nonconventional metric, where the "distance" between any two points is the number of points between them, plus one. This claim makes sense to the extent that one is concerned with C_1, that is, with the question of whether a relation is definable or logically constructible from a basic set. Grünbaum's arguments do not seem well suited for showing that such a metric would be C_2 nonconventional, since nothing about the meaning of 'distance' constrains the adoption of this as opposed to infinitely other metrics.

Another worry that might be raised concerns Grünbaum's insistence that distance functions defined in terms of physical bodies and their behaviors under transport are extrinsic to a manifold. If the manifold is a purely mathematical object, then such a view seems more plausible. But if one is concerned with a physical manifold of space-time points, it is less obvious that physical bodies should be treated as extrinsic. More importantly, even if one grants Grünbaum that physical bodies and their behaviors are extrinsic to the physical space-time manifold, one might want to grant a scientist the right to specify which features of the structures *that are being posited* are to count as basic or intrinsic. For example, Grünbaum proposes that Sir Isaac Newton in effect made a conceptual error in claiming that whether the temporal interval between one pair of instants is the same as that between another pair of instants is a factual matter, determined by the structure of time itself. Since instants form a temporal continuum in Newton's picture, there is, according to Grünbaum, no intrinsic temporal metric, contrary to Newton's claims. However, one can imagine a Newtonian responding with bewilderment. Does Newton not get to say what relations are intrinsic to the structures that he is positing?

It is difficult to see how to make room for a notion of intrinsicness that can do all of the philosophical work that Grünbaum requires of it, as Stein (1977) and others argue at length. On the other hand, Grünbaum's conventionalism about simultaneity within SR remains defensible (in particular, it escapes the trivialization charge as well as Malament's attempted refutation) if his conclusions are interpreted in the sense of C_2 conventionalism described above.

Mathematical and Logical Conventionalism

Up to now this discussion has focused on conventionalist claims concerning principles of an empirical science, physics. Other propositions that have attracted conventionalist treatment are those from the nonempirical sciences of mathematics and logic. Conventionalism seems especially attractive here, particularly to empiricists and others who find the notion of special faculties of mathematical intuition dubious but nevertheless wish to treat known mathematical or logical propositions as nonempirical.

The axiomatic methods that had motivated Poincaré's geometric conventionalism discussed above received further support with the publication of David Hilbert's *Foundations of Geometry* ([1921] 1962). Hilbert disregarded the intuitive or ordinary meanings of such constituent terms of Euclidean geometry as 'point,' 'line,' and 'plane,' and instead proposed regarding the axioms as purely formal posits (see Hilbert, David). In other words, the axioms function as generalizations about whatever set of things happens to satisfy them. In addition to allowing Hilbert to demonstrate a number of significant results in pure geometry, this method of abstracting from particular applications had immediate philosophical consequences. For instance, in response to Gottlob Frege's objection that without fixing a reference for primitive expressions like 'between,' Hilbert's axiomatic geometry would be equivocal, Hilbert wrote:

> But surely it is self-evident that every theory is merely a framework or schema of concepts together with their necessary relations to one another, and that the basic elements can be construed as one pleases. If I think of my points as some system or other of things, e.g., the system of love, of law, or of chimney sweeps ... and then conceive of all my axioms as relations between these things, then my theorems, e.g., the Pythagorean one, will hold of these things as well. (Hilbert 1971, 13)

Schlick was quick to find in Hilbert's conclusions the possibility of generalizing conventionalism beyond geometry (see Schlick, Moritz).

If the reference of the primitive concepts of an axiomatic system could be left undetermined, as Hilbert had apparently demonstrated, then the axioms would be empty of empirical content, and so knowledge of them would appear to be a priori. Like Poincaré, Schlick rejected a Kantian explanation of how such knowledge is possible, and he saw in Hilbert's approach a demonstration of how an implicit definition of concepts could be obtained through axioms whose validity had been guaranteed (Schlick [1925] 1985, 33). In his *General Theory of Knowledge* Schlick distinguished explicitly between "ordinary concepts," which are defined by ostension, and implicit definitions. Of the latter, he wrote that

> [a] system of truths created with the aid of implicit definitions does not at any point rest on the ground of reality. On the contrary, it floats freely, so to speak, and like the solar system bears within itself the guarantee of its own stability. (37)

The guaranteed stability was to be provided by a consistency proof for a given system of axioms, which Schlick, like Poincaré and Hilbert before him, claimed was a necessary condition in a system of symbolism. Schlick followed Poincaré in treating the axioms as conventions, and hence as known a priori through stipulation (Schlick [1925] 1985, 71). But he diverged markedly from Poincaré in extending this account to mathematics and logic as well. His basis for doing so was Hilbert's formalized theory of arithmetic, which Schlick hoped (wrongly, as it turned out) would eventuate in a consistency proof for arithmetic ([1925] 1985, 357).

The conventionalism that emerged was thus one in which true propositions known a priori were regarded as components of autonomous symbol games that, while perhaps constructed with an eye to an application, are not themselves answerable to an independent reality (Schlick [1925] 1985, 37–38). Schlick thought that the laws of logic, such as the principles of identity, noncontradiction, and the excluded middle, "say nothing at all about the *behavior of reality*. They simply regulate how we *designate the real*" ([1925] 1985, 337). Schlick acknowledged that the negation of logical principles like noncontradiction was unthinkable, but he suggested that this fact was *itself* a convention of symbolism (concerning the notion 'unthinkable'), claiming that "anything which contradicts the principle is termed *unthinkable*" ([1925] 1985, 337).

Although Schlick thought that the structure of a symbol system, including inference systems such as logic, was autonomous and established by a set of implicit definitions, he also recognized the possibility that some of its primitive terms could be coordinated by ordinary definitions with actual objects and properties. In this way, a conventionally established symbol system could be used to describe the empirical world, and an object designated by a primitive term might be empirically discovered to have previously unknown features. Nonetheless, some of the properties and relations had by a primitive term would continue to be governed by the system conventions through which the term is implicitly defined (Schlick [1925] 1985, 48ff). These properties and relations would thus be knowable a priori.

The conventionalism of Schlick's *General Theory of Knowledge* anticipated many of the conventionalist elements of the position advanced by members of the Vienna Circle (see Vienna Circle). Philosophers such as Hahn, Ayer, and Carnap saw in conventionalism the possibility of acknowledging the existence of necessary truths known a priori while simultaneously denying such propositions any metaphysical significance. Inspired by Wittgenstein's analysis of necessary truths in the *Tractatus*, Vienna Circle members identified the truths of mathematics and logic with *tautologies*—statements void of content in virtue of the fact that they hold no matter what the facts of the world may be (Hahn 1980; Ayer 1952). Tautologies, in turn, were equated with analytic statements, and Circle members regarded all analytic statements as either vacuous conventions of symbolism known a priori through stipulation or as derivable consequences of such conventions knowable a priori through proof (see Analyticity).

An especially noteworthy outgrowth of the conventionalism of the Vienna Circle was Carnap's *The Logical Syntax of Language*. In this book Carnap advanced a conventionalist treatment of the truths of mathematics and logic through his espousal of the principle of tolerance, which states:

> *In logic, there are no morals.* Everyone is at liberty to build up his own logic, i.e., his own form of language, as he wishes. All that is required of him is that ... he must state his methods clearly, and give syntactical rules instead of philosophical arguments. (Carnap 1937, 52)

Carnap regarded a language as a linguistic framework that specified logical relations of consequence among propositions. These logical relations are a precondition of description and investigation. In this view, there can be no question of justifying the selection of an ideal or even unique language with reference to facts, since any such justification (including a specification of facts) would presuppose a language. Mathematics is no exception to this. As a system of "framework truths," mathematical truths do not describe any convention-independent fact but are consequences of the decision to adopt one linguistic framework over another, in Carnap's account.

Conventionalism about mathematics and logic has faced a number of important objections. A significant early objection was given by Poincaré (1952, 166–172), who rejected Hilbert's idea that the principle of mathematical induction might merely be an implicit definition of the natural numbers (see Hilbert [1935] 1965, 193). Mathematical induction should not be regarded in this way, Poincaré argued, for it is presupposed by any demonstration of the consistency of the definition of number, since consistency proofs require a mathematical induction on the length of formulas. Treating the induction principle as itself conventional while using it to prove the consistency of the conventions of which it is a part would thus involve a *petitio*.

Carnap's *Logical Syntax of Language* sidestepped such problems by removing the demand that a system of conventions be consistent; the principle of tolerance made no such demand, and Carnap regarded the absence of consistency proofs as of limited significance (1937, 134). It appears that he would have regarded a contradictory language system to be pragmatically useless but not impossible.

Kurt Gödel, however, raised an important objection to this strategy. Gödel claimed that if mathematics is to be "merely" a system of syntactical conventions, then the conventions must be *known* not to entail the truth or falsity of any sentence involving a matter of extralinguistic empirical fact; if it does have such entailments, in Gödel's view, the truth of such sentences is not merely a syntactical, conventional matter (Gödel 1995, 339). If the system contains a contradiction, then it will imply every empirical sentence. But according to Gödel's second incompleteness theorem, a consistency proof required to assure the independence of the system cannot emerge from within the system itself if that system is consistent (1995, 346).

Recent commentators (Ricketts 1994; Goldfarb 1995) have argued that a response to this objection is available from within Carnap's conventionalist framework. Gödel's argument requires acceptance of an extralinguistic domain of empirical facts, which a stipulated language system may or may not imply. But it is doubtful that Carnap would have accepted such a domain, for Carnap considered linguistic conventions, including the conventions constitutive of mathematics, as antecedent to any characterization of the facts, including facts of the empirical world. The issues involved in Gödel's objection are complex and interesting, and a thorough and satisfactory conventionalist response remains to be formulated.

Another objection to conventionalism about logic was advanced by Willard Van Quine (see Quine, Willard Van). Quine objected against Carnap that treating logical laws and inference rules as conventional truths implicitly presupposed the logic that such conventions were intended to establish. Consider for instance a proposed logical convention *MP* of the form "Let all results of putting a statement for p and a statement for q in the expression 'If p, then q and p, then q' be true." In order to apply this convention to particular statements A and B, it seems that one must reason as follows: *MP* and if *MP*, then (A and if A, then B imply B); therefore, A and if A, then B imply B). But this requires that one use *modus ponens* in applying the very convention that stipulates the soundness of this inference. Quine concluded that "if logic is to proceed *mediately* from conventions, logic is needed for inferring logic from the conventions" (Quine 1966, 97). Quine at one point suggested that similar reasoning might undermine the intelligibility of the notion of a linguistic convention in general (98–99). However, after David Lewis' analysis of tacit conventions (Lewis 1969), Quine acknowledged the possibility of at least some such conventions (Quine, in Lewis 1969, xii).

Quine's original objection suggests that the conventionalist about logical truth must have an account of how something recognizable as a convention can intelligibly be established "prior to logic." As with Gödel's objection, the issues are difficult and have not yet been given a fully satisfying conventionalist account. However, there are a variety of strategies that a conventionalist might explore. One will be mentioned here. Consider an analogous case, that of rules of grammar. On one hand, one cannot specify rules of grammar without employing a language (which has its grammatical rules). Thus one cannot acquire one's first language by, say, reading its rules in a book. On the other hand, this fact does not seem to rule out the possibility that any given rules of grammar are to some extent arbitrary, and conventionally adopted. Whether the conventionalist can extend this line of thought to a defensible form of conventionalism about logic remains to be conclusively demonstrated.

Cory Juhl
Eric Loomis

The authors acknowledge the helpful input of Samet Bagce, Hans-Johan Glock, Sahotra Sarkar, and David Sosa.

References

Ayer, Alfred J. (1952), *Language, Truth and Logic*. New York: Dover.

Carnap, Rudolph (1937), *The Logical Syntax of Language*. London: Routledge and Kegan Paul.

Duhem, Pierre ([1906] 1954), *The Aim and Structure of Physical Theory*. Translated by Philip P. Wiener. Princeton, NJ: Princeton University Press.

Eddington, A. S (1953), *Space, Time and Gravitation*. Cambridge: Cambridge University Press.

Gergonne, Joseph Diaz (1818), "Essai sur la théorie de la définition," Annales de mathématiques pures et appliquées 9:1–35.

Giedymin, Jerzy (1982), *Science and Convention: Essays on Henri Poincaré's Philosophy of Science and the Conventionalist Tradition*. Oxford: Pergamon.

Gödel, Kurt (1995), "Is Mathematics Syntax of Language?" in Solomon Feferman (ed.), *Kurt Gödel: Collected Works*, vol. 3. Oxford: Oxford University Press, 334–362.

Goldfarb, Warren (1995), "Introductory Note to *1953/9," in Solomon Feferman (ed), *Kurt Gödel: Collected Works*, vol. 3. Oxford: Oxford University Press, 324–334.

Grünbaum, Adolph (1968), *Geometry and Chronometery in Philosophical Perspective*. Minneapolis: University of Minnesota Press.

—— (1973), *Philosophical Problems of Space and Time*, 2nd ed. Dordrecht, Netherlands: Reidel.

Hahn, Hans (1980), *Empiricism, Logic, and Mathematics*. Translated by Brian McGuinness. Dordrecht, Netherlands: Reidel.

Hilbert, David ([1935] 1965), "Die Grundlegung der elementaren Zahlenlehre," in *Gesammelte Abhandlungen*, vol. 3. New York: Chelsea Publishing, 192–195.

—— ([1921] 1962), *Grundlagen der Geometrie*. Stuttgart: B. G. Teubner.

—— (1971), "Hilbert's Reply to Frege," in E. H. W. Kluge (trans., ed.), *On the Foundations of Geometry and Formal Theories of Arithmetic*. London: Yale University Press, 10–14.

Lewis, David (1969), *Convention: A Philosophical Study*. Oxford: Blackwell.

Lie, Sophus (1959), "On a Class of Geometric Transformations," in D. E. Smith (ed.), *A Source Book in Mathematics*. New York: Dover.

Malament, David (1977), "Causal Theories of Time and the Conventionality of Simultaneity," *Nous* 11: 293–300.

Norton, John (1992), "Philosophy of Space and Time," in M. H. Salmon et. al. (eds), *Introduction to the Philosophy of Science*. Indianapolis: Hackett.

Poincaré, Henri ([1905] 1952), *Science and Hypothesis*. Translated by W. J. Greestreet. New York: Dover Publications.

—— (1952), *Science and Method*. Translated by Francis Maitland. New York: Dover Publications.

Putnam, Hilary (1975a), "An examination of Grunbaum's Philosophy of Geometry," in *Mathematics, Matter and Method Philosophical Papers*, vol. 1. New York: Cambridge University Press.

—— (1975b), "Reply to Gerald Massey," in *Mind, Language and Reality, Philosophical Papers*, vol. 2. New York: Cambridge UP.

Quine, Willard Van (1966), "Truth By Convention," in *The Ways of Paradox and Other Essays*. New York: Random House, 70–99.

Ricketts, Thomas (1994), "Carnap's Principle of Tolerance, Empricism, and Conventionalism," in Peter Clark and

CORROBORATION

Karl R. Popper (1959) observed that insofar as natural laws have the force of prohibitions, they cannot be adequately formalized as unrestrictedly general material conditionals but incorporate a modal element of natural necessity (Laws of Nature). To distinguish between possible laws and mere correlations, empirical scientists must therefore subject them to severe tests by serious attempts to refute them, where the only evidence that can count in favor of the existence of a law arises from unsuccessful attempts to refute it. He therefore insisted upon a distinction between 'confirmation' and 'corroboration' (see Popper, Karl Raimund).

To appreciate Popper's position, it is essential to consider the nature of natural laws as the objects of inquiry. When laws of nature are taken to have the logical form of unrestrictedly general material conditionals, such as $(x)(Rx \rightarrow Bx)$ for "All ravens are black," using the obvious predicate letters Rx and Bx and $\rightarrow$, as the material conditional, then they have many logically equivalent formulations, such as $(x)(\neg Bx \rightarrow \neg Rx)$, using the same notation, which stands for "Every nonblack thing is a nonraven." As Carl G. Hempel (1965) observed, if it is assumed that confirming a hypothesis requires satisfying its antecedent and then ascertaining whether or not its consequent is satisfied, then if logically equivalent hypotheses are confirmed or disconfirmed by the same evidence, a white shoe as an instance of $\neg Bx$ that is also $\neg Rx$ confirms the hypothesis "All ravens are black" (see Confirmation Theory; Induction, Problem of).

Popper argued that there are no "paradoxes of falsification" parallel to the "paradoxes of confirmation." And, indeed, the only way in which even a material conditional can be falsified is by things satisfying the antecedent but not satisfying the consequent. Emphasis on falsification therefore implies that serious tests of hypotheses *presuppose satisfying their antecedents*, thereby suggesting a methodological maxim of deliberately searching for examples that should be most likely to reveal the falsity of a hypothesis if it is false, such as altering the diet or the habitat of ravens to ascertain whether that would have any effect on their color. But Popper's conception of laws as prohibition was an even more far reaching insight relative to his falsificationist methodology.

Popper's work on *natural necessity* distinguishes it from logical necessity, where the notion of projectible predicates as a pragmatic condition is displaced by the notion of dispositional predicates as a semantic condition. It reflects a conception of laws as relations that cannot be violated or changed and require no enforcement. Fetzer (1981) has pursued this approach, where lawlike sentences take the form of subjunctive conditionals, such as $(x)(Rx \Rightarrow Bx)$, where $\Rightarrow$ stands for the subjunctive conditional. This asserts of everything, "If it were a raven, then it would be black." The truth of this claim, which is logically contingent, depends on a difference between permanent and transient properties. It does not imply the counterpart, "If anything were nonblack, then it would be a nonraven."

CORROBORATION

Among Popper's most important contributions were his demonstration of how his falsificationist methodology could be extended to encompass statistical laws and his propensity interpretation of probability (see Probability). Distributions of outcomes that have very low probabilities in hypotheses are regarded, by convention, as methodologically falsifying those hypotheses, tentatively and fallibilistically. Popper entertained the prospect of probabilistic measures of corroboration, but likelihood measures provide a far better fit, since universal hypotheses as infinite conjunctions have zero probability. Indeed, Popper holds that the appropriate hypothesis for scientific acceptance is the one that has the greatest content and has withstood the best attempts at its falsification, which turns out to be the least probable among the unfalsified alternatives.

The likelihood of hypothesis H, given evidence E, is simply the probability of evidence E, if hypothesis H is true. While probabilistic measures have to satisfy axioms of summation and multiplication—for example, where mutually exclusive and jointly exhaustive hypotheses must have probabilities that sum to 1—likelihood measures are consistent with arbitrarily many hypotheses of high value. This approach can incorporate the laws of likelihood advanced by Ian Hacking (1965) and a distinction between *preferability* for hypotheses with a higher likelihood on the available evidence and *acceptability* for those that are preferable when sufficient evidence becomes available. Acceptability is partially determined by relative likelihoods, even when likelihoods are low. Popper (1968, 360–91) proposed that where E describes the outcome of a new test and B our background knowledge prior to that test, *the severity of* E *as a test of* H, *relative to* B, might be measured by

$$P(E \mid H \wedge B) - P(E \mid B), \quad (1)$$

that is, as the probability of E, given H and B, minus the probability of E, given B alone. The intent of equation 1 may be more suitably captured by a formulation that employs a symmetrical—and therefore absolute—measure reflecting differences in expectations with respect to outcome distributions over sets of relevant trials, such as

$$|P(E \mid H) - P(E \mid B)|, \quad (2)$$

which ascribe degrees of nomic expectability to relative frequency data, for example. When B entails E, then $P(E|B) = 1$, and if $P(E|H) = 1$ as well, the severity of any such test is minimal, that is, 0. When H entails E, while B entails $\neg E$, the severity of such a test is maximal, i.e., 1. The acceptance of H may require the revision of B to preserve consistency.

A plausible measure of *the degree of corroboration* C *of* H, *given* E, *relative to* B, would be

$$c(H \mid E \wedge B) = L(H \mid E)[\,|P(E \mid H) - P(E \mid B)|\,], \quad (3)$$

that is, as the product of the likelihood of H, given E, times the severity of E as a test of H, relative to B. This is a Popperian measure, but not necessarily Popper's. Popper suggests (as one possibility)

$$C(H \mid E \wedge B) = P(E \mid H) - P(E \mid B) / P(E \mid H) + P(E \mid B), \quad (4)$$

which even he does not find to be entirely satisfactory and which Imre Lakatos severely criticizes (Lakatos 1968, especially 408–16). When alternative hypotheses are available, the appropriate comparative measures appear to be corroboration ratios

$$\frac{C(H_2 \mid E)}{C(H_1 \mid E)} = \frac{L(H_2 \mid E)[\,|P(E \mid H_2) - P(E \mid H_1)|\,]}{L(H_1 \mid E)[\,|P(E \mid H_1) - P(E \mid H_2)|\,]} \quad (5)$$

that reduce to the corresponding likelihood ratios of $L(H_2|E)$ divided by $L(H_1|E)$ and assume increasing significance as a function of the severity of those tests, as Fetzer (1981, 222–230) explains.

Popper also proposed the propensity interpretation of physical probabilities as probabilistic dispositions (Popper 1957 and 1959). In a revised formulation, the single-case propensity interpretation supports a theory of lawlike sentences, logically contingent subjunctive conditionals ascribing permanent dispositional properties (of varying strength) to everything possessing a reference property (Fetzer 1981 and 1993). Propensity hypotheses are testable on the basis of the frequencies they generate across sequences of trials. Long runs are infinite and short runs are finite sequences of single trials. Propensities predict frequencies but also explain them. Frequencies are evidence for the strength of propensities.

Popper (1968) promoted the conception of science as a process of conjectures and (attempted) refutations. While he rejected the conception of science as a process of inductive confirmation exemplified by the work of Hans Reichenbach (1949) and Wesley C. Salmon (1967), his commitments to deductive procedures tended to obscure the role of ampliative reasoning in his own position. Popper rejected a narrow conception of induction, according to which the basic rule of inference is "If m/n observed As are Bs, then infer that m/n As are Bs, provided a suitable number of As are tested under a wide variety of conditions." And, indeed, the rule he rejects restricts scientific hypotheses to those couched in observational language and cannot separate *bona fide* laws from correlations.

Although it was not always clear, Popper was not thereby rejecting induction in the broad sense of ampliative reasoning. He sometimes tried to formalize his conception of corroboration using the notion of absolute (or prior) probability of the evidence E, $P(E)$, which is typically supposed to be a subjective probability. Grover Maxwell (1974) even develops Popper's approach using Bayes's theorem. However, appeals to priors are inessential to formalizations of Popper's measures (Fetzer 1981), and it would be a mistake to suppose that Popper's account of severe tests, which is a pragmatic conception, could be completely formalizable.

Indeed, Popper's notion of accepting hypotheses on the basis of severe tests, no matter how tentatively and fallibilistically, implies ampliative reasoning. Its implementation for probabilistic hypotheses thereby requires large numbers of trials over a wide variety of conditions, which parallels the narrow inductivist conception. These results, however, must be subjected to severe tests to make sure the frequencies generated are robust and stable under variable conditions. When Volkswagens were first imported into the United States, for example, they were all gray. The narrow inductivist rule of inference justified inferring that all Volkswagens were gray, a conclusion that could not withstand severe tests.

JAMES H. FETZER

References

Fetzer, James H. (1981), *Scientific Knowledge: Causation, Explanation, and Corroboration*. Dordrecht, Netherlands: D. Reidel.
——— (1993), *Philosophy of Science*. New York: Paragon House.
Hacking, Ian (1965), *Logic of Statistical Inference*. Cambridge: Cambridge University Press.
Hempel, Carl G. (1965), *Aspects of Scientific Explanation*. New York: Free Press.
Lakatos, Imre (1968), "Changes in the Problem of Inductive Logic," in *The Problem of Inductive Logic*. Amsterdam: North-Holland, 315–417.
Maxwell, Grover (1974), "Corroboration without Demarcation," in Paul A. Schilpp (ed.), *The Philosophy of Karl R. Popper*, part 1. LaSalle, IL: Open Court, 292–321.
Popper, Karl R. (1959), *The Logic of Scientific Discovery*. New York: Harper & Row.
——— (1957), "The Propensity Interpretation of the Calculus of Probability, and the Quantum Theory," in Stephan Korner (ed.), *Observation and Interpretation in the Philosophy of Physics*. New York: Dover Publications, 65–70.
——— (1959), "The Propensity Interpretation of Probability," *British Journal for the Philosophy of Science* 10: 25–42.
——— (1968), *Conjectures and Refutations*. New York: Harper & Row.
Reichenbach, Hans (1949), *The Theory of Probability*. Berkeley and Los Angeles: University of California.
Salmon, Wesley C. (1967), *The Foundations of Scientific Inference*. Pittsburgh: University of Pittsburgh Press.

See also **Confirmation Theory; Hempel, Carl Gustav; Induction, Problem of; Popper, Karl Raimund; Verisimilitude**

COSMOLOGY

See **Anthropic Principle**

COUNTERFACTUALS

See **Causality**

DECISION THEORY

Decision theory seeks to provide a normative account of rational decision making and to determine the extent to which human agents succeed in living up to the rational ideal. Though many decision theories have been proposed, the version of *expected utility theory* developed in L. J. Savage's ([1954] 1972) classic *Foundations of Statistics* remains the best developed and most influential. It will serve as the principal focus of this article. Savage established a general framework for thinking about decision problems. He codified core tenets of the theory of rational preference and argued cogently for them. Most important, he proved a *representation theorem* that helps to legitimize subjective Bayesian approaches to epistemology and to justify *subjective expected utility maximization* as the foundation of rational decision making (see Bayesianism). Indeed, Savage's contributions are so seminal that the best way to approach the topic of decision theory is by treating his theory as a kind of "gold standard" and discussing other views as reactions or additions to it. This is the approach taken here. This article has three sections. The first discusses the general notion of a decision problem. The second introduces the expected utility hypothesis and explains Savage's representation theorem. The third presents the standard theory of rational preference and discusses objections to it.

Decision Problems

Savage's model assumes a rational decision maker, hereafter the *agent*, who uses beliefs about possible *states of the world* to choose *actions* that can be expected to produce desirable *consequences*. The states describe all relevant contingencies that lie beyond the agent's direct control. Any uncertainty that figures into the agent's choice is portrayed as ignorance about which state obtains. *Events* are disjunctions of states that provide less specific descriptions of the circumstances under which choices are made than states do. Consequences serve as objects of *noninstrumental* desire. Each specifies a possible course of events that is sufficiently detailed to settle every matter about which the agent intrinsically cares. Acts are objects of *instrumental* desire; the agent values them only insofar as they provide a means to the end of securing desirable consequences. When there are only finitely many acts and states, the person's choice can be described using a *decision matrix*:

	S_1	S_2	$S_3\ldots$	S_n
A_1	$C_{1,1}$	$C_{1,2}$	$C_{1,3}\ldots$	$C_{1,n}$
A_2	$C_{2,1}$	$C_{2,2}$	$C_{2,3}\ldots$	$C_{2,n}$
A_3	$C_{3,1}$	$C_{3,2}$	$C_{3,3}\ldots$	$C_{3,n}$
$\vdots$	$\vdots$	$\vdots$	$\vdots$	$\vdots$
A_m	$C_{m,1}$	$C_{m,2}$	$C_{m,3}\ldots$	$C_{m,n}$

where S_j = states, A_i = acts, and $C_{i,j}$ = the outcome that A_i will produce when S_j obtains.

This model of decision problems applies to "one-choice" decisions made at a specific time. Though early decision theorists, like Savage, believed that sequences of decisions could be represented by one-shot choices among contingency plans, or *strategies*, this view now has few adherents. The topic of dynamic decision making lies beyond the scope of this article. (For relevant discussions, see Hammond 1988; McClennen 1990; and Levi 1991.)

A decision problem is counted as rational only if the following conditions hold:

- The value of each consequence C is independent of the act and state that bring it about.
- Each act/state pair (A, S) determines a unique consequence $C_{A,S}$.
- The agent cannot causally influence that which the state obtains.

Many misguided objections to expected utility theory involve decision problems that violate these requirements. For example, the following is a central tenet of the theory:

- *Comparative Probability (CP)*: For any event E and consequences C and C^* with C preferred to C^*, the agent prefers an act that produces C when E and C^* when $\neg E$ to an act that produces C^* when E and C when $\neg E$ if and only if the agent is more confident in E than in $\neg E$.

This seems susceptible to counterexample. Imagine a person who is convinced that the average annual inflation rate over the next decade will be either 10% or 1% but that 10% is far more likely:

	E = The rate of inflation will be high over the next decade.	The rate of inflation will not be high over the next decade.
A	Be paid $1,000 in ten years.	Be paid $0 in ten years.
A^*	Be paid $0 in ten years.	Be paid $1,000 in ten years.

Despite confidence in E, such a person may prefer A^* to A on the grounds that $1,000 will be worth more if inflation is low than if it is high. This preference does not refute CP, which applies only to preferences in well-formed decision problems, and this problem is *ill-formed*, since the consequence "Be paid $1,000 in ten years" is worth less when it appears in the upper left than when it appears in the lower right. (Proponents of *state-dependent utility theory* relax this requirement by allowing utilities of outcomes to vary with states. See Karni 1985.) To fix the problem, one needs to rewrite outcomes as follows:

	The rate of inflation will be high over the next decade.	The rate of inflation will not be high over the next decade.
A	Be paid $1,000 after ten years of high inflation.	Be paid $0 after ten years of low inflation.
A^*	Be paid $0 after ten years of low inflation.	Be paid $1,000 after ten, years of high inflation.

When the problem is redescribed this way, the preference for A over A^* does not violate **CP**.

Similar problems arise when the decision maker can influence states of the world. Another core tenet of expected utility theory is:

- *Dominance*. If the agent prefers A's consequences to A^*'s consequences in every possible state of the world, then the agent prefers A to A^*.

Dominance sometimes seems to make absurd recommendations:

	One will contract influenza this winter.	One will not contract influenza this winter.
Get a flu shot.	Get the flu, and suffer the minor pain of a shot.	Avoid the flu, but suffer the minor pain of a shot.
Do not get a flu shot.	Get the flu, but avoid the minor pain of a shot.	Avoid the flu, and avoid the minor pain of a shot.

Here it seems as if Dominance requires one to forgo the shot to avoid the pain, which is terrible advice given that the chances of getting the flu are markedly less with the shot than without it. Again, this is not an objection to expected utility theory, but an ill-posed decision problem. To properly reformulate the problem, one must use states like these:

- One will contract the flu whether or not one gets the shot.
- One will not contract the flu whether or not one gets the shot.
- One will contract the flu if one gets the shot, but not otherwise.
- One will not contract the flu if one gets the shot, but will otherwise.

Dominance reasoning always holds good for states that are independent of the agent's acts, as these states are.

The debate between *causal* and *evidential* decision theorists has to do with the sort of independence that is required here. Evidentialists believe that states must be *evidentially* independent of acts, so that no act provides a sign or signal of the occurrence of any state. Causal decision theorists adopt the stronger requirement that states must be *causally* independent of acts, so that nothing the agent can do will change the probabilities of states (for details, see Jeffrey 1983; Gibbard and Harper 1978; Skyrms 1980; Joyce 1999).

Expected Utility Representations of Preference

Following Savage, it is standard in decision theory to assume that after due deliberation, a rational agent will be able to order acts with respect to their effectiveness as instruments for producing desirable outcomes. This generates a *weak preference ranking* $A \geq A^*$ that holds between acts A and A^* just when, all things considered, the agent strictly prefers A to A^* or is indifferent between them. It is important to understand that this preference ranking among acts is *not* what the agent *starts out* with when making her decision. It is the *end result* of the deliberative process.

Decision theorists have historically understood preferences behavioristically, so that $A > A^*$ means that the agent would choose A over A^* if given the chance. Though some social scientists still adhere to this interpretation, it has been widely and effectively criticized. The alternative is to take preferences as representing one's *all-things-considered judgments* about which acts will best serve one's interests. In this reading, saying that one prefers A to A^* means that the balance of one's reasons favors realizing A rather than A^*; whether one can or will choose A is another matter.

An act A is *choiceworthy* when it is weakly preferred to every alternative. According to the expected utility hypothesis, rationally choiceworthy acts maximize the decision maker's *subjective expected utility*. In Savage's framework, an act's expected utility is defined as

$$Exp_{P,U}(A) = P(S) \times U(C_{A,S})$$

where P is a *probability function* defined over events, and U is a real-valued *utility function* defined over consequences. To show that rationally choiceworthy acts maximize expected utility, Savage imposed a system of axiomatic rationality constraints on preference rankings and then proved that any ranking satisfying his axioms would be consistent with the hypothesis that acts are ranked according to expected utility. (The first result of this type is found in Ramsey 1931.) One can think of Savage as seeking to establish the following two claims:

1. *Theory of Practical Rationality*. Any practically rational agent will have a preference ranking among acts that obeys Savage's axioms.
2. *Existence of Subjective Expected Utility Representations*. For any preference ranking that obeys Savage's axioms, there will be at least one probability/utility pair (P, U) such that:
   - **P** *represents* the agent's beliefs: One event E is taken to be at least as likely as another E^* only if $P(E) \geq P(E^*)$.
   - **U** represents the agent's (intrinsic) desires for consequences: C is weakly preferred to C^* only if $U(C) \geq U(C^*)$.
   - $Exp_{P,U}$ accurately represents the agent's (instrumental) desires for actions: A is weakly preferred to A^* only if $Exp_{P,U}(A) > Exp_{P,U}(A^*)$.

It follows directly that one whose preferences satisfy Savage's axioms will always behave *as if* one were choosing acts based on their expected utility (though it is in no way required that one actually use this method in making the decision).

Savage also proves that this representation is unique once a unit and zero point for measuring utility have been fixed. To establish uniqueness, Savage was forced to assume that the agent has determinate preferences over an extremely rich set of options. Many decision theorists reject these "richness assumptions" and so believe that an agent's beliefs and desires should be represented by *sets* of probability/utility pairs, as in Joyce (1999, 102–105).

The Theory of Rational Preference

Since there is no question about the validity of Savage's representation theorem, his case for expected utility maximization rests on the plausibility of his axioms as requirements of practical rationality. Rather than trying to formulate things as Savage does, it will be better to discuss informal versions of those of his axioms and auxiliary assumptions that are components of every expected utility theory.

Frame Invariance and Value Independence

It will serve to begin by considering two principles that are left implicit in most formulations of expected utility theory:

- *Frame Invariance (INV).* A rational agent's preferences among acts should depend only on the consequences the acts produce in various states of the world, and not on the ways in which these consequences, or the acts themselves, happen to be described.
- *Value Independence (VALUE).* A rational agent endows each act with a value that is independent of the decisions in which it figures.

While **INV**'s credentials as a requirement of rationality have never been seriously questioned, a great deal of empirical research suggests that people's preferences do often depend on the way in which decisions are "framed." For example, when presented with the following two decisions (see Tversky and Kahneman 1986),

1. One will be paid $300 to choose between (a) getting another $100 for sure or (b) getting another $200 with probability $\frac{1}{2}$ and $0 with probability $\frac{1}{2}$, and
2. One will be paid $500 to make a choice between either (a*) paying back $100 for sure or (b*) paying back $0 with probability $\frac{1}{2}$ and paying back $200 with probability $\frac{1}{2}$,

a surprising number of people prefer (a) in the first choice and (b*) in the second, thus violating **INV**. The different descriptions of the same options lead people to view their choices from different perspectives. In the first case, they see themselves as having $300 and try to *improve* their lot by choosing between (a) and (b); while in the second, they (wrongly) see themselves having $500 and try to *preserve* their fortune by choosing between (a*) and (b*). This generates problems when conjoined with the following facts about human behavior (see Kahneman and Tversky 1979; Shafir and Tversky 1995):

- *Divergence from the Status Quo.* People are more concerned with gains and losses, seen as additions or subtractions from the status quo, than with total well-being or overall happiness.
- *Asymmetrical Risk Aversion.* People tend to be risk averse when pursuing gains, but risk seeking when avoiding losses.

People who choose between (a) and (a*) see $300 as the status quo, and thus prefer the less risky (a), since they are risk averse when pursuing gains. When choosing between (b) and (b*), they see $500 as the status quo and prefer the more risky (b*) because they are risk seeking when aiming to avoid losses.

VALUE has a number of important implications. First, it entails that one should be able to experimentally "elicit" a person's preference between A and A^* using any of the following methods:

- *Fair Prices.* Have an agent put a "fair price" on each action, and conclude that the higher-priced act is preferred.
- *Choice.* Have an agent choose between A and A^*, and conclude that the chosen act is not dispreferred,
- *Rejection.* Have an agent reject A or A^*, and conclude that the rejected act is not preferred.
- *Exchange.* Award an agent A^*, offer to trade A for A^* plus a small fee, and conclude that A is preferred if the agent makes the trade.

Surprisingly, these procedures can all yield different results, a fact that creates havoc for behaviorist analyses of preference. In cases of *preference reversal*, an agent who sets a higher price on A also selects A^* in a straight choice. Shafir (1993) gives examples in which subjects choose A over A^* and yet reject A for A^*. This happens because they focus more on comparisons among "positive" features of options when choosing, but more on negative features when rejecting. When A has both more pronounced positive features and more pronounced negative features, it can be both chosen and rejected. There are even cases in which an agent will refuse to trade A for A^* and refuse to trade A^* for A. The mere fact that the person 'owns' a prospect seems to make it more valuable to her. (This is referred to as *loss aversion*.)

VALUE also says that an agent's preferences among options should not depend on what other options happen to be available. This entails the following two principles (Sen 1971) (note that these do *not* apply when the addition or deletion of A^{**} provides relevant information about the desirability of A or A^*):

- *Principle-α*: If the agent will choose A over A^* and A^{**}, then the agent will choose A over A^* even when A^{**} is not available.

- *Principle-β*: If the agent will choose A in a straight choice against A^*, then the agent will also choose A in a choice between A, A^*, and some third option that is inferior to A^*.

Actual choosers often violate these principles. As salespeople have long known, one can more easily convince a person to buy a product by offering an inferior product at a higher price. Likewise, offering too many good options can lead a person to refrain from buying products he would have purchased had the list of options been smaller. A disconcerting violation of both principles is the finding of Redelmeier and Shafir (1995) that physicians are *less* likely to prescribe pain medication to patients when they can choose between ibuprofen and the inferior piroxicam than when they can choose only ibuprofen. While none of these empirical results have led decision theorists to question the normative standing of **VALUE**, they clearly show that expected utility theory is not an accurate *description* of human behavior.

Completeness

Along with Dominance and Comparative Probability, expected utility theorists also generally state as axioms:

- *Transitivity*. If the agent (strictly or weakly) prefers A to A^* and A^* to A^{**}, then the agent prefers A to A^{**}.
- *Completeness (COM)*. The agent either strictly prefers A to A^*, strictly prefers A^* to A, or is indifferent between them.
- *The "Sure-Thing" Principle (STP)*. If A and A^* produce the same consequences in every state consistent with an event $\neg E$, then the agent's preference between A and A^* depends exclusively only on their consequences when E obtains.

Though all five axioms are controversial in various ways, COM and STP have been the most contentious.

Completeness can fail in two ways. First, an agent might have no definite preference between two prospects either because the agent's intrinsic desires are vague or indeterminate or because the agent has insufficient information to judge which act will better promote desirable consequences. Being indifferent is not the same as having no preference. One who is indifferent between two options judges that they are equally desirable, but one who lacks a clear preference is unable to judge that one option is better than the other, or even that they are equally good—the person simply has no view about their relative desirabilities. Most decision theorists now admit that there is nothing irrational about having a "gappy" preference ranking, and it is becoming standard to treat COM (and any other axiom that requires the *existence* of preferences) as a *requirement of coherent extendibility* (see Jeffrey 1983; Kaplan 1983; Joyce 1999, 103–5). In this reading, it is irrational to have an incomplete preference ranking only if it cannot be extended to a complete ranking that obeys all the other axioms.

A more serious objection to COM comes from those who hold that rational agents may be unable to compare acts or consequences not because of any vagueness in their beliefs or desires, but because they regard the values of these prospects as genuinely *incommensurable* (Raz 1986; Anderson 1993). In one version of the view, a rational agent might regard distinctive standards of evaluation as appropriate to different sorts of prospects and so see prospects that do not fall under a common standard as incomparable. For example, a person might see it as perfectly appropriate to set a monetary price on a share of stock but also regard it as improper to put a price on spending an afternoon with one's children. If this is so, then a person's preference ranking will not compare these prospects, and no extension of it consistent with that person's values will do so. The incommensurability debate is too involved to pursue here. The heart of the issue has to do with the ability of rational agents to "balance off" reasons for and against an option so as to come to an all-things considered judgment about its desirability. Utility theorists think that such a balancing of reasons is always possible. Proponents of incommensurability deny this.

The Sure-Thing Principle

The Sure-Thing Principle forces preferences to be *separable across events*, so that a rational agent's preference between A and A^* depends *only* on what happens in states of the world in which these prospects produce *different* outcomes. When there are three states to be considered, STP requires an agent facing the following decision to prefer A over A^* if and only if the agent also prefers B over B^*:

	S_1	S_2	S_3
A	C_1	C_2	C_3
A^*	C_1^*	C_2^*	C_3
B	C_1	C_2	D_3
B^*	C_1^*	C_2^*	D_3

In deciding between A and A^* or between B and B^*, STP tells the agent to *ignore* what happens when S_3 holds, since the same result occurs under S_3 whichever option is chosen. In effect, the requirement is that the agent be able to form a preference between the following two *act types* whether or not the value of x is known:

	S_1	S_2	S_3
X	C_1	C_2	x
X^*	C_1^*	C_2^*	x

STP has generated more controversy than any other tenet of expected utility theory. Much of the discussion concerns two putative counterexamples, the *Allais* and *Ellsberg paradoxes*, which seem to show that an important component of rational preference—the amount of risk or uncertainty involved in an option—is nonseparable in the sense required by STP. In the jargon of economists, a prospect involves *risk* when the agent knows the objective probability of each state of the world. It involves *uncertainty* when the agent does not have sufficient information to assign objective probabilities to states. STP entails that an agent's attitudes toward risk and uncertainty can be fully captured by the combination of the agent utility function for outcomes and the probabilistic averaging involved in the computation of expected utilities. The Allais and Ellsberg paradoxes appear to show that the theory is wrong about this.

In *Allais' paradox* (Allais 1990) agents choose between A and A^* and then between B and B^* (where known probabilities of states are listed):

	0.33	0.01	0.66
A	$2,500	$0	$2,400
A^*	$2,400	$2,400	$2,400
B	$2,500	$0	$0
B^*	$2,400	$2,400	$0

Most people violate STP by preferring A^* to A and B to B^*, and these preferences remain stable upon reflection. The thinking seems to be that in the first choice one should play it safe and take the sure $2,400, since a 0.33 chance at an extra $100 does not compensate for a 0.01 risk of ending up with nothing. On the other hand, since one will probably end up with nothing in the second choice, the chance of getting an extra $100 makes the risk worth taking. Thus, Allais choosers think (a) that there is more risk involved in choosing in A over A^* than in choosing B over B^*, and (b) that this added risk justifies their nonseparable preferences.

In *Ellsberg's paradox* (Ellsberg 1961), a ball is drawn at random from an urn that is known to contain 30 red balls and 60 balls that are either white or blue but in unknown proportion. The agent is asked to choose between A and A^* and then between B and B^*.

	Red	White	Blue
A	$100	$0	$0
A^*	$0	$100	$0
B	$100	$0	$100
B^*	$0	$100	$100

Most people prefer A to A^* and B^* to B. People tend to prefer risk to equivalent levels of uncertainty when they have something to gain and to prefer uncertainty to risk when they have something to lose. Thus, A is preferred to A^* because it has $100 riding on a prospect of known risk 0.33, while A^* has that same sum riding on an uncertainty (ranging between risk 0 and risk 0.66). Likewise, B^* is preferred to B because it offers a definite 0.66 risk of $100 where B offers only an uncertainty (ranging between risk 0.33 and risk 1.0).

Some decision theorists take the Allais and Ellsberg paradoxes to show that expected utility theory is incapable of capturing rational attitudes toward risk. The problem with STP, they say, is that a rational agent need not be able to form any definite preference between the act types X and X^* because information about x's value might provide information about the relative *risk* of options, and this information can be relevant to the agent's preferences.

Many expected utility theorists respond to this objection by arguing that the Allais and Ellsberg paradoxes are *underdescribed* (Broome 1991, 95–115). One can render the usual Allais preferences consistent with STP by rewriting outcomes as follows:

	0.33	0.01		0.66
A	$2,500	$0 instead of a sure $2,400 with A^*		$2,400
A^*	$2,400	$2,400		$2,400
B	$2,500	$0 instead of a probable $0 with B^*		$0
B^*	$2,400	$2,400		$0

If the agent prefers the second outcome in A to the second outcome in B, then there is no violation of STP. Moreover, there is a plausible psychological explanation for this preference. A person who ends up with $0 instead of a sure $2,400 might experience pangs of regret that would not be felt if that person thought that ending up with $0 was likely anyhow. Thus, the person's decision really looks like this:

	0.33	0.01	0.66
A	$2,500	$2,400 and pangs of regret	$0
A*	$2,400.	$2,400	$2,400
B	$2,500	$0 with little regret	$0
B*	$2,400	$0	$2,400

The Ellsberg paradox can be handled similarly. If the agent feels a special sort of discomfort when gains ride on uncertain prospects (or losses ride on risky prospects), then the correct description of this problem might really be:

	Red	White	Blue
A	$100	$0	$0
A*	$0	$100 and discomfort	$100
B	$100 and discomfort	$0	$100 and discomfort
B*	$0	$100	$100

Again, there is no violation of STP here.

This way of eliminating counterexamples to STP worries many people, since it looks like an expected utility theorist can *always* use it. The fact that one can *always* explain away any seeming counterexamples to expected utility theory by redescribing outcomes and postulating the necessary beliefs and desires seems to show that the theory is contentless. This objection is especially effective against behaviorist interpretations of preference. Since behaviorists can appeal to only overt choices to isolate preferences, they have no principled way of distinguishing legitimate from ad hoc redescriptions of decision problems. Nothing in an Allais chooser's behavior, for example, indicates whether the agent is seeking to avoid some (unobservable) feeling of regret or is acting on the basis of nonseparable attitudes toward risk.

The objection is less effective when preferences are understood as all-things-considered judgments, for it is then possible to argue that certain redescriptions are correct because they *best explain* the totality of the person's behavior. If the hypothesis that people experience regret explains a great deal of human behavior, aside from violations of STP, then it is legitimate to use it to explain the common

DECISION THEORY

Allais preferences. Consider an analogy: It could be claimed that Newtonian mechanics is empty because (as is true) any pattern of observable motions can be made consistent with Newton's laws by positing the right constellation of forces. What makes this objection unconvincing is the fact that Sir Isaac Newton was able to account for a vast array of distinct motions using the single force of gravity. The same might be true in decision theory. If it can be shown that a small number of relatively simple psychological mechanisms, including feelings of regret or discomfort in situations of risk, explain a great deal of human behavior, then the *best explanation* for the Allais and Ellsberg choices might be among those proposed by the expected utility theorists. Of course, this places a burden on these theorists to show that by standard canons of scientific reasoning, their explanation is indeed the best available.

An alternative response is to take the description of the Allais and Ellsberg paradoxes at face value and to argue that the common preferences are irrational. To see how the argument might go for the former, note that the common rationale for the Allais choices assumes that the difference in risk between A and A* exceeds the difference in risk between B and B*. Proponents of expected utility theory will argue that this is mistaken. The best way to determine how much two options differ in risk, they will claim, is to ask how one might ensure against the increased chances of loss that one assumes in exchanging the less risky option for the more risky one. Someone who switches from A* to A in the Allais paradox can ensure against the risk of loss by buying an insurance policy that pays out $2,400 contingent on the 0.01 probability event. Moreover, the person can ensure against the incurred risk of switching from B* to B by purchasing *the same policy*. Since a single policy does both jobs, the actual change in risk must be the same in each case. Allais choosers, who perceive a greater risk in the first switch, are committed to paying more for the policy when using it as insurance against the A*-to-A risk than when using it as insurance against the B*-to-B risk. This difference in "risk premiums" shows that the Allais choosers' perceptions of risk do not track the actual risks of prospects. Similar things can be said about Ellsberg choosers.

Opponents of expected utility theory may deny that it is appropriate to measure risk by the costs of ensuring against it. In the end, the issue will be settled by the development of a convincing method for measuring the *actual* risks involved in prospects. (Ideally, this theory would be augmented by a plausible psychological account of *perceived* risks that explains the common Allais choices.) While there is a well-developed model of risk *aversion* within expected utility theory, this model does not seek to measure risk itself, only an agent's *attitudes* toward risk. While some progress has been made in the measurement of risk, a great deal remains to be done. It is known that no simple measure (standard deviation, mean absolute deviation, entropy) will do the job. Building on the classic paper by Michael Rothschild and Joseph Stiglitz (1970), economists have made great strides toward providing a definition of the "riskier than" relation. This work strongly suggests that risk is indeed a separable quantity, and thus that the Allais and Ellsberg choosers are irrational. Still, there is no universally accepted way of measuring the amount of risk that prospects involve. Until such a measure is found, the proper interpretation of the Allais and Ellsberg paradoxes is likely to remain controversial, as will expected utility theory itself.

JAMES M. JOYCE

References

Allais, Maurice (1990), "Allais Paradox," in J. Eatwell, M. Millgate, P. Newman (eds.), *The New Palgrave: Utility and Probability*. New York: Norton, 3–9.

Anderson, Elizabeth (1993), *Value in Ethics and in Economics*. Cambridge, MA: Harvard University Press.

Broome, John (1991), *Weighing Goods*. Oxford: Blackwell Publishers.

Ellsberg, Daniel (1961), "Risk, Ambiguity and the Savage Axioms," *Quarterly Journal of Economics* 75: 643–669.

Gibbard, Allan, and William Harper (1978), "Counterfactuals and Two Kinds of Expected Utility," in C. Hooker, J. Leach, and E. McClennen (eds.), *Foundations and Applications of Decision Theory*. Dordrecht, Netherlands: Reidel, 125–162.

Hammond, Peter (1988), "Consequentialist Foundations for Expected Utility Theory," *Theory and Decision* 25: 25–78.

Jeffrey, Richard (1983), *The Logic of Decision*, 2nd rev. ed. Chicago: University of Chicago Press.

Joyce, James M. (1999), *The Foundations of Causal Decision Theory*. New York: Cambridge University Press.

Kahneman, Daniel, and Amos Tversky (1979), "Prospect Theory: An Analysis of Decision Under Risk," *Econometrika* 47: 263–291.

Kaplan, Mark (1983), "Decision Theory as Philosophy," *Philosophy of Science* 50: 549–577.

Karni, Edi (1985), *Decision Making Under Uncertainty: The Case of State-Dependent Preferences*. Boston: Harvard University Press.

Levi, Isaac (1991), "Consequentialism and Sequential Choice," in M. Bacharach and S. Hurley (eds.), *Foundations of Decision Theory*. Oxford: Blackwell, 92–112.

McClennen, Edward (1990), *Rationality and Dynamic Choice: Foundational Explorations*. Cambridge: Cambridge University Press.

Ramsey, Frank (1931), "Truth and Probability," in R. Braithwaite (ed.), *The Foundations of Mathematics and Other Logical Essays*. London: Kegan Paul, 156–198.

Raz, Joseph (1986), *The Morality of Freedom*. Oxford: Clarendon Press.

Redelmeier, Donald, and Eldar Shafir (1995), "Medical Decision Making in Situations That Offer Multiple Alternatives," *Journal of the American Medical Association* 273: 302–305.

Rothschild, Michael, and Joseph Stiglitz (1970), "Increasing Risk: I. A Definition," *Journal of Economic Theory* 2: 225–243.

Savage, L. J. ([1954] 1972), *The Foundations of Statistics*, 2nd rev. ed. New York: Dover Press. Originally published by John Wiley and Sons.

Sen, Amartya K. (1971), "Choice Functions and Revealed Preference," *Review of Economic Studies* 38: 307–317.

Shafir, Eldar (1993), "Choosing Versus Rejecting: Why Some Options Are Both Better and Worse Than Others," *Memory and Cognition* 21: 546–556.

Shafir, Eldar, and Amos Tversky (1995), "Decision Making," in E. Smith and D. Osherson (eds.), *An Invitation to Cognitive Science*, vol. 3: *Thinking*, 2nd ed. Cambridge, MA: MIT Press, 77–100.

Skyrms, Brian (1980), *Causal Necessity*. New Haven, CT: Yale University Press.

Tversky, Amos, and Daniel Kahneman (1986), "Rational Choice and the Framing of Decisions," *Journal of Business* 59: S251–S278.

See also **Bayesianism; Game Theory**

PROBLEM OF DEMARCATION

"The problem of demarcation" is Karl Popper's label for the task of discriminating science from nonscience (see Popper, Karl Raimund). Nonscience includes pseudoscience and metaphysics but also logic and pure mathematics, philosophy (including value theory), religion, and politics (Popper 1959,

34; 1963, Ch. 1). Pseudoscience and metaphysics in turn include meaningless language, speculative theory, ad hoc conjectures, and some of what is today called junk science. Given this wide range of targets, the great diversity of legitimate sciences and philosophies of science, and human gullibility, it is not surprising that there is no agreement on whether there is an adequate decision procedure or criterion of demarcation and, if so, what it is. Meanwhile, the logical empiricists formulated their own criterion of demarcation in terms of their empiricist philosophy of language, an approach that Popper rejected (see below).

Traditional solutions to the problem of demarcation have been attempts to answer such questions as, What is science? What is special about science? What is (empirical) knowledge? and, by implication, Why is science important? There is much at stake for a society in the answers to such questions insofar as science enjoys cultural authority in that society. Ironically, in recent decades, the problem of demarcation has lost visibility in philosophical circles even as science and technology have gained unparalleled authority and even though creationists and various postmodernist groups now increasingly challenge that authority, not to mention the legal and political difficulties in identifying "sound science" (see Social Constructionism).

The distinction between science and nonscience is not always invidious. In his *Tractatus Logico-Philosophicus* of 1919, Wittgenstein drew the distinction in part to protect ethics from the incursions of science. The logical empiricists used it to distinguish philosophy from empirical science (see Cognitive Significance; Logical Empiricism).

Problems of demarcation arise at two different levels. One is the public level. Given the centrality of science and technology to modern societies, those societies do in fact demarcate allegedly good science from bad science and from nonscience in various ways. The question is, How *ought* they to accomplish such demarcation? At the other level, the same question arises within specialist scientific communities, although here the basis for discrimination will normally be more technical.

Historical Background

From the ancient Greeks on, Western methodologists have attempted to solve the problem by specifying a *criterion* of demarcation in the form of necessary and sufficient conditions for *epistēmē*, *scientia*, or good science. Historically prominent criteria of demarcation draw upon virtually all the main areas of philosophy. Criteria have been couched in terms of the ontological status of the objects of knowledge (e.g., Platonic Forms, Aristotelian essences), the semantic status of the products of research (science as a body of true or at least meaningful claims about the universe), the epistemological status of the products of research (science as a body of certain or necessary or reliable or appropriately warranted claims), the logical form of those claims (universal or particular, derivability of predictions from them), and value theory (the normative method that produces the claims, e.g., inductive or hypothetico-deductive, or comparison of a field with a model discipline such as physics).

For Aristotle, a claim is scientific if it is

- general or universal,
- absolutely certain, and
- causal-explanatory.

Here "general" means that the claim is of the form "All As are essentially Bs," and "causal-explanatory" means that the argument "All As are (essentially) Bs, and such-and-such is A; therefore, it is B" explains why it is B by attributing A as the cause. The possessor of genuine scientific knowledge has a demonstrative understanding of the first causes or essences of all things of a given kind. The logic or methodology of science and the investigative process itself are distinct from science proper. Aristotle stated his demarcation criteria in terms of the qualities of the products, not the process of producing them.

Two thousand years later, Galileo, Descartes, Newton, and other seventeenth-century natural philosophers still required virtual certainty for a claim in order to include it in the corpus of scientific knowledge. These early scientists also required causal-explanatory power of a sort; witness Newton's goal of finding true causes (*verae causae*) in his first rule of reasoning in the *Principia*. However, many of the natural philosophers abandoned as impossible Aristotle's demand for first causes and real essences. Newtonian mechanics could demonstrate the motion of the planets in terms of the laws of motion and gravitation, but it failed to find either the cause or the essence of gravity itself. Thus it could not provide a demonstrative chain of reasoning back to first causes (McMullin 2001).

In the wake of the English Civil War, the Royal Society of London expressly excluded religion and politics from its discussions and insisted that scientific discourse be conducted in plain (nonmetaphorical) language. Although Descartes had previously rejected rhetoric and the other humanities as bases for science, the secular saint of the Royal Society was Francis Bacon, who was usually interpreted as a simple inductivist. In this view, to be scientific, a claim must be induced from a body of previously gathered experimental or observational facts. Nature must be allowed to speak first as well as last.

The thinkers of the scientific Enlightenment shaped the modern concern with demarcation. If science is to be the supreme expression of human reason and the broom that sweeps away the cobwebs of tradition and folk wisdom, then it is crucial to distinguish science from pretenders (Amsterdamski 1975, 29). The Enlightenment legacy is that science and representative government are the two sacred institutions of modern society and that the special status of both must be preserved. Somewhat ironically, then, demarcation became a conservative exercise in exclusion. In its strongest versions, the demarcation project is associated with foundationist epistemologies. In particular, strong empiricists have regarded any claim with a suspicion proportional to its distance from experimental observation. (They have legitimized mathematics in a different way.)

In the eighteenth century, Hume insisted that natural science must be thoroughly empirical, since pure reason cannot say anything about the natural world (see Empiricism). Moreover, any meaningful expression must be traceable back to an origin in experience. Kant also insisted that it is philosophy's job to demarcate science from nonscience—but on *a priori* grounds—and also to adjudicate disputes among the sciences. His philosophy became especially influential because it was incorporated within the newly reformed German university system.

In the nineteenth century there was widespread agreement that "Baconian" induction was an overly restrictive method, and that the hypothetico-deductive method was not only legitimate but also superior, given that certainty is unattainable in science. The hypothetico-deductive method cannot achieve certainty, because of the fallacy of affirming the consequent, but neither can the inductive method that it largely supplanted. Some nineteenth-century and virtually all twentieth-century methodologists responded to the clarified logical situation by becoming fallibilists and by adopting self-correcting or successive-approximation methodologies of science in place of the old foundationist ones. Since these methodologists could no longer appeal to fail-safe epistemic status as the mark of substantive scientific claims, some retreated to the *process* that produced them: A claim is scientific if and only if it is produced by a proper application of the scientific *method*, and a discipline is scientific if and only if it is governed by the scientific method. This view enjoys considerable currency today, especially among textbook writers and school administrators. Of course, process or method had been part of the Baconian criterion all along, but the new dispensation considerably broadened what counted as a legitimate process as well as dropped the near-certainty of the final product.

One difficulty with this retreat from substance to method is that it becomes harder to defend the view that science, unlike other institutions, cumulatively produces objectively correct results. Another difficulty of the liberalized conception of process is that there was and is no agreement about whether there is a special scientific method at all, and if so, what that method is, what justifies its use, and whether it has changed historically (see Feyerabend, Paul). After all, how can anyone prove that a particular candidate for "the" scientific method is bound to produce results epistemically superior to those of any other method? Indeed, how can one know in advance whether or not a given method will be fruitful in this world or in any given domain of it? Ironically, prior to the Darwinian era, many methodologists would have said that the ultimate justification of method is theological. Third, John Herschel, William Whewell, Auguste Comte, W. S. Jevons, and others minimized the process of discovery in favor of the testability of the products of that process. Reversing the Bacon-Hume emphasis on antecedents, they asserted that it is observable consequences—predictions—that count, and that novel predictions count most (see Prediction). John Stuart Mill denied that novel predictions have special weight and remained a more conservative inductive-empiricist than Whewell and the hypotheticalists, who increasingly stressed the importance of conceptual and theoretical innovation as well as novel prediction. Popper and the logical empiricists would later reconstruct this divorce of antecedent exploration from logical consequences as involving a distinction of the psychological "context of discovery" from the logical "context of justification or corroboration," thereby reducing scientific method (the "logic" of science) to a logical minimum. As a criterion of demarcation, testability drastically weakens Aristotle's standard, for now a scientific statement need not even be supported by evidence, let alone be established as true (Laudan 1981, Chs. 9–11; Nickles 1987a).

Twentieth-Century Developments

The problem of demarcation was a central feature of the dominant philosophies of science—logical empiricism and Popperianism—at the time when philosophy of science emerged as a professional specialty area within philosophy, namely the period 1925–1970 (see Logical Empiricism; Popper, Karl Raimund; Vienna Circle). In *Tractatus*, §4.11, Wittgenstein had written, "The totality of true propositions is the whole of natural science (or the whole corpus of the natural sciences)." Inspired in part by Wittgenstein, some leading logical empiricists adopted not actual truth but empirical verifiability as their criterion of demarcation of science from metaphysics and pseudoscience. The verifiability criterion served as a criterion of both empirical meaningfulness and meaning itself. Given that a statement is meaningful (verifiable), what exactly does it mean? Roughly, the meaning is given by the method of verification. However, the logical empiricists soon realized that the unrestricted principle is untenable. It excludes mathematics and logic from science, and abstract theoretical and lawlike claims as well. Besides, its own status is unclear, since it itself is not verifiable (see Verifiability).

Pierre Duhem ([1914] 1954) had already shown, contrary to the logical empiricists and to Popper, that theories such as classical mechanics, Maxwellian electromagnetic theory, and relativity theory are not falsifiable in isolation (see Duhem Thesis). Only the larger complexes that include numerous and diverse auxiliary assumptions yield predictions, a point that Willard Van Quine expanded into a controversial, full-blown holism concerning the relation of science to experience, which in turn encouraged Thomas Kuhn (1962) to emphasize the underdetermination of theory by the facts plus logic (see Kuhn, Thomas; Quine, Willard Van). Paul Feyerabend (1981, Chs. 6 and 8) contended that some of the empirical content of a deep theory could be discovered only from the vantage point of another deep theory (see Feyerabend, Paul).

Meanwhile, the operationalism of the physicist Percy Bridgman and of several behavioral scientists required that the *smallest* linguistic units—namely, individual terms—receive operational definitions (see Behaviorism; Bridgman, Percy). Roughly, the meaning of a term is given by the operations that determine whether or not it applies. For example, the scratch test for minerals indicates the meaning of "Mineral *X is harder than* mineral *Y*." Moreover, in a reversal of the consequentialist tendency of methodological thinking, all the terms were to be defined in advance of any theorizing and were to provide a permanent conceptual basis for future science (see Scientific Revolutions).

Carl Hempel's (1965, Chs. 4–5) influential review of the literature summarized the failures of the various criteria of meaning and demarcation proposed by the logical empiricists and operationalists: They are both too restrictive and too permissive (see Cognitive Significance; Hempel, Carl Gustav). In agreement with Quine's attack on the analytic-synthetic distinction, Hempel (1965) concluded: "Theory formation and concept formation go hand in hand; neither can be carried on successfully in isolation from the other" (113). Thereafter, these programs faded in importance (see Analyticity).

The problem of demarcation is most closely associated with Popper, for whom it and the related problem of the growth of knowledge were the two central problems of philosophy (Popper 1959, 34; 1963, Ch. 1). Although both parties regarded empirical testability as the mark of a scientific statement, Popper disagreed with the leading logical empiricists on two main points. First, falsifiability alone counted as testability for Popper (a view that he had to soften in order to allow statistical-probabilistic statements). Second, refusing to take the linguistic turn, Popper rejected the logical empiricists' attempt to derive a demarcation criterion from a theory of meaning. Rudolf Carnap, for example, characterized philosophy of science as *Wissenschaftslogik*, the logic of (the language of) science, and required that all scientific statements meet the above-mentioned empiricist criterion of cognitive significance (see Carnap, Rudolf). All nonempirical statements are "metaphysical" and cognitively meaningless. Popper agreed that metaphysical statements are not scientific, but he insisted that they may be meaningful. He observed that the deepest scientific problems have their roots in metaphysical problems, which have served as a heuristic for modern science.

Popper also rejected inductive criteria of demarcation. A good hypothesis does not have to be winnowed from the empirical facts; nor can it be. Noting the logical asymmetry between verification and falsification (a single counterinstance can refute a universal claim, but no number of positive instances can verify it), Popper proposed falsifiability (empirical refutability) as the criterion of demarcation. For him a statement is scientific if and only if it is falsifiable in principle, that is, if it can fail an empirical test. This is equivalent to saying that there must be some *possible* observation statement (true or false) that logically contradicts the claim in question. Thus Newton's and Einstein's bold theories are scientific because they make risky empirical claims that can fail; but Marxist and Freudian theories are not scientific, despite their claims to wide explanatory power, because their advocates allow nothing to count as a refutation.

DEMARCATION, PROBLEM OF

Far from making strong claims about reality, these theories actually exclude nothing.

Popper criticized the age-old effort of philosophy to justify scientific claims positively. According to Popper, Hume had shown that scientific claims can never be justified, even to a probabilistic degree. For this reason he spoke of successful tests as "corroborating" rather than confirming hypotheses (see Corroboration). Scientific claims can never escape conjectural status. By contrast, Carnap and other logical empiricists responded to Hume with *probabilism,* the view that the facts can confer probability but not certainty upon universal empirical claims (see Confirmation, Theories of; Inductive Logic).

In another respect, Popper retained the view that science is a special institution, the one that best exemplifies rational and empirical inquiry into new knowledge, or what Popper called "the critical approach." The scientific community is an open society and a model for the others. But since almost any discipline can pursue a critical approach, why does demarcation remain the central problem of his epistemology? Popper's answer was that his criterion of demarcation is the key to solving the problem of the growth of knowledge, that is, the problem of how people learn from experience. The solution to this problem is that they learn from their mistakes by identifying and eliminating error, not the traditionally offered solution that they learn by induction from experience. It is falsification that enables them to learn from experience without induction—the only possible way of learning, he thought. Thus falsifiability is the crucial feature that makes learning and hence science possible. Disciplines that do not promote their own rational and empirical criticism not only are intellectually dishonest but also obstruct the advance of knowledge.

Popper's intuitively appealing criterion is widely cited in public fora. However, it too runs afoul of the objections reviewed by Hempel. For example, Popper's criterion fails utterly for singular statements such as "There are black holes," for unrestricted existential statements are not falsifiable. Moreover, it is not always clear whether Popper is criticizing Marxists and Freudians themselves or the theories they hold. Whether or not Marxists are personally responsive to reasons and evidence is one question; the testability of Marxist theory is another. A theory cannot be dishonest.

Imre Lakatos (1970) provided a thorough critique of the entire spectrum of falsificationist positions and, on this basis, arrived at his own "methodology of scientific research programs," which, in effect, makes demarcation of good from bad science and nonscience a matter of degree (see Lakatos, Imre; Research Programs). Because of Duhem's problem and other complications, what are appraised, Lakatos said, are not theories in isolation but entire series of theories generated appropriately by ongoing research programs (see Duhem Thesis). Competing programs fight long battles of attrition. A research program progresses insofar as

- it makes novel theoretical predictions in heuristically motivated (non–ad hoc) ways;
- some of these predictions are confirmed; and
- successor theories in the program can explain why their predecessors worked as well as they did.

A program degenerates insofar as it lags in these respects. Lakatos did not apply his approach with Popperian ruthlessness, for Lakatos held that it is not necessarily irrational to retain allegiance to a degenerating program for an indefinite period of time. His account implies that the predicates "is scientific" and "is good science" are relative to historical context. Phlogiston, caloric, and ether theories may have been the best available in their day, but anyone defending them today is surely unscientific. Like Popper, Lakatos and successors such as Worrall and Zahar attempt to purify science of ad hoc statements, roughly, theory modifications that are heuristically unmotivated and that lead to no new predictions. But they disagree in detail on what counts as ad hoc and why ad hoc science is bad science (Nickles 1987b).

For Kuhn (1962 and 1970) the mark of a mature science is that it supports a routine problem-solving tradition with a disciplined determination of which problems are fruitful to pursue (see Kuhn, Thomas). In his terms, this will be normal science under a paradigm. Kuhn contested Popper's treatment of falsification and falsifiability. Astrology was (and is) a pseudoscience not because it was unfalsifiable but because it could not sustain a normal-scientific puzzle-solving tradition. Kuhn's view of creation science was similar. Interestingly, Kuhnian normal science fails to be scientific by Popper's criterion, for it is uncritical and convergent rather than divergent. It does not seek major novelty. Kuhnian paradigms are not falsifiable, for their constitutive principles and practices retain a virtual synthetic a priori status for the corresponding scientific communities. Popperian falsifications become Kuhnian anomalies. Its convergent nature enables normal science to build on itself more directly than Popper allowed, yet, in the longer run, given the inevitability of scientific revolutions, not as cumulatively as even Popper wanted. And to Popper's prohibition of ad hoc adjustments in the face of threatened falsification, Kuhn replied that it is often by just such adjustments that scientific knowledge grows. While Popper minimized the importance of scientists' *believing* in scientific propositions, Kuhn emphasized commitment, but to a tradition of problem-solving practices more than to a set of specific beliefs and methodological rules (Rouse 2003).

It is largely on the basis of their implicit, practical knowledge that scientists within a specialist community agree so readily on what is good and bad science. However, this harmony of mutual comprehension goes out the window during a scientific revolution. Insofar as Kuhnian scientific revolutions actually occur in the history of science, they make solving the problem of demarcation more difficult; for by their nature, they radically undermine not only entrenched theories but also the goals, standards, and methodological practices that characterize the previous periods of normal science under a paradigm. They represent discontinuities of scientific development.

Demarcation Difficulties and Reasons for the Demise of the Philosophical Problem

Aside from the particular difficulties faced by each specific proposal, the demarcation enterprise as a whole faces some obstacles. First, the general epistemological problem of demarcating knowledge from nonknowledge is essentially an application of *the problem of the criterion,* formulated in ancient times by Sextus Empiricus. As such, the problem generates a well-known destructive dilemma. How does one decide whether a proposed criterion is correct?—for, either the candidate is self-certifying or else it is justified by a deeper criterion. The first option leads to vicious circularity and the second to vicious regress. (Simply to identify the demarcation and criterion problems would be to classify all genuine knowledge as scientific knowledge, a version of scientism.) Moreover, if the criterion of demarcation itself is an empirical claim, then how can it also be normative, and how can it be justified by appeal to empirical science without begging the question? Both the logical empiricists and the Popperians attempted to evade these difficulties by assigning the criterion of demarcation to methodology or philosophy of science rather than to empirical science itself. With some exceptions (notably Otto Neurath), they held that *philosophy* of science is an a priori or conventional and normative discipline (see Neurath Otto). But if the criterion is merely conventional, a matter of social agreement, then how can it escape being historically situated or historically relativized to competing research programs, and why is everyone obligated to accept the convention?

For Popper, the criterion of demarcation, as one of the "rules of the game of science," regulates science as a whole. Kuhn denied that there are any such rules. Larry Laudan (1981) observed that methodologists have typically used demarcation criteria and other methodological principles as *machines de guerre* in specific historical battles for control of some science or program. Moreover, since technical judgments as to what constitutes good science in a particular subspecialty are usually highly field- and problem-specific, they are not credible unless made by respected members of the research community. Philosophers and general methodologists rarely occupy this position.

The new history of science brought a historical sensitivity to philosophy of science that problematizes the entire project of demarcation. (The logical empiricists had discontinued Ernest Mach's and Duhem's practice of studying the history of science with care.) If a criterion of demarcation is supposed to answer the question, What is science? by delineating what is common to all sciences at all times, an essentialist answer is almost inevitable. Yet the diversity of past science is already so great that any criterion that encompasses all of it is bound to be too weak to be interesting or useful. Ironically, the problem of demarcation for modern science became urgent only when the sciences began to diversify sufficiently that no simple criterion was likely to succeed. These observations raise a further difficulty. It must first be decided which enterprises to include among the sciences to be compared, and this already begs the demarcation question (Amsterdamski 1975).

One reason why candidate criteria of demarcation are too narrow is that they are often backward-looking, attempting to capture what all successful sciences up to now have in common, while pretending to be suprahistorical. Yet science continues to evolve, to ramify, to diversify, to redefine itself; and this is true of methodology and goals as well as content. Insofar as scientific change is occasionally revolutionary, the difficulty may be even worse. On what basis could a criterion of demarcation formulated today presume to legislate for all future science? A criterion with any bite is likely to harm science more than help it. In giving up the search for ultimate essences, both Ptolemaic astronomy and Galilean mechanics failed to be science by Aristotle's lights (Laudan 1983).

DEMARCATION, PROBLEM OF

Many philosophers of science (as well as most science studies experts) have rejected or minimized the problem of demarcation for one reason or another. Pragmatists such as Quine tend to blur dichotomous distinctions, and this one is no exception (see Quine, Willard Van). For Quine, philosophy is continuous with science, which is in turn continuous with common sense. There is no sharp distinction between purely analytical activity and empirical investigation. And today, in the so-called postmodern era, there is a premium on discriminating the *differences* among the wide variety of seemingly legitimate scientific pursuits rather than upon identifying characteristics that they possess in common. Many science studies experts and culture theorists contest the cultural authority of science and the traditional claims that the scientific enterprise possesses a unique epistemic status (see Social Constructionism). Yet one could say that the very success of the Enlightenment project has made demarcation more difficult and less necessary today, since all major institutions strive to be more rational and scientific and less arbitrary in their practices, including some approaches to theology (Murphy 1990).

Reflecting on the steady weakening of proposed criteria of demarcation, Laudan (1983) concludes that demarcation is no longer an important philosophical problem. Popper's falsifiability criterion, he says, weakens the criterion almost beyond recognition. No longer does the criterion of demarcation mark out a body of belief-worthy claims about the world, let alone demonstrably true claims, let alone claims about ultimate causal essences—for, in Popper's criterion every empirically *false* statement is automatically scientific! Popper completely abandons the traditional attempt to characterize science in terms of either the epistemic or the ontological status of its products.

Laudan's view is that it is wrong to make invidious, holistic distinctions in advance about whether or not something is scientific. Scientists typically proceed piecemeal, he says, willing to consider anything and everything on its merits. They dismiss as bad, marginal, or fringe science much of what they encounter and keep the rest. There is no need for a separate category of pseudoscience. It is enough to reject something as bad science.

This pragmatic move deliberately blurs the distinction between the form and the content of science, i.e., between the logic or method of science and empirical claims themselves. The move rejects the traditional demarcation problem only to raise another, at least equally difficult one: How can philosophers of science (and other members of society) reliably discriminate good science from bad science? Laudan (and Kuhn) would answer that philosophers do not need to. That is a job for contemporary practicing scientists who have demonstrated their expertise. Sometimes the answers will be obvious, but often enough they will be both piecemeal and highly technical. This response may be correct, but how does it play in the sociopolitical arena? And of course it raises yet another question: How should society determine who is an expert?

Demarcation as a Social Problem

Laudan ([1982] 1996, Ch. 12) applies his position to the Arkansas trial of 1981–1982 (*McLean v. Arkansas*) over the teaching of creationism in public school biology classes. He agrees with the decision that creationism should not be taught as biology, but he is severely critical of every point of Judge Overton's philosophical justification of his decision. For example, Overton appeals to Popper's falsifiability criterion to show that creationism is not science. Laudan replies that creationist doctrine itself *is* science by that criterion, since it has been empirically falsified, however unscientific may be the behavior of some of its advocates. The reason it should not be taught is simply that it is *bad* science. Michael Ruse, who had invoked Popper's criterion in court testimony, responded that given the complexities of the legal and social situation, Judge Overton's reasoning was correct, for that is the only practical way to stop the teaching of "creation science" as a serious alternative to biological evolution (Ruse 1982).

One complication is that terms such as "bad science" and "pseudoscience" cover a variety of different sins, including incompetent but honest work, potentially good work that is difficult to test or that has utterly failed to find empirical support, and deliberately dishonest scientific pretensions. There are any number of ways in which science can be bad and many labels for bad or pretended science. "Pseudoscience" is an old term for claims that are (or are treated as) untestable. The chemist Henry Bauer (2001) prefers "anomalistics" to "pseudoscience," since all the standard criteria for the latter have failed and pseudoscience occasionally develops into real science, as Popper acknowledged. "Fringe science" includes sometimes-testable claims widely ignored by the scientific community because they violate the best naturalistic understanding of the cosmos. "Pathological science" is the name that Nobel chemist Irving Langmuir gave to those cases in which the scientists are honestly deceiving themselves, as he claimed was the case with J. B. Rhine's work on extrasensory perception (Park 2000, 40ff). "Junk science" involves deliberately exploiting scientific uncertainty to confuse and mislead judges, juries, and politicians, usually by substituting mere possibility for known probability (Huber 1991). It falls just short of "fraudulent science," in which scientists fudge their results or expert witnesses lie about the current state of knowledge. The physicist Robert Park (2000) lumps all these cases together as "voodoo science." He is especially concerned about claims with public currency that escape full scientific scrutiny because of official secrecy, political intervention, the legal adversary system, and the de facto adversary system employed by the media. The latter results in what Toumey (1996, 76) calls "the pseudosymmetry of scientific authority." That is, "unbiased reporting," like the use of expert witnesses in a courtroom, sometimes pretends that for every expert there is an equal and opposite expert.

This leads to another complication—that philosophers and scientists must make their cases to lay audiences. There is little opportunity to present esoteric detail in a court of law or the popular media. In the case of creationism, given the political, religious, and legal situation in the United States, Ruse can argue rather convincingly that labeling creationism religion rather than science is the only way to keep it from being taught as science in public school classrooms. It is doubtful whether Laudan's more nuanced treatment of the issue would have the same practical effect. Should Judge Overton have ruled that creationism cannot be taught because it is bad science, or that it can *only* be taught as an example of bad science? Surely it would be a bad precedent for sitting judges to rule on what is good or bad science. And yet in a lesser sense they must, for the U.S. Supreme Court's 1993 decision in *Daubert v. Merrill-Dow Pharmaceuticals* makes judges the gatekeepers for keeping unsound science out of the courtroom. *Daubert* requires judges to consider, in addition to their error rate, whether the alleged scientific claims have been tested, whether the claims have been subjected to peer review, and whether the relevant scientific community accepts the claims (full consensus is not required).

A related complication is that legal (and political and public policy) reasoning differs in important ways from scientific reasoning, so one should not expect full convergence between scientific and legal modes of thought and action. For example, scientific conclusions are typically guarded and open to future revision in a way that legal decisions are not. Legal judgments are final (except for appeal) and must be made within a short time span on the basis of the evidence and arguments adduced within that time, whether or not sufficient scientific knowledge is available (Foster, Bernstein, and Huber 1993). The value of a scientific claim or technique often resides in its heuristic potential, not in its known truth or correctness, whereas the judicial system wants the truth now. Scientists seek general understanding of phenomena, whereas judges and attorneys must achieve rapid closure of particular disputes. Scientific conclusions are often statistical (with margins of error given) and not explicitly causal, whereas legal decisions are typically causal and normative (assigning blame), individual, and nonstatistical, although cases involving smoking and cancer have recently broadened the law's conception of scientific reasoning. In the United States and elsewhere, many legal proceedings, both criminal and civil, are explicitly adversarial, whereas scientific competition is adversarial only in a *de facto* way. Scientists rely most heavily on evidential reasons, whereas law courts require all evidence to be introduced via testimony and accepted (or not) on that authority. Consideration of the evidence itself is beyond the purview of the court. The rules of evidence also differ. Judges must decide, in binary fashion, whether or not a given piece of evidence is admissible at all and whether a given witness is admissible as a scientific expert. When there is a jury, the judge instructs it as to what it may and may not take into consideration. In some respects, legal reasoning is more conservative than "pure" scientific reasoning, since lives may be immediately at stake; whereas in science, as Popper says, "our theories die in our stead (Popper 1985, 83).

The current situation is further complicated by the shifting use of the terms "junk science" and "sound science." In the highly litigious context of the United States, "junk science" originally meant dubious claims defended by hired expert witnesses in liability lawsuits, especially against wealthy corporations. While the increasing number of scientifically frivolous lawsuits does indeed threaten the financial stability and the innovative risk-taking of corporations, corporate executives and powerful politicians have corrupted the terminology by labeling as junk science any scientific claim or methodology that challenges their position, and as sound science any claim that favors it (Rampton and Stauber 2001).

Moreover, writing on boundary formation and maintenance in science, the sociologist Thomas Gieryn (1999) contends that the epistemic authority of science derives not from the application of philosophical criteria of demarcation, nor from empirical testing, good practices at the laboratory bench, or competent experimental design. Rather, it is generated downstream where science meets the rest of society: in schools, the media, law courts, etc. It is here where the cultural maps are drawn, he says, with sets of boundaries that confer authority, power, and prestige (or their absence) upon science and other cultural institutions. Of course, when one looks at the detailed ways in which this is accomplished, one finds such things as philosophers' criteria of demarcation being used as weapons, and so, to a degree, the issue comes full circle. From this sociological perspective, Laudan is correct to characterize criteria of demarcation as local *machines de guerre* rather than timeless principles, but the demarcation problem remains important for all that; for if Gieryn is correct, it is precisely the constellation of such maneuvers that establishes cultural boundaries. That is one reason why clashes over science in the courtroom and clashes between the epistemic standards of science and law are worrisome to those who wish to preserve the epistemic autonomy of science (see Lynch and Jasanoff 1998.)

Since the general public frequently confuses science and technology, demarcation issues carry over to technological debates. Fallibilism with respect to technological development presents more problems than fallibilism with respect to basic science. At the turn of the twenty-first century, one of the sharpest disagreements among policymakers concerns the so-called precautionary principle, the idea that scientists and technologists should proceed with caution in areas in which they are ignorant. In its strong form, the principle states that no technology shall be introduced until its safety can be assured—a measure that would curtail innovation. Defenders of the strong form apparently assume that "science" consists of fail-safe knowledge. A weaker form of the principle requires caution when the stakes (utilities) are sufficiently high, even when the probabilities are relatively small or uncertain (e.g., greenhouse gases and global warming). Equally clearly, forging ahead until there is proof of danger is a foolish policy. The problem is where and how to strike the balance, given that these are usually decisions made under large uncertainty.

Conclusion

There is no one simple distinction that marks off science (and its potential technological applications) from pseudoscience, or good science from bad. This is the conclusion of the last two generations of philosophers of science, reinforced by modern science studies. The problem is still more complex than many writers have realized, for these are no context-free distinctions. What sort of demarcation is appropriate within science depends upon subtleties of historical and technical context, and what sort of demarcation is appropriate in public policy contexts will likewise depend upon contextual details, including the particular interests at stake. So what began as a metaphysical or logical issue ends up being a concern modulated by pragmatic reasons (Resnik 2000). While there is some truth to the reported demise of the traditional demarcation problem, context-specific demarcation issues abound and are more important than ever, both within science and in the public arena, a domain that urgently needs more philosophical participation. Before emigrating to the United States and England, the logical empiricists, Popper, and Lakatos were deeply engaged in sociopolitical issues.

Clearly, the demarcation problem cannot be solved by simply identifying science with the body of currently accepted "truths," nor is it possible simply to retreat from substance to method if this implies a commitment to "the" scientific method as a set of rules. Current emphasis on future promise rather than past results, and on scientific practices rather than belief systems and universal logical criteria, may offer a more feasible approach to the problem. Rouse (2003, 119) notes the irony that despite Kuhn's and science studies' challenge to "textbook science," the leading philosophical models of science remain representational and hence lend encouragement to the creationists' fideistic conception of science and of science education. Education continues to emphasize correct beliefs (scientific facts) over productive practices and future-oriented attitudes. Half a century after the decline of logical empiricism, philosophers still tend to view science in terms of a retrospective theory of justification rather than problem-solving productivity and future promise. Although it is not absolute, the belief/practice distinction can also eliminate a persistent confusion over who is an expert. At the frontier of knowledge, where no one knows what lies beyond, there are no experts in the sense of those who know *that* such-and-such is true, but there clearly are experts in the sense of those who know *how* to proceed with frontier research and are able to furnish comparative heuristic appraisals of the competing proposals.

THOMAS NICKLES

References

Amsterdamski, Stefan (1975), *Between Experience and Metaphysics*. Dordrecht, Netherlands: Reidel.

Bauer, Henry (2001), *Science or Pseudoscience?* Urbana: University of Illinois Press.

Duhem, Pierre ([1914] 1954), *The Aim and Structure of Physical Theory*. Princeton, NJ: Princeton University Press.

Feyerabend, Paul (1981), *Realism, Rationalism and Scientific Method: Philosophical Papers* (Vol. 1). Cambridge: Cambridge University Press, 239–245.

Foster, Kenneth, David Bernstein, and Peter Huber (eds.) (1993), *Phantom Risk: Scientific Inference and the Law*. Cambridge, MA: MIT Press.

Gieryn, Thomas (1999), *Cultural Boundaries of Science: Credibility on the Line*. Chicago: University of Chicago Press.

Hempel, Carl G. (1965), *Aspects of Scientific Explanation*. New York: Free Press.

Huber, Peter (1991), *Galileo's Revenge: Junk Science in the Courtroom*. New York: Basic Books.

Kuhn, Thomas (1962), *The Structure of Scientific Revolutions*. Chicago: University of Chicago Press.

——— (1970), "Logic of Discovery or Psychology of Research?" in I. Lakatos and A. Musgrave (eds.), *Criticism and the Growth of Knowledge*. Cambridge: Cambridge University Press, 1–23.

Lakatos, Imre (1970), "Falsification and the Methodology of Scientific Research Programmes." in Lakatos and A. Musgrave (eds.), *Criticism and the Growth of Knowledge*. Cambridge: Cambridge University Press, 91–195.

Laudan, Larry (1981), *Science and Hypothesis*. Dordrecht, Netherlands: Reidel.

——— (1983), "The Demise of the Demarcation Problem," in R. S. Cohen and L. Laudan (eds.), *Essays in Honor of Adolf Grünbaum*. Dordrecht, Netherlands: Reidel, 111–127.

——— ([1982] 1996) *Beyond Positivism and Relativism*. Boulder, CO: Westview Press, 1996.

Lynch, Michael, and Sheila Jasanoff (eds.) (1998), *Social Studies of Science* 28: 5–6. Special issue on "Contested Identities: Science, Law and Forensic Practice."

McMullin, Ernan (2001), "The Impact of Newton's *Principia* on the Philosophy of Science," *Philosophy of Science* 68: 279–310.

Murphy, Nancey (1990), *Theology in the Age of Scientific Reasoning*. Ithaca, NY: Cornell University Press.

Nickles, Thomas (1987a), "From Natural Philosophy to Metaphilosophy of Science," in Robert Kargon and Peter Achinstein (eds.), *Kelvin's Baltimore Lectures and Modern Theoretical Physics: Historical and Philosophical Perspectives*. Cambridge, MA: MIT Press, 507–541.

——— (1987b), "Lakatosian Heuristics and Epistemic Support," *British Journal for the Philosophy of Science* 38: 181–205.

Park, Robert (2000), *Voodoo Science: The Road from Foolishness to Fraud*. Oxford: Oxford University Press.

Popper, Karl (1959), *The Logic of Scientific Discovery*. London: Hutchinson. This is a translation and expansion of Popper's *Logik der Forschung* of 1934.

——— (1963), *Conjectures and Refutations*. New York: Basic Books.

——— (1985), *Popper Selections*. Edited by David Miller. Princeton, NJ: Princeton University Press.

Rampton, Sheldon, and John Stauber (2001), *Trust Us, We're Experts!* New York: Tarcher/Putnam.

Resnik, David (2000), "A Pragmatic Approach to the Demarcation Problem," *Studies in History and Philosophy of Science* 31: 249–267.

Rouse, Joseph (2003), "Kuhn's Philosophy of Scientific Practice," in T. Nickles (ed.), *Thomas Kuhn*. Cambridge: Cambridge University Press, 101–121.

Ruse, Michael (1982), "Pro Judice," *Science, Technology, and Human Values* 7: 19–23.

Toumey, C. (1996), *Conjuring Science*. New Brunswick: Rutgers University Press.

See also **Carnap, Rudolf; Cognitive Significance; Duhem Thesis; Empiricism; Feyerabend, Paul; Lakatos, Imre; Logical Empiricism; Kuhn, Thomas; Popper, Karl Raimund; Prediction; Quine, Willard Van; Research Programs; Social Constructionism**

DETERMINISM

Determinism is a topic of broad interest in philosophy, with important connections to issues in metaphysics, epistemology, ethics, and philosophy of action (see e.g., O'Connor 1995; Belnap 2001). Within the philosophy of biology, there has been some discussion of determinism in evolutionary theory (e.g., Brandon and Carson 1996; Graves, Horan, and Rosenberg 1999) as well as in genetics, where a consensus has emerged that no interesting thesis of determinism can be sustained (Sarkar 1998; Kitcher 2001). Most discussion of determinism in the philosophy of science has focused on the issue as it arises in physics, and this will be the primary focus of this article.

DETERMINISM

Formulations of Determinism

The thesis of determinism has been defined in numerous ways. The basic idea is that one part of the world's history determines another part of the world's history. In order to extend this intuitive but vague idea into a crisp thesis, it is necessary to make some choices. Which part of the world's history does the determining, and which part gets determined? How should the kind of determination at issue be understood? Should determinism be thought of as a characteristic of a world, a theory, a set of laws, or something else? Is determinism an all-or-nothing affair, or are there useful notions of degrees of determinism, or "partial determinism"? This section will survey a number of different ways in which these questions have been answered.

Perhaps the most famous exposition of the doctrine of determinism in the context of modern science is due to Pierre Laplace:

> We ought to regard the present state of the universe as the effect of its antecedent state and as the cause of the state that is to follow. An intelligence knowing all the forces acting in nature at a given instant, as well as the momentary positions of all things in the universe, would be able to comprehend in one single formula the motions of the largest bodies as well as the lightest atoms in the world, provided that its intellect were sufficiently powerful to subject all data to analysis; to it nothing would be uncertain, the future as well as the past would be present to its eyes. (Laplace [1814] 1951, 282)

Laplace's formulation makes a claim about the nature of the universe, *viz.*, that the total state at one time causally determines the state at later times and is causally determined by the state at past times. It also makes a claim about predictability, namely that an idealized intelligence would be able to predict (or retrodict) the total state of the universe at any time given a specification of the state at any other time. The first claim is (broadly speaking) metaphysical or ontic, while the second seems to be epistemological. Presumably, Laplace thought that each claim followed from the other, but as will be seen, there are many physical contexts in which this is not so. There has been some controversy about whether determinism should be thought of primarily as an epistemological thesis or an ontic one.

Laplace's 'intelligence' is an extremely unrealistic idealization, and in order to understand his characterization of determinism, one must understand how this idealization is supposed to function. Karl Popper in effect takes Laplace's intelligence to be a limiting case of an actual observer and so takes determinism to be a thesis about the kind of prediction available in principle to actual observers. Actual observers, in addition to having more limited powers of calculation than Laplace's intelligence, must gather the information they use for making predictions from empirical observations. Accordingly, Popper defines *scientific determinism* as

> the doctrine that the state of any closed physical system at any given future instant of time can be predicted, even from within the system, with any specified degree of precision, by deducing the prediction from theories, in conjunction with initial conditions whose required degree of precision can always be calculated . . . if the prediction task is given. (Popper 1982, 36)

Popper thus takes seriously the link between determinism and predictability, where prediction must be done by an agent who is part of the total system the agent wishes to predict; the agent makes allowances for the fact that such predictions can never be expected to be perfectly precise (see Popper, Karl Raimund; Prediction).

Popper goes on to argue (41–85) that scientific determinism is false, even if classical mechanics, which has been traditionally regarded as a deterministic theory, is true. Popper's formulation thus makes determinism an epistemological thesis. Many other contemporary philosophers of science who write about determinism, for example, John Earman (1986, 6–10) and Patrick Suppes (1993), insist that determinism is an ontic or physical thesis and should not be analyzed in terms of epistemological concepts such as predictability. In their view, Laplace's characterization is acceptable only if the reference to the idealized intelligence is understood as an aid to the imagination; what is really important is that the laws of nature, together with the state of the universe at a given time, suffice to determine the state of the world at other times, whether it is in principle possible to exploit this fact to make predictions or not. Thus, it is crucially important to distinguish between determinism proper and predictability.

One may of course accommodate both views about determinism by distinguishing between a predictability sense of determinism and an ontic/physical sense of determinism. But advocates of the two views see more than mere terminological disagreement here. Popper (1982, 8) argues that such nonepistemological conceptions of determinism are not falsifiable and are thus metaphysical rather than scientific. (For criticism of this argument, see Earman 1986, 10.) One argument for the importance of distinguishing between predictability and determinism, given by Suppes (1993, 245–246), concerns Turing machines. A Turing machine is, in an intuitive sense, an outstanding paradigm of a deterministic system. But it is known that there is no algorithm for determining whether an arbitrary Turing machine in an arbitrary configuration will ever halt. Suppes argues that this shows that it is possible for a 'good sense' deterministic system to be in an initial state such that there is no method available for any epistemic subject to predict its future behavior.

If one decides to formulate determinism in a way that distinguishes it from predictability, one still has options. A typical logical-empiricist formulation makes determinism a property of theories and defines deterministic theories syntactically: A theory is deterministic if and only if a set of sentences specifying the state of the world at one time, together with lawlike sentences drawn from the theory, deductively entail sentences characterizing the state of the world at any other time (Nagel 1953, 420–423) (see Nagel, Ernest). The problem with this definition of determinism, pointed out by Richard Montague (1974, 303–304) is that if the set of possible states has the cardinality of the continuum, there will not be enough sentences to describe the state of the world at a given time sufficiently precisely for the required deductions to go through. Hence, the syntactic formulation makes determinism exceedingly fragile.

Montague (1974) proposes an alternative, semantic characterization of deterministic theories. A theory is associated with a class of *models*, in the sense used in formal semantics (see Scientific Models; Theories). Montague then defines a number of senses of determinism. The general pattern is that a theory is deterministic in a given sense if and only if all models of the theory that agree on the state of the world at one time also agree at certain other times. Following the lead of Earman (1986, 12–14), it is possible to strip away much of the formal semantic apparatus employed by Montague and characterize these senses of determinism in terms of physically possible worlds. Let W be the set of physically possible worlds allowed by the theory T. Then:

- T is *futuristically deterministic* if and only if: For any w_1, w_2, W, and any time t, if w_1 and w_2 agree on the complete physical state at t, then they agree on the complete physical state at any time $t^* > t$.
- T is *historically deterministic* if and only if: For any w_1, w_2, W, and any time t, if w_1 and w_2 agree on the complete physical state at t, then they agree on the complete physical state at any time $t^* < t$.
- T is *Laplacian deterministic* (or simply *deterministic*) if and only if T is both futuristically and historically deterministic.

These definitions provide a straightforward explication of the intuitive idea that one part of the world's history determines another part of that history: The physical possibilities allowed by the former leave no room for variation in the latter. These definitions establish a general pattern that can be used to generate further varieties of determinism:

- T is *(X, Y) deterministic* if and only if: For any w_1, w_2, W, if w_1 and w_2 agree on the complete physical state in spatiotemporal region X, then they agree on the complete physical state in spatiotemporal region Y (cf. Earman 1986, 17).

This pattern is useful when one turns to relativistic physics, where there is generally no well-defined sense of 'the state of the world at a time t' (see Space-Time).

Further variations on this theme are available. For example, one might hold that determinism is not an all-or-nothing affair. A theory may be *conditionally deterministic* in the sense that any two physically possible worlds allowed by that theory that agree at all times on a certain range of magnitudes or properties, and agree at a given time about everything, also agree at all times about everything. Alternatively, one may hold that some aspects of the world are deterministic (say, the properties in the set P) and others are not. One way to capture this intuitive idea is as follows:

- T is *Laplacian deterministic in the properties in the set* P if and only if: For any w_1, w_2, W, and any time t, if w_1 and w_2 agree on all of the properties in P at t, then they agree on all of the properties in P at all times.

In this way, one can formulate the idea that, for example, the world is deterministic in its physical aspects but not in its mental aspects. But as many authors have pointed out (Popper 1982, 25–26; Earman 1986, 13–14), determinism with respect to P coupled with the failure of determinism for other properties leads to the consequence that the properties in P must be nomologically independent of those outside of P. In the case where P is the set of physical properties, assumed not to include the mental properties, this leads to epiphenomenalism about the mental. (Other senses of less-than-complete determinism are defined by Montague 1974, 321–324; Earman 1986, 13–14.)

The senses of determinism just discussed take the basic notion to be that of a *deterministic theory*, where the property of determinism is defined by quantifying over all the physically possible worlds allowed by the theory. Alternatively, one can define determinism as a property of a set of laws, proceeding as above, but quantifying over all the possible worlds allowed by that set of laws. Determinism can also be defined as a property of a world. One straightforward way of doing this is to define a world as deterministic just in case the laws of nature of that world are deterministic. But Earman (1986, 12–13) provides a more sensitive way of defining a deterministic world, which may be stated as follows:

- Let W be the class of physically possible worlds relative to w. Then w is (futuristically, historically, Laplacian) deterministic if and only if: For any $w^* \in W$ and any time t, if w and w^* agree on the complete physical state at t, then they agree on the complete physical state (at all times $t' > t$, at all times $t' < t$, at all times t).

This definition allows, for example, that a world may be Laplacian deterministic even if the laws of that world are not Laplacian deterministic. (This means that it could be that the laws alone do not guarantee that the present state of the universe determines its state for all times, but given the actual state of the universe, its entire history is settled by the laws.) But it entails that a sufficient condition for the Laplacian determinism of a set of laws is that each physically possible world allowed by those laws is Laplacian deterministic.

In the case of a theory whose laws take the form of differential equations or partial differential equations, these definitions can be reformulated in terms of boundary value problems and their solutions, eliminating reference to physically possible worlds. So, for example, Laplacian determinism can be reformulated:

- T is *Laplacian deterministic* just in case a physically possible specification of the physical magnitudes at a time t, together with the laws of T, define a well-posed boundary value problem: There exists a unique solution for all time satisfying the boundary values provided by the specified physical magnitudes at t.

As will be seen later, space-time theories allow for additional versions of this formulation, with different characterizations of the boundary conditions.

In the following sections, the question of whether various theories of modern physics are deterministic will be examined. Of primary concern will be the ontic versions of determinism favored by Earman (1986) and Suppes (1993), but occasional reference will be made to the predictability conception of determinism favored by Popper (1982).

Determinism in Classical Physics

Classical physics is traditionally viewed as the very paradigm of a deterministic physical theory (see Classical Mechanics). This is probably in large part due to Laplace's influential formulation, which was produced in the context of a discussion of classical mechanics. However, it is now known that classical physics is not deterministic, in either the predictability sense or the ontic sense.

Classical physics does not satisfy any very interesting requirement of predictability. One counterexample is provided by the three-body problem in Newtonian gravitation theory: No closed analytic solution of the general problem exists, and methods of numerical approximation give predictions that are accurate only for limited periods of time (Suppes 1993, 244–245). Further, many classical systems are known in which future evolution depends so sensitively on small differences in initial data that reliable predictions for arbitrary future times are impossible. One of the most famous is Lorenz's model of the weather (Earman 1986; Suppes 1993). More generally, many classical systems exhibit the feature known as *chaos*, which rules out the possibility of predictability, though not that of Laplacian determinism as defined above. Chaotic systems are for this reason an excellent case for illustrating the way in which the two senses of determinism can come apart. Chaos in classical systems will be discussed in a later section.

Though the failure of predictability in classical physics is now widely appreciated, it is still widely but falsely believed that classical physics does satisfy the ontic formulation of Laplacian determinism. A counterexample to Laplacian determinism in classical physics is provided by any complete, consistent specification of the history of a physically possible world at a given time where this specification together with the laws of classical physics do not determine a unique future (or past) for that time. There are several known such counterexamples, and they come in a variety of kinds.

One interesting class of counterexamples is provided by collisions between perfectly elastic bodies. The most straightforward counterexample of this class is simply a collision of three or more such bodies. For a collision of three such bodies, the standard classical laws of elastic impact determine four equations, concerning conservation of each component of the total momentum and conservation of the kinetic energy. A system of three particles has nine degrees of freedom, which may be reducible to six by means of selecting a frame of reference in which the center of mass of the system is at rest. This leaves more variables than equations, so a unique solution representing the evolution of the system after the collision is not determined (cf. Earman 1986, 38–39).

More exotic elastic-collision counterexamples can be constructed if the possibility of an infinite number of particles is allowed. One such counterexample, discovered by Jon Perez Laraudogoitia (1996), goes as follows. Consider a countable infinite series of perfectly elastic balls of unit mass, laid out in a straight line of unit length, with each ball half the diameter of the preceding one, and the distance between balls decreasing by half with each successive ball (see Figure 1a). Now suppose that the first ball is struck by a cue ball of unit mass moving with unit speed. In one unit of time, the motion will have been communicated to each of the infinitely many balls. Since the balls are all perfectly elastic, each will come to rest when it strikes its successor, so that at the end of one unit of time, all of the balls are at rest, with the n-th ball occupying the initial position of the $(n+1)$-th ball and the cue ball occupying the initial position of the first ball (see Figure 1b). The important thing to notice is that no ball pops out on the right-hand side; every ball has a successor, so every ball gets stopped. The evolution just described is a solution of the classical equations of elastic impact. But those equations are time-reversal invariant, so the time-reversed process, leading from Figure 1b to the time-reversal of Figure 1a in a unit of time, is also a solution. Now consider a physically possible world in which the balls are in the situation depicted in Figure 1b at all times $t < t^*$. The laws leave open the possibility that the balls will remain in this configuration forever, and they also leave open the possibility that the time-reversal of the original process will occur, starting at any time later than t^*. This shows that the classical mechanics of elastic particles is not a deterministic theory (in any of the senses discussed above).

A similar counterexample, which uses elastic balls of the same size but initially distributed over an infinite region of space, is given by Oscar Lanford (1975). A curious feature of such counterexamples is that they involve violations of global conservation of momentum and energy; for example, in the Laraudogoitia example, the total momentum and energy of the system is zero before t^* but nonzero after t^*. But there is no violation of *local* conservation of energy or momentum, for the conservation laws are satisfied by each of the elastic collisions. Whether the violation of the global conservation laws is sufficient to show that these are not genuine counterexamples to determinism in classical physics is a point that will be addressed at the end of this section.

A second class of counterexamples to determinism in classical physics involves systems of massive particles governed by the classical law of gravitation. It is possible for such systems to exhibit *singularities* in which some quantity of motion figuring in a law of classical physics becomes arbitrarily large in a finite amount of time (see Classical

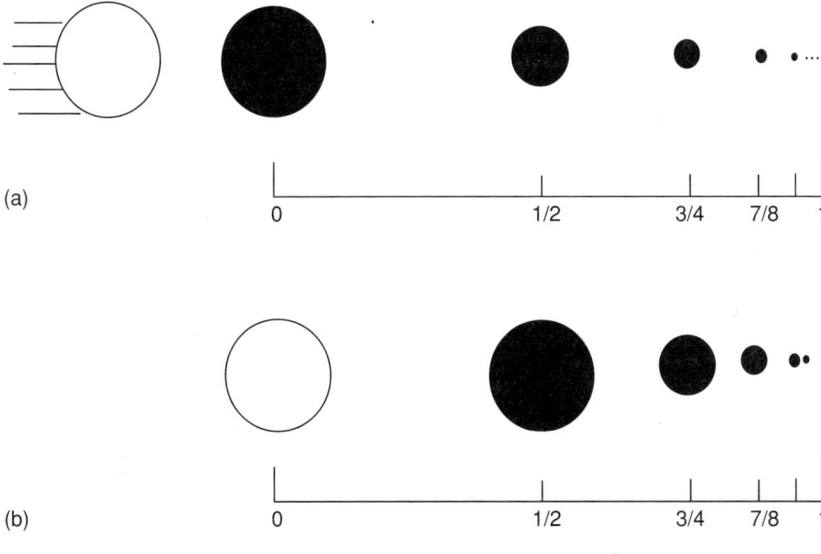

Fig. 1. (a) The Laraudogoitia example: Initial state. (b) The Laraudogoitia example: Final state.

Mechanics). The simplest kind of singularity occurs when two point particles, accelerated by their mutual gravitation, collide, so that the denominator in the inverse-square law blows up. In such cases, there will be a finite time t^* such that no solution of the classical equations up to time t^* is extendable to times later than t^*, which is to say that the laws of classical physics do not determine what happens after t^*. This problem can perhaps be dealt with by saying that, really, point particles are just an idealization, useful only because they approximate the behavior of extended bodies, whose centers of mass will never actually coincide. But this move makes it necessary to provide a theory of what happens during particle collisions, and the problem of determinism for triple-body elastic collisions mentioned above looms large.

Furthermore, it is known that there exist solutions to the classical dynamical equations with noncollision singularities. (For details and references to the physics literature, see Earman 1986, 33–39.) Such noncollision singularities entail the physical possibility of a quite bizarre species of counterexample to determinism. What typically happens in the case of such a noncollision singularity is that one or more bodies will be accelerated to an indefinitely large velocity in a finite period of time; the form of its trajectory is depicted in Figure 2a, in which a particle reaches arbitrarily high speeds before the finite time t^*. Since, again, the laws of classical mechanics are time reversible, the time reverse of this process, depicted in Figure 2b, is physically possible relative to classical mechanics. Figure 2b shows a particle suddenly "coming in from infinity," making it somewhat appropriate to call it a "space invader." Clearly, in Figure 2b, nothing about the state of the world prior to time $-t^*$ can determine whether a space invader will appear at $-t^*$ or not. So here is another striking violation of determinism in classical physics, which, unlike the Laraudogoitia and Lanford examples, does not depend on an improbable initial arrangement of an infinite number of elastic balls.

The Lanford example and the space-invader example depend on the fact that in classical physics, there is no upper bound on the speed with which an influence can propagate. This is a feature shared by classical nonrelativistic theories of fields (rather than particles), and this fact can be exploited to construct violations of determinism for classical field theories. The basic idea is that the field equations permit solutions involving a field-theoretic analog of space invaders: a wave or other disturbance in the field propagating "in from infinity" (Earman 1986, 40–45 and 48–52).

As noted above, the Laraudogoitia and Lanford examples involve violations of global conservation laws, and the space-invader example does as well, since the mass, energy, and momentum of the space invader will be added to that of the whole universe after it appears. It may be thought that this is sufficient to rule out such examples as genuine violations of determinism in classical physics, since the laws of classical physics include the global conservation laws. But there are reasons to be dubious of this move (Earman 1986, 37–39), for none of the examples just mentioned involves any violation of any *local* conservation law. (For example, the space-invader example does not violate the principle that a particle's mass must remain constant along its entire world line, which is a plausible candidate for the local principle of the conservation of mass.) Further, the global conservation laws have at best a dubious status as *fundamental* laws of classical physics. They can be derived from more fundamental laws, given certain assumptions—for example, that the universe as a whole is a closed system (which rules out space invaders) or that there are only finitely many particles (ruling out Laraudogoitia's example). One could rule out the examples of indeterminism in question by defining classical physics in such a way that it includes such assumptions. But what the examples in question

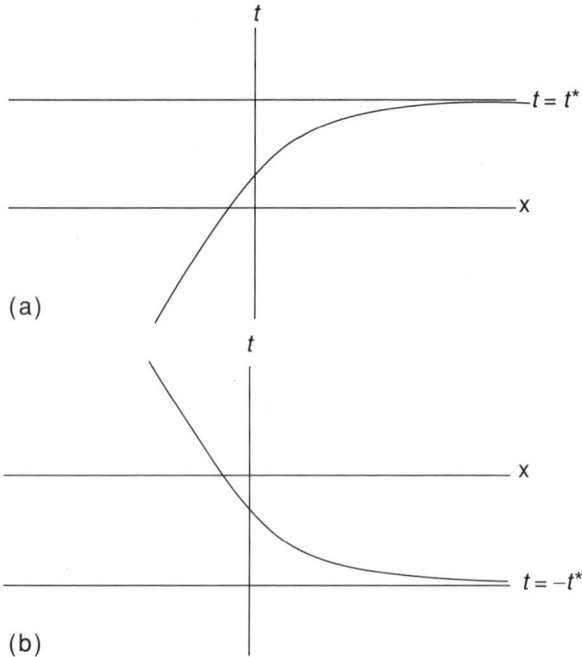

Fig. 2. (a) World-line of a "space fugitive" which may result from a non-collision singularity in classical gravitation theory. (b) A "space invader." (The time-reversal of the process depicted in Figure 2a.)

show is that the fundamental laws of classical physics do not by themselves guarantee that such assumptions hold, for theses examples provide solutions of the classical equations of motion in which these assumptions are violated. To stipulate that classical physics must be understood as essentially including these assumptions (for example, that the universe as a whole is a closed system) is arguably tantamount to trying to make classical physics deterministic by fiat. Moreover, it would still not address the problem of elastic collisions involving three or more bodies.

Chaos and Unpredictability in Classical Physics

Unfortunately there is no general definition of 'chaos' that is widely agreed upon (Belot and Earman 1997, 150), but there are a variety of conditions that are generally thought to characterize chaotic systems. These conditions come in two categories: those that amount to a kind of unpredictability and those that amount to a kind of highly sensitive dependence on initial conditions. Both kinds of condition seem to be essential to the concept of chaos.

These conditions can be made precise by making use of the concept of an *abstract dynamical system*, which can be defined as an ordered triple $<X, T, \mu>$ where X is a mathematical space (which can be thought of a state space or a phase space), T is an invertible mapping of X onto itself that represents the unit time evolution of the system, and μ is a probability measure on X, which can be thought of as representing either the state of one's information about the system or a statistical measure on an ensemble of systems (Belot and Earman 1997, 151).

The notion of highly sensitive dependence on initial conditions can be made precise by means of Liapunov exponents. Liapunov exponents are quantities defined for trajectories through the state space, and they characterize the rate of exponential divergence of nearby trajectories (see Lichtenberg and Lieberman 1992, 296, for the technical details). So a given trajectory t through the state space X has positive (nonzero) Liapunov exponents just in case trajectories that start out very close to t will diverge from t exponentially. In simpler terms, if the Liapunov exponents are greater than zero, then initial conditions that are very close together will lead to later states that differ greatly, and the amount of this difference increases exponentially with time.

The unpredictability of a dynamical system can be made precise in more than one way. First, one can use the algorithmic concepts of complexity and randomness. The *conditional complexity* of finite sequence S given information I is defined as the length of the shortest program that, when fed to a universal Turing machine together with input I, will yield S as output. For an infinite sequence S, the complexity of S, $K(S)$, is defined as:

$$\lim_{n \to \infty} \frac{1}{n} K(S_n | n)$$

where S_n is the initial sequence of S having length n, and the last n in this formula should be read as the information that the sequence S_n is of length n. An infinite sequence is random if and only if its complexity is greater than zero. Roughly speaking, an infinite sequence is random just in case no matter how long an initial segment one takes, the shortest computer program that will deliver that initial segment as output is comparable in length to the segment itself. This means that there is no way to compress the information contained in the sequence. This concept of randomness can be extended to the trajectories of an abstract dynamical system by partitioning the states in X into cells and then considering the sequence of cells in which a trajectory is found at a given time, at one unit time later, at one unit time later than that, and so on. This can be thought of as a way of coding a trajectory, by using an infinite sequence of cells. A trajectory is random just in case there exists some partition of the state space into cells such that the sequence of cells that codes the trajectory is random in the sense defined above (see Belot and Earman 1997, 152, for a more rigorous exposition).

A second way to characterize unpredictability makes use of a hierarchy of statistical properties of dynamical systems. The weakest property in this hierarchy is ergodicity: The system $<X, T, \mu>$ is *ergodic* if and only if for every function f of the state space X, for every point x in X, the time average of $f(x)$ equals the average of f over the whole of X (weighted by μ). A stronger property is mixing: The system $<X, T, \mu>$ is mixing if and only if it converges to equilibrium, in the sense that for any function f defined over X, for every point x in X, $f(x)$ approaches the average value of f over X (weighted by μ) as the time gets arbitrarily large. An even stronger property is that of being a K-system. K-systems are dynamical systems with positive, nonzero metric entropy (see Lichtenberg and Lieberman 1992, 304, for details). K-systems conform to the 0–1 law, which says that a complete specification of the system's entire past history does not enable one to predict with certainty any future event except those whose probability of occurrence is 1 independently of the past history

(Batterman 1992, 60). *K*-systems thus exhibit a very strong form of unpredictability. An even stronger form is exhibited by Bernoulli systems, in which it is true at each time that any future event that does not have a probability of 1 independently of the past history is completely statistically independent of the entire past history. There are classical mechanical systems exhibiting each of the properties in this hierarchy (Lichtenberg and Lieberman 1992).

These two ways of characterizing the unpredictability of a dynamical system—by using (1) algorithmic concepts of complexity and randomness and (2) a hierarchy of statistical properties—are related by Brudno's theorem, which says that if an abstract dynamical system $<X, T, \mu>$ is ergodic, X is compact, and T is a homeomorphism, then for almost all trajectories, the complexity of the trajectory is equal to the metric entropy of the system. Hence, almost all trajectories are random if and only if the system is a *K*-system (Batterman 1992, 61; Belot and Earman 1997, 157).

Unpredictability is related to the concept of extreme sensitivity to initial conditions by Pesin's theorem, which says that under certain conditions common among classical systems, the metric entropy of an abstract dynamical system $<X, T, \mu>$ is equal to the average value of the sum of all positive Liapunov exponents (where this average is weighted by μ) (Lichtenberg and Lieberman 1992, 304; Belot and Earman 1997, 157). Thus, for systems that satisfy these conditions, having nonzero Liapunov exponents throughout a region of X of nonzero measure is equivalent to being a *K*-system; in other words, roughly speaking, having very many trajectories that diverge exponentially from very similar initial conditions is equivalent to being unpredictable in the sense in which a *K*-system is.

Mixing, being a *K*-system, and having positive Liapunov exponents have all been proposed as necessary conditions, sufficient conditions, or necessary-and-sufficient conditions for chaos (Belot and Earman 1997). Again, there is no universally accepted definition of chaos. It is sometimes assumed that unpredictability and sensitive dependence on initial conditions are so intimately related that chaos can be defined simply in terms of one or the other. For example, Joseph Ford (1989, 350) defines chaos as "a synonym for randomness" in the algorithmic sense. But as Robert Batterman (1992, 62–63) points out, behavior that is random in this sense can be generated by systems that exhibit no exponential divergence of trajectories at all (such as a spinning roulette wheel) and are therefore poor candidates for chaos. What does seem clear is that a very strong form of unpredictability is a consequence of chaos (Batterman 1992, 63). Yet, chaos is perfectly compatible with Laplacian determinism as defined above. (In fact, the preceding discussion of dynamical systems has presumed throughout that the systems in question are deterministic, in that there is a unique, invertible mapping T that represents dynamical evolution.) This shows how important it is to distinguish between the predictability and ontic senses of determinism.

Determinism in Quantum Physics

The standard formulation of quantum mechanics posits two dynamical processes (see Quantum Mechanics). The first of these is evolution of the quantum state according to the Schrödinger equation, which is linear, continuous, and deterministic. In fact, there is a clear sense in which this evolution is more deterministic than is evolution in classical mechanics, for Schrödinger evolution does not exhibit the sensitivity to small changes in initial conditions allowed in classical mechanics (Belot and Earman 1997). Schrödinger evolution is supposed to take place in any system not being observed. When an observation takes place, the second dynamical process, called "state reduction" or "collapse," kicks in. State reduction is discontinuous: The system being observed typically jumps from a state in which it is in a superposition of values of the quantity being measured to one in which it has some definite value, with the probabilities of the various possible outcomes given by the Born rule. The probabilistic nature of state reduction entails that the standard formulation of quantum mechanics is indeterministic in all of the senses discussed above.

It is generally considered problematic that the standard formulation of the theory uses 'observation' as a primitive concept and that the standard dynamics discriminates on the basis of whether an observation is taking place (cf. Albert 1992). There are a variety of contemporary approaches to dealing with this problem (the "measurement problem"), some of which seek to provide a fuller account of state reduction and some of which seek to eliminate state reduction altogether (see Quantum Measurement Problem). It remains an open question which approach is preferable, along with whether the best approach will preserve or eliminate the indeterminism of the standard formulation of the theory.

Among the interesting current options for dealing with the measurement problem are a family of interpretations called *modal interpretations*. According to modal interpretations, all evolution of the quantum state takes place according to the Schrödinger equation without state reduction, and the physical state of a system is not completely characterized by its quantum state. Some, but not all, quantities that characterize a system have determinate values at a given moment in time, and a precise rule is supplied for determining which do and which do not. The quantum state of a system determines at most probabilistic information about the values of the quantities that do have definite values (Bub 1997, 173–80). Modal interpretations are indeterministic in the predictability sense; some modal interpretations, but not all, are indeterministic in the ontic sense (235–236).

Another approach that has received a great deal of attention is David Bohm's (1952) alternative to the standard quantum theory. In Bohm's theory, the world consists of particles with definite positions at all times and wave functions that exert a nonlocal and stochastic influence on the motions of particles. Bohm's theory, too, preserves the indeterminism of the standard theory.

Hugh Everett (1957) proposed an alternative to the standard version of quantum mechanics according to which there is no state reduction and the physical quantities that characterize physical systems do not in general have determinate values, but only values relative to a given state of the rest of the universe. In this account, when an observer measures the value of some physical quantity, the observer's state typically "branches" into a number of different "relative states" such that every physically possible outcome is observed in one of these branches, and all branches are ontologically on a par. This revised version of quantum mechanics is not deterministic in the predictability sense, because it permits an observer to make only probabilistic predictions about the results of future measurements. But it satisfies Laplacian determinism, since the total state of the universe at one time determines the state at any other time. Precisely because of this, many critics have argued that Everett deprived the quantum probabilities of any intelligible meaning (Barrett 1997). What could it mean to say that the probability of getting a certain result is, say, 0.75, when every possible result is bound to be realized on some branch or other? Critics have also taken aim at the very notion of "branching observer states," arguing that it is unintelligible or at least stands in need of interpretation (e.g., Albert and Loewer 1988; Barrett 1997). Attempts to fill in the needed interpretation include assorted versions of the "many-worlds" interpretation (e.g., Albert 1992, 112–116) and the "many-minds" interpretation (e.g., Lockwood 1996).

Determinism in Relativistic Physics

Many of the failures of determinism in classical physics are due to the absence of any limit on the speed with which causal influence can propagate. In relativistic physics (excluding the possibility of tachyons), a speed limit is imposed, suggesting that relativistic physics may be more friendly to determinism than is classical physics. As will emerge, this is the case. But matters are complicated by the fact that in relativistic space-times, absolute time is not in general definable. The definitions of determinism presented in the first section referred to the complete physical state of the world at a given time. So these definitions will have to be modified before they will be applicable in relativistic physics.

To this end, it is useful to introduce some terminology (see Space-Time). A *relativistic space-time* $<M, g>$ is a four-dimensional manifold M and a metric of Lorentz signature g defined everywhere on M. In special relativity, M is topologically equivalent to $\mathbf{R}^4$, and g is the constant Minkowski metric. In general relativity, M can take any of a variety of topological structures, and g can vary from point to point. A model of general relativity is a triple $<M, g, T>$ with $<M, g>$ a space-time and T a stress-energy tensor, where g and T jointly satisfy Albert Einstein's field equations. Henceforth, it will be assumed that space-times have a temporal orientation, allowing one to distinguish light cones and timelike curves as either future directed or past directed. (There are models of general relativity in which this is not the case, such as the Gödel space-time model [Earman 1986]. If such models cannot be excluded, then there seems little hope of formulating any very interesting form of determinism satisfied by general relativity.)

An *achronal hypersurface* in a space-time $<M, g>$ is a three-dimensional surface no two points of which may be joined by a timelike curve. Let $<M, g>$ be a space-time, and let S be any achronal surface in $<M, g>$. Then *the future domain of dependence of S*, $D^+(S)$, is the set of all points p in M such that every future-directed timelike curve in M, with future endpoint p and no past endpoint, intersects S. The *past domain of dependence of S*, $D^-(S)$, is defined analogously. The *domain of dependence of S*, $D(S)$, may be defined as the union of $D^+(S)$ and $D^-(S)$ (Geroch 1977, 83–84). Figure 3 illustrates a domain of dependence in Minkowski space-time.

A family of theorems implies that a complete specification of all physical magnitudes over the S suffices to determine, up to a diffeomorphism, all such magnitudes throughout $D(S)$. (See Geroch

DETERMINISM

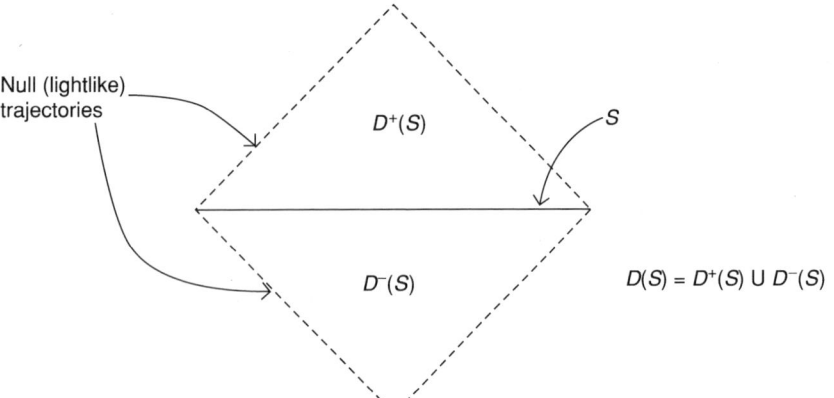

Fig. 3. Domain of dependence of a surface S in Minkowski space-time.

1977, 84; Geroch restricts his discussion to $D^+(S)$, but symmetry considerations allow these results to be extended to $D(S)$.) This shows that every model that agrees with a given model on the physical magnitudes is related to S by a diffeomorphism the restriction of which to S is the identity map. In special relativity, it is assumed that space-time has the fixed structure of Minkowski space-time, and in this setting any such diffeomorphism is just the identity map. So special relativity has the property that a complete specification of the physical magnitudes over a surface S suffices to determine the physical magnitudes over that surface's domain of dependence $D(S)$. This in itself is an interesting, albeit local, form of determinism. But special relativity also has a property that is a natural analog of Laplacian determinism. The natural special-relativistic analog of 'the complete state of the world at a given time' is the specification of all physical magnitudes over a spacelike hyperplane, which can be thought of as all of space at a given time relative to some inertial reference frame. For such a surface S, $D(S)$ is the entire space-time. So a specification of all physical magnitudes over a spacelike hypersurface suffices, in special relativity, to determine the complete physical history of the world (cf. Earman 1986, 58–60).

Things are more complicated in the case of general relativity. Assume for the moment that two models of general relativity that are related by a diffeomorphism represent the same physical situation, or the same physically possible world. Then, general relativity exhibits a weak form of determinism, in that the complete physical state over the surface S suffices to determine the physical state throughout $D(S)$.

But it is not so clear that general relativity satisfies an analog of Laplacian determinism. In general relativity, a maximally extended spacelike hyperplane (i.e., a global time-slice) need not be such that its domain of dependence is the entire space-time, in marked contrast to the situation in special relativity. A spacelike surface S in a space-time $<M, g>$ whose domain of dependence $D(S)$ does include all of $<M, g>$ is called a Cauchy surface (Earman 1986, 176–177.) It seems clear that no natural analog of Laplacian determinism can be true of a general-relativistic world that does not have a Cauchy surface, for there is no analog of the complete state of the world at a given time that suffices to determine what is going on throughout the space-time. Further, there are many models of general relativity that lack Cauchy surfaces.

One kind of example, shown in Figure 4, can be generated by starting with a space-time $<M, g>$, where M has the topology of $\mathbf{R}^4$, and deleting a compact region from it. For example, the space-time of Figure 4 has no Cauchy surface: The point p_1 is not in the domain of dependence of the surface S_1 because the timelike curve C_1 has no past endpoint and does not intersect S_1; and the point p_2 is not in the domain of dependence of the surface S_2 because of the timelike curve C_2. Clearly, the existence of a Cauchy surface requires the absence of such "holes" in the manifold. But the absence of such holes is not a sufficient condition for the existence of a Cauchy surface; there are less contrived examples of general relativistic space-times lacking Cauchy surfaces. One example is anti–de Sitter space-time; others are space-times containing singularities (though not all space-times with singularities lack Cauchy surfaces). Perhaps the general theory of relativity can be strengthened by adding conditions that guarantee the existence of Cauchy surfaces. But there are important difficulties facing this task (see Earman 1986, 177–183, for a detailed discussion).

Thus far, this section has dealt with nonpredictability senses of determinism. It turns out that relativistic physics is far more hostile to the

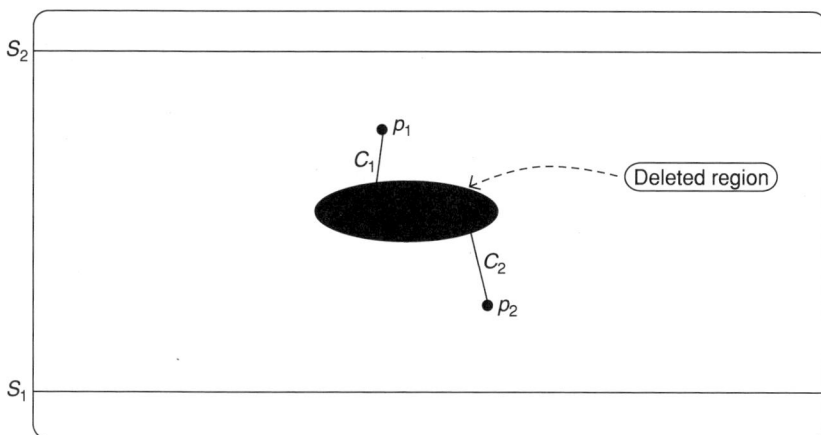

Fig. 4. A "hole-laden" general-relativistic space-time that lacks a Cauchy surface.

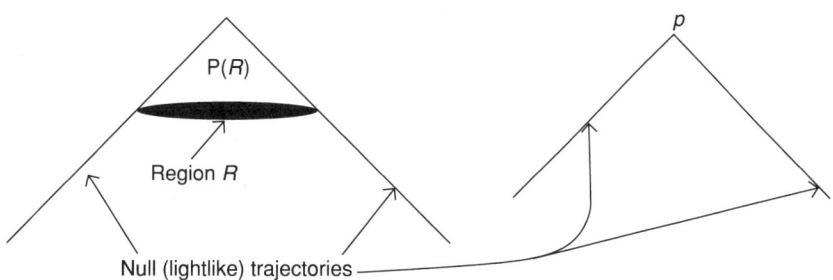

Fig. 5. Domains of prediction in special relativity (Minkowski space-time). P(R), the domain of prediction of the region R, is the only region to the future of R where the physical state is determined by the physical state in regions in R's causal past; hence it is the only region where the physical state can be predicted on the basis of information in principle available to an observer occupying R. For a point p, the domain of prediction is null.

predictability conception of determinism. If it is assumed that predictions must be made on the basis of the laws together with data drawn from empirical observations, then what an observer can predict is limited to what is determined by the laws together with the physical state throughout the region of space-time contained in one's own past-directed light cone. In special relativity, this means that a pointlike observer can reliably predict nothing at all, and an extended observer can make reliable predictions concerning only a very limited spatiotemporal region (see Figure 5) (Popper 1982, 57–62; Earman 1986, 63–6). In general relativity, matters are more complicated but not much more hospitable to predictability (see Geroch 1977 and Hogarth 1993 for details).

Is It Possible to Learn Whether the World Is Deterministic on the Basis of Empirical Evidence?

The characterization of determinism as a property of a physical theory suggests that empirical evidence can show whether determinism is true, since it can show which theory or theories have a chance of being true, while analysis of a theory can show whether that theory is deterministic. But an argument of Suppes (1993) suggests that, in fact, it may be impossible to determine whether our world is deterministic.

Suppes (1993, 254) cites a theorem due to Ornstein to the effect that there exist processes that can equally well be analyzed as deterministic classical-mechanical systems or as indeterministic semi-Markov processes. Suppes goes on to argue that it is plausible that this result applies to a great many processes found in the actual world. The conclusion Suppes draws is that the issue of determinism is "transcendental," not capable of being settled by empirical research. Be that as it may, the concept of determinism continues to serve as a useful tool for probing the foundations of a variety of physical theories.

JOHN T. ROBERTS

References

Albert, David (1992), *Quantum Mechanics and Experience*. Cambridge, MA: Harvard University Press.

Albert, David, and Barry Loewer (1988), "Interpreting the Many-Worlds Interpretation," *Synthese* 77: 195–213.

Barrett, Jeffrey A. (1997), "On Everett's Formulation of Quantum Mechanics," *Monist* 80: 70–96.

Batterman, Robert W. (1992), "Defining Chaos," *Philosophy of Science* 60: 43–66.

Belnap, Nuel D. (2001), *Facing the Future: Agents and Choices in Our Indeterminist World*. Oxford: Oxford University Press.

Belot, Gordon, and John Earman (1997), "Chaos Out of Order: Quantum Mechanics, the Correspondence Principle and Chaos," *Studies in History and Philosophy of Modern Physics* 28: 147–182.

Bohm, David (1952), "A Suggested Interpretation of Quantum Theory in Terms of 'Hidden Variables,'" *Physical Review* 85: 166–193.

Brandon, Robert, and Scott Carson (1996), "The Indeterministic Character of Evolutionary Theory: No 'No Hidden Variables Proof' but No Room for Determinism Either," *Philosophy of Science* 63: 315–337.

Bub, Jeffrey (1997), *Interpreting the Quantum World*. Cambridge: Cambridge University Press.

Earman, John (1986), *A Primer on Determinism*. Dordrecht, Netherlands: Reidel.

Everett, Hugh M. (1957), "'Relative-State' Formulation of Quantum Mechanics," *Reviews of Modern Physics* 29: 454–462.

Ford, Joseph (1989), "What Is Chaos, That We Should Be Mindful of It?," in Paul Davies (ed.), *The New Physics*. Cambridge: Cambridge University Press, 348–371.

Geroch, Robert (1977), "Prediction in General Relativity," in John Earman, Clark Glymour, and John Stachel (eds.), *Foundations of Space-Time Theories: Minnesota Studies in the Philosophy of Science*, vol. 8. Minneapolis: University of Minnesota Press, 81–93.

Graves, Leslie, Barbara L. Horan, and Alexander Rosenberg (1999), "Is Indeterminism the Source of the Statistical Character of Evolutionary Theory?" *Philosophy of Science* 66: 140–157.

Hogarth, Mark (1993), "Predicting the Future in Relativistic Spacetimes," *Studies in History and Philosophy of Science* 24: 721–739.

Kitcher, Philip (2001), "Battling the Undead: How (and How Not) to Resist Genetic Determinism," in Rama S. Singh, Costas B. Krimbas, Diane B. Paul, and John Beatty (eds.), *Thinking About Evolution: Historical, Philosophical, and Political Perspectives*. Cambridge: Cambridge University Press, 396–414.

Lanford, Oscar E. (1975), "Time Evolution of Large Classical Systems," in J. Moser (ed.), *Dynamical Systems, Theory and Applications*. New York: Springer-Verlag, 1–111.

Laplace, Pierre Simon ([1814] 1951), *A Philosophical Essay on Probabilities*. Translated by Frederick Wilson Truscott and Frederick Lincoln Emory. New York: Dover Publications, 1951. Originally published as *Essai Philosophique sur les Probabilités* (Paris: Courcier).

Laraudogoitia, Jon Perez (1996), "A Beautiful Supertask," *Mind* 105: 81–83.

Lichtenberg, A. J., and M. A. Lieberman (1992), *Regular and Chaotic Dynamics*, 2nd ed. New York: Springer-Verlag.

Lockwood, Michael (1996), "'Many Minds' Interpretations of Quantum Mechanics," *British Journal for the Philosophy of Science* 47: 159–188.

Montague, Richard (1974), "Deterministic Theories," in Richmond H. Thomason (ed.), *Formal Philosophy*. New Haven, CT: Yale University Press, 303–359.

Nagel, Ernest (1953), "The Causal Character of Modern Physical Theory," in Herbert Feigl and May Brodbeck (eds.), *Readings in the Philosophy of Science*. New York: Appleton-Century-Crofts, 419–437.

O'Connor, Timothy (ed.) (1995), *Agents, Causes, and Events: Essays on Indeterminism and Free Will*. Oxford: Oxford University Press.

Popper, Karl R. (1982), *The Open Universe: An Argument for Indeterminism*. Totowa, NJ: Rowman and Littlefield.

Sarkar, Sahotra (1998), *Genetics and Reductionism*. Cambridge: Cambridge University Press.

Suppes, Patrick (1993), "The Transcendental Character of Determinism," in P. A. French, T. E. Uehling, and H. K. Wettstein (eds.), *Midwest Studies in Philosophy Volume 18*. Notre Dame, Indiana: University of Notre Dame Press, 242–257.

See also **Prediction; Quantum Mechanics; Space-Time**

DUHEM THESIS

The Duhem thesis holds that scientific hypotheses are not tested against experimental data in isolation but as part of a larger body of beliefs. This holistic epistemological doctrine was first put forward by Pierre Duhem in his *Aim and Structure of Physical Theory* in 1906 (Duhem [1906] 1954). It runs counter to the view that the rational acceptability of any scientific hypothesis can be unambiguously determined by empirical data. In particular it challenges the possibility that a hypothesis can be conclusively falsified by data. Karl Popper ([1935] 1951) subsequently held such a possibility as distinctive of the demarcation criterion of scientific belief.

Duhem developed his position through a critique of a staple of traditional doctrine of scientific method: the Baconian and Newtonian idea of the crucial experiment, or *experimentum crucis*. In order to derive a prediction with which a hypothesis or theory may be tested, the prediction must be itself testable; but the latter is possible only if we introduce into the derivation assumptions about the functioning of the experimental apparatus or measurement instrument. For example, one way in which the phenomenon of superconductivity was established experimentally was by deriving the mathematical expression of physical effects in terms of values of the magnetic field in and around a superconducting crystal—the Meissner effect. Changes in the distribution of such values could be measured and mapped out with a galvanometer, which was assumed to convert magnetic variations into measurable variations in an electric current according to Ampere's law. Assumptions of this sort are typically known as auxiliary hypotheses.

The holistic thesis led Duhem to a form of conventionalism in scientific methodology that required 'good sense' on the part of the scientist in the choice of hypotheses or theories and an act of faith in the belief in their truth. This kind of conventionalism is different from the one advocated by Henri Poincaré. The latter emphasizes the different possible definitions of geometrical and mechanical systems compatible with our experience of the world. They are neither a priori nor empirical; instead, they are conventions, in this case convenient preconditions of experimental physics, and hence not amenable to empirical testing. Poincaré's conventionalism differs from Duhem's holism. The Duhem thesis entails a form of underdetermination of theory by data that differs from the thesis that indefinitely many alternative hypotheses are compatible and can be deductively connected to a given body of empirical data. It entails that for any successful theoretical hypothesis, all of its rivals—that is, any falsified, or incompatible, hypothesis—can be made to fit the data with a suitable modification in the auxiliary hypotheses.

In the subsequent three decades, the Duhem thesis played an important role in the formulation of views associated with logical empiricism and the members of the Vienna Circle. Thus, Moritz Schlick argued in 1915 that geometry was nonempirical, since the non-Euclidean geometry used in general relativity is justified as part of the simplest total system of natural laws consistent with the empirical data, so that Euclidean geometry remains a genuine possibility as part of a different, more complicated total system of natural laws (Schlick [1915] 1979).

For Otto Neurath, in the same decade, the thesis legitimized the possibility of retaining a hypothesis in the face of conflict with data. It suggested the possibility of imagining an indefinite number of theoretical possibilities; it opened the door for pragmatic considerations, especially in the world of natural and social phenomena; and it motivated the methodological desirability of a unification of the sciences, allowing for exact predictions of the behavior of complex phenomena involving factors studied by different sciences. Neurath's holism was more radical than Duhem had envisioned. In the early 1930s Neurath stressed that within the evolving complexes of beliefs that make up science and culture, the distinction between analytic and synthetic statements, just like the validity of logical principles, would be a contingent historical matter (Neurath 1983). Similarly, Rudolf Carnap (1937) argued, in *The Logical Syntax of Language*, that the Duhem thesis was compatible with a distinction between meaning-constitutive (analytic) and factual (synthetic) statements, and defended further the more general type of holism that allowed for the revisability of both types of statements. This appreciation led to a defense of a spirit of tolerance—Carnap's *principle of tolerance*—and of pragmatic considerations that included logic and mathematics, as well as the choice of linguistic frameworks (Carnap 1937).

In the 1950s, Willard Van Quine pointed to an alleged failure of Carnap's attempts to articulate an account of analyticity and revived Duhem's arguments as part of his rejection of the distinction between analytic and synthetic propositions. Quinean, more radical holism entails equality of status of empirical (physical) and theoretical (logical or mathematical) statements. Moreover, Quine's additional criticism of accounts of analyticity prompted his epistemological naturalism, shifting the emphasis to synthetic empirical statements originating in sensory stimuli at the periphery of our "web of beliefs." In the holistic web of science, Quine distinguished beliefs only by their different degrees of centrality and entrenchment, a naturalistic counterpart to the more traditional reference to degrees of certainty.

This view has been criticized most recently by Friedman (2001), who draws attention instead to different functions or roles of beliefs within the evolving and unpredictable whole of science. One such role is played by constitutive principles, as relativized a priori in the post-Kantian tradition of Reichenbach's and Kuhn's ideas. Such principles made other beliefs within the constituted framework empirically testable. For instance, Sir Isaac Newton's universal law of gravitation gets empirical sense and application from his laws of motion linking forces to mass and acceleration and defining an inertial frame of reference. A similar idea appears in Poincaré's formulation of conventionalism.

Another sort of assumption involved in the application and testing of hypotheses concerns the complete description of the situation in terms of the absence of interfering factors or disturbing influences, known as provisos (following Hempel), *ceteris paribus* ("other things being equal") clauses (following John Stuart Mill), or more radically, *ceteris absentibus* clauses ("other things being absent"). The assumption that *ceteris paribus* clauses cannot be dismissed raised a debate about the contents and conditions of applicability of natural laws. For Cartwright (1989), their presence has important philosophical implications. She argues that the causal nature of the contents of *ceteris paribus* conditions renders the reduction of causal laws to statements of regularities about manifest behavior impossible. They imply also that in cases of composition, laws about components cannot satisfy Hempel's criteria of scientific explanation; instead, they can only literally describe either fictions (i.e., models) or causal capacities, but not manifest behavior. Their descriptive contents have been defended also in terms of counterfactual statements about the isolation of systems, as in the description of dispositions (Suppe 2000) or in terms of additional explanatory commitments subject to empirical testability (Pietroski and Rey).

In the 1980s Hempel (1988) raised the so-called problem of provisos: The acceptable determinate formulation of a law in a theory as well as the derivation of empirical consequences require the explicit statement of an indefinitely large number of relevant provisos, but this requirement is impossible to satisfy. The problem can be solved by assuming that a theory need not ground its isolation conditions. Such grounding can be either a pragmatic and contextual question (Lange 1993) or else a question of independent causal knowledge about experimental settings (Suppe 2000).

JORDI CAT

References

Carnap, R. (1937), *The Logical Syntax of Language*. London: Kegan, Paul, Trechner, Teuben & Co.
Cartwright, N. (1989), *Nature's Capacities and Their Measurement*. Oxford: Oxford University Press.
Duhem, P. ([1906] 1954), *The Aim and Structure of Physical Theory*. Princeton, NJ: Princeton University Press.
Friedman, M. (2001), *The Dynamics of Reason: The 1999 Kant Lectures at Stanford University*. Stanford, CA: CSLI Publications.
Hempel, C. G. (1988), "Provisoes: A Problem Concerning the Inferential Function of Scientific Theories," *Erkenntnis* 28: 147–164.
Lange, M. (1993) "Natural Laws and the Problem of Provisos," *Erkenntnis* 38: 233–248.
Neurath. O. (1983), *Philosophical Papers, 1913–1946*. Dordrecht, Netherlands: Reidel.
Popper, K. R.([1935] 1951), *Logic of Scientific Discovery*. London: Unwin.
Schlick, M. ([1915] 1979), "The Philosophical Significance of the Principle of Relativity," in *Philosophical Papers*, vol. 1: *1909–1922*. Dordrecht, Netherlands: Reidel, 153–189.
Suppe, F. (2000), "Hempel and the Problem of Provisos," in J. Fetzer (ed.), *Science, Explanation and Rationality: The Philosophy of C. G. Hempel*. Oxford: Oxford University Press, 187–213.

See also **Neurath, Otto; Poincaré, Henri; Popper, Karl; Quine, Willard Van; Underdetermination of Theories**

DUTCH BOOK ARGUMENT

The Dutch Book argument was first presented by Frank Ramsey in his 1926 paper "Truth and Probability" (Ramsey 1926) (see Ramsey, Frank Plumpton). The argument purports to show that an agent's degrees of belief, or degrees of confidence, should satisfy the Kolmogorov axioms of probability (termed coherence) (see Probability). It is often cited by Bayesians, who take degrees of belief to be probabilities and endorse a probabilistic approach to theory confirmation (see Bayesianism; Confirmation Theory). If $P(H)$ represents the probability assigned to the statement (or sentence or proposition) H, then the axioms require that:

1. $0 \leq P(H) \leq 1$;
2. if H is a tautology, then $P(H) = 1$; and
3. if H_1 and H_2 are mutually exclusive, then $P(H_1 \vee H_2) = P(H_1) + P(H_2)$.

The Dutch Book argument (DBA) assumes that an agent's degrees of belief are linked to a set of betting quotients, so that if an agent's degrees of belief are incoherent (i.e., fail to satisfy the axioms), then the agent will possess an incoherent set of betting quotients. The argument then appeals to a mathematical theorem, the Dutch Book theorem, which states that if a set of betting quotients fails to satisfy the axioms of probability, there will be a series of bets, each of which is individually fair according to that set of betting quotients but which taken together will produce a net loss (a "Dutch Book"). The fact that incoherent degrees of belief are linked to a sure loss is taken as reflecting a defect in those beliefs. The mechanics of making a Dutch Book against an agent whose betting quotients fail to satisfy the axioms will be considered first, and will be followed by a discussion of what the possibility of constructing such a series of bets reveals about a person's beliefs.

To convey the content of the Dutch Book theorem, it will be shown how a bookie can exploit a bettor whose betting quotients violate the probability axioms. With de Finetti (1937), it is here assumed that a bet on a proposition H is an arrangement that has the following canonical form:

H	Payoff
T	$S - qS$
F	$-qS$

If H is true, the bettor on H collects the amount $S - qS$, but if H is false the bettor loses qS. The quantity S is called the stake and q is the betting quotient. S is the amount won if H is true and qS is the cost of the bet. Here it is assumed that an agent will bet either for or against H, provided that the betting quotient $Q(H)$ equals q. These are presumed to be fair bets by the agent's lights, or as having an expected value of zero. It can now be shown that if the agent's betting quotients fail to satisfy the axioms, a Dutch Book can be made against the agent by a clever bookie, provided that the agent will take either side of a bet for which $Q(H)$ equals q. For simplicity, the stakes for each bet will be set at $1.

- **Axiom 1:** Suppose that $Q(H) < 0$. In this case, the bookie buys the bet that pays $1 if H is true and 0 otherwise, for the negative price $Q(H)$. This is equivalent to betting against H for the agent and so the payoff table is as follows:

H	Payoff
T	$-[1 - Q(H)]$
F	$Q(H)$

  Since $Q(H)$ is negative, the agent will suffer a net loss regardless of the truth value of H. Suppose on the other hand that $Q(H) > 1$. In this case the bookie sells the bet that pays $1 if H is true and 0 otherwise, for $Q(H)$. Of course, this means that the agent will pay more for the bet than it is possible to gain and so end up with a net loss to the bookie.

- **Axiom 2:** Suppose that an agent's betting quotient for a tautology H is not equal to 1. The case where $Q(H) > 1$ was included above, so assume that $Q(H) < 1$. Here the bookie will buy the bet in which the agent pays the bookie $1 if H is true, and nothing if H is false, for $Q(H)$. The payoff table for the agent will be:

H	Payoff
T	$-[1 - Q(H)]$
F	$Q(H)$

  Notice that the agent is bound to lose the amount $[1 - Q(H)]$, since H is a tautology and hence must be true.

- **Axiom 3** (additivity): Assume that H_1 and H_2 are mutually exclusive and that $Q(H_1 \vee H_2) \neq Q(H_1) + Q(H_2)$. There are two cases,

  a) $Q(H_1 \vee H_2) > Q(H_1) + Q(H_2)$
  and
  b) $Q(H_1 \vee H_2) < Q(H_1) + Q(H_2)$.

  Suppose that $Q(H_1 \vee H_2) < Q(H_1) + Q(H_2)$, then the bookie will offer the agent the bet that pays $1 if H_1 and 0 otherwise for $Q(H_1)$ and the bet which pays $1 if H_2 is true and 0 otherwise for $Q(H_2)$. The bookie then buys the bet which will lead to a gain of $1, if $(H_1 \vee H_2)$ is true and 0 otherwise, for the price of $Q(H_1 \vee H_2)$.

The possible payoffs to the agent are summed up in the following table:

H_1	H_2	Net Payoff
T	F	$[1 - Q(H_1) - Q(H_2) + Q(H_1 \vee H_2) - 1]$
F	T	$[1 - Q(H_1) - Q(H_2) + Q(H_1 \vee H_2) - 1]$
F	F	$[-Q(H_1) - Q(H_2) + Q(H_1 \vee H_2)]$

Since $Q(H_1 \vee H_2) < Q(H_1) + Q(H_2)$, the agent is assured of a loss in each case. If $Q(H_1 \vee H_2) > Q(H_1) + Q(H_2)$, then the bookie simply reverses the direction of the bets.

It has been demonstrated above how an incoherent set of betting quotients can be exploited to produce a sure loss. It can also be shown that the bookie is not guaranteed a net profit if the agent's betting quotients are coherent. It is now time to examine what such betting quotients show about an agent's degrees of belief. Ramsey understood a person's degree of belief in a proposition as reflecting the person's willingness to act on it, and maintained that betting quotients are at least an approximate measure of a person's degrees of belief. He argued from what is in effect the Dutch Book theorem that

> If anyone's mental condition violated (the laws of probability), his choice would depend on the precise form in which the options were offered him, which would be absurd. He could have a book made against him by a cunning bettor and would then stand to lose in any event. (Ramsey 1926, 80)

This has been interpreted as the claim that incoherence is irrational because it leaves the agent vulnerable to bad consequences, and so is a kind of pragmatic defect. If the argument is understood in this way, it is open to serious objections. The main problem is that the link between having incoherent degrees of belief and suffering a loss is a weak one. First an agent's degrees of belief need not correspond to the bets that the agent will consider fair, or be willing to take. A person might be highly confident in a proposition, yet be unwilling to bet at the corresponding odds, because of risk aversion or a view of gambling as inappropriate. Even if the agent's degrees of belief and betting quotients match for individual bets, the agent is not compelled to regard the corresponding bets as jointly fair, as is needed to show that violation of the additive law involves Dutch Book vulnerability. Putting aside objections that an incoherent agent need not regard bets involved in producing a Dutch Book as fair given corresponding degrees of belief, the connection between the existence of such bets and suffering a bad outcome is rather tenuous. Whether an incoherent agent suffers an actual loss depends on whether the necessary bets will actually be made and collected. Furthermore, the agent could simply refuse to bet, and thus avoid any potential loss. Finally, the assumption that it is irrational to put oneself in a situation that could lead to a sure loss seems simply wrong, for putting oneself in such a situation could be the best available (if not a terribly desirable) option.

It has been argued that the DBA is misunderstood if it is thought to force compliance with the probability axioms as a means of avoiding bad outcomes. The suggestion is that it is not the threat of a sure loss that is the problem, but rather that having choice guide beliefs that are tied to a sure loss signals an inconsistency in those beliefs. Indeed, this interpretation fits well with Ramsey's claim that "any definite set of degrees of belief which broke (the probability axioms) would be inconsistent in the sense that it violated the laws of preference between options" (Ramsey 1926, 80). In this reading, having degrees of belief can be reduced to having certain preferences, with incoherence then being inconsistency of preference. Moreover, given appropriate constraints on a set of preferences, a utility function can be defined relative to which the value of bets is additive. It can thus be argued that by appealing to the theory of preference, a crucial assumption of the DBA can be defended. Further, within the theory of preference it is possible to define both utility and probability functions such that an agent's preferences can be represented as maximizing expected utility relative to those utility and probability functions. Such representation theorems yield a direct argument for probabilism by showing that given rational preferences, an agent can be interpreted as having degrees of belief that satisfy the probability axioms. There is controversy over the attempt to justify the DBA by appealing to principles of utility theory. Some view it as at best irrelevant, given the representation theorems. Others raise questions about representation arguments and see the DBA as providing important motivation for the Bayesian constraints on rational belief (see discussion, see Armendt 1993).

Some philosophers have objected to Ramsey's idea that incoherence reduces to inconsistency of preference, as well as to the idea that the force of the DBA derives from fundamental assumptions about preference and decision. Several attempts have been made to "depragmatize" the DBA and to show clearly that incoherence involves a type of inconsistency that is essentially epistemic rather than pragmatic in nature (Howson and Urbach 1993; Christensen 1996). Instead of reducing degrees of belief to preferences, they are reduced to evaluations of the fairness of bets, or are understood as justifying certain bets as fair. In each case, the Dutch Book theorem is invoked to establish that incoherence involves believing bets to be fair that cannot be, or as justifying bets as fair that cannot be fair. Here incoherence bears a clearer resemblance to inconsistency for full belief than on Ramsey's preference interpretation, but it is doubtful that the argument can be made to work without the resources of decision theory.

The attempts to depragmatize the DBA have foundered in providing a noncircular, and genuinely nonpragmatic, account of fairness according to which violation of the probability axioms *always* involves the appropriate sort of unfairness (see Maher 1997). Further, it is just as implausible that degrees of belief reduce to judgments of fairness, or that they alone justify certain beliefs as fair, as is the claim that degrees of belief can be reduced to preferences. Still the underlying idea that incoherence is an epistemic, and not essentially a pragmatic, defect is surely correct. It is the notion of valuation, together with the underlying logic of propositions, that yields the Dutch Book theorem and suggests that coherence is, at least, an epistemic ideal.

Dutch Book arguments have also been devised in support of the principles of conditionalization, Jeffrey conditionalization, and reflection. (For discussion, see Teller 1973; Armendt 1980; and van Fraassen 1995.)

Susan Vineberg

References

Armendt, B. (1980), "Is There a Dutch Book Argument for Probability Kinematics?" *Philosophy of Science* 47: 583–588.

——— (1993), "Dutch Books, Additivity and Utility Theory," *Philosophical Topics* 21: 1–20.

Christensen, D. (1996), "Dutch-Book Arguments Depragmatized: Epistemic Consistency for Partial Believers," *Journal of Philosophy* 93: 450–479.

Christensen, D. (2001), "Preference-Based Arguments for Probabilism," *Philosophy of Science* 68: 356–376.

de Finetti, B. (1937), "Foresight: Its Logical Laws, Its Subjective Sources," in H. E. Kyburg and H. E. K. Smokler (eds.), *Studies in Subjective Probability*. Huntington, NY: Robert E. Kreiger Publishing Co.

Howson, C., and P. Urbach (1993), *Scientific Reasoning: The Bayesian Approach*. La Salle, IL: Open Court.

Kaplan, M. (1996), *Decision Theory as Philosophy*. Cambridge: Cambridge University Press.

Kennedy, R., and C. Chihara (1979), "The Dutch Book Argument: Its Logical Flaws, Its Subjective Sources," *Philosophical Studies* 36: 19–33.

Maher, P. (1993), *Betting on Theories*. Cambridge: Cambridge University Press.

——— (1997), "Depragmatized Dutch Book Arguments," *Philosophy of Science* 64: 291–305.

Ramsey, P. F. (1926), "Truth and Probability," in H. E. Kyburg and H. E. K. Smokler (eds.), *Studies in Subjective Probability*. Huntington, NY: Robert E. Kreiger Publishing Co.

Skyrms, B. (1975), *Choice and Chance*. Belmont, CA: Wadsworth.

——— (1987), in N. Rescher (ed.), Coherence. Scientific Inquiry in Philosophical Perspective. Pittsburgh: University of Pittsburgh Press, 225–242.

Teller, P. (1973), "Conditionalization and Observation," *Synthese* 26: 218–258.

van Fraassen, B. (1995). "Belief and the Problem of Ulysses and the Sirens," *Philosophical Studies* 77: 7–37.

Vineberg, S. (2001), "The Notion of Consistency for Partial Belief," *Philosophical Studies* 102: 281–296.

See also **Bayesianism; Confirmation Theory; Decision Theory; Probability**

E

ECOLOGY

Ecology is composed of a remarkably diverse set of scientific disciplines, and many different subfields can be distinguished, such as physiological, behavioral, evolutionary, population, community, ecosystem, and landscape ecology. Clearly, no summary will do them all justice. However, for the present context, ecology as a science can be divided into three basic areas—population, community, and ecosystem ecology. This article will introduce some of the fundamental philosophical issues raised by these three disciplines.

The first order of business is to ask, What is the science of ecology? and more importantly, What is it *not?* (see Brennan 1988). Sometimes the term "ecology" is treated as synonymous or coextensive with three different concepts or sets of concepts:

- *The science of ecology*: the study of organisms, their groups, and their relation to their environment.
- *Environmentalism*: a set of sociopolitical views about the right relationship between humans and nature.
- *The ecology of an organism, population, or community*: in the case of organisms, roughly the life history of that organism.

This essay will focus only on ecology in the first of these senses—as a set of scientific disciplines. It should be noted, however, that there are important questions about how the science of ecology is related to environmental ethics and public policy (see Ludwig, Mangel, and Haddad 2001).

Metaphysics and Ecological Communities

One of the standard topics of ecology is succession. Succession concerns the structural and compositional changes that occur in communities and ecosystems as populations and species replace each other. Traditionally, succession is broken into three stages. The first is the *pioneer* stage, when the first colonizers arrive in an area; each subsequent stage is called a sere; up to the final, relatively stable stage, called the *climax*. Succession is either *primary* or *secondary*. Primary succession involves the colonization of bare ground where no ecosystem has been present. Examples of areas where primary succession occurs are sand dunes, volcanic flows, mud flats, and glacial tills. Secondary succession involves the replacement of communities after some disturbance that may involve abandoned fields, wind-blown gaps in forests, or wildfires. An example of temperate terrestrial secondary succession is the sequence of annual weeds, perennial weeds, shrubs, young pine forest, and oak forest with a well-defined canopy.

One of the foundational controversies in community ecology arose between Frederic Clements (1916) and Henry Gleason (1917) concerning succession and the nature of communities. Clements argued that communities follow a very specific sequence of stages that can be characterized in terms of nutrient cycling, species diversity, and biomass. He claimed that there is a single climax community that is self-perpetuating and tightly integrated as the result of biotic interactions among species. Clements considered communities to be "superorganisms":

> The developmental study of vegetation necessarily rests upon the assumption that the unit or climax formation is an organic entity. As an organism the formation arises, grows, matures, and dies. . . . The life-history of a formation is a complex but definite process, comparable in its chief features with the life-history of an individual plant. (1916, 16).

Gleason considered Clements' views to be without empirical support, and argued that succession results from individual species' physiological requirements and local climatic conditions:

> [I]t may be said that every species of plant is a law unto itself, the distribution of which in space depends upon its individual peculiarities of migration and environmental requirements. (1917, 26)

Hence, Gleason's views were oriented more to the individual. Likewise, he did not think that there was a final climax community, but rather communities were continually changing and nonequilibrial. These two approaches to succession and communities continue to be of influence in ecology (see Levins and Lewontin 1985; Simberloff 1980).

Nonetheless, one is still left wondering what communities are. Ecologists have conceived of communities in roughly three different ways (Shrader-Frechette and McCoy 1993, 11–31):

- Communities are groups of species at particular places and times and nothing more.
- Communities are functionally interrelated groups of species.
- Communities are groups of species that are organism-like entities.

Biologists grant that an ecological community is minimally a set of species. However, what else, if anything, is required? As Kristin Shrader-Frechette and Earl McCoy (1993, 13) ask: "Envision a group of species occurring in the same place at the same time. Conceptually, what attributes might be used to link these species together, such that they could be distinguished from other similar groups?" Better sense can be made of the three different concepts of communities by considering some metaphysics.

All objects (except possibly the very simplest) are composed of parts. Those parts may or may not be related to each other. Objects can be classified by the relations that exist between their parts, as either aggregates, wholes, or individuals. If an object is an *aggregate*, its parts bear little or no causal relations to one another. Thus, aggregates are not causally integrated at a time or over time. If an object is a *whole*, then certain causal relations exist among its parts. Wholes exist as causally structured entities that are minimally integrated at a time and through time. Finally, an *individual* is an object whose parts bear causal relations to one another such that the object is highly structured and integrated. The differences between aggregates, wholes, and individuals concern the causal relations amongst their parts and the strength of those relations. These objects form a continuum, and the differences between them are of degree. Communities can exist as aggregates, wholes, or individuals.

Now consider the sort of community that Gleason (1926) had in mind:

> Are we not justified in coming to the general conclusion, far removed from the prevailing opinion, that an association [i.e., community] is not an organism, scarcely even a vegetation unit, but merely a *coincidence*? (16)

Communities, according to Gleason, are composed of whatever species coexist in space and time. This is a Gleasonian community: a group of species in a particular area at a particular time. In effect, this type of community consists of aggregates—objects whose parts bear few if any causal relations to one another.

Now consider those communities that exist as wholes. Here there is a set of species that exists as a structured entity—there are causal relations that at least weakly integrate the species at a time and through time. This type of community concept is sometimes associated with George Evelyn Hutchinson. Hutchinson thought of communities as having "feedback loops" that assure their self-regulation and persistence.

What sorts of causal relations or feedback loops might bind species in a community? One possibility is that various interspecific interactions exist amongst organisms and populations. Between any two species, these interactions can be classified as either positive (+), negative (−), or nonexistent (0), depending on how they affect the growth or abundance of the respective species. These relations include competition [−,−], predator-prey [−,+], mutualisms [+,+], amensalisms [−,0], and commensalism [0,+]. Likewise, some interactions take place between more than two species. These *indirect effects* occur when a donor species' influence is transmitted through at least one transmitter species to a receiver species. Finally, pairwise interactions themselves may be *nonadditive* if the interaction between the pair changes as the number of species in the community changes. If there are interspecific interactions between species that integrate the species into something more than an aggregate—a whole—then this community will be called a Hutchinsonian community: a group of species that at least weakly interact with one another and no others. The community exists as a group of species structured by various interspecific causal relations. One can also see why some ecologists are skeptical of the existence of plant communities and animal communities, since they leave out causally salient parts.

Finally, a community may be a tightly integrated group of species that bear various causal relations among their component species. The community forms an individual, as if it were a multicellular organism. This is a Clementsian community: a group of species that strongly interact with one another.

It is an empirical issue whether any of these community concepts apply to any group of species. Nonetheless, some progress has been made in understanding what ecological communities *might* be. Next, several arguments will be considered for why one might think ecological communities do not exist. Here is one such argument. Communities are real only if they have distinct boundaries. However, many purported communities do not have distinct boundaries. Hence, many purported communities are not real (see Simberloff 1980, 16–17; Levins and Lewontin).

There are several general points to be made about this argument. By all accounts, a community consists of a group of species. Moreover, the community exists wherever those species exist. Thus, its boundary consists of its outermost species. So, though it may be difficult, a community's boundary can be determined by its species' boundaries. However, putatively different communities blend continuously into one another unless there is some ecotone—a relatively discrete zone of transition (Figure 1). If they blend continuously,

then it is not clear where the communities begin and end.

This may be an epistemological problem for ecologists, but from a metaphysical point of view, it need not be. For example, if two Hutchinsonian or Clementsian communities share a common habitat, they still are distinct in virtue of the causal interactions between their respective species. As Richard Levins and Richard Lewontin write, "The question of boundaries of communities is really secondary to the issues of interaction among species" (1985, 138). Hence, the problem of continuous overlap need not be a particular problem for the Hutchinsonian and Clementsian approaches.

There does seem to be reason for denying the existence of Gleasonian communities. Recall that a Gleasonian community is a set of species in a particular area at a particular time. Suppose there is a group of n species at a particular place at a particular time. If the group is a Gleasonian community, then it can properly be asked why some other $(n + 1)$-th species is *not* a member of the community. In one of the other approaches, the answer would be given by the causal interactions of the n species. The $(n + 1)$-th species would be excluded from such interactions. However, in the Gleasonian approach, it appears that the membership of the community is not secured by mind-independent

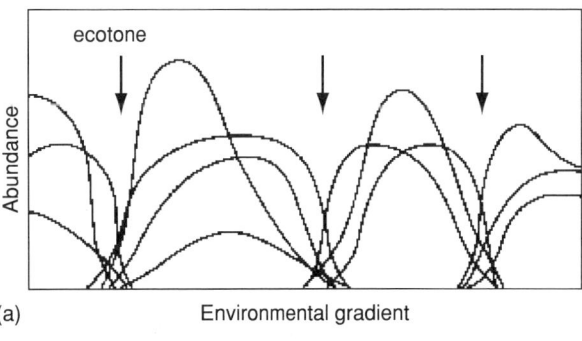
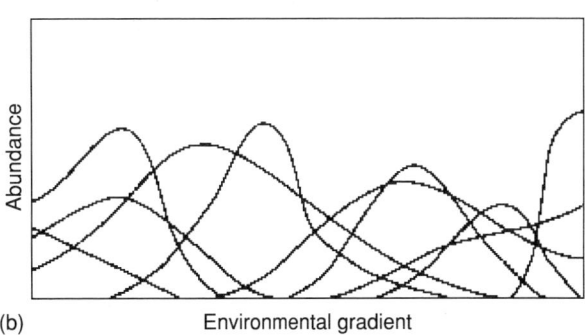

Fig. 1. (a) Ecotones generating discrete boundaries between groups of species.
(b) No ecotones generating discrete boundaries between species.
[*From Ricklefs and Miller 2000, 524*]

causal interactions but rather by the ecologist's choice about spatial and temporal boundaries. However, if Gleasonian communities objectively exist, they must exist independently of mind. The communities depend on ecologists' decisions—arbitrary or not—as to what species to consider members of them. Hence, they do not objectively exist. This in effect was the view of the ecologist Robert MacArthur (1962):

> Irrespective of how other ecologists use the term "community"—and there are almost as many uses as there are ecologists—I use it here to mean any set of organisms currently living near each other and about which it is interesting to talk. (189–190)

Likewise, something is *natural* only if it does not depend on human activities. Hence, even if Gleasonian communities exist, they would be nonnatural in this sense. Hence, they either do not objectively exist or are nonnatural.

This discussion has considered the nature of ecological communities and has only skimmed some of the issues. There are, however, many conceptual and metaphysical problems concerning ecological entities. *Token* ecological communities may exist—however, what about *types* of communities, or biomes, such as temperate grassland, chaparral, savanna, deserts, taiga, and tropical rain forests? Much of early community ecology consisted of classifying communities, and traditional accounts of succession seem to depend on such classifications. Do other ecological entities like ecosystems or guilds exist? If ecosystems exist, do they possess fashionable properties like *health* and *integrity*?

A Balance of Nature?

One might consider it "folk wisdom" that flora and fauna exhibit a "balance of nature" (Egerton 1973; Pimm 1993). Ecologists have often thought that the more diverse or complex a community or ecosystem is, the more stable it would be. This section will consider the diversity/complexity/stability hypothesis conceptually.

As was mentioned in the introductory section, ecologists have debated the meanings of 'community' for some time. Similarly, stability has been construed as the return of species abundances to pre-perturbation equilibrium values, the resistance of invasion by exotics, and the persistence of species composition of the community after a disturbance. At first glance, one might conclude that ecology is in conceptual disarray, since ecologists cannot even agree on what they are theorizing about (Shrader-Frechette and McCoy 1993).

To formulate diversity/stability hypotheses carefully, ecologists have provided precise notions of ecological stability. This hypothesis can be understood as the following claim:

- As the diversity or complexity of a community increases, so does the stability of the community.

However, this is really a schema for a variety of hypotheses depending on how one characterizes stability, diversity, and complexity. In order to understand the concept(s) of stability, it is useful to begin with an examination of the work of Stuart Pimm (1984a, 1984b, and 1993).

Pimm distinguishes among complexity, stability, and variables of interest. The complexity of a community can be defined in terms of species richness, connectance, interaction strength, or evenness. *Species richness* is the number of species in a community. *Connectance* is the number of interspecific interactions divided by those possible. *Interaction strength* is the mean magnitude of interspecific interaction, i.e., the size of the effect of one species' density on the growth rate of another species. Species *evenness* is the variance of the species abundance distribution. The variables of interest are individual species abundances, species taxonomic composition, and trophic level abundance. One important issue to note is that diversity (species richness and evenness) forms a proper part of the complexity concept. Hence, as ecologists have moved from evaluating diversity/stability to evaluating complexity/stability, they have broadened the very nature of their hypotheses.

The stability of a community is characterized in one of the following ways (Pimm 1984b, 322):

- *Stability*: just in case all the variables in a system return to their initial equilibrium values following a perturbation
- *Resilience*: how fast the variables return to their equilibrium following a perturbation
- *Persistence*: how long the value of a variable lasts before it changes to a new value
- *Resistance*: the degree to which a variable is changed following a perturbation
- *Variability*: the degree to which a variable varies over time

Thus, there are four definitions of complexity, five of stability, and three variables of interest. Consequently, there are an extremely large number of contenders for the complexity/stability hypothesis.

Robert May (1973) was one of the first to explore precisely the connections between complexity and stability with what are called *local stability analyses*. May assumes that there is a community of species described by a set of nonlinear first-order differential equations. Let $N_i(t)$ be the density of the species i at time t. To determine the joint equilibrium density N_i^* of the species, their growth rates, $dN_i(t)/dt$, are set equal to zero and the equations are solved. One must then determine whether the joint equilibrium density is stable or not. That is, if the species were perturbed in a relatively small way from that joint density, would it return in the limit? If the community would return, then it is asymptotically locally stable and is not locally stable otherwise.

May constructed his model communities with S species by choosing the interaction coefficients a_{ij}, a parameter measuring the effect of species j on species i, at random. Thus, some species interaction coefficients were greater than, less than, or equal to zero, and hence some species pairs interacted as competitors, predator and prey, and mutualists. He defined the connectance C of the community as the proportion of interspecific interactions a_{ij} not equal to zero. The intensity I of the interspecific interaction a_{ij} was a random variable with a mean of 0 and a variance of I^2. May infamously demonstrated that a community is qualitatively stable if and only if

$$I(SC)^{1/2} < 1.$$

Hence, *increases* in the number of species, connectance, or interaction strength all lead to a *decrease* in the stability of a community. May's result has not gone uncriticized. Nonetheless, more realistic models lead to the same general result, that stability decreases with increasing complexity.

Pimm (1984b) investigated larger perturbations of a different kind than the arbitrarily small demographic ones of May's analysis. Pimm's larger perturbation was the deletion of single species from the community. Informally, a community is species-deletion stable if and only if following the removal of a species from the community, all of the remaining species are maintained at a new locally stable equilibrium. Pimm found with qualifications that the number of interactions decreases the community's species-deletion stability.

Empirically oriented ecologists have not always looked favorably on the work of May and his mathematical cohorts. For example, J. S. McNaughton (1977) argued that the truth of the diversity/stability hypothesis depends on empirical tests and that all else are "acts of faith, not science" (516). One study he and his colleagues conducted was on the grasslands in the Serengeti-Mara ecosystem in Tanzania and Kenya. McNaughton examined the effect of the grazing African buffalo *Syncerus caffer* on the grasslands. He found that species diversity in the more diverse stands decreased more than in the less diverse stands because of grazing. Amazingly, though, the more diverse community suffered less of a reduction in primary production (biomass) than the less diverse community. McNaughton concluded from his study that "[t]he weight of evidence resulting from explicit tests of the diversity-stability hypothesis ... suggests, not that the hypothesis is invalid, but that it is correct" (1977, 522). It thus seemed that species diversity stabilizes ecosystem properties like primary production, and so the diversity/stability hypothesis is true and the recent models incorrect.

In 1983, Anthony King and Stuart Pimm replied to McNaughton's work, attempting to "resolve this apparent contradiction between theory and empiricism" (King and Pimm 1983). King and Pimm devised grazing food web models with n plant species and one herbivore. They examined the models with respect to three types of complexity—species richness, connectance, and species diversity. They found that each type of complexity increased relative to biomass stability, which is the ratio of the total plant biomass without the herbivore to the total plant biomass with the herbivore. They also found that if stability is determined by species composition of the community, then stability decreases with increasing complexity. So King and Pimm's and McNaughton's results generally coincide. King and Pimm (1983) argue that McNaughton was incorrect in supposing that either the field ecologists or the mathematical modelers were right. There are different types of stability, which increasing complexity can increase or decrease independently of each other. The conflict between the work of McNaughton and that of the modelers was only apparent.

Since 1982, David Tilman has continued the work of May, Pimm, and McNaughton by conducting large-scale experiments at Cedar Creek Natural History Area in Minnesota. These experiments have shown that species richness is positively correlated with plant community stability—there is a decreased coefficient of variability of plant community biomass with increasing numbers of species. However, diversity does not seem to have much effect on the variability of the component populations. There is still much controversy over whether increasing diversity *causes* decreasing plant biomass variability.

Lastly, Shrader-Frechette and McCoy (1993) argue that the terms "stability" (and "community") are "ambiguous, imprecise, and inconsistent." They claim that if community ecology is to produce predictive, general theories that are adequate for environmental applications, the foundational concepts of ecology must be clear and precise. Otherwise, there will be conceptual confusion, and different interpretations of those concepts will lead to different conservation strategies (54, 57–8). They conclude that the theories of community ecology are not well equipped for conservation purposes (for a response, see Odenbaugh 2001).

Ecological Theories: Contingency, Predictive Accuracy, and Explanation

This section will consider various methodological problems that have haunted ecological theory. Ecology has not suffered from a lack of theories. However, these models and the practice from which they arise have been heavily criticized (Peters 1991; Shrader-Frechette and McCoy 1993). As Simberloff (1981) writes,

> Ecology is awash in all manner of untested (and often untestable) models, most claiming to be heuristic, many simple elaborations of earlier untested models. Entire journals are devoted to such work, and are as remote from biological reality as are faith-healers. (3)

Critics have been skeptical of the construction of theory for these and other reasons. Whatever its merits, this skepticism does force one to wonder how the success of model building, if it is successful, *could* arise. Put dramatically, how is it distinct from a sort of numerology? This section will consider three questions:

- Can ecologists build *successful* theories and models?
- How should ecologists *evaluate* their theories and models?
- Can ecological theories or models be *explanatory*?

For want of space, the treatment here will consider some characteristic answers and is not intended to be exhaustive.

Can Ecologists Build Successful Models and Theories?

Ecologists have long desired general theories that account for the behavior of populations, communities, and ecosystems. More than any other ecologist, MacArthur has been associated with the building of such theories, often in mathematical form. He argues that ecologists are in the business of finding and explaining general patterns in the distribution and abundance of organisms. They should seek theories that minimize history and emphasize the equilibria so dear to their hearts.

However, if a research program like MacArthur's is to succeed, the biological world must cooperate. Philosophers and ecologists have suggested two problems with such theorizing. First, there is the problem of *contingency* (Sterelny 2001). One can argue that there simply are no general patterns about which ecologists can theorize. For example, historical accidents of distribution involving geographic barriers can play important causal roles in determining which species occur where. Australia, for instance, has bats and marsupials but very few other mammals. As Kim Sterelny (2001) writes:

> The worry posed by extreme versions of the contingency hypothesis is that there are no patterns at all. The thought here is that membership and abundance within a community is sensitive to so many causal factors that we cannot project from one community to another. (158–9)

More generally, ecological systems can be sensitively dependent on their prior states. This means that if the system's state at time t had been otherwise, then the system at $t + \Delta t$ would be significantly different. However, if ecological systems exhibit this "sensitive dependence" or if history matters, then ecologists should provide narratives, not mathematical models. At least in part, ecology would consist in labor-intensive case studies (Shrader-Frechette and McCoy 1993).

The second problem is that of *complexity*. Ecological systems can be exceedingly complex. They have large numbers of parts that usually interact in nonlinear ways. Moreover, ecologists themselves are limited cognitively. First, there are limitations that arise from the inability to manipulate experimentally the systems as directly as is desirable. In the field, there are multifarious factors at work and only some of them are recognized at any given time. Second, there are limitations in the ability to use mathematical representations of the systems of interest. Present capacities for storing and retrieving information, carrying out various inferences, and abstracting from details make it difficult to use certain types of mathematical formalisms. Hence, ecological modeling may be too labor intensive and mathematically intractable to be of any use for prediction (Levins 1966). There may be general ecological patterns that cannot be discerned or explained.

In light of the problems of contingency and complexity, many ecologists have accepted theoretical pluralism (McIntosh 1987). First, metaphysically, ecologists must grant that there is no single biotic or abiotic process that is responsible for ecological patterns. Second, models must be built with differing degrees of realism, generality, and precision. Some models should be more mechanistic and some more phenomenological. Moreover, one may have to trade these desiderata off, as Levins (1966) has long suggested. Finally, methodologically there must be a dynamic division of labor amongst modelers, laboratory experimentalists, and field workers.

How Should Ecologists Evaluate Their Theories and Models?

There are two issues to consider here concerning the role of prediction in modeling:

1. Should ecological models be evaluated on the basis of their predictive accuracy and that alone?
2. Provided that some models make predictions that can be tested, how should those predictions be evaluated?

Critics of ecological modeling offer the following argument. If models are going to be epistemically successful, then these theoretical hypotheses must be empirically testable. However, models are not straightforwardly testable. They make either *no* predictions, no *testable* predictions, or testable *false* predictions. Ecologist R. H. Peters (1991) writes,

> Ecology seeks to predict the abundances, distributions and other characteristics of organisms in nature. . . . This book contends that much of contemporary ecology predicts neither the characteristics of organisms nor much of anything else. Therefore it represents neither ecological nor more general scientific knowledge. (17)

Therefore, theoretical models are not a successful part of ecology.

Different critics recommend different ways of coping with the predictive failure of models. Even without delving into those proposals, serious problems can be seen with the preceding argument. First, some ecological models can accurately represent some empirical systems. Second, the argument assumes that predictive accuracy is the most important function of models. However, models, even empirically inaccurate ones, can be used for a variety of purposes. For example, they allow ecologists to explore possibilities, clarify ecological concepts, and provide conceptual frameworks for experimentation and fieldwork. As theoretical ecologist Hal Caswell (1988) argues, it is false to think that the only important thing to do with theories is to test them, that refuted theories must be abandoned, and that idealizations are a "methodological evil."

Models must be evaluated for performing the tasks for which they are designed. Theoretical ecologists like Caswell have argued that theories are tools: theoretical instruments that allow biologists to further their cognitive goals, which include predictive accuracy, but not exclusively so. As William Wimsatt (1987) has suggested, "False models can lead to truer theories." However, ecologists and philosophers have been slow to explain how the heuristics of model building work and what the standards are.

These pragmatists have also suggested that model building is *inescapable* for ecologists. For example, Charles Elton, without mathematics, suggested that communities that are more complex are more stable. Through the work of May and Pimm, it can be seen that there are many different ways of characterizing stability, complexity, and variables of interest. As Caswell writes:

> None of these distinctions were, or could have been, drawn by Elton. Their importance became apparent only as the original verbal theory was studied using mathematical models. (1988, 35)

The same is true in more applied matters. One trend in applied ecology is population viability analysis (Boyce 1992). Ecologists utilize simple logistic equations, Leslie projection matrices, and probabilistic models of demographic and environmental stochasticity to simulate the expected time to extinction of various species. These tools are needed because the relevant autoecological data are lacking and, if conservation is at stake, it is impossible to perturb experimentally these demographically depressed populations. There is no choice but to predict the fates of endangered species with mathematical models even if they are not especially accurate. Thus, model building is an essential part of theoretical and applied ecology.

Turning to the question of how the predictions of models should be tested, one of the most cantankerous debates in ecology concerns the *null hypotheses* (Gotelli and Graves 1996). In essence, the debate arose over the importance of interspecific competition in structuring properties of organisms such as body size and resource use. This debate also led to more general issues surrounding how ecological theories should be tested and evaluated. In 1975, Jared Diamond published a study on the distribution of

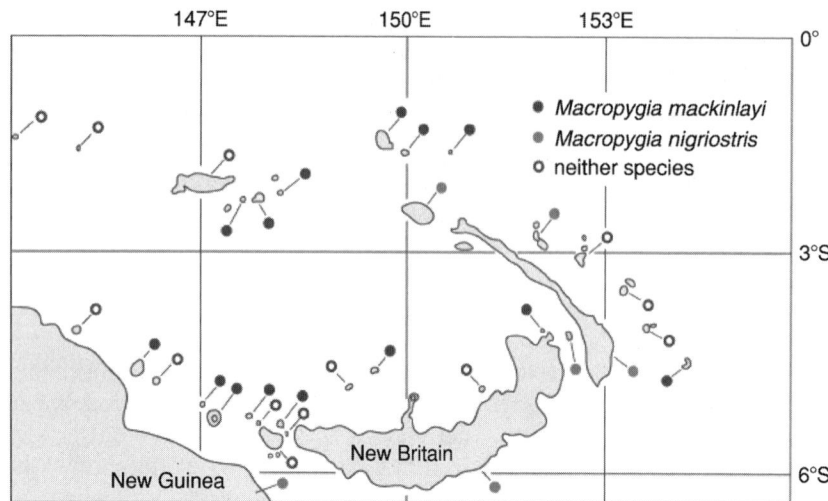

Fig. 2. The distribution of *Macropygia* species in the Bismarck Archipelago.
[*From Ricklefs and Miller 2000, 613*]

bird species among fifty islands in the Bismarck Archipelago near New Guinea (Diamond 1975). Diamond recognized that certain combinations of species were never found together in the archipelago. For example, two species of cuckoo dove, *Macropygia nigriostris* and *M. mackinlayi*, occurred on six and fourteen islands, respectively (Figure 2). However, they never co-occurred on any island.

This "checkerboard pattern," or complementary distribution, suggested that interspecific competition was at work through niche differentiation.

In the late 1970s, Edward Connor and Daniel Simberloff (1979) argued that Diamond's work was seriously flawed. They suggested that the checkerboard distribution could have resulted from random colonization rather than competition. They devised *neutral* or *null models* of communities that retained certain features, such as the number of species per island, the relative abundances of species, and their incidence functions (the probability of a species occurring on an island given the total number of species on that island), but they reassembled the rest at random excluding the effects of competition. If the actual data differ in statistically significant ways from the null hypothesis, then the null is rejected and the interaction is strongly suggested. Connor and Simberloff claimed that null hypotheses were more parsimonious and "logically prior" to competitionist hypotheses. Contrary to the Popperian philosophy adopted by Connor and Simberloff, Diamond looked for confirming evidence as opposed to first trying to refute a null hypothesis.

The work of Simberloff and his colleagues has been heavily criticized. First, in traditional Neyman-Pearson statistical testing, one formulates two mutually exclusive and exhaustive hypotheses, the null and the alternative. However, the null hypotheses articulated by the Florida group were not always logically inconsistent with competitionist hypotheses according to Michael Gilpin and Diamond. Key features of the null models—the species pools, dispersal abilities of species, and "incidence functions" of species—could be affected by competition (Gilpin and Diamond 1983). Hence, the "ghost of competition past" might be built into the null model itself, and thus it might have a "hidden structure." Second, Connor and Simberloff (1979) performed their analyses using sets of species that were not restricted to guilds (groups of species that utilize similar resources in similar ways). Competition is to be expected between two species only if they occupy the same guild. One would thus bury the effects of competition in a morass of irrelevant data (Gilpin and Diamond 1983).

It should be noted that Connor and Simberloff argued that even if one could delineate guild assignments with good evidence and had the checkerboard pattern of Figure 2, one still could not conclude that interspecific competition had been in operation. Likewise, they argued (Strong et al. 1984) that Gilpin and Diamond had not provided independent evidence for their hidden-structure claims.

The null model controversy continued in paper after paper and forced ecologists to address subtle issues concerning how predictions of ecological theory should be evaluated. It has invigorated hypothesis testing in ecology and led to more refined statistical tools for judging theory.

Can Ecological Theories or Models Be Explanatory?

Theories and models in ecology presumably provide scientific understanding of the systems they represent. A common philosophical supposition is that a theory or model explains some event or regularity only if it is true. Generally, though, ecological models are highly idealized—whatever their virtues, truth is not one of them. Hence, they cannot be explanatory. However, it does appear that ecological models do explain some events and regularities. Thus, models in ecology are not explanatory, or truth is not a necessity for successful scientific explanation.

As an example, consider the following *why*-question: Why is omnivory [feeding on more than one trophic level] rare in vertebrate food webs rather than common? Pimm and Lawton (1978), using Lotka-Volterra community models, gave one possible explanation for this. They demonstrated by computer simulations that food webs with omnivores were generally dynamically unstable. Either they were locally unstable or, if locally stable, their return time was excessively long. Hence, vertebrate food webs with omnivores would be unlikely to persist. A possible answer to this *why*-question is that vertebrate food webs with omnivores are dynamically fragile and hence do not persist.

The Lotka-Volterra community model is a caricature of empirical food webs. Some of the idealizations of the model include assuming that there is no migration and no age or genetic structure in the populations and that the density dependence is linear. Nonetheless, Pimm and Lawton argued that dynamical models explain various patterns of food webs, including the infrequency of omnivory. The fact that the model is severely idealized does not render it unexplanatory.

There are several ways to deal with this problem, one of which will be discussed here (see Cartwright 1983; Wimsatt 1987). Philosopher Gregory J. Cooper (2003) has offered a position that countenances the possibility that false ecological theories and models are explanatory. He argues, following Nancy Cartwright, that ecological models represent the *capacities* or *tendencies* of objects, which is how they would behave if there were no interfering forces. The dynamical equations of the models are true only of these dispositions or propensities. So, for example, the Lotka-Volterra model is false of most, if not all, actual food webs, though true of "interference-free" food webs. Cooper's proposal requires that capacities and tendencies exist that may sound implausible to empiricists. However, like Cartwright, he believes that much of science cannot be accounted for without them. Nonetheless, even if the existence of capacities and tendencies is accepted, how idealized models explain ecological dynamics when there are interfering forces still needs to be understood.

Conclusion

Ecology presents philosophy with several conceptual and methodological problems. These issues are not just of an abstract bent, but speak to how one should understand the role of science in society (see Conservation Biology). Many issues of importance are connected to the empirical studies and theoretical analyses that ecologists perform. These include determining the status of invasive species, considering whether a population is threatened or endangered, and estimating the risks in losing many of the communities of plants and animals across the globe. To make sense of the roles these ecologists play in policy formation, their scientific activities must also be considered. These issues are enveloped in political and ethical issues of some complexity—all the more reason to exercise philosophical care. After all, how the science of ecology is understood affects both human lives and the environment.

JAY ODENBAUGH

References

Boyce, Mark (1992), "Population Viability Analysis," *Annual Review of Ecology and Systematics* 23: 481–506.

Brennan, Andrew (1988), *Thinking About Nature: An Investigation of Nature, Value and Ecology*. Athens: University of Georgia Press.

Cartwright, Nancy (1983), *How the Laws of Physics Lie*. Cambridge: Cambridge University Press.

Caswell, Hal (1988), "Theory and Models in Ecology: A Different Perspective," *Ecological Modelling* 43: 33–44.

Clements, Frederic (1916), *Plant Succession*. Washington, DC: Carnegie Institution of Washington, Publication no. 242.

Connor, E. F., and Daniel Simberloff (1979), "The Assembly of Species Communities: Chance or Competition?" *Ecology* 60: 1132–1140.

Cooper, Greg (2003), *The Science of the Struggle for Existence: On the Foundations of Ecology*. Cambridge: Cambridge University Press.

Diamond, Jared (1975), "Assembly of Species Communities," in M. L. Cody and J. M. Diamond (eds.), *Ecology and Evolution of Communities*. Cambridge, MA: Belknap Press of Harvard University Press.

Egerton, Frank (1973), "Changing Concepts of the Balance of Nature," *Quarterly Review of Biology* 48: 322–350.

Gilpin, Michael, and Jared Diamond (1983), "Are Species Co-occurrences on Islands Non-Random, and Are Null Hypotheses Useful in Community Ecology?" in Donald Strong, Daniel Simberloff, L. G. Abele, and A. B. Thistle (eds.), *Ecological Communities: Conceptual Issues and the Evidence*. Princeton, NJ: Princeton University Press.

Gleason, Henry (1926), "The Individualistic Concept of the Plant Association," *Bulletin of the Torrey Botanical Club* 53: 7–26.

- (1917), "The Structure and Development of the Plant Association," *Bulletin of the Torrey Botanical Club* 44: 463–481.
Gotelli, Nicholas, and Gary Graves (1996), *Null Models in Ecology*. Washington, DC: Smithsonian Institution.
King, Anthony, and Stuart Pimm (1983), "Complexity and Stability: A Reconciliation of Theoretical and Experimental Results," *American Naturalist* 122: 229–239.
Levins, Richard (1966), "The Strategy of Model Building in Population Biology," *American Scientist* 54: 421–431.
Levins, Richard, and Richard Lewontin (1985), "Dialectics and Reductionism in Ecology," in *The Dialectical Biologist*. Cambridge: Harvard University Press.
Ludwig, Donald, Mark Mangel, and Brent Haddad (2001), "Ecology, Conservation, and Public Policy," *Annual Review of Ecology and Systematics* 32: 481–517.
MacArthur, Robert (1962), "Patterns of Terrestrial Bird Communities," in D. Farner, J. King, and K. Parkes (eds.), *Avian Biology*, vol. 1. New York: Academic Press.
May, Robert (1973), *The Stability and Complexity of Model Ecosystems*. Princeton, NJ: Princeton University Press.
McIntosh, Robert (1987), "Pluralism in Ecology," *Annual Review of Ecology and Systematics* 18: 321–341.
McNaughton, John (1977), "Diversity and Stability of Ecological Communities: A Comment on the Role of Empiricism in Ecology," *American Naturalist* 111: 515–525.
Odenbaugh, Jay (2001), "Ecological Stability, Model Building, and Environmental Policy: A Reply to Some of the Pessimism," *Philosophy of Science* 68 (Proceedings), S493–S505.
Peters, Robert (1991), *A Critique for Ecology*. Cambridge: Cambridge University Press.
Pimm, Stuart (1984a), *Food Webs*. London: Chapman and Hall.
- (1984b), "The Complexity and Stability of Ecosystems," *Nature* 307: 321–326.
- (1993), *The Balance of Nature?* Chicago: University of Chicago Press.
Pimm, Stuart, and John Lawton (1978), "On Feeding on More Than One Trophic Level," *Nature* 275: 542–544.
Ricklefs, Robert, and Gary Miller (2000), *Ecology*. New York: W. H. Freeman and Co., 2000.
Shrader-Frechette, Kristin, and Earl McCoy (1993), *Method in Ecology*. Cambridge: Cambridge University Press.
Simberloff, Daniel (1980), "A Succession of Paradigms in Ecology: Essentialism to Probabilism to Materialism and Probabilism," *Synthese* 43: 3–39.
Sterelny, Kim (2001), "Darwin's Tangled Bank," in *The Evolution of Agency and Other Essays*. Cambridge: Cambridge University Press.
Strong, Donald, Daniel Simberloff, L. G. Abele, and A. B. Thistle (eds.) (1984), *Ecological Communities: Conceptual Issues and the Evidence*. Princeton, NJ: Princeton University Press, 1984.
Wimsatt, William (1987), "False Models as Means to Truer Theories," in M. Nitecki and A. Hoffman (eds.), *Neutral Models in Biology*. London: Oxford University Press.

PHILOSOPHY OF ECONOMICS

The philosophy of economics concerns itself with conceptual, methodological, and ethical issues that arise within the scientific discipline of economics. The philosophy of economics is now a well-established subdiscipline within philosophy. Significant contributions to the discipline include Allen Buchanan (1985), Daniel Hausman (1984 and 1992), Hausman and Michael McPherson (1996), Daniel Little (1995), Amartya Sen (1987), and Alexander Rosenberg (1992). The primary focus is on issues of methodology and epistemology—the methods, concepts, and theories through which economists attempt to arrive at knowledge about economic processes. Philosophy of economics is also concerned with the ways in which ethical values are involved in economic reasoning—the values of human welfare, social justice, and the trade-offs among priorities that economic choices require. Economic reasoning has implications for justice and human welfare; more importantly, economic reasoning often makes inexplicit but significant ethical assumptions that philosophers of economics have found it worthwhile to scrutinize. Finally, the philosophy of economics is concerned with the concrete social assumptions that are made by economists. Philosophers have given attention to the institutions and structures through which economic activity and change take place. What is a "market"? Are there alternative institutions through which modern economic activity can proceed? What are some of the institutional variants that exist within the general framework of a market economy? What are some of the roles that the state can play within economic development so as to promote efficiency, equity, human well-being, productivity, or growth?

The dimension of the philosophy of economics that falls within the philosophy of science has to do with the status of economic analysis as a body of empirical knowledge. Primary questions include: What is economic knowledge *about*? What kind of knowledge is provided by the discipline of economics? How does it relate to other social sciences and the bodies of knowledge contained in those disciplines? How is economic knowledge justified or evaluated? Does economic theory purport to offer abstract theories of real social processes—their mechanisms, dynamics, and institutions? What is the nature of economic explanation? What is the relationship between abstract mathematical models and theorems, on the one hand, and the empirical reality of economic behavior and institutions, on the other? What is the nature of the concepts and theories in terms of which economic beliefs are formulated? Are there lawlike regularities among economic phenomena? What is the status of predictions in economics?

The Intellectual Role of the Philosophy of Economics

Philosophers are not empirical researchers, and on the whole they are not formal theory builders. So what constructive role does philosophy have to play in economics? There are several. First, philosophers are well prepared to examine the logical and rational features of an empirical discipline. How do theoretical claims in the discipline relate to empirical evidence? How do pragmatic features of theories such as simplicity, ease of computation, and the like play a role in the rational appraisal of a theory? How do presuppositions and traditions of research serve to structure the forward development of the theories and hypotheses of the discipline? Second, philosophers are well equipped to consider topics having to do with the concepts and theories that economists employ—for example, economic rationality, Nash equilibrium, perfect competition, transaction costs, or asymmetric information. Philosophers can offer useful analysis of the strengths and weaknesses of such concepts and theories—thereby helping practicing economists to further refine the theoretical foundations of their discipline. In this role the philosopher serves as a conceptual clarifier for the discipline, working in partnership with the practitioners to bring about more successful economic theories and explanations.

This account describes the position of the philosopher as the "underlaborer" of the economist. But in fact, the line between criticism and theory formation is not a sharp one. Economists such as Sen (1999, 1973, and 1987) and philosophers such as Hausman (1992; Hausman and McPherson 1996) have demonstrated that there is a very constructive crossing of the frontier that is possible between philosophy and economics; philosophical expertise can result in significant substantive progress with regard to important theoretical or empirical problems within the discipline of economics. The cumulative contents of the journal *Economics and Philosophy* provide clear evidence of the productive engagements that are possible when philosophy meets economics.

Important Questions in the Philosophy of Economics

Are There Laws in Economics?

The concept of a "law of nature" has been central to the modern understanding of the natural sciences. The intellectual power of classical physics derived from the fact that it was capable of advancing statements of physical laws that were simple and universal—laws of gravitation and planetary motion, optics, electricity and magnetism, and so on. Is this an essential feature of a successful empirical science? And does economics possess such laws? Several authors are affirmative on both points (Kincaid 1996). However, several points have emerged in recent discussions of the social sciences—including economics—that lead to doubt about the centrality of laws to them. First, there are significant differences between natural and social phenomena that should raise doubts about the availability of strong "laws of nature" describing social phenomena. Second, it is clear that there are regularities within the discipline of empirical economics—consumption usually rises when prices fall, trade increases when transport costs fall, and infant mortality usually falls when states devote more resources to public health. But these are fairly humdrum empirical regularities, exception laden and obvious. Are there strong "economic laws" that have the force of Maxwell's laws of electromagnetic propagation? Nothing in current economic theory provides reason to think that there are such laws. The foundational assumptions of economic theory plainly do not fall in the category of laws of nature. And as will be shown below, the assumption of economic rationality does not constitute a universal generalization about individual behavior. Here, as is the case in other areas of social science, it is more justifiable to seek out causal mechanisms rather than social laws (see Laws of Nature; Explanation).

ECONOMICS, PHILOSOPHY OF

How Do Economic Laws Relate to Individual Behavior?

Economists generally assume some form of the doctrine of methodological individualism (see Methodological Individualism). This principle maintains that social facts, social entities, and social laws are constituted by facts about individuals and their behavior—defined prior to the social properties that individuals possess (Kincaid 1986; Miller 1978; Watkins 1968). The doctrine of methodological individualism is an expression of the perspective of reductionism, that is, the view within scientific metaphysics that insists that higher-level structures must be explained by reference to the properties of the lower-level entities that make them up. According to the reductionist, there are no "emergent" properties of complex entities or structures. A more satisfactory perspective on this issue has been developed on the basis of the theory of supervenience—the view that differences in higher-level properties depend upon differences in lower-level properties (Kim 1984). Extreme versions of methodological individualism and reductionism make the task of social or economic explanation very difficult, since explaining or describing individual behavior seems to require referring to social entities (rules, structures, institutions).

Are the Assumptions of Economics "Realistic?"

Do economic theories and hypotheses serve to describe unobservable economic mechanisms and structures? Milton Friedman (1953) set the stage for one answer to this question by arguing for an instrumentalist interpretation of economic assumptions. In Friedman's view, the value of a theory is entirely expressed in its ability to predict observable phenomena; the theory is an instrument of prediction (see Instrumentalism). Instrumentalism, however, has generally faced strong criticism from philosophers of science (Leplin 1984). This doctrine makes the empirical success of a theory a source of mystery. The best explanation of a theory's having generally reliable predictions about a range of phenomena is that the mechanisms that it postulates are in fact true. So it is a deficiency in a theory that the mechanisms it postulates are implausible or false. And economic theory would be substantially undermined if its premises were judged to be profoundly inconsistent with the real underlying causal processes that constitute a working economy. Against this instrumentalist framework, Hausman puts forward a realist approach to economic theory (Hausman 1992) (see Realism). Within this approach, the goal for the economist is to arrive at assumptions that are approximately true. (This methodological principle suggests that economists ought to pay greater attention to economic institutions, comparative economic analysis, and economic history; see below.)

Are Economic Theories Testable or Falsifiable?

Before asking whether a theory is testable, it is necessary to have a clear specification of its empirical content. This requires asking the question, What is the theory intended to describe, predict, or explain? A theory has empirical content if it makes assertions about causal processes underlying a domain of phenomena and those assertions have consequences for observable states of the world. Under these circumstances it is possible to engage in a variety of forms of empirical test of the hypothesis. The investigator can

- perform experiments (arrange the world in a certain way, observe the outcome, and compare this with the theory's predicted outcome);
- undertake a protocol of controlled observation (collect "before/after" cases and compare the outcomes with the theory's predictions);
- perform process-tracing observation (examine elements of a process in order to assess whether the postulated causal processes did in fact occur); and so on.

Through these efforts it is possible to bring empirical evidence to bear on the task of assessing the truth of the hypothesis. So the question is this: Does economic theory contain substantive assumptions about the causal workings of the economic world that are intended to have implications for future observable states of the economic world? And is it possible to perform observations of states of the world that confirm or falsify the theory (Hands 1992)? In principle, it is clear that the answer to this question is affirmative. Consider a range of theories of specific economic processes—economic growth, trade, unemployment, wages, or discrimination. Such theories have predictive consequences, and it is not especially difficult to describe the observations that would test these theories. The epistemic difficulty comes later: Most theories of complex phenomena are in fact falsified—without necessarily being far from the mark in their description of the underlying processes. So, how is it possible to distinguish among "falsified" theories to single out the more likely from the less likely (Lakatos 1974)? Are economic theories simply formal mathematical systems, without empirical relevance?

Rosenberg makes a case for the formalist view of economic theory, having concluded that economists have not succeeded in producing empirical theories or explanations of real empirical phenomena (Rosenberg 1992, Ch. 8). Rosenberg likens microeconomics to Euclidean geometry rather than classical physics or evolutionary biology; the "theory" is a set of abstract and nonempirical axioms, and the exercise of "doing economics" is one of deriving theorems from these axioms. However, this is not a satisfactory way of understanding the intellectual program of economics. The intellectual charge for the discipline of economics—not always or successfully achieved—is to provide a social-scientific basis for understanding, explaining, and, perhaps, predicting economic phenomena. Why do interest rates affect investment levels? Why are inflation and unemployment related? Why is economic growth more rapid in the context of one set of institutions than another? What are the causal links that secure connections among economic variables? These are the sorts of questions that economists are charged to answer. And the approach to economic theorizing that stipulates that the discipline is purely formal will not aid in shedding light on these real, though unobservable, economic mechanisms. In this line of thought, the persistent mathematization of economics ought to be construed as a means to an end rather than the end itself. The formal or mathematical machinery of economics is intellectually valuable only insofar as it contributes to a better understanding of real, empirically given economic processes, causes, and systems.

What Is the Status of the Concept of Economic Rationality?

The concept of economic rationality is foundational within economic theory, and especially so within neoclassical economics. Economists are concerned with analyzing individual rational choice in the context of reward, risk, and uncertainty, where the individual's outcome depends on the probabilities and rewards associated with the various options available for choice (see Decision Theory). And they are concerned with rationality in the context of strategic interactions among two or more agents, where the individual's reward depends on the rational choices made by other agents (see Game Theory). What is the nature of the "decisions rules" that constitute rational behavior in these two stylized contexts? A special concern for philosophers of economics has been to provide critical examination of the theory of economic rationality. Taken together, these criticisms have led to a substantial enhancement in our understanding of the concept of rationality. First, philosophers have devoted a great deal of attention to the gap between a theory of utility and a theory of individual preference. Second, they have taken issue with the assumption of egoism or rational self-interest that is presupposed in the pure theory (Sen 1987; Anderson 2000). Third, philosophers and others have pointed out that real psychological actors reason in ways that are at odds with the pure theory of economic rationality (Kahneman, Slovic, and Tversky 1982; Simon 1983). Fourth, philosophers and others have devoted significant attention to the assumptions underlying game theory. Finally, some philosophers have undertaken to study the characteristics of "economic rationality" in real human persons through experiment (Schmidtz 1991). For example, Axelrod (1984) has used experimental settings to examine how real human reasoners deal with Prisoners' Dilemmas; he finds that experimental subjects are frequently able to achieve cooperation rather than defection, contrary to the prediction of two-person game theory. The results of this research suggest that real reasoners behave intelligently—but differently from the axioms of the theory of pure economic rationality.

What Is the Role of Ethical Values in Economics?

Economists often portray their science as "value free"—as a technical analysis of the demands of rationality in the allocation of resources rather than a specific set of value or policy commitments. In this interpretation, the economist wishes to be understood in an analogy with the civil engineer rather than the transportation policymaker: The engineer can prescribe how to build a stable bridge, but not where, when, or why to do so. It is for citizens and policymakers to make the judgments about the public goods that are needed in order to decide whether a given road or bridge is socially desirable; it is for the technical specialist to provide design and estimate of costs. This description of the discipline of economics fails in several important respects, however. Economic theory contains a family of substantive presuppositions about the nature of the good—individual and social—that directly influence the policy recommendations to which economic theory gives rise. For example, the assumption of rational egoism is inconsistent with several of the values of communitarianism; the assumption that equity is subordinate to efficiency is inconsistent with an egalitarian political philosophy; and the assumption that a bundle of commodities constitutes individual "well-being" is inconsistent with a more Aristotelian conception of the good human life (Nussbaum 2000). So the premises and assumptions of economics are substantially intertwined with normative assumptions about the good human life and the good society. This is not a deficiency, but it needs to be recognized in order to observe the workings of the unstated value assumptions. And it certainly invalidates the assumption of value-free social science. In general, it seems fair to say that the ethical assumptions that neoclassical economics presupposes fall together into a family of normative ideals that privilege individualism, inequality, and the minimal exercise of public policy.

Is Distributive Justice a Topic for Economists?

Once it is recognized that economics has ethical content, it becomes apparent that it is necessary to examine the content of these ethical premises in detail and offer a critique when these assumptions are found wanting. In particular, economics is obliged to confront issues of distributive justice much more explicitly than it has to date. A market economy implies some degree of inequality, in terms of outcomes (wealth and income), opportunity, power and influence, and levels of well-being (health, longevity, education). What sorts of inequalities are morally acceptable in a just society? How extensive can inequalities be before they create differences among citizens that interfere with their human dignity and the preconditions of democracy? Throughout the past 30 years philosophers have made substantial contributions to current understanding of these issues of distributive justice and the moral status of inequality (Nozick 1974; Elster 1992; Rawls 2001). There is more to be done.

Is There a Basis for Rational Debate About Economic Institutions?

What sort of social world does economic theory presuppose? In considering this type of question, philosophers begin to move into substantive debates about the nature of the empirical phenomena under study. The discussion falls under the rubric of "criticism," in that it focuses on blind spots that can be discerned within the visual field of economic theorizing. Economists make assumptions about the institutions that constitute the framework of economic transactions, and these assumptions are sometimes inflexible and unrealistic. It is therefore worthwhile for philosophers to devote attention to the shortcomings of the social-institutional assumptions that economists often make. The new institutionalism in the social sciences has focused substantial interest on the specifics of the institutions within which social activity takes place (Powell and DiMaggio 1991; Brinton and Nee 1998). Institutions matter; so a more refined account of the economic institutions of a particular market economy may lead to better understanding of the phenomena. For example, incorporation of transaction costs and asymmetric information between buyer and seller has significantly changed the current understanding of market institutions. One strand of philosophical criticism comes from the level of abstractness of typical economic theories. Greater empirical detail may well change the inferences that can be drawn about the workings of the institution. Market "imperfections" may be the rule rather than the exception—so it is important to incorporate some of these empirical characteristics into accepted theories of economic institutions.

Are There Alternative Economic Institutions That Can Work in a Modern Economy?

Economic activity within a modern society requires institutions that define the use, management, and enjoyment of resources; the deployment and management of labor; and the management of enterprises. Neoclassical economics presupposes private ownership of capital, "free" workers who do not own property, and states that have minimal economic influence. Are there other institutions through which economic activity might be conducted within a modern and productive society (Elster and Moene 1989)? For example, what is the economic logic of workers' cooperatives? How could worker-controlled pension funds be used to enhance democratic equality? Is there more to be learned from the experience of market socialism, state ownership, or workers' control of industrial processes? Are alternative institutions feasible? Are they efficient? Are they equitable?

What Can Be Learned from Comparative Economic Analysis?

Economic development has proceeded in very different ways in different nations and regions since the emergence of modern technologies and economic institutions. Market institutions developed very differently in Britain, France, and the United States during the nineteenth and twentieth centuries. Collectivized economies followed different institutional trajectories in Yugoslavia, the Soviet Union, and China. What can be learned about economic processes and dynamics by studying and comparing national economies in significant detail? For example, what do the parallel yet different experiences of China and India since 1945 teach about alternative pathways of economic development (Drèze and Sen 1989)? Does this sort of comparative economic research provide a "post–Cold War" basis for analyzing the political economy of development? As economists come to confront the intellectual challenge of providing realistic causal accounts of economic systems, they will be able to arrive at significant new insights through comparative economic analysis.

What Is the Intellectual Relevance of the History of Western Industrial Capitalism for Economic Theory?

Reexamination of the history of European capitalism suggests that there were feasible alternative paths of economic development besides mass manufacture and specialized production (Sabel and Zeitlin 1997). Mass manufacture and mass unskilled labor represented one important alternative, but there were others that were historically feasible. As Sabel and Zeitlin demonstrate, another feasible system of industrial production involves highly skilled workers, flexible production, and flexible tools and production processes (Sabel and Zeitlin 1985). Once again, the moral for the discipline of economics is an important one: It is possible to arrive at more empirically satisfactory economic theories in the context of considering the range of institutions through which economic activity and growth have taken place.

Conclusions

The philosophy of economics serves as a source of sympathetic yet rigorous critique of the science of economics, broadly construed. It raises familiar questions about the epistemology of this branch of the social sciences—questions about theory structure, theory confirmation, explanatory adequacy, and the like. It questions the implicit normative assumptions that economics contains. It raises some of the ethical questions that economics is almost forced to confront, but rarely does. And it suggests the value of a broader and more eclectic approach to economic theorizing: making more extensive use of alternative theoretical approaches, incorporating more study of economic institutions, paying more attention to comparative economic trajectories, and giving more rigorous attention to economic history. Economics will be a more successful social science when it embraces more of the role it often played in the nineteenth century as a seminal social science—an area of social inquiry that was equally interested in the concrete social and economic institutions that constituted a "modern" economy, interested in the ethical implications of the social phenomena with which it was concerned and willing to consider a variety of theoretical models in aspiring to the goal of achieving a scientific understanding of economic processes, institutions, and outcomes.

DANIEL LITTLE

References

Anderson, Elizabeth (2000), "Beyond Homo Economicus," *Philosophy and Public Affairs* 29: 170–200.
Axelrod, Robert M. (1984), *The Evolution of Cooperation*. New York: Basic Books.
Brinton, Mary C., and Victor Nee (eds.) (1998), *New Institutionalism in Sociology*. New York: Russell Sage Foundation.
Buchanan, Allen E. (1985), *Ethics, Efficiency, and the Market*. Totowa, NJ: Rowman & Allanheld.
Drèze, Jean, and Amartya Kumar Sen (1989), *Hunger and Public Action*. Oxford: Clarendon Press.
Elster, Jon (1992), *Local Justice*. New York: Russell Sage Foundation.
Elster, Jon, and Karl Ove Moene (eds.) (1989), *Alternatives to Capitalism*. Cambridge and New York: Cambridge University Press.
Friedman, Milton (1953), *Essays in Positive Economics*. Chicago: University of Chicago Press.
Hands, D. Wade (1992), *Testing, Rationality, and Progress: Essays on the Popperian Tradition in Economic Methodology, Worldly Philosophy*. Lanham, MD: Rowman & Littlefield Publishers.
Hausman, Daniel M. (ed.) (1984), *The Philosophy of Economics: An Anthology*. Cambridge and New York: Cambridge University Press.
——— (1992), *The Inexact and Separate Science of Economics*. Cambridge: Cambridge University Press.
Hausman, Daniel M., and Michael S. McPherson (1996), *Economic Analysis and Moral Philosophy*. Cambridge and New York: Cambridge University Press.
Kahneman, D., P. Slovic, and A. Tversky (1982), *Judgment under Uncertainty: Heuristics and Biases*. Cambridge: Cambridge University Press.
Kim, Jaegwon (1984), "Supervenience and Supervenient Causation," *Southern Journal of Philosophy* 22(suppl): S45–S56).
Kincaid, Harold (1996), *Philosophical Foundations of the Social Sciences: Analyzing Controversies in Social Research*. Cambridge and New York: Cambridge University Press.
——— (1986), "Reduction, Explanation, and Individualism," *Philosophy of Science* 54: 492–513.
Lakatos, Imre (1974), "Methodology of Scientific Research Programmes," in Lakatos and Alan Musgrave (eds.), *Criticism and the Growth of Knowledge*. Cambridge: Cambridge University Press.
Leplin, Jarrett (1984), *Scientific Realism*. Berkeley and LA: University of California Press.

Little, Daniel (1995), *On the Reliability of Economic Models: Essays in the Philosophy of Economics*. Recent Economic Thought Series. Boston: Kluwer Academic Publishers.

Miller, Richard (1978), "Methodological Individualism and Social Explanation," *Philosophy of Science* 45: 387–414.

Nozick, Robert (1974), *Anarchy, State, and Utopia*. New York: Basic Books.

Nussbaum, Martha Craven (2000), *Women and Human Development: The Capabilities Approach*. Cambridge and New York: Cambridge University Press.

Powell, Walter, and Paul J. DiMaggio (eds.) (1991), *The New Institutionalism in Organizational Analysis*. Chicago: University of Chicago Press.

Rawls, John (2001), *Justice as Fairness: A Restatement*. Cambridge, MA: Harvard University Press.

Rosenberg, Alexander (1992), *Economics: Mathematical Politics or Science of Diminishing Returns?* Chicago: University of Chicago Press.

Sabel, Charles F., and Jonathan Zeitlin (1985), "Historical Alternatives to Mass Production: Politics, Markets and Technology in Nineteenth Century Industrialization," *Past and Present* 108: 133–176.

——— (eds.) (1997), *Worlds of Possibility: Flexibility and Mass Production in Western Industrialization*. Cambridge and New York: Cambridge University Press.

Schmidtz, David (1991), *The Limits of Government: An Essay on the Public Goods Argument*. Boulder, CO: Westview Press.

Sen, Amartya (1987), *On Ethics and Economics*. New York: Basil Blackwell.

——— (1973). *On Economic Inequality*. Oxford: Oxford University Press.

——— (1999). *Development as Freedom*. New York: Knopf.

Simon, Herbert A. (1983), *Reason in Human Affairs*. Stanford, CA: Stanford University Press.

Watkins, J. W. N. (1968), "Methodological Individualism and Social Tendencies," in May Brodbeck (ed.), *Readings in the Philosophy of the Social Sciences*. New York: Macmillan.

See also **Behaviorism; Game Theory; Methodological Individualism; Social Sciences, Philosophy of**

EMERGENCE

The concept of emergence stems from a family of related doctrines known collectively as emergentism. Regardless of variation in its formulation, emergentists generally hold an ontological premise and an epistemological one. The *ontological premise* is that (i) there are properties (or laws) that obtain of certain complex physical entities that do not obtain of any of the individual parts or lower-level constituents of those entities. The *epistemological premise* is that (ii) the instantiation of those properties cannot be derived from an exhaustive knowledge of the nonrelational properties of the parts, in addition to any laws of composition that obtain among lower-level entities (e.g., additivity, fundamental forces) and statements of definition. (In the following, it will generally be assumed that fundamentally, properties are emergent and that laws are emergent in a derivative sense; thus "properties" will be used instead of "properties and laws.") If one allows a broad reading to include nonmaterial parts (e.g., *élan vital*, mind substance), then it follows from (i) and (ii) that emergentism rejects the use of any sort of substance dualism for the purpose of explaining the appearance of a higher-level property. Furthermore, if a "reductionist" explanation is understood as one that explains a property of an entity in terms of the nonrelational properties of its parts, in addition to the lower-level laws that obtain over these properties, then it also follows from (i) and (ii) that the instantiation of some properties cannot be given a reductionist explanation (see Reductionism). Hence emergentism takes its place in contemporary philosophical parlance as a variety of nonreductionist physicalism (see Physicalism).

Paradigmatic examples of emergence often touted by early emergentists were taken from chemical bonding—for instance, the production of common salt from sodium and chloride. At the time, little was known about the microstructure of atoms, and consequently it seemed plausible that the phenomenon in question could be conceptualized only in terms of a *de novo* or fundamental physical law relating atomic interactions to macro-level phenomena, rather than one derivable on the basis of the atomic structure of sodium and chloride and generic laws of chemical bonding. (In fact, the success of quantum chemistry in providing reductions of this type remains disputed; see final section.) Although, given such a fundamental law, the instantiation of a macro-level property would be *determined* by the instantiation of a set of micro-level properties, this property would not be reductively explainable on the basis of the latter.

A stronger form of emergentism includes a third premise, that (iii) these properties that hold of certain complex physical entities are not *determined* by the nonrelational properties that hold of their individual parts and the relation of these parts to one another. This view is sometimes characterized as involving a failure of "microdeterminism" (following Klee 1984). Karl Popper (see Popper and Eccles 1977) argues for such a view on the basis of the probabilistic character of quantum mechanical laws. Finally, some emergentists hold a fourth premise, that (iv) the emergent properties of a whole system can affect the behavior of that system's own parts. This view, referred to as "downward causation" (Campbell 1974), "hierarchical downward control" (Sperry 1969), or "macrodeterminacy" (Weiss 1968), has played a significant role in the emergentist literature since the 1960s, although an intimation of this idea can be found in C. Lloyd Morgan's notion of the "dependence" of lower-level properties on those of the higher level (Lloyd Morgan 1923, 16).

Moreover, there are regional variants of these views. In Anglo-American philosophy of science and philosophy of mind, most references to the history of emergentism implicitly or explicitly refer to the British emergentist tradition, which traces its intellectual roots back to John Stuart Mill in the mid-nineteenth century and the doctrines of "emergent evolution" that were prominent among philosophers such as Scott Alexander and biologists such as Lloyd Morgan and W. M. Wheeler in the early twentieth century. However, some writers on emergentism also refer to the German organicist tradition, which traces its intellectual roots back to Immanuel Kant in the late eighteenth century and was prominent among German biologists and psychologists in the early twentieth century.

History of the Concept of Emergence

Mill's Heteropathic Effects

What is now referred to as British emergentism (McLaughlin 1992) begins with Mill's *A System of Logic* of 1843, in which he describes two classes of phenomena. The first is that class of effects produced by the mechanical mode of causation, or according to the "composition of causes" (1874, 267), that is, the principle that the effect produced by the joint action of several causes can be inferred by summating the effects that would have been produced by each of the agents acting separately. The paradigmatic example of the mechanical mode of causation is the composition of forces, in which the velocity and direction of a given particle at time t can be derived by summating over the velocities and directions of each of the particles that strike it at a prior time t' by vector addition.

This is contrasted with that class of effects produced by the chemical mode of causation. The laws governing this mode are referred to as *heteropathic laws* (269). By this expression Mill refers to the new uniformities that arise in those cases in which the combination of different substances produces a substance with new properties not possessed by any of the parts. For Mill, the production of heteropathic laws is a sufficient condition for the origination of higher levels of the scientific hierarchy. (Lewes [1875] is credited with coining the term "emergents"—to be distinguished from "resultants"—to refer to Mill's heteropathic effects.)

Mill is often ambiguous concerning whether the notion of a heteropathic effect should be understood primarily ontologically or epistemologically. His paradigmatic example of a heteropathic effect is chemical bonding, in which some of the properties of a compound are qualitatively different from those of its constituents and hence cannot be derived by mere summation of some known property of those constituents. It would appear from this example that, for Mill, the failure of additivity for the higher-level property is a consequence of the qualitative *novelty* of the property in question, and hence his form of emergentism would amount to an ontological doctrine that presupposes some account of the individuation of properties independent of their extension. This ontological doctrine would then presumably explain the failure of derivability by summation. However, Mill's view can also be interpreted epistemologically, in that what is to be considered a heteropathic effect is relative to a given state of knowledge at a particular time:

> There most of the uniformities to which the causes conformed when separate, cease altogether when they are conjoined; and we are not, at least in the present state of our knowledge, able to foresee what result will follow from any new combination, until we have tried the specific experiment. (1874, 267)

The notion of 'novelty' might then be understood in terms of the unexpectedness of the result of a specific experiment.

EMERGENCE

Some contemporary philosophers have questioned Mill's apparent restriction of emergence to *nonadditive* properties (Wimsatt 2001; McLaughlin 1992; Kim 1999). For example, in the case of probabilities, multiplication rather than addition is the appropriate operator for deriving the probability of the joint occurrence of independent events. Some philosophers have suggested that virtually *no* constraints be imposed upon the sort of mathematical function by which the value of a system-level property is derived from those of its constituents; so long as it is so derivable, it is not an emergent property (e.g., Kim 1999, 7). But this is too strong a criterion for an emergent property, since a nonlinear function mapping a system's input to its output suggests that the relation between the two is highly dependent upon the specific nature of the interactions that take place between the system components. Hence, if the nature of such interactions cannot be discerned, then the applicability of the function remains completely mysterious. These considerations suggest that the peculiarity that Mill attempted to conceptualize as the failure of "additivity" may be more accurately conceptualized in terms of an essential interaction between the parts of a system, or an interaction that is not discernible given the best available theoretical account of the parts and their relations.

Emergent Evolution

Mill's notion of heteropathic effects gave rise to the doctrine of *emergent evolution*, which gained some currency in the early twentieth century amongst philosophers and biologists such as Lloyd Morgan (Lloyd Morgan 1923), Scott Alexander (1920), and C. D. Broad (1925). According to these emergentists, over a cosmological time scale, certain forms of complexity are successively brought into existence that are novel, in the sense that they represent qualitatively new properties that are not the resultants of the properties that existed previously and are unpredictable in the sense that an exhaustive knowledge of the properties of the previous existents would not allow a prediction or derivation of the new properties before the fact of their appearance. In short, the emergent evolutionists took the two basic characteristics of Mill's heteropathic effects—novelty and unpredictability—and transposed them more explicitly into a cosmological time frame.

However, the emergentists of this period typically did not reject the idea that the appearance of emergent properties is causally *determined* by the lower-level properties (see Alexander 1920, 330; Broad 1925, 67), in the sense that anytime certain constituents P_i are arranged in relation R under the same conditions C, a higher-level property Q will be instantiated by the whole. Consequently, according to this classical viewpoint, an emergent property is one that can also be said to supervene upon the properties of its constituents (see Supervenience). However, the idea that emergence embodies a supervenience assumption has in more recent times been placed into question (see final section).

German Organicism

German organicism was an independent philosophical and scientific tradition with distinct roots, which slowly merged with the emergentist tradition during the mid-twentieth century. The organicist tradition often traces its roots to Immanuel Kant's *Critique of the Power of Judgment*. According to Kant, biological forms appear to the mind as inherently purposive rather than as the products of blind mechanism; the reciprocity of part and whole in the organism appeals to the faculty of judgment and provides transcendental justification for the use of teleological reasoning in biology (see especially §65). For Kant, such forms cannot be understood by analyzing the operation of each of its parts in isolation and then inferring the activity of the whole; rather, the existence and operation of each part can be understood only in terms of its contribution to the functioning of the whole. Like emergentism, organicism traces a middle path between mechanism and vitalism.

Within biology, organicism was adopted by such figures as Paul Weiss, Joseph Needham, and Ludwig von Bertalanffy. It found expression, for instance, in Weiss's (1968) concept of the morphogenetic field, as it did in psychology in the notion of the gestalt. The concept of the morphogenetic field was introduced to explain cell differentiation and specialization on the basis of the relative location of whole groups of cells within the organism, rather than on the basis of the intrinsic properties of individual cells and their relations to immediately adjacent cells. Gestalt psychologists such as Köhler held that, for instance, the perceptual patterns that organize the visual field *are* explainable on the basis of the patterns of electrical activity within the visual cortex but that the formation of these electrical patterns is governed by fundamental physical laws that do not involve reference to the individual neurons and their relations. Hence both of these examples satisfy premises (i) and (ii) of emergentism (see the introductory paragraph). (For a brief conceptual overview of the organicist tradition, see Gilbert and Sarkar 2000.) From this perspective it should not be surprising that the partial revival of interest in emergentism in Anglo-American philosophy of science was initiated less by philosophers than by philosophically oriented scientists entrenched in the organicist tradition, such as Michael Polanyi (1968), Weiss (1968), and Roger Sperry (e.g., 1969).

Criticism of Emergence

Criticism of the concept of emergence typically falls under three categories. The first type of criticism holds that "emergent properties" exist, but in some trivial or uninteresting sense, and that whatever philosophical interest it seems to possess is largely a product of conceptual confusion that can be resolved by proper explication of the concept. The second type of criticism (espoused most recently by Kim [1999]) holds that the concept of emergence is conceptually clear but that emergent properties cannot exist for a priori reasons. A third line of criticism, proposed by McLaughlin (1992), holds that emergence is both conceptually interesting and a priori possible but that emergent properties do not in fact exist, and their nonexistence is attested to by the overwhelming historical success of reductionist explanation.

The most important of the critiques that belong to the first category are those of Hempel ([1948] 1965) and Nagel (1961), both of whom argued on the basis of the deductive-nomological model of explanation that the failure of predictability (Hempel) or the failure of reducibility (Nagel) is *logically* trivial and hence does not warrant any important ontological conclusions. According to Nagel (1961, 369), emergence, like reduction, should be seen as a relation between theories, rather than properties. Specifically, it involves a relation between a (suitably axiomatized) "higher-level" theory, T_A (e.g., biology), and a "lower-level" theory, T_B (e.g., chemistry). To say, "P is an emergent property" is to say that the law statements of T_A that contain the predicate P cannot be derived from the law statements of T_B. But clearly, if the predicate P does not appear in T_B—as is often the case in theories at different levels—then nonderivability is a point of logic and is therefore trivial. In order for the derivation to go through, one must specify "bridge laws" or "translation rules" that connect the predicates of T_A with those of T_B via a series of conditionals, where such bridge laws are either lawlike generalizations or definitional stipulations. But once emergence claims are relativized to a given pair of theories and a given set of bridge laws, the very notion of *in principle* irreducibility or unpredictability appears to be incoherent. The most that such nonderivability allows one to infer is the (relatively uninteresting) claim that a given scientific theory is currently unable to explain a certain phenomenon.

However, this criticism is misguided to the extent that it centers upon formal rather than substantive facets of explanation. Nothing in Mill's account, or in that of the later emergentists, prohibits the postulation of bridge laws. Mill's view is that if P is an emergent property, then such bridge laws will find only a purely inductive justification; they will not be derivable in turn from a more fundamental theory, along with the relevant statements of definition. As such, these laws would appear to represent ultimate, inexplicable synthetic facts about the world, and this inexplicability would vitiate the purpose of the putative reduction (Broad 1925, 55). In other words, the emergentist would question the ontological status of the bridge laws themselves, rather than the success or failure of the formal derivation, in evaluating whether P is an emergent property.

Kim's (1999, 32) argument against the possibility of emergence falls under the second type of criticism, in that it seeks to expose an inherent tension between the *novelty* of emergent properties and their supervenient status. According to Kim, in order for a whole system to possess a "novel" property, that property must possess novel causal powers, or powers to bring about changes that cannot be attributed to the *emergence base* of that system, which consists of the nonrelational properties of the parts and the relations of those parts to one another. He also holds that the emergent properties of a system would supervene upon this emergence base, in the sense that the instantiation of the emergence base would determine the instantiation of the emergent property. But, because of this supervenience, it would appear that any putative scientific law L that refers to an emergent property can be replaced by a law L^* that refers not to the emergent property itself but to the emergence base upon which it supervenes. As a consequence, L^* would not lack any explanatory power possessed by L, and hence the reference to the emergent property would be superfluous in a scientific law. One way of countering Kim's argument, then, would be to reject the claim that emergent properties are supervenient, which is what some contemporary advocates of emergentism do, as will be discussed in the next section.

Contemporary Emergentism in the Philosophy of Science

Emergentist approaches in the philosophy of science today are oriented toward the evaluation of reductionist claims that appear within specific scientific contexts (see Primas 1998 for an overview of several problematic claims). Philosophers of science have also drawn upon examples from physics, chemistry, and biology in order to provide new explications of emergence. Three fairly recent explications of emergence are described below.

In the philosophy of physics, one significant area of attention involves nonseparable systems. In quantum mechanics, a nonseparable state is one the state vector of which cannot be represented as a tensor product of component state vectors (see Shimony 1987, which relates nonseparability in quantum mechanics to holism). Hence the behavior of a nonseparable system cannot be explained in terms of the behavior of its components, in addition to a fundamental compositional principle. Additionally, nonseparability in quantum mechanics leads to limitations in the reduction of chemistry to physics, insofar as it entails the use of approximations in the derivation of molecular structure on the basis of quantum mechanics (Woolley 1991; Jaeger and Sarkar 2003).

Some philosophers have taken quantum nonseparability as a model for constructing novel conceptions of emergence. According to one emergentist interpretation (Humphreys 1997), one should not speak of the "components," or "component properties," of a nonseparable system at all; rather one should say that the properties of the constituents have undergone a "fusion" such that they can no longer be meaningfully individuated. Generalizing this example, Humphreys (1997) proposes an abstract fusion operator as a way of explicating the concept of emergence; this, he argues, can also be used to explicate the slippery notion of downward causation. This explication of emergence can also be used to counter Kim (1999), since, in the case of nonseparability, there is no way of independently characterizing the emergence base in terms of the nonrelational properties of the parts of the system and their relations to one another.

A second interpretation (Teller 1986) holds that the components of a nonseparable system *can* be meaningfully individuated but that they stand in "inherent relations" to one another, that is, relations that do not supervene on the nonrelational properties of each part. In contrast to the first interpretation, this conception accepts the independent ontological status of relations; hence Teller (1986) refers to this view as "relational holism."

A third recent approach to elaborating the concept of emergence focuses upon the relation between theories and phenomena, rather than between parts and wholes of systems. Hence it is not, strictly speaking, necessary that an emergentist accept the ontological and epistemological premises ([i] and [ii]) outlined above. Batterman (2002) argues at length that emergence is best understood as involving the failure of a "smooth" or regular asymptotic limiting relation between two theories. As Nickles (1973) shows, in some exemplary cases of theory reduction, one theory or formula is shown to be a special case of another theory or formula when some parameter of the latter approaches a limiting value. An example is classical mechanics, which is a limiting case of the special theory of relativity when velocities approach zero. In some cases, this limiting relation between theories is "singular" rather than smooth or regular; that is, the behavior of the formula becomes highly irregular as the value of a parameter approaches a limit, or the limit itself is undefined. Berry (1994) and Batterman (2002) describe how certain natural phenomena, such as the state of a fluid at its critical point, or the appearance of supernumerary bows of a rainbow, are best described by the irregular behavior of formulae as they approach some limiting value. Supernumerary bows, for example, appear (or "emerge") only when the wavelength of light becomes very small; theoretically, this can be described as a singularity of certain wave-optical formulae as the wavelength parameter approaches zero (that is, as wave optics approximates ray optics). Such phenomena may be called emergent, although they do not appear to involve a special part/whole relation. Additionally, they call into question the putative reduction of ray optics to wave optics (Berry 1994).

Justin Garson

References

Alexander, S. (1920), *Space, Time, and Deity*, vol. II. London: Macmillan.

Batterman, R. W. (2002), *The Devil in the Details: Asymptotic Reasoning in Explanation, Reduction, and Emergence*. Oxford: Oxford University Press.

Berry, M. (1994), "Asymptotics, Singularities, and the Reduction of Theories," in D. Prawitz, B. Skyrms, and D. Westerstahl (eds.), *Logic, Methodology and Philosophy of Science IX: Proceedings of the Ninth International Congress of Logic, Methodology and Philosophy of Science, Uppsala 1991*. Amsterdam: Elsevier, North-Holland, 597–607.

Broad, C. D. (1925), *The Mind and Its Place in Nature*. New York: Harcourt, Brace, and Company.
Campbell, D. (1974), "'Downward Causation' in Hierarchically Organised Biological Systems," in F. J. Alaya and T. Dobzhansky (eds.), *Studies in the Philosophy of Biology*. Berkeley and Los Angeles: University of California Press, 179–186.
Gilbert, S., and S. Sarkar (2000), "Embracing Complexity: Organicism for the 21st Century," *Developmental Dynamics* 219: 1–9.
Hempel, C. ([1948] 1965), "Studies in the Logic of Explanation," in *Aspects of Scientific Explanation and Other Essays in the Philosophy of Science*. New York: Free Press, 245–290.
Humphreys, P. (1997), "How Properties Emerge," *Philosophy of Science* 64: 1–17.
Jaeger, G., and S. Sarkar (2003), "Coherence, Entanglement, and Reductionist Explanation in Quantum Physics," in A. Ashtekar, R. S. Cohen, D. Howard, J. Renn, S. Sarkar, and A. Shimony (eds.), *Revisiting the Foundations of Relativistic Physics: Festschrift in Honor of John Stachel*. Dordrecht, Netherlands: Kluwer, 523–542.
Kim, J. (1999), "Making Sense of Emergence," *Philosophical Studies* 95: 3–36.
Klee, R. (1984), "Micro-Determinism and Concepts of Emergence," *Philosophy of Science* 51: 44–63.
Lewes, G. H. (1875), *Problems of Life and Mind*, vol. II. London: Kegan, Paul, Trench, Trubner, and Co.
Lloyd Morgan, C. (1923), *Emergent Evolution*. London: Williams and Norgate.
McLaughlin, B. (1992), "The Rise and Fall of British Emergentism," in A. Beckermann, H. Flohr, and J. Kim (eds.), *Emergence or Reduction: Essays on the Prospects of Nonreductive Physicalism*. Berlin: Walter de Gruyter, 49–93.
Mill, J. S. (1874), *A System of Logic*, 8th ed. New York: Harper and Brothers.
Nagel, E. (1961), *The Structure of Science*. New York: Harcourt, Brace and World.
Nickles, T. (1973), "Two Concepts of Intertheoretic Reduction," *Journal of Philosophy* 70: 181–201.
Polanyi, M. (1968), "Life's Irreducible Structure," *Science* 160: 1308–1312.
Popper, K., and J. Eccles (1977), *The Self and Its Brain*. Berlin: Springer.
Primas, H. (1998), "Emergence in the Exact Natural Sciences," *Acta Polytechnica Scandinavica* 91: 83–98.
Shimony, A. (1987), "The Methodology of Synthesis: Parts and Wholes in Low-Energy Physics," in R. Kargon and P. Achinstein (eds.), *Kelvin's Baltimore Lectures and Modern Theoretical Physics: Historical and Philosophical Perspectives*. Cambridge, MA: MIT Press, 399–423.
Sperry, R. W. (1969), "A Modified Concept of Consciousness," *Psychological Reviews* 76: 532–536.
Teller, P. (1986), "Relational Holism and Quantum Mechanics," *British Journal for the Philosophy of Science* 37: 71–81.
Weiss, P. (1968), "The Living System: Determinism Stratified," in A. Koestler and J. R. Smythies (eds.), *Beyond Reductionism*. London: Hutchinson, 3–42.
Wimsatt, W. C. (2001), "Emergence as Non-Aggregativity and the Biases of Reductionisms," *Foundations of Science* 5: 269–297.
Woolley, R. G. (1991), "Quantum Chemistry Beyond the Born-Oppenheimer Approximation," *Journal of Molecular Structure* 230: 17–46.

EMPIRICISM

Empiricism is the position according to which experience is the only source of warrant for one's claims about the world. Having assigned experience this exclusive role in justification, empiricists then have a range of views concerning the character of experience, the semantics of claims about unobservable entities, the nature of empirical confirmation, and the possibility of nonempirical warrant for some further class of claims, such as those accepted on the basis of linguistic or logical rules. Given the definitive principle of their position, empiricists can allow that one can have knowledge independent of experience only where what is known is not some objective fact about the world, but something about the way it is conceptualized or described. Some empiricists say that one can have knowledge of verbal equivalences or trivialities; some argue that any nonempirical tenets are not even properly called 'knowledge,' but should be seen as notions accepted on pragmatic rather than properly epistemic grounds. What no empiricist will allow is substantive a priori knowledge: According to empiricism, one can have no pure rational insight into real necessities or the inner structure of nature, but must rely on the deliverances of the senses for all information about external reality. Some versions of empiricism argue against the very notion of real necessities or metaphysical structure behind the phenomena; other versions take a more agnostic approach, arguing that if there is a metaphysical structure behind the phenomena, it is either out of epistemic reach or known only to the extent that it can be grasped through experience, rather than through rational reflection.

Early Modern Background

First published in 1689, John Locke's *Essay Concerning Human Understanding* sets out a version of empiricism whose basic framework remains an inspiration to contemporary advocates of the position. Expressing admiration for the accomplishments of Sir Isaac Newton and Boyle, Locke aims to show a similar respect for observation and theoretical simplicity in his investigation of the powers of the human mind. In contrast to the rationalist project of searching for the essence of the mind or the metaphysical principles behind the way it ought to work, Locke promises to pursue the "historical, plain method" of describing the type of process that would result in the ordinary formation of human knowledge (Locke [1689] 1975, 44). Locke contends that in this sequence of events one begins with a blank slate, a mind empty of ideas. The contrary postulation of innate ideas or principles is incompatible with what is observed in children and the dull-witted, Locke maintains, and in any event superfluous. Human cognition can be explained without helping oneself to the rationalist notion that some truths are built into the mind from the start; the positive task of providing such an explanation becomes the main project of Locke's *Essay*.

Locke maintains that all thought can be analyzed into ideas whose ultimate origin is in experience, broadly conceived to include both sensation (the passive reception of ideas from external objects through the senses) and reflection (the passive reception of ideas from the mind's introspective access to its own workings). Experience provides simple ideas (like those of 'blue,' 'sweet,' or 'pain'); the mind then manipulates and conjoins these simple ideas to form complex ideas (like the ideas of particular individual objects, modes, and relations). Because the mind is able to combine its ideas, the acquisition of knowledge is not restricted to the passive ingestion of ideas in experience; in fact the highest grade of certainty comes from assessing the internal structure of, and the relations among, complex ideas that are one's own construction. A lower degree of certainty accrues to knowledge of the external world, made possible in part by noting that certain ideas reliably come in clusters, which is presumed to indicate the presence of external substances, and also by one's consciousness of one's passivity in receiving ideas of sensation. While it may be the case that certain perceivable qualities necessarily coexist in certain substances (e.g., ductility and weight in gold) in virtue of the microscopic constitution of that substance, one's powers of perception are such that it is impossible to have the same kind of direct knowledge of this necessary coexistence as one has of the perceivable qualities themselves.

In Locke's theory, ideas received from experience are the only ingredients of thought, but many entities other than ideas get postulated during the course of the theory: the external objects causing ideas, powers inherent in those objects and causal relations among them, and the mind itself. Advanced some 50 years later, David Hume's version of empiricism exposes some of the difficulties of attempting to maintain this kind of mixed ontology within the empiricist framework (Hume [1739–1740] 1978). Hume is more careful than Locke to extract evidence for his theory of human cognition only from perceivable phenomena and to refrain from positing the kind of physical and metaphysical entities access to which would be unaccountable from an empiricist perspective. In the first wave of reaction to Locke, George Berkeley had already shown that even the apparently straightforward claim that one's ideas of sensation are caused by external objects could prove difficult for an empiricist to defend: If one were directly conscious only of one's ideas, with what right could one claim that these ideas resemble, and have their origin in, things of an entirely different kind that are not themselves directly present to the mind? Berkeley argues for a phenomenalist understanding of objects: The objects of which one is conscious are not independent matter but in fact collections of perceptions. Hume agrees with Berkeley that given the empiricist premise that one is only aware of one's perceptions, the postulation of independent matter is unjustifiable, but Hume also notes that people have a tendency to conceive of objects as having an independence and continuity that is not ascribed to perceptions. From the perspective of a consistent empiricist, Hume suggests that this tendency can be seen only as a blind instinct toward fabrication: Experience never delivers anything other than fleeting perceptions, so the sense of permanence is nothing more than an illusion, which Hume explains by pointing to the near resemblance of successive perceptions and the ease with which people can confuse resembling particulars for the same thing.

Causation receives a similar treatment: Where Locke had helped himself to a realist understanding of causation, Hume points out that causation is not itself perceived and cannot be construed as a pure relation of ideas. That no purely conceptual connection links a cause to an effect can be seen by reflecting on the ability one has to imagine a change in the course of nature. Like the stability of external objects, objective necessary connections among objects are an illusion, generated in this case by one's consciousness of one's instinctive (as opposed to rational) habit of expecting past patterns to continue. Where one has seen many events of type A followed by events of type B, one develops a mental custom of associating these ideas, and with this custom in place, the sight of type A compels the mind to think of B. The subjective sense of being pushed in this way gives rise to the idea of necessary connection, which then mistakenly projects onto nature and imagines it to be an objective relation among events.

If Hume's analysis is aimed at showing that such fundamental components of the commonsense worldview as enduring external objects and causation are illusory, he does not suggest that this philosophical result will overthrow that worldview; indeed, he argues that observation of the natural tendencies of the human mind shows that people will naturally continue in their instinctive patterns of thinking in terms of objective things and causes, however unjustified these instincts may seem from a philosophical standpoint. It is a difficult interpretive question to what extent this skeptical outcome should be read as a philosophical condemnation of ordinary claims to knowledge or as a demonstration of the shortcomings of either philosophical analysis in general or empiricism in particular.

One influential response to Hume was to see his skepticism as pointing to the inadequacy of the empiricist starting point. Immanuel Kant argued that one's thought about matters such as causation could not be understood without the postulation of something more than mere sensory perceptions as available to the mind; he maintained that one can make sense of empirical knowledge only if one sees sensory perceptions as entering a mind already possessed of a priori knowledge of the underlying causal structure of nature and the geometrical form of time and space. The exact nature and status of the metaphysical and geometrical commitments Kant envisaged is a matter of some controversy. For later advocates of the broadly Kantian style of response to empiricism (e.g., Reichenbach 1965), the complexity of the task of articulating a reasonable such set of a priori constraints was made particularly evident by such developments as the emergence of the theory of relativity (see Causality).

Early-Twentieth-Century Background

Until the twentieth century, geometry, or the study of the pure structure of space, had typically been seen as the paradigmatic example of an a priori discipline, and as an obstacle for empiricist accounts of knowledge. Einstein's use of non-Euclidean geometry in the theory of relativity made it hard to resist the conclusion that if geometry was a priori at all, it had this status only when considered as an uninterpreted deductive enterprise: The study of the structure of space itself could now be taken as either an empirical matter or a matter of the postulation of conventions rather than the discovery of objective facts. The reexamination of the status of questions once considered intuitive or rational was a significant source of inspiration for logical positivism, originating in Germany and Austria in the 1920s. Positivism drew inspiration from the development of Frege and Russell's symbolic logic and the new clarity it brought to the problem of the foundations of mathematics; at the same time, the legacy of Ernest Mach's eliminative empiricism was also a powerful if short-lived force behind the movement.

The relation between positivism and empiricism is a complex matter. It is clear that the positivists thought that all substantive questions about the world were to be answered by empirical science, but it is less clear that their conception of empirical science was straightforwardly empiricist. Some examination of the details of positivism is in order here.

The positivists conceived of philosophy as an enterprise of clarifying and making explicit the conceptual, linguistic, and logical structures of science, rather than as a means of discovering further characteristics of reality at a deeper metaphysical level than the empirical phenomena. The positivists hoped for a clean divide between the material questions about nature to be answered by the empirical sciences and the formal questions about science to be answered by philosophy. Given this formal approach, it is not surprising that the positivists cast the central problems of epistemology in linguistic terms. Locke's causal picture of sensation saddled him with a metaphysics not easily defended from within empiricism; the positivists aimed to avoid metaphysics altogether and take the question of the relation between experience and theory as a question about the proper form of observation reports and their formal relations to other sentences in the language of science. So Moritz Schlick writes in "The Foundation of Knowledge": "I think it a great improvement in method to try to aim at the basis of knowledge by looking not for the primary *facts* but for the primary *sentences*" (Schlick 1959, 212). These primary or protocol sentences are seen as idealized records of basic experience, cast in a vocabulary of observational terms and separated sharply from the higher-level theoretical claims whose confirmation they supply. Positivists divided into several factions over the question of the form of these statements. On Schlick's "foundationalist" side of the debate, a protocol statement aims to capture the content of what Schlick called a "confirmation," or decisive moment, of experience, whose certainty is beyond doubt; other parts of the system of science are ultimately justified by their relations to these confirmations, but the confirmations themselves are justified by the character of experience itself, and not by anything further within the system of science (see Schlick, Moritz). In opposition to Schlick, Otto Neurath proposed a fallibilist approach to protocol statements: A protocol statement is, like any other statement in the system of science, subject to rejection in light of considerations of overall coherence (see Neurath, Otto). Schlick has difficulty explaining the relation between basic confirmations and their linguistic expressions in protocol statements without recourse to metaphysics. Neurath has difficulty explaining how his solution maintains a special role for experience, or how he maintains empiricism without leaving himself open to the charge that science and fantasy could be equally well grounded given just sufficient internal consistency.

While Rudolf Carnap's original position was closer to Schlick's, he soon moved to adopt what he took to be a neutral stance, declaring that the question of the form of protocol sentences is "not answered by assertions but rather by postulations. . . . [T]he task consists in investigating the consequences of these various possible postulations and in testing their practical utility" (Carnap [1932] 1987, 458). Rather than supposing that something in the nature of reality determines the correct syntactical form and role of observation statements in science, Carnap now maintained that this is not a question of fact with a single correct answer, but a question about which postulation will be found most convenient for one's purposes. Carnap's work went on to exhibit an increasing emphasis on conventions adopted for pragmatic reasons.

The exact extent of Carnap's allegiance to empiricism is subject to debate (see Friedman 1999; Sarkar 2001). Carnap does not start from the position that the justification of empirical science is in doubt until science can be shown to be derived from the contents of experience, nor does he think that the immediately given has a specially certain or unproblematic epistemic status. In *The Logical Structure of the World*, Carnap tries to show how scientific concepts could be reduced to relations among moments of experience, but he claims that he could have taken other basic elements, like space-time points or even physical entities such as subatomic particles, as his starting point: His aim is strictly to analyze the internal logical structure of science rather than to justify science by appeal to something better grounded. By his own admission, Carnap's analysis of the internal logical structure of science was incomplete, most crucially in its failure to exhibit the dispensability of the basic relation of recollected similarity. Carnap's work in the decade after *Logical Structure* shows further departures from the verificationist empiricism of early positivism. While early positivists had claimed that every scientific term could be explicitly defined in terms of observable properties, in his "Testability and Meaning" (1936 and 1937) Carnap argues that some theoretical terms have a less direct relation with observation. Because dispositional concepts such as 'solubility,' for example, need to be understood in terms of relationships among various possible test conditions and observable outcomes, sentences involving terms of this sort cannot just be translated into sentences using the original observational vocabulary (see Carnap, Rudolf).

A form of positivism that lies squarely in the empiricist tradition is presented in A. J. Ayer's (1936) *Language, Truth and Logic*. Ayer insists on a phenomenalist account of external objects and a verificationist theory of meaning. According to Ayer, only two kinds of statements have literal significance and the possibility of truth or falsity: synthetic statements, identified as those statements that can be rendered more or less probable by some specifiable course of experience, and analytic statements, whose acceptability is wholly determined by the syntactic rules for the symbols they contain. All other statements, and in particular the statements of traditional metaphysics, are not even false, but meaningless. Philosophy itself is seen as falling on the analytic side of the line: Epistemology is concerned with the rules governing the use of symbols, and aims to identify the formal relations between the various strings of symbols that constitute observational and theoretical statements in the language of science (see Ayer, Alfred Jules).

Ayer's version of empiricism was one of the first targets of a wave of arguments that led to the decline of positivism by the middle of the twentieth century. Phenomenalism was attacked as incoherent (see Chisholm 1948); Nelson Goodman ([1954] 1979) argued that confirmation could not be explained in syntactic terms; Wilfred Sellars (1956) urged that empiricism's view of what is given in experience made experience an inadequate basis for knowledge of the world; and Willard Van Quine (1953) argued that the positivists had no acceptable way of drawing their distinction between analytic and synthetic statements. Quine's criticism proved particularly influential in the subsequent development of empiricism.

Empiricism after Positivism

Quine's "Two Dogmas of Empiricism" attacked both the positivist notion of a sharp distinction between analytic and synthetic sentences and what Quine calls the doctrine of reductionism, according to which each synthetic sentence is associated with a fixed set of actual or possible experiences tending to confirm or discredit that sentence. On the first front, Quine argues that various positivist efforts to identify the distinctive features of analytic sentences have either been inadequate to distinguish the set of sentences the positivists needed to take as analytic or slipped into an empty circularity, in which, for example, analyticity is understood with the help of the notion of cognitive synonymy, and cognitive synonymy is either left unexplained or itself defined in terms of what is analytically true. On the question of reductionism, Quine finds a lesson in Carnap's failure to reduce individual statements about the physical world to statements about immediate experience, and recalls Duhem's claim that one is always free to maintain a theory in the face of apparently contrary evidence by amending an auxiliary hypothesis. According to the slogan that has become known as the Quine-Duhem thesis, "Our statements about the external world face the tribunal of sense experience not individually but only as a corporate body" (Quine 1953, 41) (see Duhem Thesis; Quine, Willard Van; Underdetermination of Theories).

Quine intended his essay strictly as an attack on the positivist version of empiricism, and not on empiricism itself. In the final section, "Empiricism Without the Dogmas," experience is clearly identified as the only source of information for theories about the world, but the relation between experience and theory is not as the positivists had thought. Beliefs about everything from general physical laws to mundane claims about particular objects form a single system, the parts of which are amended in response to recalcitrant experience and kept in line with each other in accordance with rules of logic that are themselves part of the web. Nothing is immune to revision, and everything is revised on the same basis of accommodating experience, so that there is no difference in principle between changing a logical law to simplify quantum mechanics and changing from a geocentric to a heliocentric cosmology, or revising any other empirical claim. In place of the formal positivist approach to confirmation, Quine introduces a relation of 'germaneness' in his account of the relation between sensory evidence and the theory it supports. A body of sensory experience is more germane to one claim than to others when this experience will leave one more likely in practice to revise this particular claim. Rather than engaging in the study of how an ideal scientific language would be formulated, or how one ought to reform one's thinking, the epistemologist is directed to engage in an empirical study of the relationship between the actual input of sensory stimulation and the output of theoretical utterances. Following through with this program would require epistemology ultimately to become a chapter of psychology.

Quine insisted throughout his career that this naturalist position counted as a form of empiricism, but this classification is controversial. Indeed, Donald Davidson (1973–1974) argues that a natural extension of Quine's argument will do away with the contrast between form and content and leave nothing recognizable as empiricism. Also, while Quine contends that there is a normative element in his position, insofar as it leaves room for people to be criticized for having beliefs that accommodate their sensory experience poorly, it is clear that Quine's naturalism does not have the same normative ambition of traditional empiricism. Traditional empiricism was concerned with the question of what one ought to believe, or how common ways of thinking might be reformed to respect the limits of warrant; Quine's naturalism aims to take cognition as a given object of empirical inquiry, and does away with the traditional conception of warrant (see Hookway 1994). For Quine, the question is always about what sentences people *do* revise in practice, and not about what sentences *would be right* to revise, whether people actually do so or not. Whether Quine is an "empiricist" will depend in part on how one wants to use the term. If one emphasizes Quine's advocacy of empirical methods for the study of knowledge itself, then it may seem appropriate to classify his epistemological naturalism as a development continuous with the main thrust of empiricism; indeed, Quine is sometimes faulted for not having gone far enough in using the empirical data he recommends as useful in epistemology. On the other hand, if one sees epistemology as an enterprise that is aimed at figuring out what justifies beliefs, then it is hard not to see Quine's naturalism as constituting a change of topic rather than a development of earlier empiricism.

The version of empiricism that constitutes the most influential contribution to traditional epistemology since the collapse of positivism has been put forward by Bas van Fraassen, in support of the view of science he calls "constructive empiricism." According to van Fraassen, the positivists were mistaken in assuming that once empiricists take experience as the sole source of warrant, they are required to reduce everything to experience or to reinterpret statements about unobservable entities as abbreviations for more complex statements about observable phenomena. Empiricism does set limits on what one can see oneself as rationally obliged to believe, but by invoking a distinction between acceptance and belief, van Fraassen is able to defend an empiricist approach to science without requiring a positivist reformulation of the language of theories. When one accepts a theory, and commits oneself to a certain research program, one must believe what the theory says about observables, that is, one must believe that the theory is empirically adequate; but one does not have to believe the whole theory, including what it says about unobservables. Allowing this agnosticism about the unobservable makes accepting less committal than believing, but van Fraassen argues that science can be understood without the stronger realist stance; nothing that matters is lost by seeing science as aiming just at empirical adequacy, rather than full-blown truth. Equally, nothing is gained by the stronger realist position if van Fraassen is right, other than the need to contend with, and explain epistemic access to, various items of metaphysical baggage like causes and laws, realistically construed.

Van Fraassen allows that theories may have virtues that go beyond empirical adequacy (perhaps simplicity or explanatory power), but such informative virtues do not make the theory more likely to be true. Indeed, the more informative a theory is, the more risk it runs of being false; if one chooses informative theories over their less committal counterparts, it can only be for pragmatic reasons and not because these theories are more likely to be true. In van Fraassen's empiricism, scientists need never accept ampliative rules of inference (like inference to the best explanation [IBE]) as forcing them to go beyond the limits of observation: If positing the real existence of electrons would explain some observable phenomenon, this is not in itself a reason to take the step of believing that the unobservable electrons exist. Respecting the limits of his warrant, scientists may rationally stick to the more modest position that all observable phenomena are *as they would be if* the electron theory were true.

Van Fraassen shares with the positivists a sense of the epistemic significance of the line between what is observable and what is not, but instead of aiming to find a syntactical way of drawing the line, say, by developing a purely observational vocabulary, he argues that the problem can be naturalized: Scientific theories themselves can show how the realm of the observable is delimited. According to constructive empiricism, a scientific theory shows a picture of how the world could be, giving a set of models corresponding to various initial conditions. The theory itself can then specify parts of these models (the "empirical substructures") as potentially representing observable phenomena. A theory is empirically adequate if it has a model in which the observable phenomena can be embedded.

Van Fraassen himself notes that while belief in a theory's empirical adequacy is weaker and therefore safer than belief in its truth, it is not without risk: In claiming that a theory is empirically adequate, one is still going out on a limb and committing oneself to the truth of claims about states of affairs that are not observed by oneself, or have not yet been observed, or will never actually be observed, and so on. If one's motivation were just to maintain the weakest possible beliefs compatible with the evidence, one should shrink in the direction of a solipsism of the present moment rather than adopting the scientific rationality of constructive empiricism. So van Fraassen's position does not enable one to be maximally certain of one's beliefs. He has argued that his aim is rather to develop a characterization of the aim of science, or the standards for what counts as success or failure in that enterprise; if scientists do not restrict admissible evidence to what is observed by themselves alone, then no adequate account of science can give supreme epistemic significance to that special class of evidence.

This is not to suggest that van Fraassen sees his constructive empiricism as a sociological summary of the attitudes of working scientists. In particular, van Fraassen is ready to acknowledge that scientists may often believe that their theories are not merely empirically adequate but true, even with respect to unobservables. Because of the way van Fraassen defines rationality, he does not have to classify such thinking as irrational: His conception of rationality is permissive, rather than prescriptive. In this view, the scientist does not need to be rationally compelled to believe something in order for the scientist's belief to count as rational; rather, she may believe anything as long as she is not rationally compelled to believe otherwise. Rationality requires one to maintain logical consistency and accept the testimony of the senses, but if such minimal limits are respected, it neither requires nor forbids one from making conjectures about what lies beyond sensory evidence. In this view, then, the main upshot of an empiricist conception of rationality is negative: If warrant comes only from experience, rationality can never require belief in entities and characteristics of reality to which one lacks empirical access.

Criticisms of Constructive Empiricism

The most direct way to attack van Fraassen's empiricist view of science would be to identify a properly epistemic (as opposed to merely pragmatic) reason to believe in the claims that science makes about entities that lie below the threshold of observation. Many critics of van Fraassen have attempted to defend the rationality (as opposed to the mere practical convenience) of abduction or IBE. The best-known move here is Hilary Putnam and Richard Boyd's "no miracle argument" (NMA), according to which it is only by taking scientific theories to be true or approximately true that the success of science will be anything other than miraculous. It would be a tremendously strange coincidence, they argue, if all observable phenomena were just as though quarks existed and yet in fact they did not exist. This argument would have more force against an eliminative empiricist who would actually forbid belief in the unobservable. Against van Fraassen, the realists need to establish not just that belief in quarks is rationally permissible (he already grants this) but that it is rationally required. The main difficulty the NMA faces in establishing that conclusion is that it appears to be an argument with the very same abductive form as is in question (see Fine 1991). The argument urges that the truth of scientific theories is the best explanation for the phenomenon of their success; but even if that is so, unless one is already convinced that one is entitled to infer that whatever is the best explanation of a phenomenon is for that reason likely to be true, then one has no reason to accept the realist conclusion (see Putnam, Hilary).

A number of empiricist positions are intended to suggest that sound arguments in support of IBE are unlikely to be forthcoming. According to the "pessimistic induction," it is a mistake to infer the truth of a scientific theory from its acceptability as an explanation of the known phenomena, because of the many historical examples of theories that were explanatory successes in their day but have since been shown to be false. From the past course of events, there is no reason to believe that the theories now found persuasive as explanations of the phenomena are in fact true descriptions of things seen and unseen. In response to this argument, realists have noted that doubts about whether a current theory is exactly right may not provide a reason to withhold belief in the entities posited by that theory. Many theories that are shown to be false are superseded by theories that continue to use the same basic framework of entities, although there is some question about whether the realist can present a historical argument about the reasons for past predictive successes without presupposing the legitimacy of abduction (for a detailed historical discussion, see Psillos 2000). In addition, there is a more abstract and general form of the pessimistic induction available to the empiricist. According to the *argument from the bad lot*, the label "inference to the best explanation" is misleading, because there is no guarantee that one is in a position to choose the *best* explanation: One's choice is from among the explanations scientists have in fact been able to concoct so far, a range of alternatives that might in fact fail to include the true story. One can at most think of oneself as choosing the best available story, rationally weighing various rival theories only on the basis of evidence about observable phenomena.

The "conjunction objection" to constructive empiricism constitutes a quite independent move (Boyd 1973; Putnam 1978; Friedman 1983). It may be correct that in terms of vulnerability to recalcitrant evidence, a single theory's truth is never more credible than its empirical adequacy, but by taking theories to be true, one logically has the right to conjoin them, and the conjoined theory $(T_1 + T_2)$ can have richer empirical consequences, which can give additional confirmation to its each of its conjuncts T_1 and T_2 taken separately. In addition, the larger unified theory can give the kind of integrated explanation of phenomena at which science (arguably) must aim. Meanwhile, accepting that two theories are empirically adequate does not automatically give one the right to conjoin them (they may, for example, include contradictory statements about unobservables), and even where they can be conjoined, the claim that "T_1 is empirically adequate and T_2 is empirically adequate" will have fewer observational consequences than $(T_1 + T_2)$. It is open to the empiricist to challenge the realist idea that science aims at such unified explanations rather than unifying, where it does, as a pure consequence of the search for empirical adequacy; it is also possible to challenge the extent to which science does in fact engage in this kind of unification, or whether in fact later theories are used to correct earlier ones, rather than being straightforwardly conjoined with them (see van Fraassen 1980, Ch. 4).

Other points in the empiricist program that have attracted critical attention include the issue of modal concepts of possibility and necessity, even as they figure in van Fraassen's own statement of his position (Rosen 1994; Ladyman 2000), and the question of whether empiricism can give an adequate characterization of experience (Nagel 2000).

In raising doubts about whether the truth might always lie outside of the range of theories available, van Fraassen is sometimes seen as risking a collapse into skepticism. If warrant is so restricted that one can never have rational grounds to believe in any unobservable entity, no matter how well it would explain observations, then it may seem that by similar reasoning one will never be rationally compelled to believe anything as strong as the empirical adequacy of a theory, or even anything at all beyond the present testimony of the senses. Conversely, if van Fraassen wants to support the rationality of believing that certain theories are empirically adequate (true in all they say about the observable, and not just about what is presently observed), or even that perceived objects continue to exist after one leaves the room, then perhaps he is already committed to the admissibility of ampliative rational rules. Against the idea that continuously existing tables and trees are posited as the best explanation of given sense data, van Fraassen (1989) argues that philosophers have given ample arguments to show that one's awareness of the world cannot be a matter of making inferences from a body of raw sense data. What are perceived are not sense data but the observable parts of an objective world: "[W]e can and do see the truth about many things: ourselves, trees and animals, clouds and rivers—in the immediacy of experience"(178). Experience itself can be understood only "in the framework of observable phenomena ordinarily recognized" (1980, 72). This marks a reversal from the earlier empiricist strategy of attempting to show how the framework of observable phenomena could be constructed out of the ideas of experience.

In this version, empiricism is insulated from skepticism by setting its focus on the manner in which beliefs are updated, and not on their initial formation. According to van Fraassen (1989), "It is possible to remain an empiricist without sliding into skepticism, exactly by rejecting the skeptics' pious demands for justification where none is to be had" (178). Once one is committed to the general framework of observable phenomena, one will be in a position to examine critically the ways in which beliefs are changed, but there is no useful prospect of a critical examination of one's initial commitments. Critics of empiricism can wonder whether this pessimism about the scope of epistemology is justified and whether van Fraassen is right to characterize people's initial position as involving no commitments other than those to observables. It has also been suggested that what is in dispute between empiricism and realism may not be decidable on the basis of considerations acceptable to both sides, and this has generated some skepticism about the legitimacy of this conflict.

Skepticism About Empiricism

Both the empiricist and the realist are committed to the project of giving a philosophical analysis of the aim of science. But Arthur Fine argues that there is something wrong with that project. According to Fine (1986), realists and empiricists are mistaken in supposing that science has a single essence amenable to philosophical examination. There is nothing in scientific practice itself, Fine argues, that requires the possession of a philosophical theory of the point of science, and nothing in the deliverances of scientific inquiry yields an answer to whether empiricism or realism is correct. As an alternative, Fine advocates what he calls the natural ontological attitude, according to which one allows science to "speak for itself" and refrains from attempting to construct a notion of truth that goes beyond that "already in use in science." Of course, both realists and empiricists take themselves to be articulating exactly that conception of truth that is already in use in science. But Fine's contention is that they do not have any neutral or unprejudiced perspective from which to pass judgment on what science involves.

One of Fine's central criticisms of empiricism is that the empiricist's effort to create a special epistemic status for claims about observables could be based only on a priori commitments that do not square well with the basic orientation of empiricism. Observations alone do not force upon one any particular epistemic attitude to observation. If Fine is right about that, then the empiricist has some reason to resist the naturalist's suggestion that the claims advanced in epistemology are, like the claims of empirical science, themselves warranted only by experience (see van Fraassen 1995 for an argument along these lines). Empiricism is then a theory about what claims are warranted within science; the separate question of what claims are warranted within epistemology would lie beyond the scope of empiricism itself.

JENNIFER NAGEL

The author acknowledges the helpful input of Anjan Chakravartty, University of Toronto.

References

Ayer, A. J. (1936), *Language, Truth and Logic*. London: Gollancz, 1936.
Boyd, Richard (1973), "Realism, Underdetermination, and a Causal Theory of Evidence," *Noûs* 7: 1–12.
——— (1984), "The Current Status of Scientific Realism," in Jarrett Leplin (ed.), *Scientific Realism*. Berkeley and Los Angeles: University of California Press, 41–82.
Carnap, Rudolf ([1928] 1967), *Der logische Aufbau der Welt*. Translated by Rolf George as *The Logical Structure of the World*. Berkeley and Los Angeles: University of California Press.
——— ([1932] 1987), "On Protocol Sentences" (translated by R. Creath and R. Nollan), *Noûs* 21 (1987): 457–470. Originally published as "Über Protokollsätze," *Erkenntnis* 3: 107–142.
——— (1936),"Testability and Meaning," *Philosophy of Science* 3: 419–471.
——— (1937),"Testability and Meaning—Continued," *Philosophy of Science* 4: 1–40.
Chisholm, Roderick (1948), "The Problem of Empiricism," *Journal of Philosophy* 45: 512–517.
Davidson, Donald (1973–4), "On the Very Idea of a Conceptual Scheme," *Proceedings and Addresses of the American Philosophical Association* 67: 5–20.
Fine, Arthur (1986), "Unnatural Attitudes: Realist and Instrumentalist Attachments to Science," *Mind* 95: 149–179.
——— (1991), "Piecemeal Realism," *Philosophical Studies* 61: 79–96.
Friedman, Michael (1999), *Reconsidering Logical Positivism*. New York: Cambridge University Press.
——— (1983), *Foundations of Space-Time Theories*. Princeton, NJ: Princeton University Press.
Goodman, Nelson ([1954] 1979), *Fact, Fiction and Forecast*, 4th ed. Cambridge, MA: Harvard University Press.
Hookway, Christopher (1994), "Naturalized Epistemology and Epistemic Evaluation," *Inquiry* 37: 465–485.
Hume, David ([1739–40] 1978), *A Treatise of Human Nature*, 2nd ed. Edited by L. A. Selby-Bigge, revised P. H. Nidditch. Oxford: Clarendon.
Ladyman, James (2000), "What's Really Wrong with Constructive Empiricism: Van Fraassen and the Metaphysics of Modality," *British Journal for the Philosophy of Science* 51: 837–856.
Locke, John ([1689] 1975), *An Essay Concerning Human Understanding*. Edited by P.H. Nidditch. Oxford: Clarendon.
Nagel, Jennifer (2000), "The Empiricist Conception of Experience," *Philosophy* 75: 345–376.
Psillos, Stathis (1999), *Scientific Realism: How Science Tracks Truth*. New York: Routledge.
Putnam, Hilary (1978), *Meaning and the Moral Sciences*. Boston: Routledge and Kegan Paul.
Quine, Willard Van (1953), "Two Dogmas of Empiricism," in *From a Logical Point of View*, 2nd ed. Cambridge, MA: Harvard University Press.
Reichenbach, Hans (1965), *The Theory of Relativity and A Priori Knowledge*. Berkeley and Los Angeles: University of California Press.
Rosen, Gideon (1994), "What Is Constructive Empiricism?" *Philosophical Studies* 74: 143–178.
Sarkar, Sahotra (2001), "Rudolf Carnap," in *A Companion to Analytic Philosophy*. London: Blackwell, 94–109.
Schlick, Moritz (1959), "The Foundation of Knowledge," in A.J. Ayer (ed.), *Logical Positivism*. New York: Free Press, 209–227.
Sellars, Wilfrid (1956), "Empiricism and the Philosophy of Mind," in Herbert Feigl and Michael Scriven (eds), *Minnesota Studies in the Philosophy of Science*, vol. 1. Minneapolis: University of Minnesota Press, 253–329.
van Fraassen, Bas (1980), *The Scientific Image*. Oxford: Clarendon Press.
——— (1985), "Empiricism in the Philosophy of Science," in P. M. Churchland and C. A. Hooker (eds.), *Images of Science: Essays on Realism and Empiricism*. Chicago: University of Chicago Press.
——— (1989), *Laws and Symmetry*. Oxford: Clarendon.
——— (1995), "Against Naturalized Epistemology," in Paolo Leonardi and Marco Santambrogio (eds.), *On Quine: New Essays*. Cambridge: Cambridge University Press.

See also **Epistemology; Instrumentalism; Logical Empiricism; Realism**

EPISTEMOLOGY

Epistemology is often identified as the theory of knowledge, and epistemologists have often taken this to mean the giving of an analysis of the concept of knowledge. This is motivated, in large part, by skeptical worries about the possibility of knowledge. The project of answering the skeptic has once again emerged as an important theme in epistemology, largely due to the impact of contextualist analyses of knowledge (Cohen 1988; DeRose 1995; Lewis 1996).

While analyses of knowledge are certainly a major theme of much of contemporary epistemology, they by no means exhaust its central concerns. Of particular relevance for the philosophy of science are high-level theories about the general structure of epistemic justification and theories of epistemic change (belief revision or theory change).

The Gettier Problem

Conventional wisdom, pre-1963, had it that the concept of knowledge admitted of a straightforward analysis: A subject S knows that p iff ("if and only if") S believes that p, S is justified in so believing, and p is true. This justified true belief (JTB) analysis, however, is far from adequate. Gettier (1963) offered alarmingly simple counterexamples to the JTB analysis, establishing that justified true belief cannot be sufficient for knowledge. Consider the case of Jones. Jones has impeccable evidence that Smith owns a Ford—he sees Smith driving a Ford, Smith is always talking about his great Ford, and so on. So Jones justifiedly believes that Smith owns a Ford. Jones has also recently taken a logic class and has done very well. He realizes that "Smith owns a Ford" entails that Smith owns a Ford or Brown is in Barcelona. So he adopts this latter, disjunctive belief on the basis of his belief about Smith and the Ford. It seems as though Jones is justified in this disjunctive belief if he is justified in believing that Smith owns a Ford. Now, unbeknownst to Jones, Smith actually does *not* own a Ford, but by sheer coincidence Brown *is* in Barcelona. Hence, Jones has a justified true belief—namely, that Smith owns a Ford or Brown is in Barcelona—that does not seem to count as knowledge. His justification for the belief is not aligned in the "right way" with the facts that make the content of the belief true. The Gettier problem is that of giving an analysis of knowledge (or an analysis of the locution "S knows that p") that makes clear what the right way is.

An immediate thought is that what has gone wrong in the case of Jones is that he is reasoning from a false belief. The natural suggestion, then, is to add to the JTB account a fourth condition to the effect that S knows that p only if S's justification for p does not rely on any false beliefs. This simple fix, however, is not adequate, and this fact was apparently known to Bertrand Russell much earlier (see Russell, Bertrand). Adapting an example from Russell (1912), suppose that Jones walks through the town square every morning on his way to work. Every morning Jones looks at the clock tower, forms the relevant belief about the current time, and adjusts his pace to make sure he stays on schedule. The clock is a remarkably reliable timepiece, as everyone in town knows. Now, on a particular morning, Jones walks through the square, sees that the clock reads 8:15, and forms the belief that it is 8:15 A.M. In fact, it *is* 8:15 A.M., but unbeknownst to Jones, the clock has stopped! There was an electrical storm the night before, and lightning struck the clock precisely at 8:15 P.M. Jones seems to be justified in his belief that it is 8:15 A.M. and his belief is true, yet it does not seem that Jones knows that it is 8:15 A.M. His belief that it is 8:15 A.M. is not (or not obviously) inferred from any false belief, nevertheless this does not count as knowledge, since his justification for the belief does not connect up with its truth in the so-called right way.

The Gettier problem is seductive for its simplicity, but devilishly hard to solve. Gettier-type counterexamples tend to proliferate, so that given a particular candidate theory constructed with an eye to avoiding the original cases, it succumbs to a close variant. And patched to handle the variant, the theory typically falls prey to a variant of the variant. (See Plantinga 1993 for a detailed survey of attempts at providing a fourth condition to the JTB analysis.) A related project attempts to solve the Gettier problem (and also skeptical worries) by analyses of the locution "S knows that p" in which justification plays no essential role. So-called tracking accounts of knowledge (Nozick 1981), certain forms of contextualism (Lewis 1996), and what might be called "neighborhood reliabilism" (Williamson 2000) fit more or less uncontroversially into this broad category.

Theories of Justification

Theories of epistemic justification have traditionally been at the center of work in epistemology. This is largely because of the double duty that justification is asked to play. On the one hand, it is thought by many to be a necessary condition of knowledge, and many theories of justification aim at answering the Gettier problem. But "justification" is also a term of epistemic appraisal, signifying the positive epistemic status of a belief, and so the contours of this concept are of independent epistemological interest. Pretheoretically and from an epistemic point of view, justified beliefs are those that it is permissible to hold. Theories of epistemic justification aim at codifying and systematizing this rough-and-ready intuitive gloss by specifying the structure and nature of justification—delivering predictions about how exactly a belief is justified. The class of theories of justification can be neatly categorized in a number of ways. One way is by a series of distinctions (Pollock 1979; Pollock and Cruz 1999). The first distinction is between doxastic and nondoxastic theories. Justification of a particular belief is at least in part a function of what else an agent believes. Doxastic theories insist that in addition, justification is only a function of the agent's beliefs. Nondoxastic theories deny this. The second distinction is the internalist–externalist divide. This is a difference over the sorts of things that can be justifiers of beliefs. Internalist theories insist that justifiers must be "internal states," in some sense internally accessible to the cognitive agent in question. Externalist theories deny this. All doxastic theories are internalist, but not all internal states are doxastic, since it is possible for such states to be nevertheless relevant to the justification of a belief. Alternatively, one can describe the space of theories of justification as being carved according to the sources of the justifiers and the structure that those justifiers must have when related to other beliefs (Alston 1986). That is, the space can be carved along the dimensions of the internalist–externalist divide (the source of justifiers) and the structure of the justification relationship.

Foundationalism

Doxastic theories start from the assumption that whether an agent S's belief is justified is a function of, and only a function of, what else S believes—S's doxastic state at the time. After all, an agent's information about the world (at least the information about the world that is relevant to the concerns of epistemic justification) is codified by the beliefs that the agent has about it. Now, the project of a theory of justification is to say what beliefs one ought to hold. But then what information could be relevant, aside from an agent's beliefs, in deciding what one should believe? The assumption that nothing else could be relevant is just the assumption that epistemic justification is a function exclusively of one's doxastic state.

Foundationalist theories start from the intuition that some beliefs serve as reasons for others and that justification is a function of this "reason-for" relation. Now, belief in candidate belief p is justified only if the beliefs upon which p depends (in S's doxastic state) are also justified. But, the foundationalist argues, this tracing of reasons must either come to an end or continue ad infinitum (perhaps by running in a circle). Such infinite chains would be of no help in developing a theory of justification for finite beings (or else would be viciously circular), so the tracing of reasons must come to an end with beliefs that are self-justifying, or *basic*. The foundationalist maintains that there is a proper subset of an agent's beliefs that are basic in this way, epistemologically privileged and standing in no need of (exogenous) justification. There is a recursive structure to the nature of epistemic justification, with the basic beliefs laying the foundation: S's belief that p is justified iff p is a basic belief, or else it is properly based on or legitimately inferred from other justified beliefs. This schema can be turned into a full-blown theory once the foundationalist gives an account of what beliefs are basic (Are there enough of them? In what sense are they self-justifying?) and what sorts of relations count as "proper" inference relations between beliefs.

The foundationalist theories of the early twentieth century foundered on the nature of the reason-for relation. It was thought that the only good reasons were either deductive or inductive. But if that is right, then it seems impossible to give a foundationalist account of justified perceptual beliefs. Suppose S enters a room, sees an object that appears red, and forms the belief that there is a red object. This belief is intuitively justified but does not seem to be entailed by her perceptual information, the agent's other beliefs, or any combination of the two. Attempts at analyzing perceptual content in such a way as to guarantee such an entailment while staying within the doxastic framework (the phenomenalist project of [the early] Rudolf Carnap, [the early] Willard Van Quine, Goodman, and others) met with insuperable difficulties and were abandoned (see Carnap, Rudolf; Quine, Willard Van Orman). Similarly, S's belief cannot, in general, be justified by appeal to inductive reasons. The suggestion would be that S's belief that there is a red object is justified because S believes that she is in circumstances relevantly similar to other circumstances in which S has had similar perceptual evidence that turned out to be justified-belief-about-red-object circumstances. But this will not in general do, since in those cases S's justified belief could not have been inductively justified. But clearly S's belief is justified, and arguably this is a function of some reasons S has for it. So there must be epistemically good, noninductive, defeasible reasons. Establishing the existence of general defeasible reasons and their role in a theory of justification is one of the most important contributions to epistemology, independently discovered by Chisholm (1966) and Pollock (1967, 1974).

There are, in addition to doxastic foundationalist theories, nondoxastic relatives. Pollock and Cruz (1999) propose such a theory, which they call "direct realism." Their theory is largely foundationalist in structure, with defeasible reasoning at its core. While the details of the theory of defeasible reasoning have evolved significantly, the general framework is closely related to the foundationalist theory developed, for example, in Pollock (1974). However, they reject the doxastic assumption. They argue that any belief (even appearance beliefs) can be held for bad reasons without the agent in question realizing that the relevant belief is being held for a bad reason. A theory of justification ought to predict that in such cases those beliefs are unjustified. If this is right, then it follows that there can be no epistemically privileged class of beliefs. Their solution to this problem is to allow subdoxastic states to enter (as antecedents) into the reason-for relation—for instance, the having of a perceptual image is a defeasible reason for an agent to believe. And, of course, the reason can be defeated. Justifiers, according to this view, are still internal states, but they are not all beliefs.

Coherence Theories

While foundationalist theories assert that some nonempty proper subset of an agent's beliefs are basic, coherence theories deny this. (Thus a doxastic theory is a coherence theory iff it is not a foundationalist theory.) Coherence theories deny the existence of an epistemologically privileged class of beliefs, insisting instead that all of an agent's beliefs have the same fundamental status. Justification is then a function of how well a given belief fits in, or "coheres," with the rest of an agent's doxastic state. The main task for a coherence theorist is to give an account of the coherence relation.

One can taxonomize the class of coherence theories according to distinctions along two dimensions. Along one dimension, an analysis of the coherence relation will specify the structure of the relation and how it relates to other beliefs: For a candidate belief p, is it a linear relationship between p and some other belief(s) that matters (much like the foundationalist's tracing of reasons), or is it a holistic relationship between p and all of the agent's other beliefs? Along a second dimension, analyses of the coherence relation must specify its role, which may be positive (in which case a belief cohering with an agent's doxastic state constitutes positive grounds for the belief to be justified) or relevant only in a negative way (in which case the absence of coherence is a reason to get rid of some beliefs). This yields four types of coherence theory in logical space: positive linear, positive holistic, negative linear, and negative holistic.

Positive linear theories insist that coherence is a linear relationship between a candidate belief p and another belief q (or perhaps a smallish subset of an agent's beliefs). This linear relationship is just a version of the reason-for relation. Note that since such theories deny that linear appeals between beliefs must always come to an end (else they would be foundationalist theories), positive linear coherence theories must admit either that such an application of reasons can go on infinitely or else that the reason-for relation may be nontrivially cyclic. On the face of it, this seems an unpalatable dilemma: Either some legitimate justificatory chains go on infinitely or some legitimate justificatory chains run in a circle. The first horn connects with the intuition that reasons seem to be only a conduit through which justification flows—reasons cannot, in other words, generate justification, but only allow beliefs to be justified conditional on good starting points. The second horn connects with the intuition that circular reasoning can never lead to a justified belief. Sequences like the following seem clearly bad from an epistemic point of view: Suppose one's only reason for thinking that p is that q_1 is a reason for it, and one's only reason for thinking that q_1 is that q_2 is a reason for it, . . . , and one's only reason for thinking that q_n is that p is a reason for it. In such cases, how could S's belief that p be justified? These considerations are often taken together and labeled as the "regress problem" for coherence theories of justification (Sosa 1980). It is important to note that they properly apply only to (some) positive linear coherence theories, and not to the class of coherence theories in general.

Lehrer (2000) proposes a positive linear coherence theory of justification. In his theory, a belief p coheres with a background doxastic system just in case it is more reasonable to accept p than any of its "competitors" (roughly, propositions negatively relevant to p). One can conclude that accepting p is reasonable in this way by appeal to what he calls the "trustworthiness argument": Suppose S is trustworthy in what she accepts (for purely epistemic goals). Call this belief (T). Then S is trustworthy in accepting the candidate acceptance p (for purely epistemic goals). But then it must be reasonable to accept that p (otherwise S would not be trustworthy in S's acceptance of it). But how does one accept (T), the first premise of this argument? Lehrer says it is a "keystone belief," made reasonable by applying the trustworthiness argument to (T). So the picture is that coherence reduces to reasonableness, which is a linear relation. All justified beliefs end up appealing to (T). But (T) is not a basic belief, and itself is linearly supported by appeal to the trustworthiness argument and itself. This is what Lehrer calls the "virtuous loop."

There are a number of positive holistic coherence theories, notably those advocated by Lehrer (1974) and BonJour (1985). Positive holistic coherence theories still insist that coherence is positively relevant to justification but claim that the coherence relation is not linear. The picture is that an agent's doxastic system is a "web" of interrelated beliefs with coherence between p and a doxastic system being a function of the entire system (without p, of course). This function might be a measure of how well p would be explained by the truth of the other beliefs or how well the other beliefs "mutually support" p or some function of the agent's subjective probability assignment with respect to p. The most general sort of difficulty for this class of theories is making sense of the difference between a belief being justified and a belief being (merely) justifiable. For instance, S's doxastic state may exhibit the proper coherence between p and the rest of S's beliefs without S being able to tell, perhaps due to complexity considerations, that this is so. Is S's belief justified? More worrisome are cases in which S's belief in p is spontaneous or ungrounded (due, say, to an epistemically serendipitous brain lesion) but nevertheless fits in well with S's other beliefs. Such a belief could be justified, since it *ex hypothesi* exhibits the right properties to fit in well with S's other beliefs, but it is far from clear whether S believes p in an epistemically permissible way.

Negative linear and holistic coherence theories insist that coherence plays only a negative role in assessing whether an agent is justified in believing some candidate belief p. Such theories are motivated by the Neurath metaphor: Doxastic states are raftlike and free-floating without an anchor, each plank held in place by its relationship to all the others, with repairs and maintenance taking place at sea (Sosa 1980) (see Neurath, Otto). The picture is that doxastic agents find themselves with a plethora of existing beliefs, and any such belief is automatically *prima facie* justified. The role of coherence is a maintenance tool to shape and refine one's doxastic state by weeding out beliefs that have lost this *prima facie* status. Though a conceptual possibility, there are in fact no extant negative linear coherence theories. Harman (1986) offers the clearest example of a negative holistic coherence theory. According to this view, coherence is overall explanatory coherence. So suppose an agent S believes that p. This makes S automatically *prima facie* justified in believing that p. S should stop believing that p if and when p no longer coheres with the rest of S's doxastic state—that is, if and when S's other beliefs make it hard to explain how p could be true.

A difficulty facing negative coherence theories, and so negative holistic coherence theories, is that they insist that coherence (and, more broadly, reasoning in general) plays exclusively a negative, undermining role with respect to epistemic justification. As Pollock and Cruz (1999) point out, this is equivalent to the thesis that all of an agent's beliefs are *prima facie* justified, no matter how the agent acquired them. But that seems counterintuitive in cases in which an agent holds a belief, say p, on the basis of wishful thinking. S may have no specific reason for thinking that belief in p is no good, and so may have no negative, undermining lack of coherence that forces S to withdraw the belief. Nevertheless, such a belief seems unjustified, in which case it cannot be an undefeated *prima facie* justified belief, as the negative holistic coherence theory appears to predict. In principle, just as there are nondoxastic foundationalist theories, there is room in logical space for nondoxastic coherence theories of justification. However, this presently remains unexplored territory.

Externalist Theories of Justification

Foundationalist theories, both doxastic and nondoxastic, and their coherence counterparts take justification to be a function exclusively of the internal states of the epistemic agent in question. Externalist theories of justification deny this. Some external states—states of the world or modal facts about such states—are relevant to justification. Put another way: Internalists claim that justification supervenes on internal states of the cognizer in question (and so justification is invariant across manipulations in which the internal states are fixed but the facts about the outside world vary), and externalists deny this claimed supervenience.

A common motivation for externalism is the pretheoretic connection between epistemic justification and knowledge. Why would S want justified beliefs, as opposed to unjustified beliefs? The immediate thought is that, *qua* epistemic agent, S wants to believe what is true and avoid believing what is false. The point of having justified beliefs, then, is instrumental to attaining truth while avoiding error. And so an analysis of justification should have as a consequence that if a belief is justified, it is in some sense likely to be true (where this likelihood is a measure of actual success, not subjective probability). Justification is not merely a matter of a belief's chance of being true, understood as a result of a stochastic process. Rather, the view is that beliefs are formed and generated by cognitive processes, and (external, perhaps modal) properties of these processes determine the justificatory status of the beliefs they output.

The externalist position picks out a large region of logical space, but most of the extant externalist theories of justification, as a matter of fact, are versions of reliabilism. Broadly, reliabilist theories take belief to be justified just in case they (or their producers) are reliable indicators that the belief in question is true. The most straightforward reliabilist theory is process reliabilism (Goldman 1979). The analysis (schema) is simple: S's belief that p, produced by belief-forming process M, is justified iff M is a reliable process. To turn this into a full account, the process reliabilist must offer a criterion for reliable belief-forming processes. Different process reliabilist accounts then differ (in complexity and scope) depending on the complexity and scope of the analyses of reliable belief-forming processes. The basic analysis is simply that a belief-forming process M is reliable iff the actual ratio of true beliefs to false beliefs produced by M is sufficiently high. In such an account of reliability, S's perceptual belief that she sees a red apple is justified iff the module forming S's perceptual beliefs tends to produce true beliefs significantly more often than false ones.

As appealing as this simple form of process reliabilism is, it faces major difficulties, which (from the reliabilist's point of view) point toward more sophisticated accounts of reliability. First, tying justification to actual truth ratios does not rule out what might be called "accidental" reliability. For example, suppose S has the following belief-forming process: On a certain occasion t, S decides to believe that it is sunny in a distant location at $t + S$, say Amsterdam, if the toss of a certain coin at t turns up heads. S flips the coin and it turns up heads, whereupon S believes that it is sunny in Amsterdam. In fact, suppose it *is* sunny in Amsterdam. Then S's belief-formation process has maximal reliability, but clearly it is not a justified belief. BonJour (1985) raises similar worries by applying the basic process reliabilist account to the case of a reliable clairvoyant. Accidental reliability, in other words, seems just as bad as accidental truth. Some more subtle account of reliability seems to be in order, one in which modal stability of the truth ratio plays an important role. The "normal-worlds reliabilism" in Goldman (1986) aims, at least in part, at providing such an account. In this more sophisticated version, actual truth ratios are replaced by truth ratios across so-called normal worlds, defined as compatible with the agent's general beliefs about the general presumed structure of the actual world.

Another difficulty for reliabilist accounts as a class, which has proved to be the impetus for much further research, is the generality problem. Take the case of process reliabilism. For any belief, there are a multitude of ways of circumscribing which belief-forming process generated it. For instance, a perceptual belief about the color of an object (say, a red apple) can be described as the output of a color-vision module, color-vision-normally-situated-on-Earth module, color-vision-any-way-situated-on-Earth module, and so on indefinitely, with ever more general descriptions of the process. But it can equally be described as S's-color-vision module, S's-color-vision-since-1983 module, and so on, down to S's-color-vision-module-used-on-this-occasion-to-view-this-apple. Which of these processes is the right one by which to judge the justificatory status of S's belief? Since reliability is being treated as an objective conditional probability (i.e., the probability that a belief is true given it is produced by process M), the answer matters a great deal.

A total evidence requirement seems to compel the description to be as specific as possible, but then this conditional probability goes to 1 (if the apple is the color S believes it to be) or to 0 (if not), trivializing the account, since it would render a belief justified iff true. Determining the appropriate reference class of worlds raises essentially the same issue for normal-worlds reliabilism. Indeed any account that relativizes justificatory status of a belief to circumstances in the local environment must face this problem.

A final puzzle for reliabilism, and for externalist theories of justification more generally, is the problem posed by good epistemic agents who find themselves in epistemically unfortunate circumstances (Cohen 1984). Consider two epistemic agents S_1 and S_2. S_1 is a very careful reasoner, always forming beliefs on the basis of excellent evidential support and in short is as epistemically responsible as any normal cognizer. S_2, on the other hand, is a very sloppy epistemic agent, always forming beliefs on the basis of wishful thinking, fancy, and faulty reasoning. As a happenstance, S_1 and S_2 have a lot of beliefs in common—that is, there are many beliefs about the physical world that they share, which form the set C. Now, as it happens, S_1 and S_2 inhabit (unbeknownst to them) an Evil Demon world—their beliefs about the world are systematically and radically false, and in particular all beliefs in C are false. The facts are being manipulated by the whimsy of an Evil Demon whose aim is to deceive them. For any belief in C, can there be any difference in justificatory status between the two agents? Reliabilist theories, insofar as the two cognizers *ex hypothesi* have identical actual and counterfactual truth ratios (namely, 0), treat any two beliefs of the respective agents as unjustified. But this seems implausible, since there is an important epistemological difference between S_1's belief that, say, most swans are white (formed on the basis of a standard induction with a large inductive base) and S_2's belief that most swans are white (a randomly occurring thought to S_2). The difference seems to be a difference in justification.

The extent to which such worries upset the externalist project is an open question. Some externalists just deny the purported intuitive difference above. It is also unclear to what extent this example depends on particular incarnations of externalist theories like process reliabilism. Goldman (1986), in fact, seems to advocate exploring both of these themes, remarking that many do not find the process reliabilist prediction obviously counterintuitive (113, n. 32), but that perhaps the best way to understand reliability is in terms of normal worlds (113).

Epistemic Change

While theories of justification aim at a characterization construed as a property of beliefs in a given epistemic state, there is a related project that has received much attention, that of characterizing justification construed as a property of transitions between sets of beliefs, or, more generally, as a property of transitions between epistemic states. The main desideratum for such theories is to say in a precise way how an epistemic state ought to change under the impact of some new information—that is, to provide a characterization of rational or justified belief change. From a structural point of view, this is just the problem of theory change in science: How ought a theory to be changed in light of new, perhaps connecting, information? (See Gärdenfors 1988 and Hansson 1999 for introductions, surveys, and references to belief dynamics research.)

From a logical point of view, a main task in belief dynamics research is to specify a model of the revision process. In general terms, a model specifies four things: a set of epistemic states, a language L suitable for expressing epistemic commitments, a canonical relation of epistemic commitment (relating states to formulas of L), a language L_0 (possibly different from L) of possible epistemic inputs (these represent the impetus for change), and a revision function over the set of states. Given a set of beliefs, say, the deductive closure of $\{p, p \to q\}$, logic alone will not in general tell one how to revise it. If one in such a state learns, to one's surprise, that not-q, then one has to either stop believing p or stop believing $p \to q$. The task, then, is to attempt to codify in very general terms what one should do in all such cases—what the rational mandates are on the transitions between possible epistemic states.

The most well known model of belief revision is the AGM model, named after its trio of developers Alchourrón, Gärdenfors, and Makinson (1985). The AGM model takes epistemic states to be deductively closed subsets of a language of classical propositional logic. They then place constraints on rational revision functions by listing postulates that such functions ought to obey. The postulates are largely driven by codifying the intuition that rational changes of belief should be minimal changes of belief. It turns out that the functions that meet the AGM postulates are exactly those that can be described in the following way. Fix a space W of possible worlds. Think of an epistemic state as a subset of that space—intuitively, just those worlds consistent with what an agent believes. Suppose that the agent can assign a system of spheres around such a subset. The idea is that given a state X, an agent assigns relative implausibility of the worlds in W; the farther out the sphere containing a world, the more implausible the agent finds that world relative to X. If one has to revise X to reflect the new information that $°$, then one should adopt as candidates for the actual world the set of least implausible $°$-worlds. So one adopts as one's new epistemic state the closest $°$-worlds in terms of the system of spheres centered around X. In order to be prepared to repeat the process, one would now also need to adopt a new system of spheres, one that is centered around the new epistemic state.

Theories of belief revision differ in at least four important ways from theories that lie within the Bayesian tradition (see Bayesianism). First, revision models tend to treat 'belief' as flat footed and full, whereas in the Bayesian tradition, *degrees* of belief are of primary interest. Second, the representation here is entirely "qualitative" in the sense that an epistemic state is represented by a set of possible worlds (or, what amounts to the same thing, a logically closed set of sentences) plus a comparative ranking. Epistemic states in the Bayesian tradition, on the other hand, assume a much richer representation of states, full probability distributions over the language. In this sense the revision models discussed here are more general than their Bayesian counterparts, since they require less structure to be assumed on the agents' states. Third, sets of beliefs are deductively closed in belief revision models, whereas this is true in the Bayesian tradition only for special (and one assumes rare) subsets of beliefs in which all members have probability 1. In this sense, Bayesian models are the more general of the two. And, fourth, in belief revision models, there is no special problem of revising by some fact that an agent previously had ruled out, whereas this is a notoriously difficult problem in the Bayesian tradition.

From a philosophical point of view, there are many unsettled debates surrounding theories of belief revision. One such debate is borrowed and adapted from the literature on justification taken as a property of beliefs: Is rational belief change constrained by coherence principles or foundationalist principles? Foundationalist belief change takes seriously, in one way or another, the foundationalist intuition: Agents hold some beliefs just because they hold others, and this difference makes a difference to the landscape of belief revision. For, if S believes q just because she believes p, then if she is forced to give up her belief in p, she should also give up her belief in q. Coherence theories of belief revision deny that such asymmetrical relations play an interesting role in belief change (they typically try to explain away the problematic examples); instead they insist that the guiding aim in rational belief change is information conservation.

The AGM model is a coherence theory of belief dynamics in this sense. Unlike the case for classical theories of justification, there is at present no clear consensus on just what counts as a foundationalist model (this seems to be a much wider class than that of coherence models), or on the characterization of theories that take the foundationalist intuition seriously, or on whether coherence theories or their foundationalist counterparts are to be preferred (Harman 1984; Hansson 1999; Pollock and Gillies 2000; Gillies 2004). Other open questions in belief dynamics include the status of conditionals and epistemic modalities in belief revision models, how revision relates to questions in qualitative decision theory, the relationship between revising an epistemic state (to reflect an agent learning it was mistaken about some fact) and updating it (to reflect that the world has changed), and generally how the rational dynamics of epistemic states relates to dynamics in other cognitive domains.

ANTHONY S. GILLIES

References

Alchourrón, C. E., P. Gärdenfors, and D. Makinson (1985), "On the Logic of Theory Change: Partial Meet Functions for Contraction and Revision," *Journal of Symbolic Logic* 50: 510–530.

Alston, W. (1986), "Internalism and Externalism in Epistemology," *Philosophical Topics* 14: 179–221.

BonJour, L. (1985), *The Structure of Empirical Knowledge*. Cambridge, MA: Harvard University Press.

Chisholm, R. (1966), *Theory of Knowledge*. Englewood Cliffs, NJ: Prentice-Hall.

Cohen, S. (1984), "Justification and Truth," *Philosophical Studies* 46: 279–296.

——— (1988), "How to Be a Fallibilist," *Philosophical Perspectives* 2: 91–123.

DeRose, K. (1995), "Solving the Skeptical Problem," *Philosophical Review* 104: 1–52.

Gärdenfors, P. (1988). *Knowledge in Flux: Modeling the Dynamics of Epistemic States*. Cambridge, MA: MIT Press.

Gettier, E. (1963), "Is Justified True Belief Knowledge?" *Analysis* 23: 121–123.

Gillies, A. S. (2004), "New Foundations for Epistemic Change," *Synthese* 138: 1–48.

Goldman, A. (1979), "What Is Justified Belief?" in G. Pappas (ed.), *Justification and Knowledge*. Dordrecht, Netherlands: D. Reidel.

——— (1986), *Epistemology and Cognition*. Cambridge, MA: Harvard University Press.

Hansson, S. O. (1999), *A Textbook of Belief Dynamics*. Boston: Kluwer Academic Publishers.

Harman, G. (1984), "Positive versus Negative Undermining in Belief Revision," *Noûs* 18: 39–49.

——— (1986), *Change in View*. Cambridge, MA: MIT Press.
Lehrer, K. (1974), *Knowledge*. Oxford: Oxford University Press.
——— (2000), *Theory of Knowledge*, 2nd ed. Boulder, CO: Westview Press.
Lewis, D. (1996), "Elusive Knowledge," *Australasian Journal of Philosophy* 74: 549–567.
Nozick, R. (1981), *Philosophical Explanations*. Oxford: Oxford University Press.
Plantinga, A. (1993), *Warrant: The Current Debate*. Oxford: Oxford University Press.
Pollock, J. L. (1967), "Criteria and Our Knowledge of the Material World," *Philosophical Review* 76: 28–62.
——— (1974), *Knowledge and Justification*. Princeton, NJ: Princeton University Press.
——— (1979), "A Plethora of Epistemological Theories," in G. Pappas (ed.), *Justification and Knowledge*. Dordrecht, Netherlands: D. Reidel.
Pollock, J. L., and J. Cruz (1999), *Contemporary Theories of Knowledge*, 2nd ed. New York: Rowman and Littlefield.
Pollock, J. L., and A. S. Gillies (2000), "Belief Revision and Epistemology," *Synthese* 122: 69–92.
Russell, B. (1912), *The Problems of Philosophy*. Oxford: Oxford University Press.
Sosa, E. (1980), "The Raft and the Pyramid: Coherence versus Foundations in the Theory of Knowledge" [in French], in T. E. Uehling and H. K. Wettstein (eds.), *Midwest Studies in Philosophy*, vol. 5: *Studies in Epistemology*. Minneapolis: University of Minnesota Press.
Spohn, W. (2001), "A Brief Comparison of Pollock's Defeasible Reasoning and Ranking Functions," *Synthese* 131: 39–56.
Williamson, T. (2000), *Knowledge and Its Limits*. Oxford: Oxford University Press.

ERROR

See **Statistics, Philosophy of**

EVOLUTION

The term "evolution" is used to describe both the fact of common descent—that all organisms living today have descended from common ancestry—and the process of descent, or the ways in which species have diversified over time. The question of whether evolution is fact or theory trades on this dual use of the term; it is both a fact that evolution has gone forward, and there is a theory—evolutionary theory—that describes the process of change over time. Those who raise the fact-or-theory question may have in mind another question, however: Is the evidence sufficient to support the claim of common descent? Over 150 years of natural history, paleontology, biogeography, developmental biology, and molecular genetics have provided ample evidence for evolution in this sense (for a review, see Ridley 1996). Biologists do not argue about the fact of common descent; they do, however, argue about which mechanisms have operated in specific cases and which patterns and processes occur most frequently. Some of these debates will be discussed below. The structure of this article will be as follows. First, there will be a very brief overview of the history of evolutionary theory since Darwin; then a discussion of evolutionary theory today; and finally a review of debates among philosophers of biology, many of which originated in historical debates among biologists about evolutionary theory and its interpretation.

EVOLUTION

History of Evolutionary Theory

The term "evolution," from the Latin *evolutio* (*evolvere*, to unroll or unfold), was first used in the scientific context to describe the process of embryological development in an individual organism. The term was chosen in light of some early embryologists' views that development was simply the unfolding of preexisting parts. Lyell first applied the English word "evolution" as early as 1832 to theories of species change. Only by the late nineteenth century did Herbert Spencer use the term to characterize Darwin's theory of the origin of species (Bowler 1975).

The idea of evolution, that diversity of life may have arisen by a natural process, is much older than the 1830s. The germ of this idea was arguably nascent in Lucretius, Descartes, and Hume. For example, in Hume's *Dialogues Concerning Natural Religion* ([1779] 1947, 185), Cleanthes attributes to Philo the view that

> No form, [of plant or animal] . . . can subsist, unless it possess those powers and organs requisite for its subsistence: some new order or economy must be tried, and so on, without intermission; till at last some order, which can support and maintain itself, is fallen upon.

This characterization of origins of new varieties is arguably a predecessor to the idea of natural selection: Options are tried until something works. The novelty in Darwin's contribution to evolutionary biology was not the suggestion that natural, rather than supernatural, causes explained the diversity and adaptation of species. Rather, it consisted in the discovery of a very simple mechanism by which this process moved forward: natural selection.

Natural selection is the differential success in survival and reproduction of some entity (gene, organism, population) due to differences in adaptation to environmental conditions (see Natural Selection). Darwin ([1859] 1964) introduced this idea in *On the Origin of Species*. There, he summarizes evidence for the following empirical generalizations:

1. There exists variation among members of a species.
2. There are more organisms born than survive and reproduce.
3. Certain traits are correlated with an individual's propensity to survive and reproduce.
4. Many such traits are heritable.

From this set of generalizations, Darwin concludes the following: "Every slight modification, which in any way favored the individuals of any species, by better adapting them, would tend to be preserved." Darwin ([1859] 1964) called this his "hypothesis of natural selection" (61). In Darwin's work, the individual is the primary unit of selection. However, he does appeal to competition among groups in attempting to explain the evolution of altruism in human populations. Darwin argued that this simple process could yield not only novel adaptations within species but, over time, the diversity of species seen today. All species, thus, share common ancestry. The idea of common descent was not original to Darwin, nor was it (at least in the scientific community) altogether controversial when proposed by him. Lamarck, St. Hilaire, and many others, including Darwin's grandfather, Erasmus Darwin, had speculated on what was then called the "transmutation" of species. What was novel to Darwin was the view that natural selection was the main mechanism of descent. In later editions of the *Origin*, however, Darwin placed greater emphasis on other mechanisms, through which changes were somehow induced by "effects of the environment," taken up, and passed on to offspring.

Darwin spent five years traveling and collecting specimens of living and fossil animals and plants in South America as a companion to the captain on the *Beagle*. The *Origin* was an attempt to address the following questions, in part inspired by Darwin's travels: Why do animals and plants vary as they do relative to different geographies and climates? Why do species on islands, such as the Galápagos, seem so closely related to species on the mainland, and yet vary in certain ways? And finally, what explains the fact that fossil species share so many characteristics with living species at the same locations? Darwin deliberated on these questions for several years after returning to London. There, he was surrounded by other scientists (several of whom he put to work in cataloguing and identifying his specimens), such as Lyell, Grant, Owen, and Hooker, who were interested in similar questions about the history of the Earth and the diversity of animal life and its causes. Darwin spent a number of years publishing scientific work as well as popular work on his travels; however, he was reluctant to publish his work on the origin of species. Many of his colleagues were extremely skeptical of evolutionary hypotheses, some of which were published in popular presses and motivated by political, and not always scientific, purposes. However, after reading a manuscript on the topic by Wallace (1858), "On the Tendency of Varieties to Depart Indefinitely from the Original Type," Darwin was prompted to put forward the theory of evolution by natural selection. Wallace had independently arrived at a very similar hypothesis to Darwin's, that diverse types of organisms have greater or lesser success at survival and reproduction and that cumulatively, this leads to change in the constitution of populations.

One of the main difficulties for Darwin's theory is that he did not have an adequate view of heredity. Darwin mistakenly thought that heredity was a process of "blending," thus opening himself to the objection that the effects of selection would be "swamped," or reversed, in every generation (see Population Genetics). The rediscovery of Mendel and subsequent development of genetics supplied a mechanism for inheritance that treated traits as discrete, rather than blending, thus resolving the difficulties for Darwin's theory of inheritance (see Genetics). Further, the consequences of Mendelism and the notion of heritability were given precise quantitative formulations with the development of population and quantitative genetics. Haldane, Fisher, and Wright developed mathematical models to represent evolutionary change as occurring in genotypic frequencies, due to mutation, migration, selection, drift, and assortative mating. The early population geneticists demonstrated how, even with very small selective differences, over time, large evolutionary changes could be effected. The mathematical models of evolutionary genetics provided evolutionary biology with a firm theoretical foundation.

This mathematization of evolution, along with the collective efforts of paleontologists, systematists, and geneticists, was the basis of a new "synthetic" theory of evolution (for an overview, see Mayr and Provine 1980). Beginning in the 1930s and 1940s, biologists developed a consensus on the main mechanisms of evolution. Essentially, the evolutionary synthesis was an agreement that independent evidence from diverse fields showed that the pattern of diversity today and in the fossil record could be explained as a result of gradual change, due primarily to selection in response to environmental changes over the course of geological history. This view has been called "neo-Darwinian," insofar as it can be traced back to Darwin's view that the gradual changes seen accumulating in species, such as in domestic races, would gradually lead to new varieties and, eventually, to the diversity of life seen today.

The evolutionary synthesis was in part a reaction to challenges from two different sectors. First, because the gaps between species seemed so significant, some biologists (notably Goldschmidt 1940) argued that macroevolution, or change among species, was a fundamentally different process than microevolution, or change within species. Goldschmidt claimed that macroevolution required a change in the genetic makeup of organisms that was different in *kind* from those changes resulting from simple mutation, selection, migration, assortative mating, and drift. He wrote:

> Microevolution by means of micromutation leads only to diversification within the species.... [T]he large step from species to species is neither demonstrated nor conceivable on the basis of micromutations. (396–397)

According to Goldschmidt, there is a "bridgeless gap" between species. The geographic races are thus not incipient species, as Darwin had argued. The road to novel species was not via simple mutation and selection, but via what Goldschmidt called "systemic mutations."

In response to Goldschmidt and other "macromutationists," the authors of the synthesis held that novel species arose by the very same mechanisms observed to cause change within species (Dobzhansky 1937; Mayr 1942; Simpson 1944). As evidence for this perspective, Dobzhansky demonstrated that genetic variation among species is not different in kind than variation within species. The very same mutations, ranging from base-pair changes to transversions and translocations, may be observed both within and among species lineages. Further, Mayr (1964) documented in great detail how species varied in ways that correlated with features of their environment, such as soil type, vegetation, microclimate, and topography, and how these variations gave rise to geographic races. These races may sometimes (though not always) give rise to incipient species, by the very same mechanism (selection) that yields changes within species. Geographic isolation of a group of individuals by a feature of the environment (e.g., mountain range, river) or a catastrophic event (e.g., a flood) will change a population over time in such a way that it becomes reproductively isolated, or unable to interbreed with its parent population. Mayr's preferred mechanism of speciation is called allopatry.

The founders of the evolutionary synthesis were also reacting to the development of molecular biology (see Dietrich 1998; Beatty 1990). Many biologists, particularly Mayr and Dobzhansky, felt that the growth of molecular biology challenged the work of "classical" evolutionists. In the late 1950s, some molecular biologists argued that the major questions of phylogeny and taxonomy could be solved by molecular methods; "chemical phylogenetics" and "protein taxonomy" would replace classical methods of natural history, biogeography, and classical systematics. What some have called the dogmatism of the synthesis was thus arguably prompted in part by a battle for institutional power and resources between the "young Turks" (molecular biologists such as Watson, Zuckerkandl, Pauling, Margoliash, and Fitch) and the old guard (Mayr and Dobzhansky). Mayr argued that the new molecular methods, while useful, could not supersede or serve to answer the same questions about the diversification of species that natural history could. In order to understand the pattern and process of speciation, one needed to investigate not only species' biochemical or genetic makeup, but also how climate and biogeography could yield different patterns of species diversity.

Critics of the synthesis argued that proponents were too narrow in their views about the pattern and process of evolutionary change (Gould 1983). According to Gould (2002), the founders of the synthesis emphasized gradual change, as opposed to punctuation followed by stasis, and ignored theoretical and empirical contributions from embryologists and developmental biologists. Punctuated equilibrium is the view that there are long periods of very little change in the fossil record, followed by rapid transformation of species (Gould and Eldredge 1977). However, some have argued that Gould's criticism rests on a false account of the history of the synthesis (see Charlesworth, Lande, and Slatkin 1982). Nonetheless, there have been perceived challenges to the "synthetic" view of evolution from molecular biology, paleontology, systematics, and developmental biology in the past 50 years. For instance, Kimura (1968) and King and Jukes (1969) argued that many changes at the molecular level are neutral with respect to selection. Kimura's claim was not just that most changes are selectively neutral, but that most evolutionary changes are due to the random fixation of neutral (or nearly neutral) alleles. At first, this was perceived as a challenge to what many felt was a consensus that selection was the main force driving evolutionary change. However, many of these apparent challenges have been integrated into mainstream evolutionary theory.

Evolutionary Theory Today

There has effectively been a new synthesis between classical evolutionary biology and molecular evolution. Molecular biologists, systematists, paleontologists, and developmental biologists work together to understand the pattern and process of evolutionary change at the molecular level and above. Since the 1960s, new technologies have enabled evolutionary biologists to study evolution at the molecular level. Molecular evolutionary biologists may observe and quantify the amount of genetic divergence within and among species (Nei and Kumar 2000). Molecular methods are used by biologists of every stripe, from ecologists to developmental biologists. Today, systematists and paleontologists routinely sequence samples of genetic materials from their specimens to determine rates of change and time since the most recent common ancestor. (However, most think that such inferences cannot be based purely on sequence data.)

The synthetic theory has been moderated by new evidence and research. For example, it is no longer universally agreed that speciation is primarily a gradual process, that the origin of species requires changes in many genes, or that speciation in sympatry, or within the range of the ancestral population, is an extremely infrequent occurrence. There are examples of species differences due to as little as two genes (e.g., mimicry in butterflies [Sheppard et al. 1985]). Modes of speciation other than allopatry have been recognized and are gaining acceptance—notably, speciation in sympatry (Coyne and Orr 2004). Some argue against what they claim is one of the central theses of the synthesis, that rates of macroevolutionary change are uniform rather than punctuated by stasis followed by rapid bursts of change (Gould and Eldredge 1977).

Finally, the investigation of evolution has come to employ not only the tools of genetics, ecology, systematics, and paleontology, but also developmental biology. Evolutionary developmental biology, or the study of the evolution of developmental systems, has become an active area of investigation (see Carroll, Grenier, and Weatherbee 2001). Goldschmidt's challenge has thus been addressed insofar as biologists are now more attentive to the effects of major developmental changes, such as heterochrony (changes in rates of development), as playing a role in the generation of novel species.

Philosophical Issues

What is the structure of evolutionary theory? Theoretical population genetics uses the laws of probability and idealized mathematical models to generalize about evolving populations. Theoretical population genetics has thus been described as the "dynamics" of the theory (Sober 1984), akin to Newton's dynamical theory of physics. In this view, the mathematical models of theoretical population genetics describe the main "forces" effecting change in the genetic constitution of populations over time, or how migration, selection, drift, assortative mating, and mutation effect the genetic constitution of populations from one generation to the next. So, the mathematical theory of evolution is constituted by a family of models describing how different types of causal factors effect evolutionary change.

What does it mean to speak of selection as a force, a mechanism, or a cause? It is clear that selection, for instance, is not a cause of genetic changes in populations in the way that a baseball striking a window causes it to shatter. It is a cause of change at the level of populations rather than individuals. But what does it mean to say that there are "population-level causes"? What is the relationship between causal processes at the individual and at the population level? These sorts of concerns about the interpretation of causal terms and concepts in biology have generated some controversy in the philosophical literature. Some have argued that given the statistical character of evolutionary theory, the language of "forces" is inappropriate.

However, deployment of the force metaphor is not inconsistent with the recognition that evolution is a statistical theory. On behalf of Sober's "forces" view, Wright's classic paper ([1931] 1986) is a reminder of how it is a useful shorthand to speak of evolutionary theory as a theory of forces. Wright recognized that selection and drift were statistical processes. Yet, he argued that they act "deterministically," in the sense that they both deterministically decrease genetic heterogeneity. Yet, they are also "statistical" or "indeterministic" processes, insofar as possession of this or that selectively advantageous trait does not guarantee survival or reproduction, but selection coefficients describe an "average" survivorship or fecundity relative to one's cohort. Nonetheless, Wright describes selection, drift, and linkage as forces that decrease genetic variability in a population, and he describes recombination, migration, and large population size as forces that tend to increase genetic variability. This is not to say that these are 'forces' in the Newtonian sense, but rather that on average, selection systematically increases homozygosity, as does decreased population size. Smaller populations are, on average, less genetically diverse than larger populations. Wright took it to be the case that there needs to be an appropriate "balance" of forces leading to both genetic homogeneity and heterogeneity in order for populations to be "plastic" enough to evolve: "A balance between factors of homogeneity and heterogeneity may provide a more favorable condition for evolution than either factor by itself" (Wright [1931] 1986, 146). Of course, Wright was well aware that, as noted earlier, large population size is not a 'force' in the Newtonian sense; rather, it is a condition that makes it the case that chance events are less likely to change the genetic constitution of a population. And selection is not, strictly speaking, a force that *directs* evolution, but a consequence of the differential survival of individual organisms due to their differences in adaptation relative to their local environments. However, Wright usefully deployed the analogy with Newtonian physics to illuminate how adaptive evolution required a combination of factors or forces operating in combination. Specifically, Wright believed that a combination of isolation, drift, and intra- and interdemic selection was an optimal balance of forces for generating adaptation.

This appeal to the notion of 'force' did not prohibit Wright's simultaneous conviction that evolution was a statistical process. He was particularly sensitive to the fact that given populations of interbreeding organisms are finite and that chance, or sampling error, must play a key role in changes in gene frequency in a population from one generation to the next. In populations of small size, drift (or sampling error) will govern changes in genetic constitution to a greater extent than will selection. (More precisely, the strength of the selection pressure determines how small the population would need to be for drift to be the primary factor.) In larger populations and over the long term, even a very small difference in fitness between organisms possessing genotype x and genotype y may yield dramatic changes in the constitution of the population. These generalizations have something like the structure of a law. However, one might argue that this particular generalization is not a law of nature, but at bottom a fact about probability, that is, it is reducible to something like the claim that when one flips a coin biased toward heads ten times, one is not as likely to be able to determine that it is biased as when one flips the coin a hundred times. Any finite population will be subject to drift; or chance or "sampling" error play an inevitable role in the change in the genetic constitution of populations. This is what some people mean when they say that evolution is subject to "chance," or that evolution is irreducibly "probabilistic" in character. It is an interesting philosophical question whether and in what sense drift is a cause of evolutionary change. Yet, to concede this is not to deny that Wright's talk of forces is an effective, albeit metaphorical, way of describing the balance of factors that contribute to genetic variability in populations. Wright's work deserves praise as an example of how one can usefully combine multiple metaphorical ways of describing and explaining evolution.

Finally, to return to the question that opened this essay: What is the relationship between evolutionary theory and the evidence in support of it? The support for common descent is indirect and depends on several lines of evidence. Nevertheless, the fact of common descent is not a matter of controversy, nor is the thesis that natural selection has been a major factor at work in descent with modification. Ultimately, the most well supported theory is the one that, all evidence taken into consideration, best explains the phenomena than any alternative. It is uncontroversial (at least in the scientific community) that this is true for evolution. The evolution of life is the most likely explanation of a wide range of data—not simply the diversity and adaptation of life, but also the uniformity of life from the molecular level on up. Consider the following. Heredity is controlled by DNA and RNA in organisms as diverse as viruses and humans. The genetic code, or the codons that determine amino acids that make up proteins, which themselves control all the functions in living cells, is uniform across species (with very few exceptions). The same genes (cytochrome C, hemoglobin) can be found across species, and the relatedness among species (time since most recent common ancestor) is correlated with the number of substitutions of nucleotides in these sequences. The embryonic stages of development of diverse vertebrates such as chickens, dogs, and chimpanzees parallel almost exactly. All of these observations are not only well explained by the hypothesis of evolution, but evolutionary theory also entails precise predictions—for instance, about rates of change in various sequences—that are borne out by the evidence. Sequences of DNA and RNA with important functions are strongly conserved, whereas nonfunctional portions of the genome have a relatively quick rate of turnover.

Moreover, many lineages, and natural populations, have been so well studied over so many generations that biologists have been able to describe the ecological conditions affecting change in populations, the rates of change, and the relative significance of selection and drift. An example is Darwin's finches (Geospizinae) on the Galápagos Islands. Gibbs and Grant (1987), for instance, have demonstrated that large adult size is favored under drought conditions and that smaller sizes tend to increase when there is an increase in rainfall. Thus, there should be no question whether Darwin's theory has been borne out by the evidence to date.

The pattern and process of evolution, and how to study it, is more than a useful case study for classic questions in the philosophy of science about theory structure, confirmation, and explanation. Evolution is also the explanation for how *Homo sapiens* came to exist, and so is potentially relevant to questions in ethics, moral psychology, philosophy of mind, and epistemology (see Evolutionary Epistemology; Evolutionary Psychology). In addition, questions that arise within the science of evolutionary biology itself are not simply empirical, but also conceptual, and so repay philosophical examination. For instance, debates in evolutionary biology about the possibility of the evolution of altruism have moved forward in part via critical philosophical examination of the models in question (Sober and Wilson 2001) (see Altruism). In addition, the question of how to define species, the problem of reconstructing the history of species lineages, determining the relative significance of drift and selection in evolving lineages, examining the relationship between micro- and macroevolution, as well as the relationship between molecular and evolutionary biology, are all active areas of investigation in biology that may be usefully served by philosophical inquiry (see Molecular Biology; Species). More generally, evolutionary biology and its sister disciplines of genetics and ecology are important case studies for broader questions in the philosophy and history of science about modeling and idealization, determinism, reduction, explanation, theoretical unification, instrumentalism and realism, and scientific progress.

ANYA PLUTYNSKI

The author acknowledges the helpful input of Gary Hatfield.

References

Beatty, John (1990), "Evolutionary Anti-Reductionism: Historical Reflections," *Biology and Philosophy* 5: 199–210.
Bowler, Peter J. (1975), "The Changing Meaning of 'Evolution,'" *Journal of the History of Ideas* 36: 95–114.
Carroll, Sean B., Jennifer K. Grenier, and Scott D. Weatherbee (2001), *From DNA to Diversity: Molecular Genetics and the Evolution of Animal Design*. Cambridge, MA: Blackwell Science.
Charlesworth, Brian, Russell Lande, and Montgomery Slatkin (1982), "A Neo-Darwinian Commentary on Macroevolution," *Evolution* 36: 474–498.
Coyne, Jerry, and H. Allen Orr (2004), *Speciation*. Sinauer Associates Publishing.
Darwin, C. ([1859] 1964), *On the Origin of Species*. Edited by Ernst Mayr. Cambridge, MA: Harvard University Press.
Dietrich, Michael (1998), "Paradox and Persuasion: Negotiating the Place of Molecular Evolution within Evolutionary Biology," *Journal of the History of Biology* 31, 85–111.

Dobzhansky, T. (1937), *Genetics and the Origin of Species*. New York: Columbia University Press.

Gibbs, Leslie, and Peter Grant (1987), "Oscillating Selection on Darwin's Finches," *Nature* 327: 511–513.

Goldschmidt, Richard (1940), *The Material Basis of Evolution*. New Haven, CT: Yale University Press.

Gould, Stephen Jay (1983), "The Hardening of the Modern Synthesis," in M. Grene (ed.), *Dimensions of Darwinism*. Cambridge: Cambridge University Press, 71–93.

——— (2002), *The Structure of Evolutionary Theory*. Cambridge, MA: Harvard University Press.

Gould, Stephen J., and Niles Eldredge (1977), "Punctuated Equilibria and the Tempo and Mode of Evolution Reconsidered," *Paleobiology* 3: 115–151.

Hume, David ([1779] 1947), *Dialogues Concerning Natural Religion*. Edited by Norman Kemp Smith. London and New York: Thomas Nelson and Sons.

Kimura, Motoo (1968), "Evolutionary Rate at the Molecular Level," *Nature* 217: 624–626.

King, Jack, and Thomas Jukes (1969), "Non-Darwinian Evolution," *Science* 164: 788–798.

Mayr, E. (1942), *Systematics and the Origin of Species*. New York: Columbia University Press.

——— (1964), *Animal Species and Evolution*. Cambridge, MA: Harvard University Press.

Mayr, Ernest, and William Provine (1980), *The Evolutionary Synthesis: Perspectives on the Unification of Biology*. Cambridge, MA: Harvard University Press.

Nei, Masatoshi, and Sudhir Kumar (2000), *Molecular Evolution and Phylogenetics*. Oxford: Oxford University Press.

Oyama, Griffiths, and R. D. Gray (2001), *Cycles of Contingency: Developmental Systems and Evolution*. Cambridge, MA: MIT Press.

Richards, Robert (1994), "Evolution," in E. Fox Keller and E. Lloyd (eds.), *Key Words in Evolutionary Biology*. Cambridge, MA: Harvard University Press.

Ridley, Mark (1996), *Evolution*. Cambridge, MA: Blackwell Science.

Sheppard, Paul, J. R. Turner, K. S. Brown, W. W. Benson, and M. C. Singer (1985), "Genetics and the Evolution of Muellerian Mimicry in Heliconius Butterflies," *Philosophical Transactions of the Royal Society of London, B: Biological Sciences* 308: 433–610.

Simpson, G. C. (1944), *Tempo and Mode in Evolution*. New York: Columbia University Press.

Sober, Elliott (1984), *The Nature of Selection: Evolutionary Theory in Philosophical Focus*. Cambridge, MA: MIT Press.

Sober, E., and D. S. Wilson (1998), *Unto Others: The Evolution and Psychology of Unselfish Behavior*. Cambridge, MA: Harvard University Press.

Wallace, A. R. (1858), "On the Tendency of Varieties to Depart Indefinitely from the Original Type," *Journal of the Proceedings of the Linnaean Society, Zoology* 3: 53–62.

Wright, Sewall ([1931] 1986), "Evolution in Mendelian Populations," in W. B. Provine (ed.), *Evolution: Selected Papers by Sewall Wright*. Chicago: University of Chicago Press. Originally published in *Genetics* 16: 97–159.

See also **Adaptation and Adaptionism; Biological Information; Ecology; Evolutionary Psychology; Fitness; Natural Selection; Population Genetics; Species**

EVOLUTIONARY EPISTEMOLOGY

Evolutionary epistemology refers to a variety of approaches to the theory of knowledge that emphasize the evolutionary dynamic of knowledge acquisition and evaluation. It therefore has its roots in Charles Darwin's *On the Origin of Species* and *The Descent of Man* and Herbert Spencer's independently developed evolutionary theory of just about everything. The motivating idea is that the capacity to know and knowledge itself, human or otherwise, is a product of evolutionary forces. Any attempts to analyze or understand the nature of knowledge must take this fact into account.

Almost immediately upon the publication of Darwin's work, others began to extend the Darwinian insights to the problem of knowledge. Chief among these were the American pragmatists Charles Peirce, William James, Chauncey Wright, and John Dewey, the psychologist James Mark Baldwin, and many others, including Friedrich Nietzsche and the author Samuel Butler. Much of the work on evolutionary epistemology in the twentieth century derives from the work of Konrad Lorenz, Donald Campbell, Karl Popper, and Jean Piaget (for historical references, see Campbell 1974; Bradie 1986).

Evolutionary insights, models, and metaphors have been brought to bear on a bewildering array of diverse issues, from the evolution of organisms themselves to the development of scientific knowledge. Some have urged that biological evolution in and of itself is a knowledge process. Others have urged that conceptual change is an evolutionary process that mimics the process of natural selection. The literature is rife with analogies and metaphors. Some are drawn from biology to characterize the growth of knowledge. Some are drawn from models of knowing to inform our understanding of biological evolution. Some see biological natural selection and the evolution of scientific understanding as two examples of a single process. Others see selective processes everywhere and promote a view that has come to be labeled universal Darwinism (Plotkin 1993; Cziko 1995; Blackmore 1999).

Evolutionary epistemologies are often seen to be related to or a variant of so-called naturalized epistemologies.

Two Programs

It is useful to distinguish between two interrelated yet arguably distinct projects. On the one hand, there are projects that aim to explain or understand the development of the physical and psychological mechanisms by means of which animals and humans come to acquire and process information about the world. These have been labeled evolution of epistemic (or epistemological) mechanisms (EEM). On the other hand, there are projects that aim to understand the nature and development of the content, norms, and methods of information systems, knowledge corpuses, and scientific theories or traditions. These have been labeled evolution of epistemic theories (EET) (Bradie 1986). Not everyone accepts this division, but it is useful to keep the two projects distinct. For one thing, they do not stand or fall together. EEM programs involve the application of evolutionary biological methods to the study of the development of brains, sensory organs, nervous systems, motor systems, and the like, which are, as far as is known, the *sine qua non* for sentient and sapient creatures. As such, they share the cachet of the success and general acceptance of a broadly Darwinian view of the evolution of characters and traits. EET programs, on the other hand, trade on analogies or metaphors drawn from evolutionary biology and may well turn out to be false or unfruitful characterizations of the development of knowledge. Even those, like David Hull, who argue that the two processes are exemplars of a single overriding model admit that the details of the mechanisms promoting change in the two cases are not identical (Hull 1988). Therefore, the one could turn out to be essentially right and the other essentially wrong. At the moment, it is clear that EEM programs are probably basically right, though filling in the details is fraught with all the problems and more that plague phylogenetic reconstructions.

Brains and their cultural products, unlike bones, do not fossilize easily. If one includes the problem of reconstructing the phylogeny of the evolution of mental capacities, the difficulties become formidable indeed. The verdict is not yet in with respect to the various attempts to reconstruct the development of human knowledge in terms of evolutionary models (see Evolutionary Psychology).

In addition to the distinction between the EEM and EET programs, there is the distinction between *phylogenetic evolution* and *ontogenetic development*. In order to understand the structure of the human brain, for instance, two separate though related questions can be asked, both of which can be couched as, Why do human beings have the kind of brains they do? Such a question, as Ernest Mayr (1961) pointed out, can be given either proximate or ultimate answers. The ultimate, or phylogenetic answer, will turn on the contingencies of the evolution of the brain in the human lineage. The proximate, or ontogenetic answer, will turn on the details of the interaction between the genetic makeup of particular human beings and the ambient environment in which they develop. Both questions are part of the EEM program. EET questions can be similarly partitioned. One can ask, for example, about the development of human understanding of the nature of motion from Aristotle through Descartes, Newton, Einstein, to the present. This question, in effect, is asking about the phylogeny of a particular strand of human understanding about the nature of the universe. On the other hand, one may inquire into the development of a given individual's knowledge and understanding of the nature of motion as he or she develops from child to adult. Such questions, in effect, are asking about the ontogeny of a particular strand of human understanding in particular individuals. With these distinctions in hand, it is time to turn to an examination of some representative examples of each.

EEM Phylogenetic Projects

These include attempts to reconstruct the emergence of the biological substrate that serves as the basis for sentience, cognition, and knowledge. A philosophical example of this is Popper's notorious view of three world stages (Popper 1972; Popper and Eccles 1977). Popper correctly pointed out that according to the best modern theories, the early universe was composed of matter, energy, and radiation. Before life could emerge, suitable planets had to be formed. When life did finally emerge on the planet Earth, natural selection kicked into gear, and lineages began to proliferate and diversify. At some point, organismic brains evolved the capacity for sentience. At some later point, consciousness emerged. At this point, Popper claims, there were two worlds. World 1 was purely physical. World 2 was the world of consciousness. How did consciousness evolve? No one knows, but Popper's evolutionary hypothesis is that it emerged as an adaptation that conferred selective advantages on those that possessed it. Another long period ensued until the emergence of sapient creatures capable of knowing. At this point, a World 3, the world of objective knowledge, came into being (Popper 1972; Popper and Eccles 1977). The rationale remained the same: Creatures who could command objective knowledge had a selective advantage over those who could not. The metaphysics of this view are extravagant but the sentiment is clear. Popper went on to argue that the evolutionary process in all three independent but mutually interacting worlds was the same. In the view argued for here, the evolution of life and minded creatures is a result of Darwinian evolution broadly construed. These are EEM processes. Once humans evolved to the point where they could codify and develop their knowledge of the physical world, the evolution of that understanding was no longer a matter of biological selection but involved other mechanisms that may or may not have had relevant structural similarities to evolution by natural selection. This is EET territory. A less extravagant picture along similar lines is drawn by Daniel Dennett (1995).

EEM Ontogenetic Projects

EEM ontogenetic projects are concerned with the development, in the individual organism, of the physical structures that support cognitive and epistemic activities. For much of the twentieth century, developmental biology took a back seat to evolutionary biology. The developmental pathways were too complicated to sustain fruitful investigation. In any case, the processes underlying development seemed quite different from those underlying phylogenetic evolution. This thinking has changed in recent years with the emergence of neural Darwinism as developed by G. M. Edelman, Jean-Pierre Changeaux, and their colleagues (Changeaux 1985; Edelman 1987). The basic idea involves modeling ontogenetic development using the variational and population models characteristic of evolutionary biology. In particular, the neuronal structure of the brain is no longer construed as a lockstep unfolding of instructions hardwired in the genes. Rather, a variety of neural pathways are constructed and then some are selected and others atrophy. The result is that the brain structures of different individuals, even those with identical genetic endowments, turn out to be unique.

The signature of research on EEM projects is a focus on the development of the physical and psychological mechanisms that enable organisms to gain information and knowledge about the world they live in. What they do with this information, at least in the case of human beings, is to construct bodies of knowledge that are then acquired by individuals as they mature, are communicated among individuals, and are transmitted from one generation to another. There have been a number of proposals on how to construct evolutionary models of what may be called the dynamics and kinematics of conceptual change. This leads to the arena of EET. These projects can also be divided into phylogenetic investigations into the transmission of information from one generation to another and ontogenetic investigations into the means by which individuals come to acquire and process information.

EET Phylogenetic Projects

The phylogenetic models of the growth of knowledge have tended to focus on the growth of scientific knowledge. More recently, "universal Darwinism" and the so-called science of memetics have been postulated as models of conceptual development in general (Cziko 1995). Evolutionary models of scientific change run up against a formidable objection at the onset. The growth of scientific knowledge appears to be progressive, directed, and converging on truth. Biological evolution appears to be nonprogressive, nondirected, and focused on survival and reproductive fitness. Some evolutionary models of science, notably Thomas Kuhn's and Stephen Toulmin's, bite the bullet and opt for a nonconvergent theory of the growth of science (Kuhn 1973; Toulmin 1972). Other models, notably those proposed by Popper, Campbell, and David Hull, seek to finesse these worries in various ways.

Kuhn's model portrays science as a series of periods of "normal science" punctuated by periods of scientific "revolutions." In the revolutionary stages, many of the specific methods, theories, and norms associated with the previous stage of normal science are called into question. What happens during these revolutions is something Kuhn likened to a "gestalt shift," as sometimes radically new perspectives are tried out and adopted. Near the end of *The Structure of Scientific Revolutions*, Kuhn notes the similarity to the competition among varieties that characterizes the selective processes of biological evolution (Kuhn 1973, 172f). The winning scientific perspectives are analogous to the survivors of selection. He draws the obvious conclusion: Just as biological evolution is not seen as progressing toward some global goal, so perhaps the view of scientific progress as a series of stages leading to a "permanent fixed scientific truth" should be reexamined. This was basically a throwaway line at the end of his book, but it raised a firestorm of criticism from those who saw Kuhn as a defender of an invidious relativism. Kuhn complained that he had been misunderstood. Despite the turmoil created by revolutionary stages that threatened to upend the standards and practices of normal-science traditions, there still were general scientific values such as predictive accuracy and problem-solving capacity that served as transcendent guidelines for evaluating research programs separated by a revolutionary chasm. Kuhn denied he was a relativist but did not renege on his rejection of a global sense of progress for science.

Toulmin's evolutionary model of science also rejects the unidirectionality of scientific change and the notion of global progress. Toulmin's (1972) book *Human Understanding*, the first of a projected trilogy, lays out an ambitious project for interpreting the history of ideas in terms of a form of epistemological Darwinism. The general Darwinian model of variation within populations as the material on which selection acts is, for Toulmin, just "one illustration of a more general form of historical explanation; and . . . this same pattern is applicable also, on appropriate conditions, to historical entities and populations of other kinds" (Toulmin 1972, 135). Science, in this view, develops in a two-step process, with the same structure as evolution by natural selection. At each stage in the historical development of science, a pool of intellectual variants—theories, laws, techniques/procedures, and norms—exist along with a selection process that determines which variants survive and which die out (Toulmin 1967, 465). The constraints on theory development imposed by nature are only one selective factor among many in the evolution of scientific knowledge. The net result is a picture of the evolution of scientific knowledge that provides no promise of 'progress.' The details concerning the nature of the selective forces and an explanation of how they worked were left for further volumes, which never appeared.

The roots of Popper's version of evolutionary epistemology can be found in his 1935 classic, *The Logic of Scientific Discovery*, which first appeared in English in 1959. There one finds the first glimmerings of what came to be codified as the method of "conjectures and refutations," explicitly couched in Darwinian terms. In laying out the demarcation criterion based on the falsifiability of scientific conjectures, Popper (1961) writes:

> What characterizes the empirical method is its manner of exposing to falsification, in every conceivable way, the system to be tested. Its aim is not to save the lives of untenable systems but, on the contrary, to select the one which is by comparison the fittest, by exposing them all to the fiercest struggle for survival. (42)

Later, he argues that our choice of one theory over another is a reflection of our choice of that "theory which best holds its own in competition with other theories; the one which by natural selection, proves itself the fittest to survive" (Popper 1961, 108). In 1974, he still took Darwinism to be a "metaphysical research program" (Schilpp 1974, 133–43). He later recanted biological Darwinism and cemented his Darwinian approach to epistemology (Popper 1984), arguing that "the evolution of scientific knowledge is, in the main, . . . a Darwinian process. The theories become better adapted through natural selection: they give us better and better information about reality (they get nearer and nearer to the truth)" (Popper 1984, 239). It is not clear whether he was now construing the evolutionary model of scientific change to be itself a testable and potentially falsifiable hypothesis. If so, he would be proposing a 'science' of science in the sense advocated by Hull (1988). But this reading would make Popper more of an epistemological naturalist than the textual evidence warrants.

Campbell (1974) coined the term *evolutionary epistemology* in his influential review of the literature. Campbell developed a model he called blind variation and selective retention (BVSR), which was designed to cover both biological evolution and conceptual change. Popper was very sympathetic to Campbell's view and held it to be in almost complete agreement with his own. Both views intertwine elements of the evolution of cognitive capacities (an EEM project) and the evolution of science (an EET project). The heart of Campbell's view, developed in a series of papers that began before he was aware of Popper's views, is the construal of the evolution of organisms as the result of a nested hierarchy of levels of biological and conceptual development. The key to this process is the subsumption of organismic evolution and conceptual change under the rubric of *problem solving*. So, the earliest forms of life and most basic organisms first must develop techniques for finding nourishment and sustenance. The organisms move about randomly in search of food. At the next stage, various "vicarious" sensory modalities evolve that allow for exploration of the environment without the organisms having to move into potential danger. The more advanced modalities, overlapping to an extent, include the development of habits, instincts, visually supported thought, mnemonically supported thought, "socially vicarious exploration" (including the development of observational learning and imitation), and the development of language (Munz 1993). This in turn allows for the emergence and accumulation of culture, of which the development of science is one aspect. There are several aspects of Campbell's picture that have appeared problematic to his critics. For one thing, empirical confirmation of his nested hierarchy view is yet forthcoming, as Campbell himself admitted. For another, Campbell insists that conjectures or tentative solutions to the problems faced by organisms be "blind." Some have seen this as inconsistent with the apparent intentionality of scientific research.

Hull (1988) takes a Darwinian approach to scientific change very seriously. He proposes, in effect, to develop an empirical hypothesis about the development of knowledge along selectionist lines. Rather than interpreting scientific change as *merely* analogous to biological evolution, he argues that both biological evolution and conceptual development are examples of a common selectionist structure. For his analysis, Hull borrowed a useful distinction, first introduced by Richard Dawkins (1976), between "replicators" and "vehicles." Replicators are what get handed down from one generation to the next, and vehicles are what serve as the packages containing the replicators in any given generation. Hull replaced the term "vehicle" with "interactor" to emphasize the fact that the selection forces that pick out the fittest variants work in virtue of the interaction between vehicles and their environments. For Hull, the interactors in science are the scientists themselves, who compete with one another for success. Success is measured by inclusive conceptual fitness or the measure of how widespread one's views become. What get replicated are their ideas, theories, conjectures, and methods. In contemporary science, most workers are members of a research group, or *deme*, and these groups compete with one another as well.

Hull's model, unlike Popper's, is clearly constructed along naturalistic lines. In addition, Popper's three-world view and his rejection of the *justified-true-belief* picture of knowledge leads him, unlike Hull, to downplay the role of scientific agents (the scientists) in the growth of objective knowledge. The process of conjectures and refutations that Popper sees as the core of the scientific method involves competition among *hypotheses* or *conjectures*, not among scientists. Another virtue, then, of Hull's approach over Popper's is the emphasis Hull places on the social dimension of science.

The Darwinian models of science proposed by Campbell, Popper, Toulmin, and Hull all share a commitment to a selectionist model of scientific theory change. Not all those who invoke Darwin, however, see a corresponding commitment to such a model. Nicholas Rescher argues for what he calls "methodological Darwinism," as opposed to "thesis Darwinism" (Rescher 1977 and 1990). In his view, it is methods that compete with one another for acceptance, not theses or theories. When Michael Ruse, an early skeptic about the virtues of evolutionary epistemology, changed his mind, he too rejected the idea that embracing epistemological Darwinism entails embracing a natural selection model of theory change (Ruse 1996).

EET Ontogenetic Projects

Evolutionary models of the ontogenesis of knowledge in individual organisms have also been constructed. Briefly, as Campbell (1974) has noted, these views have their roots in nineteenth-century philosophers and psychologists. B. F. Skinner's theory of operant conditioning has obvious affinities to the theory of natural selection, as he himself has noted (Skinner 1981).

Jean Piaget's extensive writings on "genetic epistemology" develop themes with clear connections to the concerns and interests of evolutionary epistemologists. His 1971 book *Biology and Knowledge: An Essay on the Relations Between Organic Regulations and Cognitive Processes* provides a useful introduction to his ideas. As the title suggests, Piaget argues that cognitive processes are rooted in and are extensions of the fundamental organic "autoregulatory" feedback processes that are the basis of organismic existence. Living organisms, he argues, are basically interactive systems that adapt to their environments by means of the "assimilation" and "accommodation" of new elements into their structural organization. These autoregulatory systems operate at all organic and psychological levels—evolutionary, ontogenetic, physiological, cognitive, and psychological. Piaget suggests that the central problem about knowledge is the relationship between subjects and objects. This relationship, in his view, corresponds directly with the relationship between organism and environment (Piaget 1971, 99). The relationship of organism to environment is an interactive one. Transposed to the realm of knowledge, this undercuts, in Piaget's view, any "copy theory" of knowledge, whereby whole bodies of knowledge are acquired, communicated, and transmitted intergenerationally (see "EEM Ontogenetic Projects" above). Knowledge, for Piaget, is active and regulatory. He thus urges the development of appropriate cybernetic models that incorporate both Darwinian and Lamarckian elements. Piaget's work in evolutionary epistemology has not received the attention it deserves.

More recently, Henry Plotkin, Susan Blackmore, and others have begun to argue for a "science of memetics," based on the idea that conceptual change can be modeled as the differential replication of cultural units, or "memes" (Plotkin 1993; Blackmore 1999). Blackmore, in particular, argues that memetic evolution has become decoupled from genetic evolution. Memes were created to enhance the fitness of these vehicles, but once generated they take on a life of their own and become replicators in their own right. Blackmore sees implications for both the evolution of culture and the development of the big brains necessary to create culture in the first place. These programs cut across all the distinctions drawn here and have implications for the ontogeny and phylogeny of both epistemic mechanisms and the conceptual systems that they produce. It is too early to pass judgment on this project, but should it prove fruitful, it has implications for a general analysis of the evolution of culture.

Evolutionary Epistemology and the Tradition

The evolutionary approach to epistemology is most closely allied with naturalistic approaches to epistemology. The focus is on the biological conditions of knowing and the dynamics of conceptual change. Therefore, it has seemed to some critics to be "epistemology" in name only. Critics have charged that evolutionary epistemology fails to address the traditional normative issues, such as the nature of justification and the reliability of evidence. Thus, it is beside the point or involves changing the question. There is no doubt that evolutionary approaches to epistemology entail a radical reevaluation of what it means to do "proper epistemology." It is appropriate to note here that John Dewey argued that one of the consequences of taking Darwin seriously would be to restructure the kinds of questions that philosophers ask and the kinds of answers they deem appropriate (Dewey 1910). Not all are prepared to be so accommodating. In Jaegwon Kim's (1988) view, if epistemologists abandon the task of providing justifications, they have abandoned epistemology. Campbell's approach was to argue that evolutionary epistemology was "descriptive" and hence complementary to the traditional normative approach. Others stand ready to abandon the tradition and the search for justifications altogether (Radnitsky 1987). Hull concurs in part, although he allows a role for contextually articulated epistemic norms that arise from the practice of science itself. If, on the other hand, norms are construed as instrumental procedural and methodological rules, then a selectionist account can be given of them. In the marketplace of ideas, those rules that promote the development of successful strategies for coping with the environment, including the development of successful scientific theories and inferential practices, will be at a selective advantage over those that do not. The norms that emerge are justified by the fact that their deployment does lead to successful practices. From a naturalistic and evolutionary standpoint, one can ask for nothing more.

Students of the human epistemic condition stand at a fork in the road. In one direction lies the tradition that denies the relevance to "real" epistemology of any or most of the considerations discussed above. In the other lie research projects that seek to integrate the latest work on evolutionary biology, psychology, and computer modeling into a philosophically sophisticated understanding of the nature of knowledge, how it is acquired, and how it is transmitted. (Those wishing to pursue these latter issues should see the extensive bibliography in Cziko and Campbell 1997.)

MICHAEL BRADIE

References

Barkow, Jerome H., Leda Cosmides, and John Tooby (eds.) (1992), *The Adapted Mind: Evolutionary Psychology and the Generation of Culture*. New York: Oxford University Press.

Blackmore, Susan (1999), *The Meme Machine*. Oxford: Oxford University Press.

Bradie, Michael (1986), "Assessing Evolutionary Epistemology," *Biology and Philosophy* 4: 401–459.

——— (1994), "Epistemology from an Evolutionary Point of View," in Elliott Sober (ed.), *Conceptual Issues in Evolutionary Biology*. Cambridge, MA: MIT Press, 453–475.

Campbell, D. T. (1974), "Evolutionary Epistemology," in P. A. Schilpp (ed.), *The Philosophy of Karl Popper*. LaSalle, IL: Open Court.

Changeux, Jean-Pierre (1985), *Neuronal Man: The Biology of Mind*. New York: Pantheon Books.

Cziko, Gary (1995), *Without Miracles: Universal Selection Theory and the Second Darwinian Revolution*. Cambridge, MA: MIT Press.

Cziko, Gary, and D. T. Campbell (1997), *Selection Theory Bibliography*. http://faculty.ed.uiuc.edu/g-cziko/stb/Default.asp?bhcd2=1033275860

Dawkins, Richard (1976), *The Selfish Gene*. Oxford: Oxford University Press.

Dennett, Daniel (1995), *Darwin's Dangerous Idea*. New York: Simon & Schuster.

Dewey, John (1910), *The Influence of Darwin on Philosophy and Other Essays in Contemporary Thought*. New York: Henry Holt & Co.

Edelman, Gerald M. (1987), *Neural Darwinism: The Theory of Neuronal Group Selection*. New York: Basic Books.

Hull, David (1988), *Science As a Process: An Evolutionary Account of the Social and Conceptual Development of Science*. Chicago: University of Chicago Press.

Kim, Jaegwon (1988), "What Is 'Naturalized Epistemology'?" in James E. Tomberlin (ed.), *Philosophical Perspectives 2*. Atascadero, CA: Ridgeview Publishing, 381–405.

Kuhn, Thomas S. (1973), *The Structure of Scientific Revolutions*. Chicago: University of Chicago Press.

Mayr, Ernest (1961), "Cause and Effect in Biology," *Science* 134: 1501–1506.

Munz, Peter (1993), *Philosophical Darwinism: On the Origin of Knowledge by Means of Natural Selection*. London: Routledge.

Piaget, Jean (1971), *Biology and Knowledge: An Essay on the Relations Between Organic Regulations and Cognitive Processes*. Translated by Beatrix Walsh. Chicago: University of Chicago Press.

Plotkin, Henry (1993), *Darwin Machines and the Nature of Knowledge*. Cambridge, MA: Harvard University Press.

Popper, Karl (1961), *The Logic of Scientific Discovery*. New York: Science Editions.

——— (1972), *Objective Knowledge: An Evolutionary Approach*. Oxford: Clarendon Press.

——— (1984), "Evolutionary Epistemology," in J. W. Pollard (ed.), *Evolutionary Theory: Paths into the Future*. London: John Wiley & Sons Ltd.

Popper, Karl, and John Eccles (1977), *The Self and Its Brain*. New York: Springer International.

Radnitzky, Gerard, and W. W. Bartley (1987), *Evolutionary Epistemology, Rationality, and the Sociology of Knowledge*. La Salle, IL: Open Court.

Rescher, Nicholas (1977), *Methodological Pragmatism*. Oxford: Basil Blackwell.

——— (1990), *A Useful Inheritance: Evolutionary Aspects of the Theory of Knowledge*. Savage, MD: Rowman & Littlefield.

Ruse, Michael (1996), *Taking Darwin Seriously*. Oxford: Basil Blackwell.

Skinner, B. F. (1981), "Selection by Consequences," *Science* 213: 501–504.

Toulmin, Stephen (1967), "The Evolutionary Development of Natural Science," *American Scientist* 55: 456–471.

——— (1972), *Human Understanding: The Collective Use and Evolution of Concepts*. Princeton, NJ: Princeton University Press.

See also **Evolutionary Psychology; Naturalism**

EVOLUTIONARY PSYCHOLOGY

The evolutionary study of the mind in the twentieth century has been marked by three self-conscious movements: classical ethology, sociobiology, and Evolutionary Psychology (capitalized to indicate that it functions here as a proper name). Classical ethology was established in the years immediately before the Second World War, primarily by Konrad Lorenz and Niko Tinbergen (Burckhardt 1983). Interrupted by the war, the movement blossomed in the early 1950s, when ethologists established major research institutes in most developed countries and had a major impact on the broader culture through popular science writing. From the outset, ethology sought to apply its methods for the comparative study of animal behavior to human beings, something that was especially prominent in popular works written by ethologists. Lorenz's *On Aggression* (1966a) is perhaps the best known of these, but several other ethologists wrote books advocating the application of the new evolutionary science of the mind to problems of international conflict and social unrest. The ethologist who focused most on human beings in his empirical research was Lorenz's student Irenaus Eibl-Eibesfeldt, who throughout the 1960s and 1970s sought to document innate, universal behavior patterns in *Homo sapiens* through photography and film (Eibl-Eibesfeldt 1989).

Classical ethology was largely displaced in the 1970s by sociobiology, a movement that sought to apply to humans a set of new mathematical techniques for the study of animal behavior (Wilson 1975). During the 1960s behavioral ecologists had come to view animal behaviors primarily as strategies adopted in competitions among and within species. Models of these competitive interactions could be constructed using evolutionary game theory, and the predictions of these models could be tested against actual behavior. Animal behaviors were expected to correspond to "evolutionarily stable strategies," that is, to equilibria in the relevant game-theoretic models. The game-theoretic approach had the advantage that it did not require knowledge of the neural mechanisms underlying behavior (or the genetic mechanisms underlying its transmission). The early ethologists' "hydraulic model" of neural mechanisms had collapsed during the 1950s as it became clear that the prewar neuroscience on which the model was based had not been borne out by further investigation. The hydraulic model also failed to accommodate many of the new behavioral phenomena uncovered as ethology matured (Hinde 1956). No new model of similar generality was available to replace the hydraulic model and its relatives, making a method that dealt directly with behavior highly desirable.

Sociobiologists also argued that their approach was intrinsically more scientific than classical ethology because it made predictions about behavior and tested them, rather than merely describing behavior and explaining it. This led to the hope that evolutionary models could guide psychological research and point it toward important phenomena that would otherwise be misunderstood or overlooked, an idea that remains central to today's Evolutionary Psychology, which includes advocates of sociobiology among its leading figures. Among them is Jerome Barkow (1979), who expressed this viewpoint succinctly in his title "Classical Ethology: Empirical Wealth, Theoretical Dearth." But despite such oppositional rhetoric, there was considerable continuity of practice and personnel between ethology and sociobiology. Richard Dawkins and Desmond Morris, for example, key figures in the popularization of sociobiology, were students of Niko Tinbergen and regarded sociobiology as a continuation of the tradition he had established (see the introduction to Dawkins, Halliday, and Dawkins 1991).

At the end of the 1980s sociobiology itself came under attack from a new movement calling itself Evolutionary Psychology (Barkow, Cosmides, and Tooby 1992; Crawford, Smith, and Krebs 1987). The Evolutionary Psychologists argued that the whole project of explaining contemporary human behaviors as a direct result of adaptive evolution was misguided (Symons 1992). The contemporary environment is so different from that in which human beings evolved that their behavior probably bears no resemblance to the behavior that was important in evolution. This problem had been identified by many of the best-known critics of sociobiology (e.g., Kitcher 1985), but Evolutionary Psychology followed it up with a positive proposal. Evolutionary theory should be used to predict which behaviors *would have been* selected in postulated ancestral environments. Human behavior today can be explained as the output of mechanisms that evolved to produce those ancestral behaviors when these mechanisms operate in their very different, modern environment. Furthermore, the diverse behaviors seen in different cultures may all be manifestations of a single, evolved psychological mechanism operating under a range of local conditions, an idea that originated in an offshoot of sociobiology known as Darwinian anthropology (Alexander 1979). Refocusing research on the Darwinian algorithms that underlie observed behavior, rather than on the behavior itself, lets the Evolutionary Psychologist "see through" the interfering effects of environmental change and cultural difference to an underlying human nature (see Natural Selection).

Adherents of today's Evolutionary Psychology normally present their approach as something very novel, typically describing it as "the new science of the mind" (Cosmides and Tooby 2001). They allege that the social and behavioral sciences have until recently been dominated by the "standard social science model" (SSSM), which denies the existence of any evolved features of the mind. The SSSM grew out of the liberal political agendas of the 1960s, which aimed to change traditional social behavior: "'Not so long ago jealousy was considered a pointless, archaic institution in need of reform. But like other denials of human nature from the 1960s, this bromide has not aged well,'" as Stephen Pinker puts it on the dustjacket of a work of Evolutionary Psychology (Buss 2000). But the claim that Evolutionary Psychology is a rebellion against an antibiological consensus in the social and behavioral sciences is at best a considerable exaggeration. Instead, Evolutionary Psychology represents the latest stage of a tradition of evolutionary psychology dating back at least to Lorenz. Nor is the public prominence of Evolutionary Psychology entirely new. Lorenz was as successful a popular author in the 1950s and 60s as Richard Dawkins was in the 1970s and 80s. Furthermore, in some important respects, Evolutionary Psychology actually represents a return to the positions of classical ethology. Classical ethologists thought that modern human behavior was the (often maladaptive) result of ancient, evolved mechanisms operating in radically new environments. They also shared the "modular" conception of the mind, described below. Most importantly, classical ethology and Evolutionary Psychology offer very similar critiques of conventional psychology. Lorenz's complaints were directed against those he liked to call "American behaviorists." The laboratory-based search for general laws of learning seemed to him as misguided as dropping automobiles from buildings under controlled conditions and writing down the results. Without an evolutionary perspective, he argued, psychology does not know what it is looking for, and when it finds something, it does not know what it is looking at (e.g., Lorenz 1966b, 274). In the same way, advocates of Evolutionary Psychology argue that empirical psychology without an evolutionary perspective has no way to determine whether it is studying meaningful units of behavior or mental functioning:

> Cognitive scientists will make far more rapid progress in mapping this evolved architecture if they begin to seriously incorporate knowledge from evolutionary biology and its related disciplines . . . into their repertoire of theoretical tools, and use theories of adaptive function to guide their empirical investigations. (Tooby and Cosmides 1998, 195)

Evolutionary Psychology and Cognitive Science

The classical ethologists based their ideas about mental mechanisms on the neuroscience of the interwar years. Similarly, Evolutionary Psychology has turned to "classical" cognitive science, with its guiding idea that the mind is computer software implemented in neural hardware (Fodor 1983; Marr 1982). Evolutionary Psychology argues that the representational, information-processing language of classical cognitive science is ideal for describing the evolved features of the mind. Behavioral descriptions of what the mind does are useless because of the problem of changing environments, described above. Neurophysiological descriptions are inappropriate, because behavioral ecology does not predict anything about the specific neural structures that underlie behavior. Models in behavioral ecology predict which behaviors would have been selected in the ancestral environment, but they cannot distinguish between different mechanisms that produce the same behavioral output. Hence, if one accepts the conventional view in cognitive science that indefinitely many different neural mechanisms could potentially support the same behavior, it follows that behavioral ecology predicts nothing about the brain except which information-processing functions it must be able to perform:

When applied to behavior, natural selection theory is more closely allied with the cognitive level of explanation than with any other level of proximate causation. This is because the cognitive level seeks to specify a psychological mechanism's function, and natural selection theory is a theory of function (Cosmides and Tooby 1987, 284). It is thus slightly confusing that Evolutionary Psychologists talk of discovering psychological "mechanisms," a term that suggests theories at the neurological level. What 'mechanism' actually refers to in this context is a performance profile—an account of what output the mind will produce given a certain range of inputs (see Cognitive Science).

The fact that evolutionary reasoning yields expectations about the performance profile of the mind fits neatly with the explanatory framework of classical cognitive science. According to the influential account by David Marr (1982), explanation in cognitive science works at three mutually illuminating levels. The highest level concerns the tasks that the cognitive system accomplishes—for example, recovering the shape and position of objects from stimulation of the retina. The lowest level concerns the neurophysiological mechanisms that accomplish that task—the neurobiology of the visual system. The intermediate level concerns the functional profile of those mechanisms, or as it is more usually described, the computational process that is implemented in the neurophysiology. Hypotheses about the neural realization of the computational level constrain hypotheses about computational processes: Psychologists should propose only computational models that can be realized by neural systems. Conversely, hypotheses about computational processes guide the interpretation of neural structure: Neuroscience should look for structures that can implement the required computations. Similar relations of mutual constraint hold between the level of task description and the level of computational processes. But there remains something of a puzzle as to how the highest level—the task description—is to be specified other than by stipulation. It seems obvious that the task of vision is to represent things around us, but what makes this true? According to Evolutionary Psychology, claims about task descriptions are really claims about evolution. The overall task of the mind is survival and reproduction in the ancestral environment, and the subtasks performed by parts of the mind correspond to separate adaptive challenges posed by the ancestral environment. For example, it would have been useful for the ancestors of humans to be able to see, so it is predictable that humans will have a visual system. This kind of thinking becomes useful when the function of a psychological mechanism is not as blindingly obvious as in the case of vision. What, for example, is the task description for the emotional system, or for individual emotions such as jealousy or grief? Evolutionary Psychology argues that in such cases it should be evolutionary thinking that sets the agenda for cognitive science, telling it what to look for and how to interpret what it finds.

The Massive Modularity Thesis

One of the best-known features of Evolutionary Psychology is the massive modularity thesis, or the "Swiss army knife" model, according to which the mind contains few if any general-purpose cognitive mechanisms. The mind is a collection of separate modules, each designed to solve a specific adaptive problem, such as mate recognition or the enforcement of female sexual fidelity. The flagship example of a mental module is the *language acquisition device*, the mechanism that allows human infants to acquire a language in a way that, it is widely believed, would not be possible using any general-purpose learning rules (Pinker 1994). Other well-known examples include the perceptual input devices for which the modularity concept was originally introduced (Fodor 1983). The massive modularity thesis is an example of the kind of evolutionary guidance for cognitive science described in the last section. Evolutionary Psychology argues that evolution would favor multiple modules over domain-general cognitive mechanisms because each module can be fine-tuned for a specific adaptive problem. So, cognitive scientists should look for domain-specific effects in cognition and conceptualize their work as the search for and characterization of mental modules.

The Monomorphic Mind Thesis

The leading Evolutionary Psychologists John Tooby and Leda Cosmides have argued strongly for the monomorphic mind thesis, or "psychic unity of humankind" (Cosmides, Tooby, and Barkow 1992, 72). This thesis states that any differences that exist in the cognitive adaptations of individual humans or human groups are not due to genetic differences. Psychological differences are always, or almost always, due to environmental factors that trigger different aspects of the same developmental program. If true, this would make cognitive adaptations highly atypical, since most human traits display considerable individual variation related to differences in genotype. All human beings have eyes, but these eyes exhibit differences in color, size, shape, acuity, and susceptibility to various forms of degeneration over time, all due to differences in genotype. It has been known for half a century that wild populations of most species contain substantial genetic variation, and humans are no exception.

Tooby and Cosmides (1990) offer one main argument for the conclusion that the genes involved in producing cognitive adaptations will be the same in all human individuals:

> Complex adaptations necessarily require many genes to regulate their development, and sexual recombination makes it combinatorially improbable that all the necessary genes for a complex adaptation would be together at once in the same individual, if genes coding for complex adaptations varied substantially between individuals. Selection, interacting with sexual recombination, enforces a powerful tendency towards unity in the genetic architecture underlying complex functional design at the population level and usually the species level as well. (393)

The authors apply this argument to only psychological adaptations, but its logic extends to all traits with many genes involved in their etiology. The argument fails because it assumes that development is a mechanical consequence of the exact sequence of genes on each chromosome. What Cosmides and Tooby seem to have overlooked is the phenomenon described by C. H. Waddington in the 1940s as "developmental canalization": Development is buffered against genetic variation, as well as against environmental variation (Waddington 1959). This is why surprisingly many gene knock-out experiments produce negative results. Disabling a gene known to be involved in a developmental pathway frequently produces no effect (a null phenotype), because development contains positive and negative feedback mechanisms that increase transcription of the required gene product from the other allele, initiate transcription from another gene copy, or initiate transcription of a different gene product that can produce the same outcome (Freeman 2000). On a larger scale, it has become a commonplace amongst evolutionary developmental biologists that complex phenotypic features of organisms can be conserved over evolutionary time despite changes in the specific genes used to construct them and even in the general form of the developmental pathway by which they are constructed (Raff 1996; Wagner 1994) (see Developmental Biology).

One reason for the popularity of the doctrine of the monomorphic mind is probably as a bulwark against racism. If all human beings have substantially the same genes, then racial differences are superficial and modifiable. But no such bulwark is necessary. If it is assumed that variation in evolved human phenotypes roughly mirrors the known variation in human genotypes, then it follows that the vast majority of traits are pancultural and that the differences among human groups are dwarfed by the differences among individuals within those groups (Cavalli-Sforza, Menozzi, and Piazza 1994).

Alternatives to Evolutionary Psychology

The Evolutionary Psychology movement has been as controversial as it has been successful. Jerry Fodor, one of the originators of Evolutionary Psychology's preferred framework for cognitive science, rejects the massive modularity thesis and has expressed considerable skepticism about the value of evolutionary thinking as a heuristic for cognitive science (Fodor 2000). Many researchers accept that evolutionary thinking can and should transform psychology and cognitive science but disagree, often radically, with Evolutionary Psychology's specific program for accomplishing this transformation. Several recent collections of papers present the views of such evolutionary psychologists (Heyes and Huber 2000; Holcomb 2001; Scher and Rauscher 2002). Finally, a "developmentalist" tradition in animal behavior research with its roots in classical ethology and comparative psychology has criticized both sociobiology and Evolutionary Psychology for failing to integrate the Darwinian study of behavior with the study of how behavior develops (Gottlieb 1997; Bjorklund and Pellegrini 2002). Accessible introductions to this tradition are provided by Patrick Bateson and Paul Martin (1999) and David Moore (2001).

PAUL E. GRIFFITHS

References

Alexander, R. (1979), *Darwinism and Human Affairs*. Seattle: Washington University Press.
Barkow, J. H. (1979), "Human Ethology: Empirical Wealth, Theoretical Dearth," *Behavioral and Brain Sciences* 2: 27.
Barkow, J. H., L. Cosmides, and J. Tooby (eds.) (1992), *The Adapted Mind: Evolutionary Psychology and the Generation of Culture*. Oxford: Oxford University Press.
Bateson, P. P. G., and P. Martin (1999), *Design for a Life: How Behavior and Personality Develop*. London: Jonathan Cape.
Bjorklund, D. F., and A. D. Pellegrini (2002), *The Origins of Human Nature: Evolutionary Developmental Psychology*. Washington DC: American Psychological Association.
Burckhardt, R. W. (1983), "The Development of an Evolutionary Ethology," in D. S. Bendall (ed.), *Evolution: From Molecules to Men*. Cambridge: Cambridge University Press, 429–444.
Buss, D. M. (2000), *The Dangerous Passion: Why Jealousy Is As Essential As Love and Sex*. New York: Simon and Schuster.
Cavalli-Sforza, L. L., P. Menozzi, and A. Piazza (1994), *The History and Geography of Human Genes*. Princeton, NJ: Princeton University Press.
Cosmides, L., and J. Tooby (1987), "From Evolution to Behaviour: Evolutionary Psychology As the Missing Link," in J. Dupré (ed.), *The Latest on the Best: Essays on Optimality and Evolution*. Cambridge, MA: MIT Press, 277–307.
——— (2001), *What Is Evolutionary Psychology? Explaining the New Science of the Mind*. New Haven, CT: Yale University Press.
Cosmides, L., J. Tooby, and J. H. Barkow (1992), "Introduction: Evolutionary Psychology and Conceptual Integration," in J. H. Barkow, L. Cosmides, and J. Tooby (eds.), *The Adapted Mind: Evolutionary Psychology and the Generation of Culture*. Oxford and New York: Oxford University Press, 3–15.
Crawford, C., M. Smith, and D. Krebs (eds.) (1987), *Sociobiology and Psychology: Ideas, Issues and Applications*. Hillsdale, NJ: Lawrence Erlbaum Associates.
Dawkins, M. S., T. R. Halliday, and R. Dawkins (eds.) (1991), *The Tinbergen Legacy*. London: Chapman and Hall.
Eibl-Eibesfeldt, I. (1989), *Human Ethology*. New York: Aldine de Gruyter.
Fodor, J. A. (1983), *The Modularity of Mind: An Essay in Faculty Psychology*. Cambridge, MA: Bradford Books/MIT Press.
——— (2000), *The Mind Doesn't Work That Way: The Scope and Limits of Computational Psychology*. Cambridge, MA: MIT Press.
Freeman, M. (2000), "Feedback Control of Intercellular Signaling in Development," *Nature* 408: 313–319.
Gottlieb, G. (1997), *Synthesizing Nature-Nurture: Prenatal Roots of Instinctive Behavior*. Hillsdale, NJ: Lawrence Erlbaum Associates.
Heyes, C. M., and L. Huber (2000), *The Evolution of Cognition*. Cambridge, MA: MIT Press.
Hinde, R. A. (1956), "Ethological Models and the Concept of 'Drive,'" *British Journal for the Philosophy of Science* 6: 321–331.
Holcomb, H. H. III. (ed.) (2001), *The Evolution of Minds: Psychological and Philosophical Perspectives*. Dordrecht, Netherlands: Kluwer.
Kitcher, P. (1985), *Vaulting Ambition*. Cambridge, MA: MIT Press.
Lorenz, K. (1966a), *On Aggression*. Translated by M. K. Wilson. New York: Harcourt, Brace and World.
——— (1966b), "Evolution of Ritualisation in the Biological and Cultural Spheres," *Philosophical Transactions of the Royal Society of London* 251: 273–284.
Marr, D. (1982), *Vision*. New York: W. H. Freeman.
Moore, D. S. (2001), *The Dependent Gene: The Fallacy of "Nature versus Nurture"*. New York: W. H. Freeman/Times Books.
Pinker, S. (1994), *The Language Instinct: The New Science of Language and Mind*. New York: William Morrow.
Raff, R. (1996), *The Shape of Life: Genes, Development and the Evolution of Animal Form*. Chicago: University of Chicago Press.
Scher, S., and M. Rauscher (2002) (eds.), *Evolutionary Psychology: Alternative Approaches*. Dordrecht, Netherlands: Kluwer.

Symons, D. (1992), "On the Use and Misuse of Darwinism in the Study of Human Behavior," in J. H. Barkow, L. Cosmides, and J. Tooby (eds.), *The Adapted Mind: Evolutionary Psychology and the Generation of Culture.* Oxford: Oxford University Press, 137–159.

Tooby, J., and Cosmides, L. (1990), "The Past Explains the Present: Emotional Adaptations and the Structure of Ancestral Environments," *Ethology and Sociobiology* 11: 375–424.

——— (1998), "Evolutionizing the Cognitive Sciences: A Reply to Shapiro and Epstein," *Mind and Language* 13: 195–204.

Waddington, C. H. (1959), "Canalisation of Development and the Genetic Assimilation of Acquired Characters," *Nature* 183: 1654–1655.

Wagner, G. P. (1994), "Homology and the Mechanisms of Development," in B. K. Hall (ed.), *Homology: The Hierarchical Basis of Comparative Biology.* New York: Academic Press, 273–299.

Wilson, E. O. (1975), *Sociobiology: The New Synthesis.* Cambridge, MA: Harvard University Press.

See also **Adaptation and Adaptionism; Cognitive Science; Evolution; Evolutionary Epistemology; Natural Selection; Naturalism**

EXPERIMENT

Over the past two decades the historical development of experimental science has been studied in detail. One focus has been on the nature and role of experiment during the rise of the natural sciences in the sixteenth and seventeenth centuries. Earlier accounts of this so-called Scientific Revolution emphasized the universalization of the mathematical method or the mechanization of the worldview as the decisive achievement. In contrast, the more recent studies of sixteenth- and seventeenth-century science stress the great significance of a new experimental practice and a new experimental knowledge. Major figures were Bacon, Galileo, and Boyle. The story of the controversy of the latter with Hobbes has been made a paradigm of the recent history of scientific experimentation by Shapin and Schaffer (1985). In this controversy, the legitimacy of experiment as a way to knowledge was at issue. While Hobbes defended the "old" axiomatic-deductive style of the geometric tradition, Boyle advocated the more modest acquisition of probable knowledge of experimental "matters of fact." According to Shapin and Schaffer, what was at stake were simultaneously the technical details of Boyle's air-pump experiments, the epistemological justification of experimental knowledge, and the social legitimacy of the new experimental style of doing science. While Shapin and Schaffer emphasize the novelty of sixteenth- and seventeenth-century experimentation, some others have questioned this claim by arguing that Hellenistic antiquity, the Arab world, and the Middle Ages all have significant experimental traditions—for instance, in the areas of optics and alchemy.

A more wide-ranging account of the role of experimentation in the emerging natural sciences has been proposed by Thomas Kuhn (see Kuhn, Thomas). He argues that the rise of modern physical science resulted from two simultaneous developments (Kuhn 1977). On the one hand, a radical conceptual and worldview change occurred in what he calls the classical, or mathematical, sciences, such as astronomy, statics, and optics. On the other, the novel type of Baconian, or experimental, sciences emerged, dealing with the study of light, heat, magnetism, and electricity, among other things (see Scientific Change). An important additional claim put forward by Kuhn is that it was not before the second half of the nineteenth century that a systematic interaction and merging of the experimental and mathematical traditions took place. An example is the transformation of the Baconian science of heat into an experimental-mathematical thermodynamics during the first half of the nineteenth century. At about the same time, the interactions between (at first, mainly experimental) science and technology increased substantially. Important results of this scientification of technology were chemical dye stuffs and artificial fertilizers.

Starting in the second half of the nineteenth century, systematic experimentation also took root in various other sciences. This happened in medicine, in particular in physiology, somewhat later in psychology, and still later in the social sciences. A characteristic feature of many experiments in those sciences is a strong reliance on statistical methods. Thus far, most philosophers of science have focused on experimentation in physics, chemistry, and biochemistry, while many analyses of statistical experiments can be found in the methodological literature of the medical, psychological, and social sciences (see e.g., Campbell and Stanley 1963). In this article, the focus will be on the philosophical approach to experimentation.

The Philosophy of Scientific Experimentation

A central feature of experimentation is the manipulation of, and the interference with, material things. Historians and philosophers, however, have focused on science as mostly a theoretical activity, a matter of thinking and reasoning. Thus, although the logical empiricists acknowledged the importance of observation and experiment, they took these activities mostly for granted and concentrated their studies on the philosophical problems of theories and theoretical knowledge (see Logical Empiricism).

Yet, historically some authors—including scientists and philosophers—did write about the nature and function of scientific experimentation. Among the better known examples are Bacon's and Galileo's advocacy of the experimental method. Mill (around the middle of the nineteenth century) and Mach (late nineteenth and early twentieth centuries) provided some methodological and epistemological analyses of experimentation. Bernard promoted and analyzed the use of the experimental method in medicine. His *Introduction to the Study of Experimental Medicine* (Bernard [1865] 1957) influenced a number of twentieth-century French writers, including Duhem, Bachelard, and Canguilhem. While those authors addressed some aspects of experimentation in their accounts of science, a substantial and coherent tradition in the philosophy of scientific experimentation did not yet arise.

Such a tradition did spring up in Germany, in the second half of the twentieth century. Within this German tradition, two approaches may be distinguished. One developed the pioneering work of Dingler (1928), who emphasized the action and production character of experimentation, and hence its kinship to technology. One of the aims of his operationalist approach (see Bridgman, Percy) was to show how the basic theoretical concepts of physics, such as length and mass, could be grounded in concrete experimental actions. This part of Dingler's philosophy was taken up and systematically developed by a number of other German philosophers, including Lorenzen, Holzkamp, and Janich. More recently, the emphasis on the methodical construction of theoretical concepts in terms of experimental actions has given way to more self-contained accounts of scientific experimentation and a more culturalistic interpretation of its results (see Janich 1996).

A second approach within the German tradition took its departure even more directly from the kinship between experiment and technology. The major figure here is the early Habermas. In his work from the 1960s, Habermas ([1968] 1978) conceived of (empirical-analytical) science as "anticipated technology," the crucial link being experimental action. In the spirit of Marx, Heidegger, and Marcuse, Habermas's aim was to develop not merely a theory of (scientific) knowledge but rather a critique of technocratic reason. More recently, attempts have been made to connect this German tradition to Anglo-Saxon philosophy of experiment (Radder 1996, Ch. 2) and to contemporary social studies of science and technology (Feenberg 1999). Recent work on science as technology by Lelas (2000) can be characterized as broadly inspired by this second branch of the German tradition.

In the English-speaking world, a substantial number of studies of scientific experimentation have been written since the mid-1970s. They resulted from the Kuhnian "programs in history and philosophy of science" (see Kuhn, Thomas). In their studies of (historical or contemporary) scientific controversies, sociologists of scientific knowledge often focused on experimental work (e.g., Collins 1985), while so-called laboratory studies addressed the ordinary practices of experimental scientists (e.g., Latour and Woolgar 1979). An approach that remained more faithful to the history and philosophy of science started with Hacking's argument for the relative autonomy of experimentation and his plea for a philosophical study of experiment as a topic in its own right (Hacking 1983). It includes work by Franklin, Galison, Gooding, and Rheinberger, among many others (see the volumes edited by Gooding, Pinch, and Schaffer 1989; Buchwald 1995; Heidelberger and Steinle 1998).

More recently, some philosophers argue that a further step should be taken by combining the results of the empirical and historical study of experiment with more developed theoretical-philosophical analyses (see Radder 2003). A mature philosophy of experiment, they claim, should not be limited to summing up its empirical features but should attempt to provide a systematic analysis of experimental practice and experimental knowledge. The latter is often lacking in the sociological and historical literature on scientific experimentation.

Action and Production and Their Philosophical Implications

Looking at the role of experiments within the overall practice of science, there is one feature that stands out. In order to perform experiments, whether they are large-scale or small-scale, experimenters have to intervene actively in the material world; moreover, in doing so they produce all kinds of new objects, substances, phenomena, and processes. More precisely, experimentation involves the material realization of the experimental system (that is to say, the object[s] of study, the apparatus, and their interaction) as well as an active intervention in the environment of this system. In this respect, experiment contrasts with theory even if theoretical work is always attended with material acts (such as the typing or writing down of a mathematical formula). Hence, a central issue for a philosophy of experiment is the question of the nature of experimental action and production, and their ontological, epistemological, and methodological implications.

Clearly, not just any kind of intervention in the material world counts as a scientific experiment. Quite generally, one may say that successful experiments require, at least, a certain stability and reproducibility, and meeting this requirement presupposes a measure of control of the experimental system and its environment as well as a measure of discipline of the experimenters and the other people involved in realizing the experiment.

Experimenters employ a variety of strategies for producing stable and reproducible experiments (see e.g., Bhaskar 1978; Franklin 1986; Janich 1996; Radder 1996). One such strategy is to attempt to realize "pure cases" of experimental effects. For example, in some early electromagnetic experiments carried out in the 1820s, Ampère investigated the interaction between an electric current and a freely suspended magnetic needle. He systematically varied a number of factors of his experimental system and examined whether or not they were relevant, that is to say, whether they had a destabilizing impact on the experimental process.

Furthermore, realizing a stable object/apparatus system requires knowledge and control of the (actual and potential) interactions between this system and its environment. Depending on the aim and design of the experiment, these interactions may be *necessary* (and hence required), *permitted* (but irrelevant), or *undesirable* (because disturbing).

Thus, in his experiments on electromagnetism, Ampère anticipated a potential disturbance exerted by the magnetism of the Earth. In response, he designed his experiment in such a way that terrestrial magnetism constituted a permitted rather than a disturbing interaction.

A further aspect of experimental stability is implied by the notion of reproducibility. Investigating the questions of *what* should be reproducible and *by whom* leads to different types of experimental reproducibility, which can be observed to play different roles in experimental practice. A successful application of the strategy of reproducing an experiment is an achievement that may depend on certain idiosyncratic aspects of a local situation. Yet, a purely local experiment that cannot be carried out by other experimenters and in other experimental contexts will, in the end, be unproductive for science.

Laboratory experiments in physics, chemistry, and biochemistry often allow one to control the objects under investigation to such an extent that the relevant objects in successive experiments may be assumed to be in identical states. Hence, when statistical methods are employed, it is primarily to further analyze or process the data (see e.g., the error-statistical approach in Mayo 1996) (see Statistics, Philosophy of). In contrast, in field biology, medicine, psychology, and social science such a strict experimental control is often not feasible. To compensate for this, statistical methods in these areas are used directly to construct groups of experimental subjects that are presumed to possess identical average characteristics. It is only after such groups have been constructed that one can start the investigation of hypotheses about the research subjects. One can phrase this contrast in a different way by saying that in the former sciences, statistical considerations bear mostly upon linking experimental data and theoretical hypotheses, while in the latter, it is often the case that statistics play a role already at the stage of producing the actual individual data (see Psychology, Philosophy of; Social Sciences, Philosophy of the).

The action and production aspect of scientific experimentation carries implications for ontological and epistemological questions. A general ontological lesson, already drawn by Bachelard, appears to be this: The action and production character of experimentation entails that the actual objects and phenomena themselves are, at least in part, materially realized through human intervention. Hence, it is not just the knowledge of experimental objects and phenomena but also their actual existence and occurrence that prove to be dependent on specific, productive interventions by the experimenters. This fact gives rise to a number of important philosophical issues. If experimental objects and phenomena have to be realized through active human intervention, does it still make sense to speak of a "natural" nature, or does one merely deal with artificially produced laboratory worlds? If one does not want to endorse a full-fledged constructivism, according to which the experimental objects and phenomena are nothing but artificial, human creations (see Social Constructionism), one needs to go beyond an actualist ontology and introduce more differentiated ontological categorizations (see Realism). In this spirit, various authors (e.g., Bhaskar 1978) have argued that an adequate ontological interpretation of experimental science needs some kind of dispositional concepts, such as powers, potentialities, or tendencies. These human-independent dispositions would then enable the human construction of particular experimental processes.

Next to such ontological problems, the interventionist character of experimentation engenders a number of epistemological questions. At least for those who assume the ontological independence of nature, a further important question is whether scientists, on the basis of artificial experimental intervention, can acquire knowledge of a human-independent nature (see also Epistemology; Realism). Some philosophers claim that at least in a number of philosophically significant cases, such "back inferences" from the artificial laboratory experiments to their natural counterparts can be justified. Another approach accepts the constructed nature of much experimental science but stresses the fact that its results acquire a certain endurance and autonomy with respect to both the context in which they have been realized in the first place and later developments. In this vein, Baird (2004) offers a neo-Popperian account of "objective thing knowledge," the knowledge encapsulated in material things, such as Watson and Crick's material double-helix model or the indicator of Watt and Southern's steam engine (see Popper, Karl Raimund).

Another epistemologically relevant feature of experimental science is the distinction between the working of an apparatus and its theoretical accounts. In actual practice it is often the case that experimental devices work well, even if scientists disagree on how they do so. This fact supports the claim that variety and variability at the theoretical and ontological levels may well go together with a considerable stability at the level of the material realization of experiments. This claim can then be exploited for philosophical purposes—for example, to vindicate entity realism (Hacking 1983) or referential realism (Radder 1996).

At times, scientists devise and discuss so-called thought experiments (see Brown 1991). Such experiments—in which the crucial aspect of action and production is missing—are better conceived as not being experiments at all but rather as particular types of theoretical argument, which may or may not be materially realizable in experimental practice. Furthermore, recent scientific practice shows an ever-increasing use of "computer experiments." These involve various sorts of hybrids of material intervention, computer simulation, and theoretical and mathematical modeling techniques. Often, more traditional experimental approaches are challenged and replaced by approaches based fully or primarily on computer simulations (sometimes this replacement is based on budgetary considerations only). This development raises important questions for the philosophy of scientific experimentation. Prominently, there is the epistemological question of the justifiability of the results of the new approaches. Should experiments that involve a substantial material component remain the standard, or are simulated experiments equally reliable and useful?

A further issue prompted by these computer experiments concerns the nature of philosophy of science itself. Apparently, practicing scientists do not mind calling such computational procedures "experiments." This raises the question of how philosophers' notions of 'experiment' should relate to scientists' usages? Of course, this is just one example of a quite general hermeneutical issue: To what extent should philosophers take into account the concepts and interpretations of the people who are being studied (in this case, scientists)? Answers to this question will depend on the conception of philosophy one adheres to. Those philosophers who advocate a more descriptive approach will tend to follow the scientists' usages, while those who favor a more theoretical or normative approach will emphasize the legitimacy of employing their own terminology.

The Relationship Between (Experimental) Science and Technology

Traditionally, philosophers of science have defined the aim of science as, roughly, the generation of reliable knowledge of the world. Moreover, as a consequence of explicit or implicit empiricist influences, there has been a strong tendency to take the production of experimental knowledge for granted and to focus on theoretical knowledge. However, if one takes a more empirical look at the sciences, at both their historical development and their current condition, this approach must be qualified as one-sided. After all, from Archimedes' lever-and-pulley systems to the cloned sheep Dolly, the development of (experimental) science has been intricately interwoven with the development of technology (see Tiles and Oberdiek 1995). Experiments make essential use of (often specifically designed) technological devices, and conversely, experimental research often contributes to technological innovations. Moreover, there are substantial conceptual similarities between the realization of experimental and of technological processes, most significantly the implied possibility and necessity of the manipulation and control of nature. Taken together, these facts justify the claim that the science/technology relationship ought to be a central topic for the philosophy of (experimental) science.

One obvious way to study the role of technology in science is to focus on the instruments and equipment employed in experimental practice. Many studies have shown that the investigation of scientific instruments is a rich source of insights for a philosophy of scientific experimentation (Gooding, Pinch, and Schaffer 1989; Heidelberger and Steinle 1998; Radder 2003). One may, for example, focus on the role of visual images in experimental design and explore the wider problem of the relationship between thought and vision (see Visual Representation). Or one may investigate the problem of how the cognitive function of an intended experiment can be materially realized, and what this implies for the relationship between technological functions and material structures. Or one may study the modes of representation of instrumentally mediated experimental outcomes and discuss the question of the epistemic or social appraisal of qualitative versus quantitative results.

In addition to such studies, several authors have proposed classifications of scientific instruments or apparatus. One suggested distinction is between instruments that represent a property by measuring its value (e.g., a device that registers blood pressure), instruments that create phenomena that do not exist in nature (e.g., a laser), and instruments that closely imitate natural processes in the laboratory (e.g., an Atwood machine).

Such classifications form an excellent starting point for investigating further philosophical questions on the nature and function of scientific instrumentation. They demonstrate, for example, the inadequacy of the empiricist view of instruments as mere enhancers of human sensory capacities. Yet, an exclusive focus on the instruments as such may tend to ignore two things. First, an experimental setup often includes various "devices," such as a concrete wall to shield off dangerous radiation, a support to hold a thermometer, a spoon to stir a liquid, curtains to darken a room, and so on. Such devices are usually not called instruments, but they are equally crucial to a successful performance and interpretation of the experiment and hence should be taken into account. Second, a strong emphasis on instruments may lead to a neglect of the environment of the experimental system, especially of the requirement to control the interactions between the experimental system and its environment. Thus, a comprehensive view of scientific experimentation needs to go beyond an analysis of the instrument as such by taking full account of the specific setting in which the instrument needs to function.

Finally, there is the issue of the general philosophical significance of the experiment/technology relationship. Some of the philosophers who emphasize the importance of technology for science endorse a "science-as-technology" account. That is to say, they advocate an overall interpretation in which the nature of science—not just experimental but also theoretical science—is seen as basically or primarily technological (see e.g., Dingler 1928; Habermas [1968] 1978; Lelas 2000). Other authors, however, take a less radical view by criticizing the implied reduction of science to technology and by arguing for the *sui generis* character of theoretical-conceptual and formal-mathematical work. Thus, while stressing the significance of the technological (or perhaps more precisely, the action and production dimension of science), these views nevertheless see this dimension as complementary to a theoretical dimension.

The Role of Theory in Experimentation

This brings us to a further central theme in the philosophy of scientific experimentation: the relationship between experiment and theory (see Theories). The theme can be approached in two ways. One approach addresses the question of how theories or theoretical knowledge may arise from experimental practices. Thus, Franklin (1986) has developed an epistemology of experiment by arguing that following established strategies for producing stable and reproducible experiments provides a good reason for believing in the validity of the experimental results. Hon (2003) has put forward a classification of experimental error and argued that the notion of error may be exploited to elucidate the transition from the material, experimental processes to propositional, theoretical knowledge.

A second approach to the experiment/theory relationship examines the question of the role of existing theories, or theoretical knowledge, within experimental practices. Over the last two decades, this question has been debated in detail. Are experiments, factually or logically, dependent on prior theories, and if so, in which respects and to what extent? The remainder of this section reviews some of the debates on this question.

The strongest version of the claim that experimentation is theory dependent says that all experiments are planned, designed, performed, and used from the perspective of one or more theories about the objects under investigation. In this spirit, von Liebig and Popper, among others, claimed that all experiments are explicit tests of existing theories. This view completely subordinates experimental research to theoretical inquiry. However, on the basis of many studies of experimentation published during the last two decades, it can be safely concluded that this claim is most certainly false. For one thing, quite frequently the aim of experiments is just to realize a stable phenomenon or a working device. Yet, the fact that experimentation involves much more than theory testing does not, of course, mean that testing a theory may not be an important goal in particular scientific settings.

At the other extreme, there is the claim that experimentation is basically theory free. The older German school of "methodical constructivism" (see Janich 1996) came close to this position. A somewhat more moderate view is that in important cases, theory-free experiments are possible and do occur in scientific practice. This view admits that performing such "exploratory" experiments does require some ideas about nature and apparatus, but not a well-developed theory about the phenomena under scrutiny. Hacking (1983) and Steinle (1998) make this claim primarily on the basis of case studies from the history of experimental science. Heidelberger (2003) aims at a more systematic underpinning of this view. He distinguishes between theory-laden and causally based instruments and claims that experiments employing the latter type of instruments are basically theory free.

Another view admits that not all concrete activities that can be observed in scientific practice are guided by theories. Yet, according to this view, if certain activities are to count as constituting a genuine experiment, they require a theoretical interpretation (Morrison 1990; Radder 1996; Hon 2003). More specifically, performing and understanding an experiment depends on a theoretical interpretation of what happens in materially realizing the experimental process. In general, quite different kinds of theory may be involved, such as general background theories, theories of the (material, mathematical, or computational) instruments, and theories of the phenomena under investigation.

One argument for such claims derives from the fact that an experiment aims to realize a reproducible correlation between an observable feature of the apparatus and a feature of the object under investigation. The point is that materially realizing this correlation and knowing what can be learned about the object from inspecting the apparatus depends on theoretical insights about the experimental system and its environment. Thus, these insights pertain to those aspects of the experiment that are relevant to obtaining a reproducible correlation. It is not necessary, and in practice it will usually not be the case, that the theoretical interpretation offers a full understanding of any detail of the experimental process.

A further argument for the significance of theory in experimentation notes that a single experimental run is not enough to establish a stable result. A set of different runs, however, will almost always produce values that are, more or less, variable. The questions then are: What does this fact tell us about the nature of the property that has been measured? Does the property vary within the fixed interval? Is it a probabilistic property? Is its real value constant, and are the variations due to random fluctuations? In experimental practice, answers to such questions are based on an antecedent theoretical interpretation of the nature of the property that has been measured.

Regarding these claims, it is important to note that in actual practice, the theoretical interpretation of an experiment will not always be explicit and the experimenters will not always be aware of its use and significance. Once the performance of a particular experiment or experimental procedure becomes routine, the theoretical assumptions drop out of sight: They become like an (invisible) window to the world. Yet, in a context of learning to perform and understand the experiment or in a situation where its result is very consequential or controversial, the implicit interpretation will be made explicit and subjected to empirical and theoretical scrutiny. This means that the primary locus of the theoretical interpretation is the relevant epistemic community and not the individual experimenter.

EXPERIMENT

Further Issues for the Philosophy of Scientific Experimentation

As was explained before, the systematic philosophical study of scientific experimentation is a relatively recent phenomenon. Hence, there are a number of further issues that have received some attention but merit a much more detailed account. In concluding this article, three such issues will be briefly discussed.

The first bears upon the notion of (scientific) experience. If one takes into account the fact that in many cases scientific experience is experimentally realized experience, the empiricist view that reduces scientific experience to sense perception or even to visual sensation needs to be revised. Of course, this view has already been challenged by the claim that all observation is theory laden (see Empiricism; Observation; Perception). Yet, a systematic study of the action and production aspects of experimentation will lead to a more radical criticism. The studies that have been done so far suggest that gaining scientific experience depends not just on intersubjectively communicable language but also on human agency and that it requires particular skills that cannot be supposed to be simply universally available.

A second subject that merits more attention from philosophers of science is the nature and role of experimentation in the social and human sciences, such as economics, sociology, medicine, and psychology. Practitioners of those sciences often label substantial, or even large, parts of their activities "experimental." So far, this fact is not reflected in the philosophical literature on experimentation, which has focused primarily on the natural sciences. Thus, a challenge for future research is to connect the primarily methodological literature on experimentation in economics, sociology, medicine, and psychology with the philosophy of science literature on experimentation in natural science.

One subject that will naturally arise in philosophical reflection upon the similarities and dissimilarities of natural and social or human sciences is the problem of the double hermeneutic. Although it is true that the nature of this problem has been transformed by the more recent philosophical accounts of the practices of the natural sciences, the problem has by no means been resolved. The point is this: In experiments on human beings, the experimental subjects, in addition to the scientists, will often have their own interpretation of what is going on in these trials, and this interpretation may influence their responses over and above the behavior intended by the experimenters. As a methodological problem (of how to avoid "biased" responses), this is of course well known to practitioners of the human and social sciences. However, from a broader philosophical or sociocultural perspective, the problem is not necessarily one of bias. It may also reflect a clash between a scientific and a commonsense interpretation of human beings. In the case of such a clash, social and ethical issues are at stake, since the basic question is, Who is entitled to define the nature of human beings: the scientists or the people themselves? In this form, the methodological, ethical, and social problems of the double hermeneutic will continue to be a significant theme for the study of experimentation in the human and social sciences.

This brings us to a last issue. The older German tradition explicitly addressed wider normative questions surrounding experimental science and technology. The views of Habermas, for example, have had a big impact on broader conceptualizations of the position of science and technology in society. Thus far, the more recent Anglophone approaches within the philosophy of scientific experimentation have dealt primarily with more narrowly circumscribed scholarly topics. Insofar as normative questions have been taken into account, they have been mostly limited to epistemic normativity—for instance, to questions of the proper functioning of instruments or the justification of experimental evidence. Questions regarding the connections between epistemic and social or ethical normativity are hardly addressed.

Yet, posing such questions is not far-fetched, and they often relate to ontological, epistemological, or methodological concerns quite directly. For instance, those experiments that use animals or humans as experimental subjects are confronted with a variety of normative issues, often in the form of a tension between methodological and ethical requirements. Also normatively relevant are (1) the ontological issue of the artificial and the natural in experimental science and (2) science-based technology. Consider, for example, the question of whether experimentally isolated genes are natural or artificial entities. This question is often discussed in environmental philosophy, and different answers to it entail different environmental ethics and politics. More specifically, the issue of the contrast between the artificial and the natural is crucial to debates about patenting, in particular the patenting of genes and other parts of organisms. The reason is that discoveries of natural phenomena are not patentable, while inventions of artificial phenomena are (see Sterckx 2000).

Although philosophers of experiment cannot be expected to solve all of those broader social and normative problems, they may be legitimately asked to contribute to the debate on possible approaches and solutions. In this respect, the philosophy of scientific experimentation could profit from its kinship to the philosophy of technology, which has always shown a keen sensitivity to the interconnectedness between technological and social or normative issues.

HANS RADDER

References

Baird, Davis (2004), *Thing Knowledge: A Philosophy of Scientific Instruments*. Berkeley and Los Angeles: University of California Press.
Bernard, Claude ([1865] 1957), *An Introduction to the Study of Experimental Medicine*. New York: Dover Publications.
Bhaskar, Roy (1978), *A Realist Theory of Science*. Hassocks, UK: Harvester Press.
Brown, James R. (1991), *The Laboratory of the Mind: Thought Experiments in the Natural Sciences*. London: Routledge.
Buchwald, Jed Z. (ed.) (1995), *Scientific Practice: Theories and Stories of Doing Physics*. Chicago: University of Chicago Press.
Campbell, Donald T., and Julian C. Stanley (1963), *Experimental and Quasi-Experimental Designs for Research*. Chicago: Rand McNally College Publishing.
Collins, H. M. (1985), *Changing Order: Replication and Induction in Scientific Practice*. London: Sage.
Dingler, Hugo (1928), *Das Experiment. Sein Wesen und seine Geschichte*. Munich: Verlag Ernest Reinhardt.
Feenberg, Andrew (1999), *Questioning Technology*. London: Routledge.
Franklin, Allan (1986), *The Neglect of Experiment*. Cambridge: Cambridge University Press.
Gooding, David, Trevor Pinch, and Simon Schaffer (eds.) (1989), *The Uses of Experiment*. Cambridge: Cambridge University Press.
Habermas, Jürgen ([1968] 1978), *Knowledge and Human Interests*, 2nd ed. London: Heinemann.
Hacking, Ian (1983), *Representing and Intervening*. Cambridge: Cambridge University Press.
Heidelberger, Michael (2003), "Theory-Ladenness and Scientific Instruments in Experimentation," in Hans Radder (ed.), *The Philosophy of Scientific Experimentation*. Pittsburgh: University of Pittsburgh Press, 138–151.
Heidelberger, Michael, and Friedrich Steinle (eds.) (1998), *Experimental Essays—Versuche zum Experiment*. Baden-Baden: Nomos Verlagsgesellschaft.
Hon, Giora (2003), "The Idols of Experiment: Transcending the 'Etc. List,'" in Hans Radder (ed.), *The Philosophy of Scientific Experimentation*. Pittsburgh: University of Pittsburgh Press, 174–197.
Janich, Peter (1996), *Konstruktivismus und Naturerkenntnis*. Frankfurt am Main: Suhrkamp.
Kuhn, Thomas S. (1977), "Mathematical versus Experimental Traditions in the Development of Physical Science," in *The Essential Tension*. Chicago: University of Chicago Press, 31–65.
Latour, Bruno, and Steve Woolgar (1979), *Laboratory Life: The Social Construction of Scientific Facts*. London: Sage.
Lelas, Srðan (2000), *Science and Modernity: Toward an Integral Theory of Science*. Dordrecht, Netherlands: Kluwer.
Mayo, Deborah G. (1996), *Error and the Growth of Experimental Knowledge*. Chicago: University of Chicago Press.
Morrison, Margaret (1990), "Theory, Intervention and Realism," *Synthese* 82: 1–22.
Radder, Hans (1996), *In and About the World*. Albany: State University of New York Press.
——— (ed.) (2003), *The Philosophy of Scientific Experimentation*. Pittsburgh: University of Pittsburgh Press.
Shapin, Steven, and Simon Schaffer (1985), *Leviathan and the Air-Pump: Hobbes, Boyle and Experimental Life*. Princeton, NJ: Princeton University Press.
Steinle, Friedrich (1998), "Exploratives vs. theoriebestimmtes Experimentieren: Ampères erste Arbeiten zum Elektromagnetismus," in Michael Heidelberger and Friedrich Steinle (eds.), *Experimental Essays—Versuche zum Experiment*. Baden-Baden: Nomos Verlagsgesellschaft, 272–297.
Sterckx, Sigrid (ed.) (2000), *Biotechnology, Patents and Morality*, 2nd ed. Aldershot, UK: Ashgate.
Tiles, Mary, and Hans Oberdiek (1995), *Living in a Technological Culture*. London: Routledge.

See also **Bridgman, Percy; Empiricism; Epistemology; Kuhn, Thomas; Logical Empiricism; Observation; Perception; Popper, Karl Raimund; Psychology, Philosophy of; Scientific Change; Scientific Realism; Social Constructionism; Social Sciences, Philosophy of the; Statistics, Philosophy of; Theories; Visual Representation**

EXPLANATION

One of the most important aims of science is to provide explanations of natural phenomena. Consequently, philosophers have devoted much attention to the nature of scientific explanation. The twentieth century's most influential model of scientific explanation is known as the covering-law model, which has been articulated most fully in the work of Carl Hempel (1965). According to this model, an explanation is an argument whose conclusion is a statement of some fact to be explained (the explanandum) and whose premises (the explanans) comprise a set of statements that include at least one natural law and collectively provide either inductive or deductive support for the explanandum. The covering-law model suggests that the explanandum is explained by rendering it *nomically expectable*. The covering-law model is part of the legacy of logical empiricism. While this model is rejected by most contemporary philosophers of science, the majority of the literature in the last 50 years has been concerned with either defending or correcting problems with the covering-law model.

Both exponents and critics of the covering-law model have emphasized explanation in the physical sciences. Philosophers concerned with explanation in the biological and social sciences have sometimes argued that within these domains different kinds of explanations—notably *functional explanations* and *reductive explanations*—play some special role. Another major issue in the contemporary literature concerns the nature of these kinds of explanation and their relation to the covering-law model.

The Covering-Law Model of Explanation

A covering-law explanation is any explanation in which the explanandum is the conclusion of an argument whose premises contain at least one natural law. Hempel recognizes three varieties of covering-law explanation:

1. *Deductive-nomological* (D-N): These are deductive arguments whose premises include universal (deterministic) laws along with statements of particular conditions.
2. *Inductive-statistical* (I-S): These are inductive arguments whose premises include statistical laws.
3. *Deductive-statistical* (D-S): These are deductive arguments in which statistical laws are entailed by more comprehensive statistical laws.

Explanations may be further subdivided according to the logical character of the explanandum statement. The explanandum statement may be either a singular statement, a universal statement, or a statistical generalization.

The Deductive-Nomological Model

As an example of a D-N explanation, consider why a partially submerged oar appears to bend at the point where it enters the water. The phenomenon is a consequence of refraction, and a derivation of the observed angle can be given using Snell's law together with the indices of refraction of air and water. Note that the derivation will rely both upon general law statements (Snell's law together with statements of the indices of refraction for air and water) and particular statements (in particular, the angle of incidence of the oar with the water). Schematically, one can represent the D-N explanation as an argument:

$$\frac{L_1, L_2 \ldots L_k \quad C_1, C_2 \ldots C_n}{E}$$

Explanans (top), Explanandum (bottom)

where $L_1, L_2 \ldots L_k$ are laws, $C_1, C_2 \ldots C_n$ are statements of antecedent particular conditions, and E is a statement of the explanandum. In their original statement of the D-N model, Hempel and Oppenheim (1948) stipulated that for an argument to be a D-N explanation, it must meet three logical conditions of adequacy:

1. The explanandum must be a logical consequence of the explanans.
2. The explanans must include at least one law.
3. The explanans must have empirical content, in the sense that (at least some of) its component statements must be susceptible to empirical test.

Hempel and Oppenheim take the second requirement to be a logical condition, because they hold that the distinction between laws and nonlaws is essentially syntactic (see Hempel, Carl). Following a Humean analysis, they take laws simply to be universal generalizations of unrestricted scope, with no designations of particular objects and containing only purely qualitative predicates. They do not require that the explanation contain any singular statements, because they wish to allow D-N explanations of laws, in which case no singular statements should be required. A fourth condition, not mentioned, but clearly in the spirit of their model, is that the laws are essential to the explanation in the sense that omitting them from the explanans will make the argument invalid.

To the three logical conditions, Hempel and Oppenheim add an empirical condition of adequacy, which is that the statements in the explanans be true. It is useful to distinguish between a potential explanation, which meets the logical criteria, and an actual explanation, which meets both the logical and empirical criteria. Thus, for instance, Descartes offered a potential explanation of Snell's law, but it is not an actual explanation because certain laws of the corpuscular theory of light on which his derivation was based are in fact false.

An important consequence of the D-N model, and more generally of the thesis that explanations are arguments, is that explanations are logically indistinguishable from predictions. This consequence, often called the *structural identity thesis*, suggests that every explanation is a potential prediction and every prediction a potential explanation. According to Hempel, the distinction between explanation and prediction is essentially pragmatic. An explanatory argument can serve as a prediction when an explanandum statement is not antecedently known but the explanans statements are.

While many scientific explanations seem to fit the D-N model, critics have presented a number of counterexamples showing that the D-N model is either too restrictive, in the sense that its requirements rule out genuine explanations, or too permissive, in the sense that there are pseudo-explanations that meet the logical and empirical requirements of the model.

The claim that the D-N requirements are too restrictive has chiefly been made on the grounds that some genuine explanations do not involve any laws. Scriven (1962) considers the example of how to explain an ink stain on his carpet. If asked for such an explanation, Scriven might truthfully claim that the carpet was stained when he hit his writing table with his knee, overturning an ink bottle on the table, causing the ink to drip onto the carpet. According to Scriven, the assertion of these (singular) facts constitutes a complete explanation of the explanandum event. He does not know, and it is impossible to supply, universal laws that together with particular conditions entail the explanandum statement.

Hempel (1965) responds that such an explanation is not complete but is rather an enthymeme. One can have confidence that the (partial) explanation is a good one only if one believes that there exists some as yet unknown law or laws that together with the particulars cited in the explanans would entail the explanandum. Scriven points out that people quite often, as in this case, have causal knowledge without having knowledge of laws, but Hempel argues that this practical ability is not inconsistent with the claim that the existence of a causal connection implies the existence of a law. Scriven's putative counterexamples highlight the fact that scientific explanations typically cite causes of the explanandum event. Whether the D-N model is too restrictive depends, then, on whether all causally related events are instances of lawful regularities.

The D-N model is also subject to several well-known counterexamples that apparently show that the D-N requirements are too permissive. Four important ones are as follows:

1. It is possible to predict the length of a shadow cast by a flagpole using some elementary trigonometry plus measurements of the height of the flagpole and the angle of the sun. This calculation provides a D-N explanation of the length of the flagpole's shadow. However, it is equally possible to use the length of the shadow and the angle of sun to calculate the height of the flagpole. Such a calculation satisfies the requirements of the D-N model, but clearly the length of the shadow cannot be used to explain the height of the flagpole. This example, due to Bromberger (1966), calls into question the structural identity thesis, because the length of the shadow can be used to predict the height of the flagpole, but it cannot explain it.

2. Consider the following argument:

    Whenever the barometer drops, a storm occurs.
    <u>The barometer drops.</u>
    A storm occurs.

    Supposing that the first premise is a law, this argument meets the requirement for a D-N explanation, and yet, while the falling barometer may serve to predict the storm, it cannot explain it.

3. Suppose Smith ingests a lethal dose of a slow-acting poison that kills everyone who takes it within 24 hours. Suppose that immediately after ingesting the poison, Smith steps into the street and is run over by a bus. Given that Smith has ingested the poison, one can predict Smith's death, but the fact that Smith died is not explained by that fact.

4. Joe Jones (a male), though sexually active, regularly takes birth control pills. If the pill is 100% effective, then the following is a D-N explanation:

    Whenever a person takes birth control pills, that person avoids pregnancy.
    <u>Joe Jones regularly takes birth control pills.</u>
    Joe Jones avoids pregnancy.

While the argument is sound, Jones's failure to get pregnant is not explained by his use of birth control pills, but by his gender.

Each of these counterexamples offers an argument that purportedly meets the logical stipulations on D-N explanations but is not genuinely explanatory. In each case, however, the argument is predictive. Thus the counterexamples raise doubts both for the adequacy of the D-N model and for the correctness of the structural identity thesis. In each of these cases, the account of why the explanation is spurious has to do with the failure of the putative explanation to cite the causes of the explanandum. The explanatory asymmetry in the flagpole case arises from the asymmetry of cause and effect. It is possible both to explain and to predict effects from causes, but one can predict but not explain causes from effects. In the second case, the law connecting storms to falling barometers does not display a direct causal relationship but is explained by the operation of a common cause. Both the barometer's drop and the storm's occurrence are caused by falling atmospheric pressure. One effect of a common cause may be used to predict the other, but it cannot explain it. The third counterexample is a case of causal preemption. The ingestion of the poison initiates a causal process that will lead to Smith's death, but the process is preempted by the bus, which has the same effect. Again, the moral is that the D-N argument allows for prediction but not explanation. The fourth is a kind of overdetermination. The law cited is predictive, not explanatorily relevant.

The last major difficulty with the D-N model is that it is not possible to assess the correctness of a D-N explanation without an adequate understanding of what constitutes a natural law (see Laws of Nature). Hempel understood laws to be universal generalizations of unrestricted scope. For instance, it is a law that no object travels faster than the speed of light. Such laws can be represented in first-order predicate logic by a universal generalization. The problem with such a characterization is that not all universal generalizations express laws. Many universal generalizations are only accidentally true. For instance, it may be the case that all of the students in my classroom are under 20 years of age, but this universal generalization would be only accidentally true. It does not support counterfactuals, in the sense that it does not imply the claim that if a student were in my class, she would be under 20 years of age. The idea that a law is a universal generalization that supports counterfactuals is one that has considerable plausibility, but it is difficult to explicate the semantics of counterfactuals within a framework acceptable to empiricists.

Various philosophers, including Hempel (1965) and Nagel (1961), have tried instead to stipulate extra syntactic and semantic conditions that would distinguish lawful from accidental generalizations. In particular they have suggested that laws are exceptionless universal generalizations that include no reference to particulars and involve purely qualitative predicates. One consequence of this definition is that many general claims that scientists call laws would not be considered such. For instance, Kepler's and Mendel's laws would not be genuine laws, since they refer implicitly or explicitly to particulars of Earth's solar system and life on Earth. In fact, it may well be the case that no science except physics has any laws in the required sense. Hempel may not have been concerned with this, since he assumed that in principle, laws of the special sciences could be derived from more general physical laws in combination with statements about particulars. This assumption has been widely challenged by critics of reduction. Perhaps more telling is Goodman's critique of the concept of a purely qualitative predicate. Goodman's (1956) new riddle of induction seems to imply that there is no empirically respectable way to characterize purely qualitative predicates.

Another way to approach this problem, advocated by Woodward (2000), is to replace the appeal to laws in D-N explanations with appeals to invariant generalizations. In Woodward's view, explanations can be made by subsuming explananda under generalizations of varying degrees of invariance. True laws, which are strictly invariant, are a rare but limiting case. While Woodward's approach is broadly in the spirit of Hempel's covering-law models, Woodward's analysis of invariance requires appeals to counterfactuals of a kind inconsistent with stricter versions of empiricism.

Deductive-Nomological and Deductive-Statistical Explanations of General Laws

In the examples considered so far, the explananda have been singular events or states of affairs. Hempel and Oppenheim (1948) also envisioned using the D-N model to analyze explanations of laws and regularities. A D-N explanation of a law is simply a derivation of that law according to the D-N model. If the explanandum is a statistical law, Hempel called the explanatory argument a D-S explanation, but D-S explanations are essentially a species of D-N explanation.

The idea that less fundamental laws can be explained by more fundamental ones is appealing. The case of Newton's explanation of Kepler's laws in terms of his laws of motion and gravitation seems to fit this mold. There are, however, problems with Hempel and Oppenheim's explication. Perhaps the most important difficulty is illustrated by the following example: Consider an argument whose single premise is the conjunction of the law of universal gravitation (UG) with Snell's law and whose conclusion is UG alone. This apparently meets the formal requirements for a D-N explanation of UG, and the conjunctive law is "more general" than UG in the sense that both UG and Snell's law are derivable from it. Nonetheless, this is clearly not a genuine explanation of UG. The example shows that it is unclear how to use derivability relations to distinguish more and less fundamental laws.

The Inductive-Statistical Model

In "Aspects of Scientific Explanation," Hempel (1965) extends the covering-law model to statistical explanations with his I-S model. As an example of an I-S explanation, Hempel considers the treatment of streptococcus with penicillin. Suppose that a patient Jones is suffering from streptococcus, takes penicillin, and subsequently recovers from the infection. The following inductive argument serves as an I-S explanation of Jones's recovery:

> A patient with streptococcus who takes penicillin has a high probability of recovery.
> <u>Jones has streptococcus and takes penicillin (with high probability)</u>
> Jones recovers.

The formal requirements for an I-S explanation are identical to those of a D-N explanation, except that (1) the explanans must contain a statistical law and (2) the relationship between the premises and the conclusion of the explanatory argument is one of inductive strength rather than deductive validity. The simplest I-S explanations will have the following form:

$$P(G|H) = r$$
$$\underline{Hj} \qquad [r]$$
$$Gj$$

Hempel understood the probability used in the statistical law as a relative frequency, while the bracketed probability was understood as an inductive (logical) probability. The use of these probabilities raises questions about I-S explanation. First, the concept of inductive probability is very difficult to explicate, and it has so far proved impossible to establish a definitive measure of inductive probability (see Probability). Second, it is unlikely that the statistical probability concept Hempel uses in his characterization of statistical laws is adequate for distinguishing statistical laws from accidental statistical associations. The problems are in many respects analogous to those of distinguishing lawful from accidental universal generalizations, but it is in other respects worse (cf. Dupré and Cartwright 1988).

A more immediate difficulty is what Hempel calls the problem of the ambiguity of I-S explanation. The problem is that it may be possible to formulate two inductively strong arguments with true premises that support opposite conclusions. Suppose, for instance, that the strain of streptococcus with which Jones is infected is known to be penicillin resistant. One then has the I-S explanation:

> A patient with penicillin-resistant streptococcus who takes penicillin has a low probability of recovery.
> <u>Jones has penicillin-resistant streptococcus and takes penicillin (with high probability)</u>
> Jones does not recover.

The premises of both this and the previous argument are true, and yet the arguments support opposite conclusions. This situation does not arise with D-N explanations because of an important difference between inductive and deductive arguments. While deductively valid arguments can never be made invalid by addition of further premises, inductively strong arguments can be weakened by additional premises. Here, the original argument explaining Jones's recovery is weakened by the additional information that the strain with which Jones was infected is penicillin resistant.

This general problem with inductive inference motivated Rudolf Carnap (1950, 211) to stipulate that correct measures of inductive support of a hypothesis can be made only in light of total evidence. Applying this requirement naively to the I-S model would suggest that the appropriate explanans for any I-S explanation is the entire knowledge base K. Hempel (1965, 399–400) suggested a refined version of Carnap's principle, which he called the *principle of maximal specificity*.

In the simple version of Hempel's model, the explanandum statement is an assertion that a particular individual j is a member of a class G. The statistical law in the explanans is an assertion that the relative frequency of individuals in H who are in G is r. The ambiguity of I-S explanation arises because different choices of H lead to different values of r. Hempel's solution is to demand that the choice of H be maximally specific. If j is a member of a class H and of a class H' that is a proper subclass of H, the explanans should contain the probabilistic law $P(G|H') = r'$, rather than $P(G|H) = r$. For instance, since the class of persons with penicillin-resistant streptococcus infections is a subclass of the class of persons with streptococcus infections, the explanatory argument should be based on the former class. So if Jones has a penicillin-resistant infection, his recovery is not explained by his having taken penicillin.

A major problem with Hempel's proposal concerns what classes may legitimately be taken as reference classes. If one is allowed to take any class as a reference class, one may simply take the reference class H' to be the intersection of H and G. This generates a trivial and nonexplanatory I-S explanation. Hempel does not have a satisfactory formal way out of this difficulty, but clearly his intention is that reference classes are the extensions of observable or theoretical predicates used in characterizing total scientific knowledge K. A given body of knowledge will entail a particular set of reference classes with respect to which statistical probabilities should be measured. For instance, in a given knowledge situation, when one wishes to explain the cause of lung cancer, it is recognized that reference classes should be partitioned by such properties as age, gender, weight, smoking habits, presence of environmental pollutants, etc. The consequence of this proposal, as Hempel recognized, is that the concept of statistical explanation is essentially relativized to a knowledge situation. This means that, in Hempel's account, there is an essential pragmatic element of statistical explanation that is not present in D-N explanation.

While it is possible to construct statistical analogs of the problems facing D-N explanation, the I-S model has certain difficulties that are peculiar to it. The most widely discussed of these concerns Hempel's high-probability requirement. Hempel believed that all I-S explanations must show that the explanandum is to be expected. Recognizing that what counts as high probability is a pragmatic issue, Hempel's requirement rules out the probabilistic explanation of unlikely events. This is in keeping with his defense of the structural identity thesis.

The difficulty with this requirement is illustrated by Scriven's example of syphilis and paresis. Paresis is a form of tertiary syphilis that can be contracted only by persons who have originally contracted syphilis. However, paresis is relatively rare even among syphilitics, so the probability of contracting paresis given that one has syphilis is low. Nonetheless, Scriven argues, it is reasonable to cite a person's syphilis as an explanation for his paresis. About this case, Hempel insists that paresis is not explained by syphilis, arguing that necessary conditions are not generally explanatory and that, given the fact that most syphilitics do not contract paresis, other factors must be cited in the explanation.

The high-probability requirement is also open to the objection that it is sometimes too lax. Suppose, for instance, one seeks to explain teenagers' interest in sex by reference to the TV programs they watch. It is clearly spurious to argue that teenagers' interest in sex is explained by their TV-viewing habits, just because most teenagers who watch TV are interested in sex. Most teenagers appear to be interested in sex quite independently of TV.

What both of these examples suggest is that the central issue in statistical explanation is not whether the probability of the explanandum given the explanans is high, but whether the factors cited in the explanans make a difference to the probability of the explanandum. This is the intuition behind the statistical relevance approach discussed below.

Alternatives to Covering-Law Models

The problems for Hempel's D-N and I-S models described above have led to a number of alternative theories of explanation. Four alternative approaches will be considered: the statistical relevance model (Salmon, Greeno, Jeffrey), the causal/mechanical approach (Railton, Salmon), the pragmatic approach (Bromberger, van Fraassen), and explanatory unification (Friedman, Kitcher).

The Statistical Relevance Model

The fundamental intuition behind Hempel's I-S model is that a statistical explanation explains its explanandum by providing an argument that renders the explanandum probable. Scriven's example of the syphilitic man calls this intuition into question. The man's paresis is explained by the fact that he has syphilis, even though paresis is unusual among syphilitics. The statistical relevance (SR) approach, developed chiefly by Wesley Salmon (Salmon, Greeno, and Jeffrey 1971; Salmon 1984), provides a model of explanation that shows how such factors can be explanatory. In general, a factor C is statistically relevant to a factor B just in case $P(B|C) \neq P(B)$. The intuition behind the SR model is that the presence of B is explained in a particular case by finding factors C that are positively relevant to B. The factors may be explanatory even if the probability of B given C is small.

Suppose that one is interested in explaining why an adolescent female Jenny became pregnant. The explanatory query can be phrased in this way: Why is it that Jenny, who is an adolescent female, is also pregnant? The reference class A, the class of adolescent females, serves as a baseline from which one calculates probabilities (construed as relative frequencies). One then partitions this class by a mutually exclusive and exhaustive set of factors B_i.

In this case, there are only two—B_1 refers to the class of those who have become pregnant and B_2 refers to the class of those who have not. This partition is called the *explanandum partition*. One then partitions the reference class by a set of explanatory factors into classes C_i that are mutually exclusive and exhaustive. In this case, factors might include the parents' education level, family income, ethnic group, religious affiliation, etc. Values for C_i represent various conjunctions of these factors. An SR explanation would show that a set of factors are explanatory by finding the partition C_a to which Jenny belongs and showing that $P(B_1|A \& C_a) > P(B_1|A)$. For instance, one might find that Jenny is a Polish Catholic woman of high school educated parents with a family income of less than $30,000 and that women within this group are more likely to become pregnant than teenage women as a whole. The SR model is a formalization of an approach to statistical explanation familiar from the social sciences. Social scientists interested in explaining the occurrence of a property within a population will partition that population according to some set of potentially relevant factors and collect data to discover in which partitions the property is most frequent.

This procedure is open to a number of objections. One of the most familiar is connected with a statistical phenomenon called *Simpson's paradox* (Cartwright 1979). The problem raised by Simpson's paradox is that a partition C that is positively relevant to B may sometimes be partitioned into subpartitions D, in which each D is negatively relevant to B:

$$P(B \mid A \wedge C) > P(B \mid A),$$

but for all i,

$$P(B \mid A \wedge D_i) < P(B \mid A).$$

Cartwright discusses a famous example of this case concerning admissions to graduate school at the University of California, Berkeley. Questions had been raised about whether Berkeley was discriminating against women in admissions, because it turned out that women had a lower admission rate than men. A closer look dispelled this concern. It turned out that on a department-by-department basis, women were admitted at rates equal to or higher than men's. The lower admission rate for women was caused by the fact that women applied disproportionately to departments with lower overall admission rates.

The trick to avoiding spurious inferences is to create reference-class partitions that are so fine-grained that their members are homogeneous with respect to all causally relevant properties. If one can be sure that members of a reference-class partition are *objectively homogeneous* with respect to causally relevant factors, then one can justifiably say that the set of factors defining the partition is positively or negatively relevant to the explanandum. Unfortunately, the concept of an objectively homogeneous reference class is fraught with conceptual and epistemic difficulties (cf. Salmon 1984, Ch. 3).

A second problem with the SR approach is that it admits explanatory factors that are correlated with the explanandum but that are not *causally* relevant. This problem can arise when the explanatory factor is correlated with the explanandum due to a common cause. In the earlier counterexample given, the correlation of the barometer and the storm to the D-N model is a case in point, as the barometer's falling is statistically relevant to the occurrence of the storm. To meet this objection, one can amend the SR account by stipulating that a statistically relevant factor is not explanatory if it can be *screened off* by another factor. A factor D screens off a factor C from B if

$$P(B \mid C \& D \& A) = P(B \mid D \& A)$$

but

$$P(B \mid C \& D \& A) = P(B \mid C \& A)$$

Applying this to the barometer case, actual change in atmospheric pressure screens off barometer readings, because once the atmospheric pressure is fixed, variations in barometer readings (say, due to barometer malfunctions) become irrelevant. While this is intuitively plausible, there are again conceptual and empirical problems with applying screening-off criteria in such a way as to completely eliminate spurious causes.

Salmon himself ultimately became convinced that it was not possible to solve all of the problems associated with the SR approach. Statistical relevance relations provide an evidential basis for making judgments of causal relevance, but causal (and hence explanatory) relevance must be understood independently of statistical relations. Salmon's mechanistic account of causation, described below, was meant to provide this missing ingredient.

The Causal Mechanical Approach

Peter Railton (1978) first introduced the concept of mechanism into the contemporary literature on explanation. His deductive-nomothetic model of probabilistic explanation (the D-NP model) was meant as an alternative to Hempel's I-S model. Railton was concerned with Hempel's requirement that the explanans of an I-S explanation rendered the explanandum probable or nomically expectable. Railton argued that explanations describe causes, and sometimes the causal sequence of events leading up to the event to be explained may be improbable. According to Railton, while an explanation of some event may include a reference to a law that renders the event nomically expectable, the account must be supplemented by "an account of the mechanism(s) at work" (1978, 208). Railton is vague on just what a mechanism is, indicating only that an "account of the mechanism(s)" is "a more or less complete filling-in of the links in the causal chains" (ibid).

Salmon's work on causal-mechanical explanation, beginning with his seminal *Scientific Explanation and the Causal Structure of the World* (1984), elaborates Railton's earlier account of mechanistic explanation. Though he dubs his theory "mechanistic," his actual analysis is not of the concept of mechanism. Rather, he argues that explanations must refer to what he calls the "causal nexus," which he takes to be a vast network of interacting causal processes. Salmon defines a *process* to be an entity that maintains a persistent structure through space-time, a *causal process* to be a process capable of transmitting changes in its structure, and a *causal interaction* to be an intersection between causal processes in which an alteration of the persistent properties of those processes occurs. In Salmon's original formulation, interactions were defined in terms of a counterfactual criterion of mark transmission. In response to criticisms of this criterion, he has eliminated the reference to counterfactuals (Salmon 1994), relying instead on a definition in which causal interactions involve exchanges of conserved quantities.

Both versions of Salmon's theory seem vulnerable to a criticism raised by Hitchcock (1995), whose concern is that there can be events causally connected to but irrelevant to the explanation of an explanandum event. Salmon's example of Jones and the birth control pills illustrates the point. Jones's ingestion of birth control pills counts as a causal interaction under either the counterfactual or the conserved quantity account. It is part of the causal nexus preceding Jones's failure to get pregnant. How is this interaction to be excluded as explanatorily irrelevant? The obvious answer is that the counterfactual claim that Jones would have gotten pregnant if he had not taken birth control pills is false; but nothing in Salmon's theory appears to require that this claim be true.

Glennan (2002) has argued that this and other difficulties with the mechanistic approach to explanation arise from an incorrect analysis of the concept of mechanism. Salmon and Railton both conceive of mechanisms as constituting a nexus of intersecting causal processes. An alternative view of mechanisms, advocated by Glennan and by Machamer, Darden, and Carver (2000), among others, suggests that mechanisms are complex systems—ensembles of interacting parts. While a causal process in Salmon's sense may involve the operation of a mechanism, the mechanism is not identified with the single instance of this process but is rather the system that reliably underlies processes of a certain type. The spurious explanation of Jones's failure to get pregnant is rejected both because the only reliable mechanism for preventing pregnancy by birth control pills involves the female reproductive system and because there is no reliable mechanism for the production of male pregnancy, so there is no need to explain the failure of the mechanism.

Pragmatic Accounts of Explanation

From a linguistic point of view, an explanation can be viewed as an answer to a *why*-question. Aristotle, in his theory of the four causes, already recognizes that the same *why*-question can be correctly answered in a number of ways, depending upon the beliefs and interests of the questioner. Pragmatic theories of explanation seek to explicate the relationship between the context in which a *why*-question is asked and the kinds of answers that can be appropriately given.

Hempel was certainly aware that explanation had a pragmatic dimension, but at least in the case of D-N explanation, the essential part is semantic. A D-N explanation is a valid argument, and the validity of an argument is independent of pragmatic factors. Pragmatic factors will explain which questions are asked as well as how the argument is presented (e.g., which premises are treated as implied), and not much more.

The pragmatic response to Hempel's account begins with Scriven but has been more fully developed in the work of Bromberger (1966) and, especially, van Fraassen (1980, Ch. 5). Van Fraassen claims that his pragmatic theory of explanation has the resources to resolve the problems of asymmetry and irrelevance that plague the D-N model.

An explanation is an answer to a *why*-question, "Why *P*?," where *P* is some true proposition. *P* is, in Hempel's terminology, the explanandum. Van

Fraassen argues, however, that there is more to the question than P itself. First, when one asks "Why P?," one is implicitly contrasting the state of affairs expressed by P to an alternative set of states of affairs, called the *contrast class*. To take one of van Fraassen's examples, the question "Why did Adam eat the apple?" can be understood variously as asking (1) why *he* ate the apple (as opposed to the serpent or other creatures in the garden), (2) why he *ate* the apple (as opposed to refusing it, or perhaps throwing it at Eve), or (3) why he ate *the apple* (as opposed to some other fruit in the garden). Besides the contrast class, van Fraassen suggests that the context of the question includes a *relevance relation*, which specifies the kinds of answers that are considered relevant to the question. For instance, if one were to ask why primates have opposable thumbs (in contrast to the pattern of fingers in other mammals), one could either be interested in an evolutionary explanation, in which case the relevance relation would relate selectively relevant features of environments of ancestral species to various morphological traits. Or one could be interested in a developmental explanation, in which case the relevance relation would relate combinations of genetic and environmental factors to traits that these combinations cause to develop.

Van Fraassen's (1980) formal account incorporates the following definitions:

1. A question Q is a triple $<P_K, X, R>$, where P_K is the topic (or explanandum), X is the contrast class, which is a set of propositions that includes the topic, and R is a relevance relation of propositions to ordered pairs of topic propositions and contrast classes.
2. The *presupposition* of Q is the conjunction of the claims
    a. that the topic P_K is true,
    b. that every other proposition in the contrast class X is false, and
    c. that there is at least one proposition that bears the relation R to $<P_K, X>$ that is true.
3. A *direct answer* to Q is the conjunction of the presupposition and a proposition A that bears relation R to $<P_K, X>$.
4. A is called the *core* of the answer to Q. (143)

Most commentators agree that van Fraassen's account goes a long way toward elucidating explanatory practices. It shows, for instance, why certain *why*-questions can be rejected (e.g., if the topic is false or if the other elements of the contrast class are not all false) and why a verbally identical *why*-question can admit of different answers (e.g., because of different implied but unstated relevance relations). Van Fraassen, however, claims that his theory is sufficient to solve the explanatory asymmetry problems that plague the D-N model. He supports his claim by means of an amusing parable regarding a tower and its shadow (van Fraassen 1980, 132–134). As with Bromberger's flagpole, one would expect that the length of the shadow can be explained in terms of the height of the tower and the altitude of the sun, but not vice versa. In van Fraassen's parable, however, a wealthy chevalier has constructed a tower of a certain height in order that it should cast a shadow over the spot where he proclaimed his love to a woman he subsequently killed. Thus, in this case, it really would be appropriate to explain the tower's height in terms of the length of the shadow it cast. Van Fraassen's point is that a particular context will fix a particular relevance relation, and this relevance relation will specify the direction of explanation.

While van Fraassen shows that a change in context is sufficient to reverse the direction of explanation, this fact does not by itself show that pragmatic constraints can eliminate spurious explanations generated by the symmetries. What makes the reversal of direction legitimate in the case of the chevalier's tower is that there is an objective relevance relation connecting the chevalier's mental states to his actions (including building the tower). But, as Kitcher and Salmon (1987) point out, van Fraassen's theory places no substantive constraints on the choice of relevance relations. They show how to construct gerrymandered relevance relations meeting van Fraassen's formal criteria but giving rise to spurious explanations like Bromberger's flagpole. One can grant van Fraassen's point that there can be other legitimate relevance relations that give rise to different answers while still maintaining that a central task of a theory of explanation is to describe the kinds of relevance relations that are objectively legitimate. The various explanatory accounts, including Hempel's D-N and I-S models, the statistical relevance model, and Salmon's causal theory, can be seen as attempts to accomplish this task.

Explanatory Unification

The explanatory unification approach, introduced by Friedman (1974) and developed by Kitcher (1981 and 1989), represents another attempt to remedy the inadequacies in the covering-law approach to explanation. While distinctly different from Hempel's version of the covering-law model, explanatory unification is probably more in its spirit than are any of the other models discussed in this essay. Explanations are arguments. What the explanatory unification model adds is the requirement that the arguments used to explain explananda be instances of unifying explanatory patterns. The counterexamples to covering laws (and the D-N model in particular) are thought to be ruled out on the grounds that the spurious explanatory arguments are not unifying.

To understand what is meant by a unifying explanatory pattern, it is useful to consider some examples of unification. Perhaps the greatest of Newton's achievements was to unify celestial and terrestrial mechanics. In practice, what this achievement amounted to was the discovery that the motion of celestial and terrestrial bodies could be explained by deriving the trajectory of that motion from a common set of laws. For instance, the same explanatory pattern can be used both to derive the trajectory of a satellite around the Earth and to derive the trajectory of a projectile like a ballistic missile. As a second example, consider Mendelian explanations of the distribution of traits in successive generations of populations. Mendelian genetics is unifying because it allows for a diverse set of facts about distributions of traits in populations of different species to be explained in terms of common patterns like dominance and recessiveness.

Explanation, according to Kitcher, begins with a set of accepted beliefs K. Given this set, the problem is to identify a set of argument patterns, called the "explanatory store," $E(K)$, from which explananda are derived. The explanatory unification model suggests that $E(K)$ is the set of argument patterns that maximally unify K. Most of Kitcher's work is devoted to spelling out what counts as an argument pattern. For Kitcher, an argument pattern consists of three parts: (1) a set of *schematic sentences* containing dummy letters for some nonlogical vocabulary; (2) a set of *filling instructions* regarding the sorts of terms that can be substituted for the dummy letters, and (3) a *classification* of instructions regarding which sentences are to be regarded as premises and of the rules of inference that may be used to derive conclusions from those premises. Kitcher's argument patterns are similar in some respects to metalinguistic argument schemata familiar from formal logic. The essential difference is that the filling instructions are semantic rather than syntactic. The terms that replace a particular dummy letter need not have precisely the same logical form, but they must belong to a similar semantic category. For instance, a filling instruction for a Newtonian pattern of explanation might specify that the term to replace a dummy variable refer to a body or to a position within a Cartesian coordinate system.

It might seem that the unification approach is open to counterexamples of spurious explanation very similar to those that confront the D-N model. For instance, the explanatory store for all of science could contain just one argument pattern $\frac{P}{P}$ because every statement in K is derivable from itself. Kitcher's answer to this objection is that the unifying power of the explanatory store is judged not just on the numbers of argument patterns in it, but on the *stringency* of those patterns. A stringent argument pattern is one whose schematic sentences and filling instructions limit the number of possible ways in which the pattern can be instantiated. The argument pattern above is clearly not stringent. While stringency requirements are a plausible solution to the problem of spurious unification, a major problem for the explanatory unification approach is to find an adequate account of stringency, as well as of the way in which to assess the trade-off between the size of the explanatory store and the stringency of its argument patterns.

Critics of the unification approach have also raised more general concerns. For one thing, it should be noted that what counts as a good explanation is relativized to the current set of accepted beliefs K. As those beliefs change, so will the explanations. While it is certainly the case that the arguments *accepted* as explanatory will change with changes in K, many philosophers will argue that the true explanation of some fact should remain constant over time and that as beliefs change, formerly accepted explanations are regarded as spurious. A second puzzling feature of the unification approach is what can be called the *nonlocality* of explanation. According to the unification approach, whether something counts as a correct explanation depends upon whether the argument pattern used to explain it is also useful elsewhere. This means that one cannot judge the adequacy of an explanation of a particular event simply by reference to claims about other local events. Such a requirement is at odds with causal approaches to explanation, which suggest that explanations consist in describing events and processes causally relevant to the explanandum event. Whether similar causal patterns actually occur elsewhere is immaterial. Kitcher's (1989) reply to this sort of objection is that the processes identified as causal are just those that can serve in maximally unifying explanatory patterns. Advocates of the causal approach respond that Kitcher has confused ontological and epistemological issues. While considerations like simplicity and scope enter into epistemological judgments about the correctness of theories, what makes an explanation correct is that it describes an actual causal process.

Reductive Explanation

Reduction can be understood as a kind of explanation and can be analyzed using the general models of explanation so far discussed (see Reductionism). Historically, most discussion of reduction has focused on three general areas. The first concerns the attempts of logical empiricists to provide a reduction of theoretical terms to observational terms. The second, generally called successional reduction, involves the study of the relationship between succeeding theories of the same domain, such as the relationship between Newtonian mechanics and the general theory of relativity. The third, generally called interlevel reduction, involves the study of the relationship between theories of different levels of organization, such as the relationship between neurobiological and psychological theories. The focus here will be on the third, which is most clearly a kind of scientific explanation.

The classical model of theoretical reduction is due to Nagel (1961). According to this model, a theory is understood as a collection of laws and other statements. Reduction is accomplished by discovering a set of bridge principles, which are universal biconditional statements that identify terms of the reduced theory with terms of the reducing theory. For instance, if one were to attempt a reduction of classical to molecular genetics, the bridge principle would say "For all x, x is a gene if, and only if, $\phi(x)$," where $\phi(x)$ is some (presumably complicated) formula of molecular biology. In a successful reduction, the laws of the reduced theory should be derivable from the laws (and perhaps other statements) of the reducing theory. A reduction of this sort would satisfy the conditions for a D-N explanation of a general law.

Nagel's account has been criticized by Fodor, Putnam, Wimsatt, and Kitcher, among others (see Sarkar 1992 for a review). Perhaps the most influential of these criticisms is Fodor's (1974) *multiple realizability argument*. Suppose one attempts to give a reduction of a higher-level science like economics to physics. Economic laws (if there are any) will describe relations between theoretical terms like money, debt, interest, etc. What Fodor points out is that the property of, for instance, being money is in fact a functional property. What makes something money is that it plays a certain causal role in an economic system. Many things, from gold to dollar bills to digitized numbers on magnetic media, can perform this causal role, and these things—the realizations of money—have very little in common in terms of their physical properties. It would, consequently, be difficult or impossible to formulate bridge principles. At best one would identify economic predicates with a large disjunction of physical predicates. Even if this were possible, the disjunction would not form a physical kind and the bridge principle would not be lawlike.

The purport of arguments like Fodor's is to establish the explanatory autonomy of special sciences. For instance, it would seem to justify the view that psychological theory can be used to explain psychological events without reference to the physical substrate of psychological agents. However, critics of Fodor and the "antireductionist consensus" point to the various ways in which lower-level theories can increase understanding of a higher level of phenomena. They argue that the explanatory significance of interlevel relations suggests that the problem is not with reductionism per se, but with Nagel's model of reduction. An example of this approach can be found in the work of Kim (2000), who suggests that the properties in higher-level sciences are functional. In Kim's account, a reductive explanation consists in the specification of the mechanism that realizes this function. It is true that a given function may have different realizers in different contexts, but it is nonetheless explanatory to consider how a function is realized in a particular case. Moreover, it is often the case that many or all actual instances of a function may be realized in the same general way. So even if, for instance, having a pain is a functional property, that property is realized by similar mechanisms among all human beings, and to some degree among many other species. If Kim's account is correct, then reductive explanation is possible, but it is closely connected both to functional explanation and to mechanical explanation.

Functional and Teleological Explanation

The ideas of functional and teleological explanation originated with Aristotle's concept of the final cause, the reason or purpose of the existence of a thing. Although the metaphysical assumption that everything in nature has a final cause has generally been rejected since the seventeenth century, teleological explanation is still considered legitimate in areas like biology, where systems are products of design or selection processes. Thus, for instance, if one explains the structure of the human hand in terms of its adaptive function, one is providing a legitimate teleological explanation (see Function).

Hempel (1959) considered how explanations of this kind could be integrated into the framework of the D-N model. He saw the task of functional explanation as that of explaining the presence of a certain part within a complex system by showing that it contributed to that system's functioning. For instance, one could explain the presence of the heart in the human body by showing that its pumping of blood contributes to the proper functioning of the body. If, however, one attempts to frame this explanation as a D-N argument for a particular human body (say Joe's), one gets something like this:

> Joe's body functions properly.
> Pumping blood is an essential activity in the proper functioning of a body, and <u>a heart is a body component that functions as a blood pump.</u>
> Thus, Joe's body has a heart.

Hempel, however, points out that the fact that a component plays an indispensable role in the functioning of a system does not allow one to infer that that component must be present. Because functions are multiply realizable, the same role could be played by something different than a heart (e.g., an artificial heart). The point is even clearer in the case of opposable thumbs. Although opposable thumbs play a certain role in human bodies that is clearly adaptive, one could not *predict* the development of opposable thumbs, because other morphologically distinct traits could perform the same function. Hempel concluded that functional arguments are, at best, very weak explanations, because all that can legitimately be inferred from premises like those in the argument above is that one of an indefinitely large range of realizers will be present in the system.

Hempel's conviction that functional explanations are weak is a consequence of his attempt to fit them within the D-N model, and in particular of his belief in the structural identity of explanation and prediction. Other philosophers, especially Cummins (1975), have argued that these assumptions mistake the distinctive character of functional explanation. According to Cummins, the point of functional explanation is to show not that the presence of a certain component in a system was predictable, but rather how that component contributes to the functioning of the system in which it is contained. Functional analysis involves identifying a certain capacity of a system and showing how more basic capacities of the system or its components give the system that capacity. For instance, functional analysis shows how the heart, arteries, veins, lungs, etc., give the human body its capacity to transport oxygen and other products to its various areas. As Craver (2001) has pointed out, when these more basic capacities are associated with components of a system and when the organization of these components is specified, functional analysis leads to a pattern of explanation very similar to mechanistic explanation in the complex-systems sense. Whatever the success of functional analysis as an explanatory strategy, many philosophers of science still believe that there is a kind of functional explanation in which the adaptive value of a trait explains the presence of a trait.

STUART GLENNAN

References

Bromberger, Sylvain (1966), "Why Questions," in Robert Colodny (ed.), *Mind and Cosmos*. Pittsburgh: University of Pittsburgh Press, 86–111.
Carnap, Rudolf (1950), *Logical Foundations of Probability*. Chicago: University of Chicago Press.
Cartwright, Nancy (1979), "Causal Laws and Effective Strategies," *Noûs* 13: 419–437.
Craver, Carl (2001), "Role Functions, Mechanisms and Hierarchy," *Philosophy of Science* 68: 53–74.
Cummins, Robert (1975), "Functional Analysis," *Journal of Philosophy* 72: 741–765.
Dupré, John, and Nancy Cartwright (1988), "Probability and Causality: Why Hume and Indeterminism Don't Mix," *Noûs* 22: 521–536.
Fodor, Jerry (1974), "Special Sciences, or the Disunity of Sciences as a Working Hypothesis," *Synthese* 28: 97–115.
Friedman, Michael (1974), "Explanation and Scientific Understanding," *Journal of Philosophy* 71: 5–19.
Glennan, Stuart (2002), "Rethinking Mechanistic Explanation," *Philosophy of Science* 69: S342–S353.
Goodman, Nelson (1956), *Fact, Fiction and Forecast*. Indianapolis: Bobbs-Merrill.
Hempel, Carl (1965), "Aspects of Scientific Explanation," in *Aspects of Scientific Explanation and Other Essays in the Philosophy of Science*. New York: Free Press, 331–496.
——— (1959), "The Logic of Functional Analysis," in Llewellyn Gross (ed.), *Symposium on Social Theory*. New York: Harper & Row.
Hempel, Carl, and Paul Oppenheim (1948), "Studies in the Logic of Explanation," *Philosophy of Science* 15: 135–175.
Hitchcock, Christopher (1995), "Discussion: Salmon on Explanatory Relevance," *Philosophy of Science* 62: 305–320.
Kim, Jaegwon (2000), *Mind in a Physical World: An Essay on the Mind–Body Problem and Mental Causation*. Cambridge, MA: Bradford Books.
Kitcher, Phillip (1981), "Explanatory Unification," *Philosophy of Science* 48: 507–531.
——— (1989), "Explanatory Unification and the Causal Structure of the World," in Philip Kitcher and Wesley Salmon (eds.), *Scientific Explanation: Minnesota Studies in the Philosophy of Science*, vol. 13. Minneapolis: University of Minnesota Press, 410–505.
Kitcher, Phillip, and Wesley Salmon, "Van Fraassen on Explanation," *Journal of Philosophy* 84: 315–330.

Machamer, P., Darden, L., and Carver, C. (2000), "Thinking about Mechanisms," *Philosophy of Science* 67: 1–25.
Nagel, Ernest (1961), *The Structure of Science: Problems in the Logic of Scientific Explanation*. New York: Harcourt, Brace and World.
Railton, Peter (1978), "A Deductive-Nomological Model of Probabilistic Explanation," *Philosophy of Science* 45: 206–226.
——— (1981), "Probability, Explanation, and Information," *Synthese* 48: 233–256.
Salmon, Wesley (1984), *Scientific Explanation and the Causal Structure of the World*. Princeton, NJ: Princeton University Press.
——— (1994), "Causality without Counterfactuals," *Philosophy of Science* 61: 297–312.
Salmon, Wesley, Richard Greeno, and Richard Jeffrey (1971), *Statistical Explanation and Statistical Relevance*. Pittsburgh: University of Pittsburgh Press.
Sarkar, Sahotra (1992), "Models of Reduction and Categories of Reductionism," *Synthese* 91: 167–194.
Scriven, Michael (1962), "Explanations, Predictions and Laws," in Herbert Feigl and Grover Maxwell (eds.), *Scientific Explanation, Space and Time: Minnesota Studies in the Philosophy of Science*, vol. 3. Minneapolis: University of Minnesota Press.
van Fraassen, Bas (1980), *The Scientific Image*. Oxford: Clarendon Press.
Woodward, James (2000), "Explanation and Invariance in the Special Science," *British Journal for the Philosophy of Science* 51: 197–254.

See also **Carnap, Rudolf; Causality; Function; Hempel, Carl Gustav; Laws of Nature; Mechanism; Prediction; Reductionism; Scientific Models; Theories**

EXPLICATION

Explication is a form of conceptual clarification developed by the philosopher Rudolf Carnap. The purpose of explication is to diminish scientific and philosophical vagueness and confusion. Although the development and advocacy of Carnap's notion of explication was largely due to Carnap himself, many of the ideas and methods presented in his explication project have been incorporated into contemporary analytic philosophy and linguistics (see Carnap, Rudolf).

Carnap on Explication

Carnap's most detailed exposition of his notion of explication appeared in the first chapter of his *Logical Foundations of Probability* (Carnap 1950). There he described explication as the following procedure:

> [E]xplication consists in transforming a given more or less inexact concept into an exact one or, rather, in replacing the first by the second. We call the given concept (or the term used for it) the *explicandum*, and the exact concept proposed to take the place of the first (or the term proposed for it) the *explicatum*. The explicandum may belong to everyday language or to a previous stage in the development of scientific language. The explicatum must be given by explicit rules for its use, for example, by a definition which incorporates it into a well-constructed system of scientific either logicomathematical or empirical concepts. (Carnap 1950, 3)

(A similar description appears in Carnap 1947, 7–9). Explication thus involves the replacement of an inexact concept by another, more exact one. Since the explicandum in an explication is replaced rather than elucidated or elaborated upon, explication is distinct from lexical definition, and more closely related to stipulative definition.

Explication is also distinct from the analysis of a concept, where 'analysis' is understood either as the breaking down of a concept into its constituent parts (as in Kant 1965, 48) or as the substitution of an ordinary concept with a formally more precise one for the purposes of clarifying the ordinary concept's ontological commitments (as in Russell 1956). Both of these notions of analysis appear to require that the analysans preserve the meaning of the analysandum in such a way that the former can be viewed as the definiens for the latter (cf. Orilia and Varzi 1998, 107). Carnap imposed no such constraint on explication. While he thought that the explicatum should be similar to the explicandum, he did not require that it function as a definiens for it, and in fact explicitly allowed for the possibility that some loss of the meaning of the explicandum could occur in explication (Carnap 1950, 7). Furthermore, in analysis the existence of distinct and nonequivalent analysanda for the same concept would almost certainly signal an ambiguity in that concept. But in explicating a concept, Carnap thought it possible and at times even desirable to have distinct and nonequivalent explicata for it, even if the concept were unambiguous. This is discussed further below.

Carnap (1950) identifies a simple example of explication in the replacement of the "prescientific" concept 'fish' by the concept *piscis* within a systematic zoology (5–6). The concept 'fish' is vague and broad. It arguably includes, for instance, tadpoles, seals, whales ("Wal*fische*" in German) and possibly other aquatic animals that are not cold-blooded or that do not have gills throughout life. The concept *piscis*, on the other hand, was stipulated to denote just those aquatic animals having the characteristics of being cold-blooded and having gills throughout life. This stipulation was introduced, Carnap thinks, because it was more *fruitful*, within zoology, to classify these animals together. For example, the concept has proved more fruitful with respect to its appearance in laws or useful generalizations: More true and informative generalizations involve *piscis* than *fish* (in the older sense of 'fish'). Within a systematic zoology, *piscis* would function as the explicatum of fish, the explicandum.

Carnap suggested (1950, 3) that his notion of explication was informed by Immanuel Kant's notion of an explicative judgment, in which the concept of the predicate is analyzed within the subject (see Kant 1965, 48), and by Edmund Husserl's notion of an *Explikat*, or the distinct, articulated outcome of an analysis (cf. Husserl 1973, 112ff). These connections are superficial, however, for Carnap's notion of an explication was embedded within the project of a systematic axiomatization of knowledge and discourse that has no correlate in Kant or Husserl. From the perspective of axiomatics and language construction, Carnap's explication project has a rather greater affinity with Leibniz's notion of a "universal language," which was a proposal for a constructed language that would precisely specify and define key philosophical and scientific terms. Unlike Leibniz, however, Carnap would have rejected the existence of a single "correct" language. A more contemporary and direct influence on Carnap's explication project was David Hilbert's pioneering work in formal axiom systems, which Carnap made extensive use of in his later work (see Hilbert, David).

While the explicatum is a new concept that replaces the explicandum, the explicandum nonetheless guides the choice of its replacement, in that Carnap (1950) makes it a requirement on explication that the explicatum be "similar" to the explicandum in the sense that the former can be used in most cases in which the latter is used, although he emphasizes that a "close similarity" is not required (7). Similarity is the first of four conditions that Carnap places on an explicatum (7–8), *viz*.:

1. The explicatum should be *similar* to the explicandum.
2. The explicatum should be given an *exact* specification within a rule-governed system of scientific concepts.
3. The explicatum should be a *fruitful* concept, and in particular allow for the formulation of many universal statements.
4. The explicatum should be as *simple* as possible. (This condition Carnap makes subsidiary to the first three [8]).

Carnap attempted explications of a variety of concepts throughout his later works. Examples of explicanda/explicata pairs proposed by him include: denotation/extension, meaning/intension, logical truth/L-truth, logical implication/L-implication, empirical truth/semantic truth (all given in Carnap 1947), verification/confirmation, inductive inference/logical probability, and estimation/degree of confirmation (Carnap 1950). As these examples illustrate, Carnap thought that explications could be performed not just on the concepts of empirical science, but on concepts from philosophy or from formal sciences such as set theory.

Consistent with the constraints guiding the process of explication, Carnap sometimes proposed different explicata for the same explicandum. For example, Carnap (1950) offered two distinct explicata for the explicandum "probability" (23f.). For instance, probability is explicated as the degree of confirmation of a hypothesis H with respect to an evidence statement. According to another explication, probability is the relative frequency (in the long run) of one property of events or things with respect to another. Carnap recognized that both conceptions of probability were important (25), and he did not regard the existence of different and even incompatible explicata for the same explicandum as an inconsistency or defect by itself. He did, however, regard it as essential that distinct probability concepts be recognized as such, and saw his explication of probability as removing various confusions that had been generated by a failure to clearly identify distinct probability concepts (Carnap 1950, 35).

The exactness condition on explication in particular calls for further discussion, since it leads to Carnap's philosophical framework for the activity of explication. This framework in turn helps to illuminate the remaining conditions and gives some indication of why there are not more than these four.

Exactness in an explication is, for Carnap, ideally accomplished through the *axiomatization* of an area of knowledge. It was Carnap's lifelong conviction that progress within philosophy and science was hampered by the vagueness and imprecision of concepts and theories, which he believed was often manifested in the form of "sterile and useless" metaphysical disputes (Carnap 1963a, 44–5). The construction of axiomatic systems with precisely specified rules was his proposed solution. If a set of basic axioms can be stipulated for some domain of knowledge, and if such axioms can be conjoined with clear definitions of key concepts and with rules governing inferential relations and relations of justification or confirmation, disputes arising from vagueness or imprecision could, Carnap thought, be eliminated.

Consider a simple example axiom system for a theory of the thermal expansion of iron rods (Carnap 1938, 199f). One might lay down a series of syntactical rules that specify sets of typed signs and permissible concatenations of those signs, and then "translate" certain fundamental empirical laws into that language by including unary predicates such as *Sol* and *Fe* and function symbols like $te(x,t)$, $lg(x, t)$, and $th(x)$. A pair of axioms in such a (quantified) logical language might appear as:

A1: $(\forall x)(\forall t)(\forall l)(\forall T)[\text{IF }(((((Sol x \text{ and } lg(x,t_1) = l_1) \text{ and } lg(x,t_2) = l_2)) \text{ and } te(x,t_1) = T_1)) \text{ and } te(x,t_2) = T_2) \text{ and } th(x) = \beta) \text{ THEN } (l_2 = l_1 \times (1 + \beta(T_2 - T_1)))]$.

A2: $(\forall x)(\text{IF }(Sol x \text{ and } Fe x) \text{ THEN } th(x) = 0.000012)$.

Here x, β, and subscripted formulae are real number variables. From this basis semantical rules may be introduced. These rules assign to signs of the primitive, or "ground," type a class of material objects. Such rules would further assign to the predicates *Sol* and *Fe* the properties of being solid and ferrous, respectively, and assign to the functions *te*, *lg*, and *th* the values of temperature in degrees centigrade (T), length in centimeters, and coefficient of thermal expansion, respectively, for bodies x at times t. Under this interpretation, axiom 1 (A1) is the quantitative law of thermal expansion, and axiom 2 (A2) gives the coefficient of thermal expansion (in the appropriate units) for iron.

Carnap (1938) illustrates how even simple interpreted axiom systems like this can, when conjoined with a mathematical calculus, be used to derive predictions about the changes in length that an iron body will undergo when heated (201–2). Hence, unlike a set of axioms for a formal science such as logic, an interpreted set of axioms for an area of empirical science contains statements that have empirical content, and so allows the derivation of "factual" statements whose truth can be empirically determined, as well as of theorems. This raises questions concerning how the truth of such axioms is to be understood. For instance, is A2 akin to a stipulative definition? If so, what explains the fact that it has merely contingent empirical applicability and seems to have been discovered rather than stipulated? If on the other hand A2 is just a formal expression of an experimentally determined result, what does it mean to regard it as an axiom, as opposed to a true inductive generalization? To the extent that axioms for Carnap are to be treated as akin to stipulative definitions, one can see how philosophers such as Willard Van Quine found it natural to question Carnap's distinction between axioms and other truths of an empirical theory (cf. Quine 1953 and 1966) (see Quine, Willard Van). Yet, Carnap was not oblivious to such concerns as these, as a look at his research into axiomatics reveals.

Explication and Axiomatics

The nature of the relationship of axiom systems to empirical reality was something that Carnap worked throughout his career to clarify. His position changed over time in response to developments in axiomatics in the first half of the twentieth century. One of his major concerns was the status of the concepts implicitly defined by an axiom system. Like his teacher Frege, Carnap regarded implicitly defined concepts, such as 'point' or 'line' in axiomatic geometry, as problematic on the grounds that the law of the excluded middle does not typically hold for them (Carnap 1927, 364–366). It is now evident from his unpublished work on axiomatics that in the late 1920s Carnap (1927) thought that the specification of a consistent set of axioms for a complete and decidable theory T would guarantee the categoricity of T, and conversely (364–365). He (wrongly) believed himself to have a proof of this result, which he called the *Gabelbarkeitssatz* (the results of this unpublished work are presented in Awodey and Carus 2001). Carnap believed that the *Gabelbarkeitssatz* allowed him to "ground" an axiom system *A* in empirical reality in the following sense: Given a demonstration of the categoricity of the theory generated by *A*, it would follow by Carnap's proof that *A* was decidable. This in turn means that the concepts implicitly defined by *A* would be such that the law of the excluded middle held for them. On the basis of this result, Carnap set to work on a general theory of axiomatics.

In 1930, however, this project was abandoned after Carnap became persuaded by Alfred Tarski and Kurt Gödel that the *Gabelbarkeitssatz* was incorrect. Tarski persuaded Carnap to distinguish more clearly between statements framed in the formal language used *in* the axiom system and statements framed in the language used to talk *about* the axiom system (cf. Carnap 1963a, 53–54). A confusion of this distinction arguably lies at the basis of the defective *Gabelbarkeitssatz* proof (Awodey and Carus 2001, 159). And Gödel's incompleteness theorems showed Carnap that there exist categorical axiom systems (such as the Peano axioms formulated in second-order logic) that are not decidable.

Carnap's response to these developments seems to have been to abandon his earlier concerns about implicit definitions and liberalize the constraints on philosophically suitable axiom systems. No longer able to specify a formal feature internal to a system of axioms that would guarantee an application for concepts defined by those axioms, Carnap in 1934 introduced his principle of tolerance, according to which there are no "morals" for logic to obey beyond the constraints of fruitfulness and exactness in specifying a language and axioms formulated in it (Carnap 1937, 51–52) (see Conventionalism). The principle of tolerance was to guide Carnap throughout the remainder of his career, and it illustrates his considered attitude toward the formal systems in terms of which explications are to be performed.

After 1934, the principle of tolerance informed Carnap's treatment of the truth of axioms with empirical content (such as A2). When such axioms were interpreted in such a way that their nonlogical descriptive signs had empirical designata, Carnap came to simply regard them as true only insofar as their truth had been established inductively by observation and experiment (Carnap 1930, 203; see also 1963a, 60). What then renders such statements axioms? His answer relied upon the flexibility enabled by the principle of tolerance. If, as Carnap thought, there are no "facts of the matter" in logic, that is, if there is no single true or correct logic (or language system) that must be accepted, but rather a plurality of logics and language systems, each of which may be engineered to suit particular purposes, then nothing prevents the construction of systems in which certain statements are stipulated to have a privileged role. If making something like A2 an axiom leads to greater simplicity and fruitfulness in the application of mechanics, then it may be made into an axiom, despite the fact that it is regarded as true in virtue of empirical data (and not, say, a priori intuitions). The conceptual framework that one operates with—the language and formalizations of knowledge that are constructed in it—is under human control.

Thus understood as a kind of *linguistic engineering*, explication was justified in part by the very flexibility and plurality of languages and logics that Carnap—having failed to uphold a doctrine of a single "correct" logic—now believed to exist. Nothing prohibits a philosophically minded scientist from modifying a part of language by axiomatizing and precisely defining certain concepts within those axiomatizations to serve as explicata for less clearly defined ordinary concepts, provided that it is useful to do so. Over time, some explicatum might completely displace its explicandum, even in ordinary, nonsystematic linguistic usage. The replacement in everyday parlance of the vague concept *germ* by more precise and scientifically delimited concepts such as *bacterium* and *virus* might provide an example of this tendency.

Although few philosophers, and still fewer scientists, make explicit mention or use of Carnap's explication project, the general idea of providing a formal systematization of a body of knowledge and explicating concepts within it has formed a significant component of research in contemporary linguistics, mathematics, and the philosophy of language. As suggested above, Carnap's own explication projects focused largely on probability and induction (Carnap 1950) and the philosophy of language (Carnap 1947). Both the nature of explication itself and the results of these individual projects have raised philosophical questions, some of which are examined below.

Strawson's Objections

Carnap's treatment of explication as a form of *philosophical* clarification was criticized by P. F. Strawson in an exchange with Carnap (Schilpp 1963). Strawson's critiques, and Carnap's reply to them, help to illuminate aspects of Carnapian explication, as well as to suggest potential weaknesses.

Strawson and Carnap seem to agree that Carnap's method of explication is intended as a way of clarifying philosophical concepts. Strawson thinks of Carnap's method as the construction of a formal system employing concepts that are precisely defined and then comparing the concepts within the constructed system with those that were to be explicated or clarified. Strawson objects that this method is neither the only nor the best way to clear up philosophical perplexities and confusions concerning concepts of ordinary language. He thinks that most philosophical perplexities arise within ordinary discourse, to which certain basic concepts are *essential*. Carnap's approach of constructing a formal system to clarify problems arising within ordinary discourse is "utterly irrelevant" for what is needed:

> It seems *prima facie* evident that to offer formal explanations of key terms of scientific theories to one who seeks philosophical illumination of essential concepts of nonscientific discourse, is to do something utterly irrelevant—is a sheer misunderstanding, like offering a textbook on physiology to someone who says (with a sigh) that he wished he understood the workings of the human heart. (Strawson 1963, 505)

Strawson's charge here thus appears to be that Carnap's explications will miss analyzing those concepts that he thinks are essential for ordinary discourse, and thereby fail to resolve the philosophical problems that they may give rise to. If our ordinary concepts lead to puzzles, then it would seem to be *these* concepts that one ought to investigate and clarify, and not some other, more or less homologous ones. Carnap responds to this by drawing on his considerably more "tolerant" conception of language, according to which all ordinary concepts are "dispensable" and hence can be replaced when the need (e.g., philosophical perplexity) arises. Carnap thus provides a different analogy:

> A natural language is like a crude, primitive pocketknife, very useful for a hundred different purposes. But for certain specific purposes, special tools are more efficient, e.g., chisels, cutting-machines, and finally the microtome. If we find that the pocketknife is too crude for a given purpose and creates defective products, we shall try to discover the cause of the failure, and then either use the knife more skillfully, or replace it for this special purpose by a more suitable tool, or even invent a new one. The naturalist's thesis is like saying that by using a special tool we evade the problem of the correct use of the crude tool. But would anyone criticize the bacteriologist for using a microtome, and assert that he is evading the problem of correctly using a pocketknife? (Carnap 1963b, 939)

The "linguistic naturalist"—Carnap's term for advocates of Strawson's position—could presumably agree with Carnap that no one is to be criticized for employing concepts or tools as needed. But Carnap's analogy allows for the obvious rejoinders that (1) pocketknives are not replaceable by microtomes for most ordinary uses and (2) someone who was having trouble using a pocketknife in an ordinary circumstance would not be helped in the least by being shown the workings of a microtome. So it is not obvious that Carnap's analogy adequately answers Strawson's charge of the irrelevance of explication for unraveling perplexity involving ordinary notions.

In addition, it may be misleading to treat an entire language as analogous to a tool, as Carnap does here. Just as tools can be used in tandem with each other, so can concepts. Just as it would be a terrible impediment to restrict oneself to a single tool for everything, so too would it be disastrous to get by with a single concept. However, if one thinks of individual concepts as akin to tools rather than whole languages, Strawson's worries about the irrelevance of some tools to the workings of other tools return to salience.

Carnap provides another example, involving the concepts of warmth and temperature, to help clarify the role of explication. Two different people, or one person in different circumstances, might describe the same thing as warm and as not warm. This might lead to questions such as whether warmth is a feature that things can have independently of perceivers. Carnap writes:

> In order to solve this puzzle, we have first to distinguish between the following two concepts: (1) "the thing x feels warm to the person y" and (2) "the thing x is warm", and then to clarify the relation between them. The method and terminology used for this clarification depends upon the specific purpose we may have in mind. First it is indeed possible to clarify the distinction in a simple way in ordinary language. But if we require a more thorough clarification, we must search for explications of the two concepts. The explication of concept (1) may be given in an improved version of the ordinary language concerning perceptions and the like. If a still more exact explication is desired, we may go to the scientific language of psychology. The explication of concept (2) must use an objective language, which may be a carefully selected qualitative part of the ordinary language. If we wish the explicatum to be more precise, then we use the quantitative term "temperature" either as a term of the developed ordinary language, or as a scientific term of the language of physics. (Carnap 1963b, 934)

EXPLICATION

Although he does not say so explicitly, it seems that Carnap may here be identifying ever greater precision with ever "more thorough clarification." If he is, then Strawson's worries persist. Substituting more and more precise concepts need not yield any clarification of issues pertaining to other (even if less precise) concepts, any more than understanding the workings of a microtome will tell us how to skillfully use a pocketknife. At the very least, Carnap needs to show how clarity is supposed to emerge in such cases. In particular, recall the question of whether warmth is an objective feature of objects, independent of perceivers. Why should we assume as a matter of course that this question becomes uninteresting or obviously answered by pointing out that temperature is a reasonably precisely defined feature of things? Later in his reply to Strawson, Carnap (1963b) adds:

> The process of the acquisition of knowledge begins with common sense knowledge; gradually the methods become more refined and systematic, and thus more scientific.... Suppose the statement "it will probably be very hot tomorrow at noon" is made for the purpose of communicating a future state to be expected, perhaps with regard to practical consequences. The use of the explicatum "temperature" instead of "very hot" in the above statement makes it possible to fulfill the same purpose in a more efficient way: "the temperature tomorrow at noon will probably be about so and so much". (934, 936)

Carnap thus thinks of improvements in precision as signs of scientific progress and clarification. But just how invoking temperature has made anything "more efficient" in this context, he does not explain.

Perhaps a better example for Carnap (1963b) is one that he mentions later on in his response to Strawson (939): the solution of Zeno's paradoxes. Zeno, the ancient Eleatic philosopher, raised some well-known perplexities about how motion is possible. For example, the traversal of any finite distance or spatial interval involves crossing infinitely many subintervals and thereby completing an incompletable (infinite) process. But since completing the incompletable is impossible, motion is impossible. Carnap takes it that Zeno's paradoxes of motion are definitively resolved by appeal to technical advances in mathematics involving limits, the real numbers, and other notions. To the extent that this is correct (an interesting issue that cannot be addressed here), it would appear that the construction of a new relatively formal system of concepts has provided a solution or "conceptual clarification" of paradoxes involving ordinary concepts such as motion from one location to another. Strawson does not discuss this example in his paper, but a possible response that he might give leads to another strand of his argument.

Strawson grants that the construction of formal systems of precisely defined concepts, when coupled with the comparison of such constructed concepts to ordinary ones that lead to perplexities, can yield illuminating results. In order to illuminate the ordinary landscape, one might introduce an artificial one with which to compare it, and this sort of comparison and contrast is arguably what helps to resolve Zeno's paradoxes. However, Strawson argues, such applications *presuppose* the sort of analysis of the relations between ordinary concepts of the sort that he takes to be primary and essential. A sketch of the "ordinary" landscape is required before it can be compared with another.

To the extent that commonsense concepts, troubling as they are, are *essential* to ordinary language and action, then it may be that Carnap's "explication" does not help with certain problems involving such concepts that will necessarily arise. To the extent that one thinks that commonsense concepts are dispensable in favor of "more precise" reconstructed concepts, then Strawson's concerns will seem uninteresting. One will naturally focus instead on the construction of new and improved schemes, rather than muddle about with confused ones.

In his concluding remarks in response to Strawson, Carnap (1963b) writes as though there were a well-defined notion of success that could be used to evaluate his program as well as Strawson's, so that one can, as it were, inductively decide which one is better according to the standard:

> We all agree that it is important that good analytic work on philosophical problems be performed. Everyone may do this according to the method which seems the most promising to him. The future will show which of the two methods, or which of the many varieties of each, or which combinations of both, furnishes the best results. (940)

But however laudatory Carnap's tolerance might be in other settings, it is not so clearly appropriate here. Either the philosophical goal is conceptual clarification of ordinary concepts, in which case Strawson's argument (that the examination of those concepts is the most essential component) seems sound; or the goal is the construction of conceptual systems that optimize our capacity to predict and control "experience," in which case Carnap's "constructive" process of "precisification" seems appropriate.

Quine's Objections

Carnap's explication project received criticism from another philosopher, Willard Van Quine. Quine argued at length that there could be no philosophically useful definition of "analyticity" and that the definitions that had been provided for this notion, including Carnap's, suffered from defects such as circularity and an empty extension (see Analyticity). Yet Carnap's explication project seems to require that some statements, such as the axioms and "meaning postulates" of a formalized system in which explications are conducted, have a privileged role, and this role seems to make such statements functionally very similar to the analytic statements that Quine rejected. Indeed, Carnap himself emphasized that the difference between the logical and mathematical formulae of an (interpreted) axiomatized theory, on the one hand, and the physical propositions of that theory, on the other hand, was essential both to the theory and to the clarification that the theory might provide (see e.g., Carnap 1938, 202). Quine found this distinction unintelligible or at best unhelpful.

Quine's criticism of analyticity can be seen to pose a challenge to Carnap's explication project on two fronts. One front concerns the question of whether Carnap was correct in invoking and assigning philosophical importance to a distinction between statements that are true by stipulation and those that are true in virtue of matters of fact. Here the difference between Quine and Carnap needs to be carefully identified. Quine did not deny the possibility that certain statements, such as meaning postulates, could be stipulated to be true (cf. Quine 1953, 34). And as has been noted, Carnap did not deny that a statement that is functioning as an axiom, such as A2 above, could not also be regarded as an empirical generalization. What appeared to separate the two positions was rather each philosopher's appraisal of the significance of elevating empirical generalizations to the status of axioms within the systematization of an area of discourse. Carnap regarded this elevation as an essential component of explication, and thereby of philosophical clarification, while Quine regarded it as a useless and singularly unhelpful bit of stipulation.

Resolving this dispute is difficult, as it was for the original disputants themselves (see, for instance, the correspondence that Quine and Carnap exchanged over the issue in Creath 1990). It is noteworthy that even if Quine's attack on analyticity is judged a failure, its very presence exposes a second front on which Carnap's explication project may be criticized. For Carnap's attempts to defend his use of the term "analytic" against Quine's objections itself involved recourse to explication. Thus in replying to Quine, Carnap wrote that "it is not clear whether [Quine] is asking about the elucidation explicandum, 'analytic,' or about an explicatum. If he means the latter, then it is given in the rules of a semantical system" (Carnap, in Creath 1990, 430).

The problem here is that an appeal to "rules of a semantical system" to clarify "analytic" was exactly the kind of thing that Quine found objectionable. As Quine put it, "the explanation 'true according to the semantical rules of L' is unavailing; for the relative term 'semantical rule of' is as much in need of clarification, at least, as 'analytic for'" (Quine 1953, 34). It is thus understandable that Carnap's claim that philosophical disputes about "analytic" would be resolved by explication failed to placate Quine. Indeed, it appears that Carnap's desire to resolve Quine's concerns about the analytic/synthetic by means of explications risked begging the question in Carnap's favor. On the other hand, there is an argument to be made that Quine's own position at this point is problematic (see Analyticity).

There is some irony in the fact that the very project that Carnap had hoped would lead to the resolution of apparently intractable and seemingly interminable philosophical disputes appeared to be constitutionally incapable of being applied (in a non-question-begging way) to the very dispute with Quine in which Carnap found himself increasingly enmeshed. So at the very least, Quine's objections may expose the presence of philosophical problems that arguably fall outside the purview of explication. In a further irony, Quine and Strawson, who were at loggerheads over the analyticity issue, found themselves allied in their rejection of Carnap's explication project, for both regarded this project as inadequate to the task of philosophical and conceptual clarification, although they did so for very different reasons.

ERIC LOOMIS
CORY JUHL

References

Awodey, S., and A. W. Carus (2001), "Carnap, Completeness, and Categoricity," *Erkenntnis* 54: 145–172.

Carnap, Rudolf (1927), "Eigentliche und Uneigentliche Begriffe," *Symposion* 1: 355–374.

——— (1930), "Bericht über Untersuchungen zur allgemeinen Axiomatik," *Erkenntnis* 1: 303–307.

——— (1938), "Foundations of Logic and Mathematics," in Otto Neurath, Rudolf Carnap, and Charles Morris (eds.), *International Encyclopedia of Unified Science*. Chicago: University of Chicago Press.

EXPLICATION

——— (1963a), "Intellectual Autobiography," in Paul A. Schilpp (ed.), *The Philosophy of Rudolph Carnap*. La Salle, IL: Open Court, 3–85.

——— (1963b), "P. F. Strawson on Linguistic Naturalism," in Paul A. Schilpp (ed.), *The Philosophy of Rudolph Carnap*. La Salle, IL: Open Court, 933–940.

——— (1950), *Logical Foundations of Probability*. Chicago: University of Chicago Press.

——— (1937), *The Logical Syntax of Language*. London: Routledge and Kegan Paul.

——— (1947), *Meaning and Necessity: A Study in Semantics and Modal Logic*. Chicago: University of Chicago Press.

Creath, R. (ed.) (1990), *Dear Carnap, Dear Van: The Quine-Carnap correspondence and Other Material*. Berkeley and Los Angeles: University of California Press.

Husserl, Edmund (1973), *Experience and Judgment: Investigations in the Genealogy of Logic*. Translated by James Churchill and Karl Ameriks. Evanston, IL: Northwestern University Press.

Kant, Immanuel (1965), *Critique of Pure Reason*. Translated by Norman Kemp Smith. New York: St. Martin's Press.

Orilia, F., and A. C. Varzi (1998), "A Note on Analysis and Circular Definitions," *Grazer Philosophische Studien* 54: 107–115.

Quine, Willard Van (1966), "Carnap and Logical Truth," in *The Ways of Paradox and Other Essays*. New York: Random House.

——— (1953), "Two Dogmas of Empiricism," in *From a Logical Point of View*. Cambridge, MA: Harvard University Press, 21–46.

Russell, Bertrand (1956), "On Denoting," in R. Marsh (ed.), *Logic and Knowledge*. London: Allen and Unwin, 39–55.

Schilpp, Paul (ed.) (1963), *The Philosophy of Rudolph Carnap*. La Salle, IL: Open Court.

Strawson, P. F. (1963), "Carnap's Views on Constructed Systems versus Natural Languages in Analytic Philosophy," in Paul A. Schilpp (ed.), *The Philosophy of Rudolph Carnap*. La Salle, IL: Open Court, 503–518.

See also **Analyticity; Carnap, Rudolf; Quine, Willard Van**

F

FALSIFICATION

See **Cognitive Significance; Demarcation, Problem of; Popper, Karl Raimund**

FEMINIST PHILOSOPHY OF SCIENCE

This article will encompass philosophical analyses of science undertaken from feminist perspectives. The tradition is part of the larger field of feminist science studies, which includes feminist internal science critique, feminist work in the history of science, and feminist engagements in social studies of science.

Feminist philosophy of science is a dynamic research tradition, loosely delineated by its origins, research questions, and history. Its origins include the emergence in the 1970s of feminist scholarship, in which academics, including scientists, brought the analytic category of gender to bear on research questions, methods, and theories in their fields. In philosophy, feminists analyzed relationships between gender, on the one hand, and theories of ethics, social and political theory, metaphysics, and epistemology, on the other. Although the details varied by area, historical period, and specific theory, feminists found that assumptions about gender informed many philosophical theories, including those about knowledge and science. Some associated men and women with what were argued to be opposing characteristics or categories—respectively, for example, mind and body, reason and emotion, objectivity and subjectivity, culture and nature, and activity and passivity—and took the first characteristic of each pair to be rightfully dominant in relation to or superior to the second. Many incorporated symbolic gender associations in which valued traits or characteristics (e.g., rationality) were closely aligned with traits associated with (stereotypical) masculinity. And many took men, or more accurately some subset of men, as their only or primary subject.

During this period, feminists in a number of sciences found that androcentrism, or "male centeredness," as well as sexism, informed research questions, methods, and hypotheses in their fields. They identified and criticized the emphasis on male behavior and activities in psychology, the social sciences, biobehavioral science, animal sociology, and various fields in the biological sciences. They criticized biological explanations for what were alleged to be differences in behavior, cognitive abilities, and temperament between the sexes. They also explored symbolic gender associations in general views about science, including how traits closely aligned with it, such as detachment and objectivity, were also strongly aligned with (stereotypical) masculinity. Finally, they detailed informal barriers women continued to confront in entering the sciences or succeeding in them. Both bodies of research would contribute to the emergence of feminist philosophy of science in the 1980s.

Developments in the philosophy of science in the 1970s have also influenced the emphases and methods of this tradition. Challenges in the preceding decades to what Suppe (1972) called "the received view" led not only to the abandonment of specific traditional positions, but also to changes in how those in the discipline viewed it. The challenges included arguments against the plausibility of an extrascientific "foundation" for science, and for the theory-ladenness of observation, underdetermination, and versions of holism (see Theories). Together with increased interest in the history of science and the details of scientific practice, both in part a response to the work of Kuhn, these arguments contributed to the emergence of more contextualist approaches in mainstream philosophy of science (see Kuhn, Thomas), to finely focused studies of the special sciences, of the role of community-specific standards in research, and of the role of other contingent and contextual (or external) factors in scientific practice. Like their colleagues, feminist philosophers have explored the implications of these developments, including how they might provide insights into the relationships between gender and science, and their implications for interpretive notions such as social constructivism and realism (see Scientific Realism; Social Constructionism).

Research in feminist philosophy of science has also developed apace with work in other traditions in science studies, including anthropology and sociology of science. Some feminists have found resources in one or more of these traditions (e.g., Barad 1996). Others maintain that there are important differences between their own methods and goals and those characterizing one or more of them (e.g., Harding 1986). In particular, although feminists are interested in understanding the role of social factors in scientific theorizing, many reject the epistemic relativism espoused in early work in the sociology and anthropology of science. Increasingly, however, work in these traditions recognizes the role of the material world in scientific practice, and further engagements between them and feminist philosophy of science are likely.

The traditions and developments noted do not receive equal treatment in this article, but each has influenced the trajectories of feminist philosophy of science.

Research Questions

Feminist philosophy of science encompasses a variety of methods and emphases, and these continue to evolve. But several questions, albeit differently formulated and pursued in the last two decades, inform much of the work undertaken in it. There is the general question:

What are the relationships between social relations (e.g., gender, race, class, culture) and methods (directions and/or content) of the sciences?

Feminist philosophers of science have explored several kinds of relationship between the sciences and their social contexts. As noted earlier, they study the ways in which assumptions concerning gender contribute to scientific questions, methods, hypotheses, and other aspects of scientific practice. In addition, there is increasing (though some would argue still insufficient) attention to the ways in which race, class, and other social relations impact the directions or content of science (e.g., Harding 1991; Schiebinger 2004; Weasel 2004). Conversely, feminists analyze the impact of scientific hypotheses and technologies on cultures as a whole and/or on specific groups, including the recurrent interest in establishing differences in abilities or behavior along the politically salient axes of gender and race (e.g., Schiebinger 2004; Weasel 2004). They also explore the nature and consequences of divisions in cognitive authority between women and men, scientists and laypersons, specialties within science, White persons and persons of color, and Western science and knowledge of other cultures (e.g., Addelson 1983 and 2003; Harding 1991).

These investigations differ in several respects from those that characterized mid-twentieth-century philosophy of science. For one thing, many feminists do not assume that the social or cultural identities of scientists, or divisions in cognitive authority and labor along the lines of gender or other social relations, are of no epistemological consequence. For another, many do not assume that all relationships between social interests and research compromise the science in question. These, they argue, are empirical issues to be investigated on a case-by-case basis (e.g., Anderson 2004; Harding 1986; Keller 1985; Longino 1990 and 1996; Nelson 1990 and 1996; Wylie 2004). Finally, many feminist analyses are empirically based, focusing on specific episodes in the history of science or contemporary research. The emphasis on case studies, also common in naturalized philosophy of science and philosophy of the special sciences, is in part a function of the developments in the philosophy of science noted above. But these methodological approaches and the empirical hypotheses underlying them also trace their roots to the internal science critiques leveled by feminist scientists in the 1970s and 1980s, which are summarized in the next section.

The role of epistemic values in scientific practice, such as simplicity, generality of scope, and conservatism, are of long-standing interest to philosophers of science, including feminists. Among feminists and others, there is now substantial interest in the question, What role(s) do *nonepistemic* values have in scientific practice and what role(s) should such values have?

Again reflecting a break with traditional approaches in the philosophy of science, feminists are among those who explore how nonepistemic values might have a positive role in science and how, whether positively or negatively, they can influence the directions or content of well-regarded research yet remain unrecognized (e.g., Longino 1990; Potter 1989). These two lines of empirical investigation yield another question: *If* relationships are found between social relations and science, and/or between science and so-called nonepistemic values, what are the *normative implications* for scientific practice and for the philosophy of science?

In the 1970s and 1980s, many feminist scientists took it to be an obvious implication of the research they analyzed that science was a more human and culturally bound activity than previously acknowledged in much mainstream philosophy of science, although recognized by previous work in other traditions (e.g., neo-Marxism). So stated, the empirical content of this view is quite vague and its normative implications are unclear. In the intervening years, feminist philosophers of science and scientists have worked to understand both.

The questions just outlined are not unique to feminist philosophy of science, so one might well ask, What work is done by *feminist* in "feminist philosophy of science"? In this article, 'feminist' locates a dynamic research tradition in relation to its history and a now extensive tradition of feminist science studies (cf. Alcoff and Potter 1993). This essay also follows Longino in understanding feminist philosophy of science as a *way of doing* the philosophy of science, not as a specific theory about science. As Longino puts this point, it is a way of engaging in the philosophy of science that reflects a commitment to not let gender "be disappeared," of studying its relationships to science and, as needed, working to change them (Longino 1994). Again, reflecting developments in a range of disciplines and approaches, many feminists have come to recognize that gender is an insufficient variable for understanding women's experiences, including those of women scientists, or for understanding the impact that the sciences have on women. Accordingly, recent work in feminist philosophy of science analyzes the role of other social relations, such as race and culture, in addressing the questions earlier outlined (e.g., the essays in Nelson and Wylie 2004) and makes increasing use of the work done in postcolonial science studies (e.g., Harding 1996 and Schiebinger 2004).

Its origins, core questions, and internal history delineate feminist philosophy of science as a recognizably distinct tradition. But the following discussion also reveals its strong relationships to the broader tradition of the philosophy of science. Feminists have appealed to and built on a number of contextualist approaches and positions, particularly neo-empiricism and naturalism, in the broader discipline. The contrasts earlier emphasized distinguish feminist approaches from mid-twentieth-century philosophy of science and from some still quite traditional approaches in philosophical epistemology.

Feminist Internal Science Critiques: Critical and Constructive

As noted earlier, feminist philosophy of science emerged in part in response to the analyses undertaken by feminist scientists. A continuing focus of feminist internal science critique is the relative underrepresentation of women and minorities in the sciences, and the informal barriers to full participation in the sciences that members of these groups have faced. It first arose as an issue of fairness. But feminist scientists would soon investigate the epistemic consequences of inequities in access to and opportunities in the sciences—that is, their potential consequences for the directions and content of science. It is to such consequences, and the questions they raised for traditional understandings of science, that this section is devoted.

In the 1970s and 1980s, feminists in the social sciences identified several levels of androcentrism in the research questions, methods, and theories of their fields—perhaps of most significance, in research not concerned to identify or explain sex or gender differences. They criticized methodological approaches to and accounts of social life that emphasized men's activities as defining the so-called public sphere and "culture" and associated women with the "private" sphere and reproductive activities, in turn treated as "natural" and without need of explanation. They also criticized the lack of attention to issues of concern to women, including gender discrimination in the workplace and violence against women. These issues, they argued, were of epistemic consequence. For one thing, such accounts of social life were at least incomplete because they ignored the productive and diverse nature of women's activities in specific cultural contexts, as well as phenomena such as rape and domestic violence. For another, they argued, the association of men with culture and production, and of women with nature and reproduction, obscured basic relationships between the domains so dichotomized. Analyses detailing these problems and proposing constructive alternatives were offered in economics (e.g., Hartmann 1981), sociology (e.g., Smith 1987), history (e.g., Kelly-Gadol 1976), anthropology (e.g., Rosaldo and Lamphere 1974), and human evolution (e.g., Tanner and Zihlman 1976). Finally, many argued that the fact that most scientists were White males was somehow implicated in the androcentrism and other biases they were identifying, although few maintained that the bulk of the problems were purposeful.

These critiques and alternatives were paralleled in other sciences. In psychology, feminists criticized models of psychological development and maturity based solely on research involving boys and men, and hypotheses that women's trajectory was "truncated" when it did not fit such models. They maintained not only that the models were likely to be empirically inadequate, but that the alternative trajectories that became visible when the subject pool was enlarged to include women suggested that social factors had more of a role in development than earlier models recognized. They also developed alternative models based on empirical research devoted to women and girls (e.g., Gilligan 1982).

In empirical psychology and developmental biology, feminists argued that laboratory-animal investigations into the effects of prenatal sex hormones on brain development, behavior, and temperament were characterized by circular reasoning and androcentric assumptions. For example, much research began from assumptions linking males with "aggressivity" and "spatial abilities," and females with "receptivity," assumptions that feminist scientists argued unduly influenced the nature of research tests and the interpretation of results (e.g., Bleier 1984). In animal sociology and biobehavioral science, feminists argued that stereotypical gender associations, such as that of males with aggression and dominance hierarchies and of females with passivity and reproduction, shaped organizing principles, observations, and hypotheses. They argued that the models generated not only were incomplete, but distorted relevant phenomena, and offered alternative observations, methods, and hypotheses (e.g., Haraway 1978).

In the 1980s, feminist biologists expanded their critiques beyond research concerned with a biological "origin" for alleged sex differences, to more subtle ways that androcentrism and other cultural assumptions informed the biological sciences. In embryology, they criticized a then exclusive emphasis on androgens and other features of male fetal development in models of "human" fetal development (e.g., Fausto-Sterling 1985). They criticized the imposition of gender connotations and sexual dimorphism on objects that are not sexed, including hormones, the nucleus and cytoplasm, and bacteria (e.g., Biology and Gender Study Group 1988). In evolutionary biology, they criticized the lack of attention to the selection pressures on females (e.g., Hubbard 1982). Feminist biologists (e.g., Bleier 1984) and biophysicist Keller (e.g., Keller 1985) also criticized linear and hierarchical models of biological processes that posited discrete entities and linear trajectories, including models of cellular protein synthesis that posited DNA as the "executive" of the process. Such models, they argued, reflect unwarranted assumptions about the ubiquity of single and dominant causes for complex processes and, in so doing, oversimplify the relevant processes. Some, including Keller, argued that a preference for linear, hierarchical models of biological processes was linked to masculine experience and self-identity issues in cultures fostering strong sex differences in temperament and behavior (Keller 1985). More plausibly, many biologists argued that such models functioned to support empirically inadequate and determinist explanations of alleged psychological and behavioral sex differences (e.g., as offered in human sociobiology in the 1970s and 1980s). As alternatives, they proposed multifactor and nonlinear models of a number of processes, including cellular protein synthesis, fetal brain development, and the relationships between genes and traits (e.g., Bleier 1984; Fausto-Sterling 1985; Keller 1985). These, they argued, make biological determinist explanations of traits and capacities highly implausible. Feminist scientists were not alone in relating linear models to biological determinism. Other scientists concerned with both the empirical adequacy and the political import of such models include Gould (e.g., 1981) and Lewontin (e.g., 1992).

In arguments both critical and constructive, feminist scientists appealed to the epistemic virtues of empirical adequacy, explanatory power, and generality of scope. They also explored social and political issues: how scientific theories can reinforce cultural beliefs and values, and the ways in which such beliefs and values can inform scientific theorizing. And they often identified epistemological questions. For instance, Were the cases involving androcentrism "bad" science and/or idiosyncratic, and thus without implication for "science as usual"? It seemed to many that the answer to this question was no. Social and cultural assumptions had been found to inform mainstream and credible research, much of it not concerned with sex differences, and few thought the androcentrism they were uncovering was conscious. In addition, many feminists recognized a relationship between their own feminist commitments and their ability to recognize problems that many of their colleagues had not (e.g., Bleier 1984; Hubbard 1982; Wylie 1996).

In considering the philosophical import of these internal science critiques, some emphasized social constructivist notions, citing "scientific facts" as human constructions, and scientific theories as "self-fulfilling prophecies" (e.g., Hubbard 1982, 7). But some also recognized the apparent tensions between their dual emphasis on and concern with the empirical and the political. Representative is Keller's (1985) argument that concluding that science is "just politics" would undermine the empirical force of feminist science critiques and that determining "how things are" is crucial in choosing effective courses of action for improving women's lives.

Themes in Feminist Philosophy of Science

The work of feminist philosophers of science is continuous with that of feminist scientists in two ways. Feminist philosophers also explore the role of androcentrism and other contextual factors in the sciences and, reflecting the emphasis in the broader field of philosophy of science, often do so through focused case studies. These include studies of the relative ignorance about female sexuality and anatomy (Tuana 2004), androcentric hypotheses in archaeology and related fields (Wylie 1996), and androcentrism in investigations into relationships between prenatal hormones and sex differences in behavior and/or cognitive abilities (Longino 1990; Nelson 1990). Feminist philosophers have also offered analyses of historical episodes, including those not overtly concerned with gender. For example, Potter has analyzed the role of then current debates concerning gender and Boyle's choice of one formulation of the ideal gas law over another formulation, equally compatible with available data, the metaphysics of which was aligned with liberal political positions concerning gender to which Boyle was opposed (Potter 1989).

Second, feminist philosophers of science often engage epistemological questions initially identified by feminist scientists—for example, What are the ways in which androcentrism can and does inform research and is it only bad science that is so informed? But they also engage the empirical questions cited at the outset of this essay, concerning (1) how precisely social relations and beliefs, such as androcentric and feminist perspectives, *can* come to inform scientific practice and (2) the normative implications for scientific practice and the philosophy of science of the findings of such investigations.

As will become clear, the approaches of feminist philosophers to both the empirical and normative issues often reflect developments in the philosophy of science in the second half of the twentieth century. Many are also keenly aware of the dangers noted by Keller of embracing a thoroughgoing social constructivism. Accordingly, they have sought to develop models of scientific practice and reasoning that encompass the role, suggested by feminist science critiques and developments in the philosophy of science, of both the natural/material world and contextual factors. Feminist models of science, many have argued, must take seriously constructivist insights into the historically and culturally contingent aspects of science "without sacrificing the ability to explain and justify its existence as a reliable, though not foolproof, process" (Alcoff 1989, 122). After all, these philosophers argue, feminist scientists appealed to evidence and to cognitive values such as empirical adequacy in both their critical and constructive engagements with science (e.g., Nelson 1990). Many argued that feminist models of science must also be able to reconceptualize objectivity in ways that disentangle epistemic adequacy from unattainable ideals of value freedom (Harding 1986; Longino 1990; Nelson 1990;

Wylie 2004). So understood, the challenge is to explain "with some precision" how science that is "good on all traditional criteria" can nonetheless be influenced by contextual values such as androcentrism or ethnocentrism (Potter 1989, 132), and how science informed by feminist values may be even better science. Feminist engagements with these issues are extensive, and only a representative sample is mentioned.

Harding's *The Science Question in Feminism* (1986) had a substantial impact on feminist philosophy of science in the 1980s and early 1990s. Harding identified three epistemological frameworks emerging in feminist science studies: feminist empiricism, feminist standpoint theory, and feminist postmodernism. She noted that each represented an effort to revise an earlier, nonfeminist tradition in light of the emergence of feminist science critiques. (The origins of standpoint theory are in Marxism.) Harding urged ambivalence toward the frameworks, predicting their further development in light of one another and feminist science studies. But her analysis was widely understood to suggest that feminist empiricism was the least promising. Harding (1986) argued that unlike the other two frameworks, central tenets of empiricism (her list included individualism and the distinction between contexts of discovery and justification) ruled out any relationship between social movements and scientific progress (24). If correct—that is, if such tenets are inseparable from empiricism, a claim other feminists would dispute (e.g., Longino 1990; Nelson 1990)—then feminist empiricists would be limited to claiming that androcentrism represents a failure to uphold traditional norms, would contribute no new insights into science, and would be unable to cite feminism as enabling feminist science critiques.

In contrast, Harding argued, feminist standpoint theory provides a framework for understanding the emergence of feminist science critiques precisely because it insists on relationships between power and knowledge. As Harding would later formulate this argument, those in dominant positions in political hierarchies (in this case, men) are at an epistemic disadvantage because their specific locations "organize and set limits on what [they] can understand about themselves and the world around them." From such perspectives, "the real relations of humans with each other and with the natural world are not visible" (Harding 1991, 54). Conversely, the activities and experiences of those disadvantaged (in this case, women) can provide a standpoint from which contradictions between reality and the dominant ideology could come to be recognized. This is not to say that such standpoints are inevitable. Given divisions in experiences and labor by gender, feminist standpoints are possible but are achieved through social movements. Harding argued that standpoint theory's hypothesis that some locations allowed for "better" knowledge than others could more reasonably explain both androcentrism in the sciences and feminist scientists' ability to recognize it, and avoid relativism.

In her 1986 analysis, Harding also argued that feminist postmodernist critiques of epistemology indicated that aspects of both feminist empiricism and feminist standpoint theory were problematic. She cited the work of Haraway, Flax, and others as constituting important challenges to "universalizing claims" about the power of reason, science, and the "subject/self" (Harding 1986, 28), which she and they saw as implicit in the other two frameworks.

In the intervening years, feminist philosophers of science have devoted considerable attention to understanding the ways in which scientists and other knowers are "situated" in socially and historically specific contexts, and the epistemic and normative implications of this hypothesis. Most have worked to develop understandings of science that recognize both the epistemic limits of any given location (political, disciplinary, historical, and so forth) *and* the constraints the world imposes. Most work in the tradition has also been committed to a symmetry thesis: that gender analysis is relevant to understanding not just bad science, but *good* science—indeed, even the best. In this respect, their work parallels the *methodological* relativism advocated in the Strong Programme in the Sociology of Knowledge (see Social Constructivism). But, for reasons cited above, many strongly reject the epistemic relativism earlier espoused by its advocates. In what follows, feminists' analyses are grouped according to Harding's three categories in cases in which their authors use them. It will be clear, however, that divisions between feminist empiricism and standpoint theory are increasingly less definitive and that although few feminist philosophers of science have wholeheartedly embraced postmodernism, the latter's arguments against universalizing claims have been influential.

Feminist standpoint theory has developed in response to criticism that early versions presupposed gender essentialism and did not adequately address the differences in women's lives along the axes of race, class, ethnicity, and culture, problems that Harding herself identified in her initial analysis. Some have incorporated divisions by race (e.g., Collins 1990) and class (e.g., Harstock 1983) into their analyses of women's standpoints. Harding and others have also worked to develop the implications of the deep divisions in perspectives and knowledge it suggests and whether, in this and other ways, it entails relativism. Two developments are significant. Harding has developed an argument for how individuals can "reinvent themselves as others," that is, learn and understand the standpoints of those differently situated. And she and others have developed the notion of strong objectivity, which calls not for "nonsituated" perspectives (unattainable according to standpoint theory) but for "reflexivity," for seeking to understand the limits of one's own perspective through active efforts to understand alternatives. Such reflexivity, standpoint theorists argue, can lead to theories that are "less false" (Harding 1991).

Feminist philosophers interested in developing empiricist models challenged Harding's account of empiricism, particularly her arguments that tenets such as those earlier noted are inseparable from empiricism and her account of the limitations of feminist empiricism (see Empiricism). Feminist empiricists have worked to understand and accommodate the evidence for situatedness and for contingent and discipline-specific features of scientific practice suggested by both recent research in the philosophy of science and feminist internal science critiques (e.g., Longino 1990; Nelson 1990). They too reject relativism, and many have adopted or developed holistic models of evidential relations. These assume the general thesis, advanced by Duhem, Hesse, Kuhn, and Quine, that individual hypotheses are constrained by data and convey evidential status on data only as part of larger bodies of theories (see Duhem Thesis; Kuhn, Thomas; Quine, Willard Van Orman). Feminist empiricists have used such models, together with Quine's thesis of underdetermination (see Theories, Underdetermination of), to engage the empirical questions earlier outlined: how androcentric and feminist assumptions can mediate the inferences drawn between data and hypotheses, the interpretation of research results, the hypotheses entertained, and the categories and methods of well-respected research (e.g., Alcoff 1989; Campbell 1998; Longino 1990; Nelson 1990; Potter 1989). In these efforts, they have argued that the bodies of theories within which hypotheses emerge and are accepted are not limited to those of science proper but include social and cultural beliefs, which can operate (often unrecognized) as background assumptions in specific research programs. Feminist empiricists have also explored how underdetermination can enable a role for nonepistemic values in scientific practice (Alcoff 1989; Longino 1990; Nelson 1990). At a more abstract level, some have explored how, in specific cases, nonepistemic values can contribute to the weight scientists attribute to epistemic values that, as Kuhn (1977) argued, cannot be realized simultaneously—for example, simplicity and empirical adequacy (e.g., Anderson 2004; Longino 1996; Ruetsche 2004).

As earlier noted, feminist standpoint theory emphasizes the role of social movements in enabling less distorted "angles of vision" (Wylie 1996). An emphasis on the social nature of science and the normative implications of this feature also characterize work in feminist empiricism. Some use holism, arguments against foundationalism, and the emergence of feminist science critiques to argue against reconstructions and explanations of scientific practice that focus on the reasoning capacities and methods of scientists *qua* individuals. As alternatives, they have developed models that take shared theories, standards, and practices of science communities as their primary focus and that understand the weighting of evidential warrant as a social (and contingent) rather than an individual achievement solely driven by logic and data (e.g., Longino 1990 and 1996; Nelson 1990 and 1996). Such approaches parallel those in so-called mainstream philosophy of science taken by "social empiricists" (e.g., Solomon 2001). Although few advocate scientific realism (Campbell 1998 is an exception), feminist empiricists contend that research assumptions and methods, hypotheses, and the interpretation of data can and should be assessed on the basis of empirical adequacy and other cognitive values.

At the same time, because they recognize that different background assumptions, theories, and interests are at work in contemporary science (of which androcentric and feminist perspectives are but two examples), recent models in feminist empiricism, like others in the philosophy of science, understand the sciences to be more "disunified" than early versions of holism recognized (see Unity and Disunity of Science). Indeed, proposals for ways to enable reflexivity on the part of scientists draw on the perception of such disunities. Some propose more diverse science communities and the development of norms that encourage conceptual discussion and debate, on the grounds that they would lead to more empirically adequate theories (e.g., Longino 1990; Nelson 1990), while others study how disunities have led to advances in specific research programs (e.g., Nelson 1996; Wylie 2004).

Important work in feminist philosophy of science is neither self-identified nor easily categorized in terms of the three frameworks Harding identified.

Barad's and Wylie's work are representative. Barad has developed "agential realism" as a model of the epistemology of science that is both realist and social constructivist (see Scientific Realism; Social Constructionism). She builds from the epistemology she attributes to Bohr to argue that the "phenomena" that are the actual objects of scientific study are simultaneously the products of human construction and of nature. Wylie's work incorporates features of feminist empiricism as well as of feminist standpoint theory (e.g., Wylie 2004). Her model of evidential warrant is like that advocated by feminist empiricists in assuming underdetermination, taking "data" to be the least defeasible aspect, and in seeing the relationship between data and hypotheses as mediated by background assumptions. But Wylie also builds on Hacking's work to argue that the more evidence supporting a hypothesis enjoys a degree of horizontal and/or vertical independence from it, the stronger the hypothesis. In these arguments, Wylie makes use of the standpoint hypothesis that different angles of vision yield different and, in some cases, better hypotheses, and supports her account with case studies from archaeology (Wylie 1996).

Recent work in feminist empiricism is characterized by an understanding that situatedness, holism, and underdetermination entail the contingency of all knowledge claims, including those of science. This view, common also to feminist standpoint theory, is generally taken to call for an understanding of objectivity different from the traditional "the view from nowhere" as well as from "the view from everywhere" associated with relativism. Not unlike the strong objectivity advocated by standpoint theorists, feminist empiricists and some feminist postmodernists have argued that reflexivity is a necessary condition for objectivity in doing science and in the philosophy of science. Representative is Haraway's notion of "partial vision," a metaphor for situatedness and contingency, in which she makes use of both standpoint and postmodern insights:

> Not so perversely, objectivity turns out to be about particular and specific embodiment, and definitely not about the false vision promising transcendence of all limits and responsibility.... Feminist objectivity is about limited location and situated knowledge, not about transcendence and splitting of subject and object. (Haraway 1988, 190)

Directions

A special issue of *Hypatia*, a journal of feminist philosophy, was published in 2004 devoted to feminist science studies (Nelson and Wylie 2004). It reflected many of the themes that have characterized feminist engagements with science and the philosophy of science. A number of its authors were scientists, and the collection as a whole contained numerous cross-disciplinary engagements. The symmetry thesis prominent in the mid- and late 1980s and the interest in carving out a middle ground between oppositional interpretive positions such as scientific realism and social constructivism remained prominent. Notably, however, few contributors felt the need to "defend" feminist science studies from charges of relativism or irrationality, something that feminists did feel the need to do in the 1980s and early 1990s, or to offer full-blown arguments for situatedness, contingency, and constraint. Developments in science studies disciplines, including the philosophy of science, are seen to have made such arguments unnecessary. There were also analyses that advanced the study of relationships among race, gender, culture, and science (Schiebinger 2004; Weasel 2004).

The collection suggested new directions. Some authors extended and revised earlier feminist arguments. For example, some offered more substantive analyses of the nature of noncognitive values and of ways in which they could inform scientific practice than had earlier analyses (Anderson 2004; Ruetsche 2004), and some provided more substance to earlier feminist arguments that science was inherently "social" (Sobstyl 2004). One author (Okruhlik 2004) called on feminist philosophers of science to reassess the Vienna Circle and the usefulness to feminist science studies of the arguments and positions developed by some of its members, such as Neurath (see Neurath, Otto; Vienna Circle). Perhaps also predictive, few contributors identified their methods or approaches as "empiricist," "standpoint," or "postmodernist," although they often incorporated one or more views earlier associated with these approaches. Finally, reflexivity was emphasized by many authors, who discussed what it meant to practice science and/or to engage in the history or philosophy of science as a feminist.

LYNN HANKINSON NELSON

References

Addelson, Kathryn Pyne (1983), "The Man of Professional Wisdom," in S. Harding and M. Hintikka (eds.), *Discovering Reality*. Dordrecht, Netherlands: Kluwer, 165–186.

——— (2003), "Naturalizing Quine," in L. H. Nelson and J. Nelson (eds.), *Feminist Interpretations of Willard Van Quine*. University Park: Penn State University Press, 241–268.

Anderson, Elizabeth (2004), "Uses of Value Judgments in Science," *Hypatia* 19: 1–24.

Alcoff, Linda Martin (1989), "Justifying Feminist Social Science," in N. Tuana (ed.), *Feminism and Science*. Bloomington: Indiana University Press, 85–103.

Alcoff, L. M., and Elizabeth Potter (eds.) (1993), *Feminist Epistemologies*. New York and London: Routledge.

Barad, Karen (1996), "Meeting the Universe Halfway: Realism and Social Constructivism without Contradiction," in L. H. Nelson and J. Nelson (eds.), *Feminism, Science, and the Philosophy of Science*. Dordrecht, Netherlands: Kluwer, 161–194.

Biology and Gender Study Group (1988), "The Importance of Feminist Critique for Contemporary Cell Biology," *Hypatia* 3: 61–76.

Bleier, Ruth (1984), *Science and Gender*. New York: Pergamon Press.

Campbell, Richmond (1998), *Illusions of Paradox: A Feminist Epistemology Naturalized*. Ithaca, NY: Cornell University Press.

Collins, Patricia Hill (1990), *Black Feminist Thought*. Boston: Unwin Hyman.

Fausto-Sterling, Anne (1985), *Myths of Gender*. New York: Basic Books.

Gilligan, Carol (1982), *In a Different Voice*. Cambridge, MA: Harvard University Press.

Gould, S. J. (1981), *The Mismeasure of Man*. New York: Norton.

Haraway, Donna (1978), "Animal Sociology and a Natural Economy of the Body Politic, Part I," *Signs* 4: 21–36.

——— (1988), "Situated Knowledges," *Feminist Studies* 14: 575–599.

Harding, Sandra G. (1986), *The Science Question in Feminism*. Ithaca, NY: Cornell University Press.

——— (1991), *Whose Science? Whose Knowledge? Thinking from Women's Lives*. Ithaca, NY: Cornell University Press.

——— (1996), "Multicultural and Global Feminist Philosophies of Science: Resources and Challenges," in L. H. Nelson and J. Nelson (eds.), *Feminism, Science, and the Philosophy of Science*, Dordrecht, Netherlands: Kluwer.

Harstock, Nancy (1983), "The Feminist Standpoint," in S. Harding and M. Hintikka (eds.), *Discovering Reality*. Dordrecht, Netherlands: Kluwer Academic Press, 283–310.

Hartmann, Heidi (1981), "The Family As the Locus of Gender, Class, and Political Struggle," *Signs* 6: 366–394.

Hubbard, Ruth (1982), "Have Only Men Evolved?" in R. Hubbard, M. Henifin, and B. Fried (eds.), *Biological Woman—The Convenient Myth*. Cambridge, MA: Schenkman, 7–36.

Keller, Evelyn Fox (1985), *Reflections on Gender and Science*. New Haven, CT: Yale University Press.

Kelly-Gadol, J. (1976), "The Social Relations of the Sexes: Methodological Implications of Women's History," *Signs* 1: 809–823.

Kuhn, Thomas (1977), *The Essential Tension: Selected Studies in Scientific Tradition and Change*. Chicago: University of Chicago Press.

Lewontin, R. C. (1992), *Biology as Ideology: The Doctrine of DNA*. New York: HarperPerennial.

Longino, Helen E. (1990), *Science as Social Knowledge*. Princeton, NJ: Princeton University Press.

——— (1994), "In Search of Feminist Epistemology," *The Monist* 77: 472–485.

——— (1996), "Cognitive and Non-Cognitive Values in Science: Rethinking the Dichotomy," in L. H. Nelson and J. Nelson (eds.), *Feminism, Science, and the Philosophy of Science*. Dordrecht, Netherlands: Kluwer, 39–58.

Nelson, Lynn Hankinson (1990), *Who Knows: From Quine to Feminist Empiricism*. Philadelphia, PA: Temple University Press.

——— (1996), "Empiricism without Dogmas," in L. H. Nelson and J. Nelson (eds.), *Feminism, Science, and the Philosophy of Science*. Dordrecht, Netherlands: Kluwer, 95–120.

Nelson, Lynn Hankinson, and Alison Wylie (eds.) (2004), Special Issue of *Hypatia: Feminist Science Studies* 19(1).

Okruhlik, Kathleen (2004), "Logical Empiricism, Feminism, and Neurath's Auxiliary Motive," *Hypatia* 19: 48–72.

Potter, Elizabeth (1989), "Modeling the Gender Politics in Science," in N. Tuana (ed.), *Feminism and Science*. Bloomington: Indiana University Press, 132–146.

Ruetsche, Laura (2004), "Virtue and Contingent History: Possibilities for Feminist Epistemology," *Hypatia* 19: 73–101.

Rosaldo, Michelle Z., and Louise Lamphere (eds.) (1974), *Woman, Culture, and Society*. Stanford, CA: Stanford University Press.

Schiebinger, Londa (2004), "Feminist History of Colonial Science," *Hypatia* 19: 233–254.

Smith, Dorothy (1987), *The Everyday World As Problematic: A Feminist Sociology*. Boston: Northeastern University Press.

Sobstyl, Edrie (2004), "Re-Radicalizing Nelson's Feminist Empiricism," *Hypatia* 19: 119–141.

Solomon, Miriam (2001), *Social Empiricism*. Cambridge, MA: Bradford Books/MIT Press.

Suppe, Frederick (1972), "What's Wrong with the Received View on the Structure of Scientific Theories?" *Philosophy of Science* 39: 1–19.

Tanner, Nancy, and A. Zihlman (1976), "Women in Evolution," *Signs* 1: 585–608.

Tuana, Nancy (ed.) (2004), "Coming to Understand: Orgasm and the Epistemology of Ignorance," *Hypatia* 19: 194–232.

Weasel, Lisa H. (2004), "Feminist Intersections in Science: Race, Gender and Sexuality through the Microscope," *Hypatia* 19: 183–193.

Wylie, Alison (1996), "The Constitution of Archaeological Evidence: Gender Politics and Science," in P. Galison and D.J. Stump (eds.), *The Disunity of Science*. Stanford, CA: Stanford University Press, 311–343.

——— (2004), "Why Standpoint Matters," in R. Figueroa and S. Harding (eds.), *Science and Other Cultures*. New York: Routledge, 26–48.

See also **Empiricism; Instrumentalism; Logical Empiricism; Scientific Realism; Social Constructivism; Unity and Disunity of Science**

PAUL KARL FEYERABEND

(13 January 1924–11 February 1994)

A Viennese émigré, Paul Feyerabend taught philosophy of science wherever his restless nature brought him—especially Berkeley, California; London, Auckland, Berlin, and Zürich. His views on methodology and the politics of science established him as one of the most controversial, eccentric, and outrageous figures in contemporary philosophy. Allegedly an irrational thinker, Feyerabend was in fact a skeptical master and iconoclast about the sciences and their philosophy. He denounced the gap between abstract normative philosophical accounts of science and actual, complex, and context-dependent scientific practice. He argued against the hegemony of any intellectual or ideological vision to promote the advantages of tolerance and pluralism in science as well as in society. His anarchistic theory of knowledge and the willingness to question the supremacy of Western scientific rationality vis-à-vis other "forms of life" made him famous beyond the boundaries of the philosophy of science.

A Philosophical Life Spent "Killing Time" and Scientific Idols

Paul Karl Feyerabend was born in Vienna in 1924. As a young man he was attracted to physics, mathematics, and astronomy (a passionate observer through the telescope he built with his father), as well as to drama, cinema, singing, and opera. Four years after the anschluss of Austria by the Third Reich in 1938, he was drafted into the Nazi work service and later entered the German army.

Posted to battle on the Russian front, he was awarded the Iron Cross. The end of the war saw him recovering from a bullet wound in his spine, which was to leave him crippled. He was granted state funding to study singing and stage management, and also cultivated Italian, harmony, piano, and diction. He then decided to study history and sociology in Vienna, but soon changed to theoretical physics and generally adhered to a positivistic scientism, which regarded science as an empirical activity and the basis of all knowledge.

During the following years Feyerabend received his Ph.D. in Philosophy with a dissertation on "basic statements" supervised by Viktor Kraft, and crossed Karl Popper's path for the first time (see Popper, Karl Raimund). He also met Bertholt Brecht, turning down an offer to work as his production assistant ("one of the greatest mistakes of my life," he would later say, adding, however, that as with Marxism and the army, he would probably not have enjoyed the gregarious group mentality prevalent in Brecht's circle).

In 1952, Feyerabend left for Cambridge, England, hoping to study under Wittgenstein; when the latter died, Feyerabend turned to the London School of Economics, where he was supervised by Popper, and genuinely embraced falsificationism. His adherence to it, however, was fairly unorthodox, combining realism and the view that all (observational) terms are theoretical with the principle of tenacity (the idea that it is rational to keep working on a theory despite empirical anomalies) and theoretical pluralism. A year later, he declined the offer of a job as Popper's assistant and left for Vienna.

In 1955, the University of Bristol, England, granted him his first academic post as lecturer in philosophy of science. During the following years Feyerabend confirmed his decision to cut all ties with what he later called the "Popperian Church," a group of scholars who preached but did not practice the critical attitude that plays a central role in Popper's philosophy. From 1958 to 1990 (the year he tendered his official resignation), Feyerabend was lecturer and then professor at the University of California–Berkeley, spending much time both in the United States (Yale University and Minnesota) and abroad (London, Berlin, Auckland, Brighton, Kassel), wherever his restlessness and growing fame took him. During the 1980s, Feyerabend accepted a chair at the Zürich Polytechnic ("ten wonderful years of half-Berkeley, half-Switzerland"). Struck by a brain tumor, he died on February 11, 1994, in Grenolier, Switzerland.

Explanation, Reduction, and Empiricism (Feyerabend 1962) marks both Feyerabend's departure from a foundationalist conception of experience and his endorsement of some of Wittgenstein's later views. Feyerabend argues against the logical empiricist accounts of explanation, theoretical reduction, and meaning invariance (see Explanation; Logical Empiricism; Reductionism). He also derives the methodological implications of his "contextual theory of meaning" and "incommensurability thesis" based on detailed historical examples. During his frequent visits to the London School of Economics, Feyerabend met Imre Lakatos, who encouraged him to collect the impertinent ideas expounded in his lectures about the nonexistence of scientific method. Lakatos was supposed to reply and defend rationality, but their joint project—provisionally titled *For and Against Method*—was never completed. Lakatos unexpectedly died in 1974, and Feyerabend's part of the project, *Against Method*, was ultimately published as a collection of essays (Feyerabend 1975). The publication of the long correspondence between Feyerabend and Lakatos (Feyerabend and Lakatos 1999) partially filled this gap and fully acknowledged the dialectical exchange of ideas between the two friends that helped sharpen Feyerabend's attack on the rationalist position.

While *Against Method* denounces the dichotomous and enigmatic relation between philosophical theories and scientific practice and advocates the freedom of science from the interference of philosophy, *Science in a Free Society* (Feyerabend 1978) argues for the freedom of all "forms of life" from the interference of science. In this book, Feyerabend complains about the "illiteracy" with which his previous book was received, but also elaborates on the political consequences of his epistemological anarchism, argues for the separation of science and state and for the equal right to survival and access to power of all traditions (including those in conflict with accepted "scientific truths"). Feyerabend's corrosive skepticism is here directed toward the uncontrolled and uncritical, yet all-powerful, authority of "scientific expertise." Feyerabend claims that in order to defend society against science, the latter has to be placed under the supervision of democratic councils of laymen—with the aim of assessing and counterbalancing experts' judgments and decisions.

Feyerabend's attempt to dethrone science from its privileged position within Western culture is also carried on in his later writings collected in *Farewell to Reason* (Feyerabend 1987), a *sui generis* apology for cultural relativism. Reviving John Stuart Mill's argument on the means of cultivating human flourishing, Feyerabend argues that the freedom of a society increases as the restrictions imposed on its traditions are removed. Moreover, societies that contain many traditions side by side and stimulate cultural diversity have a better chance to enhance both the quality of the traditions and the maturity of their citizens. The citizens, in turn, should be prepared to use the standards of the traditions to which they belong to judge and supervise the institutions. In Feyerabend's view, this constitutes the best antidote to cultural and political totalitarianism.

The Refutation of Classical and Logical Empiricism, or How to Be a Good Empiricist

Feyerabend's first iconoclastic enterprise is directed against philosophical empiricism: the view that what is to be believed is what experiences establish, and no more. In fact, Feyerabend's line of attack is broad and applies to any foundationalist epistemology (see Epistemology). A naïve appeal to experience assumes that the meaning of observational terms is unequivocally determined by the procedures of observation such as looking, listening, and the like, and that scientific theories can be grounded in independently meaningful facts thus established. To Feyerabend, this view is at variance with actual scientific practice. Moreover, empiricism in the form theorized by logical empiricist philosophers cannot contribute to the growth of knowledge; on the contrary, it is bound to lead to "a dogmatic petrifaction" of theories and "the establishment of a rigid metaphysics" (Feyerabend 1999a, 82).

Feyerabend's argument moves from the consideration that theories are all-pervading conceptualizations of the world and determine the vocabulary that is used in building up "facts." This is in particular the case with the observation-language reputed to ground scientific theories (see Observation; Theories). Feyerabend's first main thesis is that "the interpretation of an observation-language is determined by the theories we use to explain what we observe, and it changes as soon as those theories change" (1981, 31).

In principle, according to Feyerabend, all observational terms are fully theoretical, and there is no semantic difference between theoretical terms and observational terms. Thus, observational terms are neither certain nor stable but share the hypothetical and changing nature of theoretical terms. The consequences for the relation between theory and experience are radical. Crucially, if meanings of observational terms depend on the universal principles of the theory in which they are used, terms that depend on different universal principles will not share the same meaning. Feyerabend then, anticipating some of Kuhn's ideas, argues that theory testing cannot be a matter of confrontation of theory and (theory-laden) empirical data; rather it is a matter of competition between theories that are in part mutually exclusive, or *incommensurable* (see Incommensurability; Kuhn,Thomas).

Theories are incommensurable when the universal principles used to determine the concepts within one theory "suspend" the universal principles of the other, and thus all its facts and concepts. Classical Newtonian mechanics, for example, is said to be incommensurable with relativistic mechanics on the basis that the latter rejects a universal principle of the former "that shapes, masses, periods are changed only by physical interactions" (Feyerabend 1975, 269–271). Consider, in particular, the concept of 'length.' In classical mechanics, length is a relation that is independent of signal velocity, gravitational fields, and the motion of the observers; whereas in relativistic mechanics the value of length depends on these very concepts. The switch from classical mechanics to relativity entails a change of meaning of spatio-temporal concepts (see Classical Mechanics; Space-Time). Classical length and relativistic length are incommensurable notions, and classical mechanics is not explained by, or "reducible to," Einstein's relativity theory (Feyerabend 1981, 76–81). In general, according to Feyerabend, any attempt to derive the universal principles of an old theory from those of a new one necessarily leads to a change of the meanings in the old theory's terms. And this is why the "theoretical reduction" fostered by the orthodox account of explanation is not viable. Feyerabend's second main thesis is thus that there is not any reduction of a theory to another in actual science, but rather a replacement of one theory and its "ontology" with another (1999a, 86–87).

The question now is raised of "how to be a good empiricist." For Feyerabend a good empiricist is a *critical metaphysician:*

> His first step will be the formulation of fairly general assumptions which are not yet directly connected with observations; this means that his first step will be the invention of a new metaphysics. This metaphysics must then be elaborated in sufficient detail in order to be able to compete [with] the theory to be investigated as regards generality, details of prediction, precision of formulation.... Elimination of all metaphysics, far from increasing the empirical content of the remaining theories, is liable to turn these theories into dogmas. (1999a, 102)

However, it should be noticed that contrary to what many critics have claimed, Feyerabend's incommensurability thesis should not be interpreted as maintaining that competing theories cannot be compared. What his thesis entails is that theories cannot be compared in the ways in which many philosophical accounts of scientific explanation and reduction have thought that such comparisons should occur. To reject these accounts is to raise problems about certain philosophical theories of science; it is not to raise any difficulties for scientific practice itself (1981, xi).

Against (Too Much) Method

Against Method aims at demystifying another philosophical idol: the existence of a strictly binding system of rules for (good) scientific practice. Feyerabend highlights the huge gap between the "real thing" (science) and the various images of science. His therapy for philosophers' schizophrenic detachment from scientific reality is methodological anarchism. The therapy is the result of historical analyses. In particular, careful historical investigation supports the thesis that

> [t]here is not a single rule, however plausible, and however firmly grounded in epistemology, that is not violated some time or another.... Such violations are not accidental events. On the contrary we see they are necessary for progress.... The Copernican Revolution, the rise of modern atomism, the gradual emergence of [the] wave theory of life, occurred because some thinkers either *decided* not to be bound by certain "obvious" methodological rules, or because they *unwittingly broke* them. (1993, 14)

If this is the case, then any attempt to reform science by bringing it closer to the abstract image philosophers have of *the* scientific method is bound to damage science. On the contrary, "the only principle that does not inhibit progress is: *anything goes*"(1993, 5). Anything goes (perhaps paradoxically) is also the only general principle to which the coherent rationalist can be committed if looking for a rule valid in all given historical situations. But at the same time—at least in Feyerabend's intention—it is not introduced to replace one set of general rules by another set, but rather "to convince the reader that all methodologies, even the most obvious ones, have their own limits" (1993, 23). Consider for example the application of a clear, well-defined, and well-regarded rule like the consistency condition. According to it, the new hypotheses should agree with the accepted theories. But for Feyerabend it is not a reasonable condition at all. In fact, instead of being of help in obtaining better theories, it is just a factor for preservation of the old ones. Hypotheses contradicting well-confirmed theories should proliferate and not be restricted, because they help provide (theory-laden) evidence that cannot be obtained in any other way. Consider also the rule that a theory that contradicts experience should be excluded from science. This rule, Feyerabend claims, is violated at every run:

> [T]heories are refuted in every moment of their existence ... *ad hoc* hypotheses patch up gaps in the proofs and cracks in the connection of facts. And internal contradictions are almost never avoided. We do not have proud cathedrals standing before us, instead we have dilapidated ruins, architectural monstrosities whose precarious existence is laboriously prolonged through ugly patch-work by their constructors. This is scientific reality. (1981, 156)

Scientific reality is always too rich in content, too varied, too many-sided, too lively and subtle to be captured by the simple-minded rules of even the best philosophers or historians. Scientists are not rule-followers but opportunists. In the construction of their conceptual world, they cannot be restricted by the adherence to any epistemological system; rather they rely "now on one trick now on the other" (1993, 1). Galileo Galilei's cunning defense of the heliocentric cosmology is paradigmatic in this respect. According to Feyerabend, not only did Galileo develop a research program in striking contrast to the Aristotelian standards and the accepted observation of the time, he was also prepared to defend it by substituting a "natural" interpretation of motion (motion can be expressed only in terms of observable changes) with an "unnatural" and highly theoretical concept, which introduced into the phenomenon of motion some components (such as circular inertia) that cannot be observed. In this way Galileo was able to "defuse a mine" placed under the Copernican system by explaining away the objection regarding the motion of the Earth. This move was possible because people see a phenomenon and interpret it in what they regard as a natural way according with their beliefs. So it is the *interpretation* of the phenomenon and *not* the phenomenon itself that is in contradiction with a given belief. Galileo then resolved the contradiction between empirical observation and the Copernican view by providing a new and highly abstract observational language and thus a newly constructed empirical basis. This, in turn, was a new theory of interpretation (containing the idea of the relativity of motion and the law of circular inertia) fitting the Copernican system (Feyerabend 1993, 55–85).

Galileo also changed the "sensory core" of observational statements that seemed to contradict Copernicus. He claimed to have removed them with the help of a 'superior and better sense' for astronomical matters, the telescope. However, Feyerabend points out, Galileo had no theoretical reasons to support the conclusion that the telescopic phenomena are more veridical than observations by the unaided eye. Once again, behind the clashes of the senses, there was a clash of theoretical assumptions, explicit or not. Galileo chose the research program that promised him the most exciting discoveries and adopted propaganda strategies in which reason was not enough to defend it against the widely accepted methodological canons:

> We see that Galileo's view of the origin of Copernicanism differs markedly from the more familiar historical accounts. He neither points to new facts which offer inductive support [for] the idea of a moving earth, nor does he mention any observations that would refute the geocentric point of view but be accounted for by Copernicanism. On the contrary, he emphasizes that not only Ptolemy, but Copernicus as well, is refuted by the facts, and he praises Aristarchus and Copernicus for not having given up in the face of such tremendous difficulties. He praises them for having proceeded *counterinductively*. (Feyerabend 1993, 80–81)

That is, Galileo wins the battle against the Ptolemaic system by subverting the most carefully established observational results and challenging the most plausible theoretical principles.

The Value of Theoretical Pluralism

Counterinduction can be beneficial to the advancement of science. Even Feyerabend's anarchism, then, provides some positive prescriptions. In particular, counterinductive hypotheses are valuable because they provide a means of criticizing accepted theories in a manner that goes beyond the comparison of the theories with the "facts." He says that

> the only way of arriving at a useful judgment of what is supposed to be the truth, or the correct procedure, is to become acquainted with the widest possible range of alternatives.... The reasons were explained by John Stuart Mill in his immortal essay *On Liberty*. It is not possible to improve upon his arguments. (Feyerabend 1978, 86)

One of the arguments Feyerabend is referring to is that silencing the expression of an opinion robs the human race by reducing the opportunity to ascertain truth. The role of tolerant controversy in grounding knowledge is so important that, according to Mill, "if opponents of all important truth do not exist, it is indispensable to imagine them, and supply them with the strongest arguments which the most skilful devil's advocate can conjure up" (Mill [1859] 1977, 229).

Accordingly, science should be organized to generate the continuous generation of alternatives, to strengthen anomalies, and to stimulate controversies. The legacy of Mill's liberal standpoint is what Feyerabend calls the principle of proliferation: "Invent, and elaborate theories which are inconsistent with the accepted point of view, even if the latter should happen to be highly confirmed and generally accepted" (Feyerabend 1981, 105). Of course, knowledge generated by such a principle is of a peculiar sort: It is not a series of self-consistent theories that converges toward an ideal view; it is not a gradual approach to the truth. Rather,

> It is an ever increased ocean of mutually incompatible alternatives. Each single theory, each fairly-tale, each myth is part of the collection forcing others into greater articulation and all of them contributing, via this process of competition, to the development of our consciousness. (1993, 21)

As a consequence, "experts and laymen, professionals and dilettanti, truth-freaks and liars—they all are invited to participate in the contest and to make their contribution to the enrichment of our culture" (Feyerabend 1993, 21). Democratic participation in scientific matters warrants the advocacy of minority opinions and thus sustains the conditions for scientific development and human flourishing. This last consideration leads to the question of "science versus democracy," which in his later years Feyerabend (1993, 3) regarded as most important: "My main motive is humanitarian, not intellectual . . . . I want to support people, not 'advance knowledge.'" In particular, provided that there is no abstract canon ensuring success in any given field of enquiry, and that scientific achievements can be judged only *after* the event, Feyerabend claims that scientists are no better off than anybody else in these matters. The public, therefore, not only can take part in scientific decisions, but *should* do so:

> [F]irst, because it is a concerned party; secondly, because such participation is the best education the public can get—a full democratisation of science is not in conflict with science. It is in conflict with a philosophy, often called "Rationalism" that uses a frozen image of science to terrorize people with its practice. (1993, xii)

So the humanitarian motive behind Feyerabend's debunking of science is clear: Scientists should adapt their procedures and goals to the values of the people they are supposed to advise.

Feyerabend is not against science so understood—"Such a science is one of the most wonderful inventions of the human mind"—he is "against ideologies that use the name of science for cultural murder" (1993, 4).

Relativism and Beyond

Two more consequences emerge from the thesis that the sciences have no common structure, but local and distinct features. First, "the success of 'science' cannot be used as an argument for treating as yet unresolved problems in a standardized way" (Feyerabend 1993, 2); second, "'non-scientific' procedures cannot be pushed outside by arguments" (ibid). The political implication of this epistemological stand is *democratic relativism*, the view that all traditions have equal rights. Democratic relativism, in turn, denies the right of traditions to impose their "form of life" on others, and therefore recommends the protection of traditions from interference from outside, including the interferences of the tradition of Western scientific rationalism. A new question then arises: How is a citizen to judge the suggestions issuing from the institutions that surround him? It is assumed that the citizen will judge "rationally," that is, in accordance with some scientific standards. However, there are no unambiguous scientific standards. Feyerabend's answer is that in a "free society," a citizen will use the standards of the tradition to which the citizen belongs: "Hopi standards if he is Hopi; fundamentalist Protestant standards if he is Fundamentalist; ancient Jewish standards if he belongs to a group trying to revive ancient Jewish traditions" (Feyerabend 1999a, 220). To those who claim the superiority of Western achievements over other traditions, Feyerabend simply objects that such a claim needs to be backed up by comparative studies:

> The sciences, it is said, are uniformly better than all alternatives—but where is the evidence to support this claim? Where, for example, are the control groups which show the uniform (and not only the occasional) superiority of Western scientific medicine over the medicine of the *Nei Ching*? Or over Hopi medicine? Such control groups need patients that have been treated in the Hopi manner, or in the Chinese manner using Hopi experts and experts in traditional Chinese medicine. (1999a, 221)

In his later years, however, Feyerabend acknowledged that relativism can run into trouble: It reflects on traditions "from afar" in an abstract and unrealistic way. Traditions are not closed units, they are not frozen systems of thought:

Traditions not only have no well-defined boundaries, but contain ambiguities and methods of change which enable their members to think and act as if no boundaries existed: *potentially every tradition is all traditions*. Relativizing existence to a single "conceptual system" that is then closed off from the rest and presented in unambiguous details mutilates real traditions and creates a chimera. (quoted in Munévar 2000, 76)

The same, of course, applies to the tradition of scientific rationality. If scientific rationality were characterised as a well-defined, unambiguous, and "closed" system of rules, then relativism would be correct. On the contrary, scientific theories are not unified semantic domains with rigid borders; they change, they borrow from others, and they adapt to new situations. And so is the case for scientific procedures and value judgments; they are continually adapted to circumstances in an open-ended historical process. After all, Feyerabend clarifies, incommensurability is a difficulty for philosophers, not for scientists—the latter being "experts in the art of arguing across lines which philosophers regard as insuperable boundaries of the discourse" (Feyerabend 1987, 272).

Posthumous Works and the Legacy of Paulus Empiricus

At the time of his death, Feyerabend was working on *The Conquest of Abundance* (1999b), developing the theme of how different traditions, or forms of life, can learn from each other and can grow out of each other. The target here (as elsewhere) is the hegemony of any intellectual or ideological single vision—in particular the entire tradition of rationalism and its heirs. The subtitle ("A Tale of Abstraction Versus the Richness of Being") hints at the poverty of the "reality" produced by the method of abstraction typical of Western thought, compared with the abundance, richness, and boundless variety of the world around us. In a "Letter to the Reader" (quoted in Hacking 2000), Feyerabend also makes clear how to approach this text and possibly all his work, which he regarded as specially constructed plays to be performed in the theatre of ideas:

I want you to sense chaos where first you noticed an orderly arrangement of well-behaved things and processes.... This, my dear reader, is the warning I want you to remember from time to time and especially when the story seems to become so definite that it almost turns into a clearly thought-out and precisely structured point of view. (Hacking 2000, 28)

Feyerabend's (1996) fascinating autobiography shows that he often changed his mind on a variety of subjects, but it also proves that he was neither the worst enemy of science, as depicted by some of his commentators, nor the irrationalist philosopher criticized by most of the profession. He was a skeptic about the foundations of knowledge and a cunning rhetorician who knew how to use effectively all the ancient skeptical tropes. (Feyerabend used to entertain Lakatos by signing his letters and postcards as *Paulus Empiricus*.) Skepticism to him was not only a powerful rhetorical device but also highly regarded for its normative implications for the practice and the role of science in a "free society." Feyerabend's iconoclastic enterprise is against neither reason nor science. It is against the idea that there is some unique set of rules (whatever it is) that ought to be followed in order to produce good science (whatever that is). Feyerabend's favorite slogan, *anything goes,* "is a jocular summary of the predicament of the rationalists" (Feyerabend 1978, 188)—thus "anything goes" from the point of view of the rationalist who believes that only *the* scientific method is admissible. On the contrary, there are lots of ways of moving forward, including the different local and contextual methods of various sciences or traditions. (If anything goes, reason sometimes goes as well; thus Feyerabend is not guilty of any inconsistency by employing rational arguments to attack the rationalist positions he opposes.) In this respect, Feyerabend can be seen not as rejecting rationality *tout court*, but rather as urging a conception of rationality wider than that embodied in some existing version of scientific rationalism (Preston 1997, 203). Feyerabend's arguments are generally to be intended as a *reductio* against certain forms of rationalism, rather than positive arguments in favor of irrationalism (Munévar 2000, 63–64). Far from a self-defeating skepticism, Feyerabend presented an impressive challenge to the received view in the philosophy of science. He argued that its elegant but useless epistemological accounts should be substituted by a detailed study of the primary sources in the history of science:

This is the material to be analysed, and *this* is the material from which philosophical problems should arise. And such problems should not at once be blown up into formalistic tumours which grow incessantly by feeding on their own juices but they should be kept in close contact with the process of science even if this means lots of uncertainty and a low level of precision. (1999a, 137)

In this respect, Feyerabend's legacy can hardly be overestimated.

MATTEO MOTTERLINI

FEYERABEND, PAUL KARL

References

Feyerabend, P. K. (1962), "Explanation, Reduction, and Empiricism," *Minnesota Studies in the Philosophy of Science*, III. Minneapolis: University of Minnesota Press, 28–97.

——— (1975), *Against Method: Outline of an Anarchistic Theory of Knowledge*. London: New Left Books.

——— (1978), *Science in a Free Society*. London: New Left Books.

——— (1981), *Rationalism and Scientific Method: Philosophical Papers, Vol. I*. Cambridge: Cambridge University Press.

——— (1987), *Farewell to Reason*. London: Verso.

——— (1993), *Against Method* (3rd ed.). London: Verso.

——— (1996), *Killing Time: The Autobiography of Paul Feyerabend*. Chicago: University of Chicago Press.

——— (1999a), *Knowledge, Science and Relativism: Philosophical Papers, Vol. III*. Edited by J. Preston. Cambridge: Cambridge University Press.

——— (1999b), *Conquest of Abundance: A Tale of Abstraction Versus the Richness of Being*. Edited by B. Terpstra. Chicago: University of Chicago Press.

Feyerabend, P. K., and I. Lakatos (1999), *For and Against Method. Including Lakatos's Lectures on Scientific Method, and the Lakatos-Feyerabend Correspondence*. Edited by M. Motterlini. Chicago: University of Chicago Press.

Hacking, I. (2000), "'Screw You, I'm Going Home.' Review of *Conquest of Abundance*," *London Review of Books*, June 2000, 28–29.

Mill, J. S. ([1859] 1977), "On Liberty," in J. M. Robson (ed.), *Collected Works of John Stuart Mill, Vol. 18*. Toronto: Toronto University Press.

Munévar, G. (2000), "A *Réhabilitation* of Paul Feyerabend," in J. M. Preston, G. Munévar, and D. Lamb (eds.), *The Worst Enemy of Science? Essays in Memory of Paul Feyerabend*. New York: Oxford University Press, 58–79.

Preston, J. M. (1997), *Feyerabend: Philosophy, Science and Society*. Cambridge: Polity Press.

See also **Empiricism; Incommensurability; Kuhn, Thomas; Logical Empiricism; Social Constructionism; Theories**

FITNESS

Darwin's theory of evolution by natural selection is often summarized in terms first coined by Herbert Spencer as the claim that among competing organisms the fittest survive. If there is variation among the traits of organisms, and if some variant traits confer advantages on the organisms that bear them, that is, enhance their fitness, then those organisms will have a tendency to live to have more offspring, which in turn will bear the advantageous traits. The success of Darwin's theory turns on the meaning of its central explanatory concept, 'fitness.'

What is fitness and how can one tell when a trait enhances fitness, or more to the point, when one organism is fitter than another? Some opponents of the theory of evolution by natural selection have long claimed that by defining fitness in terms of actual rates of reproduction, the proponents of evolutionary theory are unknowingly condemning the principle of the survival of the fittest to triviality: If one defines fitness in terms of actual reproductive rates, one is making the claim that those organisms with higher rates of reproduction leave more offspring. This is obviously an empty, unfalsifiable tautology bereft of explanatory power.

Evolutionary theory requires a definition of fitness that will protect it from the charges of tautology, triviality, unfalsifiability, and explanatory infirmity. If no such definition is forthcoming, then what is required from the theory's advocates is an alternative account of the theory's structure and content and the theory's role in the research program of biology.

Ensemble Properties and Population Biology

Since the modern synthesis, evolution is usually described in terms of "change in gene ratios" (see Population Genetics). Population genetics is the mathematical formalism used to describe the effects of natural selection on changes in gene frequency. The subject matter of population genetics is populations, ensembles of organisms, or genes, and not individual biological organisms or pairs of them. This has suggested to more than one philosopher that the theory of natural selection is better understood solely as a theory about ensembles and not individuals. As Sterelny and Kitcher (1988, 345) put this view, "evolutionary theory, like statistical mechanics, has no use for such a fine grain of description [as the biography of each organism]: the aim is to make clear the central tendencies in the history of evolving populations."

In this view, though the word "fitness" (or w) figures in the theory, it is to be understood as exclusively expressing probabilistic reproduction rates for populations. This is generally operationalized in terms of differential success of genotypes (*genotypic fitness*). It has also been suggested that *allelic fitness* is useful, although allelic fitness will not be predictive of evolutionary change in cases such as that of dominance. Although it has been argued that cases such as heterozygote superiority could be handled at the allelic level as a case of frequency-dependent selection, others disagree (for discussion, see Sober and Lewontin 1982; Sterelny and Kitcher 1988). Whether one uses allelic frequencies, genotypic frequencies, or some other kind of census, evolutionary theory is then treated as a set of claims about how the sizes of populations and subpopulations change over time as a function of differing reproductive rates at some initial time, holding environments constant. According to that account, the theory makes no claims about the local adaptation of individual organisms to their particular environment, or for that matter about the local adaptation of populations to their environment. Thus, it does not provide a local causal explanation of these changes in organism or gene frequency (i.e., the account is silent on what specific features of environment E and the organism x lead to the changes). Explanations for these changes are to be sought elsewhere. This being said, population-level explanations do offer clear benefits: Some traits may be explainable only by stepping away from explanations that focus on the benefit to the individual organism.

Some population structures in some species (e.g., some colonies of social insects) "encourage" sterility of some of its members to increase the numbers of the overall population. As Sober (1984, 135) puts it: "According to the [Darwinian fitness notion], a sterile organism will have 0 fitness, since its chances of reproducing are nil." If the focus is shifted from the organism to its genes, and if these genes are shared by other organisms, one gets a different view of fitness. With the idea of "inclusive fitness," Hamilton (1964) explained that an organism might have a better chance of having its genes represented at a later generation if it forgoes its own reproductive success to help other bearers of its genes increase theirs. Hamilton suggested that inclusive fitness should be understood as the sum of the individual's fitness and the individual's effect on the fitness of other related individuals. Kinship is used as a tracker to "estimate" the degree to which two organisms "share the same genes." The higher the relatedness, the higher the probability that a significant proportion of genes is shared between the benefactor/donor and the recipient. The net effect in terms of the donor's genes being passed on could be greater by helping relatives reproduce than by reproducing itself. Kin selection provides an explanation for traits such as altruism that prima facie do not seem to fit into evolutionary theory.

However, kin selection has recently been somewhat demoted in favor of more general group-selection explanations (Hamilton himself shifted his views to a group-selection explanation following his reading of Price's [1970] argument on covariance). One of the advantages of group-selection explanations is that they do not presuppose that relatedness is necessary to establish traits such as altruism. More complex interactions are possible, and communities of different species can act as units of selection. However, understanding these interactions and identifying these communities demands a more ecological approach.

Solution to a Design Problem

Suppose, following Dennett (1995) and others before him, one characterizes the relation "x is fitter than y" as follows:

> x is fitter than y if and only if x's traits enable it to solve the "design problems" set by the environment more fully than y's traits do.

Call this concept *ecological fitness*. The "ecological" definition is fraught with difficulties.

What are these design problems? How many of them are there? Is there any way of measuring the degree to which x exceeds y in their solution? To begin with, the notion of "design problems" is vague and metaphorical. If treated literally, design problems will all be relative to the overarching objective of leaving more descendants: If to be fitter means that an organism offers a better solution—better fills an ecological niche—how is this differential success measured any other way than by measuring higher offspring numbers? Thus the definition may simply hide the original problem of distinguishing fitness from reproductive rates, instead of solving it.

Second, the number of design problems is equal to the number of distinct environmental features that affect survival and reproduction, and this number is probably uncountable. It is therefore no wonder that many biologists have favored defining "x is fitter than y" in terms of quantitatively measurable reproductive rates.

The Propensity Interpretation of Fitness

Among philosophers of biology, there has been a wide consensus that the solution to the problem of defining fitness is to be found in treating it as a probabilistic disposition. The most popular, or the "standard," probabilistic propensity account of fitness has been in terms of offspring contribution (note that there are other propensity accounts that aren't "offspring centric"; more on this below). As such, the propensity causally intervenes between the relationship of environments to organisms that cause it and the actual rates of reproduction, which are its effects. Thus an organism can have a probabilistic disposition to have n offspring and yet unluckily never actually reproduce (or produce a number of offspring different than n).

Comparative fitness differences are dispositions supervening on the complex of relations between the manifest properties of organisms and environments (Rosenberg, 1978) and will give rise to differential reproductive rates. Thus, definitions such as the following were advanced (Brandon 1978; Beatty and Mills 1979):

> x is fitter than y in E **iff** ["if and only if"] x has a probabilistic propensity to leave more offspring in E greater than y's probabilistic propensity to leave more offspring in E.

If fitness is a probabilistic propensity, then the fitter among competing organisms will not always leave more offspring. Fitness differences do not invariably result in reproductive differences, but only with some probability (since the theory allows for drift, this qualification on its claims will be a welcome one). However, assume, as Hume famously argued, that causes must be distinct from their effects and only contingently related to them. Then if the probability of leaving more offspring is an effect, then it will have to be distinct from its cause—the probabilistic propensities that constitute fitness differences. This is a problem that all probabilistic accounts have to face (not just propensities to leave more offspring). One possible solution is to define probability as long-run relative frequency. The idea here is that chances explain long-run frequencies: If x has a probabilistic propensity to leave more offspring than y in every generation, then the long-run relative frequency of x's having more offspring than y in any generation is greater than y's long-run relative frequency.

But there must be at least a theoretical difference between the chance and the frequency if one wishes the former to explain the latter causally. There are philosophers of science who deny that such a distinction between probabilistic propensities in general (not just biological propensities) and long-run relative frequencies is possible (Earman 1986, 149). Others have argued that, as in the case of quantum mechanics, there are such independent chancy probabilistic fitness propensities that generate the long-run relative frequencies. Proponents of the standard account in biology have envisioned two possible explanations for these propensities. One is that probabilistic propensities at the biological levels of phenomena are the result of quantum probabilities "percolating up" (Sober 1984); the second is that there are brute unexplainable probabilistic propensities at the level of organismal fitness differences (Brandon and Carson. 1996). Few doubt that quantum percolation of some kind could have a biological significance. It may well be a source of mutation (cf. Monod 1971, 111–115 for support of this idea). But the claim that it has a significant role in fitness differences is not supported by any independent evidence (cf. Millstein 2000 for discussion). Even if quantum effects could percolate up, they probably would do it so infrequently that they could not help but ground all biological propensities. The claim that there are brute probabilistic propensities at the level of organismal fitness differences is controversial, depending largely on one's acceptance of emergent autonomous propensities at the macro level.

Some qualifications of the standard propensity account will need to be offered. As Gillespie (1977) has shown, the temporal and/or spatial variance in number of offspring may have an important selective effect. In certain scenarios, it might be more beneficial for an organism to have a lower mean number of offspring with a low variance than a higher mean number of offspring with a wider variance (and with equal means, the organism with the higher variance will always be selected against). To take the example from Brandon (1990, 20): If organism A has 2 offspring each year, and organism B has 1 offspring in odd-numbered years and 3 in even-numbered ones, then, *ceteris paribus*, after 10 generations there will be 512 descendants of A and 243 descendants of B. The same holds if A and B are populations and B's offspring vary between 1 and 3 depending on location instead of period. This is not a problem for propensity accounts that reflect averages over many generations, but it would be a problem for accounts that do not in fact have access to such long-term averages.

Accordingly, the definition needs to be changed to accommodate the effects of variance in offspring number per generation. One could get something like this formulation:

x is fitter than *y* iff probably *x* will have more offspring than *y*, *unless* their average numbers of offspring are equal and the temporal and/or spatial variance in *y*'s offspring numbers is greater than the variance in *x*'s, or the average numbers of *x*'s offspring are lower than *y*'s but the difference in offspring variance is large enough to counterbalance *y*'s greater number of offspring.

It is also the case that in some biologically actual circumstances (e.g., where mean fitnesses are low), increased variance is sometimes selected for (Ekbohm, Fagerstrom, and Agren 1980). As Beatty and Finsen (1987) further showed, the definition will also have to accommodate skewness along with offspring numbers and variance, on pain of falsity. One simple way to do this is to add a *ceteris paribus* clause to the definition. But the question must then be raised as to how many different exceptions to the original definitions need to be accommodated. If the circumstances under which greater offspring numbers do not make for greater fitness are indefinitely many, then this definition will be unsatisfactory.

Some proponents of the propensity definition recognize these difficulties and are prepared to accept that at most a "schematic" definition can be provided. Thus Brandon (1990, 20) defined the "adaptedness," or expected fitness, of an organism O in an environment E as:

$$A^*(O, E) = \sum P(Q_i^{OE})Q_i^{OE} - f(E, \sigma^2)$$

where Q_i^{OE} represents a range of possible offspring numbers in generation i, $P(Q_i^{OE})$ is the probabilistic propensity to leave Q_i^{OE} in generation i, and, most importantly, $f(E, \sigma^2)$ is "some function of the variance in offspring numbers for a given type, σ^2 and of the pattern of variation" (Brandon 1990, 20), or, in other words, some function or other that is not known in advance of examining the case. Moreover, one will have to add to variance other factors that determine the function, such as Beatty and Finsen's skewnes. Thus, the final term in the definition will have to be expanded to $f(E, \sigma^2, \ldots)$, where the ellipsis indicates the additional statistical factors that sometimes combine with or cancel the variance to determine fitness levels.

But how many such factors are there, and when do they play a nonzero role in fitness? The number of such factors is probably indefinitely large. The reason for this is given by the facts about natural selection as Darwin and his successors uncovered them. The fact about selection that fates this definition to be forever schematic is the "arms-race" strategic character of evolutionary interactions. Since every strategy for enhancing reproductive fitness (including how many offspring to have in a given environment) calls forth a counterstrategy among competing organisms (which may undercut the initial reproductive strategy), the number of conditions covered by the *ceteris paribus* clause is equal to the number of strategies and counterstrategies of reproduction available in an environment. Brandon (1990) writes, "In the above definition of $A^*(O, E)$, the function $f(E, \sigma^2)$ is a dummy function in the sense that the form can be specified only after the details of the selection scenario have been specified" (20). He acknowledges that the function f will differ for different O and E and will have to be expanded to accommodate an indefinite number of further statistical terms beyond variance. Schematically, it will take the form $f(E, \sigma^2, \ldots)$. Again, adapting Brandon's notation, none of the members of the set that express his generic definition of adaptedness, or expected fitness, $[P(Q_i^{OE})Q_i^{OE} - f_1(E, \sigma^2, \ldots)$. $P(Q_i^{OE})$ $Q_i^{OE} - f_1(E, \sigma^2, \ldots), P(Q_i^{OE})$ $Q_i^{OE} - f_2(E, \sigma^2, \ldots), P(Q_i^{OE})$ $Q_i^{OE} - f_3(E, \sigma^2, \ldots)$, $\ldots]$, is in fact a definition of either term. It is the set of operational measurements of the property of comparative fitness.

It is for reasons such as these that the standard propensity account endorses a schematic view that accommodates all the *ceteris paribus* clauses that could appear (from variance, low mean fitnesses, kin selection effects, etc.). But since the number of these clauses may be indefinite, the standard propensity view does not truly offer a definition of fitness. The schematic nature of this propensity interpretation, along with other problems elaborated in Sober (2002) and Beatty and Finsen (1987), will motivate some to consider alternative approaches to the treatment of fitness.

Conclusion: Models, Ecological Fitness, and the Problem of Evolutionary Drift

Any attempt to turn a generic probabilistic schema for fitness into a complete general definition that is both applicable and adequate to the task of vindicating the truth of the theory of natural selection is problematic. These problems suggest to some philosophers that there is a need to rethink the cognitive status of the theory altogether.

Some (mainly Williams 1970; Rosenberg 1985) have argued that if one strives for an axiomatic formalization of evolutionary theory, one will have to understand fitness as being a primitive notion: A definition of fitness is not available from within the theory itself. Only operational definitions of fitness will be available, which will be only provisional characterizations to help guide investigations and not definitions in the strict sense, since fitness can be characterized only "by appeal to the phenomena that it is employed to account for" (Rosenberg 1985, 141). In other words, fitness can be characterized only through its actual causal role (analogously to the concepts of force and charge). Among other reasons, the very abstract nature of this axiomatic account made it unpalatable for most philosophers and biologists: Biologists do not use a "primitive" notion of fitness in their inquiry, but a more empirical notion better reflected in semantic accounts.

Trying to address these pragmatic concerns, others have argued that the theory of natural selection should not be viewed as a body of general laws but as the prescription for a research program (see Brandon 1990, chap. 4). As such, its central claims need not meet standards of testability, and fitness need not be defined in terms that assure the nontriviality, testability, and direct explanatory power of the theory of natural selection. Evolutionary theory remains a scientifically respectable, but nevertheless untestable, organizing principle for biological science.

Thus, in each particular selective scenario, a different specification of the schematic propensity definition figures in the antecedent of a different and highly restricted principle of natural selection that is applicable only in that scenario. The notion that there is a very large family of principles of natural selection, each with a restricted range of application, may be attractive to those biologists uncomfortable with a single principle or law of natural selection, and to those philosophers of science who treat the theory of natural selection as a class of models. In the "semantic" approach to the theory of natural selection (see Theories), each of the substitution instances of the schematic principle of natural selection generated by a particular specification of the propensity definition of fitness is treated as a definition of a different Darwinian system of population change over time (Beatty 1981). The evolutionary biologist's task is to identify which definition is instantiated by various populations in various environments.

The difficulties of the probabilistic propensity definitions of fitness (standard accounts and others) are serious enough to make the notion of ecological fitness worth revisiting. Recall that in this view "A is fitter than B in E" is defined as "A's traits result in its solving the design problems set by E more fully than do B's traits." The terms in which this definition is couched are certainly in as much need of clarification as is "fitness." There does appear to be important biological work that the ecological fitness concept can do that a definition of fitness solely in terms of differential reproductive rates (actual, expected, or dispositional) cannot.

Suppose one measures the fitness differences between population A and population B to be in the ratio of 7:3 (e.g., $w_A = 1$, $w_B = 0.428$), and suppose further that in some generation the actual offspring ratio is 5:5. There are two alternatives: (1) The fitness measure of 7:3 is right but drift explains the deviation or (2) the fitness measure of 7:3 is incorrect and drift has occurred.

In the absence of information about the initial conditions of the divergence, there is a way empirically to choose between (1) and (2), which requires that there be ecological fitness differences that are detectable. Suppose that fitness differences were matters of probabilistic differential reproductive success. Then the only access to fitness differences would be via population censuses in previous generations (since these form the bases of the probabilities). Suppose that this census did indeed show a 7:3 ratio between A and B in the recent past. In order to exclude the absence of fitness differences instead of drift as the source of the current generation's 5:5 outcome, one needs to be able to establish that the 7:3 differences in previous populations were not themselves solely the result of drift. But this is the first step in a regression, since the original problem was in discriminating drift from mismeasures of fitness. Of course, the problem does not arise if one has access to fitness differences independently of previous population censuses. And this access is available, at least in principle, if fitness is a matter of differences in the solution of identifiable design problems, that is, if there is such a thing as ecological fitness and it is (fallibly) measured by probabilistic propensities to leave offspring.

If there is access to ecological fitness differences, one can, at least in principle, decide whether the divergence from predicted long-run relative frequencies is a matter of drift or reflects an ignorance either of ecological fitness differences or the unrepresentativeness of the initial conditions of individual births, deaths, and reproductions. But given the epistemic problems related to ecological fitness highlighted earlier, can one truly say that ecological fitness differences are within reach?

The problem of defining fitness remains. Or at any rate, it does if biology cannot live with an imperfect definition of fitness in terms of an overall design-problem solution or an understanding of fitness as a research program for building models.

FRÉDÉRIC BOUCHARD

The author acknowledges the helpful input of Alex Rosenberg, Duke University.

References

Beatty, J. (1981), "What's Wrong with the Received View of Evolutionary Theory?" in P. Asquith and R. Giere (eds.), *PSA 1980*, vol. 2. East Lansing, MI: Philosophy of Science Association, 397–426.

Beatty, J., and S. Finsen (1987), "Rethinking the Propensity Interpretation," in M. Ruse (ed.), *What Philosophy of Biology Is*. Dordrecht, Netherlands: Kluwer.

Beatty, J., and S. Mills (1979), "The Propensity Interpretation of Fitness," *Philosophy of Science* 46: 263–288.

Brandon, R. (1990), *Adaptation and Environment*. Princeton, NJ: Princeton University Press.

——— (1978), "Adaptation and Evolutionary Theory," *Studies in the History and Philosophy of Science* 9: 181–206.

Brandon, R., and S. Carson (1996), "The Indeterministic Character of Evolutionary Theory," *Philosophy of Science* 63: 315–337.

Dennett, D. C. (1995), *Darwin's Dangerous Idea*. New York: Simon and Schuster.

Earman, J. (1986), *A Primer on Determinism*. Dordrecht: Reidel.

Ekbohm, G., T. Fagerstrom, and G. Agren (1980), "Natural Selection for Variation in Offspring Numbers: Comments on a Paper by J. H. Gillespie," *American Naturalist* 115: 445–447.

Gillespie, G. H. (1977), "Natural Selection for Variances in Offspring Numbers: A New Evolutionary Principle," *American Naturalist* 111: 1010–1014.

Hamilton, W. D. (1964), "The Genetic Evolution of Social Behaviour," *Journal of Theoretical Biology* 7: 1–16.

Millstein, R. L. (2000), "Is the Evolutionary Process Deterministic or Indeterministic? An Argument for Agnosticism," paper presented at the biennial meeting of the Philosophy of Science Association, Vancouver, Canada, November.

Monod, J. (1971), *Chance and Necessity*. New York: Alfred A. Knopf.

Price, G. R. (1970), "Selection and Covariance," *Nature* 227: 520–521.

Rosenberg, A. (1978), "The Supervenience of Biological Concepts," *Philosophy of Science* 45: 368–386.

——— (1985), *The Structure of Biological Science*. Cambridge: Cambridge University Press.

Sober, E. (1984), *The Nature of Selection: Evolutionary Theory in Philosophical Focus*. Cambridge, MA: MIT Press.

——— (2002), "The Two Faces of Fitness," in *Thinking about Evolution: Historical, Philosophical, and Political Perspectives*. Cambridge: Cambridge University Press.

Sober, E., and R. Lewontin (1982), "Artefact, Cause, and Genic Selection," *Philosophy of Science* 47: 157–180.

Sterelny, K., and P. Kitcher (1988), "Return of the Gene," *Journal of Philosophy* 85: 339–362.

Williams, M. B. (1970), "Deducing the Consequences of Evolution: A Mathematical Model," *Journal of Theoretical Biology* 29: 343–385.

See also **Evolution; Natural Selection; Population Genetics**

FUNCTION

Current philosophical debate concerning functions is focused around the concern that commitment to the existence of functions in nature rests uneasily on the assumption that scientific descriptions of the world should be wholly *naturalistic,* meaning that they must not involve any irreducible, mentalistic elements. Talk of 'function' is pervasive in many areas of the life sciences, and a cursory glance at how it is employed suggests that functions *are* generally taken to be real properties of the biological world. However, the concept of function carries with it all sorts of teleological and mentalistic connotations, which are seemingly in tension with an objective, scientific view of the world. For example, there is the possibility that ascription of biological function involves reference to *design* and *agent intention.* Biological functions, like artifact functions, incorporate an effect of a structure into a description of its causal capacities, thus introducing a "forward-looking" element. Thus, saying that "the function of the heart is to pump the blood" might be taken to imply that "the heart beats *so as* to pump the blood" or "the heart is designed for blood pumping." The effects that a designer *intended* an *artifact* to *achieve* clearly do play a role in determining its current causal properties. Hence, appealing to an intended effect in order to account for current properties is unproblematic. However, a conscious designer is not the sort of thing that a naturalistic biology wants to fall back on. As Allen and Bekoff (1995, 611) put it, "Successful naturalization of teleological notions in biology requires that one give an account of these notions that does not involve the goals or purposes of a psychological agent."

FUNCTION

A further problem that a naturalistic analysis of biological function must contend with is that many ascriptions of function are *normative*, meaning that the function of a structure is not just something that it *does*, but also something that it is *supposed to do*. A function of X does not *accidentally* contribute to the realization of some effect but *properly* does so. Similarly, when something fails to fulfil its function, it does not just *do something else* but rather *goes wrong* or *malfunctions*. Assuming naturalism, a tenable account of functions is required to assimilate them into a purely descriptive framework. Hence, if a concept of normative or proper function is to survive naturalization, the normative must somehow be reduced to the descriptive.

Rather than try to accommodate such concerns, why not just dispense altogether with reference to functions in biology? Why not condemn them to the fate of 'vital spirits' and 'pangenic gemmules'? Given the centrality of functional characterizations in the life sciences, this would call for dramatic revisions of contemporary scientific practices. Functions are not only invoked in describing and explaining the workings of biological structures but also play a role in the classification and individuation of many such structures (Neander 1991). The function of a wing is what makes it a wing. Similarly, a heart would not be a heart if it were not for its function; an eye would not be an eye and a leg would not be a leg. Of course, classification according to function is not exclusive. As Amundson and Lauder (1994, 453) note, biological organs can be classified both anatomically and functionally. Nevertheless, bereft of function, classifications of the biological world would be very different and, most would agree, severely impoverished. Hence, the central problem addressed by most philosophical discussions is how to analyze function in a way that keeps biological science natural and at the same time licenses continued employment of teleological language, whose elimination would make biology very difficult indeed.

Naturalizing Function

The parameters of current debates owe much to the classic accounts of Nagel (1961) and Hempel (1965). They both structure their discussions around the need to account for scientific usage of 'function' in nonteleological terms, accepting (a) that irreducible teleologies are scientifically unacceptable and (b) that some appeals to function in the life sciences are legitimate. Nagel takes it as given that functional language in the life sciences does not ultimately appeal to irreducible teleologies:

We shall ... assume that teleological (or functional) statements in biology normally neither assert nor presuppose in the materials under discussion either manifest or latent purposes, aims, objectives, or goals.... [D]espite the prima facie distinctive character of teleological (or functional) explanations, we shall first argue that they can be reformulated, without loss of asserted content, to take the form of nonteleological ones, so that in an important sense teleological and nonteleological explanations are equivalent. (402–403)

Hempel (1965) takes a similar line, conceding a historical connection between biological functions and full-blooded teleology but maintaining at the same time that any teleological explanations involving irreducible purposes or entelechies are "pseudo-explanations" (304). However, like Nagel, he assumes that certain uses of 'function' in science are legitimate, having a "definitely empirical core" (ibid). Hence, the object of both analyses is to account for function in nonteleological terms, defending the assumption of a clear distinction between function and teleology through a naturalistic analysis of the latter.

Nagel's analysis proposes that functions are contributions made by parts of a system to that system's ability to carry out characteristic processes and behaviors. As Nagel (1961) puts it, structure A has a function in cases where "[e]very system S with organization C and in environment E engages in process P; if S with organization C and in environment E does not have A, then S does not engage in P; hence, S with organization C must have A" (403). He observes that although A is never *logically* necessary for process P, structures such as brains and livers can be said to be "necessary" in a different sense, in that there are no *actual* alternatives in the biological world (404). Thus an account of the function of a systemic constituent can contribute to a robust *explanation* of a system's ability to sustain a certain behavior or process.

Functions, for Nagel (1961), are essentially causal contributions to systemic goals. A successful naturalization of 'function' therefore requires that 'goal' also be naturalized. Nagel analyzes goals in terms of the properties of *directively organized systems*, invoking criteria such as adaptability and plasticity (417–418), but concedes that any demarcation between goal-directed and non-goal-directed systems will be somewhat vague (419). Given a naturalistic account of functions and goals along the lines set out, Nagel suggests that any remaining differences between teleological and nonteleological explanations in science should be attributed to "emphasis and perspective in formulation" rather than content (422).

Hempel's (1965) account is similar to that of Nagel, in maintaining that functions are roles played by parts of a system that contribute to "keeping the given system in proper working order or maintaining it as a going concern" (305). Hempel locates proper scientific usage of 'function' within a procedure called functional analysis, whose aim is to "exhibit the contribution which the behavior pattern makes to the preservation or the development of the individual or group within which it occurs" (305). He constrains the domain of legitimate function assignment to those *constituents of self-regulating systems that contribute to the activity of self-regulation*, thus ruling out numerous possible systemic goals that would be admitted by Nagel's more permissive account. Hempel is less optimistic than Nagel with regard to the explanatory potential of functional analysis, which, he observes, cannot deductively explain the presence of a functional item and has very little predictive power (312–313). However, he does assign it a heuristic role, in "determining the respects and the degrees in which various systems are self-regulating" (330). Hempel also departs from Nagel in adding a historical dimension to his analysis. Functions are not simply *current* contributions made by parts of a system to that system's capacities. The objects of functional analysis are also "standardized" or "repetitive" items (307) that occur in a context of sustained self-regulation, implying a historical or backward-looking aspect to the concept of function (see McLaughlin 2001, part II).

This historical dimension is developed substantially and in a novel way by Wright's (1973) analysis, from which the majority of recent attempts to naturalize biological functions take their lead. Wright seeks to provide a unified account, embracing both natural and artifact functions. The two key requirements of such an account are, according to Wright, that it be able to distinguish a function of X from an accidental consequence of X and be able to capture the sense in which statements of function are intrinsically explanatory, a statement of X's function constituting an explanation of why X exists. Wright proposes a consequence etiology, which is intended to specify necessary and sufficient conditions for an entity X to have a function Z. The 'function of X is Z' means:

- X is there because it does Z.
- Z is a consequence (or result) of X's being there. (161)

The analysis is devised to accommodate both natural and artifact functions, both conscious design and natural selection. A car (X) has the function of transportation (Z) because transportation is not merely one of its effects but also what it was *designed for* and hence why it exists. A kidney (X) has the function of cleansing the blood (Z), as it was *selected for* that role and hence exists *because* of its role in cleansing the blood. Wright's analysis attempts to preserve the forward-looking element of function (by appealing to history in order to show how the effect can explain the presence of cause), the association between function and origin (the function of X is also the reason for X's existence), and the normativity of function (X is supposed to do what X was designed or selected to do) whilst also providing a naturalistic, unitary account, covering all uses of 'function.'

There are a number of possible exceptions to Wright's analysis, many of which were first pointed out by Boorse (1976). For instance, consider a break in a hose (X), which releases chlorine gas (Z). The break causes the gas to be released, but release of the gas keeps the break open, by gassing anybody who gets close enough to seal it, and so the break continues to exist because it releases the gas (72). This maps onto both of Wright's criteria for Z being a function of X. However, it is intuitively clear that this is not the kind of scenario to which one would want to assign a function.

More recent etiological accounts take Wright's basic formulation as their starting point and attempt to insulate it against various counterexamples in order to capture a sense of function as the notion is employed specifically in biology. This quest has resulted in several deviations from the goals of Wright's original analysis. For example, Millikan (1989) and Neander (1991) both depart from the project of conceptual analysis to formulate accounts that are, to varying degrees, stipulative and not intended to reflect every instance of everyday function assignment. Part of the motivation for this is the need to guard against far-fetched counterexamples such as car engines and livers materializing out of nowhere, or more plausible-sounding scenarios such as complex biological structures emerging through chance macromutations and thus owing nothing to the action of natural selection. Even if commonsense intuitions do lean toward the assignment of function in these cases, such intuitions can simply be declared irrelevant if conceptual analysis is not one of its aims.

Coupled with the departure from conceptual analysis is a renunciation of the goal of a unitary account. Recent etiological approaches tend to focus specifically on *biological* function, which they analyze in terms of natural selection and its products, rather than on the basis of some more general etiology common to both conscious design and natural selection. Hence, they treat biological functions as essentially distinct from artifact functions, given that artifact functions are not the result of a process of blind, nonconscious selection:

> It is the/a proper function of an item (*X*) of an organism (*O*) to do that which items of *X*'s type did to contribute to the inclusive fitness of *O*'s ancestors, and which caused the genotype, of which *X* is the phenotypic expression, to be selected by natural selection. (Neander 1991, 176)

Even given this restriction in scope to biological function and natural selection, further questions have been raised as to whether short-term or long-term selection history should be given priority when assigning functions. For example, a structure might originally have been selected for task *A* and have subsequently been selected for task *B*, with *A* fading out of the picture completely. Most accounts maintain that recent selection pressures are more relevant in determining current function (Godfrey-Smith 1994). Buller (1998) adds yet further fine-tuning to etiological approaches by differentiating between "strong" and "weak" etiological theories. These are distinct in that the former explicitly emphasizes selection *for* a trait, whereas the latter stresses only that the trait be a *result* of the reproduction of structures that had the same effect.

Despite the often subtle differences between contemporary etiological theories of function, all aim to provide an analysis that preserves a concept of normative, teleological, and (most importantly) *natural* function that can be legitimately employed to describe the biological world.

Etiological accounts are not the whole story, however, and causal role accounts, which, like the early analyses of Nagel (1961) and Hempel (1965), emphasize the current contribution of a structure to systemic capacities, are still viewed by many as a plausible alternative or supplement to the etiological view. A causal role account proposed by Cummins ([1975] 1984) has been particularly influential. He argues that functions are contributions to the capacities of containing systems, playing a pivotal role in an explanatory strategy that he, like Hempel, refers to as "functional analysis":

> When a capacity of a containing system is appropriately explained by analyzing it into a number of other capacities whose programmed exercise yields a manifestation of the analyzed capacity, the analyzing capacities emerge as functions. (407)

A function of *X* is a capacity of *X* that contributes to the explanation of a more complex capacity of a containing system *Y*. So functions, for Cummins, play a role in a reductive explanatory strategy that is popular in various areas of the biological sciences, requires an essentially ahistorical understanding of causal function, and, as Amundson and Lauder (1994) point out, does not require the incorporation of normativity.

Cummins' account has been criticized for being both too vague and too liberal. It is unclear precisely what is meant by "system" and, without any explicit restrictions concerning acceptable systems, just about anything can have a Cummins function in some context. In relation to some conceivable system (in the Cummins sense), the function of cancer is to cause death and the function of the nose to support spectacles (see e.g., Manning 1997, 70). Hence, the account needs to be constrained in order to isolate a more specific sense of function that can be applied to biological practice without opening the floodgates to permit all manner of bizarre functions. Certain more recent causal role accounts insulate against excessive generality by restricting function assignment to those roles that enhance the *fitness* of biological systems (Bigelow and Pargetter 1987; Mitchell 1995; Walsh 1996). In contrast to etiological approaches, these theories construe function in terms of *current* rather than *historical* propensity to contribute to biological fitness. Bigelow and Pargetter (1987) propose their propensity account as an *alternative* to etiological accounts. However, the two accounts need not be antagonists and can be thought of as working together to accommodate distinct but equally permissible uses of the notion of function. In recent years a consensus has emerged along these lines, which recognizes the legitimacy of etiological functions and Cummins functions in biology, with fitness functions composing a subfamily of Cummins functions. Etiological approaches pin down a distinctive, specific use of the term "function," whilst Cummins' account identifies a broader, more generally applicable use (Godfrey-Smith, 1993). There are still some issues left concerning the nature and extent of any relationships between Cummins functions and etiological functions. Buller (1998, 515) suggests that one has to identify a Cummins function before going on to ask whether that function is also an etiological function. That is, one has to know whether and how *X* contributes to *Z* (Cummins function) before one can determine whether *X* was selected for because it contributes to *Z* (etiological function). Griffiths

(1993) also regards the two senses as intimately connected:

> We can incorporate the etiological approach into the Cumminsesque picture of function ascription. The proper functions of a biological trait are the functions it is ascribed in a functional analysis of the capacity to survive and reproduce (fitness) which has been displayed by animals with that feature. This means that a feature will have a proper function only if it is an adaptation for that function. The trait must have been selected because it performs that function. (412)

However, Godfrey-Smith (1993) places more of an emphasis on disunity and advises philosophers not to "join what science has put asunder" (207), stressing that the two kinds of function are distinct and symptomatic of different patterns of scientific explanation.

In summary, despite a series of minor disagreements amongst its proponents, there is a currently popular consensus in philosophy to the effect that biological functions can be satisfactorily naturalized via some combination of etiological and causal role accounts, with only the former yielding a formulation of normative, teleological, or proper function. The philosophical interest of a naturalized proper function is not restricted to biology, however. The concept also has considerable potential for application in psychology and the philosophy of mind.

Function and Mind

Etiological accounts purport to provide a naturalistic reduction of normative, teleological function, which accounts for normative properties in purely descriptive terms. The resultant formulation of a proper function has proved to be a very powerful tool, whose application is becoming increasingly central to a number of different projects sharing the common goal of naturalizing the mind. For example, one possibility is in accounting for the seemingly intractable normativity of mental states in terms of the more tangible normativity of biological function. This is the aim of currently popular teleological theories of mental representation, such as that of Millikan (1993). The connection between intentional states, such as beliefs and desires, and their objects cannot be adequately characterized in terms of causation alone. The belief that there is a bird rustling in the bushes might well be caused by the foragings of a rat, a brief glimpse of which gives the mistaken impression of a bird. So what is it about the belief that specifies the content 'bird' rather than 'rat and/or bird'? This is the so-called disjunction problem: determining how the content of a belief is fixed as X rather than as X and/or Y, given that it can be present in both cases. The belief about the rustling in the bush is so not just because it is *commonly* caused by birds rather than rats, bees, or buffalo, but because it is *properly* caused only by birds and thus *mistaken* when it is caused by rats. Teleological theories of representation attempt to account for this normativity of intentional states in terms of the normativity of biological function. In its simplest form, their central claim is that a belief X is about Y rather than Z because its proper function is to represent Y and not Z:

> Just as the characteristic mark of intentionality is that intentional items can be false, unsatisfied, or seemingly 'about' what does not exist, so the characteristic mark of the purposive, of that which has a function, is that it may not in fact fulfill that purpose or serve that function. (Millikan 1993, 23)

Functional characterizations are also employed more generally in the philosophy of mind. Many philosophers hold that *all* mental processes are best characterized in terms of their *biological roles*, as opposed to a physical description of the structures that instantiate those roles. This general position, known as *teleological functionalism* (see e.g., Sober 1985), maintains that proper functions provide a means of describing what various mental processes essentially *do*, without getting sidetracked by accidental/contingent effects of those processes or excessive attention to the anatomy of the biological structures that perform them. For example, an account of the proper function(s) of consciousness will aim to tell us what consciousness *does* and why it came to *be*.

If one goes so far as to adopt the line that mental processes simply *are* what they *do*, then a comprehensive characterization of psychological states and processes in terms of their various etiological functions will amount to a comprehensive account of what the mind *is*. Some recent work in evolutionary psychology has precisely this goal. Evolutionary psychologists such as Cosmides and Tooby (1992) not only employ 'function' to *describe* psychological processes but also maintain that psychological processes are *individuated* or *defined* by their etiological functions. Cosmides and Tooby's aim is to account for the mind in terms of a set of *modules*, which are innate, domain-specific programs selected to deal with the many problems posed by the environment in which humans evolved. Modules are not *first* identified and *then* assigned a function. Instead, their function is *constitutive* of what they are. For

Cosmides and Tooby, etiological functions are utterly indispensable for a scientific understanding of the mind.

Thus, etiological functions play a major part in several projects in naturalistic philosophy of mind and psychology. These projects are structured around the assumption that if we can rid biological teleology of mind, we can then employ it to rid the rest of the world of mind, by reducing various psychological phenomena to their biological functions and thus naturalizing them. So etiological proper functions are increasingly indispensable, not so much for biology but for projects in the philosophy of mind that require a naturalistic formulation of normative, teleological function.

Function in Mind?

Employment of functions in the service of naturalizing the mind presupposes the possibility of objective, mind-independent functions. However, not all agree with Godfrey-Smith's (1993) naturalistic consensus. In his response to Wright (1973), Boorse (1976) argues that all functions are ultimately *contributions to goals* and that only the inclusion of a goal can ultimately serve to distinguish between cases where the term "function" is and is not legitimate. This sort of claim has also been employed to criticize more recent formulations of the etiological theory. For instance, Manning (1997) examines several cases that fit the etiologists' criteria but clearly do not have functions. Both "junk DNA" and "selfish DNA" (which enhances the chances of its own replication but has a detrimental effect on the organism) have the right etiology but neither is assigned a function (74–75). Manning suggests that functions are assigned only when goals are incorporated into one's description of a system or process. Thus, if naturalism requires the elimination of goal directedness, etiological accounts cannot meet the requirements of naturalism without an additional account of goals (80–81).

If functions depend upon goals or other related mentalistic notions, where do these goals come from? Neither Nagel (1961) nor Hempel (1965), both of whom claim that function assignment involves reference to systemic goals, think that the problem of goals is insurmountable for naturalism. However, an alternative response is to concede that goals come from us; we tacitly incorporate them into our conceptions of the biological world, slipping in values, ends, and intentions that have their ultimate source in human agency. Functions, if they do indeed presuppose psychological goals, turn out to be mind-dependent and so cannot be employed to naturalize the mind.

Some recent versions of this kind of view maintain that 'biological function' originates from a metaphor or analogy with human agency and artifice, which results in one's thinking of the biological world in terms of end directedness, normativity, and value. For instance, Matthen (1997) argues that "function attributions seem to be dependent on user, role, mode, use, and the utility that the user realizes from the outcome." Functions are "attributed to natural things by virtue of an analogy with instruments designed for use by or actually used by an agent for a purpose" (31).

Accepting that functions are ultimately dependent on an analogy with human artifice, one might wonder whether they can and should be eliminated from biology. Such a move would be highly contentious, given that function talk is widespread, entrenched, and apparently central to many areas of biological thinking. Matthen (1997, 37) suggests that elimination of function from biology is possible *in principle*, but he does not recommend such a course of action. Ruse (2000) argues that elimination would be undesirable, as even though functions do not correspond to properties of the mind-independent world, they still serve a central role in biological explanations. Functions need not be *out there* in order to be legitimate and useful tools to enhance "science's heuristic power" and "its predictive fertility" (231). Ratcliffe (2001, 45–47) argues that the role played by teleological language in biology renders it impossible to eliminate without eliminating most of biological practice along with it. Functions are not just dispensable metaphors that attach to biological descriptions, but rather constitutive conditions for the possibility of biology, whose absence would render much of that science impossible.

Others claim that certain understandings of biological function could and even *should* be eliminated. Davies (2001) argues that biological proper function cannot be naturalized and ought therefore to be abandoned, leaving only nonnormative causal role functions. For different reasons, Amundson (2000) argues that a "normal" function, construed as a normative ideal of biological performance, should be discarded. He claims that Darwinian evolutionary processes result in a plethora of differences rather than a few "proper" ways of doing things, deviations from which constitute *mal*function or *ab*normality. In place of the contrast between proper function and malfunction, there are a multitude of effective modes of performance, all of which get the job done, if a little differently from each other (34). Amundson (2000) goes on to suggest that the concept of normative biological function is symptomatic of an ideology or prejudice that serves to disadvantage certain people, who are classified as *dis*eased, *dis*abled or *ab*normal and excluded to varying degrees from society on the grounds that society is not obliged to accept or enable those who are biologically *dys*functional: "The disadvantages of people who are assessed as 'abnormal' derive not from biology, but from implicit social judgments about the acceptability of certain kinds of biological variation" (33). So, according to Amundson, so-called normal function is neither part of the biological world nor useful, but an ideological distortion of biology that serves to justify social attitudes that would otherwise appear morally unacceptable, a contention that adds an important social dimension to the functions debate.

Conclusion

To summarize, current debates concerning biological function center around three issues: (a) the number of distinct senses of 'function' at work inside and outside biology, (b) whether or not certain kinds of function are reducible to properties of the objective, mind-independent world, and (c) the extent to which talk of functions is useful or even indispensable in biology.

The prognosis for biological naturalism may well hinge on the outcome of these debates. Crucial to the success of attempts to naturalize the mind through biology is the possibility of a naturalistic reduction of 'normative function,' which etiological accounts claim to supply. But if one concedes, contrary to these accounts, that normative functions are in some sense irreducibly mind dependent, resting on conceptions of goals, values, design, or agency that are not part of the objective, biological world, then employment of function as an ontological component in naturalistic accounts of mind is ruled out. Employing a mind-dependent concept to rid the world of mind would constitute a viciously circular endeavor.

However, all of this is contingent on what commitments one takes to be essentially constitutive of naturalism. For instance, Bedau (1991) argues that although it is not possible to eliminate *value* from the concept of function, this does not imply that values are woven into the biological world by minds or are dependent on an analogy with values that play an integral role in human artifice. Bedau (1991) advocates a "broader naturalism" that acknowledges that "value notions apply to living things, even those which are not human" (655). If naturalism is compatible with values and goals being an intrinsic part of the natural world, then irreducible teleology in nature will threaten neither a naturalistic biology nor naturalistic accounts of psychological processes that rest upon that biology. However, as Manning (1997, 80–81) notes, any such account will "not amount to the full-scale naturalization of functional properties *if* such a naturalization requires the absolute elimination of the analysis and of all teleological, purposive or 'goal-directed' notions." And whatever its merits, "full-scale" naturalization or reduction of teleological properties is indeed the central aim of most so-called naturalistic analyses of function.

Functions are a philosophical problem precisely because of the specter of their incompatibility with a naturalism that refuses to admit teleology or normativity as ground-floor properties of the world and grants them reality only on the condition that they be reducible to more basic properties. Reductive naturalism not only seeks to accommodate function but has increasingly come to depend upon it, as an essential tool in the project of naturalizing other troublesome concepts, such as intentionality or representation. Hence, an awful lot hinges on whether normative, teleological functions can ultimately be cashed out in descriptive, nonteleological terms.

MATTHEW RATCLIFFE

References

Allen, C., and M. Bekoff (1995), "Biological Function, Adaptation and Natural Design," *Philosophy of Science* 62: 609–622.

Amundson, R. (2000), "Against Normal Function," *Studies in History and Philosophy of Biological and Biomedical Sciences* 31: 33–53.

Amundson, R., and G. Lauder (1994), "Function without Purpose: The Uses of Causal Role Function in Evolutionary Biology," *Biology and Philosophy* 9: 443–469.

Bedau, M. (1991), "Can Biological Teleology Be Naturalized?" *Journal of Philosophy* 88: 647–655.

Bigelow, J., and R. Pargetter (1987), "Functions," *Journal of Philosophy* 84: 181–196.

Boorse, C. (1976), "Wright on Functions," *Philosophical Review* 85: 70–86.

Buller, D. J. (1998), "Etiological Theories of Function: A Geographical Survey," *Biology and Philosophy* 13: 505–527.

Cosmides, L., and J. Tooby (1992), "The Psychological Foundations of Culture," in H. Barkow, L. Cosmides, and J. Tooby (eds.), *The Adapted Mind: Evolutionary Psychology and the Generation of Culture*. Oxford: Oxford University Press.

Cummins, R. ([1975] 1984), "Functional Analysis," in E. Sober (ed.), *Conceptual Issues in Evolutionary Biology*. Cambridge, MA: MIT Press, 386–407. Originally published in *Journal of Philosophy* 72: 741–765.

Davies, P. S. (2001), *Norms of Nature: Naturalism and the Nature of Functions*. Cambridge, MA: MIT Press.

Godfrey-Smith, P. (1993), "Functions: Consensus with;out Unity," *Pacific Philosophical Quarterly* 74: 196–208.
——— (1994), "A Modern History Theory of Functions," *Noûs* 28: 344–362.
Griffiths, P. (1993), "Functional Analysis and Proper Functions," *British Journal for the Philosophy of Science* 44: 409–422.
Hempel, C. (1965), *Aspects of Scientific Explanation and Other Essays in the Philosophy of Science*. New York: Free Press.
Manning, R. (1997), "Biological Function, Selection and Reduction," *British Journal for the Philosophy of Science* 48: 69–82.
Matthen, M. (1997), "Teleology and the Product Analogy," *Australasian Journal of Philosophy* 75: 21–37.
McLaughlin, P (2001), *What Functions Explain: Functional Explanation and Self-Reproducing Systems*. Cambridge: Cambridge University Press.
Millikan, R. (1989), "In Defense of Proper Functions," *Philosophy of Science* 56: 288–302.
——— (1993), *White Queen Psychology and Other Essays For Alice*. Cambridge, MA: MIT Press.
Mitchell, S. D. (1995), "Function, Fitness and Disposition," *Biology and Philosophy* 10: 39–54.
Nagel, E. (1961), *The Structure of Science: Problems in the Logic of Scientific Explanation*. London: Routledge.
Neander, K. (1991), "Functions as Selected Effects: The Conceptual Analyst's Defense," *Philosophy of Science* 58: 168–184.
Ratcliffe, M. (2001), "A Kantian Stance on the Intentional Stance," *Biology and Philosophy* 16: 29–52.
Ruse, M. (2000), "Teleology: Yesterday, Today and Tomorrow?" *Studies in History and Philosophy of Biological and Biomedical Sciences* 31: 213–232.
Sober, E. (1985), "Panglossian Functionalism and the Philosophy of Mind," *Synthese* 64: 165–193.
Walsh, D. M. (1996), "Fitness and Function," *British Journal for the Philosophy of Science* 47: 553–574.
Wright, L. (1973), "Functions," *Philosophical Review* 82: 139–168.

See also **Evolutionary Psychology; Natural Selection; Naturalism; Teleology**

G

GAME THEORY

Game theory is the branch of decision theory that analyzes interdependent decision problems between rational, strategic agents. A rational agent is one who has a consistent set of preferences defined over some set of possible outcomes and who makes choices consistent with these preferences. A strategic agent is one who, given these preferences, reasons about the best course of action to take in order to satisfy them. Interdependent decision problems arise when the outcome for any particular agent depends upon the actions chosen by all of the agents; that is, when the optimal choice for an agent A depends upon the choices made by other agents, and the optimal choice for the other agents depends in turn upon the choice made by A. It is this strategic feature that distinguishes game-theoretic problems from simpler decision problems such as parametric choice under conditions of risk or uncertainty.

The birth of modern game theory is usually attributed to von Neumann and Morgenstern (1944). However, precursors to game-theoretic analyses of strategic problems can be found in Zermelo (1913), Borel ([1921] 1953), and von Neumann ([1928] 1959), as well as in the works of Hobbes and Hume.

A Theory of Utility

One of von Neumann and Morgenstern's primary contributions was their development of a mathematical theory of utility, allowing one to define, for a given agent, an interval utility measure unique up to a strictly increasing affine transformation. The need for such a notion of utility originates in the fact that in game theory, agents often need to make decisions under conditions of risk or uncertainty, and hence one needs a measure of how strong their preferences for a given outcome are.

If an agent's preferences over outcomes satisfy certain basic coherence criteria, it is possible to define a utility function with the property that if one makes choices consistent with one's preferences, one acts *as if* one were choosing to maximize expected utility. The following axioms (from Luce and Raiffa's [1957] classic text *Games and Decisions*) formalize the coherence criteria necessary to satisfy in order to define a von Neumann–Morgenstern utility function. Let $A = \{a_1, \ldots, a_n\}$ denote the set of outcomes, and let $a_j \precsim a_i$ denote that the agent either prefers a_i over a_j or is indifferent between them. A *lottery* $L = (p_1 a_1, \ldots, p_n a_n)$ is simply a randomization over outcomes, where the

GAME THEORY

outcome a_i occurs with probability p_i. A *compound lottery* $Q = (q_1L_1, \ldots, q_mL_m)$ is a lottery over lotteries, where the chance that the lottery L_i occurs is q_i.

Ordering of Alternatives

For any outcomes a_i, a_j, and a_k either $a_i \precsim a_j$ or $a_j \precsim a_i$ (and possibly both). Moreover, the relation "$\precsim$" is *transitive*; that is, if $a_i \precsim a_j$ and $a_j \precsim a_k$, then $a_i \precsim a_k$.

Reduction of Compound Lotteries

Let $L^i = (p_1^i a_1, p_2^i a_2, \ldots, p_n^i a_n)$ be a lottery, for $i = 1, \ldots, m$. Then the agent is indifferent between the compound lottery $(q_1L^1, q_2L^2, \ldots, q_mL^m)$ and the simple lottery $(p_1 a_1, \ldots, p_n a_n)$, where $p_i = q_1 p_i^1 + q_2 p_i^2 + \ldots + q_m p_i^m$.

Continuity

Suppose that $a_n \precsim a_{n-1} \precsim \ldots \precsim a_1$. Then there exists a number u_i such that the agent is *indifferent* between a_i and the lottery $[u_i a_1, 0 \bullet a_2, \ldots, 0 \bullet a_{n-1}, (1 - u_i)a_n]$, which is denoted $\hat{a}_i$.

Substitutibility

In any lottery, $\hat{a}_i$ is substitutable for a_i.

Transitivity of Lotteries

The preference and indifference relations over lotteries are transitive relations.

Monotonicity

A lottery $(pa_1, (1-p)a_n)$ is preferred or indifferent to $(p'a_1, (1-p')a_n)$ if and only if $p \geq p'$.

If an agent's preferences satisfy the above axioms, it is possible to find a number u_i for each outcome a_i such that for any two lotteries L and L' the magnitudes of the expected values $p_1 u_1 + \ldots + p_n u_n$ and $p_1' u_1 + \ldots + p_n' u_n$ indicate the preference between the lotteries. From this assignment of utilities to the basic alternatives, one can construct a utility function f over the set of risky alternatives (the lotteries). Consequently, when an agent makes choices consistent with her preferences, she acts as if she is choosing to maximize personal utility as measured by f.

Representations of a Game

Games are most commonly represented in an extensive or a strategic form. One also finds the strategic form referred to as the *normal* form, following von Neumann and Morgenstern, who believed that normally one should reduce the extensive form of a game to the strategic form for the purpose of analysis.

The extensive form uses a game tree to represent the order of play (see Figure 1). Each node in the tree represents a *choice point* for a particular player; the player whose turn it is to move at a particular choice point is indicated by a label attached to the node. All games have a privileged node, the *root* or *initial* node where the game begins. The leaves of the tree, also known as *terminal* nodes, represent endpoints, or outcomes, of the game. Every node in the game tree except for the terminal nodes has at least one edge lying on a path between it and a terminal node; such edges represent choices available to a player at that choice point. In some games, the moves available to a player depend not only on the previous moves of other players, but on the outcome of a chance event like the roll of a die. Such games may be represented by including a fictitious player in the game tree, Chance, whose available moves at a point correspond to the possible outcomes of the random event. A player's choice at a given point is a *move* in the game, and each edge has an attached label naming the move. A path from the root node to a terminal node is one possible *play* of the game. In Figure 1, terminal nodes are labeled with W or L, meaning that Player 1 wins or loses the game, respectively.

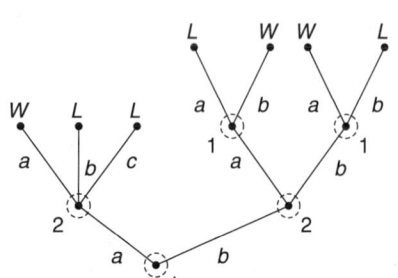

(a) Game of perfect information

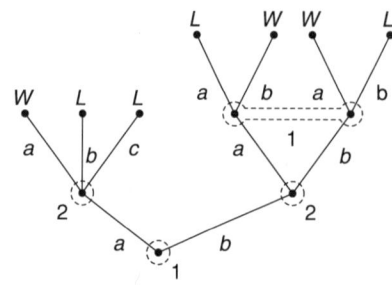

(b) Game of imperfect information

Fig. 1. A simple two-player game in extensive form. Terminal nodes labeled "W" and "L" indicate whether Player 1 wins or loses, respectively.

If all players know their exact position in the game tree at every point, the game is said to be one of *perfect information*; all other games are of *imperfect* information. Although players do not always know their exact position in the game tree in a game of imperfect information, they often know that their position is one of a limited number of possible nodes. This subset of nodes is a player's information set. In an extensive form game, a player's *strategy* specifies the choices that the player would make at each of his information sets. A player's information sets are indicated in the game tree by grouping together those nodes among which the player cannot distinguish. Thus, an alternative definition of a game of perfect information is one in which all information sets contain only a single node. Figure 2b illustrates a game of imperfect information in which Player 1 moves first but keeps the move hidden from Player 2. When it is Player 2's turn to move, he does not know whether the choice occurs at the left or the right side of the game tree.

The strategic form of a game is a minimal representation that omits all information about the game except for the relationship between strategies and payoffs. The strategic form of a game consists of a set of players $P = \{1, \ldots, N\}$, a set of pure strategies S_i for each player $i \in P$, and, for each player i, a payoff function u_i that maps pure strategy profiles $\sigma = (\sigma_1, \ldots, \sigma_N) \in S_1 \times \ldots \times S_N$ to a real number r. In a two-player game, the strategic form can be represented as a matrix, where each row corresponds to a strategy for Player 1, each column a strategy for Player 2, and each cell the resulting payoffs obtained by Players 1 and 2 when they choose those respective strategies.

In many cases, it proves convenient to allow players to adopt *mixed* strategies, where they choose a pure strategy at random according to some probability distribution defined over the set of pure strategies S_i. The payoff for a mixed strategy $\bar{\sigma}_i$ is defined to be the expected payoff $\sum_\sigma P(\sigma \mid \bar{\sigma}_i) u_i(\sigma)$, where the sum is over all strategy profiles σ and $P(\sigma \mid \bar{\sigma}_i)$ denotes the probability that the strategy profile σ occurs when player i adopts the mixed strategy $\bar{\sigma}_i$.

Although it is often said that which form one uses to represent a game is merely a practical question, on the grounds that any game represented in one form may be represented in the other, this is a topic of some debate. To begin with, it is clear that moving from the extensive form to the strategic form results in a loss of information, for it is possible for two *different* extensive games to have the *same* strategic form. In some cases, this lost information may be relevant to the analysis of the game; if so, it may not always be possible to adequately analyze a game just given its strategic form (see Harper 1988). For example, Figure 2 illustrates the strategic and extensive forms for the decision problem central to Puccini's opera *Gianni Schicchi*, in which the causal dependencies between the players' choices are lost in the normal form, yet seem crucial to the game's analysis.

Noncooperative Games

In a noncooperative game, players independently decide what strategy to adopt in the light of their knowledge of the other players and the payoff matrix. Most of the classical results in game theory have been obtained for noncooperative games, for

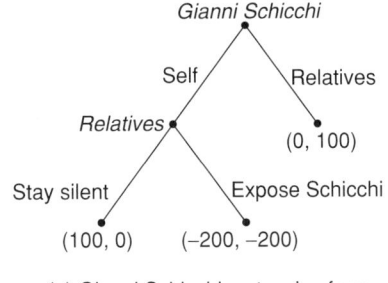

	Stay silent, if Schicchi awards fortune to self	Expose Schicchi, if Schicchi awards fortune to self
Relatives	(0, 100)	(0, 100)
Self	(100, 0)	(−200, −200)

(a) Gianni Schicchi, extensive form (b) Gianni Schicchi, strategic form

Fig. 2. Strategic and extensive form of the Gianni Schicchi. In Puccini's opera *Gianni Schicchi*, the wealthy Buoso Donati dies, and before his will is read, his relatives learn that he has willed a large portion of his fortune to friars. They conspire to have a noted mimic, Gianni Schicchi, impersonate Buoso Donati on his deathbed in order to dictate a new will. Gianni Schicchi agrees but while impersonating Buoso Donati and dictating a new will, he declares his wish to leave a large portion of his fortune to his devoted friend Gianni Schicchi. The relatives contemplate notifying the authorities but decide against it, knowing that the punishment for tampering with a will is banishment and amputation of a hand.

the inability of players to form coalitions and enter into binding agreements make noncooperative games much easier to analyze. Some game theorists (such as Nash) also have a methodological reason for concentrating on noncooperative games: These theorists hold that such games are "more basic" than cooperative games and that the appropriate way to solve a cooperative game is first to transform it into a noncooperative game. However, these views are not universally held (see Osborne and Rubinstein 1994; Binmore 1992).

A *solution* of a game is a specification of the outcomes that may be expected to occur when the game is played by rational agents. Two widely used techniques for solving noncooperative games are *dominance arguments* and *equilibrium analysis*. The goal of each of these approaches is to identify, for each player, a best-response strategy to the anticipated play of all other players. Given a strategy profile σ, the strategy σ_i is a best-response for player i if $u_i(\sigma_{-i}, \sigma_i) \geq u_i(\sigma_{-i}, \sigma_j)$ for all $\sigma_j \in S_i$, where σ_{-i} denotes the set of strategies in the profile σ for the opponents of player i.

A dominance argument rules out certain strategies for play on the grounds that those strategies are inferior to other alternatives, where an inferior strategy is one that is either weakly or strictly dominated: A strategy σ is *weakly dominated* if there exists another strategy σ' such that the payoff from σ' is never worse than the payoff from σ, and there is at least one instance in which the payoff from σ' exceeds that of σ. A strategy σ is *strongly dominated* if there exists another strategy σ^* such that the payoff given by σ^* always exceeds the payoff given by σ.

Iterated elimination of strongly dominated strategies is a procedure for transforming games into a reduced form. One eliminates the strongly dominated strategies for Player 1, transforming the game G to the game G', and then eliminates the strongly dominated strategies for Player 2 from G' to obtain G'', repeating this procedure until no strongly dominated strategies for any player remain. At the end, one obtains a reduced game G^* with the property that every remaining strategy for every player is a best-response to some possible strategy profile. In addition, the resulting game obtained does not depend on the order or the rate at which strongly dominated strategies are removed. This result does not hold for iterated elimination of weakly dominated strategies. The resulting game G^* obtained by iterated elimination of weakly dominated strategies may depend on the order in which strategies are eliminated, as shown in Figure 3. It is never rational to play a strongly dominated strategy, but there are cases where it is not irrational to play a weakly dominated strategy. Although some game theorists freely apply iterated dominance arguments to reduce the complexity of games, others caution against adopting this as a general approach toward their solution (see Binmore 1992).

Although dominance arguments are useful in analyzing a game, the primary tool of analysis in noncooperative game theory is a Nash equilibrium. A strategy profile σ is a Nash equilibrium if each player's strategy is a best-response to the strategies selected by the rest of the players; alternatively, a Nash equilibrium occurs when no player's expected payoff improves by adopting a different strategy unless another player adopts a different strategy as well. More formally, a strategy profile $\sigma = (\sigma_1, \ldots, \sigma_N)$ is a Nash equilibrium if, for $1 \leq i \leq N$, $u_i(\sigma) \geq u_i(\sigma_{-i}, s_i)$ for all $s_i \in S_i$. The wide acceptance of the Nash equilibrium for solving games derives from the fact that it is the only such concept compatible with the rules of the game, the rationality of the players, and the independent selection of strategies all being common knowledge. (For a discussion of common knowledge, see Lewis 1969.)

If players are restricted to pure strategies, not all games have a Nash equilibrium. The game of Matching Pennies, shown in Figure 4, has no Nash equilibrium when the players are restricted to playing either heads or tails. If players may adopt mixed strategies, then it can be shown that all finite games (that is, games in which each player has only finitely many strategies) have at least one Nash equilibrium (Nash 1950).

	μ_1	μ_2
σ_1	(1,1)	(0, 0)
σ_2	(1, 1)	(2, 1)
σ_3	(0, 0)	(2, 1)

Fig. 3. A game in which order matters for the iterated elimination of weakly dominated strategies.

	Heads	Tails
Heads	(1, −1)	(−1, 1)
Tails	(−1, 1)	(1, −1)

Fig. 4. Matching Pennies, a game with no Nash equilibria (in pure strategies).

GAME THEORY

Refinements of Nash Equilibrium

Although it is generally agreed that a solution to a game must be a strategy profile in a Nash equilibrium, this provides only a necessary, not a sufficient, condition. In general, Nash equilibria lack several desirable properties: They need not be unique, they need not be optimal, and they may allow players to make incredible threats or promises. The game Battle of the Sexes, shown in Figure 5a, has two Nash equilibria, (Boxing, Boxing) and (Ballet, Ballet). The well-known Prisoner's Dilemma, illustrated in Figure 5b, has (Defect, Defect) as its sole Nash equilibria, yet this outcome yields a payoff of 2 to each player, whereas the outcome (Cooperate, Cooperate) yields payoffs of 3. In the game G (Binmore 1992), (rr, LLL) is a Nash equilibrium, but note that this strategy profile requires that Player 2 commit to playing L at node N, an irrational move, as Player 2 would thereby lose the game if that node were reached, whereas Player 2 would win by playing R. Consequently, a number of refinements and extensions to the concept of a Nash equilibrium have been introduced, two of which are discussed below.

Subgame Perfect Equilibrium

Each node v in an extensive game G induces a subgame of G. A subgame is produced by keeping the node v, along with the subtree rooted at v, and deleting the rest of the game. If σ is a Nash equilibrium of G, it need not be true that σ is a Nash equilibrium for every subgame of G as well. Selten (1965) introduced a refinement of the Nash equilibrium concept known as a *subgame perfect* equilibrium, which requires that a strategy profile σ be a Nash equilbrium for every subgame as well. It has been shown that every finite extensive game of perfect information has at least one subgame perfect equilibrium. Since every subgame perfect equilibrium is also a Nash equilibrium, subgame perfection counts as a refinement of the concept of a Nash equilibrium, because it often eliminates Nash equilibria that are unlikely to be adopted by rational players, such as the strategy profile (rr, LLL) in the Prisoner's Dilemma game of Figure 5b.

Correlated Equilibrium

The definition of a Nash equilibrium assumes that the selection of strategies by players occurs independently. Aumann (1974 and 1987) defined a notion of *correlated* equilibrium for noncooperative games. By correlating on shared information about the state of the world (although the information need not be the same for all the players), it is possible for players to arrive at an equilibrium that is self-enforcing in the sense that no player would have reason to deviate from equilibrium play. The fact that correlated equilibria are self-enforcing is significant because it means that adhering to a correlated equilibrium does not require the existence of a binding agreement among the players. In many cases, adopting a strategy profile in correlated equilibrium rewards each player with a higher expected payoff than she could receive in the absence of correlation. For example, consider the game of Battle of the Sexes from Figure 5a and suppose that the players have shared information about the result of a toss of a fair coin. If both players (independently) adopt the strategy of going to a boxing match whenever the coin turns up heads and going to

	Boxing	Ballet
Boxing	(2, 1)	(0, 0)
Ballet	(0, 0)	(1, 2)

(a) Battle of the sexes

	Cooperate	Defect
Cooperate	(3, 3)	(1, 4)
Defect	(4, 1)	(2, 2)

(b) The prisoner's dilemma

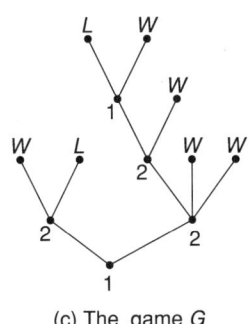

(c) The game G

Fig. 5. Games with multiple or suboptimal Nash equilibria.

the ballet whenever the coin turns up tails, each player has an expected payoff of $\frac{3}{2}$, a significant improvement upon their expected payoffs in the absence of correlating their strategies. It has been proven that the set of correlated equilibria always contains the set of Nash equilibria and hence is an extension of the concept of a Nash equilibrium.

Cooperative Games

In a cooperative game, players can enter into binding agreements in which they are committed to playing certain strategies. Whereas strategy profiles in noncooperative games need to be self-enforcing (e.g., a Nash equilibrium) in order to be plausible outcomes of play, in cooperative games the binding agreement can be used to bring about any possible outcome. Of the many possible outcomes, how should one be selected?

Nash (1950 and 1953) proposed the following approach to analyzing cooperative games: Although players *may* enter into a binding agreement, they need not. If they choose not to, then there is a noncooperative game in which each player can, adopting the appropriate mixed strategy, be assured of a certain minimum expected payoff; call this outcome the *disagreement point*. The original cooperative game can thus be conceived as a bargaining problem in which players seek to improve their situation by moving away from the disagreement point to a new, more desirable point conferring greater utility. Exactly which point is selected depends upon the particular arbitration scheme used. An arbitration scheme can be thought of as a function mapping the set of possible outcomes to a single outcome—the solution offered by the arbitrator. A cooperative game, then, can be conceived as an extensive form of a noncooperative game where the early stages of the game involve the selection of the disagreement point and the arbitration scheme. This approach, of reducing cooperative games to noncooperative games, is known as the *Nash program*.

Nash argued that a reasonable arbitration scheme for a bargaining problem should satisfy the following four conditions:

- **Pareto optimality:** It is not possible to increase any player's utility without decreasing another player's utility.
- **Independence of irrelevant alternatives:** The selection of the outcome of the bargaining problem should not depend upon alternatives which were not chosen. (One should be aware that Nash's proposed solution is not universally accepted. This axiom is generally viewed as the most controversial.)
- **Symmetry:** If the set of outcomes is symmetric, then the solution point awards the same payoff to all players.
- **Invariance:** Since utility functions are unique only up to a strictly increasing affine transformation, no player should be able to affect the solution point by rescaling his or her utility function.

The fact that there exists a unique outcome satisfying these four conditions was proved by Nash (1950) for the two-person case.

Solution concepts differing from the one suggested by Nash have been defended by Kalai and Smorodinsky (1975), Braithwaite (1954), and Gauthier (1986). The Kalai-Smorodinsky solution has a natural geometric construction that illustrates the underlying intuitions. Define the "Utopia point" as the outcome awarding each player the maximum amount of utility possible for the game under consideration. In all cases of interest, the Utopia point lies outside the set of feasible solutions. Draw a line *l* connecting the disagreement point to the Utopia point. The point of intersection between *l* and the Pareto frontier is the Kalai-Smorodinsky solution. That is, the Kalai-Smorodinsky solution is the point arrived at when each player makes "appropriate" relative concessions from the Utopia point. The solution point identified by the Kalai-Smorodinsky solution is often not the same point as that identified by the Nash axioms.

Evolutionary Game Theory

Evolutionary game theory originated as an application of game theory to biology, arising from the realization that frequency-dependent fitness introduces a strategic aspect into evolution. Evolutionary game theory has since become an object of interest to economists in part because the rationality assumptions underlying it are more appropriate for modeling strategic deliberation by real humans, who are only boundedly rational, as opposed to the perfectly rational agents modeled by traditional game theory. In addition, evolutionary game theory provides a way of modeling the dynamics of strategic interaction in a way not possible with the traditional theory of games. Recall that the only way to model the temporal aspect of a game is to use the extensive form of representation. However, methods of analyzing extensive games typically proceed by envisioning that players select

	Rock	Paper	Scissors
Rock	(1, 1)	(0, 2)	(2, 0)
Paper	(2, 0)	(1, 1)	(0, 2)
Scissors	(0, 2)	(2, 0)	(1, 1)

Fig. 6. The game of Rock–Paper–Scissors.

a strategy at the beginning of the game that specifies their course of action at each choice point, which really does not model the dynamical aspect of the game.

The primary equilibrium concept in evolutionary game theory is that of an evolutionarily stable strategy (see Maynard Smith 1982). A strategy is *evolutionarily stable* if when almost every member of the population follows it, no individual who adopts a novel strategy can successfully invade. If σ is evolutionarily stable, the fitness of an individual following σ must be greater than the fitness of an individual following μ (otherwise the individual following μ would be able to invade, and so σ would not be evolutionarily stable). Let $F(s_1, s_2)$ denote the change in fitness for an individual who plays the strategy s_1 against an opponent playing the strategy s_2. Then σ is evolutionarily stable if and only if:

$$F(\sigma, \sigma) > F(\mu, \sigma)$$

or

$$F(\sigma, \sigma) = F(\mu, \sigma) \text{ and } F(\sigma, \mu) > F(\mu, \mu).$$

If a strategy is evolutionarily stable, it must be a best reply against itself, for, if not, a mutant strategy would be able to invade. This means that all evolutionarily stable strategies are Nash equilbria when played against themselves. However, not all games have evolutionarily stable strategies, and not all Nash equilibria are evolutionarily stable. The game of Rock–Paper–Scissors, shown in Figure 6, has a unique Nash equilibrium in mixed strategies where each individual plays Rock, Paper, or Scissors with probability $\frac{1}{3}$, but no evolutionarily stable strategy.

J. McKenzie Alexander

References

Aumann, R. (1974), "Subjectivity and Correlation in Randomized Strategies," *Journal of Mathematical Economics* 1: 67–96.
——— (1987), "Correlated Equilibrium as an Expression of Bayesian Rationality," *Econometrika* 55: 1–18.
Binmore, K. (1992), *Fun and Games*. Lexington, MA: D. C. Heath and Company.
Borel, É. ([1921] 1953), "The Theory of Play and Integral Equations with Skew Symmetric Kernels" (translated by L. J. Savage), *Econometrika* 21: 97–100. Originally published as "La théorie du jeu et les équations, intégrales à noyau symmétrique gache," *Comptes Renduc de l'Académie des Sciences* 173: 1304–1308.
Braithwaite, R. B. (1954), *Theory of Games As a Tool for the Moral Philosopher*. Cambridge: Cambridge University Press.
Gauthier, David (1986), *Morals by Agreement*. Oxford: Oxford University Press.
Harper, W. (1988). "Causal Decision Theory and Game Theory: A Classic Argument for Equilibrium Solutions, a Defense of Weak Equilibria, and a New Problem for the Normal Form Representation," in W. Harper and B. Skyrms (eds.), *Causation in Decision, Belief Change, and Statistics II*. Dordrecht, Netherlands: Kluwer.
Kalai, E., and M. Smorodinsky (1975), "Other Solutions to Nash's Bargaining Problem," *Econometrika* 43: 513–518.
Kreps, D. M. (1990), *Game Theory and Economic Modelling*. Oxford: Oxford University Press.
Lewis, David (1969), *Convention: A Philosophical Study*. Oxford: Basil Blackwell.
Luce, R. D., and H. Raiffa (1957), *Games and Decisions*. New York: Wiley.
Maynard Smith, J. (1982), *Evolution and the Theory of Games*. Cambridge: Cambridge University Press.
Myerson, R. B. (1991), *Game Theory: Analysis of Conflict*. Cambridge, MA: Harvard University Press.
Nash, J. (1950), "Equilibrium Points in N-person Games," *Proceedings of the National Academy of Sciences of the United States of America* 36: 48–49.
——— (1953), "Two-Person Cooperative Games," *Econometrica* 21: 128–140.
Osborne, M. J., and A. Rubinstein (1994), *A Course in Game Theory*. Cambridge, MA: MIT Press.
Samuelson, L. (1998), *Evolutionary Games and Equilibrium Selection*. Cambridge, MA: MIT Press.
Selten, R. (1965), "Spieltheoretische Behandlung eines Oligopolmodells mit Nachfragetragheit," *Zeitschrift für die gesampte Staatswissenschaft* 121: 301–324.
von Neumann, John ([1928] 1959), "On the Theory of Games of Strategy" (translated by Sonya Bargmann), in A. W. Tucker and R. D. Luce (eds.), *Contributions to the Theory of Games*, vol. 4 (*Annals of Mathematics Studies* 40). Princeton, NJ: Princeton University Press, 13–43. Originally published as "Zur Theorie der Gesellschaftsspiele," *Mathematische Annalen* 100: 295–320.
von Neumann, John, and Oskar Morgenstern (1944), *Theory of Games and Economic Behavior*. Princeton, NJ: Princeton University Press.
Zermelo, E. (1913), "Über eine Anwendung der Mengenlehre auf die Theorie des Schachspiels," in *Proceedings of the Fifth International Congress of Mathematicians* 2: 501–504.

GENETIC INFORMATION

See **Biological Information; Molecular Biology**

GENETICS

Classical Genetics

Genetics was the name given in 1906 by William Bateson (1861–1926) to the emerging branch of biology "devoted to the elucidation of the phenomena of heredity and variation" (Bateson 1928, 1943). The founding opus of genetics is Gregor Mendel's (1822–1884) *Versuche über Planzenhybriden* [Experiments on Plant Hybrids] (Mendel 1866), read at the meetings of the Naturalist Society of Brno (Moravia) on 8 February and 8 March 1865. In 1900, while elaborating his theory of *Intracellulare Pangenesis* into the *Mutationstheorie* on the origin of species by discontinuous, rather than continuous variations (of Darwinian theory), Hugo de Vries (1848–1935) modified his model along the lines of Mendel's hypothesis of particulate inheritance of thirty-five years earlier. However, instead of Mendel's abstract notion of factors for characters, experimentally demonstrated by seven carefully selected discrete traits, de Vries introduced the preformationist notion of organisms composed of "unit characters," for each of which pangenes existed. Cell nuclei, including those of the gametes, contained the full gamut of pangenes, thus providing for continuity of intergenerational inheritance, whereas development was due to differential activation of specific pangenes farmed out to the cytoplasm of the cells of various organs. Thus, whereas de Vries adopted Mendel's insight of dealing with inheritance in terms of *factors* for *discrete traits*, he accorded these abstract factors properties of material entities, introducing into genetic theory a dialectical confrontation that has been part of it ever since.

Mendel, who was educated in the physical sciences, apparently believed that laws of nature were expressible as mathematical statements. His experiments on hybridization, mainly in garden peas, were carefully designed to establish numerical laws for the *inheritance* of selected individual traits, irrespective of the *nature* of these traits. His paper is a masterpiece of didactic presentation of experimental results as support for his theory of inheritance. Mendel posited discrete and independent factors for each trait. A maternal and a paternal factor for any given trait may combine in hybrids without losing their identity and segregate again in the gametes of the hybrids, to combine according to the laws of probability in progeny of further generations (*law of segregation*). Factors of different traits segregate independently of each other (*law of independent segregation*). Plants that produce only one kind of factor for a trait are homozygous for that trait. Those that produce two kinds of factors for a trait are heterozygotes for that trait. Heterozygotes for a trait are often indistinguishable in appearance from one of the homozygotes; the factors for that trait are considered dominant, whereas those of the trait that does not show in heterozygotes are recessive. The alternatives or complementary appearances that a unit can obtain (red or white flowers, A, B, or O blood type) are its *allelomorphs*, or alleles.

Wilhelm Johannsen (1857–1947) studied seed dimensions of bean plants. By repeated inbreeding, "pure lines" were obtained, in which selection was ineffective, since practically all variation among the progeny was due to environmental fluctuations. In 1909, he concluded that the visible or phenotypic variation of a character in a mixed population of individuals is composed of heritable, or genotypic, variance and varies due to nonhereditary fluctuations, or environmental effects. By extending the distinction between the genotypic and phenotypic components of variation to the changes during individual development and to variation in morphological and physiological traits of individuals, Johannsen extended the notion of the predestined factors of unit characters, or the preformationist link between hereditary factors and characters. Following Mendelian theory, Johannsen termed the genotypic component of a distinct character its gene. Genes are *invariant entities of inheritance* and development, which are present in the gametes and the zygotes, through which "a property of the developing organism is or may be conditioned or codetermined" (Johannsen 1911). For Johannsen the concept of the gene was merely an abstraction, an *intervening variable* that purely "summarized" characters, a quantity obtained by a specified manipulation of the values of empirical variables (see Falk 1986):

> The segregation of one sort of "gene" may have influence upon the whole organization. Hence the talk of "genes for any particular character" ought to be omitted.... It should be a principle of Mendelian workers to minimize the number of different genes as much as possible. (Johannsen 1911, 147)

Once the notion of unit characters became redundant, biologists could conceive the Mendelian theory of heredity of distinct factors or genes as providing necessary but not sufficient conditions for traits of living organisms. The discriminative trait became merely the phenotypic "marker" of a gene. Evidence from cytological observations indicated that the cell nucleus, or more precisely its chromosomes, provided the material basis of inheritance. Chromosomes maintained continuity between cell generations, and the specific functional role of each chromosome was revealed by the dysfunction of any embryo lacking a full set of them.

Sexually reproducing organisms contain a maternal and a paternal set of chromosomes in each of their cells. Before cell division, or *mitosis*, the chromosomes are duplicated, and a precise division of the duplicated nuclear content to the daughter cells is orchestrated. Before the production of the reproductive cells, or gametes, two nuclear divisions follow only one chromosome duplication. During these coupled divisions, or *meiosis*, a complex process of chromosome pairing and exchange of parts takes place. As a result, corresponding segments of paternal and maternal chromosomes segregate to different gametes, so that each gamete contains a single but full set of chromosomes, although no set is either paternal or maternal. Edmund B. Wilson (1856–1938) and his students showed that chromosome pairs segregate independently at meiosis and suggested that segregation of chromosomes at meiosis and their recombined association at fertilization is what is expected of the material bearers of Mendelian factors.

In 1910, Thomas H. Morgan (1866–1945) observed genes that segregated according to the pattern of the sex chromosomes of his new experimental organism, the fruit fly *Drosophila melanogaster*: Of the pair of sex chromosomes (X chromosomes) present in females, only one—of maternal origin—is found in males (who have a paternal Y chromosome instead of a second X chromosome). Morgan adopted an instrumental approach to genes as entities detected by function while accepting that, materially, these were entities localizable to chromosomes. Thus, rather than being abstract intervening variables, genes were envisioned by Morgan as *hypothetical constructs* to which existence properties were added that were not explicitly defined by the empirical relations. With this dialectical approach, Morgan maintained an epigenetic view of many-to-many relationships between genes and characters, in which there were "manifold effects of each gene" and "each character is the product of many genes" (Morgan 1917).

However, not all observed deviations from independent segregation of traits were explicable in terms of the same gene affecting several traits (*pleiotropy*). Such correlation was considered to be due to material dependence between different genes:

> If the materials that represent these factors are contained in the chromosomes, and if those factors that "couple" be near together in a linear series, then when the parental pairs (in the heterozygote) conjugate, like regions will stand opposed. There is good evidence that during [meiosis] homologous chromosomes twist around each other, but when the chromosomes separate (split), the split is in a single plane.... In consequence, we find coupling in certain characters and little or no evidence at all of coupling in other characters; the difference depending on the linear distance apart of the chromosomal materials that represent the factors. (Morgan 1911)

Morgan's distinction between multiple effects of genes and physical dependence between genes was elaborated by him and his students into the theory of *genetic linkage*, according to which genes are located at specific *loci* along the chromosomes. Linked genes on a given chromosome may recombine at a rate that depends on the relative distance of the loci along the chromosome. Alfred H. Sturtevant (1891–1971) provided in 1913 the first linkage (or recombination) map of a chromosome of *Drosophila melanogaster*. These maps, however, were merely of abstract intervening variables of experimental linkage data, that is, linear representations of the deviations from independent segregation of gene pairs in Mendelian hybridization experiments. Notwithstanding, Morgan's students, notably Calvin B. Bridges (1889–1938) and Hermann J. Muller (1890–1968), provided increasing evidence for the genes being discrete material entities arranged along the chromosomes: "[B]esides the ordinary proteins, carbohydrates, lipoids, and extractives, of their several types, there are present within the cell *thousands* of distinct substances—the 'genes'; these genes exist as ultramicroscopic particles" (Muller 1922, 32).

The necessary prerequisite properties of these atoms of heredity are:

- self-replication, or autocatalysis;
- involvement in physiological and developmental processes of organisms, or heterocatalysis; and
- a form of catalysis that upon a change in the structure of the gene "may become correspondingly changed, in such a way as to leave [the gene] still *auto*catalytic" (Muller 1922, 34).

The third property derives from Muller's insistence that given "inheritance *of* change," evolution would automatically follow (35). It was through this "general feature of gene construction," which was indispensable for matter evolving by a Darwinian process of trial and error, that Muller set out to investigate the genes. Such a pivotal image of the gene also led, however, to the genocentric notions that overwhelmed future discourse far beyond the image's heuristic value, although Muller himself emphasized that whatever the genes may be or do, they make sense only in the context of the living cell and its environment:

> Each of these effects, which we call a "character" of the organism, is the product of a highly complex, intricate, and delicately balanced system of reactions, caused by the interaction of countless genes, and every organic structure and activity is [liable to become altered] when the balance of the reaction system is disturbed by an alteration in the nature or the relative quantities of any of the component genes of the system. (Muller 1922, 33)

Muller developed quantitative methods of mutagenesis to investigate the physical properties of the genes, and in 1927 showed that x-rays may induce mutations and chromosomal aberrations. These studies also allowed correlations to be established between genes and their location on chromosomes. The discovery of giant "polytenic" (multistranded) chromosomes in the cells of some insect larvae finally allowed the detailed physical maps of genes on chromosomes. This confirmed the collinearity of the linkage maps and the cytological maps of chromosomes (Figure 1).

However, the dialectics of the theory of the genes did not conceive of them as necessarily particulate atomic entities of heredity. Richard Goldschmidt (1878–1958) conceived of whole chromosomes as integrative functional entities. Changes, such as breaks and rearrangements in the chromosomal continuity, cause functional deviations that may operationally be localizable as mutations in discrete genes. Induced changes in the arrangement of chromosomes that did affect function (*position effects*) supported this notion. L. J. Stadler (1896–1954) emphasized as late as 1953 that the operational tests to support the existence of genes could not prove their indivisibility (Stadler 1954). Doubts about genes as the physical entities of heredity grew when estimates of their size changed under different conditions of mutagenesis. Also, intensive experiments revealed that recombination could occur between what were considered to be alleles, alternative mutants of the same gene. Such a possibility to separate by recombination what turned out to be similar yet different adjacent functional entities, or *pseudo-alleles*, was first believed to be a property of complex loci, but it eventually allowed the experimental analysis of the gene. By 1953, when the structural organization of the hereditary material became clear, the indivisible nature of the abstract and cytological gene entities had already been replaced by a gene analyzable by intragenic recombination in organisms from Drosophila to the mold *Aspergillus nidulans*. Contrary to Muller's project to study the properties of the gene indirectly because "[a] gene can not effectively be ground in a mortar, or distilled in a retort" (Muller 1922, 36), it was eventually the physicochemical analysis of molecules that resolved the puzzle (Watson and Crick 1953a and 1953b). It did not escape the notice of Watson (1928–) or Crick (1916–2004) that their initial paper, "Genetical Implications of the Structure of Deoxyribose Nucleic Acid [DNA]," addressed exactly the three properties that Muller expected of genes as the atoms of inheritance.

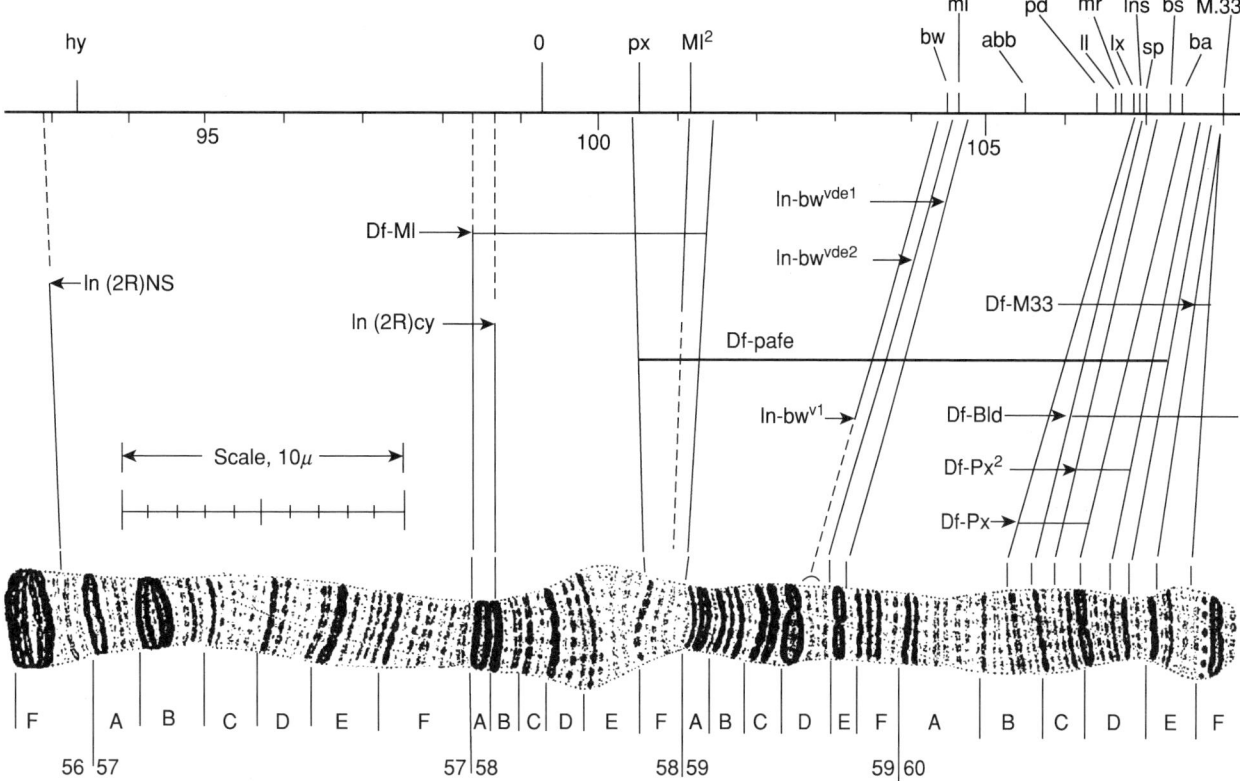

Fig. 1. *Drosophila* linkage map and polytenic chromosome map, aligned.

Thus, genetic analysis of abstract, intervening variables, which evolved in parallel with a phenomenological approach engaged with hypothetical constructs, finally landed at the physicochemical definition of molecular genetic matter. However, the Watson-Crick model of the molecular structure of DNA did not indicate any secondary organization into discrete entities, and experimental data showed chromosomes to be straightforward, continuous DNA sequences. Thus, the model of the molecular structure of DNA did not resolve the confrontation of genes as abstract entities versus that of genes as material atoms of heredity, and the dynamics of this dialectic still play a major role in genetic research. Population geneticists and breeders still may refer to genes as variables in a frequency distribution space, although the complexity of the organization of the genetic material is acknowledged. Likewise, reference to phenomenological entities as "genes for" diseases or behavioral properties are made frequently, often irrespective of available information on their detailed molecular structure.

Formal Genetics

In the early 1940s, the modern synthesis suggested that evolution should be expressed in terms of changes in gene frequencies in populations of interbreeding individuals (see Evolution). The basic law of population genetics was formulated in 1908 independently by G. H. Hardy (1877–1947) and by Wilhelm Weinberg (1862–1937) (see Population Genetics). The Hardy-Weinberg law posits that in an infinitely large, randomly mating population, where p is the frequency of allele A_1 and q is the frequency of allele A_2 at a given locus ($p + q = 1$), the frequencies of the genotypes at that locus will be p^2 A_1A_1, $2pq$ A_1A_2, and q^2 A_2A_2, as long as no other forces, like mutation, selection, or migration, affect this population. In other words, within one generation of random mating in an infinite population, in which no outer forces act on the alleles of the gene under consideration, equilibrium in genotype frequencies will be established. Population genetics is essentially the study of theoretical and empirical factors that may cause deviations from Hardy-Weinberg equilibrium.

GENETICS

An important project, especially for animal and plant breeders, was the extension of Mendelian inheritance to quantitative traits, or traits with continuous variation. The effect of numerous genes, each affecting the trait only slightly and more or less to the same extent, would give a binomial distribution of genotypes in a population, which, considering environmental fluctuations, would dissolve into a normal distribution of phenotypes. According to the instrumental reductionist approach, as many genes are allocated to the quantitative trait as are necessary to explain its distribution (Sarkar 1998). The formal analysis of quantitative traits by R. A. Fisher (1890–1962) (Fisher 1918) reconciled the biometricians' interpretation of inheritance with Mendelism. This could be used to construct efficient breeding designs. Even today, identification of quantitative trait loci (QTLs), which are identified as variables at the abstract or phenomenological level, may provide anchors for the search of the functional equivalents at the molecular segments.

Mendelian genetics was also extended to deliberations on biological impacts on human societies. As early as the end of the nineteenth century, Francis Galton (1822–1911) introduced the notion of eugenics, the application of the insights of the science of inheritance and evolution to humans, in order to prevent an anticipated "deterioration of the species." Human populations may be exposed to the effects of mutation and selection like those of any other organism. Social revolutions and advances in health services allegedly caused relaxation of selection, which had to be countered. In the first decades of the twentieth century, the eugenics movement got widespread support from geneticists, who saw it as part of their moral and social obligation to face the consequences of their scientific insights. However, persons who wished to use eugenic arguments to promote social and political aims increasingly usurped the eugenics movement. Eugenics became an important discriminatory tool in the hands of social and political conservatives as well as reformers. At the level of genetic theory, eugenic thinking has suffered too much from oversimplified reductionism, underestimating the extent of environment/genotype interactions and their flexibility, as well as the interactions of the *genome*, the complete collection of genetic information. (For details, see Paul 1995.)

Material Genetics

Although most geneticists accepted the chromosomal theory of heredity from early on, for many years few efforts were made to investigate the chemical aspects of chromosome structure or function. Following Troland, Muller believed that genes acted like enzymes. Suggesting that the newly discovered bacterial viruses might be "naked genes," he hoped that geneticists would "be able to grind genes in a mortar and cook them in a beaker after all" (Muller 1922, 48). The physician Archibald Garrod (1857–1936) recognized as early as the 1910s that gene mutations caused dysfunction or malfunction of relevant enzymes in the normal metabolic pathways of the organism, and accordingly interpreted some diseases as "inborn errors of metabolism." This idea was elaborated by Beadle (1903–1989) and Tatum (1909–1975) into the "one gene–one enzyme" concept, according to which each gene is responsible for one specific enzyme. Beadle and Tatum (1941) studied the growth capacity of the bread mold *Neurospora crassa* on well-defined media from which specific nutrients could be omitted *ad lib*. The one gene–one enzyme concept provided a major framework for genetics, although it was soon overhauled when it turned out that more than one gene may be involved in an enzyme or that genes may code for structural, nonenzymatic proteins.

Seymour Benzer's (1921–) high-resolution recombination analysis of the *r*II gene of the bacterial virus, the bacteriophage T4, indicated that the gene as a functional unit, or *cistron*, might be presented as a linear recombination map of mutated sites. When he had "run the map into the ground," it was possible "to translate linkage distances, as derived from genetic recombination experiments, into molecular units"(Benzer 1955). Similar, though less extensive, recombination experiments with bacterial genes proved that the information along cistrons was collinear with that of the sequence of amino acids in the polypeptides corresponding to those cistrons. The primary information for polypeptide structure is materially coded in the DNA molecule, as revealed by the linear recombination map of the functional units.

Although Friedrich Miescher (1844–1895) had identified nucleic acids by the end of the nineteenth century, and increasingly overwhelming evidence for their involvement in heredity had accumulated since, the role of the carrier of the genetic specificity was persistently ascribed to the protein component of the chromosomal nucleoproteins. The structure of nucleic acids was believed to be a monotonous, repetitive polymer inadequate for encoding complex hereditary specifications. What was crucial was the recruitment of bacteria and viruses as experimental systems for the elucidation of this problem. This was possible only after demonstrating that prokaryotic microorganisms, that is, organisms lacking discernible cell nuclei, obey the same rules of random mutations and selection that were accepted for Darwinian evolution in eukaryotes, those organisms with well-defined cell nuclei (Luria and Delbrück 1943). Still, even in the 1950s, the convincing evidence for the role of DNA was derived from elegant experiments with bacterial viruses rather than from the more scrupulous, straightforward chemical work with bacteria (Avery, MacLeod, and McCarty 1944).

The DNA model of Watson and Crick is that of a double helix constructed of two antipolar strands. Each strand is a string of nucelotides composed of deoxyribose (S) phosphoric acid (P) and a nitrogen base. There are four nucleotides in DNA: adenine (A), guanine (G), thymine (T), and cytosine (C). The strands are held together by weak hydrogen bonds between complementary bases. As a rule, A pairs with T, and G pairs with C. There are no structural limitations on the sequence of nucleotides along the helices (Figure 2).

The Watson-Crick model of the molecular structure of DNA was enthusiastically adopted largely because of the elegance with which it purported to resolve the three properties that Muller assigned to the entities of genetic material. Self-replication was shown to be accomplished by each strand becoming a template for a complementary new strand (*semiconservative replication* [Meselson and Stahl 1958]). The lack of constraints on the sequence of nucleotides along the strands allowed for the endless variability needed for coding genetic specificity, now conceptualized as genetic information. Finally, the consistent structure of the backbone of the helical strands allowed exchange of one base pair for another or rearrangement of whole sequences, causing changes in coding without affecting the self-replication or coding capacity of the molecule, that is, mutations. However, major physicochemical problems, such as the unwinding of the two strands at replication, or the directional specificity of replication enzymes facing the opposite polarity of the two strands, were resolved only years after the model was firmly established. (This process of connecting formal genetics with material genetics has been often controversially interpreted by philosophers of biology as a case of reduction) (Sarkar 1998) (see Reductionism).

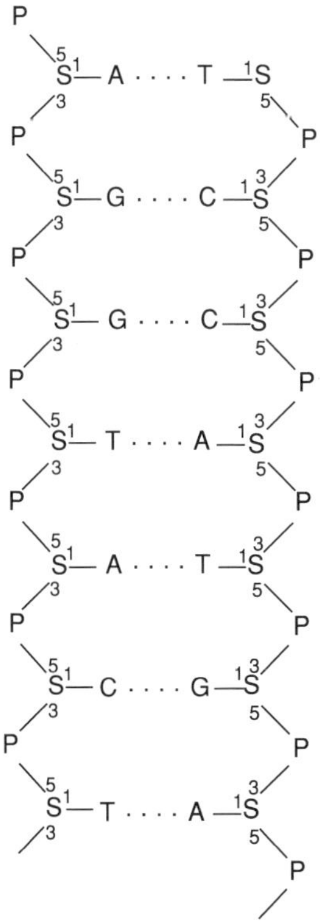

Molecular Genetics

Enzymes and many cell-structure components are proteins. Proteins are polypeptides composed of specific sequences of an array of (usually) twenty amino acids. Protein function relies on the molecules' folding into three-dimensional structures that depend on the sequence of the amino acids in the polypeptide (and on the cellular environment). The Watson-Crick model posits that the information for the sequence of amino acids in polypeptides is encoded in the sequence of the DNA base pairs (see Molecular Biology). In 1958, Crick formulated the central dogma of molecular biology (Crick 1958), according to which the flow of genetic information is unidirectional, from the nucleic acids to proteins but never from proteins to nucleic acids, proving "beyond any doubt but in a totally new way the complete independence of genetic information from events occurring outside or even inside the cell" (Judson 1979, 217). The dogma posited further that the information in the DNA is first transcribed into intermediary polynucleotide molecules, from which it is translated into amino acid sequences at the sites of protein synthesis (see Biological Information). The basic details of cellular-information reading have been elucidated mainly in prokaryotic systems. The intermediaries are molecules of ribonucleic acid (RNA), termed *messenger RNAs* (mRNA). RNA is, as a rule, a single-stranded polymer of nucleotides

Fig. 2. The Watson-Crick model of double-stranded DNA.

composed of a ribose (instead of the deoxyribose of DNA) and phosphate in the backbone and of four kinds of bases attached to the ribose residue: adenine (A), guanine (G), uracil (U), and cytosine (C). Transcription is mediated by RNA-polymerase complexes that bind at sites upstream of the sequences to be transcribed and is accomplished by nucleotide complementarity of DNA and RNA:

- G with C
- C with G
- T with A
- A with U.

Translation occurs on special cytoplasmic organelles, the ribosomes, and is catalyzed by them. To code for all twenty amino acids, sequences of three bases of nucleic acids are needed. The code was found to be redundant and comma free. Of the 64 possible triplets, or codons, 61 are sense codons, 1–6 of which code for each of the twenty amino acids; the remaining 3 are termination signals, or nonsense codons (Figure 3). On the ribosomes the code is read sequentially from the mRNA, one triplet after another, from a given starting point (Crick et al. 1961). Each codon is translated into its corresponding amino acid by a specific molecule of transfer RNA (tRNA). Amino acids are enzymatically attached to their specific tRNA, and when the specified tRNA anticodon sequence pairs with its complementary codon in the mRNA on the ribosome, the amino acid is transferred from the tRNA to the nascent polypeptide chain.

Regulation of the genes' activity occurs at transcription as well as posttranscription levels. The elaboration by Jacob (1920–) and Monod (1910–1975) of the negative feedback regulation mechanism of transcription of an (adaptive) enzyme, β-galactosidase, in the bacterium *Escherichia coli* became paradigmatic for genetic regulation (Jacob and Monod 1961). Transcription initiation is controlled by the attachment of an RNA polymerase at the *promoter* site. Numerous transcription factors must combine with the polymerase for its proper function. Various intracellular metabolites (or extracellular *ligands* that attach to cellular *receptors*) affect the formation of different transcription factor complexes, which allow the polymerase to initiate transcription at specific sequences, thus serving as cues that regulate transcription. Regulation may occur by negative-feedback as well as by positive-feedback mechanisms. Although the details of the transcription from DNA and the translation to polypeptides were elaborated in prokaryotes, the essential features were found to hold for all living cells. Furthermore, the same genetic code (with some minor but important exceptions) holds throughout the living world. This strongly endorses the Darwinian model of evolution from an early common ancestor.

However, the expectation of the early molecular geneticists that what was true for *E. coli* was also true for the elephant was exaggerated: Major cellular systems of eukaryotes and prokaryotes diverge significantly. Cells of eukaryotes usually contain orders of magnitude more DNA per nucleus than do prokaryotic cells, in spite of the fact that their basic cell maintenance functions are not much more numerous or complex. Britten and Kohne (1968) found that the nuclear DNA of most mouse cells contains highly repetitive sequences (some of these up to a million times). Such redundancy suggests that these sequences are not involved directly in coding or regulatory functions. In many eukaryotic genomes, not more than 10 percent of the DNA appears to be "meaningful." The observation that often the evolutionarily older taxonomic groups are those especially rich in this repetitive, so-called junk DNA has been described as the C-value paradox. Did birds and mammals evolve more efficient cellular household mechanisms for functions such as packing and unpacking of DNA, instead of the "primitive"

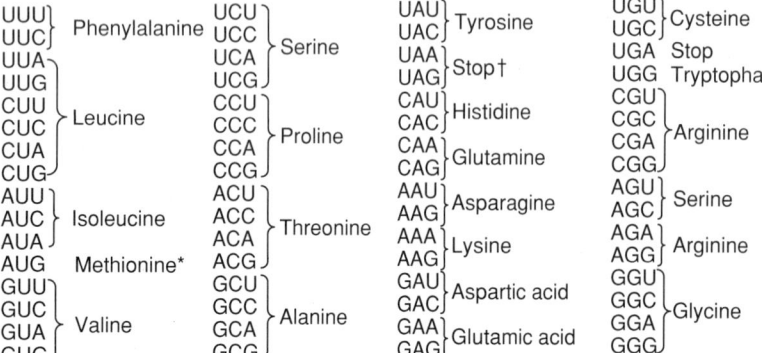

Fig. 3. The genetic code.

mechanisms that need lots of DNA in lungfish and amphibians?

Another unexpected experimental finding, inconsistent with the concept of the gene as a coherent entity of information, was that most eukaryotic coding sequences are interrupted by numerous noncoding sequences (introns). Introns vary in length and sometimes comprise sequences many times longer than the coding exons, the continuity of which they interrupt. The RNA that is transcribed from such DNA sequences is processed by splicing out the introns before the sequence of continuous exons forms a translatable mRNA. Splicing of the introns provides the cells with another level of regulatory control. This includes alternative splicing, whereby numerous different alternative assemblies of sequential exons may be spliced from a given transcription product. Alternative splicing of the same RNA stretch thus effectively codes different mRNA, hence different polypeptides.

Usually only one DNA strand, the "sense" strand, is transcribed into RNA from a given DNA sequence. Sometimes, however, coding regions overlap: Both RNAs could be transcribed from (partly) overlapping sequences of the same DNA strand or from opposite strands, one being the sense strand for one transcript, the other being the sense strand for the other transcript. *RNA editing* by enzymatically changing single nucleotides or whole stretches is another device to increase the repertoire of polypeptides translated from a given sequence of DNA. Thousands of polypeptides were experimentally shown to be referable to given DNA sequences, and more than ten thousand have been predicted for some, defying the one gene–one enzyme notion. No reduction of the classical notion of a gene to such molecular concepts seems possible (see Reductionism). Genes as material entities become merely generic terms for DNA stretches that code some information, whether structural or regulatory (see e.g., Beurton, Falk, and Rheinberger 2000).

Genetics in Context

Arguably, no new major concepts beyond that of the Watson-Crick model of DNA and Crick's central dogma have been formulated by molecular genetics because none are needed (see, however, Molecular Biology). Empirical molecular biology provided the insights that allowed phenomena of genetics to be expressed in physicochemical and physiological terms. Although no reduction of formal genetics to molecular genetics may be possible, biochemical and biophysical details replaced one by one the old concepts of abstract and phenomenological genetics (see Reductionism).

At the beginning of the 1970s, genetic research underwent a profound methodological turn with the introduction of controlled *in vitro* splicing of DNA sequences from any source and the use of appropriate vectors to insert such engineered sequences into host cells, irrespective of the donor's relationship to the host. The possible ethical and social repercussions of this development are obvious, and the scientists involved were the first to take notice (see Krimsky 1982). Genetic engineering completely revolutionized research not only in genetics and its classic sister disciplines, developmental and evolutionary biology, but also in more remote disciplines of the life sciences, such as physiology and neurobiology. With the beginning of the twenty-first century, when sequencing of the complete genome of organisms has become routine, genetic research is undergoing another major conceptual breakthrough with *genomics* and *proteomics,* focusing on the integrated study of whole genomes and on structural and functional interactions instead of the classical Mendelian concentration on one factor at a time.

Genetics has extended far beyond problems of heredity and variation of individuals. The elucidation of the principles of gene regulation allowed a molecular extension of embryological *Entwicklungsmechanik,* or as it is now called, "developmental biology," whose major mode is positive regulation of initiation of transcription. Any protein that is needed for the initiation of transcription but that is not itself part of RNA polymerase is defined as a transcription factor. Transcription factors are provided under tissue-specific control to activate a promoter or a set of promoters that contain a common target sequence upstream of the transcription initiation points. Initiation at a promoter involves a large number of factors. Some recognize the specific target sequences and, once bound to DNA, bind by protein–protein interactions to other components of the transcription apparatus. A generic promoter usually functions at a low efficiency. *Enhancer* sequences that are a major target for tissue-specific or temporal regulation are located often at considerable distance from the start point.

Genetic analysis heavily relies on deducing the normal from the deviant, whether natural or induced. A major feature of most eukaryotes is the defined life span of the organism, a property that extends to the individual somatic cells, whose growth and division is highly regulated. Genetic instability is thought to transform normal cells into cancerous ones. As a rule, growth of transformed cancer cells is less restricted or dependent on external cues than is growth of normal cells, and cancer cells appear to be immortal. Usually, multiple genetic changes are necessary to create a cancer, and the virulence of a cancer may increase as the result of progressive series of changes. One group of genes in which mutations cause transformation is the oncogenes, which have cellular counterparts: the *proto-oncogenes*, which are involved in normal regulated cell function. The generation of an oncogene is by a mutation that inappropriately activates the regulated proto-oncogene. Another tumorigenic factor is loss of function of suppressor genes that usually impose constraints on cell cycle or cell growth. Finally, tumors may result from defects in genetic checkpoint systems that should prevent further damage in cells that went astray by inducing repair mechanisms or by initiating programmed cell death, or apoptosis.

As noted, the near-perfect universality of the genetic code and of the machinery of transcription and translation strongly support the Darwinian hypothesis of evolution of life from a common primeval ancestor by a long process of trial and error of random mutations and natural selection. However, the enormous amount of genetic variability at the level of proteins and DNA suggests that not all of it could be driven or maintained by natural selection. Theoretical considerations indicate that much of this variation is due to random fluctuations of adaptively neutral or near-neutral mutants (Kimura 1968). Physical association with loci that are selected for or against may also affect genetic variability in neighboring stretches of DNA.

In vivo and *in vitro* juxtaposition of DNA sequences from different organisms and the examination of their homologies turned DNA manipulation into a central tool of evolutionary analysis. The insertion of foreign DNA sequences into cells, with or without the knockout replacement of the indigenous sequences, indicated the functional conservation of sequences. Surprisingly many genes were found to be *orthologous*, consisting of homologous, highly conserved DNA sequences in different species; the conservation is even more impressive at the level of the corresponding amino acids (the greater identity of amino acid is due to the redundancy of the genetic code). Likewise many sequences *within* the same genome were found to be homologous (*paralogous*), indicating intensive intragenomic duplication of sequences during evolution; the duplicates were usually modified and mobilized for new related or unrelated functions. A classic example is the human gene that codes the globin component of myoglobin, a structural protein of the muscle fibers, which is paralogous to the respective genes that code embryonic, fetal, and adult components of the blood hemoglobin. Orthologous genes for globin are found throughout the organic world, including in some plant species. Studies have revealed not only the evolutionary path of genes and proteins, but also the developmental constraints on evolution. *Pax*6 is a *homeotic* "master gene" for eye formation in mice as well as in *Drosophila*. A mutation in it may alter a *Drosophila* eye into a homologous structure. The sequence of many master genes contains a special domain, such as the *homeobox* in homeotic genes, which codes for an amino acid domain involved in DNA binding of transcription factors. Such homeoboxes are highly conserved in the evolution of developmental master genes, and orthologous copies are found throughout the animal kingdom, often with many paralogous copies in each.

Such detective work of relationships led to major reevaluations of accepted patterns of the evolution of species. Bacteria were split into two kingdoms. That of the *Archea*, most of the present members of which inhabit niches of high salt concentration and/or extreme temperatures, seem to be nearer relatives of the ancestors of eukaryotes than are the more common *Eubacteria*. Concomitantly, it was surmised that intensive lateral gene transfer must have been the rule in evolution, even between cells belonging to different kingdoms, especially in early phases of the evolution of life. The breakthrough in the evolution of eukaryotes was apparently facilitated by the incorporation of mitochondria and chloroplasts in their cells by symbiosis with prokaryotes, which turned obligatory. The fact that genes like *Pax*6 have similar functions in such diverse eye structures as those of arthropods, mollusks, and vertebrates suggests that much of what was considered to be convergent evolution should be regarded as divergent evolution.

The upsurge of the study of the whole genome as an entity, which depends on the development of techniques such as simultaneous screening of micro-arrays of many thousands of genes or their products, shifts the attention of genetics to multiple interactions between genes and between proteins. A significant insight from these studies is the extent of homeostatic buffering that interactions of integrated gene functions provide even at the most basic functions of living cells. Changing variables, sometimes over several orders of magnitude, may hardly affect the stability of systems in which they are involved. This could provide a new challenge to theories of evolution and development. However, such developments signify a change not only in the conceptions of genetic control of cellular and organismic development and function, but also at the practical level of their application. Technologies of *transgenic*, "genetically modified" domestic animals and plant crops have already affected various aspects of society. The impacts of gene therapy on humans appear to have to wait somewhat longer.

Philosophers of science tried for many years to establish the continuity of genetic theory by formally reducing classical or Mendelian genetics to molecular genetics (see Sarkar 1998). When this failed, it was concluded that genetic theory incorporates essentially at least two incommensurable concepts, best explicated in the central entity of genetics, the gene. Representative of this is Moss's (2003) conceptual analysis of the gene, which results in "defining and distinguishing two different genes. . . . The preformationistic gene (Gene-P) predicts phenotypes but only on an instrumental basis where immediate medical and/or economic benefits can be had. The gene of epigenesis (Gene-D), by contrast, is a developmental resource that provides possible templates for RNA and protein synthesis but has in itself no determinative relationship to organismal phenotypes"(xiv). Such an analysis, falling back on developments in the analysis of whole genomes and of "forward genetics," which purports to predict the function of sequences directly from their sequence, underestimates the fact that throughout the billions of years of organismic evolution, no novel *structurally discrete* DNA entities evolved. It has been a dialectical, philosophically loose discourse, which allowed functions to instrumentally parse DNA and refer to sequences as genes.

RAPHAEL FALK
SAHOTRA SARKAR

The authors acknowledge the helpful input of Pamela Lyon.

References

Avery, Oswald T., Colin M. MacLeod, and Maclyn McCarty (1944), "Studies on the Chemical Nature of the Substance Inducing Transformation of Pneumococcal Types," *Journal Experimental Medicine* 79: 137–158.

Beadle, George W., and Edward L. Tatum (1941), "Genetic Control of Biochemical Reaction in *Neurospora*," *Proceedings of the National Academy of Science, Washington* 27: 499–506.

Benzer, Seymour (1955), "Fine Structure of a Genetic Region in Bacteriophage," *Proceedings of the National Academy of Science, Washington* 41: 344–354.

Beurton, Peter J., Raphael Falk, and Hans-Jörg Rheinberger (2000), *The Concept of the Gene in Development and Evolution: Historical and Epistemological Perspectives*. Cambridge and New York: Cambridge University Press.

Britten, Roy J., and D. E. Kohne (1968), "Repeated Sequences in DNA," *Science* 161: 529–540.

Crick, Francis H. C. (1958), "On Protein Synthesis," in *Symposium of the Society for Experimental Biology: The Biological Replication of Macromolecules*, 138–163. Cambridge: Cambridge University Press.

Crick, Francis, H. C., Leslie Barnett, S. Brenner, and R. J. Watts-Tobin (1961), "General Nature of the Genetic Code for Proteins," *Nature* 192: 1227–1232.

Falk, Raphael (1986), "What Is a Gene?" *Studies in the History and Philosophy of Science* 17: 133–173.

Fisher, Ronald A. (1918), "The Correlation Between Relatives on the Supposition of Mendelian Inheritance," *Transactions of the Royal Society, Edinburgh* 52: 399–433.

Jacob, François, and Jacques Monod (1961), "Genetic Regulatory Mechanisms in the Synthesis of Proteins," *Journal of Molecular Biology* 3: 318–356.

Johannsen, Wilhelm (1911), "The Genotype Conception of Heredity," *American Naturalist* 45: 129–159.

Judson, Horace Freeland (1979), *The Eighth Day of Creation: Makers of Revolution in Biology*. New York: Simon and Schuster.

Kimura, Motoo (1968), "Genetic Variability Maintained in a Finite Population Due to Mutational Production of Neutral and Nearly Neutral Isoalleles," *Genetical Research* 11: 247–269.

Krimsky, S. (1982), *Genetic Alchemy: The Social History of the Recombinant DNA Controversy*. Cambridge, MA: MIT Press.

Luria, Salvator E., and Max Delbrück (1943), "Mutations of Bacteria from Virus Sensitivity to Virus Resistance," *Genetics* 28: 491–511.

Mendel, Gregor (1866), "Versuche über Pflanzenhybriden," *Verhundlungen Naturforscher Verein, Brunn* 4: 3–47.

Meselson, Matthew, and Franklin W. Stahl (1958), "The Replication of DNA in *Escherichia coli*," *Proceedings of the National Academy of Science, Washington* 44: 671–682.

Morgan, Thomas. H. (1911), "Random Segregation versus Coupling in Mendelian Inheritance," *Science* 34: 384.

——— (1917), "The Theory of the Gene," *American Naturalist* 51: 513–544.

Moss, Lenny (2003), *What Genes Can't Do*. Cambridge, MA: MIT Press.

Muller, Hermann J. (1922), "Variation Due to Change in the Individual Gene," *American Naturalist* 56: 32–50.

Paul, D. B. (1995), *Controlling Human Heredity: 1865 to the Present*. Atlantic Highlands, NJ: Humanities Press.

Sarkar, Sahotra (1998), *Genetics and Reductionism*. New York: Cambridge University Press.

Stadler, Lewis J. (1954), "The Gene," *Science* 120: 811–819.

Watson, James D., and Francis H. C. Crick (1953a), "Molecular Structure of Nucleic Acids," *Nature* 171: 737–738.

——— (1953b), "Genetical Implications of the Structure of Deoxyribose Nucleic Acid," *Nature* 171: 964–967.

See also **Biological Information; Evolution; Molecular Biology; Population Genetics; Reductionism**

HANS HAHN

(27 September 1879–24 July 1934)

Hahn was a mathematician whose contributions to analysis and topology were outstanding. In addition, he had a remarkable influence on twentieth-century philosophy—less through his writings (a mere handful of essays) than by bringing together and stimulating other thinkers. Hahn was instrumental in founding and running the Vienna Circle (see Vienna Circle). He was the thesis adviser of Kurt Gödel and a mentor of Karl Popper (see Popper, Karl Raimund). In addition, he had a hand in the chain of events that brought Ludwig Wittgenstein back to philosophy. He was both a front-seat witness and a catalyst of the great foundational debate on mathematics and logic that took place in the first third of the twentieth century.

Hahn was born on September 27, 1879, in Vienna. His father, a former music critic, eventually became one of the highest-ranking officials in the Austro-Hungarian empire. Hans Hahn grew up in the center of the fervid Viennese *fin de-siècle* atmosphere that produced Freud, Mahler, and Kokoschka. The philosophical giants of his youth were Mach and Boltzmann, who, while strongly conflicting in most of their views, were both eminent physicists with a positivistic worldview.

Both Hahn's philosophical inclinations and his networking style became apparent already before the First World War. As a young mathematician, he belonged to a group of intellectuals that included the social scientist Otto Neurath, the applied mathematician Richard von Mises, and the theoretical physicist Philipp Frank (see Neurath, Otto). The group met in Viennese coffeehouses to discuss philosophical topics, influenced by the work of Bertrand Russell, Henri Poincaré, Émile Duhem, and Heinrich Hertz. In retrospect this can be seen as the forerunner of the Vienna Circle. Indeed, when in 1909 Hahn got his first appointment as a professor in far-flung Czernowicz (an outpost of the multiethnic empire of Emperor Franz Josef), he announced to his friends that on his eventual return to a chair in the capital, they would resume their discussions with the participation of a university philosopher.

In Czernowicz, Hahn intensified his philosophical studies, writing to a friend that "last year I almost became faithless to mathematics, seduced by the charms of—philosophy" (Hahn 1906). But it was only in 1921—after years of war service and a professorship in Bonn—that Hahn got his coveted appointment in Vienna and could take steps to resume the philosophical discussions. There was no university philosopher on hand: All three chairs in philosophy happened to be vacant at the time. In particular, Stöhr, successor to Boltzmann and Mach, had died in 1920. Hahn managed to persuade the appointment committee to fill the vacancy with the German, Moritz Schlick, professor at the University of Kiel. Fittingly enough, Schlick was a former physicist, a student of Planck and friend of Einstein (see Schlick, Moritz).

The "Schlick circle," which later became known as the Vienna Circle, met on every second Thursday, during term time, in a small lecture room of the mathematical seminar. The members of the group, who were personally invited by Schlick to attend, were a congenial mixture of philosophers and mathematicians, including Hahn; his sister Olga and her husband, Otto Neurath; Hahn's young colleague Kurt Reidemeister, a professor of geometry; and (later) Hahn's two brightest students, Karl Menger and Kurt Gödel. As Popper later wrote: "It was Hahn who was the founder of the Vienna Circle, and his brother-in-law Neurath who was the organiser. ... Schlick was at first, I think, a kind of honorary president ... but he became very active" (Popper 1995, 16). Popper went on to state: "What made the Vienna Circle so special, so different from any other philosophical circle was that it was founded not by philosophers but by an important and creative mathematician, who was keenly interested in fundamental problems (also in those belonging to the philosophy of mathematics) and in applications" (ibid). Philipp Frank also designated Hahn as the true founder of the Vienna Circle.

An important part of the discussions in the Circle centered on the theory of knowledge, a topic familiar to Schlick and to Carnap (who joined in 1926), and well in line with the works of Mach and Boltzmann (see Mach, Ernest). Hahn's main contribution to the agenda of the Vienna Circle was his emphasis on the foundations of mathematics. Hahn was not looking for a proof that there exists no contradiction in mathematics, or an explanation for the astonishing efficiency of mathematical tools, or a reduction of mathematical insight to some primordial intuition. What was for him the fundamental problem was the compatibility of mathematics with an empiricist position.

Hahn had encountered foundational problems in mathematics already during his early stay in Göttingen, right after completing his doctoral thesis. He had studied with Hilbert, the foremost advocate of an axiomatization of mathematics, and worked with Zermelo, a highly influential set theorist whose "axiom of choice" aroused fierce debates among mathematicians. Later, Hahn embraced Russell's logicism, the program to reduce mathematics to logic, and engaged in an in-depth study of the *Principia Mathematica* of Russell and Whitehead (see Russell, Bertrand). His sister Olga (who lost her eyesight at the age of twenty-two) had written seminal papers on formal logic. But Hahn himself did not write on mathematical logic. His interest in the foundations of mathematics was of a more philosophical nature, and was an attempt to reconcile Russell with Hume. As Hahn stressed on several occasions, "the only possible way of facing the world seems to me the empiricist position" (Hahn 1980, 31). But since it is unthinkable that an assertion like "two times two is four" is not valid tomorrow, it cannot be based on experience. How, then, "is the empiricist position compatible with the applicability of logic and mathematics to the real world?" (ibid., 32).

Hahn found his answer in a booklet by another Viennese. Following a suggestion by Reidemeister, the Vienna Circle had started reading Ludwig Wittgenstein's *Tractatus Logico-Philosophicus* and spent several semesters discussing it sentence by sentence. Not all members of the Circle were convinced that Wittgenstein had, as he claimed, essentially solved all philosophical questions. But for both Schlick and Hahn, working through the booklet became a key experience. "It was Wittgenstein," he wrote later, "who recognised the tautological character of logic, and who stressed that there exists nothing in the world that corresponds to the so-called logical constants (like 'and,' 'or,' etc.)" (Hahn 1980, 24). Members of the Circle later criticized this view on the grounds that the concept of tautology is not precisely defined except in the realm of first-order logic. But Hahn did not aim at a precise delimitation, and rather used the term to denote any sentence that is true by its logical structure, such as the analytical statement "No object is both red and non-red" (Hahn 1995, 494).

Since mathematics, for a logicist, can be reduced to logic, it also consists of tautologies. Mathematicians often object to the view that their "hard-earned theorems can be dissolved into tautologies" (Hahn 1995, 500). But this, according to Hahn, "overlooks a minor detail, namely the circumstance that we are not omniscient." And indeed, "an omniscient being needs no logic and no mathematics," and "the reason for introducing a symbolic notation which allows to say the same in different ways is that we are not omniscient" (Hahn 1980, 23). Logic, in Hahn's view, "is a set of rules for stating the same in different ways" (ibid., 33). Logic does not deal with the most general properties of objects (this would indeed present insurmountable obstacles to empiricists): "Logic does not deal with any objects at all: it only deals with the way we talk about objects; logic first comes into being by language" (Hahn 1995, 492).

Wittgenstein, at this time, had withdrawn completely from philosophy. After much wooing, he finally condescended to meet with some members of the Circle (not Hahn), on condition that philosophical topics were avoided. But when Hahn invited the celebrated Dutch topologist L. E. J. Brouwer, the founder of the intuitionist movement in the debate on the foundation of mathematics, to give a lecture at the university, Wittgenstein showed up, and started in the after-session to discuss philosophy again. Apparently, there was still something left to say, and Wittgenstein embarked on his second phase, soon leaving Vienna to become, eventually, a professor at the University of Cambridge.

Another member of Brouwer's audience had been Kurt Gödel, a student of Hahn's who first shone in the latter's seminar on the *Principia Mathematica.* That same year, Gödel solved a problem posed by Hilbert as a first step in the latter's program of basing the foundations of mathematics on a formalistic approach. Gödel's proof that first-order logic was complete was published in his doctoral dissertation. In the following year, however, Gödel effectively destroyed Hilbert's program by showing the incompleteness of any consistent mathematical theory rich enough to allow for the natural numbers. Some true statements could not be derived from the axioms and the rules. Hahn praised Gödel and the mathematical importance of his result. The fact that a proof of the consistency of mathematics is impossible—a consequence of Gödel's breakthrough—was taken by Hahn in his stride. In his few philosophical papers, Hahn does not mention Gödel. He seems to have expected the result that "on the basis of present knowledge, an absolute proof of freedom from contradiction is probably unattainable. ... For here, as in every sphere of thought, the demand for absolute certainty of knowledge is an exaggerated demand; in no field is such certainty attainable" (Hahn 1980, 121). In the same vein, Hahn appears to have anticipated the independence of the axiom of choice from the axioms of set theory, a result proved only after his death. Hahn wrote: "The question has nothing to do with the nature of reality, as the realists think, or with pure intuition, as the intuitionists think. The question is rather in which sense we decide to use the word 'set'; it is a matter of determining the syntax of that word" (ibid., 118). This approach via analyzing how language is used seems closer to Wittgenstein than to Gödel. Hahn was fifty years old before he wrote his first philosophical essay, a pamphlet named *Occam's Razor.* This and the following philosophical papers—*On Intuition,* for instance, and *What Is Infinity?*—are models of clarity, the outcome of a lifelong concern with popularizing knowledge (Hahn 1980).

Unlike the majority of his colleagues at the university, Hahn was a stalwart member of "Red Vienna," firmly supporting school reforms, free thought, and enlightenment. Some of his fellow members of the Vienna Circle distanced themselves from what they saw as an unseemly involvement in the political quarrels of the day. But after Hahn's death from cancer on July 24, 1934—a year when Austria's political prospects took a distinct turn for the worse—Menger would mourn the demise of this "tireless and effective speaker for progressive causes" (Menger 1994, 215).

Just before the onset of his illness, Hahn had been reading the proofs of Karl Popper's *Logik der Forschung.* "His opinions were as positive as I could only wish [them] to be," wrote Popper (1995, 19) in his last paper. Popper would continue his interactions with the Viennese "Mathematical Colloquium" for the few years until his emigration, but wrote later that "of all the mathematicians at the institute, Hahn was the one who seemed to me the embodiment of mathematical discipline" (Popper 1995, 13).

KARL SIGMUND

References

Hahn, H. (1906), Unpublished letter to Ehrenfest, 26 December 26. Ehrenfest archive, Boerhave Museum, Leiden, Netherlands.

——— (1980), *Empiricism, Logic and Mathematics.* Dordrecht, Holland: Kluwer.

——— (1995), *Introduction to the Collected Works of Hans Hahn.* Edited by L. Schmetterer and K. Sigmund. New York: Springer Verlag.

Menger, K. (1994), *Reminiscences of the Vienna Circle and the Mathematical Colloquium.* Boston: Kluwer Academic Publishers.

Popper, K. (1995), 'Hans Hahn—Reminiscences of a Grateful Student," in L. Schmetterer and K. Sigmund (eds.), *Introduction to the Collected Works of Hans Hahn.* Vienna: Springer Verlag, 11–19.

See also **Logical Empiricism; Neurath, Otto; Schlick, Moritz; Vienna Circle**

NORWOOD RUSSELL HANSON

(17 August 1924–18 April 1967)

After distinguishing himself as a fighter pilot in World War II, Hanson attended the University of Chicago, receiving a B.A. in philosophy in 1946. He then went on to Columbia University, where he received degrees in physics, a B.S. in 1948, and an M.S. in 1949. With the aid of a Fulbright scholarship, Hanson studied philosophy at Oxford and Cambridge, where he also lectured in the philosophy of science. He completed his graduate studies in 1956, earning a D.Phil. from Oxford and a Ph.D. from Cambridge. In 1957, Hanson joined the philosophy department of Indiana University and was the founding chair of Indiana's Graduate Program in the History and Philosophy of Science, the first department of its kind (Hanson 1960; Grau 1999). However, injuries sustained in a plane crash caused Hanson to step down as chair in 1962, and in 1963 he left Indiana for Yale. Hanson died in 1967 in a plane crash on his way to present a paper at Cornell University.

Hanson was a prolific philosopher during his brief career, writing on scientific observation, the role of concepts in accounts of scientific facts and causation, the logic of discovery, the history of discoveries in quantum mechanics and seventeenth-century physics, and the relation between history and the philosophy of science. Hanson represented a fusion of the late Wittgenstein and logical empiricism (see Logical Empiricism). He agreed with Wittgenstein that the meaning of terms, even in science, depends on their use, and he expanded Wittgenstein's account of the conceptual loading of perception to include scientific observation. Hanson, however, shared the logical empiricist view that the function of the philosophy of science is to examine and clarify the conceptual foundations of science.

In a sense, Hanson can be seen as extending the field of conceptual analysis to areas considered off-limits, such as the context of discovery and the conceptualization of perception. Philosophers inspired by logical empiricism often spoke of matters such as observation, factuality, and perhaps causation as fundamental ideas underlying all scientific thought and practice; for Hanson, however, all of these notions can be understood, at a given time, only in terms of the theoretical and notational networks in which they figure. The great revolutions in the history of science were not generally due to observing the world, collecting facts, and finding the causes. Rather, revolutions are made possible by conceptual innovations; after such a conceptual shift, the sense of what the facts are, what has been observed, and what features of phenomena require explanation change as well.

Observation

Hanson's most significant contribution to the philosophy of science was his discussion of observation (see Observation). Hanson argued that observation is "theory-laden": more precisely, in order for perceptual experience to relate to knowledge, experience must already contain some conceptual content. Drawing on Wittgenstein's *Philosophical Investigations* and the findings of psychology on "perceptual sets," Hanson attacked the logical empiricist conception of observation. The logical empiricists generally believed that the edifice of scientific knowledge had basic statements about first-person experience, "protocol sentences," at its foundations, and that these statements were connected to scientific theory via analytic connection rules (see Protocol Sentences). Theories could be confirmed by deriving predictions about observables and then verifying that the appropriate protocol statements were produced in testing the predictions. Among the many complaints Hanson had about this picture of science was that it cannot give an adequate account of scientific controversy and discovery. The deep disputes in the history of science require a better explanation than simply claiming that the disputants were clinging to different interpretations of essentially similar protocol statements.

While other critics of the logical empiricists' conception of observation, such as Feyerabend, focused on the theory-laden character of the terms in observational reports, Hanson's main concern was to show that the *process* of observation is theoretically loaded (see Feyerabend, Paul). Hanson (1961 and 1969) treated vision as illustrative of the general perceptual case. He acknowledged that there is a sense in which seeing is just the stimulation of one's sensory organs or, alternatively, the reception of sense data; however, neither of these senses of seeing are of much epistemological importance. Simply analyzing someone's retinal imprints, or the drawings produced based on sense data, gives one little idea of what is being seen, or what knowledge has been gained through the seeing. In order for seeing to be epistemologically useful, the elements of the visual field must be ordered and categorized with concepts. The logical empiricists explained this by arguing that seeing has two discrete components: acquisition of sense data and interpretation. Hanson objected to this "formula," for he argued that for something to be an interpretation, (i) one must be introspectively aware that one is interpreting, and (ii) there must be a detectible time lag between raw perception and interpretation. Hanson used ambiguous figures from Gestalt psychology to attempt to show that neither of these conditions are met in appreciating figures in an aspect shift: The "interpretation," or the concepts, are already there in the seeing. In order to see something (in the useful sense) as an *X*, one must first have a concept of an *X*. Thus all useful seeing is *seeing as*, and *seeing as* threads experience into knowledge.

The logical empiricist tradition took the reception of sense data to be the paradigmatic case of seeing, because such seeing is incorrigible and provides the foundation for knowledge. Hanson, in contrast, considered the central function of vision to be to provide knowledge about the world, rather than to provide the indubitable foundation of a system of knowledge. Hanson claimed that *seeing that* it is four o'clock or *that* a voltmeter reads 3.5 volts is the sense of seeing of interest in the study of science. Thus, he takes epistemic seeing, or *seeing that*, to be the paradigmatic case of seeing, and asserts that study of the logic of *seeing that* will illuminate the logic of perception generally. To see something as an *X* is to see that, were certain things done to it, other things, which would be expected of *X*s, would follow; more basically, the concept of *X* incorporates our prior knowledge of *X*, such as what *X*s are composed of and what types of interactions *X*s can participate in. To see something as an *X* is to see it in all the connections that the concept of *X* has to other elements of our knowledge.

Hanson's view is open to criticism on the grounds that (i) the statement following *sees that* is always taken to be true in ordinary usage (i.e., *seeing that* is a success verb) and (ii) obsolete concepts (or those that do not apply properly to anything in the world) can be used only for *seeing as*, not *seeing that* (e.g., one cannot now *see that* a bell jar has been saturated by phlogiston). Hanson, however, was very clear that one could be wrong in what one "sees that." He did not take *sees that* to be equivalent to *knows that*; rather, *sees that* is just an indicator of certain (often unconscious) psychological inferences from perception. A person can *see that* the Earth is the center of the solar system, since for Hanson this is just to say that the person's experiences of the Earth are ordered by the concept of a geocentric universe, and thus infers other things in virtue of the experience and the concept. The reason certain conceptual orderings have fallen out of fashion, such as those associated with the geocentric solar system, is that the patterns according to which they order experience render less things intelligible than their successful rivals.

While Hanson thought that observation is theory laden, he seems not to have held that one's theoretical commitments in any sense determine or alter the phenomenology of one's experience. Thus, how one "sees as," or one's conceptual repertoire, does not place absolute constraints on how one will be able to "see as" in the future. However, the production of new conceptual orderings is no trivial, transparent, or easy process, as the theoretical struggles in the history of science attest. One must build new conceptual patterns out of existing frameworks, but it can be extremely difficult to determine which elements of the older frameworks can be transferred and which cannot, and scientists are often blinded by assumptions they have inherited from previous conceptual and notational frameworks.

Hanson's goal in his discussion of observation was not to argue for the subjectivity of science, but rather to clarify the link between perception and knowledge (Hanson 1971). In clarifying this link, Hanson provides a clue to the logic of discovery, for discoveries are achieved through seeing the world differently, which involves appreciating the world through new conceptual arrangements.

Facts

Hanson was also critical of philosophical accounts of facts that attempted to define facts syntactically as phrases following *that*-clauses, arguing that this does little to help make clear the logical and epistemological status of facts. Moreover, faith in the theoretical neutrality of fact-claims obfuscates the effects that language, notation, and idioms have on the way facts are understood. Language and notation provide a sort of template through which facts can be expressed, a pattern through which the world can be understood. Hanson uses ordinary language analysis to indicate the dependence of a fact's significance on the structure of the language in which it is couched. For example, there is a difference between saying "Grass is green" and "Grass greens," for the first formulation assigns merely a property to grass, whereas the second assigns an action; similarly, there is a difference between saying "Electrons are charged" and "Electrons induce electrostatic fields."

Hanson used a historical case to show that while facts inexpressible in a given notation are not impossible to grasp, the practical obstacle such a process involves is very conceptually important for understanding the growth of science. Hanson showed, through a careful analysis of the works of Galileo, Beeckman, and Descartes, how the correct law of free fall was grasped only after a long period of confusion, even though all the requisite data, or "facts," were known from the beginning. All three persisted in thinking of velocity as a direct proportion to the space traversed, rather than (as is correct) a direct proportion of the times. Hanson attributed this apparent obtuseness among geniuses to the geometric notation with which such problems were then treated, which left no room for the expression of a time axis. Spatial properties were more easily measured and represented than temporal ones, and it took the penetrating mind of Galileo to see through this theory-laden factual representation to the correct solution.

Causes

Hanson saw science as primarily a quest for intelligibility, and only secondarily as a search for facts and causes, since these notions are definable only within the organizing conceptual framework. Thus, scientists do not search for succeeding links in the causal chain of nature. According to Hanson, causality is best thought of not as an independent feature of the world, but as a means whereby elements of a theory are bound together inferentially. What is significant, then, about causal language is not what is asserted about objects, but the inferential relations they warrant. Therefore, the adequacy of a proffered explanation cannot be appraised according to some extratheoretical notion of causation, and different scientific programs will differ in terms of what needs to be explained, that is, what causal inferences need be defined. This does not lead to a free-for-all, however, since those conceptual programs that inferentially organize the most phenomena ultimately prevail. For instance, the standard Newtonian line that gravitation requires no explanation was taken as correct until general relativity provided an explanation (see Space-Time); but Newton prevailed over the Cartesians in spite of this inferential omission, since there were so many other inferences he could supply that they could not.

Logic of Discovery

Hanson was also a critic of the logical empiricists' doctrine that discovery is a matter of mere psychology and that philosophical assessment of science should be confined to a logical analysis of the justification of theories. Hanson repeatedly urges that the great conceptual innovations that have fueled science required the genius of a Galileo, Newton, Kepler, or Einstein for their creation, whereas the business of justification is an ordinarily pedestrian affair, better suited to the talents of assistants than geniuses. The production of hypotheses in unsettled domains of inquiry is itself rationally appraisable; there are right and wrong ways of doing it. Revolutionary discoveries are a triumph of reason, and it would be most inaccurate to consider them as such only in retrospect, after they have been justified. Thus, Hanson looked to the history of science in order to adumbrate a set of informal dicta for rational hypothesis creation.

A number of criteria can be used to assess the reasonableness of suggesting a hypothesis. The hypothesis should be consistent with background knowledge, show some capacity to explain the problem at hand, offer testable consequences and an account of the constraints on testing, and have some plausibility (in light of the rest of knowledge). Hanson attempted to outline a set of informal strategies for rational hypothesis production from problematic situations. Hypothesis creation from enumerative induction, abductive reasoning, authority, or symmetry considerations are all approaches that can be used normatively. In addition, since perception and data interpretation are theory-laden activities, they aid in making inferences and offering hypotheses. Previous knowledge shapes new expectations, thus suggesting explanations and inferences about the future. Given a theoretically loaded observation, certain hypotheses, which also make reference to these same concepts, can be reasonably suggested.

Hanson was able to provide a set of only vague criteria for rational hypothesis suggestion. His account was never complete enough to determine whether a particular hypothesis suggestion was rational without some reference to the historical consequences that followed from the suggestion.

JORDI CAT
MATTHEW LUND

References

Grau, K. T. (1999), "Force and Nature: The Department of History and Philosophy of Science at Indiana University, 1960–1998," *Isis* 90: S295–S318.

Hanson, N. R. (1960), "New Discipline Joins Science to Arts," in *Arts and Sciences: The Review* (published by the Indiana University Alumni Association) 58: 3–11.

——— (1961), *Patterns of Discovery: An Inquiry into the Conceptual Foundations of Science*, Cambridge: Cambridge University Press.

Hanson, N. R. (1963), *The Concept of the Positron: A Philosophical Analysis*. Cambridge: Cambridge University Press.

Hanson, N. R. (1969), *Perception and Discovery: An Introduction to Scientific Inquiry*. Edited by W. C. Humphreys. San Francisco: Freeman, Cooper and Co.

Hanson, N. R. (1971), *What I Do Not Believe and Other Essays*. Edited by S. Toulmin and H. Woolf. Dordrecht, Holland: Reidel.

CARL GUSTAV HEMPEL

(8 January 1905–9 November 1997)

Carl Gustav Hempel was born in and educated primarily in Germany. He studied mathematics, physics, and philosophy at the universities of Berlin, Gottingen, Heidelberg, and Vienna. He completed most of his doctoral thesis on analyses of probability under Reichenbach but was compelled to find an alternative advisor to complete the project because Reichenbach was dismissed from his position in 1933 when Hitler and the National Socialist Party came to power (see Reichenbach, Hans).

Hempel was opposed to the National Socialist Party and moved to Brussels in 1934 and then to Chicago in 1937. He taught at City College and Queens College in New York from 1939 until 1948, and Yale from 1948 to 1955. Most of his subsequent career was spent at Princeton University from 1955 until mandatory retirement in 1973, and after a two-year sojourn at the University of Pittsburgh, he returned and resided in Princeton, New Jersey, until his death in 1997.

Hempel was one of the youngest members of the Berlin Circle and was in close contact with members of the Vienna Circle (see Vienna Circle; Logical Empiricism). Because of his longevity, his career spanned the rise and decline of the logical empiricist movement. Because of his emphasis on pursuit of truth and clarity, his views underwent considerable change during his career, especially after 1964, when Thomas Kuhn became a colleague in the History and Philosophy of Science program at Princeton (see also Kuhn, Thomas). Although most philosophers who knew Hempel's and Kuhn's views expected that they would be highly antagonistic because their early views were rather divergent, they became close friends and had significant influences on each other.

Hempel contributed to numerous areas of philosophy of science, most notably to explanation, confirmation, analyses of theory and observation, and questions of scientific methodology, each of which is addressed below. However, it is important to note that Hempel also made an important contribution by serving as a personal example of the possibility of combining an unwavering pursuit of truth and clarity with great kindness toward and encouragement of his fellow philosophers. (See Richard Jeffrey's "Introduction" in Hempel 2000 for further details).

Explanation

Hempel's proposals for analyzing scientific explanation were among the most fruitful part of his research (Hempel 1965). Roughly he suggested that explanation involves a relation between a set of sentences (including at least one law) and a statement to be explained. The relations were of three sorts: deductive, inductive, and deductive-statistical (see Prediction). Respectively, the explaining statements provide a deductive argument, an inductive argument, and a deductive argument for a probabilistic conclusion.

These suggestions made depend primarily on syntactic relations. Numerous criticisms and counterexamples were proposed to these suggestions. While Hempel's suggestions are generally discredited, most of the decades after his work were devoted to criticism, defense, and modification of his views, so that it is fair to say that his ideas shaped the field for a significant period of time. The alternative views incorporate psychological, social, and pragmatic factors, as well as syntactic ones, and can be seen as developments that add further factors to his syntactic approach. (For a more detailed discussion, see Explanation; Inductive Logic).

Confirmation and the Raven Paradox

The puzzle known sometimes as *Hempel's paradox* and sometimes as the Raven paradox attracted less attention initially than some of Hempel's later work, but it has proved to be an enduring topic. It was first sketched by Hempel in 1937, but the first full development came later (Hempel 1943), and his *Studies in the Logic of Confirmation* (Hempel 1945) is more accessible both in terms of content and physically as reprinted still later (Hempel 1965). Most writers (Giere 1970; Good 1967) do not see the Raven paradox as a major puzzle because they are each convinced that they have solved it; yet there are numerous conflicting solutions with no consensus (see Confirmation Theory). The puzzle is easily described.

It seems plausible that the universal generalization "All ravens are black" is confirmed by reports of black ravens. (The general statement of this principle, that a universal generalization is confirmed by a report of a positive instance, is often called *Nicod's criterion,* though this is somewhat misleading historically, since Nicod's (1930, 219) original criterion made this a *necessary* as well as a *sufficient* condition. It is also highly plausible that whatever confirms a statement confirms any statement that is logically equivalent to the first. After all, if two statements are logically equivalent, then they are guaranteed to say the same thing about the world. (For further discussion of these principles, their justification, and history, see Hempel 1945.) The standard formalization of "All ravens are black" renders it as the universal generalization of a conditional, that is, $(\forall x)(Rx \rightarrow Bx)$ (where Rx is "x is a raven" and Bx is "x is black"). By the standard formalization, "All non-black things are non-ravens" is to be formalized as $(\forall x)(\neg Bx \rightarrow \neg Rx)$. By Nicod's criterion, a report of a non-black non-raven, for example, a white shoe, confirms that all non-black things are non-ravens. But by the logical equivalence condition, since $(\forall x)(Rx \rightarrow Bx)$ and $(\forall x)(\neg Bx \rightarrow \neg Rx)$ are equivalent, observation of a white shoe confirms that all ravens are black.

Many people find this last conclusion unacceptable, but some solutions to the paradox attempt to make it palatable. Hempel propounded the paradox in the context of attempting to develop principles for a qualitative theory of confirmation, and he accepted the conclusion in spite of its counterintuitiveness. Later writers have attempted to make the conclusion more palatable by embedding the argument in a quantitative context and saying that although the report of a white shoe confirms that all ravens are black, it does so only to a very minute degree, and thus one has the illusion of irrelevance (see Induction, Problem of).

More precisely, if one accepts a Bayesian (see Bayesianism) account of confirmation, then a hypothesis H is confirmed by evidence E just in case the evidence increases the probability of H. There are various exact formulations of confirmation, but one rather natural one is that the degree of confirmation of a hypothesis H by evidence E is the extent to which the probability of H given E, $P(H|E)$, exceeds the prior probability of hypothesis $P(H)$. Thus one plausible measure is $P(H|E) - P(H)$.

Let H be the hypothesis that all ravens (R) are black (B) and consider the two evidential statements, the statement, E_1, that the observed object is a black raven, that is, $Ra \wedge Ba$, and E_2, the statement that b is a non-black non-raven, that is, $\neg Bb \wedge \neg Rb$. Since $P(H|E) = P(H \wedge E)/P(E) = [P(E|H) \times P(H)]/P(E)$, an expression for the confirmation measure is given by $\{P(E|H) \times P(H)/P(E)\} - P(H)$. Substituting the particular positive evidence $Ra \wedge Ba$ for E produces $\{[P(Ra \wedge Ba|H) \times P(H)]/P(Ra \wedge Ba)\} - P(H)$.

Using the definition of conditional probability again, $[P(Ra \wedge BaH) \times P(H)] = P(Ba|(Ra \wedge H)) \times P(Ra|H) \times P(H) = P(Ra) \times P(H)$, on the assumptions that $P(Ba|(Ra \wedge H)) = 1$ and $P(Ra|H) = P(Ra)$. The first is a theorem of probability, the second an assumption about the irrelevance of H to whether something is a raven (given no other information). Since $P(Ba \wedge Ra) = P(Ba|Ra) \times P(Ra)$, the expression for confirmation turns into

$\{P(H)/P(Ba\,|\,Ra)\} - P(H)$. If it is taken as a background assumption that black objects are relatively uncommon and that ravenhood is antecedently thought irrelevant in this situation, the term $P(Ba\,|\,Ra)$ will be small and the confirmation large.

Working through the parallel calculation for the negative evidence, $\neg Ba \wedge \neg Ra$, leads to the parallel expression $\{P(H)/P(\neg Ba\,|\,\neg Ra)\} - P(H)$. But since the relative frequency of non-black objects is presumed high, $P(\neg Ba\,|\,\neg Ra)$ is close to 1 and the confirmation is minimal, though not presumably not zero.

This analysis vindicates Hempel's qualitative claim that the non-black non-raven does confirm the hypothesis H, but is supposed to allay the sense of paradox by showing that the degree of confirmation is extremely small and is considerably less than the confirmation provided by a black raven. However, the derivation depends on some assumptions about the probabilities, that is, $P(Ra\,|\,H) = P(Ra)$ and that $P(\neg Ra)$ is very close to 1, and these assumptions can be questioned. (See Vranas 2004 for further discussion and references.)

A second family of proposed solutions to the paradox appeals to Goodman's (1983) conception of projectiblity and his arguments that generalizations are confirmed by positive instances only if the predicates they contain are projectible. Projectibility is relative to a language community and a history of actual predictions. Roughly, a predicate is projectible if it has been used successfully in making predictions by the community in the past. New predicates may be projected if, for example, they are coextensive with prior projectible predicates. Goodman presents a complicated set of rules for projectibility (Goodman 1983, ch. 4), but what is essential for present purposes is that the class of projectible predicates is not closed under negation. Thus the presumed projectibility of "raven" and "black" does not transfer to "non-raven" and "non-black" (see Induction, Problem of).

A third family of solutions changes the subject from simple confirmation of a hypothesis by evidence, and substitutes the question whether a particular piece of evidence selectively confirms H from among a set of competing hypotheses. In this context, Hempel's reasoning shows that being a positive instance of a logically equivalent hypothesis does not generally provide selective confirmation. For example, if the alternative hypotheses to "All ravens are black" are "All ravens are white," "All ravens are green," etc., then a white shoe instantiates all of these equally. The only evidence that would selectively instantiate "All ravens are black" is a black raven.

There are at least two suggestions of how the alternatives for selective confirmation are specified. According to one alternative, specification would be in linguistic terms, so that the contrast would always use expressions in the same semantic field as the term that was the focus. The other alternative is to see the competing hypotheses as those that would be seriously entertained by members of the scientific community that investigates the relevant domain. Thus the first version makes the selective confirmation relation a function whose arguments include the language, and the second makes the function depend on the community as well as the evidential statement and the candidate for confirmation.

Arguments for one as opposed to the other of these alternatives are probably not compelling when the sentences are qualitative statements such as the Raven hypotheses. But for quantitative hypotheses, it would appear that the community approach is more plausible, since it is difficult to envision linguistic grounds for preferring some equations over others.

Yet another distinct approach to solving the paradox questions a basic assumption about the logical form of the hypotheses. Consider for example, the theory of conditionals espoused by McDermott (1996), in which a conditional is true if the antecedent and consequent are both true, false if the antecedent is true and the consequent false, and has no truth value otherwise. If this conditional is symbolized by $\Rightarrow$, so that "All ravens are black" is translated as $(\forall x)(Rx \Rightarrow Bx)$ and "All non-black things are non-ravens" as $(\forall x)(\neg Bx \Rightarrow \neg Rx)$, the two sentences are not equivalent and the paradox is resolved.

In summary, Hempel's Raven paradox continues to command the attention of philosophers of science. There are four major kinds of responses to the puzzle. The first response accepts the puzzling conclusion and attempts to explain it away. The second—projectibility—appeals to a combination of language and community practices to disarm the puzzle. The third appeals to a slightly different relation and to the role of the scientific community to change the subject. And the fourth claims to solve the problem by placing the blame on the choice of the material conditional as a way of representing the hypothesis.

Problems and Changes

Hempel's earliest work dealt with truth in science and mathematics (Hempel 2000, Essays 1–5), but his focus shifted fairly soon to more accessible and less metaphysical issues such as confirmation and explanation. Truth and realism deal with the relation between sentences and the world, whereas confirmation and explanation, on the surface at least, deal with relations among sentences. One of his most valuable later papers was "Problems and Changes in the Empiricist Criterion of Meaning" (reprinted in Hempel 1965), which chronicles the shift over several decades in the attempts to make a sharp demarcation between statements that are cognitively significant and those that are not. This piece lacks the rhetorical flourishes and the metaphoric ending of Quine's (1980) *Two Dogmas of Empiricism* (see Quine, Willard Van Orman) but is perhaps a more telling argument against the attempt to make a demarcation of the boundary of the cognitively significant. In this, as in numerous other areas, Hempel's views shifted away from attempts at purely syntactic or even semantic characterizations and toward conceptions that included social, psychological, and historical elements. (For a more detailed discussion, see Cognitive Significance).

Theories and Observation

Hempel's views on theory and observation evolved as he continued to ponder basic questions of confirmation and explanation. In evaluating the extent of this evolution, it is important to read closely the formulations that Hempel provides. For example, in *The Theoretician's Dilemma,* he states: "Formally, a scientific theory may be considered as a set of sentences expressed in terms of a specific vocabulary" (Hempel [1958] 1965, 182–183). Logical empiricists are often criticized for *identifying* theories with sets of sentences in first-order logic, but note that here Hempel is not arguing for identification, but proposing that from a particular perspective, a theory may be *considered* as a set of sentences. The distinction between formal questions about artificial languages and the related questions about natural languages was clear in Hempel's mind from very early. For example, in his one venture into the topic of vagueness and logic (Hempel 1939), he argues that since vagueness is a phenomenon of natural language, it does not provide any leverage for an argument for relinquishing two-valued formal logic.

The theoretician's dilemma is the following: Divide the vocabulary of a theory into two portions, that which is observation and that which is theoretical. If the sole function of the theory is to provide derivations of observational statements from observational statements by means of the intermediary use of theoretical statements and vocabulary, then it can be shown that the theoretical statements are dispensable. It will be useful to expand on the terms of the dilemma.

Hempel's construal of "observational" at this stage is to be distinguished from one of the earlier logical empiricist notions of "observational," which was to provide an absolute epistemological foundation for science (see Logical Empiricism; Observation). Rather, the relevant criteria for observationality is that intersubjective agreement is obtainable for the statement in question: "The observational data...are...couched in terms whose applicability in a given situation different individuals can ascertain with high agreement, by means of direct observation" (Hempel [1958] 1965, 179). Notice that this characterization of the observational terms is not syntactic or semantic, but pragmatic, in the sense of the scientific community agreeing on applicability of the terms (see Observation). A closely related conception of an observation sentence is developed by Quine (1960); it too is relativized to a time and a linguistic community, as well as other parameters (see Quine, Willard Van).

Shortly thereafter, in *Philosophy of Natural Science,* Hempel (1966) took two further steps in modifying his view. First, he abandoned the characterization of the term opposed to "theoretical" as "observational," and instead proposes to distinguish between what is an understood and agreed-upon antecedent to a particular theory and that which is not. This is, of course, a distinction that shifts over time, as what were new theories become accepted and incorporated as part of the "antecedently understood." The second shift was that he recognized that being antecedently understood is not a characteristic of terms *simpliciter,* but of statements including those terms. A term may be antecedently understood within a particular range of application, but not outside that range:

> For example, a characterization of the concept of temperature by reference to the readings of a mercury thermometer affords no general definition of temperature; it assigns no temperature below the freezing point or above the boiling point of mercury. (Hempel 1966, 79)

This illustrates that a term such as "temperature" may be antecedently understood in the context of assigning temperatures in a certain range to liquids, but be highly theoretical when applied to other objects or at much higher or lower temperatures.

Even with the change from "observation" to "antecedently understood," the theoretician's dilemma can still be stated. Consider the theory and the antecedently understood expressions as part of a formal language. Divide the sentences into those that are antecedently understood and those that are not, and consider what the function might be for the sentences that are not antecedently understood. If their role is to provide appropriate deductive connections between antecedently understood sentences, that is, predictions in the broad sense that include past statements, then it can be shown on formal grounds (Hempel [1958] 1965) that an alternative formal structure is available that does not utilize the other sentences at all.

Hempel's conclusion in 1958 was not that the theoretical is dispensable on these grounds because there are nondeductive characteristics of theoretical systematizations that are of great importance: "If it is recognized that a satisfactory theory should provide possibilities also for inductive explanatory and predictive use and that it should achieve systematic economy and heuristic fertility, then it is clear that theoretical formulations cannot be replaced by expressions in terms of observables only" (Hempel [1958] 1965, 222).

Although he came to recognize the importance of nonformal characteristics of scientific theories, Hempel never entirely lost sight of the usefulness of axiomatization and formalization for some purposes. As the commentator on an exchange between Suppes and Kuhn on whether axiomatization is valuable, Hempel clearly favored a compromise position. On the one hand, "axiomatization of theories can be of value for certain philosophical or scientific purposes," but on the other hand:

> Professor Kuhn's paper is highly relevant, for it explores ways in which the requisite agreement in the understanding and the use of scientific terms may be attained by the members of the scientific community without reliance on, or even availability of, explicitly formulated criteria of application. (Hempel 1977, 257)

Later Work

Hempel's later thought continued to be innovative and influential, but while his later conclusions were also clear, they were more tentative and less definite. He had recognized from both his own work and the criticisms of others, especially Kuhn, that the logical empiricist tradition that emphasized syntactic, and to a lesser extent semantic, relations among sentences omitted a great deal of importance for the purpose of understanding how science develops and theories are evaluated, and for assessing rationality. However, his drive for clarity and precision left him dissatisfied with the formulations provided by Kuhn and others in the more pragmatic and historical traditions that emerged from the 1950s on.

Hempel (1983a, and 1983b) provides examples of his working through the issues concerning theory choice and rationality, but the most thorough is *Scientific Rationality* (Hempel [1979] 2001). He begins by noting the goal held by analytic empiricism of formulating explicit, logically precise criteria of rationality for the formulation, testing, and evaluation of scientific claims (358). However, the historical quest for such criteria has not been successful, and the arguments of Kuhn and others seem persuasive that the quest cannot be fulfilled. Scientific inquiry, at its best, does involve shared preferences that shape theory evaluation, such as a preference for quantitative theories "whose predictions show a close fit with experimental findings; for theories covering a wide variety of phenomena; for theories that correctly predict novel phenomena; for fruitful theories; for simple theories rather than complex ones" (359).

However, these preferences do not generally suffice for the unambiguous selection among competing theories. Unless such crucial terms as "close fit," "wide variety," "fruitful," and "simple" can be explicated clearly and rigorously, there remains room for disagreement in theory choice. Hempel remained optimistic that some progress could be made on these questions in specific contexts but was persuaded by the arguments from history of science and the logical problems that no general solution would be forthcoming.

Although Hempel's views evolved considerably and in many regards moved toward convergence with those of Kuhn, the two continued to differ on some important issues, including the locus of rationality. For Kuhn the rationality of theory choice resides in the relevant scientific community and is a holistic property of that group. Hempel criticizes Kuhn's analogy between evolutionary change and scientific change because evolutionary change selects from among more or less randomly produced variations. In contrast, Hempel believed that the rationality of science is indicated by the fact that later theories are superior to earlier ones, because they have been consciously designed to be better, and this is explained by the goal-directed character of scientific change at the level of individual scientists (Hempel [1979] 2001).

RICHARD E. GRANDY

References

Giere, R. (1970), "An Orthodox Statistical Resolution of the Paradox of Confirmation," *Philosophy of Science* 37: 354–362.

Good, I. J. (1967), "The White Shoe Is a Red Herring," *British Journal for the Philosophy of Science* 17: 322.

Goodman, Nelson (1983), *Fact, Fiction and Forecast* (4th ed.). Cambridge, MA: Harvard University Press.

Hempel, C. G. (1939), "Vagueness and Logic," *Philosophy of Science* 6: 163–180.

——— (1943), "A Purely Syntactical Definition of Confirmation," *The Journal of Symbolic Logic* 8: 122–143.

——— (1945), "Studies in the Logic of Confirmation," *Mind* 54: 1–26 and 97–121.

——— (1952), *Fundamentals of Concept Formation in Empirical Science: International Encyclopedia of Unified Science* (Vol. II, no. 7). Chicago: University of Chicago Press.

——— ([1958] 1965), "The Theoretician's Dilemma: A Study in the Logic of Theory Construction," in *Aspects of Scientific Explanation and Other Essays in the Philosophy of Science* New York: Free Press, 173–226.

——— (1965), *Aspects of Scientific Explanation and Other Essays in the Philosophy of Science*. New York: Free Press.

——— (1966), *Philosophy of Natural Science*. Englewood Cliffs, NJ: Prentice-Hall.

——— (1977), "Formulation and Formalization of Scientific Theories," in Suppe, F. (ed.), *The Structure of Scientific Theories*. Urbana: University of Illinois Press, 245–254.

——— ([1979] 2001), "Scientific Rationality: Normative Versus Descriptive Construals," in James Fetzer (ed.), *The Philosophy of Carl G. Hempel*. Oxford: Oxford University Press, 357–371.

——— (1983a), "Valuation and Objectivity in Science," in R. S. Cohen and L. Laudan (eds.), *Physics, Philosophy and Psychoanalysis in Honor of Adolf Grunbaum*. Dordrecht, Holland: D. Reidel Publishing, 73–100.

——— (1983b), "Kuhn and Salmon on Rationality and Theory Choice," *Journal of Philosophy* 80: 570–572.

——— (1988a), "Limits of a Deductive Construal of the Function of Scientific Theories," in Edna Ullman-Margalit (ed.), *Science in Reflection*. The Israel Colloquium, vol. 3. Dordrecht, Holland: Kluwer Academic Publishers, 1–15.

——— (2000), *Selected Philosophical Essays*. Edited by Richard Jeffrey. Cambridge: Cambridge University Press.

Maher, P. (1999), "Inductive Logic and the Ravens Paradox," *Philosophy of Science* 66: 50–70.

McDermott, M. (1996), "On the Truth Conditions of Certain 'If' Sentences," *Philosophical Review* 105: 1–38.

Nicod, J. (1930), *Foundations of Geometry and Induction*. Translated by N. Weiner. London: Routledge and Kegan Paul.

Quine, Willard Van (1960), *Word and Object*. Cambridge, MA: MIT Press.

——— (1980) "Two Dogmas of Empiricism," in *From a Logical Point of View*. Cambridge, MA: Harvard University Press, 20–46.

Vranas, P. (2004), "Hempel's Raven Paradox: A Lacuna in the Standard Bayesian Solution" *British Journal for the Philosophy of Science* 55: 545–560.

See also **Carnap, Rudolf; Cognitive Significance; Confirmation; Explanation; Induction, Problem of; Inductive Logic; Kuhn, Thomas; Logical Empiricism; Observation; Prediction; Quine, Willard Van; Reichenbach, Hans; Verifiability; Vienna Circle**

HERITABILITY

Offspring resemble their parents. This simple observation of family resemblances is one of the oldest conceptions of inheritance. It is also the foundation of the scientific study of heredity. For centuries ideas about heredity were based on this simple qualitative measurement. As might be expected, explanations of this observation changed in the course of history, and these changing explanations reflected the prevailing (scientific) attitudes within historical periods. For Aristotle the resemblance between parents and offspring was based on the fusion of male and female "fluids" and the subsequent action of the four causes (*materialis, efficiens, formalis, finalis*), with the female providing the material cause and the male providing the semen, "that which generates," the active stimulus for the developmental dynamics. For Aristotle, development was epigenetic: The new organism was not preformed in any of the parental contributions; rather it realized its own potential in the dynamic process of development. Heredity, in Aristotle's conception, was a consequence of generation (Aristotle 1979).

During the seventeenth and eighteenth centuries, two competing positions were put forward to account for heredity and generation: preformationism and epigenesis. Each of these positions progressed through several versions that reflected both new empirical observations and changing theoretical assumptions. Preformationists, such as Malebranche, Malpighi, and Bonnet, based their argument both on new microscopic observations that showed that semen and eggs had internal structures and on a rejection of ideas of spontaneous generation. Another criticism was the lack of a satisfactory mechanism that would explain generation epigenetically. Proponents of epigenesis, such as Harvey, Gassendi, and Wolff, also claimed empirical support (especially based upon observations of chick embryos). They emphasized several phenomena that could not easily be explained within a preformationist framework, such as hybridization, regeneration, or the existence of so-called monsters (teratology). Even though they did not resolve the problem of inheritance, these seventeenth- and eighteenth-century debates brought the problem of generation and heredity within the scientific focus. It soon became clear that heredity and generation were linked, even though it was not yet obvious what mechanism could account for each (see also Roe 1981; Pinto-Correia 1997; Maienschein 2004).

Ideas about the variability and transformation of species, which emerged in the late eighteenth and early nineteenth centuries, complicated things even further. While early approaches were mostly concerned with transformations between forms (Goethe, Lamarck, Oken) or the correlations between embryological transformations and morphological complexity (Meckel, Serres, von Baer), attention soon shifted to variation at the subspecies level. This shift first happened in the context of animal and plant breeding programs, the study of geographic variation (fostered by the consolidation of colonial power and the increase in global trade), and the emerging science of anthropology, with its focus on the concept of race. In the nineteenth century, an experimental approach to the problem of heredity emerged, as well as conceptual transformations of the older notion of generation. The latter can be seen most prominently in the work of Darwin (see Evolution).

Darwin's theory of the transmutation of species was predicated on the existence of variation within populations, the action of natural selection on these variants, and a principle of heredity that would guarantee that the offspring of these selected variants would also bear the same traits that enabled the survival and reproductive success of their parents (see Natural Selection). The often repeated formal requirement of Darwin's theory of natural selection consists of phenotypic variation within a population correlated with a corresponding variation in fitness that is also heritable; in other words, offspring share the same traits that helped their parents succeed in the "struggle for existence" (Lewontin 1974a). The essence of Darwin's theory of natural selection is thus based on strictly phenotypic observations. It was, at least initially, also largely a qualitative theory, based on common sense and supported by an overwhelming body of empirical observations.

In the years after the publication of *On the Origin of Species,* the nature of biological variation was the subject of intense debates. Was variation primarily continuous, as in many quantitative characteristics, such as height, or was it primarily discontinuous and discrete, as in qualitative characteristics, such as many color variants? This was an important question, as the views about the nature of biological variation corresponded to different ideas about the mechanisms of evolutionary change (see Population Genetics). Evolution, and the origin of new variants and species, was considered to have either followed the gradual path that Darwin proposed or to have happened by means of larger changes. Some even thought that different mechanisms accounted for the gradual adaptation to changing environmental conditions and for the discontinuous origin of new species.

Two problems that were left unanswered in Darwin's theory are of special interest for the discussion of the problem of heritability: (1) How can the qualitative observations of heredity by Darwin and others be made quantitative and therefore predictive? and (2) What is the material basis of heredity? The latter problem initiated a century-long research program that led from Mendel's experiments (and the factors that he postulated would represent each of the discrete variants in them), to Johannsen's distinction between genotype and phenotype and the associated concept of a pure line, to Boveri and Sutton's chromosomal theory of inheritance, to Morgan's gene maps, and finally to the discovery of the double-helical structure of DNA (see Genetics). Though this line of research (eventually) elucidated the molecular basis of heredity, it contributed very little, with the exception of the genotype/phenotype distinction, to the second problem of establishing a quantitative theory of inheritance and natural selection. For this a different approach was needed, one that was rooted in the simultaneous development of statistics.

It was Darwin's cousin, Galton, who first compared the properties of the phenotypes of parents with those of their children as well as with the rest of a population. He found that the mean value of all the offspring tended to be closer to the overall population mean than to the mean value of their parents. The analysis of this phenomenon, which he called regression to the mean, initiated the statistical analysis of the problem of inheritance. This was a separate approach to the study of inheritance, one that was focused on statistical correlations rather than on material entities. As Provine (1971) has shown, after the turn of the twentieth century, these questions became the foundation not only of theoretical population genetics, but also of quantitative genetics, which was more closely allied with traditional efforts of animal and plant breeding. It was in this context that the concept of heritability was first formulated.

Current Definitions and Problems

Today the concept of heritability is at the heart of the discipline of quantitative genetics and is increasingly also employed in medical contexts, where it is used to assess the probability for the occurrence of certain genetic conditions. Probably the best known, and also the most controversial, applications of the concept of heritability have been attempts to quantify the genetic component of the observed variance in individual values of IQ (e.g., Herrnstein and Murray 1994; Fraser 1995; Jacoby and Glauberman 1995). These debates, especially the more popular discussions, often confused several crucial components of heritability (such as broad- and narrow-sense heritability) and also unjustly equated heritability with a specific form of genetic causation. It is therefore crucial to distinguish several important dimensions and assumptions of the statistical concept of heritability as it is currently used in genetics.

In quantitative genetics, there are two definitions of heritability, broad-sense heritability and narrow-sense heritability. The former is defined simply as the ratio of the total genetic variance, V_G, and the phenotypic variance, V_P. In the case of broad-sense heritability, one is interested in quantifying the total genetic contribution, including all dominance and interaction effects, to the phenotypic variance and to distinguish those from the environmental contributions. This measurement then provides a broad estimate of the degree of genetic determination. In the simple case, one assumes that the total phenotypic variance is simply the sum of the genetic and environmental variance ($V_P = V_G + V_E$). However, this is an idealization, as it assumes that there is no variation in the interaction between a genotype and the environment, that is, that any difference in the environment has the same effect on all the different phenotypes. As this is extremely unrealistic, one should include another variance term that accounts for the variance in the genotype/environment interaction, V_{GE}. Thus we have $V_P = V_G + V_E + V_{GE}$. (Even this is an approximate relation. For a full quantitative treatment, see Sarkar 1998, Ch. 4.)

In sexually reproducing diploid species, there is the additional problem that the genotype is not passed on directly to the next generation, only gametes are. Thus, from an evolutionary point of view, the total genetic variance is not what helps an understanding of the dynamics of natural selection (see Natural Selection). The fraction of the total genetic variance V_G that is relevant for the evolutionary consequences of natural selection is called the additive genetic variance V_A. Fisher (1918) first introduced the concept of the additive effect of an allele in order to account for the additive genetic variance. He first analyzed a one-locus two-allele case, with A and a as the two alleles. If the substitution from the aa homozygote to the aA heterozygote genotype produces the same phenotypic effect as the second substitution of the a allele (from the aA to the AA genotype), then all of the genetic variance at that locus is considered additive. These ideal cases are, of course, exceptionally rare. In actuality, one has to also account for a within-locus deviation from additivity (the so-called dominance deviation), as well as for nonadditive effects between different loci (the so-called epistasis effect) that contribute to a given phenotype. The total genetic variance is thus a combination of three different genetic variance components, the *additive, dominance,* and *epistatic* (interaction):

$$V_G = V_A + V_D + V_E.$$

Based on this decomposition of the total genetic variance V_G, the narrow-sense heritability is then defined as the fraction of the additive genetic variance and the phenotypic variance:

$$h^2 = \frac{V_A}{V_P}.$$

The open question thus is, How can one estimate the additive genetic variance in order to calculate the heritability of a trait, as it is not usually possible to measure the additive effects of alleles directly? In the context of quantitative genetics, the additive genetic variance is also given by the variance of the breeding values of the genotypes within a population (Falconer and Mackay 1996). The breeding value of a genotype is defined by the mean genotypic value of its offspring. It therefore can be measured. The idea behind the concept of a breeding value is similar to Galton's regression analysis between midparent and midoffspring values. It provides a measure of how much of the phenotypic variance of the parents is passed on to their offspring. The slope of the regression line between midparent and midoffspring values is therefore another measure of heritability.

Heritability is clearly a central concept in evolutionary and quantitative genetics. It is therefore important to be aware of its assumptions and limitations. There are two main problems of special relevance for philosophers of science: the first is related to causal inferences of genetic determination is based on heritability estimates and the second related to the consequences of epistatic interactions. As has already been seen, heritability is a statistical concept, defined as a ratio between variances or as a regression coefficient. It does not imply a specific model of genetic or environmental causation, and therefore no such model or any specific interpretation of genetic causality can be inferred from a particular value of heritability. In order to establish support for a specific interpretation of genetic causality from the type of variance analysis that is part of heritability assessments, one would have to develop a rather rigorous experimental design (Lewontin 1974b; Feldman and Lewontin 1975). While this is possible in certain breeding experiments with Drosophila (see Falconer and Mackay 1996), these conditions are almost never realized in studies involving humans. The fact that there is a widespread tendency to use heritability values in support of a genetic etiology of human conditions and diseases thus points more to the existence of an underlying genetic ideology than to a well-supported understanding of the role of genes in human disease (Laubichler and Sarkar 2002).

Besides the problems related to causal inference from statistical and population-dependent values, nonlinear or epistatic interactions between genes also complicate the interpretation of heritability. The consequences of epistatic interactions for heritability are also related to the *unit of selection problem* (Wimsatt 1981; Lloyd 1988). In short, recent studies have shown that (1) in all cases of multilevel selection, there will be a certain amount of the total additive genetic variance that will be a consequence of additive effects of alleles at individual loci (Sarkar 1994) and (2) based on a physiological definition of epistasis as the effect of a gene substitution at one locus on a subsequent gene substitution at another locus (Cheverud and Routman 1995; Wagner, Laubichler, and Bagheri-Chaichian 1998), it can be shown that in the case of epistasis between loci, there will (a) always be an epistatic component to the additive genetic variance and (b) under certain conditions, such as a population that is far away from its equilibrium point, there will be irreducible higher-order additive effects that can be attributed to sets of interacting genes (or gene complexes) rather than individual genes. The latter is important for considerations of heritability, as these higher-order additive effects will also contribute to the total additive genetic variance. Consequently, there will also be covariance terms between the additive effects of individual alleles and the irreducible effects of interacting gene complexes. As these covariance terms can be negative under certain circumstances, any value of heritability that is derived from an estimate of the additive genetic variance based solely on the additive effects of individual alleles can thus potentially overestimate the actual value of heritability. As there are only a small number of diseases that are caused by a single gene, these recent studies offer some corrective to the often unrealistically high estimates of heritability for genetic disorders reported in the medical literature.

Today, the concept of heritability continues to be central to quantitative, evolutionary, and medical genetics. It represents a culmination of a century-long quest to quantify and understand the consequences and phenomena of inheritance. However, there continues to be a discordance between the technical interpretation of heritability and the many roles the concept plays in medical, popular, and philosophical discourses.

MANFRED D. LAUBICHLER

References

Aristotle (1979), *Generation of Animals*. Cambridge, MA: Harvard University Press.

Cheverud, J. M., and E. J. Routman (1995), "Epistasis and Its Contribution to Genetic Variance Components," *Genetics* 130: 1455–1461.

Falconer, D. S., and T. F. C. Mackay (1996), *Introduction to Quantitative Genetics*. London: Longmans.

Feldman, M. W., and R. Lewontin (1975), "The Heritability Hang-Up," *Science* 190: 1163–1168.

Fisher, R. A. (1918), "The Correlation Between Relatives and the Supposition of Mendelian Inheritance," *Transactions of the Royal Society, Edinburgh* 52: 399–433.

Fraser, S. (1995), *The Bell Curve Culture Wars: Race, Intelligence, and the Future of America*. New York: BasicBooks.

Herrnstein, R. J., and C. A. Murray (1994), *The Bell Curve: Intelligence and Class Structure in American Life*. New York: Free Press.

Jacoby, R., N. Glauberman (eds.) (1995), *The Bell Curve Debate: History, Documents, Opinions*. New York: Times Books.

Laubichler, M. D., and S. Sarkar (2002), "Flies, Genes, and Brains: Oskar Vogt, Nicolai Timofeeff-Ressovsky, and the Origin of the Concepts of Penetrance and Expressivity in Classical Genetics," in L. Parker and R. Ankeny (eds.), *Medical Genetics. Conceptual Foundations and Classical Questions*. Dordrecht, Netherlands: Kluwer, 63–85.

Lewontin, R. (1974a), *The Genetic Basis of Evolutionary Change*. New York: Columbia University Press.
——— (1974b), "The Analysis of Variance and the Analysis of Causes," *American Journal of Human Genetics* 26: 400–411.
Lloyd, E. A. (1988), *Structure and Confirmation of Evolutionary Theory*. Westport, CT: Greenwood.
Maienschein, J. (2004), *Whose View of Life? Embryos, Cloning, and Stem Cells*. Cambridge, MA: Harvard University Press.
Pinto-Correia, C. (1997), *The Ovary of Eve: Egg and Sperm and Preformation*. Chicago: University of Chicago Press.
Provine, W. B. (1971), *The Origins of Theoretical Population Genetics*. Chicago: University of Chicago Press.
Roe, S. A. (1981), *Matter, Life, and Generation: Eighteenth-Century Embryology and the Haller-Wolff Debate*. Cambridge: Cambridge University Press.
Sarkar, S. (1994), "The Selection of Alleles and the Additivity of Variance," in D. Hull, M. Forbes, and R. M. Burian (eds.), *PSA 1994: Proceedings of the 1994 Meeting of the Philosophy of Science Association*. East Lansing, MI: Philosophy of Science Association, 3–12.
——— (1998), *Genetics and Reductionism*. New York: Cambridge University Press.
Wagner, G. P., M. Laubichler, and H. Bagheri-Chaichian (1998), "Genetic Measurement Theory of Epistatic Effects," *Genetica* 102/103: 569–580.
Wimsatt, W. C. (1981), "Units of Selection and the Structure of the Multi-Level Genome," *PSA 1980* 2: 122–183.

See also **Adaptation and Adaptationism; Evolution; Fitness; Genetics; Natural Selection; Population Genetics**

DAVID HILBERT

(23 January 1862–14 February 1943)

Hilbert was, and still is, known as one of the greatest mathematicians of the first half of the twentieth century. Although this view is doubtlessly correct, it is one-sided and incomplete because it neglects two important aspects of his work:

1. Throughout his career, Hilbert was interested in the foundations of all exact sciences—not only mathematics, but also the natural sciences. As a truly universal scientist, he contributed to fields other than pure mathematics, in particular, to theoretical physics and its dramatic development during the first quarter of the twentieth century.
2. Hilbert consciously and deliberately transcended the borders between mathematics and the exact sciences, on the one side, and epistemology and philosophy, on the other. This situates him with other twentieth-century figures such as Einstein, Bohr, Born, Schrödinger, and Weyl, who, like Hilbert, aimed to tear down the wall between traditional philosophy and the exact sciences.

In spite of numerous accounts of Hilbert's achievements in mathematics, his work in other areas—in particular, his contributions to modern physics and its philosophical implications—remains relatively neglected. Because the present volume is an encyclopedia of the philosophy of science, this article focuses on these other aspects of Hilbert's work. However, his mathematical achievements will not be entirely ignored: It will be pointed out that there is an intimate relationship between his mathematical work and his contributions to modern physics, especially its conceptual clarification.

David Hilbert was born in Königsberg, then the capital of East Prussia. He attended the local Gymnasium and spent most of his student life in Königsberg. During these years, he spent much time with Hurwitz and Minkowski studying mathematics and physics. In 1886, he became *Privatdozent* in Königsberg with a highly regarded work on the theory of invariants. Six years later, he was appointed as *Extraordinarius* and became full professor for mathematics in Königsberg a year later. In 1895, Felix Klein brought him to Göttingen, where he remained, despite many offers from other distinguished universities, until his death. In 1925, Hilbert suffered from "pernicious anemia" but recovered soon thanks to a new medication. During his academic career in Göttingen, he established (together with Klein and Courant) the best-known and most esteemed center for mathematics in the world. The fruits of his research in the foundations of mathematics and the sciences are still significant.

The principal means by which Hilbert achieved most of his fundamental results in the foundations of mathematics and science was the so-called *axiomatic method*. Since there are many obscure and confused opinions about this method, especially about its essential role (as well as its limits) in the logical analysis of the exact sciences, the subsequent sections discuss what the "essence" of this method is (see also Hilbert 2004). A similar clarification is necessary with respect to Hilbert's so-called *formalistic* approach as an alternative to Brouwer's *intuitionistic* and Frege's *logicistic* views about the foundations of mathematics (Sieg 1990).

Foundations of Geometry and the Axiomatic Method

Hilbert's inquiries into the foundations of mathematics fall into two periods, which are separated (judged by his publications) by about fifteen years. The first period began in 1893 with a lecture on "projective geometry," reached its zenith with *Grundlagen der Geometrie* (Hilbert [1899] 2004), and ended with *Über die Grundlagen der Geometrie* (Hilbert 1902). The second period, which started in 1917–1918 with the programmatic essay "Axiomatisches Denken" (Hilbert [1918] 1932), includes most of Hilbert's investigations of the foundations of arithmetic and the establishment of a radical new program, called *proof theory,* to prove the consistency of arithmetic by finite means. This important program came to a halt (at least temporarily) in 1931, when Gödel published his famous paper showing the existence of undecidable sentences within Peano arithmetic (Gödel 1986). This happened just as Hilbert was going to retire from his position at Göttingen.

This division of Hilbert's career into two periods gives the impression that the two topics that characterized them were for Hilbert unrelated, which is not the case. In fact, the latter is the continuation of the former by other much more radical means. This becomes obvious if one considers Hilbert's early development more closely, taking into account both his published papers and his unpublished lectures. The first point that should be noted is the trivial fact that Hilbert started his research in geometry quite conventionally. He did not have the axiomatic method at his disposal. Instead this method first emerged in connection with his "meta-theoretical inquiries of the logical structure of geometry." What this phrase means will become clear in the next section.

Around 1890, when Hilbert began his studies in geometry, the intellectual situation in that discipline was rather complicated. Geometry had become torn asunder into a confusing number of different branches and competing programs, such as projective versus differential geometry, Euclidean versus non-Euclidean geometry, and synthetic versus analytic geometry. Interesting but unconnected results were being discovered, and important books appeared, such as those by von Staudt (1847) and Pasch (1882). Hilbert was not acquainted with all of them initially. But once he had read Pasch's book (in about 1893), he knew, at least in principle, what his main goal was: He intended to resuscitate Euclid's axiomatic point of view, not exactly in the same way as used by Euclid, but in a very similar form. The main difference was to do it more transparently and perfectly. This meant that the desired axiom system should have a "perspicuous" logical-deductive order, that is, it should be complete in the sense that no essential assumption is missing, and simple in the sense that it contains no superfluous assumptions.

To achieve this goal, Hilbert had to develop a device by which he could prove whether a given sentence is logically dependent on or independent of a certain set of sentences. The principal idea is the following: A given sentence S is logically independent of a set of sentences $S_1, \ldots, S_n$, if there is a structure in which all sentences $S_1, \ldots, S_n$ are fulfilled (or true, as is now said), but S is not, and, instead, $\neg S$ is. If there is no such structure, that is, if in all structures in which $S_1, \ldots, S_n$ hold, S holds, S is a *logical consequence* of $S_1, \ldots, S_n$. This idea of a structure is the core of model theory (if "fulfilled" is replaced by "true"); it is also an essential ingredient of what Hilbert called the *axiomatic method*. The latter is the deliberate change or *variation* (Hilbert's term) of an axiom system in order to study the logical dependence of a specific sentence (e.g., the axiom of parallels) from a given set of axioms by model-theoretic means. Hilbert took primarily algebraic number fields as models for his proofs. In this way he was able to prove many interesting results— for example, the independence of Archimedes' axiom of continuity from the remaining axioms of his axiomatization of Euclidean geometry, the nonprovability of the Desargues sentence in Euclidean geometry without the axioms of congruence, the nonprovability of Pascal's sentence without the axiom of Archimedes, and others (see Hilbert [1899] 2004, Chs. 2, 3, and 5, for more details).

From an epistemological point of view, more important than these particular results is another, more general, aspect of the axiomatic method. With this method Hilbert could not only answer logical questions of independence and dependence, but also analyze meta-theoretical problems like the consistency and completeness of an axiom system, which roughly means (syntactically) that it is impossible to deduce a sentence A and its negation, and furthermore that every sentence which is "intuitively" true can be deduced from the axiom system.

These questions had become particularly pressing since the consistency of "non-Euclidean geometry" (taken quite generally) was still unproven (respectively the existing "proofs" were doubted). Hence, the consistency of the different non-Euclidean systems could not simply be taken for granted, but had to be shown definitively. Hilbert's idea to prove the consistency of non-Euclidean axiom systems was strikingly simple: If arithmetic is consistent (as everyone believed), then the consistency of a geometrical axiom system can be proved by relating it to the consistency of a suitably chosen number field. (This idea was quite original; it cannot be found in Pasch's work.) Today a geometrical axiom system is considered consistent if it has a numerical model. Hence, consistency means only "relative consistency": If it is not possible to deduce a contradiction within the numerical model, then the coordinated geometrical axiom system is consistent.

Execution of this program is very tricky, but Hilbert could show that his full axiom system of Euclidean geometry (including the axioms of continuity) is "complete" in the sense that it has a "numerical" model (the real numbers) whose domain of individual elements cannot be expanded without introducing a contradiction. Hence, Hilbert's axiom system of Euclidean geometry is consistent in virtue of its completeness if the real-number field is consistent, which no one doubts seriously. Hilbert's concept of completeness is today called *categoricity*. A theory is categorical iff it has, up to isomorphism, exactly one model (Majer 1998).

New Foundations of Mathematics and the Genesis of Proof Theory

With the proof of the relative consistency of a large number of non-Euclidean geometries a new problem emerged: How could the consistency of arithmetic itself be proved? Although it was immediately clear to Hilbert that this could not be done in the same style as in geometry (otherwise, one would be trapped in an infinite regress), around 1900 he did not know how to achieve this goal. It wasn't until 1920 that his first proposals occurred, in a pair of unpublished lectures given in Göttingen, which was the beginning of modern proof theory. Its basic assumptions and procedures include:

1. Arithmetic cannot be reduced to logic, because it has a (nonlogical) content, given in intuition.
2. Contradictions can be avoided totally if all operations used in calculating and reasoning remain finite.
3. Because mathematics was and shall remain a "free" science, it cannot be restricted to the finite.
4. Consequently, to save mathematics from the danger of inconsistency (by entering the transfinite), it has to be proven by *finite means* that no contradiction can be derived from an *appropriate* system of axioms for arithmetic.
5. In order to achieve this goal, the axiom system itself has to be "formalized" so that a meta-theoretical investigation about its deductive structure is possible.

"Formalization" here means two things: (a) all axioms have to be expressed as formulas of a definite language (*Zeichensprache*), with clear syntactical rules for the formation and transformation of formulas, and (b) all logical rules of deduction such as *modus ponens* and substitution *salva veritate* have to be made explicit. Once this is done, the formalized system must be investigated to determine whether it is possible to derive a pair of formulas A and $\neg A$ from the axioms. If this is impossible, the system is consistent. The real difficulty lies in the proof that this is indeed impossible, because the proof of the impossibility has to be finite, whereas the number of possible proofs within a formal system can (and usually is) infinite.

Today (70 years after Gödel) most scientists believe that such a *direct* (nonrelative) proof of the consistency of an axiom system for arithmetic is impossible. But this is incorrect for several reasons. First, Hilbert and his school presented such proofs for certain "weak" axiom systems of arithmetic. Second, if one gives up the finitist restrictions on the meta-theoretical proofs and permits, as Gentzen did, transfinite induction, the consistency of arithmetic can be proved. Third, Gödel's proof of the existence of undecidable sentences within Peano arithmetic, and hence the infeasibility of a formal proof of consistency of Peano arithmetic by finite means, is no absolute verdict. Its correctness depends on the meaning of the phrase "formal proof by finite means." There are a number of proposals about how to "sidestep" Gödel's verdict without betraying Hilbert's finite point of view (modern proof theory, inverse arithmetic, weak arithmetic, etc.) (Simpson 1999).

Hilbert's Contributions to Physics and Its Axiomatic Foundations

Hilbert delivered his first lecture in physics in 1898, a year before the *Foundations of Geometry* appeared. During the next decade, he lectured six times on classical and continuum mechanics, including hydrodynamics, electrodynamics, and thermodynamics. During this period, mechanics was for Hilbert *the* fundamental theory of all physics, as it was for Heinrich Hertz (1895), whose book *Die Prinzipien der Mechanik* Hilbert admired as an exemplar for his own axiomatic point of view in physics. The period of classical physics (based exclusively on Galilean space-time theory) ended when Hilbert (in cooperation with Minkowski) began studying Einstein's theory of special relativity. This led to a complete revision of the mechanics lecture of 1911 in which Hilbert presented Einstein's theory of special relativity and discussed its consequences for electrodynamics and thermodynamics.

Hilbert's first publications in physics date from 1912 and are closely related to his monumental work on the theory of integral equations (Hilbert 1912). In fact, "Begründung der kinetischen Gastheorie" first appeared as Chapter 22 of that book. This gives the impression that Hilbert's interest in physics was only that of a mathematician, who is "simply looking for another possible application of his mathematical theories" (Brush 1976, 448) without any real interest in physics. A cursory examination of the unpublished lectures shows that this claim is unjustified. Although Hilbert's main concern in the paper is the search for solutions to the Maxwell-Boltzmann equation, his primary concern in the lectures on the kinetic theory of gases is different. The main question he pursued is whether a logical derivation of the Maxwell-Boltzmann equation from the time-reversible equations of mechanics is possible (see Irreversibility; Time). This is a very interesting question, and the different answers proposed so far are still controversial. Hilbert himself favored a negative answer in the sense that no strict logical deduction is possible. This leads to a new problem: What does "irreversibility" mean in an objective sense, if the irreversible Maxwell-Boltzmann equation cannot be deduced from the fundamental equations of motions? (see Majer 2002 and Scheibe 1997).

Similar points can be made regarding the reception of Hilbert's work in physics (such as his papers on radiation theory) as merely applied mathematics. As in the former case, this work was closely related to his new theory of integral equations and was thought by Pringsheim (a leading figure in radiation theory) as entailing nothing physically new about Kirchhoff's law of radiation. This is a misapprehension. Hilbert pointed out a serious logical gap in the foundation of radiation theory and the deduction of Kirchhoff's law from more fundamental theories, but his work was dismissed as of merely mathematical interest.

Better received were his two papers *The Foundations of Physics* in 1915 and 1916 (republished in Hilbert 1924), in which Hilbert presented several generalized field equations, which turned out to be equivalent to Einstein's field equations of 1915. (Therefore, they are also sometimes called the Hilbert-Einstein equations.)

Nineteen fifteen was, in retrospect, the year of the most intensive research on the theory of general relativity, in both Hilbert's and Einstein's careers. Both struggled in searching for *universal* field equations, in which two fundamental forces would be united: electromagnetism and gravitation. They approached the problem from different points of view. To Hilbert as "mathematician" it was clear from the very beginning that the field equations had to be independent of the choice of the coordinates, and, more important, it was clear what this precisely meant in physical terms. Einstein, on the other hand, had to struggle as a self-taught mathematician with the problem of invariance for several years before finding an acceptable solution.

More important than their different technical approaches are their differences in physical perspectives. Hilbert, having once abandoned the *mechanical* worldview, followed Gustav Mie, who tried to develop a universal field theory from which the existence of the electron could be explained. Einstein, however, looked for a generalized field theory in which Newton's theory of gravitation could be embedded, at least approximately. The solutions Einstein and Hilbert found toward the end of 1915 are, seen from this perspective, only contingently the same. In "essence" they are rather different (Weyl 1988). Recently, a priority debate among historians of science has emerged about whether Hilbert "acquired" his field equations from Einstein (Sauer 1999).

In his last period of active research in physics (1926–1927), Hilbert lectured on "Mathematical

Methods of Quantum Theory" (unpublished typescript). This was just after the *new* quantum mechanics had been formulated by Heisenberg, Born, and Jordan, and independently by Schrödinger and Dirac ("new" in distinction to the old "quantum theory" of Bohr) (see Quantum Mechanics). Although Hilbert did not belong to the group of physicists who created the new quantum mechanics, he is doubtless one of its intellectual progenitors, since the mathematical foundations of the new theory were precisely his theory of integral equations with infinitely many variables, which he had developed fifteen years earlier. It is therefore not accidental that in modern textbooks of quantum theory, the basic concept is the so-called *Hilbert space.* Hilbert's lectures significantly influenced the further development of quantum theory, in particular its axiomatic presentation by von Neumann. The lectures predate the publication of the famous paper by Hilbert, von Neumann, and Northeim (1928), which became the starting point of von Neumann's (1932) *Mathematische Grundlagen.*

Hilbert's "Finite Point of View" and Recursive Epistemology

Hilbert was not a "professional" philosopher, but he studied Kant's *Critique of Pure Reason* and acquired an intimate knowledge of the writings of contemporary philosophers such as Husserl, Frege, and Russell. Traces of these authors (and some minor figures) can be found in Hilbert's work. Hilbert was, however, too independent a thinker to simply adopt their philosophical views. Instead he selected only those aspects that he found acceptable in light of the extraordinary progress of mathematics and science in the late nineteenth and early twentieth centuries. He tried to "unify" these aspects into a coherent view. This was not easy, because there were conflicts in these views, which had to be resolved. The best example of such a resolution is his "finite point of view" regarding the foundations of mathematics, by which the actual infinite of Cantor's set theory should be tamed. This led to the idea of proof theory, in which the "infinite" of arithmetic should be controlled by "finite means," that is, by proof of its consistency within a "finite" fraction of itself. Although there are some doubts as to what precisely this finite fraction is, there are strong indications in Hilbert's work that he thought it was primitive recursive arithmetic.

Perhaps more important than this example from pure mathematics are the conflicting moments or centrifugal forces in Hilbert's conception of geometry. They are best understood by considering the following three statements:

1. Geometry is a natural science.
2. The task of geometry is the logical analysis of human spatial intuition.
3. Geometrical conventionalism (a la Poincarè) is an untenable position in spite of the fact that experimental results (e.g., the light deflection in the gravitational field of the sun) are easier explained by assuming a "variable metric" than by clinging to Euclidean geometry and introducing new material forces; because the introduction of such forces is quite ad hoc (see Conventionalism; Poincarè, Henri).

At first glance, the three statements seem incompatible. There are, however, several ways to make them coherent.

The easiest way is to drop statement 2 and consider geometry as a purely empirical science. This is, roughly, what the logical empiricists did. But this is not Hilbert's point of view. He insists on statement 2 (Hilbert [1899] 2004). The second option is to take Einstein's view and distinguish geometry as a formal mathematical discipline (which can be known a priori) from a physical discipline (which can be known only a posteriori). This again is not Hilbert's position, because in his view such a separation is totally arbitrary and problematic. The third option is to take Poincarè's position seriously and regard the choice of geometry as a matter of convention (like the choice of meter or yard as unit of length) and then stick to Euclidean geometry as the simplest one. Hilbert rejects this opinion as confused, because it confounds two concepts of simplicity, which have to be sharply distinguished: an old intuitive notion and Hertz's new methodical notion of simplicity. Poincarè introduces a metrical structure into geometry as simple, which is unnecessary, and this violates Hertz's principle of simplicity: Do not introduce superfluous elements into a theory.

But what is Hilbert's solution? How can he make the three statements coherent? The answer can be stated in a very much abbreviated form as this: Human beings have a spatial intuition of external objects, which they use in normal life. For the first level of conscious reflection, the most significant facts of human intuition were conceptually identified and put into an axiomatic order of deduction. This is roughly what Euclid achieved in his *Elements.* In the second stage, mathematicians began making spatial intuition the object of logical investigation. This led to a multiplicity of geometries, whose logical relations could be studied by the axiomatic method (including model theory). For a correct understanding of Hilbert's epistemology, it is important to note that this process took place without any input from the natural sciences. The application of non-Euclidean geometry to physics came after the logical analysis had been achieved. There was a second, equally important part of Hilbert's view: Euclidean geometry was not simply abandoned. It still played a decisive role, intellectually as well as practically, but not, as Poincaré supposes, because it is simpler to cling to Euclidean geometry. Hilbert rejected this view because he thought it was like an "idle wheel" in general relativity. The true reason that Euclidean geometry is used not only in daily life but also in science is that the deviations from it are so unimaginably small that it would be ridiculous in most cases to replace it by a non-Euclidean geometry. This remains true also when measuring devices are constructed, by which Einstein's or any other theory of general relativity is tested.

ULRICH MAJER

References

Brush, S. G. (1976), *The Kind of Motion We Call Heat: A History of the Kinetic Theory of Gases* (Vol. 2). Amsterdam: Elsevier.
Gödel, K. (1986), *Collected Works* (Vol 1). Edited by S. Feferman, J. W. Dawson, W. Goldfarb, C. Parsons, and R. Solovay. New York: Oxford University Press.
Hertz, H. (1895), "Die Prinzipien der Mechanik," in *Gesammelte Werke* (Vol. 3). Leipzig: J. A. Barth.
Hilbert, D. ([1899] 2004), "Grundlagen der Geometrie," in M. Hallett and U. Majer (eds.), *David Hilbert's Lectures on the Foundations of Geometry: 1891–1902*. Berlin: Springer, 72–123.
—— (1902), "Über die Grundlagen der Geometrie," *Mathematische Annalen* 56: 381–422.
—— (1912), *Grundzüge einer allgemeinen Theorie der linearen Gleichungen*. Leipzig: Teubner.
—— ([1918] 1932), "Axiomatisches Denken," in *Gesammelte Abhandlungen* (Vol. 3). Berlin: Springer, 146–156.
—— (1924), "Die Grundlagen der Physik," *Mathematische Annalen* 92: 1–32.
—— (2004), *David Hilbert's Lectures on the Foundations of Geometry: 1891–1902*. Edited by M. Hallett and U. Majer. Berlin: Springer.
Hilbert, David, J. von Neumann, and L. Northeim (1928), "Über die Grundlagen der Quantenmechanik," *Mathematische Annalen* 98: 1–30.
Majer, U. (1998), "Husserl and Hilbert on Completeness," *Synthese* 110: 37–56.
—— (2002), "Lassen sich Phänomenologische Gesetze *im Prinzip* auf mikro-physikalische Theorien reduzieren?" in M. Pauen and A. Stephan (eds.), *Phänomenales Bewusstsein: Rückkehr zur Identitätstheorie?* Paderborn, Germany: Mentis Verlag.
Pasch, M. (1882), *Vorlesungen über neuere Geometrie*. Leipzig: Teubner.
Sauer, T. (1999), "The Relativity of Discovery: Hilbert's First Note on the Foundations of Physics," *Archive for the History of the Exact Sciences* 53: 529–575.
Scheibe, E. (1997), *Die Reduktion physikalischer Theorien. Ein Beitrag zur Einheit der Physik* (Vol. 2). Berlin: Springer.
Sieg, W. (1990), "Relative Consistency and Admissible Domains," *Synthese* 84: 259–297.
Simpson, S. G. (1999), *Subsystems of Second-Order Arithmetic*. New York: Springer.
von Neumann, J. (1932), *Mathematische Grundlagen der Quantenmechanik*. Berlin: Springer.
von Staudt, K. G. C. (1847), *Geometrie der Lage*. Nürnberg: Bauer and Raspe.
Weyl, H. 1988. *Riemanns geometrische Ideen, ihre Auswirkungen und ihre Verknuepfungen mit der Gruppentheorie*. Berlin: Springer.

See also **Carnap, Rudolf; Conventionalism; Logical Empiricism; Quantum Mechanics; Physical Sciences, Philosophy of; Poincarè, Henri; Space-Time; Vienna Circle**

IDEALIZATION

See **Approximation**

IMMUNOLOGY

Because of its eclectic contributions to pathology, clinical medicine, and basic biology, immunology cannot be defined by a single, unifying experimental framework. Rather, it is (and has been) characterized by multiple, even competing, thought styles (Crist and Tauber 1997), each requiring a different methodological apparatus to order its experimental program—from receptor biology to molecular biology, from allergy to xeno-transplantation, from infectious diseases to rheumatoid arthritis and diabetes. The discipline is experimentally divided by the examination of two broad arenas of immune function:

1. Innate immunity, employing more ancient phylogenetic mechanisms, which deploys various identifying proteins (lectins and complement) to target pathogens for destruction by phagocytes, and
2. Acquired immunity (found only in vertebrates), which consists of antibodies (immunoglobulins) and lymphocytes (T cells and B cells); it is more specific in its identification capabilities and its memory of prior immune encounters.

The lymphocyte, because of its central role in contemporary clinical immunology (ranging from vaccination to transplantation to neuroendocrinology), has become the intense focus of current investigations. But underlying each branch of immunology, the concept of an identified and protected "self," a theoretical construction and fecund metaphor, has served as the central theme that integrates this diverse discipline.

During the last three decades of the twentieth century, immunology has commonly been described as the science that distinguishes 'self' from 'nonself,' and upon this distinction the means to preserve organismal integrity has defined the scope of immunity. In this formulation, the host organism, perceiving an invasion by microbial pathogens, mounts a defensive response. Contemporary immunology has broadened this agenda to include surveillance of the body for malignant, effete, damaged, or dead host constituents (altered "normal" cells), as well as autoimmune processes directed against undamaged elements—some of which may be part of ordinary physiological economy, while others are pathological. The challenges to define a basis for immune identity, within the coupled ambiguities of autoimmunity and tolerance (the reciprocal nonreactivity to host constituents), has generated debate about selfhood as an organizing concept for the discipline. The immune self, an implicit entity in the late nineteenth century (Tauber and Chernyak 1991; Mazumdar 1995) and a hotly contested one today (Langman 2000), is a rich philosophical topic, in terms of both its epistemological standing as well as its metaphysical foundations (Tauber 1994 and 1999). Note that while the immune self is rooted historically in the problematics of biological individuality (Loeb 1945; Buss 1987), its philosophical attention is distinct from those concerns (Wilson 1999) and is subsumed in the broader questions of reductionism (see Reductionism).

This article will outline in a historical context the two principal theories governing immunology's research program: the theory of immune identity and the more recent one that challenges the very notion of selfhood. In those constructions are reflected the prevailing attempts to define the concept of organism.

Historical Antecedents

The first medical use of the term "immunity" (originally a legal designation conferring exemption and distinction) appears in 1775, when van Sweiten, a Dutch physician, used *immunitas* to describe the effects induced by an early attempt at variolization (Moulin 1991, 24). But the concept did not develop until the mid-nineteenth century, when Claude Bernard set the theoretical stage for the autonomous organism (Cohen 2001). In contradistinction to an animal in humoral balance (i.e., the body conceived as composed of various "humors" that were in balance during health and unbalanced in disease with a pervasive environment), Bernard postulated the primacy of the organism's essential independence. In this view, animals provided discrete sites for methodological medical experimentation, as well as a focus for a theoretical reductive strategy based on positivist principles. Together, these two views provide medicine with its modern experimental basis.

Bernard furnished biology with a new concept of the organism, which would have wider ramifications than the establishment of physiology and biochemistry. Obviously, interchange with the environment was a necessary requirement for life, but Bernard emphasized how boundaries provided the crucial metabolic limits required for normal physiological function. With his concept of the *milieu interieur*, the body was envisioned as a demarcated, interdependent, yet autonomous entity ("corporeal atomism" [Cohen 2001, 190]), thereby establishing the theoretical grounding that became the *sine qua non* for the development of the models for infectious diseases, genetics, neurosciences, and immunology in all of its various guises. But as important as Bernard's concept proved to be for certain sciences, his construction also obfuscated certain aspects of biology's complexity. Most importantly, the ecological consciousness that emerged in the twentieth century found itself enmeshed in a conceptual struggle to promote a contextualist approach to complex biological environments populated by multiple species, against a biology dominated by the centrality of the autonomous organism. Even within the confines of biomedicine, Bernard's focus on the individual proved inadequate for the hygienic movement of his own period and later developments in public health. But Bernard introduced a revolutionary formulation, notwithstanding its limitations, and immunology became one of its defining sciences—indeed, immunity was alien to the older humoral view. By radically changing the inside/outside topology so that the organism's interior became the determining context of function, Bernard effectively isolated the organism from its environment and joined a complex cultural movement of redefining the body more generally.

Bernard's notion of the body as independent of the environment complemented Malthusian economics, liberal political philosophy, and Comtian sociology. From these and other disparate sources, the autonomous, atomistic body as a political, social, economic, and medical entity was redefined in the nineteenth century (Foucault 1973; Agamben

1998), and Bernard played a central role in providing a theoretical biological foundation for its critical nexus in various discourses. Notwithstanding that "independence" is a political term and fairly represents neither the dialectical relationships of the organism and its environment (Levins and Lewontin 1985) nor the evolutionary peculiarities of individuality itself (Buss 1987), the formulation has served as the touchstone for various cultural constructions of identity. Indeed, culture critics have seized on immunology as paradigmatic for the modern notions of identity, where boundaries are contested and the body becomes the localized site of battle between self and other (Haraway 1989; Martin 1994). The warfare metaphors— "attack," "defense," "invaders"—so prevalent in immunology's lexicon, dramatically illustrate this construction, in terms of both the self/other dichotomy and the privileged regard of individuality over community.

Origins of the Immune Self

Immunology's history is generally regarded as intimately tied to those discoveries leading to the elucidation of the bacterial etiology of infectious diseases, which draws together twin disciplines— microbiology (the study of the offenders) and immunology (the examination of host defenses). Thus, in this pathological context, immunology began as the study of how a host animal reacts to pathogenic injury and defends itself against the deleterious effects from such microbial insult. This is the typical historical account of immunology as a clinical science, a tool of medicine; and as such, it focused almost exclusively on the role of immunity as a defender of the infected. The paradigmatic host is the patient, an infected "self," which is the critical element for the power of this view. The clinical orientation, which *assumes* a given entity—the self—is obviously a dominant organizing perspective, but another perspective turns this assumption into a question or a problem: Rather than the science that seeks to discern the basis of self/nonself discrimination, immunology may also be regarded as more fundamentally concerned with the *establishment* of organismal identity.

This latter point of view was offered by Elie Metchnikoff, who came to the nascent field of immunology from an unexpected theoretical and methodological perspective—that of an embryologist—and sought to discover genealogical relationships in the context of Darwinism (Tauber and Chernyak 1991). Intrigued with the problem of how divergent cell lineages were integrated into a coherent, functioning organism, Metchnikoff was thus preoccupied with the problems of development as process, which he regarded as analogous to Darwinian interspecies struggle: Cell lineages were inherently in conflict to establish their own hegemony, but he hypothesized that unlike nature writ large, a regulatory system was required to impose order, or what he called "harmony," on the disharmonious elements of the animal. He found such an agent in the phagocyte, which retained its ancient phylogenetic eating function, to devour effete, dead, or injured cells that violated the phagocyte's sense of organismal identity. When pathogenic microbes were discovered in the 1870s, Metchnikoff soon assigned his phagocyte the new role of defending the organism against invaders. Indeed, in this context, the phagocyte became an exemplary combatant of Darwinian struggle, now occurring within the organism.

In Metchnikoff's theory, immunity was a particular case of physiological inflammation, a normal process of animal economy. But there was a more subtle message: (1) Immunity was an active process, with the phagocyte's response seemingly mounted with a sense of independent arbitration, and (2) organismal identity was a problem bequeathed from a Darwinian perspective that placed all life in an evolutionary context. In short, he combined a Darwinian sensibility to a Bernardian conceptualization of autonomy.

Metchnikoff's overall representation constituted the phagocyte as an *agent* (Crist and Tauber 2001), an actor that was the cause of its own action, as a matter of endogenously generated and directed behaviors. The portrayal of the phagocyte as autonomous is largely derived from the linked features of its capacity to sense its environment and move freely within it, and the various degrees of unpredictability and meaningfulness that characterize this behavior. Indeed, the phagocyte, as an agent, becomes a metaphorical 'self,' a primordial microcosmic expression of what later immunologists would extend into an epistemology of biological identity. But while placing the identity function at the nexus of immunology's concern, Metchnikoff failed to provide the necessary preconditions for those who would seek to demonstrate those reactions that conferred protection of such an *entity*. Much of the subsequent history of immunology may be traced to the attempts to establish a definition and experimental basis that fulfills such an identity function, an effort that may be fairly regarded as remaining unresolved, as an ambiguity at the very heart of the discipline.

IMMUNOLOGY

Twentieth-Century Constructions of the Immune Self

In the first half of the twentieth century, immunology was devoted to establishing the chemical basis of specificity, which unreflectively assumed the parameters of selfhood (Silverstein 1989; Mazumdar 1995). Paul Ehrlich, whose early scientific research concerned the chemical specificity of dyes, applied his general notions of biological affinity to immunology and thereby provided the first theory of immune specificity. Analogous to Fisher's model of "lock and key" binding of organic compounds, Ehrlich proposed that antibodies and their targets bind according to corresponding structural fittings. His postulated "side chains" were *cellular receptors* (a term he coined) for bacteria and their products. When confronted with infection, proliferation of side chains (which in solution were liberated as "anti-bodies") bound and thus neutralized pathogens and their toxic products. This mechanism, coupled with phagocytes, provided the host organism with a defense against microbes, and Ehrlich shared the Nobel Prize with Metchnikoff in 1908 in recognition of the synthesis of their respective (cellular versus immunochemical) points of view.

By World War I, Karl Landsteiner had demonstrated the extraordinary finesse of chemical recognition, and the biological mechanism that accounted for antibody generation to a seemingly infinite array of antigens (targets of antibody recognition) accommodated itself to the colloid theory of protein structure. In this view, antibodies were thought to form upon an antigenic template, which then would serve as a model for the multiplication of identical antibody molecules (Silverstein 1989). With the understanding of protein synthesis inspired by Watson and Crick in the 1950s, such template models violated DNA-directed protein synthesis, and a new theory soon followed. In 1955, Niels Jerne postulated that antibodies were "selected" from a pool of "natural antibodies" on the basis of their respective affinities for antigen. From this subpopulation of the antibody pool, an array of appropriate neutralizers would be conscripted. This suggestion was soon followed by a biological model to explain how such "natural selection" operated (Tauber 1994). David Talmage and Frank Burnet's better developed "clonal selection theory" (CST) (Burnet 1959), predicted that antibody selection occurred at the level of the antibody-producing lymphocyte, whose singular antibody receptor had high affinity for antigen. Consistent with the peptide model, they proposed that antigen binding stimulated cellular proliferation and differentiation of those cells (clones) that shared an appropriate affinity profile for those pathogens, toxins, allergens, and any other "foreign substance" that was so recognized.

Within a decade, compelling evidence confirmed these theoretical musings, and the central question then became one of the intracellular mechanism of antibody generation. By the late 1970s, this beguiling puzzle was solved when it was shown that immunoglobulin (antibody) was made up of segments that were put together like so many "cards" from a genotypic "deck," and that these synthesized proteins also underwent somatic mutation. (The elucidation of antibody generation was generally important, for it demonstrated the plasticity of the genome and the genotypic variability of individual cells.) Thus the bewildering specificity of the immune reaction could be accounted for by the shuffling of a finite number of genes (coding for the cards in the deck) and a mechanism of fine-tuning somatic mutation that gave rise to "custom" antibodies with highly specific binding characteristics (Podolsky and Tauber 1997). This breakthrough was only the most celebrated of immunology's molecularization. Indeed, the entire field of immunology was now committed to defining the molecular pathways of immune effector functions, the structure/function relationships of various mediators, and the molecular control mechanisms of what became an increasingly complex system of interactive components (Tauber 1996).

But this shift to a highly sophisticated molecular approach should not obscure the underlying theoretical questions being addressed of a new biologically oriented program. While Ehrlich's chemical perspective had dominated immunology until World War II, these new hypotheses concerning antibody generation, coupled to clinical demands in the fields of organ transplantation and autoimmunity, drove immunology toward a new biological theory of immunity that was both more comprehensive than earlier chemical models and far-reaching in its theoretical implications. Burnet not only provided a mechanism for antibody selection and biological generation, he also presented a theory of immunological "tolerance" that was to henceforth dominate the field (Burnet and Fenner 1949; Tauber 1994). From Burnet's perspective, foreign bodies are destroyed by immune cells and their products, whereas the normal constituents of the animal are ignored, that is, "tolerated." In other words, the identity of the host organism was a given within the Bernardian construct, with implicit boundaries as defined by immune reactivity. What was "attacked" was "other"; that which was regarded by immune silence became the self. What was, perhaps, implicit in pre–World War II immunology now declared its theoretical basis: Without a theory of self/nonself differentiation, immune reactivity had no biological basis for control. Indeed, the 'self' was introduced by Burnet into the immunological lexicon specifically to address immunity as an organismal phenomenon.

Unlike Metchnikoff, Burnet sought a firm definition of the immune self. Burnet's theory proposed that the animal, during prenatal development, exercises a purging function of self-reactive lymphocytes (the cells responsible for synthesizing reactive antibodies and mediating so-called cellular reactions) so that all antigens (substances that initiate immune responses) encountered during this period would attain a neutrality status. Thus, lymphocytes with reactivity against host constituents are putatively destroyed during development, and only those tolerant lymphocytes that are nonreactive are left to engage the antigens of the foreign universe. The hypothesis (first presented in 1949 and later developed into the CST [Burnet 1959]) contained two key challenges that dominated immunology: (1) How was tolerance induced and autoimmunity controlled? and (2) What was the mechanism that accounted for antibody and lymphocyte diversity? As already noted, the latter issue was solved by molecular biologists by the late-1970s (Podolsky and Tauber 1997); the former question, involving systems analysis, apparently requires a comprehensive model of the immune system as a whole and remains enigmatic.

Aside from incomplete accounts of tolerance, there were early discrepancies arising from a continuum of autoimmune reactions, ranging from normal physiological and inflammatory processes to uncontrolled disease initiated by an immune reaction gone awry. Bountiful evidence in recent years has shown that autoimmunity is also a normal finding, and in these newer views, such functions are regarded as integrated within a more complex normal physiology. Thus, immune reactivity, rather than functioning only in an "other-directed" mode, is in fact bidirectional. This position contrasts with the "one-way" definition of selfhood, where there is a genetic self whose constitutive agents see the foreign bodies, and immune reactivity arises from this polarization with attack directed only against nonself (Tauber 1998). Not unexpectedly, in this turn inward, the immune self becomes increasingly difficult to define, and the concept becomes unable to easily accommodate these new appraisals.

There are at least half a dozen different conceptions of what constitutes the immune self (Matzinger 1994, 993):

1. Everything encoded by the genome
2. Everything under the skin including/excluding immune "privileged" sites
3. The set of peptides complexed with T-lymphocyte antigen-presenting complexes, of which various subsets vie for inclusion
4. Cell surface and soluble molecules of B-lymphocytes
5. A set of bodily proteins that exist above a certain concentration
6. The immune network itself, variously conceived (detailed below)

While these versions may be situated along a continuum between a severe genetic reductionism and complex organismal view (Tauber 1998 and 1999), each shares an unsettled relationship to Burnet's original dichotomous model of self and other.

Assaults on the Immune Self

Well before the current debate about the immune self (Langman 2000), Niels Jerne attempted to dispel the many ad hoc caveats and paradoxes encumbering it by eliminating the self concept altogether. He went beyond the current notion of the immune network composed of lymphocyte subsets, secreting immunostimulatory and inhibitory substances (essentially a simple mechanical model with interlaced, first-order feedback loops) to propose a novel conception of immune regulation (Tauber 1994; Podolsky and Tauber 1997). His network theory was, from its very inception, a complex amalgam of pieces of the regulation puzzle fitted into place, with the overriding goal of understanding the immune system as a cognitive enterprise that spawned different formulations (e.g., Varela et al. 1988; Atlan and Cohen 1989; Stewart 1994a). In introducing this metaphoric construction of the immune system as analogous to the nervous system as early as 1960, Jerne set the stage for understanding newer immune metaphors (recognition, memory, learning) that built on the parallel with human cognition.

Jerne's idiotypic network theory hypothesis proposed that antibodies form a highly complex interwoven system, where the various specificities "referred" to each other (Jerne 1974). Under the general rubric of "cognition," he conceived of the immune system as self-regulating, where antibodies recognize not only foreign antigens, but self constituents as antigens (the so-called *idiotopes*). There was no essential difference between the recognized and the recognizer, since any given antibody might serve either, or both, functions. In other words, immune regulation was based on the reactivity of antibodies (and later lymphocytes) with their own repertoire forming a set of self-reactive, self-reflective, self-defining immune activities. There is no self and other for the immune system, for according to Jerne's theory, the system is complete unto itself, consisting of interlocking recognizing units: Each component reacts with certain other constituents to form a complex network or lattice structure. When the system is perturbated by the introduction of a substance that is recognized (i.e., it reacts with members of the system), this disturbance initiates immune responsiveness. Thus, foreignness per se does not exist in this formulation.

Jerne's theory presents a radically altered view of immune selfhood. In Burnet's simplified world of self/nonself discrimination, the immune system learned host/foreign distinctions, generated an army of reactive antibodies and lymphocytes, and acted accordingly when "antigen" was encountered. But Jerne coupled the simple antibody/antigen interactions to the far more complex and nondiscriminatory functions of the immune system, which built upon self-recognition. In his view, autoimmunity, instead of an aberration, became the organizational rule to explain immune function. Strikingly, there is no explicit mechanism for self/nonself discrimination, and this apparent lacuna served as the nexus of critiques (reviewed in Podolsky and Tauber 1997; Tauber 1999; Tauber 2000). But for Jerne, the need to define the self as distinct from other receded from his primary theoretical concerns, and this posture was to have important repercussions.

When the immune system is regarded as essentially self-reactive and interconnected, the "meaning" of immunogenicity, that is, reactivity, must be sought in some larger framework. Antigenicity then is only a question of degree, where self evokes one kind of response, and the foreign evokes another, not because of its intrinsic foreignness, but because the immune system sees the foreign antigen in the context of invasion or degeneracy. From the immune system's perspective, it only "knows" itself (Varela et al. 1988). Indeed, for Jerne, if a self was at all needed, it would be simply the immune system. Most importantly, the singular defensive purpose of immunity was widened to include an array of physiological functions, each of them now regarded as fully integrated within the immune system (Matzinger 1994). If eventually successful, this heralds a decisive shift in immunology's theoretical foundations, one more attuned to the diversity of immune functions that contribute to evolutionary fitness (Cohen 1992 and 1994; Stewart 1994a). While host defense is a critical function, it is hardly the only one of interest. Indeed, the immune system might be regarded as primarily fulfilling an altogether different role if its phylogeny is carefully examined. On this basis, John Stewart has provocatively suggested that the immune system became defensive only after its primordial neuroendocrine communicative capabilities (Ader, Felton, and Cohen 2001) were usurped for immunity (Stewart 1994b).

Biologists have increasingly come to appreciate that such systems are highly integrated within larger wholes and require analysis of how adjustments are made in relation to these other systems. This means, simply, that immune reactivity is determined by context (Cohen 1994; Podolsky and Tauber 1997), where agent and object play upon each other. Specific recognition of an antigen by a lymphocyte receptor is not sufficient for activation, for additional signals determine whether a cellular response or cell inactivation follows. In short, an antigen is neither self nor nonself except as it attains its meaning, so to speak, within a broader construct. Orthodox immune theory encompasses this idea in the so-called two-signal model, which does not require any of Jerne's hypotheses to fulfill its agenda. But there are more radical readings of the contextualist setting by which antigens are sensed, and debate concerning what constitutes the milieu of meaning of antigenicity and ensuing reaction have spawned certain provocative, and potentially important, models of immune regulation (reviewed in Podolsky and Tauber 1997; Tauber 2000).

In summary, immunology may be seen as structured on two major theoretical developments. The first was made by Metchnikoff in framing immunology with dual functions:

- establishment of organismal identity and
- protection of this integrity.

His immunochemical contemporaries and their direct heirs followed the second agenda to the exclusion of the first. The primacy of the identity issue was reintroduced by Burnet, and his program defined lymphocyte biology for the latter half of the twentieth century.

The second theoretical advance was made by Jerne, who moved past the identity issue altogether. No longer in service to a self, the immune system functioned within a greater whole as a cognitive faculty, perceiving only what it might know—*itself*. Patterns, context, and interlocution become organizing principles, so that the self metaphor, assuming a Jernian perspective, is eclipsed by another catchall metaphor, *cognition*. Even within such new formulations, the self still resides, reflecting a deep struggle over the character of biology, one that has its roots in Bernard's original understanding of autonomy, and now linked to our own more complex ecological views of agency and determinism.

ALFRED I. TAUBER

References

Ader, Robert, David L. Felten, and Nicholas Cohen (2001), *Psychoneuroimmunology*, 3rd ed. San Diego: Academic Press.

Agamben, Giorgio (1998), *Homo Sacer: Sovereign Power and Bare Life*. Stanford, CA: Stanford University Press.

Atlan, Henri, and Irun R. Cohen (eds.) (1989), *Theories of Immune Networks*. Berlin: Springer-Verlag.

Burnet, Frank Macfarlane (1959), *The Clonal Selection Theory of Acquired Immunity*. Nashville, TN: Vanderbilt University Press.

Burnet, Frank Macfarlane, and Frank Fenner (1949), *The Production of Antibodies*, 2nd ed. Melbourne, Australia: Macmillan and Co.

Buss, Leo (1987), *The Evolution of Individuality*. Princeton, NJ: Princeton University Press.

Cohen, Edward (2001), "Figuring Immunity: Towards the Genealogy of a Metaphor," in A. M. Moulin and A. Cambrosio (eds.), *Singular Selves: Historical Issues and Contemporary Debates in Immunology*. Amsterdam: Elsevier, 179–201.

Cohen, Irun R. (1992), "The Cognitive Paradigm and the Immunological Hommunculus," *Immunology Today* 13: 490–494.

——— (1994), "Kadishman's Tree, Escher's Angels, and Immunological Homunculus," in Antonio Coutinho and Michel D. Kazatchkine (eds.), *Autoimmunity: Physiology and Disease*. New York: Wiley-Liss, 7–18.

Crist, Eileen, and Alfred I. Tauber (1997), "Debating Humoral Immunity and Epistemology: The Rivalry of the Immunochemists Jules Bordet and Paul Ehrlich," *Journal of the History of Biology* 30: 321–356.

——— (2001), "The Phagocyte, the Antibody, and Agency: Contending Turn-of-the-Century Approaches to Immunity," in Anne Marie Moulin and Alberto Cambrosio (eds.), *Singular Selves: Historical Issues and Contemporary Debates in Immunology*. Amsterdam: Elsevier, 115–139.

Foucault, Michel (1973), *The Birth of the Clinic: An Archaeology of Medical Perception*. New York: Vintage.

Haraway, Donna (1989), "The Biopolitics of Postmodern Bodies: Determinations of Self in Immune System Discourse," *Differences* 1: 3–43.

Jerne, Niels K. (1974), "Towards a Network Theory of the Immune System," *Annals of Institute Pasteur/Immunology (Paris)* 125C: 373–389.

Langman, Rodney (ed.) (2000), *Self-Nonself Discrimination Revisited*. Special issue of *Seminars in Immunology* 12(3).

Levins, Richard, and Richard Lewontin (1985), *The Dialectical Biologist*. Cambridge, MA: Harvard University Press.

Loeb, Leo (1945), *The Biological Basis of Individuality*. Springfield, IL: C. C. Thomas.

Martin, Emily (1994), *Flexible Bodies: The Role of Immunity in American Culture from the Days of Polio to the Age of AIDS*. Boston: Beacon Press.

Matzinger, Polly (1994), "Tolerance, Danger, and the Extended Family," *Annual Review of Immunology* 12: 991–1045.

Mazumdar, Pauline M. H. (1995), *Species and Specificity. An Interpretation of the History of Immunology*. Cambridge: Cambridge University Press.

Moulin, Anne Marie (1991), *Le Dernier Langage de la Medicine: Histoire de l'Immunologie de Pasteur au Sida*. Paris: Presses Universitaires de France.

Podolsky, Scott H., and Alfred I. Tauber (1997), *The Generation of Diversity: Clonal Selection Theory and the Rise of Molecular Immunology*. Cambridge, MA: Harvard University Press.

Silverstein, Arthur (1989), *A History of Immunology*. San Diego: Academic Press.

Stewart, John (1994a), "Cognition Without Neurons: Adaptation, Learning and Memory in the Immune System," *Communication and Cognition—Artificial Intelligence* 11: 7–30.

——— (1994b), *The Primordial VRM System and the Evolution of Vertebrate Immunity*. Austin, TX: R. G. Landes.

Tauber, Alfred I. (1994), *The Immune Self: Theory or Metaphor?* New York and Cambridge: Cambridge University Press.

——— (1996), "The Molecularization of Immunology," in S. Sarkar (ed.), *The Philosophy and History of Molecular Biology: New Perspectives*. Dordrecht, Netherlands: Kluwer Academic Publishers, 125–169.

——— (1998), "Conceptual Shifts in Immunology: Comments on the 'Two-Way Paradigm,'" *Theoretical Medicine and Bioethics* 19: 457–473.

——— (1999), "The Elusive Self: A Case of Category Errors," *Perspectives in Biology and Medicine* 42: 459–474.

——— (2000), "Moving Beyond the Immune Self?" *Seminars in Immunology* 12: 241–248.

Tauber, Alfred I., and Leon Chernyak (1991), *Metchnikoff and the Origins of Immunology: From Metaphor to Theory*. New York and Oxford: Oxford University Press.

Varela, Francisco J., Antonio Coutinho, B. Dupire, and N. N. Vaz (1988), "Cognitive Networks: Immune, Neural, and Otherwise," in Alan S. Perelson (ed.), *Theoretical Immunology*, part 2. Redwood City, CA: Addison-Wesley, 359–375.

Wilson, Jack (1999), *Biological Individuality: The Identity and Persistence of Living Entities*. Cambridge: Cambridge University Press.

INCOMMENSURABILITY

Incommensurability is a relation of incomparability, or limited comparability, purported to obtain between some pairs of successive or competing scientific theories. The thesis that scientific theories may be incommensurable was proposed by Paul Feyerabend and Thomas Kuhn in separate publications in 1962 (see Feyerabend, Paul; Kuhn, Thomas). Due to perceived negative consequences of incommensurability, the thesis has been the focus of considerable controversy. Before considering the objections to it, the thesis of incommensurability will first be examined.

Feyerabend on Incommensurability

Feyerabend's claim that some theories are incommensurable derives from his critique of the empiricist idea of a theory-neutral observation language. Neither experience nor pragmatic conditions of use determine the meaning of observational terms. Instead, "the interpretation of an observation language is determined by the theories which we use to explain what we observe, and it changes as soon as those theories change" (Feyerabend [1958] 1981, 31). In contrast to the empiricist view that the meaning of observational terms is independent of theory, Feyerabend holds that the meaning of such terms varies with theory.

Feyerabend introduced the concept of incommensurability in the context of a discussion of the empiricist account of inter-theory reduction by means of deductive subsumption. Against reduction, Feyerabend argues:

> What happens... when a transition is made from a theory T' to a wider theory T (which... is capable of covering all the phenomena that have been covered by T') is something much more radical than incorporation of the *unchanged* theory T' (unchanged, that is, with respect to the meanings of its main descriptive terms as well as to the meanings of the terms of its observation language) into the context of T. What does happen is, rather, a *replacement* of the ontology (and perhaps even of the formalism) of T' by the ontology (and formalism) of T, and a corresponding change of the meanings of the descriptive elements of the formalism of T' (provided these elements and this formalism are still used). This replacement affects not only the theoretical terms of T' but also at least some of the observational terms which occurred in its test statements. (Feyerabend [1962] 1981, 44–45)

For Feyerabend, change in theoretical ontology leads to variation in the meaning of the vocabulary employed by theories. One theory cannot be deductively subsumed by the other, given differences in the meaning (due to untranslatability) of the terminology employed by the theories.

According to Feyerabend, reduction fails because of incommensurability. Theories are incommensurable due to lack of semantic equivalence between terms employed by the theories. On the one hand, the concepts of one theory cannot be defined on the basis of concepts of the other. On the other hand, no empirical statement may be formulated that correlates terms of one theory with terms of the other theory. Because no neutral observation language exists in which to express the empirical consequences of such theories, Feyerabend concludes that "incommensurable theories may not possess any comparable consequences, observational or otherwise" (Feyerabend [1962] 1981, 93). The contents of incommensurable theories are unable to be compared because no consequence of one theory may either assert or deny the same thing as any consequence of a theory with which it is incommensurable (see Feyerabend, Paul).

Incommensurability in Kuhn's *Structure of Scientific Revolutions*

In *The Structure of Scientific Revolutions*, Kuhn (1962) proposed a model of the development of science divided into periods of normal science grounded in consensus on a shared scientific paradigm. Normal science is broken at intervals by periods of extraordinary science, brought on by anomaly and crisis, which may ultimately result in revolutionary displacement of a paradigm (see Scientific Revolutions).

Once a new candidate for paradigm emerges in the midst of a crisis, debate ensues between defenders of the reigning paradigm and advocates of the candidate paradigm. This debate is characterized by failure of communication that arises because of the incommensurability of the old paradigm and the new candidate paradigm. As a result of incommensurability, debate about which paradigm to adopt is unable to be brought to closure by purely rational means.

According to Kuhn, the incommensurability of competing paradigms is due to differences that arise at three levels between paradigms. The first difference involves variation at the methodological level. Paradigms address different problem-solving agendas and employ different standards of theory appraisal:

> [P]roponents of competing paradigms will often disagree about the list of problems that any candidate for paradigm must resolve. Their standards or their definitions of science are not the same. (Kuhn 1962, 148)

The second difference is at the semantic level. There is variation in the concepts employed by paradigms, which leads to change in the meanings of the terms that express key scientific concepts:

> Within the new paradigm, old terms, concepts, and experiments fall into new relationships one with the other. ... To make the transition to Einstein's universe, the whole conceptual web whose strands are space, time, matter, force, and so on, had to be shifted and laid down again on nature whole. (149)

The third difference relates to the theory-dependence of observation. Not only may scientists observe different things, but the content of their perceptual experience when they observe the same thing depends upon the paradigm in which they work:

> [P]roponents of competing paradigms practice their trades in different worlds.... [P]racticing in different worlds, the two groups of scientists see different things when they look from the same point in the same direction. (150)

Kuhn's claim that scientists work in different worlds may be taken to suggest a stronger thesis than that which states that scientists' perceptual experience depends on paradigm. In his book *Reconstructing Scientific Revolutions*, Paul Hoyningen-Huene (1993) has argued that Kuhn's position is best understood as a neo-Kantian position in which the phenomenal world of scientists varies with paradigms, while the unknowable noumenal world remains constant (see Kuhn, Thomas).

Taxonomic Incommensurability

In later work, Kuhn continued to refine his concept of incommensurability. In contrast to Feyerabend, Kuhn's original concept of incommensurability included nonsemantic elements. Kuhn came to view incommensurability as a semantic issue distinct from methodological variation and dependence of observation on theory. Semantic issues relating to translation failure are the focus of Kuhn's later work on incommensurability, of which he proposes a taxonomic version that involves localized translation failure between subsets of the special terminology employed by theories.

In *The Road Since Structure*, Kuhn (2000) claims that scientific revolutions are characterized by changes in the taxonomic schemes by means of which theories classify entities in their domain (30). In the transition between theories, both criteria of classification and membership of taxonomic categories undergo change. At the semantic level, taxonomic change gives rise to variation in meaning of some of the preserved vocabulary, as well as to introduction of vocabulary with new meaning. Because taxonomic change involves change of interconnected categories, the meanings of the terms affected by such change are related in a holistic manner. Each theory possesses a central set of interdefined terms, which cannot be translated in piecemeal fashion into the vocabulary of a theory with a different taxonomic structure (Kuhn 2000, 43–44). Translation failure between theories is a localized phenomenon that is restricted to such central sets of interdefined terms.

Objections to Incommensurability

As indicated above, the incommensurability thesis is controversial because of negative outcomes to which it gives rise. If, as Kuhn initially suggested, there are no neutral standards of theory appraisal, and communication is obstructed, it is unclear how choice between theories may proceed on a rational basis. If, as Kuhn and Feyerabend both suggest, the content of theories may not be compared due to semantic variance, it is unclear how to conduct crucial tests between rival theories or to determine whether one theory marks an advance over another. Indeed, Dudley Shapere raises the question of whether incommensurable theories may constitute rivals at all (Shapere 1984, 73). But the two objections that have proven the most telling have been Donald Davidson's critique of untranslatability and Israel Scheffler's referential objection to incomparability.

The Incoherence of Untranslatability

Davidson raises serious doubts regarding the coherence of the idea of an untranslatable language. He notes that there is an air of paradox about incommensurability: "Kuhn is brilliant at saying what things were like before the revolution using—what else?—our post-revolutionary idiom" (Davidson 1984, 184). If one provides an example of an untranslatable concept in the language into which translation fails, the example belies the untranslatability. It is also puzzling how one might understand an untranslatable concept in the first place if it cannot be translated into a language that one understands. It is not, moreover, clear what would count as evidence of untranslatability. Failure to translate a language is indeterminate between being evidence that the language is untranslatable and evidence that it is not a language at all. Davidson suggests that the idea of an untranslatable language depends on a distinction between conceptual scheme and content, which gives substance to the idea of a language independent of translation. But, he argues, no intelligible sense can be made of the distinction between scheme and content.

Davidson's objections may be defused by noting two ways in which the incommensurability thesis is less extreme than he supposes. First, failure of translation between incommensurable theories is restricted to the vocabulary employed by theories, or to a subset of such vocabulary, rather than extending to the entirety of a natural language. Thus, the thesis of incommensurability does not require sense to be made of radically alternative conceptual schemes, but only of localized translation failure within a language. Nor need untranslatable concepts be formulated in the language into which translation fails. Untranslatability is restricted to semantically variant fragments of an embracing natural language. The latter may therefore serve as metalanguage within which semantic relations between the vocabulary of meaning-variant theories may be analyzed. Second, incommensurability need entail only failure to translate between the vocabulary of theories, rather than failure to understand the content of a meaning-variant theory. One may understand what is said in another language even if it cannot be translated into one's own language. Equally, one may understand concepts of a theory that are untranslatable into one's own theory due to incommensurability.

The Referential Objection

In his book *Science and Subjectivity*, Scheffler (1967) notes that discussion of meaning variance in relation to incommensurability runs foul of the distinction between sense and reference. Variation in theoretical context may lead to variation in the sense of a scientific term. But it does not follow that the term's reference is thereby similarly affected. Terms that differ in sense may refer to the same thing. Coreference is all that is needed for claims about the world to enter into conflict. Hence, the content of meaning-variant theories may be compared if the terms employed by the theories share common reference, regardless of variation in sense.

However, Scheffler's referential objection is not entirely successful. Meaning variance in science need not be restricted to variation in the sense of scientific terms. If reference is determined by sense, then significant variation in sense may result in variation of reference. Moreover, the reference of scientific terms may be subject to variation independent of sense as a result of changes in classification or revision of linguistic use.

The Causal Theory of Reference

Since the 1970s, the referential objection has been based on the causal theory of reference. Causal theorists argue that the reference of a term is not determined by an associated description that specifies the term's sense. Rather, reference is determined in a direct manner by ostensive introduction of a term in the presence of the referent. What determines reference is the causal relation (e.g., perception) between term-introducer and referent. Subsequent use of the term by later speakers is connected by means of a causal-historical chain to the original term-introduction.

If reference is determined independently of description, then terms employed by one theory may be employed in the context of a later theory to refer to the same things as they referred to in the earlier theory. Terms employed by successive theories may continue to refer to the same things despite variation in descriptive content associated with the terms in different theories. The claims made by such theories may be compared directly on the basis of the shared referents of the terms that the theories employ.

But application of the causal theory of reference in the context of scientific theory change is not without difficulties. First, to secure reference for a kind-term, it does not suffice to identify sample members of the kind in an ostensive manner. The sample may belong to multiple kinds. Ostension must be supplemented by a description that specifies the relevant kind by means of a sortal expression. Second, theoretical terms may fail to refer if the unobservable entities to which they purport to refer do not in fact exist. But if reference is determined by the causal relation to the real cause of an actual phenomenon, then it may be impossible for reference to fail. To allow for such failure, descriptive characterization of putative theoretical entities must enter into the reference determination of theoretical terms. Third, to allow reference change, reference must be sensitive to the use of terms on occasions subsequent to their initial introduction, rather than being permanently fixed at the outset.

In light of the need for a descriptive element in the determination of reference, recent authors propose causal descriptive accounts of reference in which either causal relation combines with description to fix reference or else reference-fixing description is cast in causal terms. Causal descriptive accounts allow that the descriptive content of scientific theories affects reference. However, such accounts provide little scope for incommensurability due to radical divergence of reference. For while description may play a role in reference determination, neither is reference fully determined by description, nor is the entirety of the descriptive content associated with a term relevant to reference. Thus, so far as variation of reference is concerned, the prospects for incommensurability are greatly diminished.

HOWARD SANKEY

References

Bird, Alexander (2000), *Thomas Kuhn*. Chesham, UK: Acumen Publishing.
Davidson, Donald (1984), *Inquiries into Truth and Interpretation*. Oxford: Oxford University Press.
Feyerabend, Paul K. ([1958] 1981), "An Attempt at a Realistic Interpretation of Experience," in *Realism, Rationalism and Scientific Method: Philosophical Papers*, vol. 1. Cambridge: Cambridge University Press, 17–36. Originally published in *Proceedings of the Aristotelian Society, New Series* 58, 143–170.
——— ([1962] 1981), "Explanation, Reduction and Empiricism," in *Realism, Rationalism and Scientific Method: Philosophical Papers*, vol. 1. Cambridge: Cambridge University Press, 44–96. Originally published in Herbert Feigl and Grover Maxwell (eds.), *Scientific Explanation, Space and Time: Minnesota Studies in the Philosophy of Science*, vol. 3. Minneapolis: University of Minnesota Press, 28–97.
Hoyningen-Huene, Paul (1993), *Reconstructing Scientific Revolutions: Thomas S. Kuhn's Philosophy of Science*. Chicago: University of Chicago Press.
Hoyningen-Huene, Paul, and Sankey, Howard (eds.) (2001), *Incommensurability and Related Matters: Boston Studies in Philosophy of Science*, vol. 216. Dordrecht, Netherlands: Kluwer Academic Publishers.
Kuhn, Thomas S. (2000), *The Road Since Structure*. Edited by James Conant and John Haugeland. Chicago: University of Chicago Press.
——— (1962), *The Structure of Scientific Revolutions*. Chicago: University of Chicago Press.
Preston, John (1997), *Feyerabend: Philosophy, Science and Society*. Cambridge: Polity Press.
Sankey, Howard (1993), "Kuhn's Changing Concept of Incommensurability," *British Journal for the Philosophy of Science* 44: 775–791.
——— (1997), *Rationality, Relativism and Incommensurability*. Aldershot, UK: Ashgate.
——— (1998), "Taxonomic Incommensurability," *International Studies in the Philosophy of Science* 2: 7–16.
——— (1994), *The Incommensurability Thesis*. Aldershot, UK: Avebury.
Scheffler, Israel (1967), *Science and Subjectivity*. Indianapolis: Bobbs-Merrill.
Shapere, Dudley (1984), *Reason and the Search for Knowledge*. Dordrecht, Netherlands: Reidel.

See also **Feyerabend, Paul; Kuhn, Thomas; Logical Empiricism; Scientific Change; Unity and Disunity of Science**

INDETERMINISM

See **Determinism; Quantum Mechanics**

INDIVIDUALISM

See **Methodological Individualism**

INDIVIDUALITY

One of the most fundamental distinctions in philosophy is between individuals (also called particulars) and such things as classes, sets, kinds, universals, or whatever. As the number of synonyms might indicate, philosophers have spent much more effort clarifying classes than they have explicating the polar notion of individuals. In the briefest form of this relationship, classes range over individuals, and individuals belong to one or more classes. For example, in the claim that Socrates is mortal, Socrates is an individual human being, while 'mortal' denotes a class of beings all of whom are born and die. Socrates is mortal, but so are all other living creatures.

Philosophers have dealt with the distinction between individuals and classes in two ways. First, some have relied on ordinary folk notions of individual and class. What ordinary people take to be individuals are individuals. What ordinary people take to be classes are classes. Socrates is an individual, and 'mortal' is a class. For most philosophers, however, ordinary notions of individual and class are not good enough. They have devised technical notions of individuals and classes to conform strictly to the needs of their technical philosophies: for instance, a bare particular and the class of all bare particulars. A bare particular is an individual that has no properties of its own, an entity shorn of all characteristics. Needless to say, bare particulars are not common sense entities.

Biological Individuals

Although most of the entities that biologists treat as individuals are common sense individuals, many are not. Ordinary people consider a Portuguese man-of-war a single organism—a single individual. Biologists do not. Like philosophers, biologists have had to develop their own technical notions of individuality to fulfill their own special needs. The analyses of the 'individual' produced by biologists have several advantages over comparable philosophical accounts. All such analyses must take place in some context or other. In philosophy these contexts are supplied by very general philosophical systems, while in biology they take place within specific scientific theories. Unfortunately for philosophers, none of their systems have gained much in the way of consensus. Scientists have the advantage that many scientific theories are widely accepted. Right now, evolutionary biology is going through some fundamental revisions (see Evolution). Even so, there is more agreement among evolutionary biologists about the evolutionary process than among philosophers with respect to their systems. As a result, biologists' notions of the individual are likely to have more content and lasting influence than such notions devised by philosophers.

A second advantage that scientists, particularly biologists, have over philosophers is the wealth of examples open to them. Philosophers have a habit of making up science fiction examples to illustrate and test their analyses. Unfortunately, such examples are highly malleable; philosophers can make them serve just about any purpose they desire. Real examples, on the contrary, can force a reexamination of the decisions that are made. They can force one to see that some of one's deepest intuitions are mistaken. Nature provides much more bizarre examples than any philosopher has ever been able to dream up, and—more importantly—it provides good reason to accept one analysis over another.

Nihilists can be found claiming that how one divides up the world simply does not matter. This claim is usually made in the context of classes. Group females of one species with males of another. It makes no difference. Ignore the distinction between sexual and asexual organisms. Comparable claims are also made with respect to individuals. Consider a clone of a single organism or thousands. It simply does not matter. For scientists, it does matter. For scientists, certain ways of dividing up the world are preferable to others. Perhaps scientists cannot always settle on one way and only one way of dividing up the world, but from this state of affairs it does not follow that anything goes. In actual fact, developing alternative classifications is quite difficult. Only a very few serious alternatives can be found.

In biology at least, biologists and philosophers have pooled their conceptual resources to deal with topics such as individuality. For example, biologists commonly use "individual" and "organism" interchangeably, but such a linguistic convention leads to all sorts of confusion. All organisms are individuals, but not all individuals are organisms. "Individual" is a much broader term, and one needs such a broad term. A running argument in the biological literature concerns the levels of organization at which selection can occur (Keller 1999; Michod 1999). Those who think that organisms are the primary units of selection are called "individual selectionists," because organisms are individuals; but another group of biologists think that genes are the primary units of selection, and they too are called "individual selectionists," because genes are just as much individuals as organisms are. Of course, genes are not organisms. To avoid such confusion, "individual" is used here in a generic sense to refer not only to organisms but also to other individuals.

The literature on individuality as this notion is used in biology is replete with all sorts of bizarre problem cases. A few years ago, several mycologists discovered a clonal population of the fungus *Armillaria bulbosa* that occupied fifteen hectares in the Upper Peninsula of Michigan. As far as these mycologists could tell, all parts of this clonal entity were attached and had the same genome. Why not think of this huge fungus as a single organism or, if not a single organism, at least a single individual (Gould 1992; Wilson 1999)? To answer this question (and others), philosophers and biologists have set out a list of criteria for an entity counting as an individual in the generic sense and the implications that these criteria have for a variety of problem cases. But before these criteria are examined, a word needs to be said about the importance of developing a clear notion of individuality. As real as the fungus is, one might wonder why deciding whether or not it is one organism or a million different organisms is of any significance (Hull 1992; Wilson 1999).

The Cost of Meiosis

Such questions are important because counting is a central activity of scientists, and they have to know what it is they are counting. Like must be counted with like. For example, one commonly hears that sexual reproduction is extremely prevalent and that this prevalence poses a problem for evolutionary biologists. In sexual reproduction, homologous chromosomes line up at meiosis and separate in the formation of germ cells. At every locus where different alleles reside, sexual reproduction has a 50 percent cost. More generally, asexual organisms can pass on all their genetic material to the next generation, whereas each sexual organism can pass on only half. A 50 percent cost is extremely high. The usual conclusion is that sexual reproduction must be doing a lot of good, or else it would not be so prevalent.

If meiosis is so costly, why is it so prevalent? Several answers to this question have been suggested, but one issue is passed over too quickly. Is sexual reproduction actually all that prevalent, compared with asexual reproduction? This difference is usually presented in terms of species. Sexual species are vastly more prevalent than asexual species. But there are some difficulties here. The most popular definition of species in the past century or so includes reference to interbreeding. Those organisms that produce a genealogical nexus by mating with each other form species. Since asexual organisms do not form such a network, they do not form species. Hence, contrasting the number of sexual species with the number of asexual "species" is illicit.

Only if one resorts to defining species in terms of character distributions do sexual and asexual organisms form the same sort of species. However, the concept of morphospecies has numerous problems. At one time systematists thought that something called "overall similarity" existed out there in nature and that one degree of overall similarity could be found that was the same across all organisms. Such is not the case. Various degrees of similarity exist, and no reason can be found for choosing one degree of similarity over any of the others as the level of species. Sexual organisms produce biologically significant units, but these units are not comparable to anything found in asexual organisms. Comparable units of overall similarity can be found in both, but these units are not biologically significant. In short, comparable units at the traditional level of species cannot be found for both sexual and asexual organisms.

One way out of this difficulty is to move down to the level of organisms. Sexual and asexual organisms may not form species of the same kind, but they are 'organisms' in the same sense. Hence, the prevalence of sexual reproduction can best be determined at the level of organisms, not species. When one makes such a conceptual shift, the relative percentages change. Sexual reproduction ceases to be so overwhelmingly prevalent. For the first half of life on earth, no sexual reproduction occurred. Since then it has become increasingly prevalent, but not as prevalent as is commonly claimed. In comparing organisms with organisms, problems with the cost of meiosis remain, but their scope is greatly reduced.

Units of Selection

Are there units of selection? and if so, what are they? The different answers to these questions result from yet another failure to make necessary distinctions. Selection is not one process but two intricately connected processes—replication and environmental interaction. In replication, information is passed on from one generation to the next. Certain entities also interact with their environment in such a way that replication is differential. If selection is to occur, both replication and environmental interaction are necessary. What are the units of replication? By and large, replication is limited to the genetic material, though in special circumstances replication can occur at higher levels of organization. What are the units of environmental interaction? Environmental interaction can occur at a variety of levels, from single genes and cells through organisms and hives to demes and possibly entire species.

The discussion above mentions no entities that are uniquely units of selection. There are units of replication and units of environmental interaction, but no units of selection. When biologists such as Dawkins (1976) claim that genes are units of selection, they are actually claiming that genes are the units of replication. When Dawkins's opponents claim that selection occurs at a variety of hierarchically organized levels, they are referring to environmental interaction. The issue here is individuality—what counts as individuals and what roles these individuals play in the evolutionary process. There are entities that function as replicators, and there are entities that function in environmental interaction; but there are no entities that function as "selectors."

Species as Individuals

Species present another example of the importance of individuality in biology. According to Ghiselin (1974) and Hull (1976), species are not classes but historical entities. As such, they exhibit all the characteristics of individuals, not of classes. For example, species can go extinct. Classes cannot. A class can temporarily have no individuals exemplifying it; that is all. For example, during the first few moments of the big bang, no heavy elements existed. Through time, elements with higher atomic numbers came into existence here and there in the universe. It is possible that at any one moment, gold could cease to exist. Then, later, additional atoms of gold could reemerge.

Such occurrences pose no problems for physicists, but what about the emergence, extinction, and reemergence of biological species? Could not dinosaurs reevolve? For species as historical entities, continuity through time is required. Once a taxon as a monophyletic unit goes extinct, it cannot reevolve; once extinct, no taxon can come into existence again. This claim depends not on empirical contingencies but on the individuation of species in terms of descent. For an atom to count as a gold atom, no historical connections are required. For evolving lineages, such connections are necessary. Descent is required because natural selection requires descent. Natural selection is not the only mechanism involved in the evolution of species, but it is certainly the main mechanism.

Reinterpreting species as historical entities has numerous important implications for an understanding of the evolutionary process. Critics of present-day biology have made much of the contention that biology, unlike physics, has no laws. As an example, they cite the millions upon millions of claims made by biologists about particular species. It is often heard that all swans are white, all ravens are black, and human beings are rational animals. All these claims inevitably have exceptions because evolution proceeds by means of variations, and these variations cannot be explained away as monsters. Variability is essential to the evolutionary process. However, if species are individuals, this is exactly as they should be. 'All swans are white' is no more a candidate for a law of nature than 'the earth is the third planet from the sun.' If biology includes any genuine laws of nature, they must be found elsewhere.

Criteria for Individuality

Outside biology, the criteria for individuality are frequently quite strong—so strong that no biological entities can meet these standards. For example:

(a) *Same substance.* Individuals must retain the same substance throughout their existence. Nothing in the living world fulfills this requirement. For example, organisms such as people exchange their substance many times over during the course of their existence. Even nerve cells are lost, gained, and reconfigured, albeit quite slowly.

(b) *Same form.* Individuals must retain the same form throughout their existence. Individuating forms is far from easy, but no matter how one deals with this knotty issue, many organisms undergo dramatic metamorphosis during the course of their development. A caterpillar and the butterfly into which it develops do not share the same form. The choice is either to consider these two as stages in the life cycle of a single organism or to consider the caterpillar a separate organism from the butterfly it produces. The problem is only magnified when a single organism produces numerous individuals at a later stage in its development.

Perhaps some individuals in physics are composed of the same substance and exhibit the same form throughout their existence, but nothing fulfills these requirements in biology. Two additional criteria do apply equally inside and outside of biology:

(c) *Spatiotemporal localization (boundedness).* Individuals have locations as well as beginnings and endings in space and time. These beginnings and endings can be sharp or fuzzy. Individuals can cease to exist terminally, but they also can cease to exist by becoming another individual. That is what happens when one cell splits into two cells or one species splits into two species.

(d) *Spatiotemporal continuity.* In criteria (a) and (b) above, no change whatsoever can take place in substance or essential form. However, all sorts of things change in the natural world. If the entity involved is to count as the same individual throughout, the change must be continuous, even if it is not gradual. An organism, when it undergoes metamorphosis, may proceed from stage to stage quite abruptly, but it remains the same organism because the change is continuous. As a result, numerically, the same individual cannot come into existence more than once.

(e) *Spatiotemporal localization and continuity* define what is commonly called a "historical entity" (Wiggins 1967). Such entities can be found both in and outside biology. For example, organisms are clear cases of historical entities, but so are planets and stars. As long as the nine solar planets keep revolving around the sun, each will remain the same historical entity. However, if a very large comet were to smash into Pluto, that event might well be the end of both of them.

Biologists are not content to limit themselves to historical entities of the more generic sort. Instead, they add more criteria, which characterize few if any individuals outside biology. For example:

(f) *Structural heterogeneity.* Biological individuals are structurally heterogeneous. Outside biology, certain entities approach structural homogeneity (e.g., the proverbial billiard ball), but even the simplest biological entity, such as a single codon, exhibits structural heterogeneity. However, the most important sort of structural heterogeneity is that which plays a role in functional organization.

(g) *Functional organization.* This facility is limited to human artifacts and evolved biological organisms. Although there is considerable latitude for fluctuations in functionally organized systems, eventually this organization can be destroyed. For highly organized systems, just about any modification results in loss of function. In others, especially those that exhibit modular organization, considerable disruption can be tolerated (Buss 1987). For example, if a lobster were torn into half a dozen pieces and the pieces thrown back into the ocean, all of these pieces would die. A lobster can regenerate a leg or two, but little more. However, if the same thing happened to a starfish, one would be likely to get six new starfish.

(h) *Genetic homogeneity.* Just as biological individuals are quite heterogeneous internally, they are quite homogeneous genetically. In order for multicellular organisms to develop, some way had to be found to allow cells to cooperate. The roadblocks that hinder entities with different genetic constitutions from cooperating are well known, especially as a result of Richard Dawkins's famous book *The Selfish Gene* (1976). One way around this roadblock is to have all the genes in a multicellular organism genetically identical.

The criterion of genetic homogeneity prevents entities more inclusive than single organisms from functioning as units in the evolutionary process. A beehive can fulfill all the criteria listed above save for (a) and (b), sameness of substance and form, but not all the individual bees living in the same hive contain the same genes. In particular, the drones that succeed in mating with the queen differ from her and from the workers. The entire hive might succeed in functioning as a unit of selection, but not one that is as efficient as it should be.

Kinds of Biological Individuals

Given these considerations, Wilson (1999, 60) distinguishes four sorts of biological historical entities:

1. *Functional individuals* are entities whose heterogeneous parts are currently so causally integrated that they tend to return the individual to the same state in the face of appreciable though not unlimited alterations.
2. *Developmental individuals* are entities that are programmed to develop through time. Whereas functional individuals return to a preferred state, developmental individuals proceed from state to state.
3. *Genetic individuals* are entities that possess the same genetic makeup derived from descent from a common ancestor. For example, all the descendants of a single zygote count as genetic individuals.
4. *Units of evolution* are entities that play important roles in evolutionary change, for example, units of replication and environmental interaction (Brandon and Burian 1984).

As a result of this analysis of biological individuality, certain "individuals" do not accord with common sense notions of individuality. For example, a clone counts as a single genetic individual even though it may be made up of hundreds of distinct developmental individuals. In such situations, growth is indistinguishable from reproduction. If the clone is considered a single individual, then the production of additional individuals counts as growth. If, however, the clone is thought of as being made up of independent organisms, then the very same changes count as asexual reproduction (see, e.g., Jackson, Buss, and Cook 1986).

This entry has concentrated on only one half of the polar concepts of 'individual' and 'class.' Several fascinating issues concerning the relation between individuals and their kinds have been ignored (Lowe 1989; Wiggins 1967). For example, can an individual remain the same individual when it changes from one kind to another? Some philosophers argue that an entity's ability to change from one kind to another proves that these were not genuine kinds in the first place. For example, in certain species of fish, a female can change into a male with only minor alterations. It would seem rather strange to consider this fish two distinct individuals in the process. Hence, it would seem that male and female are not genuine kinds, but they play this role in several areas of biology. Here is but one more instance in which real examples can make a difference.

DAVID L. HULL

References

Brandon, R., and R. Burian (eds.) (1984), *Genes, Organisms, Populations: Controversies over the Units of Selection*. Cambridge, MA: MIT Press.
Buss, L. (1987), *The Evolution of Individuality*. Princeton, NJ: Princeton University Press.
Dawkins, R. (1976), *The Selfish Gene*. Oxford: Oxford University Press.
Ghiselin, M. T. (1974), "A Radical Solution to the Species Problem," *Systematic Zoology* 23, 536–544.
Gould, S. J. (1992), "A Humungous Fungus Among Us," *Natural History*, July, 10–16.
Hull, D. L. (1976), "Are Species Really Individuals?" *Systematic Zoology* 25, 174–191.
———. (1992), "Individual," in E. Fox Keller and E. A. Lloyd (eds.), *Keywords in Evolutionary Biology*. Cambridge, MA: Harvard University Press, 180–187.
Jackson, Jeremy B. C., Leo W. Buss, and Robert E. Cook (eds.) (1986), *Population Biology and Evolution of Clonal Organisms*. New Haven, CT: Yale University Press.
Keller, L. (ed.) (1999), *Levels of Selection in Evolution*. Princeton, NJ: Princeton University Press.
Lowe, E. L. (1989), *Kinds of Being: A Study of Individuation, Identity, and the Logic of Sortal Terms*. Oxford: Basil Blackwell.
Michod, R. E. (1999), *Darwinian Dynamics: Evolutionary Transitions in Fitness and Individuality*. Princeton, NJ: Princeton University Press.
Wiggins, D. (1967), *Identity and Spatio-Temporal Continuity*. Oxford: Oxford University Press.
Wilson, J. (1999), *Biological Individuality: The Identity and Persistence of Living Entities*. Cambridge: Cambridge University Press.

See also **Biological Information; Evolution; Natural Selection; Species**

PROBLEM OF INDUCTION

The Scottish philosopher David Hume (1711–1776) first focused attention on the question of what the grounds are for believing that the future will resemble the past or, more generally, that what has not been observed will resemble what has (Hume 1965 and 1999). Inference in which one takes the past as grounds for beliefs about the future, or the observed as grounds for beliefs about the unobserved, or in general an inference that is ampliative—having more content in the conclusion than in the premises—has come to be called *induction*. ("Induction" is also sometimes used in a more specific sense to refer to induction by enumeration, the inference in which one simply generalizes from instances to all cases or to the next case.) In addition, the powerful arguments that led Hume to suppose that there was no rational ground whatsoever for inference from the observed to the unobserved are difficult to answer. Hume's question has thus come to be called the *problem* of *induction*.

Hume divided all reasoning into two mutually exclusive types, that concerning relations between ideas and that concerning matters of fact and existence. All mathematical and logical reasoning fell into the former category—also called "demonstration," what would now be called "deduction"—and was regarded as unproblematic, though also nonampliative. Reasoning about relations between ideas could not suffice for natural science, which frequently makes inferences from the observed to the unobserved in making predictions, retrodictions, and generalizations, which count as Hume's second sort of reasoning. Hume asks what these ampliative inferences are based on, and he concludes that all of them are founded on beliefs about relations of cause and effect. The question of how to justify ampliative inferences thereby becomes the question of how to justify judgments about cause and effect.

One thread of Hume's further argument depends to some extent on his empiricism and his psychological views. He is committed to the views that all ampliative knowledge comes from experience and that experience is composed entirely of impressions. Thus when he asks what one can know of cause and effect, he asks what one experiences of it and answers with what one can have an impression of. He considers one billiard ball hitting another and points out that while people seem to think that there are three things here—cause, effect, and necessary connection between the two—one has impressions only of the first two, the first ball hitting the second, and the second moving. Thus, the only possible way of knowing that this effect must follow the cause, as opposed to merely that it happens to have done so in the past, gives no grounds for believing in this necessity or connection.

Hume's main argumentation does not depend on special assumptions about causation or the psychology of experience, and this is the version of his argument that has received the most attention from philosophers of science subsequently. He claims first that the inference from the collision of the first ball with the second to the belief that this will be followed by the second ball moving cannot be made by demonstration. This is because, in a demonstration, supposing the premises true and the conclusion false is always a contradiction, but there is no contradiction in supposing that after the first billiard ball hits the second, the second ball rises one foot and levitates. There is no contradiction, that is, in supposing that the second ball will not do what it usually does. By assumption, the only other form of reasoning by which the usual inference that the second billiard ball will move could be justified is the very reasoning about matters of fact that Hume was seeking a justification of. To appeal to such reasoning at this juncture would be circular. Hume concludes that the expectation that the future will resemble the past has no rational justification and is based only on instinct and custom.

To be sure, the inference to the conclusion that the second billiard ball will move in the usual way on being hit by the first could be understood as a deduction with a suppressed premise, also known as an *enthymeme*. The suppressed premise on which one's confidence depends in this view would be a claim that nature is uniform—a "uniformity of nature" assumption—and that regularities that have been observed to hold up to now will continue to hold. However, this strategy is ineffectual unless one can justify the premise that nature is uniform. One cannot show by demonstration from the claim that what has been seen of nature is uniform that nature's as yet unseen parts are uniform, because denying the latter does not contradict the former. However, one would have to know the latter to know that nature is uniform. One cannot in this context appeal to reasoning about matters of fact to defend the claim that all of nature is uniform, and not only the parts already seen. That would be circular because the assumption about the uniformity of nature was supposed to justify all of our reasoning about matters of fact. This strategy for answering Hume's question is thus best understood as an alternative way of presenting Hume's problem.

The uniformity-of-nature assumption presented has further problems. One is that not all regularities that have been observed to hold up to now will continue to hold. One does not expect regularities that are regarded as coincidences to continue, and one recognizes at once the folly of a person who jumps out of a window on the fortieth floor and when passing the tenth floor says, "So far so good." The uniformity-of-nature assumption can be made weaker, to say that everything that happens is an instance of some exceptionless general law, a claim that resembles Kant's response to Hume about causation. It may be that such a claim could be argued for on a priori grounds, but since this principle would provide no basis for identifying which events are invariably followed by which others, it also would provide no basis for distinguishing sound from unsound inductions. If the uniformity-of-nature assumption were, on the other hand, so specific as to identify the regularities that will continue, it would gain content at the expense of being difficult to defend without circularity.

Twentieth-Century Responses

There are three popular ways of responding to Hume's question by rejecting the problem. The first, the explicit formulation of which is due to Strawson, though it was also suggested by the Ayer (1952), says that attention to the meaning of 'reasonableness' shows that the supposed problem is merely a misunderstanding (Strawson 1952) (see Ayer, Alfred Jules). Induction is part of the standards for reasonableness, so the question why *it* is reasonable does not, strictly speaking, make sense. To ask why it is reasonable to make inductive inferences is to ask why it is reasonable to be reasonable, a question to which one should not expect an answer. However, the problem is not so easily dissolved. A reasonable reply to this line of argument is that the problem of induction is not generated by asking whether induction is acceptable according to itself, but rather by asking whether the inductive standards of reasonableness that are in fact employed are likely to serve the end of making true predictions as often as possible, and what grounds one has for thinking so. Thus, one is not asking why it is reasonable to be reasonable, but why it is reasonable to think that these standards serve (one of) their purpose(s) of making true predictions.

A second way of rejecting the problem is to protest that the claim that justifying induction is a problem rests on a mistaken demand for a guarantee where, as Hume has shown, a guarantee is logically impossible. This response is commonly invoked to say that it is only by holding the bar too high that one can think there is a problem of justification about induction; it is only by expecting induction to be deduction, to yield certainty, that one sees a problem when it is not and does not. However, this line involves an erroneous understanding of what Hume's arguments purport to show—for, it is not argued merely that there is no deductive (demonstrative) guarantee that future events will behave like past events. It is argued rather persuasively that there can be no rational ground whatsoever. That is, one would be happy if one could show even that a great number of occasions on which A is associated with B make it more *likely* than it would have been without those instances, that A will be associated with B in the future (Russell 1959). Hume argues that there can be none but a question-begging justification of this probability claim.

A third complaint about Hume's problem says that he did not concern himself with the kinds of complex inferences scientists actually engage in, and skeptical conclusions about the simple induction by enumeration that he had in mind are thus irrelevant to the justification of scientific claims, whatever other significance they may have. However, whether or not Hume had enumerative induction explicitly in mind, the applicability of his main argumentation is by no means restricted to that. Hume's main argument applies to any ampliative inference, which the complex scientific inferences referred to certainly involve. For any ampliative inference, it will be the case that no demonstration can secure it, for a conclusion that goes beyond the premises in content could well be different in the content that goes beyond the premises without contradicting the premises. And ampliative inference cannot be used to justify ampliative inferences because that would be relying on the very sort of inference that is in need of justification. Appeal to the complexity of real science does not erase Hume's problem.

In the 1950s, a popular response to Hume's problem was to question easy acceptance of the claim that induction could not legitimately justify itself. Black (1954) pointed out that if one appeals to the past success of induction to justify belief that induction will be successful in a new instance, the conclusion that induction will be successful in the new instance does not appear as a premise. Such an argument thus does not commit what is called "premise circularity." However, as Salmon (1966) pointed out, the argument does commit rule circularity, since appeal is made to the rule of induction in making the transition from the premises to the conclusion that recommends the rule of induction. Rule circularity has consequences as untoward as does premise circularity in that there are rules that are patently invalid or crazy that can justify themselves in the same way. Consider the following argument:

> If affirming the consequent is valid, then grass is green.
> Grass is green.
> Therefore, affirming the consequent is valid.

The argument itself proceeds by affirming the consequent, and in so doing leads to the conclusion that affirming the consequent is a legitimate procedure. It is known independently, however, that affirming the consequent is an invalid procedure that can lead from true premises to false conclusion.

In an example closer to the present topic, if one believes that induction is reasonable, one will not believe in the legitimacy of counterinduction, in which one takes past negative instances to be a good indication of future positive instances. However, counterinduction can recommend itself in the same way that affirming the consequent did:

> Counterinduction has usually been unsuccessful in the past.
> Therefore, counterinduction is likely to be successful in the next instance.

The premise is presumably true, even according to an inductivist. The conclusion recommends the rule of counterinduction. The conclusion is inferred from the premise by that same rule of counterinduction, so counterinduction is capable of the same kind of rule-circular defense of itself that was lately recommended for induction. Showing that induction can defend itself in a rule-circular though not premise-circular fashion cannot assuage worries if a rule regarded as illegitimate can do the same (Salmon 1966; Skyrms 2000).

Feigl's distinction between two types of justification, validation and vindication, helps to clarify the so-called 'pragmatic' attempt to justify inductive behavior (see Feigl, Herbert). A principle is validated when it is derived from other, more basic, principles that one accepts, as when a theorem is derived from geometric axioms. A principle is vindicated, on the other hand, when it is shown that following its rule serves the purpose for which that rule was designed (Feigl 1950). Incidentally, this distinction shows that even if inductive rules are basic, not derivable from other principles, as Strawson suggested, that alone does not excuse one from the task of justifying them. They may still require vindication.

The pragmatic justification of inductive behavior was developed by Reichenbach (1949, 469–482) (see Reichenbach, Hans). In this strategy one does not try to show that induction is likely to succeed, but only that it is likely to succeed *if any method will*. In Reichenbach's precise treatment, the rule of induction is to infer from the fact that the frequency of As among Bs in a large sample is m/n to the claim that the limit of the relative frequency of As among Bs in an infinite number of trials is m/n. This limit either exists or it does not. It must exist if the inductive procedure is to be successful, but according to Reichenbach it need not be known whether it exists when it is asked whether the inductive procedure is justified. The inductive procedure is justified as an attempt to find the limit. There is no known sufficient condition for finding the limit of the relative frequency. However, Reichenbach argued that the attainability of the limit by means of the rule of induction is a necessary condition for the existence of the desired limit. If that limit exists, then it follows analytically that the rule of induction will yield the limit at some point, for a set degree of approximation. If the limit does not exist, then there is no probability for any method to ascertain. More vividly, if the clairvoyant's predictions will identify the limit of the relative frequency, then that limit exists. If the limit exists, the rule of induction will also identify it. Hence, one cannot do better at identifying the limit than by employing the rule of induction.

The main defect in this strategy is that it does not so far defend the standard rule of induction as uniquely qualified for the specified role. The familiar "straight rule" of induction, described above, is one of an infinite number of inductive rules that satisfy the demand of giving the limit of the relative frequency if that limit exists, a class that

Reichenbach called the "asymptotic" rules. The existence of an infinite number of such rules would not be a problem if the limits they yielded in a given case were similar, but in fact they vary arbitrarily widely from each other. For a finite sample, the class of asymptotic rules tolerates any identification of the limit of the relative frequency (Salmon 1966). However, adding to asymptoticity further natural criteria, such as speed-optimality, does narrow the class of acceptable rules (Juhl 1994).

Popper accepted Hume's skeptical conclusion but thought he solved the problem Hume created by declaring that people never use induction anyway (Popper 1972). He developed a deductivist view of science according to which all scientific inference takes the form of falsification; and positive, ampliative inferences asserting the truth or probability of theories, or claims about their future performance, are never made (see Popper, Karl Raimund). When scientists appear to take confirming instances as offering positive evidence for their theories, what they are really doing is finding *corroboration*, a term Popper used for the summation of the theory's past performance under test (see Corroboration). A theory is more highly corroborated the more, and more severe, the tests it has passed that it might have failed, but corroboration never gives grounds for positive conclusions about the theory. Corroboration is thus similar to but crucially distinct from confirmation. One problem with this line of argument is the difficulty even Popper sometimes had eschewing the disallowed positive claims for theories while maintaining a plausible view of scientists' behavior and motivation. Another problem is that, despite its association with deduction, falsification can be seen to be no more decisive than verification, since falsification must assume background claims themselves in need of defense (see Duhem Thesis). Finally, defining corroboration in such a way that theories can be compared with each other when neither is a logical consequence of the other has proved troublesome. One problem is that Popper never adequately defined the notion of severity for tests, a concept on which much depended, since the more severe the test a theory passed, the better its corroboration. Mayo (1996) has recently shed light on this topic by developing the insights inherent in error statistics.

Confirmation Theory

At some point in the early to mid-twentieth century, it was recognized that however the problem of justifying induction turned out, there was much descriptive work to do in order to characterize what exactly the rules of inference in question were. The most basic rules of deduction have been known since the time of Aristotle, but the same degree of clarity had not been achieved for the other major branch of reasoning, *viz.*, induction. (Arguably, it still has not, perhaps testifying to the difficulty of the task.) Perhaps the reason there were few complaints about the fact that the rules of deduction cannot be given a noncircular justification was that those rules were so clear. Work on the description of induction, known as "confirmation theory," might shed light on, if not entirely transform, the justificatory question (see Confirmation Theory). Inductive logic is a similarly descriptive enterprise but is based on the calculus of probability (see Inductive Logic).

However, confirmation theory encountered several paradoxes on its way, including the Raven paradox discovered by Hempel (Hempel 1937, 1943, and 1945; Earman and Salmon 1992) (see Hempel, Carl Gustav). It seems reasonable to assume that any instance lends some confirmation to its generalization, and also that, if a given piece of evidence lends confirmation to a hypothesis, it also lends confirmation to every statement logically equivalent to that hypothesis. However, the statement that all non-black things are non-ravens is logically equivalent to the statement that all ravens are black. Yet a white tennis shoe, which is an instance of the former statement, supports the hypothesis that all ravens are black only on pain of elevating indoor ornithology to the status of a science. More generally, three conditions are sufficient for this paradox. The first is the instantiation condition (also known as "Nicod's condition," after Jean Nicod), by which any instance that is both P and Q provides confirmation for the hypothesis $(x)(Px \rightarrow Qx)$, and any instance that is P and not Q provides disconfirmation of that hypothesis. Another is the equivalence condition, by which an instance that provides confirmation for a hypothesis, H, provides confirmation for any statement logically equivalent to H. The third, the irrelevance condition, says that for any hypothesis, H, there are some instances that provide neither confirming nor disconfirming evidence for H. Orthodox statistics, and some Bayesian approaches, resolve the paradox by denying the instantiation condition (Giere 1970). The most well known Bayesian approach involves denying the irrelevance condition but arguing that the confirmation that instances like the white tennis shoe provide is of negligible size (Howson and Urbach 1996, 126–130) (see Bayesianism).

The paradox of confirmation discovered by Goodman, which introduced what is called the "new riddle of induction," may present a more serious problem, since it seems to show that a purely syntactical confirmation theory is impossible (Goodman 1983; Stalker 1994). Let the predicate "grue" be true of an emerald just in case it is green and observed before January 2100 or not observed before that new year and blue. Why should this not be the predicate one projects on the basis of all of the green emeralds one has seen? Syntactically speaking, the generalization that all emeralds are grue is instantiated by, and so confirmed if anything is by, all the same instances that confirm the generalization that all emeralds are green, since those are also instances in which the emeralds are grue. Indeed, the generalization that all emeralds are grue is apparently confirmed in just the same way, to just the same degree, as the generalization that all emeralds are green. Syntactically, Goodman argues, the two predicates are symmetrical. Common sense says it surely cannot be that the two are equally confirmed by emeralds that have been observed, yet what reason can be given to project the predicate "green" rather than the predicate grue? Nothing in syntax alone, Goodman submitted, can show why the predicate grue should be shunned in favor of the predicate green.

Goodman's answer to this problem was in the tradition of Hume when he developed the notion that projectible predicates can be distinguished from nonprojectible predicates on the basis of the former's superior entrenchment, that is, on the fact that the former predicates have actually been projected more in the past. "Regularities are where you find them, and you can find them anywhere," wrote Goodman, who focused attention on the fact that Hume did not emphasize—which was touched on above with uniformity-of-nature assumptions—that there are many repeated instances that are *not* expected to continue. Goodman's view that what distinguishes projectible from nonprojectible predicates is the former's entrenchment is a descriptive account, as he intended it to be. Goodman was convinced that he had dissolved the traditional problem about the justification of induction in favor of descriptive questions and answers. However, this may not be persuasive, since one can ask about entrenched predicates by what right one's ancestors chose them over other possibilities, and of oneself what reason one has for continuing that tradition.

SHERRILYN ROUSH

References

Ayer, Alfred Jules (1952), *Language, Truth and Logic*. New York: Dover, 49–50.
Black, Max (1954), "The Inductive Support of Inductive Rules," in *Problems of Analysis*. Ithaca, NY: Cornell University Press, 191–208.
Earman, John, and Wesley C. Salmon (1992), "The Confirmation of Scientific Theories," in Merrilee Salmon et al. (eds.), *Introduction to the Philosophy of Science*. Indianapolis: Hackett Publishing Company.
Feigl, Herbert (1950), "De principiis non disputandum," in Max Black (ed.), *Philosophical Analysis*. Ithaca, NY: Cornell University Press.
Giere, Ronald N. (1970), "An Orthodox Statistical Resolution to the Paradox of Confirmation," *Philosophy of Science* 37: 354–362.
Goodman, Nelson (1983), *Fact, Fiction, and Forecast* (4th ed.). Cambridge, MA: Harvard University Press, 59–124.
Hempel, Carl G. (1937), "Le Problème de la Vérité," *Theoria* (Goteberg), vol. 3.
——— (1943), "A Purely Syntactical Definition of Confirmation," *Journal of Symbolic Logic*, vol. 8.
——— (1945), "Studies in the Logic of Confirmation," *Mind* 54: 1–26, 97–121.
——— (1981), "Turns in the Evolution of the Problem of Induction," *Synthese* 46: 389–404.
Howson, Colin (2000), *Hume's Problem: Induction and the Justification of Belief*. Oxford: Oxford University Press.
Howson, Colin, and Peter Urbach (1996), *Scientific Reasoning: The Bayesian Approach* (2nd ed.). Chicago: Open Court.
Hume, David (1965), *Treatise of Human Nature*. Edited by L. A. Selby-Bigge. Oxford: Oxford University Press.
——— (1999), *An Enquiry Concerning Human Understanding*. Edited by T. L. Beauchamp. Oxford: Oxford University Press, 108–130.
Juhl, Cory (1994), "The Speed-Optimality of Reichenbach's Straight Rule of Induction," *British Journal for the Philosophy of Science* 45: 857–863.
Mayo, Deborah G. (1996), *Error and the Growth of Experimental Knowledge*. Chicago: University of Chicago Press.
Popper, Karl R. (1972), "Conjectural Knowledge: My Solution of the Problem of Induction," in *Objective Knowledge*. Oxford: Clarendon Press, 1–31.
Reichenbach, Hans (1949), "The Justification of Induction," in *The Theory of Probability*, Section 91. Berkeley and Los Angeles: University of California Press.
Russell, Bertrand (1959), "On Induction," in *The Problems of Philosophy*. New York: Oxford University Press, 60–69.
Salmon, Wesley C. (1966), *The Foundations of Scientific Inference*. Pittsburgh, PA: University of Pittsburgh Press.
Skyrms, Brian (2000), *Choice and Chance: An Introduction to Inductive Logic*. Stamford, CT: Wadsworth, 2000.
Stalker, Douglas (ed.) (1994), *Grue! The new riddle of induction*. Chicago: Open Court.
Strawson, P. F. (1952), "Inductive Reasoning and Support," in *Introduction to Logical Theory*. London: Methuen, 233–263.

See also **Causation; Confirmation Theory; Inductive Logic; Laws of Nature**

INDUCTIVE LOGIC

The idea of inductive logic as providing a general, quantitative way of evaluating arguments is a relatively modern one. Aristotle's conception of 'induction' (επαγωγή)—which he contrasted with 'reasoning' (συλλογισμός)—involved moving only from particulars to universals (Kneale and Kneale 1962, 36). This rather narrow way of thinking about inductive reasoning seems to have held sway through the Middle Ages and into the seventeenth century, when Francis Bacon (1620) developed an elaborate account of such reasoning. During the eighteenth and nineteenth centuries, the scope of thinking about induction began to broaden considerably with the description of more sophisticated inductive techniques (e.g., those of Mill [1843]), and with precise mathematical accounts of the notion of probability. Intuitive and quasi-mathematical notions of probability had long been used to codify various aspects of uncertain reasoning in the contexts of games of chance and statistical inference (see Stigler 1986 and Dale 1999), but a more abstract and formal approach to probability theory would be necessary to formulate the general modern inductive-logical theories of nondemonstrative inference. In particular, the pioneering work in probability theory by Bayes (1764), Laplace (1812), Boole (1854), and many others in the eighteenth and nineteenth centuries laid the groundwork for a much more general framework for inductive reasoning. (Philosophical thinking about the possibility of inductive knowledge was most famously articulated by David Hume 1739–1740 and 1758) (See Problem of Induction).

The contemporary idea of inductive logic (as a general, logical theory of argument evaluation) did not begin to appear in a mature form until the late nineteenth and early twentieth centuries. Some of the most eloquent articulations of the basic ideas behind inductive logic in this modern sense appear in John Maynard Keynes's *Treatise on Probability*. Keynes (1921, 8) describes a "logical relation between two sets of propositions in cases where it is not possible to argue demonstratively from one to another." Nearly thirty years later, Rudolf Carnap (1950) published his encyclopedic work *Logical Foundations of Probability*, in which he very clearly explicates the idea of an inductive-logical relation called "confirmation," which is a quantitative generalization of deductive entailment (See Carnap, Rudolf; Confirmation Theory).

Carnap (1950) gives some insight into the modern project of inductive logic and its relation to classical deductive logic:

> Deductive logic may be regarded as the theory of the relation of logical consequence, and inductive logic as the theory of another concept ["c"] which is likewise objective and logical, viz., ... degree of confirmation. (43)

More precisely, the following three fundamental tenets have been accepted by the vast majority of proponents as desiderata of modern inductive logic:

1. Inductive logic should provide a quantitative generalization of (classical) deductive logic. That is, the relations of deductive entailment and deductive refutation should be captured as limiting (extreme) cases with cases of "partial entailment" and "partial refutation" lying somewhere on a continuum (or range) between these extremes.
2. Inductive logic should use probability (in its modern sense) as its central conceptual building block.
3. Inductive logic (i.e., the nondeductive relations between propositions that are characterized by inductive logic) should be objective and logical.

(Skyrms 2000, chap. 2, provides a contemporary overview.) In other words, the aim of inductive logic is to characterize a quantitative relation (of inductive strength or confirmation), c, which satisfies desiderata 1–3 above. The first two of these desiderata are relatively clear (or will quickly become clear below). The third is less clear. What does it mean for the quantitative relation c to be objective and logical? Carnap (1950) explains his understanding as follows:

> That c is an objective concept means this: if a certain c value holds for a certain hypothesis with respect to a certain evidence, then this value is entirely independent of what any person may happen to think about these sentences, just as the relation of logical consequence is independent in this respect. [43] ... The principal common characteristic of the statements in both fields

[deductive and inductive logic] is their independence of the contingency of facts [of nature]. This characteristic justifies the application of the common term 'logic' to both fields. [200]

This entry will examine a few of the prevailing modern theories of inductive logic and discuss how they fare with respect to these three central desiderata. The meaning and significance of these desiderata will be clarified and the received view about inductive logic critically evaluated.

Some Basic Terminology and Machinery for Inductive Logic

It is often said (e.g., in many contemporary introductory logic texts) that there are two kinds of argument: deductive and inductive, where the premises of deductive arguments are intended to guarantee the truth of their conclusions, while inductive arguments involve some risk of their conclusions being false even if all of their premises are true (see, e.g., Hurley 2003). It seems better to say that there is just one kind of argument: An argument is a set of propositions, one of which is the conclusion, the rest are premises. There are many ways of evaluating arguments. Deductive logic offers strict, qualitative standards of evaluation: the conclusion either follows from the premises or it does not, whereas inductive logic provides a finer-grained (and thereby more liberal) quantitative range of evaluation standards for arguments. One can also define comparative and/or qualitative notions of inductive support or confirmation. Carnap (1950, §8) and Hempel (1945) both provide penetrating discussions of the contrast between quantitative and comparative/qualitative notions. For simplicity, the focus here will be on quantitative approaches to inductive logic, but most of the main issues and arguments discussed below can be recast in comparative or qualitative terms.

Let $\{P_1, \ldots, P_n\}$ be a finite set of propositions constituting the premises of an (arbitrary) argument, and let C be its conclusion. Deductive logic aims to explicate the concept of *validity* (i.e., deductive-logical goodness) of arguments. Inductive logic aims to explicate a quantitative generalization of this deductive concept. This generalization is often called the "inductive strength" of an argument (Carnap 1950 uses the word "confirmation" here). Following Carnap, the notation $c(C, \{P_1, \ldots, P_n\})$ will denote the degree to which $\{P_1, \ldots, P_n\}$ jointly inductively support (or "confirm") C.

As desideratum 2 indicates, the concept of probability is central to the modern project of inductive logic. The notation $P(\bullet)$ and $P(\bullet|\bullet)$ will denote unconditional and conditional probability functions, respectively. Informally (and roughly), "$P(p)$" can be read "the probability that proposition p is true," and "$P(p|q)$" can be read "the probability that proposition p is true, given that proposition q is true." The nature of probability functions and their relation to the project of inductive logic will be a central theme in what follows.

A Naive Version of Basic Inductive Logic and the Received View

According to classical deductive propositional logic, the argument from $\{P_1, \ldots, P_n\}$ to C is *valid* iff ("if and only if") the material conditional $(P_1 \wedge, \ldots, \wedge P_n) \to C$ is (logically) necessarily true. Naively, one might try to define "inductively strong" as follows: The argument from $\{P_1, \ldots, P_n\}$ to C is *inductively strong* iff the material conditional $(P_1 \wedge \ldots \wedge P_n) \to C$ is (logically?) probably true. More formally, one can express this naive inductive logic (NIL) proposal as follows:

$$c(C, \{P_1, \ldots, P_n\}) \text{ is high iff}$$
$$P((P_1 \wedge, \ldots, \wedge P_n) \to C) \text{ is high.}$$

There are problems with this first, naive attempt to use probability to generalize deductive validity quantitatively. As Skyrms (2000, 19–22) points out, there are (intuitively) cases in which the material conditional $(P_1 \wedge, \ldots, \wedge P_n) \to C$ is probable but the argument from $\{P_1, \ldots, P_n\}$ to C is not a strong one. Skyrms (21) gives the following example:

(**P**) There is a man in Cleveland who is 1,999 years and 11 months old and in good health.
(**C**) No man will live to be 2,000 years old.

Skyrms argues that $P(\mathbf{P} \to \mathbf{C})$ is high, simply because $P(\mathbf{C})$ is high and not because there is any evidential relation between **P** and **C**. Indeed, intuitively, the argument from (**P**) to (**C**) is not strong, since (**P**) seems to disconfirm or countersupport (**C**). Thus, $P((P_1 \wedge \ldots \wedge P_n) \to C)$ being high is not sufficient for $c(C, \{P_1, \ldots, P_n\})$ being high. Note also that $P((P_1 \wedge \ldots \wedge P_n) \to C)$ cannot serve as $c(C, \{P_1, \ldots, P_n\})$, since it violates desideratum 1. If $\{P_1, \ldots, P_n\}$ refutes C, then $Pr((P_1 \wedge \ldots \wedge P_n) \to C) = Pr(\neg(P_1 \wedge \ldots \wedge P_n))$, which is not minimal, since the conjunction of the premises of an argument need not have probability one.

Skyrms suggests that the mistake that NIL makes is one of conflating the probability of the material conditional $Pr((P_1 \wedge \ldots \wedge P_n) \to C)$ with the conditional probability of C, given $P_1 \wedge \ldots \wedge P_n$, that is, $P(C|P_1 \wedge \ldots \wedge P_n)$.. According to Skyrms, it is the latter that should be used as a definition of $c(C, \{P_1, \ldots, P_n\})$. The reason for this preference is that $P((P_1 \wedge \ldots \wedge P_n) \to C)$ fails to capture the evidential relation between the premises and conclusion, since $P((P_1 \wedge \ldots \wedge P_n) \to C)$ can be high solely in virtue of the unconditional probability of (C) being high or solely in virtue of the unconditional probability of $P_1 \wedge \ldots \wedge P_n$ being low. As Skyrms (20) stresses, $c(C, \{P_1, \ldots, P_n\})$ should measure the "evidential relation between the premises and the conclusion." This leads Skyrms (and many others) to defend the following account, which might be called the received view (RV) about inductive logic:

$$c(C, \{P_1, \ldots, P_n\}) = Pr(C|P_1 \wedge \ldots \wedge P_n).$$

The idea that $c(C, \{P_1, \ldots, P_n\})$ should be identified with the conditional probability of C, given $P_1 \wedge \ldots \wedge P_n$, has been nearly universally accepted by inductive logicians since the inception of the contemporary discipline. Recent pedagogical advocates of the RV include Copi and Cohen (2001), Hurley (2003), and Layman (2002); and historical champions of various versions of the RV include Keynes (1921), Carnap (1950), Kyburg (1970), and Skyrms (2000), among many others. There are nevertheless some compelling reasons to doubt the correctness of the RV. These reasons, which are analogous to Skyrms's reasons for rejecting the NIL, will be discussed below. But before one can adequately assess the merits of the NIL, RV, and other proposals concerning inductive logic, one needs to say more about probability models and their relation to inductive logic (see Probability).

Probability: Its Interpretation and Role in Traditional Inductive Logic

The Mathematical Theory of Probability

For present purposes, assume that a probability function $P(\bullet)$ is a finitely additive measure function over a Boolean algebra of propositions (or sentences in some formal language). That is, assume that $P(\bullet)$ is a function from a Boolean algebra B of propositions (or sentences) to the unit interval $[0,1]$ satisfying the following three axioms (this is Kolmogorov's (1950) axiomatization), for all propositions X and Y in B:

i. $P(X) \geq 0$.
ii. If X is a (logically) necessary truth, then $P(X) = 1$.
iii. If X and Y are mutually exclusive, then $P(X \vee Y) = Pr(X) + Pr(Y)$.

Following Kolmogorov, define conditional probability $P(\bullet|\bullet)$ in terms of unconditional probability $P(\bullet)$, as follows:

$$Pr(X|Y) = Pr(X \wedge Y)/Pr(Y),$$
provided that $Pr(Y) \neq 0$.

A probability model $M = \langle B, P_M \rangle$ consists of a Boolean algebra B of propositions (or sentences in some language), together with a particular probability function $P_M(\bullet)$ over the elements of B.

These axioms (and the definition of conditional probability) say what the mathematical properties of probability models are, but they do not say anything about the interpretation or application of such models. The latter issue is philosophically more central and controversial than the former (but see Popper 1992, appendix *iv, Roeper and Leblanc 1999, and Hájek 2003 for dissenting views on the formal theory of conditional probability). There are various ways in which one can interpret or understand probabilities (see Probability for a thorough discussion). The two interpretations that are most commonly encountered in the context of applications to inductive logic are the so-called "epistemic" and "logical" interpretations of probability.

Epistemic Interpretations of Probability

In epistemic interpretations of probability, $P_M(H)$ is (roughly) the degree of belief that an epistemically rational agent assigns to H, according to a probability model M of the agent's epistemic state. A rational agent's background knowledge K is assumed (in orthodox theories of epistemic probability) to be "included" in any epistemic probability model M, and therefore K is assumed to have an unconditional probability of 1 in M. $P_M(H|E)$ is the degree of belief an epistemically rational agent assigns to H upon learning that E is true (or on the supposition that E is true; see Joyce 1999, chap. 6, for discussion), according to a probability model M of the agent's epistemic state. According to standard theories of epistemic probability, agents learn by conditionalizing on evidence. So, roughly speaking, the probabilistic structure of a rational agent's epistemic state evolves (in time t) through a series of probability models $\{M_t\}$, where evidence learned at time t has probability 1 in all subsequent models $\{M_{t'}\}, t' > t$.

Keynes (1921) seems to be employing an epistemic interpretation of probability in his inductive logic when he says:

Let our premises consist of any set of propositions h, and our conclusion consist of any set of propositions a, then, if a knowledge of *h* justifies a rational degree of belief in *a* of degree *x*, we say that there is a *probability-relation* of degree *x* between *a* and *h* [$P(a|h) = x$]. (4)

It is not obvious that the RV can satisfy desideratum 3—that c be logical and objective—if the probability function *P* that is used to explicate c in the RV is given an epistemic interpretation of this kind. After all, whether "a knowledge of *h* justifies a rational degree of belief in *a* of degree *x*" seems to depend on what one's background knowledge *K* is. And while this is arguably an objective fact, it also seems to be a contingent fact and not something that can be determined a priori (on the basis of *a* and *h* alone). As Keynes (1921) explains, his probability function $P(a|h)$ is not subjective, since "once the facts are given which determine our knowledge [background and *h*], what is probable or improbable [*viz.*, *a*] in these circumstances has been fixed objectively, and is independent of our opinion" (4). But he later suggests that the function is contingent on what the agent's background knowledge *K* is, in the sense that $P(a|h)$ can vary "depending upon the knowledge to which it is related."

Carnap (1950, §45B) is keenly aware of this problem. He suggests that Keynes should have characterized $P(a|h)$ as the degree of belief in *a* that is justified by knowledge of *h*—*and nothing else* (the reader may want to ponder what it might mean for an agent to "*know h and nothing else*"). As Keynes's remarks suggest (and as Maher 1996 explains), the problem is even deeper than this, since even a complete specification of an agent's background knowledge *K* may not be sufficient to pick out a unique (rational) epistemic probability model *M* for an agent. (Keynes's reaction to this was to conclude that sometimes quantitative judgments of inductive strength or degree of conditional probability are not possible and that in these cases one must settle for qualitative or comparative judgments.) The problem here is that "$P(X|K)$" ("the probability of *X*, given background knowledge *K*") will not (in general) be determined unless an epistemic probability model *M* is specified, which (*a fortiori*) gives $Pr_M(X)$, for each *X* in *M*. And, without a determination of these fundamental or a priori probabilities $P_M(X)$, a general (quantitative) theory of inductive logic based on epistemic probabilities seems all but hopeless. This raises the problem of specifying an appropriate a priori probability model *M*. Keynes (1921, chap. 4) and Carnap (see below) both look to the principle of indifference at this point, as a guide to choosing a priori probability models. Before discussing the role of the principle of indifference, logical interpretations of probability require a brief discussion.

Logical Interpretations of Probability

Philosophers who accepted the RV and were concerned about the inductive-logical ramifications (mainly, regarding the satisfaction of desideratum 3) of interpreting probabilities epistemically began to formulate logical interpretations of probability. In such interpretations, conditional probabilities $P(X|Y)$ are themselves understood as quantitative generalizations of a logical entailment (or deducibility) relation between propositions *Y* and *X*. The motivation for this should be clear—it seems like the most direct way to guarantee that an RV-type theory of inductive logic will satisfy desideratum 3. If $P(\bullet|\bullet)$ is itself logical, then $c(\bullet,\bullet)$, which is defined by the RV as $P(\bullet|\bullet)$, should also be logical, and the satisfaction of desideratum 3 (as well as the other two) seems automatic. Below it will become clear that RV + logical probability is not the only way (and not necessarily the best way) to satisfy the three desiderata for providing an adequate account of the logical relation of inductive support. In preparation, the notion of logical probability must be examined in some detail.

Typically, logical interpretations of probability attempt to define $Pr(q|p)$, where *p* and *q* are sentences in some formal first-order language *L*, in terms of the syntactical features of *p* and *q* (in *L*). The most famous logical interpretations of probability are those of Carnap. It is interesting to note that Carnap's (1950 and 1952) systems are almost identical to those described 20–30 years earlier by W. E. Johnson (1921 and 1932) (Paris 1994; Kyburg 1970, Ch. 5). His later work (Carnap 1971 and 1980) became increasingly complicated, involving two-dimensional continua, and was less tightly coupled with the syntax of *L* (Maher 2000 and 2001; Skyrms 1996 discusses some recent applications of Carnapian techniques to Bayesian statistical models involving continuous random variables; Glaister 2001 and Festa 1993 provide broad surveys of Carnapian theories of logical probability and inductive logic).

Begin with a standard first-order logical language *L* containing a finite number of monadic predicates *F*, *G*, *H*, ... and a finite or denumerable number of individual constants *a*, *b*, *c*, ... Define an unconditional probability function $P(\bullet)$ over the sentences of *L*. Finally, following the standard Kolmogorovian approach, construct a conditional probability function $P(\bullet|\bullet)$ over pairs of sentences of *L*, using the ratio definition of conditional probability given above. To fix ideas, consider a very simple toy language L with only two monadic predicates, F and G and only two individual constants a and b. In this language, there are only sixteen possible states of the world that can be described. These sixteen maximally specific descriptions are called the *state descriptions* of L, and they are as follows:

$Fa \land Ga \land Fb \land Gb$	$\neg Fa \land Ga \land Fb \land Gb$
$Fa \land Ga \land Fb \land \neg Gb$	$\neg Fa \land Ga \land Fb \land \neg Gb$
$Fa \land Ga \land \neg Fb \land Gb$	$\neg Fa \land Ga \land \neg Fb \land Gb$
$Fa \land Ga \land \neg Fb \land \neg Gb$	$\neg Fa \land Ga \land \neg Fb \land \neg Gb$
$Fa \land \neg Ga \land Fb \land Gb$	$\neg Fa \land \neg Ga \land Fb \land Gb$
$Fa \land \neg Ga \land Fb \land \neg Gb$	$\neg Fa \land \neg Ga \land Fb \land \neg Gb$
$Fa \land \neg Ga \land \neg Fb \land Gb$	$\neg Fa \land \neg Ga \land \neg Fb \land Gb$
$Fa \land \neg Ga \land \neg Fb \land \neg Gb$	$\neg Fa \land \neg Ga \land \neg Fb \land \neg Gb$

Two state descriptions S_1 and S_2 are said to be *permutations* of each other if S_1 can be obtained from S_2 by some permutation of the individual constants. For instance, $Fa \land \neg Ga \land \neg Fb \land Gb$ can be obtained from $\neg Fa \land Ga \land Fb \land \neg Gb$ by permuting a and b. Thus, $Fa \land \neg Ga \land \neg Fb \land Gb$ and $\neg Fa \land Ga \land Fb \land \neg Gb$ are permutations of each other (in L). A *structure description* in L is a disjunction of state descriptions, each of which is a permutation of the others. In the toy language L, there are the following ten structure descriptions:

$Fa \land Ga \land Fb \land Gb$	$(Fa \land \neg Ga \land \neg Fb \land Gb)$ $\lor (\neg Fa \land Ga \land Fb \land \neg Gb)$
$(Fa \land Ga \land Fb \land \neg Gb)$ $\lor (Fa \land \neg Ga \land Fb \land Gb)$	$(Fa \land \neg Ga \land \neg Fb \land \neg Gb)$ $\lor (\neg Fa \land \neg Ga \land Fb \land \neg Gb)$
$(Fa \land Ga \land \neg Fb \land Gb)$ $\lor (\neg Fa \land Ga \land Fb \land Gb)$	$\neg Fa \land Ga \land \neg Fb \land Gb$
$(Fa \land Ga \land \neg Fb \land \neg Gb)$ $\lor (\neg Fa \land \neg Ga \land Fb \land Gb)$	$(\neg Fa \land Ga \land \neg Fb \land \neg Gb)$ $\lor (\neg Fa \land \neg Ga \land \neg Fb \land Gb)$
$Fa \land \neg Ga \land Fb \land \neg Gb$	$\neg Fa \land \neg Ga \land \neg Fb \land \neg Gb$

Now assign nonnegative real numbers to the state descriptions, so that these sixteen numbers sum to 1. Any such assignment will constitute an unconditional probability function $P(\bullet)$ over the state descriptions of L. To extend $P(\bullet)$ to the entire language L, stipulate that the probability of a disjunction of mutually exclusive sentences is the sum of the probabilities of its disjuncts. Since every sentence in L is equivalent to some disjunction of state descriptions, and every pair of state descriptions is mutually exclusive, this gives a complete unconditional probability function $P(\bullet)$ over L. For instance, since $Fa \land Ga \land \neg Gb$ is equivalent to the disjunction $(Fa \land Ga \land Fb \land \neg Gb) \lor (Fa \land Ga \land \neg Fb \land \neg Gb)$, one will have:

$Pr(Fa \land Ga \land \neg Gb)$
$= Pr((Fa \land Ga \land Fb \land \neg Gb) \lor (Fa \land Ga \land \neg Fb \land \neg Gb))$
$= Pr(Fa \land Ga \land Fb \land \neg Gb) + Pr(Fa \land Ga \land \neg Fb \land \neg Gb)$.

Now, it is only a brief step to the definition of the conditional probability function $P(\bullet|\bullet)$ over pairs of sentences in L. Using the standard, Kolmogorovian ratio definition of conditional probability, for all pairs of sentences X, Y in L:

$P(X|Y) = P(X \land Y)/Pr(Y)$, provided that $P(Y) \neq 0$.

Thus, once the unconditional probability function $P(\bullet)$ is specified for the state descriptions of a language L, all probabilities both conditional and unconditional are thereby determined over L. And, this gives one a logical probability model M over the language L. The unconditional, logical probability functions so defined are typically called measure functions. Carnap (1950) discusses two "natural" measure functions.

The first Carnapian measure function is $m^\dagger$, which assumes that each of the state descriptions is equiprobable a priori: If there are N state descriptions in L, then $m^\dagger$ assigns $\frac{1}{N}$ to each state description. While this may seem like a very natural measure function, since it applies something like the principle of indifference to the state descriptions of L (see below for discussion), $m^\dagger$ has the consequence that the resulting probabilities cannot reflect learning from experience. Consider the following simple example. Assume that one adopts a logical probability function $P(\bullet)$ based on $m^\dagger$ as one's own a priori degree of belief (or credence) function. Then, one learns (by conditionalizing) that an object a is F, that is, Fa. Intuitively, one's conditional degree of credence $P(Fb|Fa)$ that a distinct object b also is F, given that a is F, should not always be the same as one's a priori degree of credence that b is F. That is, the fact that one has observed another F object should (at least in some cases) make it more probable (a posteriori) that b will also be F (i.e., more probable than Fb was a priori). More generally, if one observes that a large number of objects have been F, this should raise the probability that the next object one observes will also be F. Unfortunately, no a priori probability function based on $m^\dagger$ is consistent with learning from experience in either sense. To see this, consider the simple case $Pr(Fb|Fa)$:

$P(Fb|Fa) = m^\dagger(Fb \land Fa)/m^\dagger(Fa)$
$= \frac{1}{2} = m^\dagger(Fb) = Pr(Fb)$.

So, if one assumes an a priori probability function based on $m^\dagger$, the fact that one object has property F cannot affect the probability that any other object will also have property F. Indeed, it can be shown (Kyburg 1970, 58–59) that no matter how many objects are assumed to be F, this will be irrelevant (according to probability functions based on $m^\dagger$) to the hypothesis that a distinct object will also be F.

The fact that (on the probability functions generated by the measure $m^\dagger$) no object's having certain properties can be informative about other objects also having those same properties has been viewed as a serious shortcoming of $m^\dagger$ (Carnap 1955). As a result, Carnap formulated an alternative measure function m^*, which is defined as follows. First, assign equal probabilities to each structure description (in the toy language above, n $\frac{1}{10}$). Then, each state description belonging to a given structure description is assigned an equal portion of the probability assigned to that structure description). For instance, in the toy language, the state description $Fa \wedge Ga \wedge \neg Fb \wedge Gb$ gets assigned an a priori probability of $\frac{1}{20}$ ($\frac{1}{2}$ of $\frac{1}{10}$), but the state description $Fa \wedge Ga \wedge Fb \wedge Gb$ receives an a priori probability of $\frac{1}{10}$ ($\frac{1}{1}$ of $\frac{1}{10}$). To further illustrate the differences between $m^\dagger$ and m^*, here are some numerical values in the toy language L:

Measure function $m^\dagger$	Measure function m^*
$m^\dagger(Fa \wedge Ga \wedge \neg Fb \wedge Gb) = \frac{1}{16}$	$m^*(Fa \wedge Ga \wedge Fb \wedge Gb) = \frac{1}{10}$
$m^\dagger((Fa \wedge Ga \wedge \neg Fb \wedge Gb) \vee ((\neg Fa \wedge Ga \wedge Fb \wedge Gb))) = \frac{1}{8}$	$m^*(Fa \wedge Ga \wedge \neg Fb \wedge Gb) = \frac{1}{20}$
$m^\dagger(Fa) = \frac{1}{2}$	$m^*(Fa) = \frac{1}{2}$
$Pr^\dagger(Fa\|Fb) = \frac{1}{2} = m^*(Fa) = Pr^\dagger(Fa)$	$Pr^*(Fa\|Fb) = \frac{3}{5} > \frac{1}{2} = m^*(Fa) = Pr^*(Fa)$

Unlike $m^\dagger$, m^* can model learning from experience, since in the simple language

$$P(Fa \mid Fb) = \frac{3}{5} > \frac{1}{2} = Pr(Fa)$$

if the probability function P is defined in terms of the logical measure function m^*. Although m^* does have some advantages over $m^\dagger$, even m^* can give counterintuitive results in more complex languages (Carnap 1952).

Carnap (1952) presents a more complicated framework, which describes a more general class (or "continuum") of conditional probability functions, from which the definitions of $P(\bullet|\bullet)$ in terms of m^* and $m^\dagger$ fall out as special cases. This continuum of conditional probability functions depends on a parameter λ, which is supposed to reflect the "speed" with which learning from experience is possible. In this continuum, $\lambda = 0$ corresponds to the "straight rule" of induction, which says that the probability that the next object observed will be F, conditional upon a sequence of past observations, is simply the frequency with which F objects have been observed in the past sequence; $\lambda = +\infty$ yields a conditional probability function much like that given above by assuming the underlying logical measure $m^\dagger$ (i.e., $\lambda = +\infty$ implies that there is no learning from experience). And setting $\lambda = \kappa$ (where κ is the number of independent families of predicates in Carnap's more elaborate 1952 linguistic framework) yields a conditional probability function equivalent to that generated by the measure function m^*.

But even this λ-continuum has problems. First, none of the Carnapian systems allow universal generalizations to have nonzero probability. This problem was addressed by Hintikka (1966) and Hintikka and Niiniluoto (1980), who provided various alterations of the Carnapian framework that allow for nonzero probabilities of universal generalizations. Moreover, Carnap's early systems did not allow for the probabilistic modeling of analogical effects. That is, in his 1950–1952 systems, the fact that two objects share several properties in common is always irrelevant to whether they share any other properties in common. Carnap's more recent (and most complex) theories of logical probability (1971, 1980) include two additional adjustable parameters (γ and η), designed to provide the theory with enough flexibility to overcome these (and other) limitations. Unfortunately, no Carnapian logical theory of probability to date has successfully dealt with the problem of analogical effects (Maher 2000 and 2001). Moreover, as Putnam (1963) explains, there are further (and some say deeper) problems with Carnapian (or, more generally, syntactical) approaches to logical probability, if they are to be applied to inductive inference generally. The consensus now seems to be that the Carnapian project of characterizing an adequate logical theory of probability is (by his own standards and lights) not very promising (Putnam 1963; Festa 1993; Maher 2001).

This discussion has glossed over technical details in the development of (Carnapian) logical interpretations or theories of probability since 1950. To recapitulate, what is important for present purposes is that Carnap (along with the other advocates of logical probability) was an RV theorist about inductive logic. He identified the concept c(•,•) of inductive strength (or inductive support) with the concept of conditional probability $P(•|•)$. And he thought (partly because of the problems he saw with epistemic interpretations) that in order for an RV account to satisfy desideratum 3, it needed to presuppose a logical interpretation (or theory) of probability. This led him, initially, to develop various logical measures (e.g., the a priori logical probability functions $m^†$ and m^*), and then to define conditional logical probability $Pr(•|•)$ in terms of these underlying a priori logical measures, using the standard ratio definition. This approach ran into various problems when it came to the application of $P(•|•)$ to inductive logic. These difficulties mainly had to do with the ability of Carnap's $P(•|•)$ to undergird learning from experience and/or certain kinds of analogical reasoning (for other philosophical objections to Carnap's logical probability project, see Putnam 1963). In response to these difficulties, Carnap began to fiddle directly with the definition of $P(•|•)$. In 1952, he moved to a parameterized definition of $P(•|•)$, which contained an "index of inductive caution" (λ) that was supposed to regulate the speed with which learning from experience is reflected by $P(•|•)$. Later, Carnap (1971, 1980) added γ and η to the definition of $P(•|•)$, as noted above, in an attempt to further generalize the theory and allow for sensitivity to certain kinds of analogical effects. Ultimately, no such theory was ever viewed by Carnap (or others) as fully adequate for the purposes of grounding an RV conception of inductive logic.

At this point, it is important to ask, In what sense are Carnap's theories of logical probability (especially his later ones) *logical*? His early theories (based on the measure functions $m^†$ and m^*) applied something like the principle of indifference to the state and/or structure descriptions of the formal language L in order to determine the logical probabilities $P(•|•)$. In this sense, these early theories assume that certain sentences of L are equiprobable a priori. Why is such an assumption *logical*? Or, more to the point, how is *logic* supposed to tell one which statements are equiprobable a priori? Carnap (1955) explains that

> the statement of equiprobability to which the principle of indifference leads is, like all other statements of inductive probability, not a factual but a logical statement. If the knowledge of the observer does not favor any of the possible events, then with respect to this knowledge as evidence they *are* equiprobable. The statement assigning equal probabilities in this case does not assert anything about the facts, but merely the logical relations between the given evidence and each of the hypotheses; namely, that these relations are logically alike. These relations are obviously alike if the evidence has a symmetrical structure with respect to their possible events. The statement of equiprobability asserts nothing more than the symmetry. (22)

Carnap seems to be saying that the principle of indifference is to be applied only to possible events that exhibit certain a priori symmetries with respect to some rational agent's background evidence. But this appears no more logical than Keynes's epistemic approach to probability. It seems that the resulting probabilities $P(•|•)$ will not be logical in the sense Carnap desired (at least no more so than Keynes's epistemic probabilities were), unless Carnap can motivate—on logical grounds—the choice of an a priori probability model. To that end, Carnap's application of the principle of indifference is not very useful. Recall that the goal of Carnap's project (of inductive logic) was to explicate the confirmation relation, which is itself supposed to reflect the evidential relation between premises and conclusions (Carnap 1950 uses the locutions "degree of confirmation" and "weight of evidence" synonymously). How is one to understand what it means for evidence not to "favor any of the possible events" in a way that does not require one to already understand how to measure the degree to which the evidence confirms each of the possible events? Here, Carnap's discussion of the principle of indifference presupposes that degree of confirmation is to be identified with degree of conditional probability. In that reading, "not favoring" just means "conferring equal probability on," and Carnap's unpacking of the principle of indifference reduces directly to a mathematical truth (which, for Carnap, is good enough to render the principle *logical*). If one had independent grounds for thinking that conditional probabilities were the right way to measure confirmation (or weight of evidence), then Carnap would have a rather clever (albeit not terribly informative) way to (logically) ground his choice of a priori probability models. Unfortunately, as will be seen below, there are independent reasons to doubt Carnap's presupposition here that degree of confirmation should be identified with degree of conditional probability. Without that assumption, Carnap's principle of indifference is no longer logical (by his own lights), and the problem of the contingency (nonlogicality) of the ultimate inductive-logical probability assignments returns with a vengeance. There are independent and deep problems with any attempt to consistently apply the principle of indifference to contexts in which hypotheses and/or evidence involve continuous magnitudes (van Fraassen 1989).

Carnap's later theories of $P(\bullet|\bullet)$ introduce even further contingencies, in the form of adjustable parameters, the "proper values" of which do not seem to be determinable a priori (Carnap 1952, 1971, 1980). In particular, consider Carnap's (1952) λ-continuum. The parameter λ is supposed to indicate how sensitive $P(\bullet|\bullet)$ is to learning from experience. A higher value of λ indicates slower learning, and a lower λ indicates faster learning. As Carnap (1952) concedes, no one value of λ is best a priori. Presumably, different values of λ are appropriate for different contexts in which confirmational judgments are made (see Festa 1993 for a contextual Carnapian approach to confirmation). It seems that the same must be said for the additional parameters γ and η (Carnap 1971, 1980). The moral here seems to be that it is only relative to a particular assignment of values to λ, γ, and η that probabilistic (and/or confirmational) judgments are objectively and noncontingently determined in Carnap's later systems. This is analogous to the fact that it is only relative to a (probabilistic) characterization of the agent's background knowledge and complete epistemic state (in the form of a specific epistemic probability model M) that Keynes's epistemic probabilities (or Carnap's measure functions m^* and $m^\dagger$) have a chance of being objectively and noncontingently determined.

A pattern is developing. Both Keynes and Carnap give accounts of a priori probability functions $P(\bullet|\bullet)$ that involve certain contingencies and indeterminacies. They each feel pressure (owing to desideratum 3) to eliminate these contingencies when the time comes to use $P(\bullet|\bullet)$ as an explication of c($\bullet$,$\bullet$). The general strategy for rendering these probabilities logical is to choose some privileged, a priori probability model. Here, both Keynes and Carnap appeal to the principle of indifference to constrain the ultimate choice of model. Carnap is sensitive to the fact that the principle of indifference does not seem logical, but his attempts to render it so (and useful for grounding the choice of an a priori probability model) are both unconvincing and uninformative. There is a much easier and more direct way to guarantee the satisfaction of desideratum 3. Why not just define c from the beginning as a three-place relation that depends on premises, conclusion, and a particular probability model?

The next section describes a simple, general recipe (along the lines suggested by the preceding considerations) for formulating probabilistic inductive logics in such a way that they transparently satisfy desiderata 1–3. This section will also address the following question: Is the RV *materially* adequate as an account of inductive strength or inductive support? This will lead to a fourth material desideratum for measures of inductive support, and ultimately to a concrete alternative to the RV.

Rethinking the Received View

How to Ensure the Transparent Satisfaction of Desideratum 3

The existing attempts to use the notion of probability to explicate the concept of inductive support (or inductive strength) c have foundered on the question of their contingency (which threatened violation of desideratum 3). It may be that these contingencies can be eliminated (in general) only by making the notion of inductive support explicitly relational. To follow such a plan, in the case of the RV one should rather say:

The inductive strength of the argument from $\{P_1, \ldots, P_n\}$ to C relative to a probability model $M = <B, P_M>$ is $P_M(C|P_1 \wedge \ldots \wedge P_n)$.

Relativizing judgments of inductive support to particular probability models fully and transparently eliminates the contingency and indeterminacy of these judgments. It is clear that the revision of RV above satisfies all three desiderata, since:

1. $P_M(C \mid P_1 \wedge \ldots \wedge P_n)$ is maximal and constant when $\{P_1, \ldots, P_n\}$ entails C, and $Pr_M(C \mid P_1 \wedge \ldots \wedge P_n)$ is minimal and constant when $\{P_1, \ldots, P_n\}$ refutes C.
2. The relation of inductive support is defined in terms of the notion of probability.
3. Once the conditional probability function $P_M(\bullet|\bullet)$ is specified (as it is, a fortiori, once the probability model M has been), its values are determined objectively and in a way that is contingent on only certain mathematical facts about the probability calculus. This is, the resulting c-values are determined mathematically by the specification of a particular probability model M.

One might respond at this point by asking, Where do the probability models M come from? and how does one choose an "appropriate" probability model in a given inductive logical context? These are good questions. However, it is not clear that they must be answered by the inductive logician *qua* logician. Here it is interesting to note the analogy between the P_M-relativity of inductive logical relations (in the present approach) and the language relativity of deductive logical relations in Carnap's (early) approach to deductive logic. For the early Carnap, deductive logical (or, more generally, analytic) relations obtain only between sentences in a formal language. The deductive logician is not in the business of telling people which languages they should use, since this (presumably pragmatic) question is "external" to deductive logic. However, once a language has been specified, the deductive relations among sentences in that language are determined objectively and noncontingently, and it is up to the deductive logician to explicate these relations. In the approach to inductive logic just described, the same sort of thing can be said for the inductive logician. It is not the business of the inductive logician to tell people which probability models they should use (presumably, that is an epistemic or pragmatic question), but once a probability model is specified, the inductive logical relations in that model (viz., c) are determined objectively and noncontingently. In the present approach, the duty of the inductive logician is (simply) to explicate the c-function—not to decide which probability models should be used in which contexts.

One last analogy might be useful here. When the theory of special relativity came along, some people were afraid that it might introduce an element of subjectivity into physics, since the velocities of objects were now determined only relative to a frame of reference. There was no physical ether with respect to which objects received their absolute velocities. However, the velocities and other values were determined objectively and noncontingently once the frame of reference was specified, which is the reason Einstein originally intended to call his theory the theory of invariants. Similarly, it seems that there may be no *logical ether* with respect to which pairs of propositions (or sentences) obtain their a priori relations of inductive support. But once a probability model M is specified (and independently of how that model is interpreted), the values of c-functions defined relative to M are determined objectively and noncontingently (in precisely the sense Carnap had in mind when he used those terms).

A Fourth Material Desideratum: Relevance

Consider the following argument:

(**P**) Fred Fox (who is a male) has been taking birth control pills for the past year.
(**C**) Fred Fox is not pregnant.

Intuitively (i.e., assuming a probability model M that properly incorporates one's intuitively salient background knowledge about human biology, etc.), $P_M(\mathbf{C}|\mathbf{P})$ is very high. But does one want to say that there is a strong evidential relation between **P** and **C**? According to proponents of the RV, one should say just that. This seems wrong, because intuitively $P_M(\mathbf{C} \mid \mathbf{P}) = P_M(\mathbf{C})$. That is, $P_M(\mathbf{C}|\mathbf{P})$ is high solely because $P_M(\mathbf{C})$ is high, and not because of any evidential relation between **P** and **C**. This is the same kind of criticism that Skyrms (2000) made against the NIL proposal. And it is just as compelling here. The problem here is that **P** is irrelevant to **C**. Plausibly, it seems that if **P** is going to be counted as providing evidence in favor of **C**, then **P** should raise the probability of **C** (Popper 1954 and 1992; Salmon 1975). This leads to the following fourth material desideratum for c:

- $c(C, \{P_1, \ldots, P_n\})$ should be sensitive to the probabilistic relevance of $P_1 \wedge \ldots \wedge P_n$ to C.

In particular, desideratum 4 implies that if P_1 raises the probability of C_1, but P_2 lowers the probability of C_2, then $c(C_1, P_1) > c(C_2, P_2)$. This rules out $P(C|P_1 \wedge \ldots \wedge P_n)$ as a candidate for $c(C, \{P_1, \ldots, P_n\})$, and it is therefore inconsistent with the RV. Many nonequivalent probabilistic-relevance measures of support (or confirmation) satisfying desideratum 4 have been proposed and defended in the philosophical literature (Fitelson 1999 and 2001).

One can combine desiderata 1–4 into the following single probabilistic inductive logic. This unified desideratum gives constraints on a three-place probabilistic confirmation function $c(C, \{P_1, \ldots, P_n\}, M)$, which is the degree to which $\{P_1, \ldots, P_n\}$ inductively supports C, relative to a specified probability model $M = <B, Pr_M>$:

$c(C, \{P_1, \ldots, P_n\}, M)$ is
$\begin{cases} \text{maximal and} > 0 & \text{if } \{P_1, \ldots, P_n\} \text{ entails } C \\ > 0 & \text{if } P_M(C|P_1 \wedge \ldots \wedge P_n) > P_M(C) \\ 0 & \text{if } P_M(C|P_1 \wedge \ldots \wedge P_n) = P_M(C) \\ < 0 & \text{if } P_M(C|P_1 \wedge \ldots \wedge P_n) < P_M(C) \\ \text{minimal and} < 0 & \text{if } \{P_1, \ldots, P_n\} \text{ entails } \neg C \end{cases}$

To see that any measure satisfying probabilistic inductive logic will satisfy desiderata 1–4, note that

- the cases of entailment and refutation are at the extremes of c, with intermediate values of support and countersupport in between the extremes;
- the constraints in probabilistic inductive logic can be stated purely probabilistically, and c's values must be determined relative to a probability model M, so any measure satisfying it must use probability as a central concept in its definition;

- the measure c is defined relative to a probability model, and so its values are determined objectively and noncontingently by the values in the specified model; and
- sensitivity to *P*-relevance is built into the desideratum (probabilistic inductive logic).

Interestingly, almost all relevance measures proposed in the confirmation theory literature fail to satisfy probabilistic inductive logic (Fitelson 2001, §3.2.3). One historical measure that does satisfy probabilistic inductive logic was independently defended by Kemeny and Oppenheim (1952) as the correct measure of confirmation (in opposition to Carnap's RV c-measures) within a Carnapian framework for logical probability:

$$c(C, \{P_1, \ldots, P_n\}, M) = \frac{P_M(P_1 \wedge \ldots \wedge P_n | C) - P_M(P_1 \wedge \ldots \wedge P_n | \neg C)}{P_M(P_1 \wedge \ldots \wedge P_n | C) + P_M(P_1 \wedge \ldots \wedge P_n | \neg C)}.$$

Indeed, of all the historically proposed (probabilistic) measures of degree of confirmation (and there have been dozens), the above measure is the only one (up to ordinal equivalence) that satisfies all four of the material desiderata. The four simple desiderata are thus sufficient to (nearly uniquely) determine the desired explicandum c, or the degree of inductive strength of an argument. There are other measures in the literature, such as the log-likelihood ratio, that differ conventionally from, but are ordinally equivalent to, the above measure (for various other virtues of measures in this family, see Fitelson 2001, Good 1985, Heckerman 1988, Kemeny and Oppenheim 1952, and Schum 1994).

Historical Epilogue on the Relevance of Relevance

In the second edition of *Logical Foundations of Probability*, Carnap (1962) acknowledges that probabilistic relevance is an intuitively compelling desideratum for measures of inductive support. This acknowledgement was in response to the trenchant criticisms of Popper (1954), who was one of the first to urge relevance as a desideratum in this context (see Michalos 1971 for a thorough discussion of this important debate between Popper and Carnap). But instead of embracing relevance measures like Kemeny and Oppenheim's (1952) (and rewriting much of the first edition of *Logical Foundations of Probability*), Carnap (1962) simply postulates an ambiguity in the term "confirmation." He now argues that there are two kinds of confirmation: confirmation as firmness and confirmation as increase in firmness, where the former is properly explicated using just conditional probability (à la the RV) and does not require relevance of the premises to the conclusion, while the latter presupposes that the premises are probabilistically relevant to the conclusion. Strangely, Carnap does not even mention Kemeny and Oppenheim's measure (of which he was aware) as a proper measure of confirmation as increase in firmness. Instead, he suggests for that purpose a relevance measure that does not satisfy desideratum 1 and so is not even a proper generalization of deductive entailment. This puzzling but crucial sequence of events in the history of inductive logic may explain why relevance-based approaches (like that of Kemeny and Oppenheim) have never enjoyed as many proponents as the RV.

BRANDEN FITELSON

References

Bacon, F. (1620), *The Novum Organon*. Oxford: The University Press.
Bayes, T. (1764), "An Essay Towards Solving a Problem in the Doctrine of Chances," *Philosophical Transactions of the Royal Society of London* 53.
Boole, G. (1854), *An Investigation of the Laws of Thought, on Which Are Founded the Mathematical Theories of Logic and Probabilities*. London: Walton & Maberly.
Carnap, R. (1950), *Logical Foundations of Probability*. Chicago: University of Chicago Press.
——— (1952), *The Continuum of Inductive Methods*. Chicago: University of Chicago Press.
——— (1955), *Statistical and Inductive Probability* and *Inductive Logic and Science* (leaflet). Brooklyn, NY: Galois Institute of Mathematics and Art.
——— (1962), *Logical Foundations of Probability*, 2nd ed. Chicago: University of Chicago Press.
——— (1971), "A Basic System of Inductive Logic, I," in R. Carnap and R. Jeffrey (eds.), *Studies in Inductive Logic and Probability*, vol. 1. Berkeley and Los Angeles: University of California Press, 33–165.
——— (1980), "A Basic System of Inductive Logic, II," in R. Jeffrey (ed.), *Studies in Inductive Logic and Probability*, vol. 2. Berkeley and Los Angeles: University of California Press, 7–155.
Copi, I., and C. Cohen (2001), *Introduction To Logic*, 11th ed. New York: Prentice Hall.
Dale, A. (1999), *A History of Inverse Probability: From Thomas Bayes To Karl Pearson*, 2nd ed. New York: Springer-Verlag.
Festa, R. (1993), *Optimum Inductive Methods*. Dordrecht, Netherlands: Kluwer Academic Publishers.
Fitelson, B. (1999), "The Plurality of Bayesian Measures of Confirmation and the Problem of Measure Sensitivity," *Philosophy of Science* 66: S362–S378.
——— (2001), *Studies in Bayesian Confirmation Theory*. PhD. dissertation, University of Wisconsin–Madison (Philosophy).
Glaister, S. (2001), "Inductive Logic," in D. Jacquette (ed.), *A Companion to Philosophical Logic*. London: Blackwell.

Good, I. J. (1985), "Weight of Evidence: A Brief Survey," in J. Bernardo, M. DeGroot, D. Lindley, and A. Smith (eds.), *Bayesian Statistics, 2*. Amsterdam: North-Holland, 249–269.

Hájek, A. (2003), "What Conditional Probabilities Could Not Be," *Synthese* 137: 273–323.

Heckerman, D. (1988), "An Axiomatic Framework for Belief Updates," in L. Kanal and J. Lemmer (eds.), *Uncertainty in Artificial Intelligence 2*. New York: Elsevier Science Publishers, 11–22.

Hempel, C. (1945), "Studies in the Logic of Confirmation," parts I and II, *Mind* 54: 1–26 and 97–121.

Hintikka, J. (1966), "A Two-Dimensional Continuum of Inductive Methods," in J. Hintikka and P. Suppes (eds.), *Aspects of Inductive Logic*. Amsterdam: North-Holland.

Hintikka, J., and I. Niiniluoto (1980), "An Axiomatic Foundation for the Logic of Inductive Generalization," in R. Jeffrey, *Studies in Inductive Logic and Probability*, vol. 2. Berkeley and Los Angeles: University of California Press.

Hume, D. (1739–1740), *A Treatise of Human Nature: Being an Attempt to Introduce the Experimental Method of Reasoning into Moral Subjects*, vols. 1–3. London: John Noon (1739), Thomas Longman (1740).

——— (1758), *An Enquiry Concerning Human Understanding in Essays and Treatises on Several Subjects*. London: A. Millar.

Hurley, P. (2003), *A Concise Introduction to Logic*, 8th ed. Melbourne, Australia, and Belmont, CA: Wadsworth/Thomson Learning.

Johnson, W. E. (1921), *Logic*. Cambridge: Cambridge University Press.

——— (1932), "Probability: The Deductive and Inductive Problems," *Mind* 49: 409–423.

Joyce, J. (1999), *The Foundations of Causal Decision Theory*. Cambridge: Cambridge University Press.

Kemeny, J., and P. Oppenheim (1952), "Degrees of Factual Support," *Philosophy of Science* 19: 307–324.

Keynes, J. (1921), *A Treatise on Probability*. London: Macmillan.

Kneale, W., and M. Kneale (1962), *The Development of Logic*. Oxford: Clarendon Press.

Kolmogorov, A. (1950), *Foundations of the Theory of Probability*. New York: Chelsea.

Kyburg, H. E. (1970), *Probability and Inductive Logic*. London: Macmillan.

Laplace, P. S. M. d. (1812), *Théorie Analytique des Probabilités.*. Paris: Ve. Courcier.

Layman, C. S. (2002), *The Power of Logic*, 2nd ed. New York: McGraw-Hill.

Maher, P. (1996), "Subjective and Objective Confirmation," *Philosophy of Science* 63: 149–174.

——— (2000), "Probabilities for Two Properties," *Erkenntnis* 52: 63–91.

——— (2001), "Probabilities for Multiple Properties: The Models of Hesse and Carnap and Kemeny," *Erkenntnis* 55: 183–216.

Michalos, A. (1971), *The Popper–Carnap Controversy*. The Hague: Martinus Nijhoff.

Mill, J. (1843), *A System of Logic, Ratiocinative and Inductive, Being a Connected View of the Principles of Evidence and the Methods of Scientific Investigation*. London: Parker.

Paris, J. (1994), *The Uncertain Reasoner's Companion: A Mathematical Perspective*. Cambridge: Cambridge University Press, chap. 12.

Popper, K. (1954), "Degree of Confirmation," *British Journal for the Philosophy of Science* 5: 143–149.

——— (1992), *The Logic of Scientific Discovery*. London: Routledge.

Putnam, H. (1963), "'Degree of Confirmation' and Inductive Logic," in P. A. Schilpp (ed.), *The Philosophy of Rudolf Carnap*. La Salle, IL: Open Court Publishing, 761–784.

Roeper, P., and H. Leblanc (1999), *Probability Theory and Probability Logic*. Toronto: University of Toronto Press.

Salmon, W. C. (1975), "Confirmation and Relevance," in G. Maxwell and R. M. Anderson Jr. (eds.), *Induction, Probability, and Confirmation: Minnesota Studies in the Philosophy of Science*, vol. 6. Minneapolis: University of Minnesota Press, 3–36.

Schum, D. (1994), *The Evidential Foundations of Probabilistic Reasoning*. New York: John Wiley & Sons.

Skyrms, B. (1996), "Carnapian Inductive Logic and Bayesian Statistics," in *Statistics, Probability and Game Theory: Papers in Honor of David Blackwell*. IMS Lecture Notes, Monograph Series 30: 321–336.

——— (2000), *Choice and Chance*. Melbourne, Australia, and Belmont, CA: Wadsworth/Thomson Learning.

Stigler, S. (1986), *The History of Statistics*. Cambridge, MA: Harvard University Press.

van Fraassen, B. (1989), *Laws and Symmetry*. Oxford: Oxford University Press, chap. 12.

INNATE/ACQUIRED DISTINCTION

Arguments about innateness center on two distinct but overlapping theoretical issues. One concerns the explanation of the origin of ideas in the human mind. This ancient question was famously introduced in Plato's *Meno* and it took center stage in seventeenth- and eighteenth-century debates between rationalists and empiricists. More recently, it has seen a sophisticated revival in arguments about the status of nativism in philosophy of mind and cognitive science (see Chomsky, Noam; Cognitive Science). Nativists believe that a mind cannot learn unless it comes already furnished with certain fundamental ideas and/or representational capacities. The issue remains controversial. Clearly human cognitive architecture has some effect on ways of thinking and the sort of ideas that can be had. Some have wanted to say that this amounts to having innate ideas. Others worry that the notion of innate ideas is really just another (and more confusing) way of discussing cognitive architecture.

The second point at issue concerns the explanation of evolutionary and developmental rigidity in living systems. Evolutionarily rigid traits—such as being five-fingered amongst primates—are resistant to selection pressure. Developmentally rigid traits are resistant to environmental perturbation during ontogeny. Darwin argued that some traits are more advantageous than others and that over time the traits that contributed to more successful organisms would become more common in populations. The power of natural selection in the Darwinian sense comes from the fact that it is cumulative and occurs over very large numbers of generations (see Natural Selection). This explains the existence of complex adaptations such as eyes and brains. For cumulative selection to operate, there must be very reliable inheritance of traits from one generation to the next, otherwise fitness-enhancing traits would "leech out" of populations. Similarly, the development of individual organisms must also be reliable. That is, individuals with similar genomes reared in similar environments must be more likely to develop similar traits (see Heritability). If this were not the case, then biological inheritance could not be reliably fitness enhancing. However, it has long been recognized that traits differ in the extent to which they are rigid. This is true in the development of individuals. Different people might speak different languages or be shorter and taller in stature, but they would not have different numbers of chambers in their hearts. This difference in malleability is also evident when one looks at the evolution of traits in long lineages of organisms. Arthropoda, for example, is a clade of millions of extant species. Its members have evolved a remarkable variety of morphologies and behaviours. Yet despite this diversity, there are still traits that are ubiquitous within the clade. These *diagnostic* traits have become entrenched within the lineage, and they seem highly resistant to selection pressure.

Thus, study of evolutionary theory and developmental biology says that some traits seem to be built into individuals and some seem to be built into evolving lineages. Some biologists have wanted to call such traits innate. But this leads inevitably to the question, What exactly does it mean for a trait to be "built into" an individual or a group of individuals? Is it, for example, the same as saying that the characteristic in question is genetic? Is it the same as saying that the characteristic in question is ubiquitous? Such questions have been asked variously by ethologists, developmental biologists, evolutionary theorists, and, more recently, philosophers of biology. Finally, and most importantly, one might also ask, Is there some clear characterization of innateness that will clarify both of the theoretical issues set out above?

Where Do Thoughts Come From?

The notion that human beings are the bearers of innate ideas has been the servant of many theoretical masters. In the *Meno*, Plato seeks to demonstrate that a slave boy who has received no schooling is, in fact, possessed of knowledge of geometry. Plato's intention is to argue that such knowledge is not learned but rather recollected from a time prior to birth in which humans were in direct contact with the forms. It is thus innate, in the sense of being present at birth.

A much later version of this reasoning seeks to achieve very different programmatic aims. Whereas Plato was concerned to ground a metaphysical system, the Cambridge Platonists attempted early in the seventeenth century to use the same machinery to buttress religious faith. Their claim was that all people have innate knowledge of God's existence and his moral laws. Again, the suggestion was that to be a member of humankind is to have, born within one, a number of undeniable truths (Patrides 1970).

The zenith of this style of argument is not to be found in metaphysics or in theology, but in epistemology. Both Descartes (in the *Meditations*) and Leibniz (in the *New Essays*) employ the notion of innate ideas as a means of staving off a more radical form of skepticism than that which concerned the Cambridge Platonists. As Descartes claims, anything one thinks one knows could in fact be a deception caused by an evil demon. Therefore, if one is to know about the world, then one must do so on the basis of epistemic principles that are not themselves based on the way the world is perceived to be. The solution provided by both Descartes and Leibniz was to defend a foundationalist epistemology that rested upon a God-given bedrock of undeniable truths. These truths are not obtained via perception. They might be thought of as innate, though it should be noted that for something to be known a priori does not in itself imply that the knowledge is innate.

In this tradition, which runs from Plato through to the Rationalists, there is a common denominator. All these systems share the claim that innate ideas are the product of some external metaphysical cause (be it the realm of the forms or an undeceiving God). While this tradition is long-standing, modern philosophical and scientific inquiry has taught that such premises need not underpin arguments for innate ideas.

Nativism

In the latter half of the twentieth century, philosophers of mind and cognitive scientists employed innateness in their explanation of a variety of cognitive phenomena (see Cognitive Science). Noam Chomsky (1965) and his heirs argue that much of the capacity to decipher verbal information is innate; David Marr (1982) defends a similar position with respect to the interpretation of visual information; and Jerry Fodor (1975) argues for a broadly rationalist interpretation of concept acquisition (see Philosophy of Linguistics).

While such positions are now usually categorized as "nativist," they are in some respects direct theoretical descendents of seventeenth-century rationalist arguments (Cowie 1999). Leibniz (1981) argues that a limited modal perspective dictates that one cannot learn necessary truths from observation of the actual world. One simply is not presented with sufficient data to learn facts about the way things are in every possible world (79–80). Thus, he inferred, experience alone cannot explain knowledge of necessary truths. Chomsky also avails himself of arguments based upon poverty of the stimulus. He argues that children are not presented with sufficient input from other language users to explain their acquisition of complex natural languages. More precisely, the grammar they learn is underdetermined by the instances of grammar they encounter. Nor do they receive sufficient negative reinforcement for grammatical errors. Thus, as did Leibniz, Chomsky concludes that the phenomenon in question cannot be explained as the product of experience alone. Famously, he argues that humans are possessed of an innate universal grammar (Chomsky 1965, 47–59). Thus, for him, linguistic development consists not in learning what language is, but rather in learning which language one speaks. For Chomsky, the language faculty is a mental "organ" that develops within humans. It is innate in just the same way that various aspects of morphology are innate. As he puts it, "Language acquisition is not really something that the child does; it is something that happens to the child placed in a certain environment" (Chomsky 1990, 634).

One can draw a similar parallel between Fodor's flagship argument and those of his predecessors. For Descartes, it is impossible that the mechanical process of perception (consisting of the corporeal movement of nerves and resulting flows of vital spirits) could actually result in something as perfect as the idea of blueness. Thus, he concludes that all apparently learned ideas actually come from God. This is usually described as an impossibility argument, for gaining knowledge via perception is impossible, and thus concepts must be supplied by some other means. Fodor (1975 and 1981) also avails himself of an impossibility argument. He argues that concept acquisition is strongly dependent upon a preexisting representational capacity. It would be impossible to learn about the world unless one could form hypotheses about it. But to do that, one must have some system of representation (usually called mentalese) in which to form the hypotheses in the first place. Of course, mentalese could also be learned, but one would have to have formed hypotheses, etc. Ultimately, Fodor argues, if one is to avoid an infinite regress, one must admit the existence of some innate representational vocabulary.

In recognizing that nativism is a true heir of rationalism, it should not be presumed that little has changed since the seventeenth century. It would be unfair to suggest that Fodor's impossibility argument is much the same as those of his philosophical forebears (see the discussion in the following section). Furthermore, Chomsky makes use of a wide variety of naturalistic arguments supporting the idea that human beings inherit domain-specific cognitive capacities. So, for example, Chomsky and other nativists in linguistics also argue for an innate universal grammar on the following grounds:

- All natural languages share a variety of suboptimal features that would be explained by faults in the underlying universal grammar.
- Language learning appears to be tied to biological development in a way that other types of learning are not. Acquiring a language during early childhood is much easier than acquiring one outside this crucial developmental window.
- There is a great breadth of evidence (particularly from cases of brain trauma) suggesting that language learning is modular.
- There is clear evidence that general intelligence and linguistic ability are independent.

Against Nativism

While many see modern arguments for nativism as much more compelling than Leibniz' metaphysics or Socrates' interrogation of the uneducated, it is still the case that all the arguments in favor of nativism are controversial to some degree.

Fodor's flagship argument for nativism is one that many philosophers find unpalatable. Many worry that Fodor's view implies that if humans have born within them concepts of those things that they will later learn to name, then at least some humans must be born with innate concepts of special relativity, cellular phones, and Barbie dolls. But this seems distinctly implausible given that humans have evolved in environments in which at least cell phones and Barbie dolls made no appearance. It should be noted though that Fodor's argument does not imply that *all* the terms used in natural languages must correspond to innate concepts. Fodor admits that one might combine primitive words in mentalese to represent complex thoughts that one then attaches to words in natural languages. So one need not have innate concepts corresponding to all the things that one names using natural languages. One might accept Fodor's basic argument but maintain that relatively few concepts are innate. However, Fodor (1981) rejects that conclusion, arguing that empirical evidence says that most concepts are primitive rather than constructed out of primitives (272–275). Some reject Fodor's argument on the grounds that it wrongly assumes that language acquisition involves learning facts of the form "$X_{English}$ means the same as $Y_{mentalese}$." This is just to say that anyone who rejects the language of thought hypothesis will also reject Fodor's innateness argument.

Sterelny (1989) argues against Fodor's suggestion that humans have most of their concepts innately and are merely triggered to use them, just as a duckling is triggered to imprint on its mother. Ducklings can be triggered to imprint upon all sorts of things (animals of the wrong species, cuddly toys, and even the odd ethologist). Thus the idea of a trigger implies a certain arbitrariness about what does the triggering. However, suggests Sterelny, concept acquisition appears not to work this way. One's whale concept is caused by whales. One's doorknob concept is caused by doorknobs. Had an individual's doorknob concept been caused by whales, others would be inclined to think that the individual in question had a faulty doorknob concept. But if only doorknobs can cause a doorknob concept, then triggering begins to look a lot more like learning (see Cowie 1999, 69–139, for an extended discussion of this and other objections to Fodorian nativism; see Fodor 2001 for a response to Cowie).

There is similarly a long-standing tradition of close-fought argument against Chomskian nativism. Putnam (1992) claims that suboptimal features common to all languages need not be innate and that first-language learning is much more difficult and time-consuming than nativists assume.

The majority of criticism of Chomskian nativism has been based on empirical studies. However, in one crucial respect the foundations of nativism (Chomskian and otherwise) remain philosophically suspect. This is the suggestion that there is yet no substantive theory of exactly what it means for a concept to be innate (Cowie 1999). One such suggestion is that the putative understanding of innateness is really just a familiarity with certain metaphors (such as Leibniz' veins in marble, etc.). When one asks for a definition of what it means for a behavioral trait to be innate, one finds either a lack of a clear definition or a recognition of a confusing array of definitions.

A similar worry is that while philosophy provides a good characterization of what is meant by the term 'innate,' it does not do so in a way that leads naturally to useful scientific exploration of the putative phenomenon. So, for example, Stich (1975) argues that Descartes' suggestion should be followed to analyze innateness in terms of dispositions:

> A person has a disease innately at time t, if and only if, from the beginning of his life to t it has been true of him that if he is or were of the appropriate age (or at the appropriate stage of life) then he has or in the normal course of events would have the disease's symptoms. (6)

This characterization remains popular with some philosophers but is of little use to scientists. That is because it gives little advice as to how one might detect innate traits and no explanation as to their cause. How then might nativism respond to the charge that it fails to supply a compelling scientific theory of the nature of innateness? One possibility is to point out that the opponents of nativism are no better off. They claim that some portion of the doxastic furnishings is learned, but despite serious effort in this direction, there is still no good general theory of learning. This view is championed by Fodor (2001). Alternatively, the nativist might suggest that one does not need a general theory of innateness in order to be convinced that some things are indeed innate. After all, lack of a good theory about the nature of inheritance did not stop Darwin from employing inheritance in his theory of natural selection. Finally, the nativist might suggest that actually there already is a good theory of innateness. In this vein, some argue that cognitive science (and philosophy more generally) might avail itself of a biological notion of innateness (Fodor 2001, 102). Of course, this solution to the problem rests on there being a satisfactory biological notion of innateness.

The Biological Notion of Innateness

Innateness and inheritance are biologically distinct and yet closely linked. *Inheritance* is a relation of similarity that holds between generations by virtue of some biological process such as the passing on of genetic structure. *Innateness*, on the other hand, is a claim about a certain type of rigidity within the development of a biological individual (or within a group of individuals). Thus, congenital deformities (such as those caused by the drug thalidomide) are innate but not inherited. Conversely, at least on some accounts of innateness, regional accents in human speech are inherited but not innate.

Having said this, to accept the fact of inheritance is to acknowledge a certain rigidity in biological development, and to accept the fact of cumulative natural selection is to acknowledge a certain rigidity in evolutionary history. Thus one might infer from widespread acceptance of evolution to widespread acceptance of innateness. However, such inference has proved problematic. One stumbling block has been the great variety of ways in which innate traits can be characterized. They can be described variously as:

- Traits that develop in the absence of contact with conspecifics, such as web building in spiders;
- Traits that are characteristic of particular species, such as species-specific birdsong;
- Traits that are evolutionary adaptations, such as "dancing" in bees;
- Behaviors that are unlearned, including so-called reflex actions;
- Behaviors that develop fully formed in animals that have been prevented from practicing them.

While often in agreement, these definitions are not coextensive. Linguistic ability is very much characteristic of the human species, but it does not develop in the rare cases in which human children grow up in the absence of contact with conspecifics. So, some worry that there may be no single property that all purportedly innate characteristics have in common. Certainly there has been no agreed-upon definition in the scientific or philosophical literature despite a considerable amount of work toward this end. Much of this work has been done by ethologists (beginning with Konrad Lorenz) in the 1940s who sought to explain the peculiar character of a class of inherited behavioral characteristics in nonhuman animals. These behaviors have in common that they are performed in very stereotypical ways, are inherited, are triggered by relatively simple proximal stimuli, and can be triggered in circumstances in which their performance is disadvantageous to the animal in question. Ethologists argued that these traits are properly thought of as innate. However, they were well aware of the problem of providing a clear characterization of innateness. In light of the apparent profusion of possible definitions, Lorenz (1950, 261) argued that what these putatively innate behaviors have in common is that they are all genetic.

Are Innate Traits Genetic Traits?

Despite the *prima facie* plausibility of innate traits being the products of genetic structure, this idea has come under fire from a number of developmental biologists as well as philosophers of biology. In part this is because while it is true that all organisms inherit genetic structure from their parent or parents, it is false to suggest that this is the sum total of their inheritance (see Heritability).

Developmental systems theorists (such as Griffiths and Gray [1994]) point out many inherited characteristics are reliably passed from generation to generation via nongenetic channels. Organisms inherit taught hunting behaviors, food sources, and nest sites and even gut microfauna from their parents. None of these are transmitted via the inheritance of their parents' DNA. Given this, one cannot infer that if innate traits are inherited, they must also be genetic.

If innate traits are to be characterized as genetic, there must first be a robust theory of what it means for a trait to be genetic. However, the idea that such a robust theory will be found has been the subject of considerable skepticism, most famously from Richard Lewontin (1974), who points out that one cannot mean that a "genetic trait" so called is caused exclusively by genes. All traits require both genetic and nongenetic precursors in their development. Furthermore, genes and environment interact in the course of development, so that one cannot determine the extent to which each is a cause of the development of any particular trait. Put more technically—genes and nongenetic developmental resources are typically nonadditive contributors to phenotype, and one therefore cannot partition the variance due to each (although one can partition the additive and nonadditive portions of variance) (see Heritability).

Recent Work on Innateness

Recent work has sought to tie the idea of innateness to particular biological processes. André Ariew (1999) proposes an account of innateness based on C. H. Waddington's notion of developmental canalization. This is the process by which developmental pathways are buffered against environmental perturbation. William Wimsatt (1999) argues that innateness is caused by generative entrenchment. This is the process by which adaptations that have been historically important become locked into a lineage as later adaptations are built atop them and are thus developmentally dependent upon them. But such strategies have the disadvantage of requiring the scientific community to settle on a particular biological process (generative entrenchment, canalization, or some other) as the source of all innateness. This leads back to the original problem with the biological characterization of innateness, namely that there appear to be a variety of processes that give rise to evolutionary and developmental rigidity.

Another alternative is to avoid Lewontin's argument by characterizing genetic traits in terms of genetic information (as did Lorenz, although his description of genetic information was rather sketchy). However this strategy has proved contentious (see Biological Information). Griffiths and Gray (1994) argue that all traits are products of information from both genetic and nongenetic developmental resources. Therefore, attempting to single out sources of developmental information will not provide an explanation of the distinctive nature of innate traits. Indeed, Griffiths has recently argued that the use of the term "innate" ought to be abandoned altogether (Griffiths 2002). However, Maclaurin (2002) suggests that one can nonetheless recognize particular groups of developmental resources (genes among them) as very unequivocal sources of developmental information about particular traits. Thus, innate traits can be characterized as those that are products of information from particular developmental resources that are maintained in populations by a variety of mechanisms. But this is likely to be controversial, as it rejects the idea that innate traits are necessarily genetic and it embraces the idea that innateness is a matter of degree.

The central advantage of making this move is that it focuses the study of innateness on the existence and nature of mechanisms that serve to maintain particular traits in biological populations. In doing so, it avoids the implausible assumption that the development of some traits is entirely ruled by the presence of some particular set of genetic (and perhaps environmental) precursors. This broadening of the notion of innateness is very much in line with recent findings in genetics. A study was begun in 1972 on the lives of 1,037 newborns in the city of Dunedin in the South Island of New Zealand. The most remarkable finding of the study to date has been a gene (monoamine oxidase A) that predisposes young people to violent behavior in later life. Those with low-active versions of the gene did four times the number of rapes, robberies, and assaults as they progressed to adulthood, *but only if they had also been maltreated as children*. Remarkably, the same maltreatment produced no corresponding psychological maladjustment in individuals with high-active versions of the gene (Caspi et al. 2002). No one would have called such environmentally mediated behavior innate in the old Lorenzian sense of the word, and yet here there is a very important and complex set of mechanisms that maintain a cycle of violence in a human community.

As more is learned about such inherited interactions, the study of innateness will increasingly be focused on the enhancement or amelioration of characteristics that are currently innate.

JAMES MACLAURIN

References

Ariew, A. (1999), "Innateness Is Canalization: In Defense of a Developmental Account of Innateness," in Valerie Gray Hardcastle (ed.), *Where Biology Meets Psychology*. Cambridge, MA: MIT Press, 117–138.

Caspi A., J. McClay, T. E. Moffitt, J. Mill, J. Martin, I. W. Craig, A. Taylor, and R. Poulton (2002), "Role of Genotype in the Cycle of Violence in Maltreated Children," *Science* 297: 851–854.

Chomsky, N. (1990), "On the Nature, Use and Acquisition of Language," in W. Lycan (ed.), *Mind and Cognition*. Cambridge, MA: Blackwell, 627–646.

——— (1965), *Aspects of the Theory of Syntax*. Cambridge, MA: MIT Press.

Cowie, Fiona (1999), *What's Within: Nativism Reconsidered*. New York: Oxford University Press.

Fodor, J. (2001), "Doing Without What's Within: Fiona Cowie's Critique of Nativism," *Mind* 110: 99–148.

——— (1981), "The Present Status of the Innateness Controversy," in J. Fodor (ed.), *RePresentations: Philosophical Essays on the Foundations of Cognitive Science*. Cambridge, MA: Bradford Books/MIT Press, 257–316.

(1975), *The Language of Thought*. New York: Crowell.

Griffiths, P. (2002), "What Is Innateness?" *Monist* 85: 70–85.

Griffiths, Paul, and Russell Gray (1994), "Developmental Systems and Evolutionary Explanation," *Journal of Philosophy* XCI: 277–304.

Leibniz, G. E. (1981), *New Essays on Human Understanding*. Translated by P. Remnant and J. Bennett. Cambridge: Cambridge University Press.

Lewontin, R. C. (1974), "The Analysis of Variance and the Analysis of Causes," *American Journal of Human Genetics* 26: 400–411.

Lorenz, Konrad (1950), "The Comparative Method in Studying Innate Behaviour Patterns," *Symposia of the Society for Experimental Biology* 4: 221–268.

Maclaurin, James (2002), "The Resurrection of Innateness," *Monist* 85: 105–130.

Marr, D. (1982), *Vision*. New York: W. H. Freeman.

Patrides, C. A. (ed.) (1970), *The Cambridge Platonists*. Cambridge, MA: Harvard University Press.

Putnam, H. (1992), "What Is Innate and Why: Comments on the Debate," in B. Beakley and P. Ludlow (eds.), *The Philosophy of Mind*. Cambridge, MA: MIT Press.

Sterelny, Kim (1989), "Fodor's Nativism," *Philosophical Studies* 55: 119–141.

Stich, Stephen P. (1975), "Introduction: The Idea of Innateness," in *Innate Ideas*. Berkeley and Los Angeles: University of California Press, 1–22.

Wimsatt, W. C. (1999), "Generativity, Entrenchment, Evolution, and Innateness: Philosophy, Evolutionary Biology, and Conceptual Foundations of Science," in Valerie Gray Hardcastle (ed.), *Where Biology Meets Psychology*. Cambridge, MA: MIT Press, 139–179.

See also **Biological Information; Chomsky, Noam; Cognitive Science; Heritability**

INSTRUMENTALISM

Though John Dewey coined the term *instrumentalism* to describe an extremely broad pragmatist attitude toward ideas or concepts in general, the distinctive application of that label within the philosophy of science is to positions that regard scientific theories not as literal and/or accurate descriptions of the natural world, but instead as mere tools or "instruments" for making empirical predictions and achieving other practical ends. This general instrumentalist thesis has, however, historically been associated with a wide variety of motivations, arguments, and further commitments, most centrally concerning the semantic and/or epistemic status of theoretical discourse (see below). Unifying all these positions is the insistence that one can and should make full pragmatic *use* of scientific theories either without believing the claims they seem to make about nature (or some parts of nature) or without regarding them as actually making such claims in the first place. This entry will leave aside the question of whether the term *instrumentalism* is properly restricted to only some subset of such views, seeking instead to illustrate the historical and conceptual relations they bear to one another and to related positions in the philosophy of science.

Loci Classici

Broadly instrumentalist sentiments concerning scientific theories have a remarkably long intellectual pedigree: indeed, Popper's (1963) famous critique of the position as intellectually sterile counts Andreas Osiander (author of the unsigned preface to Copernicus' *On the Revolutions of the Celestial Spheres*), Cardinal Bellarmino, and Bishop Berkeley as notable early defenders of the view (but cf. Fine 2001), even while resisting Duhem's claim to find its historical antecedents in classical Greek thinkers. Furthermore (as Popper and others note) the influential instrumentalism of nineteenth-century physicist Ernest Mach is rooted in a critique of Newtonian mechanics (and its concepts of absolute space, time, and motion) strikingly similar to Berkeley's own. Mach (1911) also resembles Berkeley in embracing a radical phenomenalism, insisting that what is represented " behind the appearances exists *only* in our understanding, and has for us only the value of a *memoria technica* or formula" (49): He argues that laws of nature (e.g., Snell's law) and theoretical hypotheses (e.g., the atomic hypothesis) are simply conceptual devices for the systematic classification, summary, organization, and coordinated expression and prediction of innumerable particular appearances (Mach [1893] 1960, 582f). Thus, Mach insists that theoretical concepts like 'atoms' are merely "provisional helps" and are ultimately to be dispensed with not because they seek unsuccessfully to describe a reality beyond appearances, but rather because they *successfully* but only *indirectly* describe coordinated and systematized collections of experiences themselves.

The instrumentalist impetus familiar from more recent philosophy of science, however, is rooted more fundamentally in developments within physics at the turn of the century, and in the related logical, epistemic, and historical concerns about the status of scientific theories articulated by thinkers like Pierre Duhem and Henri Poincaré (Worrall 1982) (see Duhem Thesis; Poincaré, Henri). The progress of physical science had by this time begun to suggest that there might be quite genuine cases of differences between actual competing scientific theories that could not possibly be adjudicated by any straightforward appeal to empirical tests or observations: to use a famous example of Poincaré's (though not a case of actual competing theories), any set of measurements of the angles in a triangle marked out by appropriately oriented perfectly rigid rods can be accommodated by the assignment of any number of different combinations of underlying spatial geometries and compensating "congruence relations" for the rods in question; if the sum of the angles differs from 180 degrees, for instance, one may either interpret the underlying geometry as Euclidean and conclude that the distance marked out by each rod varies with its position and/or orientation or assume that the distance marked out by each rod remains constant and conclude that the underlying geometry of the relevant space is non-Euclidean. Poincaré's response to this problem of theoretical underdetermination was *conventionalism*, that is, he regarded such theoretical matters as the assignment of a particular physical geometry to space as a matter of choice or convention to be decided on grounds of greatest convenience (see Conventionalism). And this in turn implied, he suggested, the distinctively instrumentalist conclusion that the quite useful ascription of a particular geometry to space by a theory should not be construed as literally attributing anything (truly or falsely) to nature itself: "[T]he question: Is Euclidean geometry true?... has no meaning. We might as well ask if the metric system is true, and if the old weights and measures are false.... One geometry cannot be more true than another; it can only be more convenient" (Poincaré [1905] 1952, 50).

To important, distinct worries about theoretical underdetermination Duhem added a further concern about the role played by idealizations in physical theories, and both he and Poincaré noted the long history of repeated and radical discontinuities in the dominant theoretical conceptions of particular domains of nature. But both argued that this history of scientific revolution and wholesale replacement is characteristic only of efforts to "surmise realities hidden under data observable by the senses" (Duhem [1914] 1954, 274). These data name only "the images we substituted for the real objects which Nature will hide forever from our eyes" (Poincaré [1905] 1952, 161). Thus, while both Duhem and Poincaré retained full confidence in the "experimental laws" or generalizations about observable phenomena uncovered by scientific investigations, each denied that such investigations were able to penetrate the actual constitution of nature, or that mathematical theories were able to describe it. Duhem went so far as to consign the explanatory ambitions of theories to the realm of metaphysics rather than science.

Thus recognizing that scientific theories or theorists *aspire* to describe an underlying, inaccessible reality and/or explain observable events by appeal to it, both Duhem and Poincaré simply reject these ambitions as ultimately either unscientific or unsatisfiable in some way. Both ranged at different times and in different works through a wide variety of importantly divergent attitudes (and not all the same ones between them) toward the cognitive, semantic, and epistemic status of theories, including the views that extant scientific theories were not in fact making claims about inaccessible realities behind observable phenomena, that the scientific enterprise need not do so, and that it should not. Moreover, there is reasonable controversy over classifying either thinker as ultimately an instrumentalist in any of these straightforward senses: in his latest work, Poincaré (1999) wholeheartedly embraced the reality of atoms, while Duhem consistently held that scientific theories are able to establish "natural classifications" of the phenomena.

Even this brief excursion through instrumentalist themes in Mach, Duhem, and Poincaré offers some sense of the variety of distinctive further commitments with which the general claim that scientific theories should be understood simply as useful instruments rather than accurate descriptions of inaccessible domains of nature has been conjoined. Among such further commitments are the following suggestions:

- Theoretical discourse is simply a device for organizing or systematizing beliefs about observational experience and its meaning is therefore exhausted by or reducible to any implications it has concerning observable states of affairs (reductive instrumentalism);
- Theoretical discourse has no meaning, semantic content, or assertoric force at all beyond the license it provides to infer some observable states from others (syntactic instrumentalism);
- Even if such discourse is both meaningful and irreducible, it can nonetheless be eliminated from science altogether (eliminative instrumentalism); and
- Even if the literal claims of theoretical science about the natural world are neither reducible, nor meaningless, nor even eliminable, such claims are nonetheless not to be believed (epistemic instrumentalism).

The Language of Science: Reductive, Syntactic, and Eliminative Instrumentalism

It is quite striking that even some of Duhem and Poincaré's explicit reservations about scientific theories have a semantic or linguistic character: Duhem ([1914] 1954) claims that "hypotheses [are] not judgments about the nature of things, only premises intended to provide consequences conforming to experimental laws" (39) and that theoretical propositions "are neither true nor false ... only convenient or inconvenient" (334), while Poincaré ([1905] 1952) adds that the "object of mathematical theories is not to reveal to us the real nature of things," but "only ... to co-ordinate the physical laws with which experiment makes us acquainted" (211). Perhaps less surprising is the fact that such a generally linguistic or semantic strategy of analysis was appealing to logical empiricist thinkers.

The logical empiricists' efforts to effect a reduction of all scientific language to a privileged phenomenological or observational basis (a project pursued most notably by the early Carnap, but also influentially by Bridgman) quite naturally grounded an instrumentalism about scientific theories of the sort described above as reductive (see Bridgman, Percy; Carnap, Rudolf). But even after this attempted reduction came to be widely regarded as a failure and such logical empiricists had given up the notion that the semantic content of apparently theoretical discourse was "really" exhausted by its implications concerning collections of observable events or subjective experiences, the distinctively syntactic variety of instrumentalism offered a fallback position: perhaps theoretical claims carry no straightforward ontological commitments regarding unobservable entities, even if they cannot be fully reduced to claims about immediately accessible experiences or states of affairs. More specifically, some logical empiricists suggested that theoretical claims are properly regarded as devoid of any semantic content whatsoever beyond the license they provide to draw inferences from one observable state of affairs to another. In the spirit of Duhem and Poincaré, this view regarded theoretical claims as nonassertoric, that is (appearances to the contrary), as not making claims about what the world is like and not possessing truth values at all.

Of course, this somewhat counterintuitive view of the semantics of theoretical discourse might be evaded by embracing the arguably more natural view (equally in the spirit of Duhem and Poincaré) that such discourse is simply *eliminable* from science altogether. This eliminative form of instrumentalism also gained considerable currency among logical empiricist thinkers, especially following the formulation and proof of an influential theorem by William Craig. Craig's theorem showed that for any recursively axiomatized first-order theory T, given any effectively specified subvocabulary O of T (mutually exclusive of and exhaustive with the remainder of the vocabulary of T), one can effectively construct another theory, T', whose theorems are exactly those of T that contain no nonlogical expressions besides those in O. As Hempel was the first to realize, this theorem implies that if the nonlogical vocabulary of any given scientific theory is partitioned into theoretical and observational components, the theory can be replaced with a "functionally equivalent" Craig transform that preserves all the deductive relationships between observation sentences established by T itself, since (by Craig's theorem) "any chain of laws and interpretive statements establishing [definite connections among observable phenomena] should then be replaceable by a law which directly links observational antecedents to observational consequents" (Hempel [1958] 1965, 186). This implied in turn, Hempel noted, that theoretical terms could be eliminated from theories altogether without losses in the purely observable consequences (deductively) obtainable from them, creating the following "Theoretician's Dilemma":

> If the terms and principles of a theory serve their purpose [of deductively systematizing the theory's observational consequences] they are unnecessary, as just pointed out; and if they do not serve their purpose they are surely unnecessary. But given any theory, its terms and principles either serve their purpose or they do not. Hence, the terms and principles of any theory are unnecessary. (Hempel [1958] 1965, 186)

The apparent feasibility of this eliminative instrumentalist program was further advanced by a related (and earlier, though largely unrecognized at the time) innovation of Frank Ramsey's: He proposed replacing any finitely axiomatized theory with a sentence that existentially generalizes on all the theoretical predicates of that theory. This "Ramsey sentence," he argued, has the same observational consequences as the original theory and therefore captures all the "factual content" of the original.

The significance of Craig's theorem was, however, immediately controversial. Nagel (1961), for instance, famously argued (136–137) that it is of quite limited relevance to the actual eliminability of theoretical discourse from science because:

(i) there is no guarantee that the axioms of T' delivered by Craig's method will not be "so cumbersome that no effective logical use can be made of them";

(ii) in fact, the axioms of T' will be infinite in number, no matter how simple the axioms of T itself, and correspond one-to-one with all the true statements expressible in the language of T', rendering them "quite valueless for the purposes of scientific inquiry"; and

(iii) Craig's method can actually be applied only if one knows, in advance of any deductions made from them, all the true statements in the restricted observational language.

In addition, Glymour (1980, Ch. 2) offers elegant technical objections to Ramsey's proposal, most importantly that as a theory of truth it fails to respect even the most elementary forms of demonstrative inference: For example, the Ramsey sentence of a conjunction may be necessarily false while the Ramsey sentences of each conjunct is individually true.

More recently, however, the profound differences between actual scientific theories and the sorts of artificial formal systems to which tools such as Craig's theorem and Ramsey's technique apply have led these formal results to be regarded as increasingly irrelevant to the genuine prospects for instrumentalism. More specifically, philosophers of science have become increasingly convinced that

(i) there is no strict, principled, or systematic division of the vocabulary of a theory into observational and theoretical parts;

(ii) the parts of a theory bear important logical, epistemic, and cognitive relations to one another that go far beyond what is captured by mere deductive systematization; and

(iii) scientific theories may not be best regarded as axiomatic formal systems in any case.

Thus, at least part of the solution to the Theoretician's Dilemma, as Hempel himself recognized, is to reject the claim that the only function of theoretical terms is to deductively systematize a theory's observational consequences.

Credibility and Belief: Epistemic Instrumentalism

Even as the philosophical fortunes of these distinctive semantic and eliminativist theses have declined, interest has remained strong in the broader instrumentalist conception of theories as tools for pursuing practical ends rather than accurate descriptions of nature itself. The most influential recent approaches have pursued this conception by exchanging the logical empiricists' reductive, syntactic, and eliminativist commitments for epistemic alternatives, that is, more recently influential forms of instrumentalism grant both the assertoric force and the ineliminability of theoretical claims but insist that such theories should simply be *used* for prediction of experimental outcomes and other practical goals without a requirement of *belief* in the claims they in fact make about nature itself (or some parts thereof). A further recent trend has been to make the case(s) for instrumentalism piecemeal, arguing that quite specific features of a given scientific theory (e.g., quantum mechanics, evolutionary biology) either require or recommend an instrumentalist stance toward that theory.

One prominent example of general instrumentalism of this epistemic variety is constructive empiricism of Bas van Fraassen (1980). Like Duhem and Poincaré, van Fraassen appeals to the underdetermination of theories by evidence to challenge the conclusion that empirically successful scientific theories describe what inaccessible domains of nature are really like, and he insists that even a reflective endorsement of the actual inferential and other practices of science itself requires only a cognitive attitude of *acceptance* toward theories, rather than belief. He argues that it is epistemically supererogatory to believe any more of scientific theories than that they are *empirically adequate*, that is, that what they say about *observable* phenomena is true, and he insists that epistemic prudence recommends agnosticism regarding even the most successful theories' claims about unobservables. Thus, constructive empiricism regards scientific theories as reliable tools for anticipating how observables will behave, while it resists the conclusion that such theories describe what unobservable domains of nature are really like—but on epistemic rather than semantic grounds.

Of course, constructive empiricism still relies fundamentally on an extremely controversial distinction (albeit itself naturalized) between observables and unobservables, so it is important to note that the distinctively epistemic form of instrumentalism *need* not rely upon any such distinction: As Fine (1991) argues the guiding commitment of instrumentalism is simply to the *reliability* of a causal story, which "treats all entities (observable or not) perfectly on par":

> Of course if the cause happens to be observable, then the reliability of the story leads me to expect to observe it (other things being equal). If I make the observation, I then have independent grounds for thinking the cause to be real. If I do not make the observation or if the cause is not observable, then my commitment is just to the reliability of the causal story, and not to the reality of the cause. (86)

Perhaps the most fully developed form of epistemic instrumentalism that eschews any important distinction between observables and unobservables is the historically oriented variety inspired by thinkers like Thomas Kuhn and pursued more recently by Larry Laudan. Like Duhem and Poincaré, these thinkers draw centrally on the history of repeated fundamental changes over time in the descriptions of nature offered by dominant scientific theories, in support of a skeptical attitude toward the claims of the dominant scientific theories of the present day. Kuhn ([1962] 1996) not only appeals to this history to undermine the notion that contemporary science is in possession of any final theoretical truth about a stable natural world, but also famously claims that the very "notion of a match between the ontology of a theory and its 'real' counterpart in nature now seems to me illusive in principle"; nonetheless he insists that scientific theories have improved over time "as instruments for puzzle-solving" (206). Laudan (1981a) argues that the long historical record of successful but ultimately rejected scientific theories undermines any justification for inferring even the approximate truth of contemporary scientific characterizations of nature (observable or not) from their dramatic empirical successes, but insists nonetheless not only that such theories can and should be used to tackle and solve a wide variety of empirical and conceptual problems, but also that there is a clear sense in which cumulative progress in this regard has been achieved over time, by attaining with the theoretical instruments of science an ever larger and more various set of effective solutions to such problems (Laudan 1977 and 1981b) (see Kuhn, Thomas).

As these influential formulations of the view illustrate, epistemic instrumentalism seems committed to some distinction between believing a theory to be true and accepting or using it *without* believing what it says. Perhaps unsurprisingly, the cogency of this distinction has itself been the target of recent influential criticisms of epistemic instrumentalism, on the grounds that these cognitive attitudes simply cannot be distinguished in the way that one or more forms of instrumentalism require. Horwich (1991) points out, for example, that some accounts of belief itself simply identify it as the mental state responsible for use, while Blackburn (1984) argues that there is no room for a distinction between merely "accepting" a statement with a truth condition and simply believing it to be true (see Fine 1986, especially sec. 4). By contrast, Sober (2002) defends the distinction, pointing out not only that idealized models known to be false are often accepted or used as the basis for accurate predictions across a range of phenomena, but also that recent work in model selection theory shows why models (statements containing adjustable parameters) known to be false will routinely serve as the basis for more accurate predictions of new data than competitors known to have higher likelihood conferred on them by the available data or even by competitors known to be true. Thus, he argues, not only is there a genuine difference between the goal of seeking instrumental or predictive reliability and the goal of seeking truth, but this distinction is respected within scientific practice itself, which typically chooses models (with adjustable parameters) with the former and fitted models (once parameters have been adjusted) with the latter goal in mind.

In a related vein, Nagel (1961, 139) famously argues that there is a "merely verbal" difference between the instrumentalist contention that a theory offers satisfactory techniques of inference and the realist contention that it is true. More recently, Stein (1989) has argued that the dispute between realism and instrumentalism is not well joined: There would be no appreciable difference (or no difference that makes a difference) between the two positions once

(a) realism becomes sophisticated enough (as Stein suggests it must) to (i) give up its pretensions to metaphysically transcendent theorizing, (ii) eschew aspirations to noumenal truth and reference, and (iii) abandon the idea that a property of a theory might somehow explain its success in a way that does not simply point out the use that has been made of the theory; and

(b) instrumentalism becomes sophisticated enough (as Stein suggests it must) to recognize the scope of a theory's role *as an instrument* to include not just calculating experimental outcomes, but also adequately representing phenomena in detail across the entire domain of nature and providing resources for further inquiry.

Thus, Stein argues for a convergence between the appropriately restricted ambitions a sophisticated realism holds out for theories and the appropriately expanded ambitions a sophisticated instrumentalism holds out for them, and indeed, that in the work of the deepest scientists (his examples are Maxwell, Newton, and Einstein) the two attitudes are present together in such a way that the alleged contradiction between them simply vanishes. Thus, even as instrumentalism persists as a viable and influential position in the contemporary philosophy of science, its comparative merits and even the coherence of its formulation remain the subject of deservedly intense controversy.

P. KYLE STANFORD

The author acknowledges the helpful input of Arthur Fine, Bas Van Fraassen, Bill Demopoulos, Elliott Sober, Larry Laudan, David Malament, Aldo Antonelli, Jeff Barrett, Stathis Psillos, and Philip Kitcher. This material is based upon work supported by the National Science Foundation under Grant No. SES-0094001. Any opinions, findings and conclusions, or recommendations expressed in this material are those of the author and do not necessarily reflect the views of the National Science Foundation (NSF).

References

Blackburn, Simon (1984), *Spreading the Word*. Oxford: Clarendon Press.
Duhem, Pierre ([1914] 1954), *The Aim and Structure of Physical Theory*. Translated by Philip P. Wiener. Princeton, NJ: Princeton University Press.
Fine, Arthur (1986), "Unnatural Attitudes: Realist and Instrumentalist Attachments to Science," *Mind* 95: 149–179.
——— (1991), "Piecemeal Realism," *Philosophical Studies* 61: 79–96.
——— (2001), "The Scientific Image Twenty Years Later," *Philosophical Studies* 106: 107–122.
Glymour, Clark (1980), *Theory and Evidence*. Princeton, NJ: Princeton University Press.
Hempel, Carl ([1958] 1965), "The Theoretician's Dilemma: A Study in the Logic of Theory Construction," in *Aspects of Scientific Explanation and Other Essays in the Philosophy of Science*. New York: Free Press.
Horwich, Paul (1991), "On the Nature and Norms of Theoretical Commitment," *Philosophy of Science* 58: 1–14.
Kuhn, Thomas S. ([1962] 1996), *The Structure of Scientific Revolutions*, 3d ed. Chicago: University of Chicago Press.
Laudan, Larry (1977), *Progress and Its Problems*. Berkeley and Los Angeles: University of California Press.
——— (1981a), "A Confutation of Scientific Realism," *Philosophy of Science* 48: 19–49.
——— (1981b), "A Problem-Solving Approach to Scientific Progress," in Ian Hacking (ed.), *Scientific Revolutions*. New York: Oxford University Press, 144–155.
Mach, Ernest ([1893] 1960), *The Science of Mechanics*, 6th ed. Translated by T. J. McCormack. La Salle, IL: Open Court.
——— (1911), *History and Root of the Principle of the Conservation of Energy*. Translated by P. E. B. Jourdain. Chicago: Open Court.
Nagel, Ernest (1961), *The Structure of Science*. New York: Harcourt, Brace and World.
Poincaré, Henri ([1905] 1952), *Science and Hypothesis*. New York: Dover.
Popper, Karl. R. (1963), *Conjectures and Refutations*. London: Routledge and Kegan Paul, chap. 3.
Psillos, Stathis (1999), *Scientific Realism: How Science Tracks Truth*. New York: Routledge.
Sober, Elliott (2002), "Instrumentalism, Parsimony, and the Akaike Framework," *Philosophy of Science* 69: S112–S123.
Stein, Howard (1989), "Yes, But... Some Skeptical Remarks on Realism and Anti-Realism," *Dialectica* 43: 47–65.
van Fraassen, Bas C. (1980), *The Scientific Image*. Oxford: Oxford University Press.
Worrall, John (1982), "Scientific Realism and Scientific Change," *Philosophical Quarterly* 32: 201–231.

See also **Conventionalism; Empiricism; Logical Empiricism; Phenomenalism; Realism; Theories**

INTENTIONALITY

Some things are *about,* or are *directed on,* or *represent* other things. For example, the sentence 'Cats are animals' is about cats (and about animals); this entry is about intentionality; Emanuel Leutze's most famous painting is about Washington's crossing of the Delaware; lanterns hung in Boston's

North Church were about the British; and a map of Boston is about Boston. In contrast, #a$b, a blank slate, and the city of Boston are not about anything. Many mental states and events also have "aboutness": the belief that cats are animals is about cats, as is the fear of cats, the desire to have many cats, and seeing that the cats are on the mat. Arguably some mental states and events are not about anything: Sensations, like pains and itches, are often held to be examples. Actions can also be about other things: Hunting for the cat is about the cat, although tripping over the cat is not. This (rather vaguely characterized) phenomenon of aboutness is called intentionality. Something that is about (directed on, represents) something else is said to "have intentionality" or to be an "intentional mental state."

This medieval terminology was reintroduced by the Austrian philosopher, Franz Brentano ([1874] 1995), in his book *Psychology from an Empirical Standpoint*, although Brentano himself did not use the word "intentionality." (For a brief history of the terminology, and further references, see Crane 1998a; for an account of Brentano's thought, see Moran 2000, Ch. 2.) In a famous passage, Brentano ([1874] 1995) claimed that every mental state/event has intentionality:

> Every mental phenomenon is characterized by what the Scholastics of the Middle Ages called the intentional (or mental) inexistence of an object, and what we might call, though not wholly unambiguously, reference to a content, direction towards an object (which is not to be understood here as meaning a thing), or immanent objectivity. In presentation something is presented, in judgement something is affirmed or denied, in love loved, in hate hated, in desire desired, and so on. (88)

Brentano's use of "intentional inexistence" is liable to confuse. Brentano did not mean that mental states are about peculiar nonexistent objects, but was rather referring to the (admittedly obscure) sense in which the object of a mental state is "in" the mind. The terminology of intentionality can also be confusing, for at least two reasons. First, intentionality has nothing in particular to do with intending, or intentions. Intentions—for instance, the intention to adopt a cat—are just one of many types of intentional mental states. Second, inten*t*ionality must be sharply distinguished from inten*s*ionality (Searle 1983). Mental states are not intensional, only sentences (and, for some sentences, other linguistic entities). A sentence *S* is intensional, or is an *intensional context*, just in case substitution of some expression *a* in *S* with some coreferring expression *b* yields a sentence with different truth value from the truth value of *S*. So, for example, "Necessarily, the number of planets is nine" and "Hegel believed that the number of planets is seven" are intensional. Substituting "nine" for the coreferential "the number of planets" turns the first false sentence into the true sentence "Necessarily, nine is nine" and the second true sentence into the false sentence "Hegel believed that nine is seven." As the first example indicates, a sentence can be intensional and yet have nothing to do with intentionality. Conversely, sentences that report intentional mental states/events need not be intensional (Crane 1998a). For example, "Berkeley heard the coach" is (arguably) not intensional: If that sentence is true, and if "the coach" and "Locke's favorite carriage" refer to the same thing, then "Berkeley heard Locke's favorite carriage" is true.

Paradoxes of Intentionality

As informally explained above, an intentional mental state (for example) is "about" something. The belief that Brentano is Austrian is about Brentano. The object that the state is about is called the intentional object of the state. (Intentional objects are sometimes taken to include states of affairs as well as particulars like Brentano: The belief that Brentano is Austrian could be said to be about Brentano's being Austrian.) So there should be a relation of aboutness that holds between a mental state and an object just in case the state is about the object—"the intentional relation," in Brentano's terminology.

Thinking of intentionality in this way, as a relation to intentional objects, leads to three classic "paradoxes of intentionality" (Thau 2002). The first paradox is that the intentional object need not exist (at any time). The belief that the fountain of youth is in Florida bears the intentional relation to the fountain of youth, and the fountain of youth does not exist. But if *a* is related to *b*, then there is such a thing as *a* and such a thing as *b*. One rather extreme solution, famously proposed by Brentano's student Alexius Meinong, is to hold that there are objects that do not exist. In this view, there *is* a fountain of youth, and the belief that the fountain of youth is in Florida bears the intentional relation to that (nonexistent) object. (This is Meinong's view, but not his terminology. Meinong used "subsists" to mean *exists*, and "exists" to mean something like *spatiotemporally exists*. Thus, in Meinong's terminology, Mont Blanc exists, the number 7 subsists but does not exist, and the fountain of youth neither exists nor subsists.)

The second paradox is that a mental state can bear an intentional relation to something without there being any particular thing that the state bears the relation to. If one wants a cat, but has no particular cat in mind, then one's state of wanting a cat bears the intentional relation to an object—a cat, presumably—yet there is no particular cat that the state bears the intentional relation to. But if a is related to something, then there is a particular object that a is related to.

The third paradox is that a mental state can bear the intentional relation to a, but not bear the intentional relation to b, even though a is b. The belief that the first postmaster general was a United States president is about the first postmaster general, but not about the inventor of bifocals, even though the inventor of bifocals was the first postmaster general, namely Benjamin Franklin. But if a bears a certain relation to b, and $b = c$, then a is related by the same relation to c.

In *Psychology from an Empirical Standpoint* Brentano ([1874] 1995) himself did appear to think that a mental state was always related to an intentional object, but in an appendix he insisted that the "only thing which is required by mental reference is the person thinking. The terminus of the so-called relation does not need to exist in reality at all" (272).

The moral of the paradoxes of intentionality is that thinking of intentionality in terms of the intentional relation is a bad idea. A better way involves drawing a distinction between the *representational content* of a mental state (or some other thing that has intentionality) and the objects (if any) the mental state is about. So, for example, the belief that the fountain of youth is in Florida has as its content the proposition that the fountain of youth is in Florida, and there is no object that the belief is about—at any rate, not the fountain of youth (the belief is about Florida). To believe that the fountain of youth is in Florida is to stand in the belief-relation to the proposition that the fountain of youth is in Florida. This proposition exists whether or not the fountain of youth does (it does not contain the fountain of youth as a constituent), and this proposition is true just in case there is such a thing as the fountain of youth and it is located in Florida. Similarly, the desire that one have a cat has as its content the proposition that one has a cat, and there is again nothing that the belief is about—at any rate, no particular cat. Finally, the belief that the first postmaster general was a United States president and the belief that the inventor of bifocals was a United States president are both about the same object, namely Benjamin Franklin. However, there is some truth behind the original mistaken claim that the two beliefs are about different objects. This can be brought out by noting that the contents of the two beliefs are true at different possible worlds (of course, the contents are both false at the actual world). Specifically, the first proposition, but not the second, is true at a possible world in which the first postmaster general became president and the inventor of bifocals never entered politics. The truth of the first proposition at a world depends on the political fortunes of whomever is the first postmaster general at that world—whether or not that individual invented bifocals.

Brentano's Two Theses

Brentano ([1874] 1995) proposed two theses that form the basis of contemporary discussions of intentionality:

1. No "physical phenomenon" has intentionality.
2. Intentionality is *the mark of the mental:* All and only mental states/events have intentionality.

> [I]ntentional in-existence is characteristic exclusively of mental phenomena. No physical phenomenon exhibits anything like it. We can, therefore, define mental phenomena by saying that they are those phenomena which contain an object intentionally within themselves. (89)

Brentano's examples of physical phenomena were not, say, brain processes, but were (chiefly) perceptible properties, like "color, sound and warmth" (92). Nonetheless, mainly through the influence of Roderick Chisholm, Brentano came to be associated with the doctrine that intentionality is not reducible to the physical—in the contemporary sense of 'physical' (see Physicalism). Although quite dubious as an interpretation of Brentano (Moran 1996), it started a debate that continues to this day. Chisholm himself argued for the irreducibility of intentionality by first transforming this thesis into one about the sort of language adequate for psychology. Thus recast, the thesis of the irreducibility of intentionality becomes one about the ineliminability of intensional contexts, like "Revere believes that the British are coming," in the language of a scientific psychology. (Chisholm called such sentences "intentional sentences.")

Chisholm's (1957) reformulation of "a thesis resembling that of Brentano" is:

> [W]e do not need to use intentional language when we describe non-psychological phenomena; we can express all of our beliefs about what is merely 'physical' in sentences which are not intentional. But ... when we

wish to describe perceiving, assuming, believing, knowing, wanting, hoping, and other such attitudes, then either (a) we must use language which is intentional or (b) we must use terms we do not need to use when we describe nonpsychological phenomena. (172–173)

Chisholm argued for his "linguistic version" of Brentano's first thesis as opposed to various behavioristically inspired analyses of "intentional language." Chisholm did not conclude that the failure of reduction impugned the reality of intentional mental states, but Quine ([1960] 1998) famously did:

> One may accept the Brentano thesis as either showing the indispensibility of intentional idioms and the importance of an autonomous science of intention, or as showing the baselessness of intentional idioms and the emptiness of a science of intention. My attitude, unlike Brentano's, is the second. (221)

Many philosophers are not so pessimistic, and there are many suggestions for providing a physicalistic or naturalistic reduction of intentionality. This is discussed in the following section.

Brentano's first thesis is true but has been (fruitfully) misinterpreted. Brentano's second thesis, on the other hand, has been correctly interpreted but seems obviously false, because of examples given in the first paragraph of this article. However, as discussed in the final section, Brentano's second thesis is in better shape than initial appearances suggest.

Reducing Intentionality

Many philosophers hold that there must be a physicalistic/naturalistic reduction of intentionality—at least if intentionality is a genuine phenomenon. Fodor (1987) is a prominent example:

> I suppose that sooner or later the physicists will complete the catalogue they've been compiling of the ultimate and irreducible properties of things. When they do, the likes of *spin, charm,* and *charge* will perhaps appear on their list. But *aboutness* surely won't; intentionality simply doesn't go that deep. ... If the semantic and the intentional are real properties of things, it must be in virtue of their identity with (or maybe of their supervenience on?) properties that are themselves *neither* intentional *nor* semantic. If aboutness is real, it must be really something else. (97)

There are many different approaches to providing the reduction of intentionality that Fodor says we need. Most adopt some kind of divide-and-conquer strategy. First, a distinction is made between *original* intentionality and *derivative* intentionality (Haugeland 1998; see Searle 1992 for a similar distinction between *intrinsic* and *derived* intentionality). A thing has derivative intentionality just in case the fact that it represents such-and-such can be explained in terms of the intentionality of something else; otherwise it has original intentionality. Often the intentionality of language and other sorts of conventional signs is said to be derivative. Language, in this view, inherits its intentionality from that of mental states, specifically from the intentions and conventions adopted by language users (Grice 1989). This is an attractive and plausible claim, although it is not obvious, and has been denied (see e.g., many of the essays in Davidson 1985). However, if it is correct, then the problem of reducing intentionality is itself reduced to the problem of reducing the intentionality of the mental.

Theories that attempt to provide a physicalistic reduction of intentionality fall into three broad groups. The first group comprises causal covariational theories (Stampe 1977; Dretske 1981; Stalnaker 1984; Fodor 1990). The basic idea is that mental states represent in much the same way that tree rings represent. The number of rings on a tree represents the tree's age, because the fact that the tree's age is n years old causes the tree to have n rings, or (a refinement) would cause the tree to have n rings in optimal conditions. A simple example of a causal covariational theory is this: A belief state S represents that p (that is, has propositional content that p) if and only if the fact that p would cause a subject to be in S. (This formulation takes the notion of a *belief state* for granted; a physicalistically acceptable version of the theory would have to provide a further reduction of a belief state.)

The second group comprises teleological theories (Papineau 1987; Millikan 1993; Dretske 1995). The basic idea is to explain the intentionality of mental states in terms of their biological functions, which might in turn be given a reductive account in terms of evolutionary history. A simple example is this: A belief state S represents that p if and only if in conditions in which a subject's cognitive system is functioning as it is designed to function by evolution, the subject would be in S when and only when it is the case that p (see Function).

The third group comprises functional role theories. Here the basic idea is that a representation or symbol means what it does because of its *functional role*—its causal interaction with other representations. A simple example (for public language): A two-place sentence connective * means *and* if and only if the acceptance of sentence P and sentence Q is disposed to cause the acceptance of the sentence $P \wedge * \wedge Q$ (i.e., P concatenated with * concatenated with Q), and the acceptance of $P \wedge * \wedge Q$ is disposed to cause the acceptance of P and the acceptance of Q. If this is to be an account of thought rather than language, then there must be an appropriate range of neural representations—perhaps words in a "language of thought." In "long-armed" theories, functional roles are taken to include causal interactions with the environment (Harman 1999); "short-armed" functional role theories exclude such causal interactions, and for that reason are often taken to be accounts of the so-called "narrow content" of mental states (Block 1986).

Two other notable approaches should be mentioned. One is Dennett's (1987) instrumentalism, which attempts to vindicate intentional notions from a physicalistic perspective without providing explicit reductions of the sort just illustrated. The other is Brandom's (1994) inferentialism, which attempts to reduce intentionality to normativity, in particular to norms governing inferential practices.

Intentionality as the Mark of the Mental

Brentano's second thesis is independent of (the misinterpretation of) his first, that intentionality cannot be given a physicalistic reduction. The irreducibility of intentionality does not imply that all and only mental states/events are intentional. Searle is an example of a philosopher who holds that intentionality is irreducible, yet that sensations are not intentional. Neither does the converse implication hold: if intentionality is the mark of the mental, it might still be reducible. Tye and Dretske, whose views are mentioned below, think that intentionality is the mark of the mental and that it can be given a physicalistic reduction.

Brenanto's second thesis divides into two parts:

- Intentionality is *sufficient* for mentality and
- intentionality is *necessary* for mentality.

The sufficiency claim is false—at least if 'intentionality' is used in the broad and loose contemporary way, to include nonmental entities like sentences, paintings, and maps (see the beginning of this article). However, the sufficiency claim might be amended as follows: Original intentionality is sufficient for mentality. According to the revised sufficiency claim, the mental is the source of all intentionality. This revised claim still faces problems. First, if the indication of a tree's age by the number of its rings is an example of intentionality at all, then it is presumably original intentionality. And if it is original intentionality, the sufficiency claim is false. Again, the sufficiency claim is false if the intentionality of language does not derive from the intentionality of the mental. But these are controversial issues, and there is at least some prospect of defending a modified version of Brentano's sufficiency claim.

Matters might seem even less promising with the other part of Brentano's second thesis, the claim that intentionality is necessary for mentality. At any rate, some philosophers think that sensations are obviously nonintentional. However, the claim that bodily sensation is a form of perception of one's own body was defended in the 1960s by D. M. Armstrong (1968) and has been revived today by a number of philosophers including Dretske (1995), Lycan (1996), and Tye (1995). And if this thesis is correct, then because perceptions have intentionality, bodily sensations are not counterexamples to the claim that intentionality is necessary for mentality. (See Brentano [1874] 1995, 82–85, for his account of the intentionality of pain, which anticipates many modern discussions.)

More problematic cases are provided by certain "objectless" emotions, like forms of anxiety or depression, where one is hard put to say what one is anxious or depressed about (Searle 1983). For defenses of Brentano's second thesis against this sort of example, see Tye (1995, 128–131) and Crane (1998b).

Assuming that every mental state/event is intentional, a further issue arises, whether the representational content of a mental state determines "what it's like" to be in the state—the state's qualia. Dretske, Lycan, and Tye, among others, endorse this determination claim. Such an "intentional theory of qualia" is controversial and has been widely discussed in the literature on consciousness.

ALEX BYRNE

References

Armstrong, David (1968), *A Materialist Theory of the Mind*. London: Routledge.
Block, Ned (1986), "Advertisement for a Semantics for Psychology," *Midwest Studies in Philosophy* 10: 615–678.
Brandom, Robert (1994), *Making it Explicit*. Cambridge, MA: Harvard University Press.
Brentano, Franz ([1874] 1995), *Psychology from an Empirical Standpoint*. Translated by Antos C. Rancurello, D. B. Terrell, and Linda L. McAlister. London: Routledge.
Chisholm, Roderick (1957), *Perceiving: A Philosophical Study*. Ithaca, NY: Cornell University Press.
Crane, Tim (1998a), "Intentionality," in Edward Craig (ed.), *Routledge Encyclopedia of Philosophy*. London: Routledge.

——— (1998b), "Intentionality as the Mark of the Mental," in Anthony O'Hear (ed.), *Current Isssues in Philosophy of Mind* (Royal Institute of Philosophy Supplement 43). Cambridge: Cambridge University Press, 229–251.
Davidson, D. (1985), *Inquiries into Truth and Interpretation*. Oxford: Oxford University Press.
Dennett, Daniel (1987), *The Intentional Stance*. Cambridge, MA: MIT Press.
Dretske, Fred (1981), *Knowledge and the Flow of Information*. Cambridge, MA: MIT Press.
——— (1995), *Naturalizing the Mind*. Cambridge, MA: MIT Press.
Fodor, Jerry (1987), *Psychosemantics*. Cambridge, MA: MIT Press.
——— (1990), *A Theory of Content and Other Essays*. Cambridge, MA: MIT Press.
Grice, Paul (1989), *Studies in the Way of Words*. Cambridge, MA: Harvard University Press, part 1.
Harman, Gilbert (1999), "(Nonsolipsistic) Conceptual Role Semantics," in *Reasoning, Meaning, and Mind*. Oxford: Oxford University Press, 206–231.
Haugeland, John (1998), "The Intentionality All-Stars," in *Having Thought*. Cambridge, MA: Harvard University Press, 127–170.
Lycan, W. G. (1996), *Consciousness and Experience*. Cambridge, MA: MIT Press.
Millikan, Ruth (1993), *White Queen Psychology and Other Essays for Alice*. Cambridge, MA: MIT Press.
Moran, Dermot (1996), "Brentano's Thesis," *Proceedings of the Aristotelian Society, Supplementary Volume* 70: 1–27.
——— (2000), *Introduction to Phenomenology*. London: Routledge.
Papineau, David (1987), *Reality and Representation*. Oxford: Blackwell.
Quine, Willard ([1960] 1998), *Word and Object*. Cambridge, MA: Harvard University Press.
Searle, John (1983), *Intentionality*. Cambridge: Cambridge University Press, chap. 1.
——— (1992), *The Rediscovery of the Mind*. Cambridge, MA: MIT Press.
Stalnaker, Robert (1984), *Inquiry*. Cambridge, MA: MIT Press.
Stampe, Dennis (1977), "Towards a Causal Theory of Linguistic Representation," *Midwest Studies in Philosophy* 2: 42–63.
Thau, Michael (2002), *Consciousness and Cognition*. Oxford: Oxford University Press, chap. 2.
Tye, Michael (1995), *Ten Problems of Consciousness*. Cambridge, MA: MIT Press.

See also **Function; Naturalism; Physicalism; Searle, John; Teleology**

IRREVERSIBILITY

Chunks of ice melt in warm water, but warm water never spontaneously forms chunks of ice. Gases expand to fill their containers, but uniformly spread gasses never contract into one corner of their (isolated) containers. These examples illustrate an incredibly pervasive regularity. Certain types of processes proceed only in one direction: they occur, but their time reverses do not. And this appears to be a matter of law. These processes are said to be irreversible. (This notion of irreversibility should be distinguished from a similarly termed notion that appears in classical thermodynamics: A system is said to undergo a "reversible change" when it changes so slowly that it remains very near thermal equilibrium throughout [Uffink 2001].)

Irreversible processes typically involve temperature-difference equalization, diffusion, the completion of chemical reactions, and certain phase transitions. More generally, isolated systems (or, rather, systems that can be treated as isolated) tend to move toward states of equilibrium. What explains this regularity? The simplest hypothesis is that fundamental laws of nature explicitly require that isolated systems tend toward equilibrium, and the science of thermodynamics provides a system of postulates that yields just this constraint.

Thermodynamics introduces a physical quantity, *thermodynamic entropy*, which is defined for systems at equilibrium. The thermodynamic entropy of a system measures how much of the system's energy is available for conversion into useful work (the higher the entropy, the less energy is available). The second law of thermodynamics says that the entropy of an isolated system never decreases (Sklar 1993, 21).

Irreversible processes—for example, a chunk of ice melts in warm water, a gas spreads to fill its container—involve increases in entropy. The second law rules out the time reversals of these processes because this would involve *decreases* in entropy. Despite the elegance and practical indispensability of classical thermodynamics, its postulates do not have the character of fundamental dynamical laws, which are thought to govern the detailed motions of the microscopic constituents of matter.

The natural next suggestion is that the fundamental dynamical laws themselves are time asymmetric and this asymmetry helps explain irreversibility. For example, particle physics has produced evidence that there are *T-symmetry violations*—interactions that have (slightly) different chances than their time reverses. So it is believed that the dynamical laws are not time symmetric. However, these slight differences in chances do not have a significant effect on the manner in which macroscopic systems undergo thermodynamic change. So this time asymmetry in the laws does not help explain irreversibility (Sklar 1993, 248).

Thus the following question is left open: How does the sort of irreversible behavior captured by the postulates of thermodynamics arise from the fundamental dynamical laws?

Statistical Mechanics

It is simplest to introduce statistical mechanics in the context of classical mechanics. As a simple example, consider a number of billiard balls, undergoing perfectly elastic collisions on a frictionless table (Figure 1). Suppose that at some initial time the balls are concentrated in the upper-left corner of the table and that they spread out over the course of a minute. Now perform a thought experiment. At the end of the minute, stop time, reverse the velocities of the balls, and start time again. The balls will retrace their original paths and return to the corner of the table in which they began.

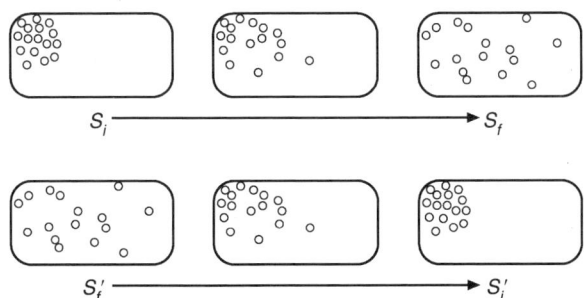

Fig. 1. The initial state of the table (S_i), in which the balls are concentrated, evolves into a final state (S_f), in which the balls are dispersed. Reversing the velocities of the balls in S_f results in a state S_f', which evolves into a state S_i' in which the balls are concentrated (Goldstein 2001).

This thought experiment illustrates a striking fact about classical mechanical laws of motion: Whenever a process is allowed by the laws, so is the time reverse of that process. So, given just these laws, there is no hope of showing that entropy-decreasing processes are downright disallowed. The best that can be hoped for is a statistical argument that entropy-decreasing processes are highly improbable. This observation is known as the reversibility objection (famously put to L. Boltzmann by J. Loschmidt [Sklar 1993, 35]).

There are many approaches to producing such an argument (Sklar 1993). One approach introduces the notion of *Gibbs entropy*. Gibbs entropy is a quantity defined not for individual systems, but for probability distributions over phase space (measuring the extent to which such distributions are spread out). This approach seeks to explain irreversibility by deriving certain facts about how such probability distributions evolve under the laws. It has been objected that this sort of result does not address the phenomenon to be explained, *viz.*, that *individual* isolated systems tend to increase in entropy (Lebowitz 1993; Goldstein 2001; Maudlin 1995). *Interventionism* attempts to avoid the reversibility objection by observing that thermodynamic systems interact with their environments. This observation is correct but does not avoid the difficulty, for one can shift attention to a larger system that is not subject to such interference.

The above difficulties are avoided by a highly influential approach to explaining irreversibility, which has its roots in the work of Ludwig Boltzmann.

Phase Space

The *phase space* of a system is a set of points, each of which is a dynamical state for the system to be in at a time. In the case of a number of classical point particles, each point of phase space determines the position and momentum of each particle. Phase space is also equipped with some additional geometric structure, including a measure that determines the volume of each of its regions. (Since the total energy of a classical system remains constant over time, attention can be restricted to that portion of phase space associated with some particular fixed total energy of the system.)

Phase space is carved up into disjoint sets, called *macrostates* (Figure 2). Points of phase space (or *microstates*) that are in the same macrostate are alike with respect to macroscopic parameters. For

IRREVERSIBILITY

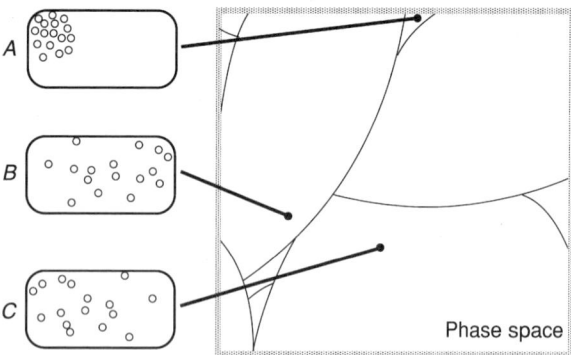

Fig. 2. Schematic representation of the phase space of a system of hard spheres. The space is divided into macrostates, which vary greatly in size. The space is dominated by equilibrium states such as *B* and *C*, which are shown in the left-hand column. Only a tiny proportion of phase space consists of far-from-equilibrium states such as *A*.

example, they have roughly the same temperature and pressure distributions.

Boltzmann noticed that for combinatorial reasons, macrostates vary greatly in size, and the variation is systematic. If all else is equal, macrostates in which the particles are spread out over physical space are bigger than ones in which they are clumped together, and macrostates in which the particles are moving with a large variety of momenta are bigger than ones in which, say, all have the same momentum.

That observation motivates the following definition: The statistical-mechanical entropy (also known as the *Boltzmann entropy*) of a point in phase space is a measure of the size of the macrostate to which it belongs—the bigger the macrostate, the greater the entropy. More precisely: the statistical-mechanical entropy of a point in phase space is proportional to the logarithm of the volume of the macrostate to which it belongs. For macroscopic systems, the imbalance in sizes of macrostates is overwhelming. Virtually all of phase space consists of points representing states in which the system is at equilibrium. This vast imbalance in size of macrostates can be used to explain why irreversible processes are so common.

A Statistical Explanation of Irreversibility

Consider a system consisting of a (nearly isolated) gas confined to a box. Take a particular low-entropy macrostate *L*, in which the gas is concentrated in the left half of the box. Notice that the phase space for this system is dominated by points whose entropies are *higher* than that of *L*. In other words, phase space is dominated by points in which the gas is *more spread out* than it is in *L*. So it is reasonable to think that practically all of the states in *L* have futures in which entropy *increases*. In other words, it is reasonable to think that practically all of the states in *L* have futures in which the gas *spreads out*.

So the following *appears* to be a statistical explanation of why gases spread out: the vast proportion of microstates compatible with a given gas-is-in-the-left-half macrostate have futures in which the gas spreads out. But as it stands, this explanation is defective. The trouble is that exactly analogous reasoning shows the following: The vast proportion of microstates compatible with a given gas-is-in-the-left-half macrostate have histories in which the gas *was more spread out in the past!*

And it is certainly *not* true that gases that are at one time concentrated in the left half of their boxes tend to have recent pasts in which they were dispersed throughout those boxes. A promising way out of this difficulty is to introduce an assumption concerning the initial state of the universe, *viz.*, that the universe started in a state of extremely low entropy. For example, one might posit an additional law of nature that has the effect of constraining the initial state of the universe in this way (Penrose 1993). Given such a law, the following modified explanation is available: almost all of the microstates compatible with a given gas-is-in-the-left-half macrostate *and compatible with the low-entropy constraint on the initial state of the universe* have futures in which the gas spreads out. Furthermore, the modified explanation (unlike the original one) does not lead to the incorrect retrodictions about the past history of the gas.

The global picture is this: judging purely by size of regions of phase space, one would expect for the universe to start (and stay) in global equilibrium. But restricting attention to just those regions of phase space compatible with a low-entropy initial condition, one would expect for global entropy to start low and increase over time. And it is reasonable to think that in a world with global increase of entropy, irreversible processes abound.

Entropy in Contemporary Physics

The explanation given above worked under the assumption of classical mechanics. Whether an explanation of the same type is consistent with contemporary physics remains to be seen. It is a reasonable (but as yet unproven) hypothesis that a definition of statistical-mechanical entropy in terms of phase space volume can be given in the context of general relativity (Bekenstein 2001 appraises this hypothesis).

It is expected that a general treatment of relativistic statistical-mechanical entropy would require a quantum-mechanical theory of gravitation (Wald 1998). Such a theory has been notoriously elusive. Nevertheless, the study of gravitational entropy is an active area of research. For example, theorists have offered statistical-mechanical measures of the entropy of black holes, from both the standpoint of quantum field theory (Sorkin 1998) and the standpoint of string theory (Horowitz 1998).

ADAM ELGA

References

Albert, David (2001), *Time and Chance*. Cambridge, MA: Harvard University Press.

Bekenstein, Jacob D. (2001), "The Limits of Information," *Studies in History and Philosophy of Modern Physics* 32: 511–524.

Callender, Craig (2001), "Taking Thermodynamics Too Seriously," *Studies in History and Philosophy of Modern Physics* 32: 539–553.

Goldstein, Sheldon (2001), "Boltzmann's Approach to Statistical Mechanics," in Jean Bricmont, Detlef Durr, Maria C. Galavotti, Giancarlo Ghirardi, Francesco Petruccione, and Nino Zanghi (eds.), *Chance in Physics: Foundations and Perspectives, Lecture Notes in Physics* 574. New York: Springer-Verlag.

Horowitz, Gary T. (1998), "Quantum States of Black Holes," in R. Wald (ed.), *Black Holes and Relativistic Stars*. Chicago: University of Chicago Press.

Lebowitz, Joel L. (1993), "Macroscopic Laws, Microscopic Dynamics, Time's Arrow and Boltzmann's Entropy," *Physica A* 194: 1–27.

Maudlin, Tim (1995), "Review of L. Sklar's Physics and Chance and Philosophy of Physics," *British Journal of the Philosophy of Science* 46: 145–149.

Penrose, Roger (1993), *The Emperor's New Mind*. Oxford: Oxford University Press.

Price, Huw (2002), "Boltzmann's Time Bomb," *British Journal for the Philosophy of Science* 53: 83–119.

Sklar, Lawrence (1993), *Physics and Chance: Philosophical Issues in the Foundations of Statistical Mechanics*. Cambridge: Cambridge University Press.

Sorkin, Rafael D. (1998), "The Statistical Mechanics of Black Hole Thermodynamics," in R. Wald (ed.), *Black Holes and Relativistic Stars*. Chicago: University of Chicago Press.

Wald, Robert M. (1998), "Black Holes and Thermodynamics," in Wald (ed.), *Black Holes and Relativistic Stars*. Chicago: University of Chicago Press.

Uffink, Jos. (2001), "Bluff Your Way in the Second Law of Thermodynamics," *Studies in History and Philosophy of Modern Physics* 32: 305–394.

See also **Classical Mechanics; Reductionism; Thermodynamics; Time**

K

KIN SELECTION

See **Natural Selection**

KINETIC THEORY

Kinetic theory explains the properties and behavior of physical systems on the basis of the hypothesis that they consist of a great number of particles (e.g., molecules or atoms) in motion (Greek: *kinesis* = motion). Its most important application is the kinetic theory of gases, but it can be applied to liquids and collections of subatomic particles as well. This discussion will be restricted to the kinetic theory of gases, as this suffices to clarify the nature of kinetic theory and to highlight its philosophical aspects.

The kinetic theory of gases proceeds from two principles:

1. *An ontological principle*: Gases are composed of freely moving particles subject to the laws of classical mechanics (atomism, mechanism); and
2. *A methodological principle*: The behavior of gases is analyzed not by tracing the trajectory of every individual particle but by applying statistical methods to the collection of particles as a whole (see Classical Mechanics).

The theory is historically and conceptually related to the theories of statistical mechanics and thermodynamics. Statistical mechanics is a generalization of kinetic theory that emerged in the course of the latter's development by Ludwig Boltzmann and J. W. Gibbs. Thermodynamics is a phenomenological theory that accounts for the behavior of

KINETIC THEORY

gases without a hypothesis about their microscopic constitution (see Thermodynamics).

Therefore, thermodynamics can be regarded as a competitor to kinetic theory, a relation that is a topic of philosophical debate. The kinetic theory of gases has played a significant role in the shaping of modern physics and has been relevant for the philosophy of science in a variety of ways, having had an impact on the development of its ideas. The kinetic theory of gases is regularly employed in philosophical discussions.

The Kinetic Theory of Gases in Historical Perspective

An essential feature of the kinetic theory of gases is the identification of temperature with molecular motion, specifically with the mean kinetic energy of molecules: $T \sim <\frac{1}{2}mv^2>$, where T is the temperature and m and v are the mass and velocity, respectively, of an individual molecule. In the second half of the nineteenth century, this kinetic view of heat replaced alternatives such as the caloric theory, which assumed that heat was a substance ("caloric"), and the wave theory of heat, which took heat to be vibrations of an ether. Before 1850, some kinetic theories of gases were advanced, which attempted to explain Boyle's law that at constant temperature the product of pressure and volume of the gas is constant (known since 1662). These theories did not have much impact because Newton had already explained Boyle's law by means of a static molecular theory of gases. The earliest kinetic theory was Daniel Bernoulli's ([1738] 1965) in which pressure is proportional to molecular velocity squared. It did not yet have a temperature scale, however. Bernoulli derived the formula (still accepted in modern kinetic theory)

$$PV = \frac{1}{3}N<mv^2>,$$

where P is the pressure, V is the volume, N is the number of molecules, and m and v are the mass and velocity, respectively, of an individual molecule. Kinetic theories were proposed by John Herapath in 1820 and J. J. Waterston in 1845 in which temperature was related to molecular velocity: Herapath (wrongly) supposed that T was proportional to v; Waterston took T to be proportional to v^2. Both theories were ignored (for historical references, see Brush 1965–1972 and 1976).

Around 1850, the scientific scene turned favorable for kinetic theories: Joule and others established the law of conservation of energy (or convertibility of heat and work), which gave essential support to the kinetic view of heat. This paved the way for the important kinetic theory of Rudolf Clausius ([1857] 1965). Like earlier theories, Clausius's statistical hypotheses were of a simple kind: calculations were based on a random variation in the *direction* of particle velocities; their *magnitude* was represented by one average value. In response to an objection by Buys Ballot, Clausius introduced the "mean free path" of molecules and calculated that this path is so small that gases mix quite slowly despite the high velocities of individual molecules.

In the 1860s, Clausius's theory was further developed by James Clerk Maxwell, who at first regarded his work on gas theory as an "exercise in statistics"; he did not (yet) believe in the atomistic view of matter. Maxwell's theory explained many properties of real gases; most important were its predictions regarding transport phenomena (heat conduction, diffusion, viscosity). On the basis of Maxwell's approach, Boltzmann later developed a general transport equation (the Boltzmann equation), which is still employed today. Maxwell refined Clausius's statistical analysis: Instead of using merely one average velocity, he introduced a statistical distribution function $f(v)$ for molecular velocities (a Gaussian curve), where v is the velocity of individual molecules. Boltzmann generalized Maxwell's distribution law for situations in which external forces are present (leading to what is now called the *Maxwell–Boltzmann distribution law*):

$$f(v) = N\sqrt{\frac{2m^3}{\pi k^3 T^3}} v^2 e^{-(\frac{1}{2}mv^2 + V[x])/kT}$$

where m and v are the mass and velocity, respectively, of an individual molecule; k is Boltzmann's constant; and $V[x]$ is a potential due to an external force depending on the position x. This law applies only to equilibrium situations; in order to explain the tendency toward equilibrium, Boltzmann advanced the H-theorem (see below).

A fundamental element of kinetic theory is the equipartition theorem, which states that every degree of freedom of the system takes up an equal part of the total kinetic energy. The theorem had an important anomalous consequence: Its prediction for the specific heat ratio gases is at odds with the experimentally obtained value for many ordinary gases (e.g., oxygen, nitrogen). This "specific heat anomaly" was discovered by Maxwell (1875), who called it "the greatest difficulty which the molecular theory has yet encountered." Boltzmann's proposed solution—the "dumbbell model" of diatomic molecules—was controversial at the time because it disregarded spectral evidence for internal atomic structure (see de Regt 1996). In his famous 1900 lecture, William Thomson (Lord Kelvin) (1904) labeled the equipartition problem as one of two "nineteenth-century clouds over the dynamical theory of heat and light." Today Boltzmann's model is accepted as an idealization: nineteenth-century objections to the model have dissolved since quantum mechanics has separated internal atomic structure (spectra) from mechanical degrees of freedom (see Quantum Mechanics).

In the early twentieth century, kinetic theory was incorporated into statistical mechanics, due to the work of J. W. Gibbs. Kinetic theory itself did not witness serious changes anymore; the most important twentieth-century contributions consisted in methods for solving the Boltzmann equation, notably by S. Chapman and D. Enskog (Brush 1976, chap. 12). Kinetic theory was fruitfully applied to other topics, such as radiation transfer, ionization, chemical reactions, evaporating liquids, and neutron transport. While kinetic theory remains a paradigm nineteenth-century theory, it has had a profound influence on twentieth-century physics, especially on the genesis and development of quantum theory: Boltzmann's work was a key element of Planck's solution of the problem of black-body radiation, marking the beginning of quantum theory in 1900.

Kinetic Theory: Atomism and Scientific Realism

Kinetic theory is based on an atomistic ontology. The unobservability of atoms led to philosophical disputes over their existence, and kinetic theory played a pivotal role in debates between scientific realists and their opponents.

Around 1850, atoms were regarded as fictional entities. However, due to the impressive successes of the kinetic theory in later years, more and more scientists took a realist stance toward the theory. But while the existence of atoms was gradually accepted by the scientific community, their structure was still open to debate and speculation. In the second half of the nineteenth century, various models of atomic structure were proposed. Maxwell treated atoms as hard elastic spheres and later as centers of force. By contrast, Kelvin's vortex theory represented atoms as spinning rings of a homogeneous, frictionless, incompressible fluid. Meanwhile, estimations of molecular sizes were made, first by J. Loschmidt (1865), who also calculated the number of molecules per volume unit (today this number, which is the same for every gas at standard temperature and pressure, is known as Avogadro's number). Subsequently, J. D. van der Waals replaced the ideal gas law ($PV = nRT$, where P is the pressure, V the volume, n the number of moles of the gas, R the ideal gas constant, and T the temperature) by an equation of state for real gases, containing correction terms depending on intermolecular forces and the size of the molecules:

$$\left(P + \frac{an^2}{V^2}\right)(V - nb) = nRT,$$

where a and b are constants depending on the type of molecule, n is the number of moles, and R is the molar gas constant.

These scientific successes firmly but only temporarily established the atomistic worldview and an accompanying realistic view of scientific theories. After 1880, kinetic theory lost momentum, and in the final decade of the nineteenth century the theory was strongly attacked by anti-atomists who based their objections on a positivist philosophy of science. A notable early example is J. B. Stallo's ([1884] 1960) criticism of kinetic theorists for having "faith in spooks" and for wasting "their efforts upon a theory so manifestly repugnant to all scientific sobriety" (151). The most important anti-atomist was Ernest Mach, whose radical empiricist movement (to which the young Max Planck adhered) influenced twentieth-century philosophy of science, particularly the logical positivism of the Vienna Circle (see Mach, Ernest; Vienna Circle).

The tide turned once again in the early twentieth century. In the course of his work on black-body theory, Planck was converted to atomism and became a staunch opponent of Mach. In 1905, Einstein published an explanation of Brownian motion—the observable irregular motion of small particles suspended in fluids, discovered by Robert Brown in 1828—on the basis of kinetic theory: Brownian motion results from the impact of surrounding molecules on the particle, and Einstein ([1905] 1965) derived a prediction of the mean displacement of Brownian particles. This prediction was successfully tested by Jean Perrin (1913), who subsequently made it into a strong case for the reality of atoms in his book *Les Atomes*. As such, Einstein's explanation (independently developed by Smoluchowski) was the final vindication of kinetic theory and atomism.

Today, the existence of atoms is uncontested among scientists, but the case of atomism and kinetic theory is still used in philosophical debates about the status of unobservable entities, particularly (not surprisingly) by realists. For example, Salmon (1984, 213–227) returns to Perrin's argument in order to defend scientific realism; he argues that from the fact that there are many independent methods of determining Avogadro's number, which all arrive at the same result, one must conclude that this number describes something real and that molecules thereby exist. (For an illuminating analysis of the debate between realists and instrumentalists from the perspective of the development of kinetic theory, see Gardner 1979.)

Kinetic Theory: Explanation and Reduction

Kinetic theory is often cited as a paradigmatic example in philosophical discussions about explanation and reduction. In the context of his deductive-nomological model of explanation, Carl Hempel cited kinetic-theoretical explanations of phenomenological gas laws (such as Boyle's law) as exemplary cases of "theoretical explanation" (see Explanation; Hempel, Carl Gustav). Crucial in Hempelian theoretical explanations (where the explanans refers to theoretical, that is, unobservable, entities) are the so-called bridge principles connecting the theoretical and the observational level. In the case of kinetic theory, the bridge principles relate macroscopic features such as temperature and diffusion rate with microscopic properties such as velocity and kinetic energy of the gas molecules (Hempel 1966, 73).

Contemporary philosophers who reject Hempel's model of explanation feel nonetheless obliged to give alternative interpretations of how kinetic theory explains gaseous behavior. Apparently, the fact that kinetic theory provides scientific explanations is undisputed: Any respectable theory of explanation should be able to account for the explanatory power of kinetic theory. Thus, the presently influential unificationist conception of explanation argues that kinetic theory provides explanations because it unifies many (sometimes seemingly unrelated) facts about nature: It accounts not only for Boyle's law, but for many other phenomenological gas laws as well, and also relates gaseous behavior to other natural phenomena governed by the laws of mechanics (Friedman 1974, 14–15). Alternatively, the causal conception of explanation claims that it is the causal-mechanical features of the kinetic theory that do the explanatory work (Salmon 1984, 227–228).

A related philosophical issue is that of intertheoretic reduction. According to Nagel (1961, 342), the relation between kinetic theory and the phenomenological theory of thermodynamics is "a classic and generally familiar instance of such a reduction" (see 338–345 for an account). *Pace* Nagel, the reduction of thermodynamics to kinetic theory is not completely unproblematic: kinetic theory appears to be unable to account for the tendency toward equilibrium described by the second law of thermodynamics. This inconsistency was already observed by Maxwell (1872), in his famous thought experiment known as Maxwell's Demon: Maxwell imagined a microscopic but "very observant and neat-fingered being" manipulating molecules in such a way as to make heat flow from a cold to a hot body, thereby contradicting the second law of thermodynamics (see Irreversibility). Boltzmann's ([1872] 1966) H-theorem was intended as a microphysical analogue to the second law. Boltzmann defined a function H on $f(v^2)$ and proved that H always decreases. As such, H can be regarded as a microphysical counterpart of entropy S. Boltzmann's proof required an extra statistical hypothesis, the *Stosszahlansatz* (molecular chaos): there is no statistical correlation between colliding molecules before and after the collisions. However, if the system behaves according to deterministic Newtonian mechanics (as kinetic theory presupposes), the *Stosszahlansatz* cannot be absolutely true.

This incompatibility between mechanical laws (read: kinetic theory) and thermodynamics was made explicit in the reversibility objection (Thomson [1874] 1966; also known as Loschmidt's [1876] *Umkehreinwand*): If one considers a process and reverses the velocity of every molecule, the resulting process will be physically possible as well. This contradicts the second law, and the experience that irreversible processes exist in nature. Boltzmann responded to the objection with his famous equation $S = k.\log W$, relating the entropy S of a macroscopic state to the number (W) of possible microscopic states corresponding with the macroscopic state in question (in other words, to its relative probability). This equation, which lies at the basis of statistical mechanics, implies that entropy decrease is not impossible but only very improbable: Because there are many more microstates corresponding with a macrostate of high entropy (disorder), the probability that the system develops into a state of higher entropy is much greater than vice versa. In contrast to Boltzmann's response, however, some authors take an antireductionist approach by claiming that the second law has absolute (ontological) validity and that attempts at reducing thermodynamics to mechanics are misguided (e.g., Prigogine 1980). A more detailed overview of philosophical issues related to kinetic theory can be found in Sklar (1993).

HENK W. DE REGT

References

Bernoulli, D. ([1738] 1965), "On the Properties and Motions of Elastic Fluids, Especially Air," in Stephen G. Brush (ed.), *Kinetic Theory*, vol. 1. Oxford: Pergamon Press, 57–65.

Boltzmann, L. ([1872] 1966), "Further Studies on the Thermal Equilibrium of Gas Molecules," in Stephen G. Brush (ed.), *Kinetic Theory*, vol. 2. Oxford: Pergamon Press, 88–175.

Brush, Stephen G. (1976), *The Kind of Motion We Call Heat*. Amsterdam: North-Holland.

——— (ed.) (1965–72), *Kinetic Theory*, 3 vols. Oxford: Pergamon Press.

Clausius, R. ([1857] 1965), "The Nature of the Motion Which We Call Heat," in Stephen G. Brush (ed.), *Kinetic Theory*, vol. 1. Oxford: Pergamon Press, 111–134.

de Regt, Henk W. (1996), "Philosophy and the Kinetic Theory of Gases," *British Journal for the Philosophy of Science* 47: 31–62.

Einstein, A. ([1905] 1956), *Investigations on the Theory of Brownian Movement*. New York: Dover.

Friedman, Michael (1974), "Explanation and Scientific Understanding," *Journal of Philosophy* 71: 5–19.

Gardner, Michael (1979), "Realism and Instrumentalism in 19th-Century Atomism," *Philosophy of Science* 46: 1–34.

Hempel, Carl G. (1966), *Philosophy of Natural Science*. Englewood Cliffs, NJ: Prentice-Hall.

Loschmidt, J. (1865), "Zur Grösse der Luftmolecüle," *Sitsungsberichte, K. Akademie der Wissenschaften in Wien, Math.-Naturwiss, Kl.* 52: 395–413.

——— (1876), "Über den Zustand des Wärmegleichgewichtes eines Systemes von Körpern mit Rucksicht auf die Schwerkraft," *Sitsungsberichte, K. Akademie der Wissenschaften in Wien, Math.-Naturwiss, Kl.* 73: 128–142.

Maxwell, J. C. (1872), *Theory of Heat*. New York: Appleton.

——— (1875), "On the Dynamical Evidence of the Molecular Constitution of Bodies," *Nature* 11: 357–359, 374–377.

Nagel, Ernest (1961), *The Structure of Science*. London: Routledge and Kegan Paul.

Perrin, Jean (1913), *Les Atomes*. Paris: Alcan. English translation: *Atoms*. New York: Van Nostrand, 1923.

Prigogine, Ilya (1980), *From Being to Becoming*. San Fransisco: Freeman.

Salmon, Wesley (1984), *Scientific Explanation and the Causal Structure of the World*. Princeton, NJ: Princeton University Press.

Sklar, Lawrence (1993), *Physics and Chance*. Cambridge: Cambridge University Press.

Stallo, John B. ([1884] 1960), *The Concepts and Theories of Modern Physics*. Cambridge, MA: Belknap Press.

Thomson, William (Lord Kelvin) ([1874] 1966), "The Kinetic Theory of the Dissipation of Energy," in Stephen G. Brush (ed.), *Kinetic Theory*, vol. 2. Oxford: Pergamon Press, 176–187.

——— (1904), "Nineteenth-Century Clouds over the Dynamical Theory of Heat and Light," in *Baltimore Lectures on Molecular Dynamics and the Wave Theory of Light*. London: C. J. Clay and Sons, 486–527.

See also **Classical Mechanics; Explanation; Irreversibility; Mechanism; Quantum Mechanics; Probability; Realism; Reductionism**

THOMAS KUHN

(18 July 1922–17 June 1996)

Thomas S. Kuhn was the most widely read, and most influential, philosopher and historian of science of the twentieth century. *The Structure of Scientific Revolutions*, first published in 1962, challenged then-dominant philosophical views of science regarding progress, rationality, observation, theories, and language. The book has been continuously in print for forty years; it has been translated into more than twenty languages, and the various editions have sold over a million copies. Unlike all other books in the history or philosophy of science, *Structure* was, and is, still widely read outside of the philosophical community.

Kuhn was also the author of *The Copernican Revolution*, and in 1978, *Black-Body Theory and the Quantum Discontinuity, 1894–1912*, as well as numerous essays, most of which were reprinted in two collections, *The Essential Tension* (Kuhn

1977) and *The Road since Structure* (Kuhn 2000). Although the monographs each made important contributions, respectively, to contemporary understanding of the Copernican revolution and of the early stages of the quantum revolution, none of the other work has attracted nearly as much attention as *Structure*, especially within philosophy of science. If Kuhn had written only these two monographs and the articles, he would have merited a minor footnote as a historian of science in the twentieth century.

Although Kuhn was the most influential philosopher of science of the twentieth century, his formal training was in physics, and his next career turn was to the history of science. He was trained as a physicist, receiving a B.S. in 1943, an M.A. in 1946, and a Ph.D. in 1949 from Harvard in that discipline. Moreover, his first teaching position, at University of California, Berkeley, from 1956 to 1964, and his second, at Princeton University (1964–1979), were in history departments and in the history and philosophy of science program at Princeton, but not in philosophy. Only when he moved to Massachusetts Institute of Technology in 1979 did he become a member of a philosophy department (though it was the *Linguistics and* Philosophy Department). He was elected president of the Philosophy of Science Association in 1989, after having been a member of that association for only a few years.

Revolutions and Two Kinds of Science

The main thesis of *Structure* is that the development of the natural sciences and their subfields proceed through alternations of two kinds of scientific development: *normal* and *revolutionary* science (see Scientific Revolutions). A period of normal science produces cumulative progress in understanding of the domain of that field and involves the application and refinement of generally accepted theories to the unresolved questions in a domain according to an agreed understanding both of what constitutes a reasonable question and on the criteria used to adjudicate answers. Revolutionary change involves rejection of a significant portion of the theories, methods, and criteria for problem solution, and their replacement by new ideas. In revolutionary change at least some of the previously "solved" problems are rejected or reopened. One of the most controversial claims of the book was that there was incommensurability between a revolutionary theory and the one it supplanted (see Incommensurability).

The book received both widespread praise and condemnation. In addition to the claims about "revolutions," which were very controversial, critics argued that Kuhn's account of science made it neither rational nor objective. Kuhn regarded this as a misinterpretation of his views. In addition to the claim about incommensurability, he claimed that when a scientific revolution occurs, "the world changes," and both of these claims provoked philosophical outrage in many critics (see Shapere 1964 for an example). An elaboration of the basic views of the book is presented first; discussion of controversies and consequences follows. For an extensive discussion of the meanings and history of the phrase 'scientific revolution' (see Scientific Revolutions).

Understanding Kuhn's account is complicated by the fact that throughout his career he was in the process of refining his positions to clarify them, to meet objections and to eliminate misunderstandings. *Structure* was only published in its actual form because Kuhn had agreed to write an entry for the Encyclopedia of Unified Science series and the editors pressed him to produce a manuscript (see Unity of Science Movement). His personal preference would have been to continue to develop the ideas and to relate them in more detail to the existing tradition in philosophy of science, with which he was then acquainting himself. In the Preface, he apologizes for leaving out many topics and more precise references because of lack of space. After the somewhat negative reception of the book, he suggested a major terminological change in the second edition (1970), which he did not incorporate in the text generally, but only in a postscript, which has been largely ignored by readers and critics. He continued to work throughout his life on a clearer and more definitive formulation of his views and he died in 1996 without having completed that project. Thus any evaluation must be of work still in progress.

Structure

In the first edition of *Structure*, Kuhn defined the two kinds of scientific development in terms of *paradigms*. Normal science involves the articulation and refinement of a paradigm that is shared by the relevant scientific community; in revolutionary scientific change, one paradigm is rejected and another takes its place. One reason for the widespread influence of the book outside of the community of philosophers and historians is that the conception of a group or community guided by a *paradigm* seemed to have explanatory value in many settings. This use of the term has become firmly entrenched as a standard expression in English and appears in cartoons and business management courses, although most of its contemporary users have no notion of its source.

However useful the term 'paradigm' has proven to be in the general culture, it was the cause of considerable criticism in the reception of the book because critical readers perceived that he was using the term very variously and loosely. One critic presented a taxonomy of twenty-two distinguishable senses of the term in *Structure* (Masterman 1974). Kuhn disagreed with the precise count, but was sufficiently persuaded by Masterman's critique and those of many other critics that clarification was required. Many of the criticisms were aired at two important conferences that focused heavily on his work, in London in 1965 and Champaign, Illinois, in 1969. The proceedings of these were eventually published as *Criticism and the Growth of Knowledge* (Lakatos and Musgrave 1974) and *The Structure of Scientific Theories* (Suppe 1977). Kuhn was also conducting graduate seminars at Princeton that were attended by philosophers of science on the faculty, as well as historians and graduate students from both disciplines. As a result of these influences and further reflection, in the postscript to the second edition of *Structure* in 1970, he expressed a desire to replace the term 'paradigm' with two new terms, 'disciplinary matrix' and 'exemplar,' which he believed expressed the two main distinct uses he had made of "paradigm." Similar qualifications of the view in *Structure* are expressed in his contributions to those two conferences.

The six elements that Kuhn intends to include in a *disciplinary matrix* are:

(i) equations or other symbolic representations,
(ii) instruments,
(iii) standards of accuracy and experimental repeatability,
(iv) metaphysical assumptions,
(v) the domain of inquiry, and
(vi) exemplars.

The *domain of inquiry* includes the problems that workers in the field regard as relevant but unsolved. An *exemplar* is, as we have seen, the second meaning of 'paradigm,' and Kuhn emphasizes that these are *concrete* examples of problem solutions. One of the most crucial points in his emphasis on exemplars is that they give guidance to future research by example; these examples are implicitly constrained by rules and are instances of a method, but neither the rules nor the method is explicit in them. A scientific field or specialty is given its coherence partly by the shared examples, but it is given its diversity of approaches by the possibility of researchers interpreting those examples somewhat differently from one another. Researchers can all agree that they want to do for their field what Newton did for his, but they may disagree fairly radically about what that was, and therefore on what they intend to achieve. Some of the confusion in interpreting *Structure* was due to the fact that one sense of paradigm, that is, exemplars, are an element of the other sense (disciplinary matrices).

Disciplinary Matrices and "The Scientific Revolution"

In the Ptolemaic/Aristotelian scheme that dominated scientific thought in the Western world for almost two millennia, one of the fundamental metaphysical assumptions is that there are unbridgeable differences between terrestrial and celestial phenomena, while the Copernican/Newtonian view assumes the uniformity of laws throughout the universe. The former emphasizes qualitative explanations; the latter, quantitative predictions. Famously, the telescope, particularly in the hands of Galileo, was a critical instrument in the arguments against the static Ptolemaic/Aristotelian view of the heavens. While Kuhn was still a graduate student, James Conant, a chemist who was then president of Harvard, asked him to assist in preparing a historically oriented physics course for non-science majors. In the process of preparing for this course, Kuhn, who had read little or no history of science previously, spent an extensive period of time reading Aristotle.

He describes how as he read Aristotle he discovered that Aristotle had known almost no mechanics, if one understands mechanics as the system discovered by Galileo, Newton, and others. This baffled Kuhn because Aristotle's contribution to logic remained of central importance at least until the twentieth century, and Kuhn believed that Aristotle's observations in biology provided models that were instrumental to the emergence of the modern biological tradition. If Aristotle had been both a keen observer and the epitome of reasoning, how could he be so mistaken (Kuhn 2000, 16)?

His conceptual difficulty led Kuhn to reflect that perhaps Aristotle's (translated) words did not mean quite the same to the modern reader as they had to Aristotle. This thought, together with continued concentrated immersion in the texts, led to an abrupt revelation:

> Suddenly the fragments in my head sorted themselves out in a new way, and fell into place together. My jaw dropped, for all at once Aristotle seemed a very good

physicist indeed, but of a sort I'd never dreamed possible. Now I could understand why he had said what he'd said, and what his authority had been. Statements that had previously seemed egregious mistakes, now seemed at worst near misses within a powerful and generally successful tradition. (Kuhn 2000, 16)

This experience initiated the intellectual development that led to *Structure* and his position on revolutionary change of worlds and worldviews, a position that raised problems that he would continue to struggle with until his death in 1996. The careful reader can detect this theme of sudden revision already in Kuhn's first book, *The Copernican Revolution*, published in 1956, although it is probably only with hindsight that one can see the importance of the idea. Writing that monograph cemented many of the major themes of Kuhn's later work in its detailed description of the complex transformation from the world as described by Ptolemaic astronomy and Aristotelian physics (or perhaps medieval neo-Aristotelian physics) to the worldview that developed through Copernicus, Kepler, and Galileo, to culminate in Newton.

In assessing this sudden transformation and the significance of the process of writing this book on Kuhn's subsequent views, it is essential to recall the unique character of the Copernican-Newtonian revolution. Indeed, for many historians and philosophers, it is called *the* Scientific Revolution. Before this revolution, humans saw themselves as situated on a motionless Earth in the center of a relatively small finite universe. Terrestrial substances were divided into four kinds, and the motions of objects depended on the substances composing them; each kind had a natural motion defined in terms of the center of the universe located at the center of the Earth. Celestial substances were a different matter—or rather were *not* matter—and followed circular paths.

By the end of the Scientific Revolution, humans were on an Earth that was not only rotating at 1,000 miles per hour but also one that was traversing an orbital path around the sun at an even greater velocity. They were not at the center of the solar system, but on the third planet from the sun. Nor is the sun at the center of the universe, for the universe is infinite and there is no center at which to be located. Terrestrial and celestial objects were now subject to the same governing laws, and those laws were abstract, quantitative, and mathematical rather than qualitative and teleological.

Notice that two abrupt changes are involved here. One is the change in the scientific worldview—the Ptolemaic/Aristotelian view that had evolved little over almost two millennia was suddenly replaced by the Copernican/Newtonian view. The second is that Kuhn's understanding of Aristotle's worldview and the relation between Aristotle's views and post-Newtonian views underwent an instantaneous change when his jaw dropped. Until much later (Kuhn 2000), Kuhn did not distinguish these two kinds of changes, one of which is a personal psychological transformation, and the second a social and epistemological change in a community. Neither the scale nor the processes of these two kinds of change are identical, and some of the lack of clarity of his earlier views is due to failure to make this distinction.

Progress

The question of scientific "progress" is a complicated one in Kuhn's thought. He clearly believed that there is directionality to scientific change, that, for example, scientific disciplines frequently split into subdisciplines that become fields in their own right, and that this process is never reversed. And at the most general level, as mentioned above, he thought that one can distinguish pre-paradigm from paradigm-driven fields. But directionality does not mean that it is movement toward some ultimate end rather than simply solutions to current problems.

As indicated above, many of Kuhn's later views are already discernible in *The Copernican Revolution*, but there are other points at which his ideas had clearly developed in the six-year period between that work and *Structure*. In the conclusion to the former book, he discusses the fact that progress in scientific concepts is not cumulative. "But though the achievements of Copernicus and Newton are permanent, the concepts that made those achievements possible are not" (Kuhn 1957, 264–265). This is clearly consistent with his more elaborated view in *Structure* that science does not make cumulative progress with respect to the underlying structure of the world, and is closely connected with his controversial views about how the "world changes" during evolutionary periods.

However, in *The Copernican Revolution* he still holds that the list of solved problems is cumulative across even the Scientific Revolution. "Only the list of explicable phenomena grows; there is no similar cumulative progress for the explanations themselves" (265). By the time he wrote *Structure*, Kuhn had ceased to think that even the list of explained phenomena was cumulative. The beginning of the evidence for this was already in *The Copernican Revolution*, but he had not seen it as such.

For example, in the Ptolemaic/Aristotelian view of the world, since the Earth was (almost) at the center of the celestial sphere of fixed stars, there was a trivial explanation of why there was no apparent parallax in the position of the fixed stars over a period of six months. On the other hand, given the Copernican view of the Earth as orbiting around the sun, one would expect to see the stars appear at slightly different angles when observed at intervals of six months because the Earth is at opposite sides of its orbit. Since no parallax was observed, this was taken by many as evidence for the Ptolemaic/Aristotelian view. The Copernican explanation of the lack of observed parallax was to infer that the distance to the fixed stars was so great that the angle of parallax was less than the limit of observation (Kuhn 1957, 159).

In this instance, what was an unproblematic observation in the Ptolemaic system was "explained" by an assumption that was seen as ad hoc by traditional astronomers when propounded. The detection of parallax remained an issue for Copernican and successive astronomical views even after the invention of the telescope, and it was not until 1838 that the first measurements demonstrating parallax were made (163).

One of the exemplars of the Copernican/Newtonian view was the pendulum. In this case the universal laws of gravitation, force, and acceleration combine to give a derivation of a precise quantitative law stating that the period of a pendulum depends only on its length. The place of the pendulum in the development of Copernican/Newtonian physics is very important. The properties of a simple ideal pendulum are easy to establish, but more complex approximations to actual physical instantiations pose important puzzles for normal science. For the Ptolemaic/Aristotelian view, a heavy object suspended from a string or chain is of no theoretical interest because it is an example of constrained motion—the string or chain prevents the object from pursuing its natural motion toward the center of the Earth and universe. The example of the pendulum was of central importance for some of Kuhn's most controversial views on world change during revolutions, a topic that will be discussed later.

Normal Science

According to Kuhn, normal science consists of periods of cumulative progress in which scientists apply generally accepted theories to the unresolved questions in a domain according to shared assumptions about what constitutes the important problems and what would count as a solution. He characterizes "normal science" as a very sophisticated form of puzzle-solving that can require great ingenuity but occurs within a stable framework of tradition.

Kuhn's characterization of "normal science" was criticized from at least two directions. Popper argued that if Kuhn were correct, then normal science was in fact not science at all, because scientists under those conditions were presupposing the cognitive elements of the disciplinary matrix and not testing them (Lakatos and Musgrave, 1974) (see Popper, Karl Raimund). According to Popper, the Kuhnian characterization of science requires that the laws and theories under consideration be falsifiable and that experiments be attempts at falsification. Kuhn did not accept this criticism—his view was that the presuppositions were necessary in order to make progress. Continual reexamination of what is taken as basic knowledge would impede the process of extending knowledge. He also held that it was only through the vigorous pursuit of normal science that further revolutions could be achieved.

A second criticism was that Kuhn's account made so-called normal science, which is quantitatively the large majority of scientific activity, uninteresting and routine. Although his use of the term "puzzle solving" to describe normal-scientific activity may have made it sound more routine, Kuhn certainly did not think that normal science was uninteresting and routine. The phrase "puzzle solving" may have been unfortunate, because it tends to suggest crossword puzzles or jigsaw puzzles, challenges that have been created by humans with explicit rules and a solution that is known in advance before the puzzle solvers enter the activity.

In contrast, the "puzzles" of normal science are posed by scientists, but the answers are not known by anyone in advance, and the rules for solution are given at best implicitly by previous exemplars. Both features—the difficulty of solution and the bridge to the next revolution—can be illustrated by many examples, but the refinement of the Newtonian theory of the solar system is a particularly clear one.

The textbook accounts say that Newton derived Kepler's laws from his own laws of motion, but this is a derivation in the sense of physicists, not of philosophers or logicians. Kepler's first law states that the planets move in elliptical orbits with the sun at one of the foci of the ellipse. However, the derivation of an elliptical orbit for one body revolving around a second body holds only when no other forces are acting on these bodies than the mutual force proportional to the inverse square of their distance apart. Thus, this derivation is possible only if the effects of all the remaining planets on the orbit are ignored, and it is clear from Newton's law of gravitation that there is a non-null effect. So the "derivation" involves a deliberate simplification, though one that was unproblematic at the time of Newton.

However, as telescopic observation improved, the discrepancy from the Keplerian model was observationally recorded, and one of the puzzles of the astronomical tradition became to provide a more exact mathematical analysis that did not oversimplify as much. For example, by 1770 it was noted that there were deviations in the motions of Jupiter and Saturn from predictions. Laplace established in 1775 that these deviations were periodic (with a period of 929 years!) and showed that the deviations followed from a mathematical analysis that included the gravitational attraction between those two planets. There is a crucial mathematical fact that intrudes here, *viz.*, that the problem of solving the differential equations for two bodies acting under the influence of gravity has a general analytic solution, but in general there is no closed analytic solution for equations involving three or more bodies. Thus significant new mathematical tools were required to solve the puzzle. And the puzzle led to the major mathematical discovery of the mathematical fact just cited, so the process influenced the development of not only astronomy but mathematics as well.

The better mathematical approximations fit well with observation until about the 1840s, when it became clear that the predictions for Uranus did not fit the data. The possibility of an undiscovered planet was one potential explanation, though others were also offered, including the possible failure of the inverse square law for gravitation. However, the standards of astronomy at that post-Keplerian point would not permit as a "solution" the mere postulation of such a planet, but required calculations to determine the location of the hypothetical planet, and observations of the planet and its positions.

When the relevant calculations were made and the appropriate portion of the sky observed, astronomers saw Neptune. But not for the first time! Once the existence of Neptune was known and its orbit calculated backward, astronomers learned that on several occasions the planet had been seen and noted, but that the observers (including Galileo) did not discern that it was a planet. This episode provided (though it was hardly necessary) more confirmation of the universal applicability and validity of the Newtonian equations.

A similar situation arose late in the nineteenth century with respect to the planet Mercury, whose orbit likewise did not fit predictions according to the standards of the time. Astronomers hypothesized another inner planet, calculated its orbit and mass, and even named it: Vulcan. However, nature was not so cooperative in this instance and Vulcan was never observed. The anomaly of Mercury's orbit later became one of the important phenomena predicted correctly by general relativity theory and was a crucial part of the overthrow of Newtonian theory. Rigorous pursuit of increasingly precise predictions within the normal-science tradition of Newtonian astronomy provided the data that gave observational leverage to the overthrow of that tradition.

Rationality

Perhaps the most common and outraged criticism of Kuhn's work was that it denies the rationality of science, because the acceptance or rejection of a new theory depends not only on "scientific" factors but also on social factors. Kuhn's response was indignation and perplexity. The first sentence of *Structure* promises (or warns) that "History, if viewed as a repository for more than anecdote or chronology, could produce a decisive transformation in the image of science by which we are now possessed." One important respect in which Kuhn wanted to correct the image of science was in the general understanding of what constitutes its rationality.

The generally accepted philosophical view of scientific rationality in the 1950s was that science rested on a basis of *observation statements*—these statements were thought to be neutral with respect to the various theories to be compared and to be unproblematic with respect to verification (see Logical Empiricism; Verifiability). While there were disagreements about the character of observation statements (whether they were subjective reports of the agent's perceptions or reports of external states that were intersubjectively agreed upon) and about whether their verification was absolute or not, the assumption of a basis of observation sentences was shared by Carnap, Hempel, and almost all others working at the time (See also Carnap, Rudolf; Hempel, Carl Gustav; Observation; Physicalism; Protocol Sentences). So the first element of scientific rationality was the observational base.

The second component of scientific rationality consisted of the method of establishing verification, confirmation or falsification of theories on the basis of observation sentences (see Confirmation Theory; Observation). Significant controversies raged among those who shared this framework over such issues as whether the basic concept should be probability (see Popper, Karl Raimund; Reichenbach, Hans), confirmation, or refutation, but all shared the overarching program of articulating exactly how the logical relation between theory and observation was to be analyzed. The guiding motivation for this program was the success that had supposedly been obtained in the foundations of mathematics. Mathematical logic had proved to be an enormously successful tool for representing mathematical statements and for analyzing the proof relations among the statements. The second stage in securing the foundations of science was to be a comparable analysis of scientific reasoning. In particular since the surprising success of mathematical logic consisted largely in showing how semantic conceptions, such as entailment, could be rendered into equivalent syntactical forms, such as derivability in a formal system, the goal was to provide a syntactic characterization of the process of theory evaluation.

Summarizing, the two principles are that the basis of evaluating scientific theories was a shared collection of observation sentences and that evaluation is to be done according to a single algorithm, which was to be proved optimal. Given these conditions, it follows that if two scientists disagree about theory choice, then either

- they are using different evidential bases; or
- (at least) one of them is not using the proper algorithm for theory evaluation.

Thus, if Carnap's original program, to prove that there is a unique correct quantitative confirmation function, could have been carried out, any scientific controversy must result from ignorance or inductive error; scientific controversies are an inefficiency in the progress of science. Furthermore, the scientific community, or at least the ideal one in which all scientists follow the algorithmic ideal, has no discernible function except to make the process of scientific development move faster by having more hands to make the tasks lighter. In the ideal scientific community, in this view, there would be no disagreements about theory evaluation. If Carnap's program does not succeed, then one must make a reevaluation of the nature of scientific controversies (see Carnap, Rudolf; Inductive Logic).

One consequence of Kuhn's work and the subsequent discussion has been to encourage alternative approaches to rationality. The Bayesian approach to theory evaluation would undoubtedly have commanded some attention in any event, but interest in this alternative to the traditional approaches was undoubtedly increased by the desire to find ways to accommodate Kuhnian insights in a more formal structure. The view that apparent incommensurability stems from scientists using very different prior probabilities is explored in interesting depth in Salmon (1990) (see Bayesianism).

Rationality and the Social

Since finding a formal analog for inductive logic to the very successful formal deductive foundation of almost all of mathematics was taken as key to demonstrating the rationality of the scientific method, it is not surprising that Kuhn was accused of advocating irrationality when he questioned the possibility of this project. On this point he was very clear in his own mind that the traditional attempts to underwrite the rationality of science misunderstood the history and character of the scientific process. For him, the process of theory choice by the scientific community was the touchstone of rationality, and his goal was to better understand that process. In other words, the scientific process is not rational because the community is embodying an independently specifiable algorithm for theory choice, but the process is rational because it is being carried out by the community.

There are two importantly different ways of understanding this claim. The first is based on an idea from information theory and signal processing. A classical engineering problem was how to improve signal detection given either imperfect detectors or noisy signal channels. Early researchers in the field showed that with imperfect signal detectors one could obtain an arbitrarily good improvement in the accuracy of signal detection by combining a sufficiently large number of the imperfect detectors and taking the result from the majority of the detectors. The application of this analogy is to think of scientists as imperfect detectors of the signals sent by nature, perhaps under experimental questioning, regarding the best of a set of available theories.

This model depends on (at least) two crucial assumptions. One assumption is that the scientists/detectors are more likely to choose the better theory or to detect the correct signal, than the alternative theory/signal. Lest the reader think that "better theory" presupposes too much, note that for these current purposes, one can take *better theory* in a limited way to mean the better theory from an available specified set for the purposes of fitting experimental results, making correct predictions, and cohering with other theories in the near future. This argument can be formulated in a way that is agnostic about metaphysical truth and realism—indeed, both realists and anti-realists can adopt this assumption. If there are more than two theories in contention, then the assumption need only be that the better theory is more likely to be chosen than any individual competitor. One does not need the stronger assumption that the better theory is more likely to be chosen than the disjunction of the competitors. And the difference in probability of the better theory being chosen does not have to be great; even small differences can be leveraged with enough agents.

Secondly, the scientists/detectors must be appropriately independent of one another. Otherwise they will all simply reproduce the same errors. Elaboration of this assumption is a complicated and subtle matter, and only an approximation can be given here. The detectors cannot be statistically independent, for if each is more likely to detect the true theory than not, then their conclusions will be at least weakly correlated. The proper statement of the assumption is that their correlation is entirely due to their propensity to detect the true theory. If the detectors are all made in the same factory, and thus are subject to the same biases and will produce the same result, true or false, then they are not independent in this sense. If the scientists are all trained by the same narrow-minded dissertation advisor and thus will produce the same result, true or false, regardless of the facts of the matter, then they are not independent in this sense.

This emphasis on the independence of judgment suggests an atomistic view of the scientists that is at odds with the emphasis Kuhn places on the weighty role played by the social community of scientists. As Kuhn discussed in *The Essential Tension* (Kuhn, 1977), the process of being trained as a scientist in a particular specialty is simultaneously a process of making the trainee conform to the standard acceptable canons of the discipline while trying to preserve the freedom and independence of judgment that will enable the new specialist to explore and evaluate alternative approaches to anomalies in the discipline. The process of training ranges inclusively from training in specific experimental or mathematical fields to absorption of the cultural tales of the field (Traweek, 1988).

This model of the scientific community is oversimplified in many ways, but it brings out important points. If not exclusively, at least largely, the community must consist of inquirers who are responsive to information about the world—they must be rational inquirers as individuals. But there are also issues of optimality at the level of the community, and one can speak of rationality at the level of the structure of the community.

Kitcher (1993) develops a version of this approach. Since there is no theory-choice algorithm, a point he takes as given, it is important to have a diversity of approaches to theory testing and evaluation. If everyone agreed on and pursued the most plausible direction, then potentially promising theories might well be ignored, often at a considerable loss to the community. One could add to his point that a diversity of cognitive styles and variability in willingness to take risks would also be beneficial. In these matters there remains the difficult task of maintaining the right balance in the essential tension Kuhn noted, the balance between having sufficient communal agreement on enough matters to define a field while leaving room for vigorous debate on others.

An alternative, bolder interpretation is that the rationality is present only at the level of the community. As Longino puts it: "Objectivity, then, is a characteristic of a community's practice of science rather than of an individual's, and the practice of science is understood in a much broader sense than most discussions of the logic of scientific method suggest" (Longino 1990, 74). In her analysis, there are four conditions that a community must satisfy in order to qualify as rationally developing scientific knowledge:

1. There must exist within the community recognized and approved forums or avenues for the criticism of theories, evidence, experiments, assumptions, and inferences.
2. The criticism must be effective in that the community at times changes its belief and practices in response to it. Criticism is not merely tolerated and ignored.
3. As a background for criticism, the community must have publicly recognized and shared standards that provide the criteria for evaluating theories, hypotheses, experiments, data analysis, etcetera. These establish standards for the quality and relevance of criticism.
4. Communities must be characterized by appropriate levels of equality of intellectual authority. This does not require that all members of the community have equal influence, but that any disparities be due to past accomplishments or training and not to political, economic, or other factors not directly

related to the epistemological task at hand (Longino 1990, 76–81).

Whether communities or individuals are the fundamental basis of rationality in science is still under debate and development, but there is no question that participants such as Kitcher and Longino are continuing a debate opened by *Structure*.

Contexts: Discovery, Justification, Development

Perhaps the most important break with the standard philosophy of science tradition was Kuhn's rejection of the distinction, due to Reichenbach, between the context of discovery and the context of justification. At the very beginning of *Experience and Prediction*, Reichenbach distinguishes the descriptive task of epistemology, *viz.*, "giving a description of knowledge as it really is," from the main evaluative task of epistemology of considering "a logical substitute rather than real processes" (Reichenbach 1938, 5) He concludes that "it will therefore never be a permissible objection to an epistemological construction that actual thinking does not conform to it" (6). Reichenbach (and many other logical empiricists) saw the task of philosophy of science as providing a "rational reconstruction" of scientific processes. These reconstructions substituted abstract logical operations for the actual psychological processes. Probably Kuhn's most important contribution was to bring both history of science and psychology into contact with philosophy of science.

At the time Kuhn was writing *Structure*, almost all of philosophy of science focused on the context of justification, understood as analyzing the relation between a theory axiomatized as a set of sentences in a formal language and evidential sentences from an agreed-on base also represented in a formal language. Kuhn's use of history was frequently seen as dealing with the context of discovery and thus as irrelevant to philosophy of science. Kuhn's insight, which is now generally accepted among philosophers of science, is that theories and evidence undergo significant transformations during the period between the time the theory is first formulated and the time it reaches the final form that is enshrined in textbooks. Most of the process of scientific development occurs between the time of discovery and the time at which a formal justification is developed, and formalization is ill-suited to represent those processes of development.

To develop Kuhn's ideas, it is essential to distinguish a third context, that of *development*. This is the stage in which an embryonic, or perhaps fetal, scientific theory is nurtured and developed so as to analyze its implications. The most famous and dramatic example is the Copernican revolution discussed above. Copernicus' theory of planetary motion described the motion of the planets in terms of circular motion with epicycles on the main cycles, and attributed the movement to a rather mystical force emanating from the sun but pushing the planets around in their orbit. It took more than a half century after Copernicus for Kepler to develop the description of the elliptical orbits, and another half century before Newton provided the dynamics to explain the approximately elliptical orbits in terms of a force attracting the planets to the sun. Although Kuhn sometimes describes scientific revolutions as sudden, he had documented in his first book the century and a half that was required for the Ptolemaic/Newtonian revolution to unfold.

The Scientific Revolution

The issue of the context of discovery is closely related to the character of normal science. As indicated earlier, in Kuhn's view, the adoption of a new disciplinary matrix by a community requires an extended period that looks instantaneous only from a rather distant perspective. For a new disciplinary matrix to be taken seriously by the community, it must solve some of the outstanding problems that have eluded solution with the older disciplinary matrix. But the new disciplinary matrix at that stage has not been developed sufficiently to resolve many of the outstanding problems or to provide new alternative solutions to previously solved problems. Thus much of the process of normal science is the development of new scientific ideas, instruments, mathematics, and experimental methods. The formalized theories on which logical empiricism focused are an end product and do not represent the character of most of the activities of scientists (see Logical Empiricism).

Gestalts and World Changes

One of the significant oversimplifications in *Structure* is that Kuhn uses the same vocabulary to describe the process of change in a scientific community and in an individual—both are characterized as "Gestalt switches," or instantaneous changes of perspective, familiar from such psychological examples as the Necker cube: a two-dimensional configuration of lines that can be seen in either of two three-dimensional orientations. (The relevance of Gestalt psychology to philosophy of science was also argued by Hanson [1958] in his *Patterns of*

Discovery. Although Hanson published these ideas earlier and Kuhn acknowledges Hanson's work in *Structure*, Kuhn had been familiar with the Gestalt examples from his time at Harvard. There is also a question of the scale of revolutions and various perspectives on them. The Copernican–Newtonian revolution took over a century, but this looks relatively abrupt compared with the millennium and a half of domination by the Aristotelian–Ptolemaic disciplinary matrix before the Scientific Revolution, and the two and a half centuries subsequently by the Copernican–Newtonian.

The contrast between the psychological processes by which an individual comes to accept one theory rather than another and the social processes by which the consensus of the scientific community shifts can be illustrated by examples. Galileo quickly adopted the Copernican framework but was part of a long and tragic struggle within (and without) the scientific community. The Gestalt character of how perception can change is illustrated by Kuhn's discussion of Planck's discovery of the constant that bears his name. In his earliest paper, Planck uses a quantity that is equal to the constant, but his conceptualization of it at that time was not that it was a constant, but was simply another quantity that emerged in accounting for the data. It was not until considerably later that Planck conceived of it as a fundamental constant (Kuhn 1978).

Two of the passages that most disturb critics of *Structure* are Kuhn's comments that before Galileo there were no pendula and that when scientists change their disciplinary matrix, as we have seen, the world changes. However, one way of interpreting Kuhn's pendulum comment less dramatically can be derived from recent work on the view that science consists primarily of modeling (Giere 1988 and 1999; Cartwright 1983). If a pendulum is to be understood as described in physics textbooks as a *point mass* suspended from a *massless*, completely *inelastic* string with *no resistance* to the medium in which it moves, then there are no pendula in the physical world, neither before nor after Galileo. However, from Galileo on, the worlds of natural philosophers and physicists include abstract pendula of all various masses and string length. And with these in the mental space of the scientist, the world looks very different because many physical objects approximate the properties of an ideal pendulum so that prediction is possible and fruitful.

Later Work

The main focus of Kuhn's later work was on language, attempting to understand and explicate precisely the nature and causes of incommensurability. This direction began in the 1960s while he was at Princeton, where Quine's ideas about radical translation were a dominant theme of discussions (see Quine, Willard Van). In the postscript to the second edition of *Structure* and subsequent work Kuhn often used the term "translation" and was interested in the parallels between language learning and scientific development. For example, he claimed repeatedly that Aristotelian "physics" cannot be translated into Newtonian terms. A frequent criticism of this claim was that it implied falsely that no one who knew Newtonian mechanics could understand Aristotle, whereas Kuhn himself explicated Aristotle. His response was that the process of learning Aristotle's science was a process of second-language learning, and the ability to speak two languages does not guarantee that what is said in one can be translated without remainder into the other.

In discussing "incommensurability" claims, it is important to bear in mind what Kuhn meant and not overinterpret the claim. He used the term in its historical mathematical sense, in which a mathematician says that the square root of 2 is incommensurable with any rational number. This means that no rational number is a square root of 2. On the other hand, one can approximate the square root of 2 as precisely as you wish by rationals. It is not clear whether Kuhn was committed to this latter consequence, that one could approximate Aristotelian claims as precisely as one wishes by Newtonian statements, but it is at least evident that his terminology did not imply that no comparison could be made between the two theories.

It is unfortunate that Kuhn did not more systematically develop his conception of disciplinary matrix, because it might have led him to recognize that in addition to the semantic incommensurability that most concerned him, other sources and kinds of incommensurability also abound. The Millikan–Ehrenhaft controversy over the electron illustrates how metaphysical and experimental commitments can produce incommensurable analyses of experimental results.

Around 1910, many physicists believed that there was a fundamental particle, now known as the electron, which had the minimal unit of negative electrical charge. In other words, all electrical charges were multiples of the charge of the electron. But many others were still not persuaded of the existence of such quantized particles. Millikan (1911), believing strongly in the existence of the electron, embarked on the research project of determining more precisely the value of the charge.

His method was to produce small oil drops that were ionized by radiation so that the drops were charged. He then observed their behavior in both the presence and the absence of an electrical field.

Meanwhile, Ehrenhaft, who was opposed to ideas of quantization, and perhaps even of atomism, was performing similar experiments to test whether the basic unit of charge existed. Ehrenhaft used small particles of metal rather than oil drops. These had the advantage that they were even smaller than the oil drops and were not susceptible to evaporation, as the oil drops were. Ehrenhaft's experimental procedure produced data that led to calculations of various charges smaller than the unit charge that Millikan reported (Ehrenhaft 1941).

Millikan, being an atomist, believed that Ehrenhaft's particles were so small that they were subject to irregularities in their motion caused by encounters with individual atoms and that Ehrenhaft's results were thus unreliable.

Ehrenhaft rejected Millikan's claims and pointed out that Millikan's process of analyzing data included omission of microscopic observations that produced data inconsistent with his atomic hypothesis. Ehrenhaft also argued that his own data were more reliable because his particles were not subject to any evaporation and could be produced as exact spheres, whereas oil drops were only approximately spherical. Ehrenhaft was never persuaded of Millikan's results, but Millikan (1965) received the Nobel Prize in 1924 for this work.

The differing metaphysical assumptions—continuity versus atomism—led to differing experimental approaches and to divergent interpretations of the reliability of data. Since those metaphysical assumptions were not only in the background, but also the subject at issue for Ehrenhaft, the result was incommensurability. (For a much more detailed discussion of the controversy and issues, see Holton 1978.)

Influence

Assessments of Kuhn's influence vary enormously, and although most philosophers of science would agree that his influence was very large, some would not; and among those who do agree, there is disagreement about whether it was positive or negative. Those who argue that his influence was slight, point to the fact that almost all of the elements of *Structure* can be found in other philosophers of science writing at the time—Hanson (1958), Hesse (1966), Feyerabend (1993), Toulmin (1961), and others. However, the constellation of ideas in *Structure* and the rhetorical tone caught readers' attention in a way that produced much more dramatic results did than any of the others. Feyerabend made more radical claims than Kuhn and was largely dismissed or ignored, and the others made less sweeping claims and attracted less attention. For complex reasons that are not fully understood, *Structure* struck a resonant chord and transformed philosophy of science.

Some of the disagreement stems from unclear aspects of the book. As discussed earlier, although Kuhn changed his mind about the central term of the book, "paradigm," he did not rewrite the book to reflect that change. This decision was the result of Kuhn's recognition that reworking *Structure* was not a very good option, since he was still in the midst of reworking many of his views, and so the "postscript" strategy was a stopgap measure until he could reach the stage where a new and more thorough book was prepared. During the 1960s and 1970s Kuhn gave frequent graduate seminars on *Structure* and his further thoughts, as well as giving lectures and publishing intermediate elaborations. In 1977, he published *The Essential Tension*, a collection of his essays ranging from reprintings of pre-*Structure* papers to items that appeared for the first time in that volume. The "essential tension" referred to is that between the desire to assimilate all data/observations within the current paradigm and the desire to find revolutionary new solutions.

The characterization of revolutionary science has attracted the most attention—both positive and negative—from readers of *Structure*. Much of the popularity of the book outside the community of philosophers and historians derived from the conceptual tool it provided to analyze change and often to attempt to bring about a "change of paradigm." The book especially attracted interest from the social sciences because, in addition to the two types of science discussed earlier, Kuhn also described the "pre-paradigm" periods of the physical sciences before they achieved maturity. Whether any changes in these sciences are due to *Structure* or whether they are positive is mostly outside the scope of this entry. In particular, the widespread use of "paradigm" to mean something like a holistic worldview has its origin in the first edition of *Structure* and is rather distinct from the official approach of the second edition, which replaces "paradigm" with the disciplinary matrix with various explicit dimensions.

Although *Structure* was one of the major final blows to the logical positivist program, Kuhn's relations with at least two of the major figures in that program were cordial. *Structure* was originally published in the Encyclopedia of Unified Science series, which was the primary publishing format for logical empiricism (see Logical Empiricism). Carnap, one of the editors, was enthusiastic about publishing *Structure*. Although it was demonstrated in a paper by English (1978) that Carnap's own account of the relation between theory and evidence leads to incommensurability, since the meaning of theoretical terms is given implicitly by the relation of the theoretical terms to observation terms, this has not been generally noted. Also, in Carnap's account of probability and changes in probability assignments, he recognizes that in addition to conditionalization on new evidence, sometimes probability assignments are made in a global way that do not rely on evidence. This distinction is not exactly that between normal-science belief revision and revolutionary belief revision, but it is also not entirely foreign to it (see Carnap, Rudolf).

Hempel was a colleague of Kuhn's at Princeton and they were frequent interlocutors. Both influenced each other's views, and by the late 1970s their positions on many issues were very similar, although they had arrived at those positions from different directions. Kuhn came to emphasize more that revolutionary changes were made for good reasons, though he continued to assert that particularly at early stages of revolutionary change there were also good reasons against the eventual successful theory. And he continued to emphasize that there was no formal algorithm for theory evaluation that could be appealed to (see Hempel, Carl Gustav).

Some examples of Kuhn's influence within philosophy of science have already been given. Kitcher (1993) refers to Kuhn as one of the two most significant influences on his work and thought (the other is Hempel). Longino (1990) does not delineate her debts so explicitly, but there are far more references in her work to Kuhn than to any other philosopher of science. Kuhn's inclusion of values in the disciplinary matrix fostered the possibility of feminist philosophy of science, which studies, among other topics, the extent to which assumptions about gender influence the choice of research topics, funding, and scientific prestige, as well as biases with regard to theory selection (see Feminist Philosophy of Science). Another consequence of note was the development by Joseph Sneed, Wolfgang Stegmuller, and others of a formalization of theories inspired by Kuhn (for more details, see Theories).

One of the effects of *Structure* was that researchers in a number of fields beyond philosophy of science cited it as justification for emerging new domains in the study of science or for transformations of old ones. Sociologists of science, which had been dominated by a model of scientific development emphasizing the rational and cumulative character of scientific knowledge, was emboldened to investigate issues of authority and power and other issues at the level of the scientific group (Crane 1972). Kuhn's introduction of Gestalt psychology into discussions of scientific change helped to encourage both psychologists and philosophers to investigate these further. Productive examples can be found in Brewer and Chinn's (1994) research on anomalous data and Giere's *Explaining Science* (Giere 1988). Anthropology of science, which did not exist in any significant way prior to *Structure*, began to study scientific communities in the same ways in which other esoteric cultures were investigated (e.g., Traweek 1988). Some of the influence, and debates about it, continue. For example, Duschl (1990) and others are currently arguing for transformation in science education on the basis of what they perceive as a Kuhnian understanding of the process of scientific development.

In conclusion, the influence of Kuhn in philosophy of science is difficult to gauge because a great deal of what he argued for is now taken as part of the underlying assumptions. Studying history of science, having more realistic accounts of scientific development, and appreciating the relevance of theories of cognition and of social processes are all accepted and valued by the mainstream of philosophy of science. Thus what is attributed to Kuhn are primarily the claims about incommensurability and the dichotomous nature of scientific change, which are the less plausible parts of his views with the hindsight of fifty years. Kuhn's work has achieved a transformation of views of science that makes his most valuable contributions invisible to many current philosophers of science.

RICHARD GRANDY

References

Brewer, W. F., and C. A. Chinn (1994), "Scientists' Responses to Anomalous Data: Evidence from Psychology, History, and Philosophy of Science," *PSA* 1: 304–313.
Cartwright, Nancy (1983), *How the Laws of Physics Lie*. Oxford: Oxford University Press, Oxford.
Crane, Diana (1972). *Invisible Colleges: Diffusion of Knowledge in Scientific Communities*. Chicago: University of Chicago Press.
Cushing, James T. (1994), *Quantum Mechanics: Historical Contingency and the Copenhagen Hegemony*. Chicago: University of Chicago Press.
Duschl, Richard A. (1990), *Restructuring Science Education: The Importance of Theories and Their Development*. New York: Teacher's College Press.

Ehrenhaft, Felix (1941), "The Microcoulomb Experiment: Charges Smaller than the Electronic Charge," *Philosophy of Science* 8: 403–457.

English, Jane (1978), "Partial Interpretation and Meaning Change," *Journal of Philosophy* 75: 57–76.

Feyerabend, Paul (1993), *Against Method* (3rd ed.). New York: Verso.

Galison, Peter (1987), *How Experiments End*. Chicago: University of Chicago Press.

Giere, Ronald N. (1988), *Explaining Science: A Cognitive Approach*. Chicago: University of Chicago Press.

——— (1999), *Science Without Laws*. Chicago: University of Chicago Press.

Hanson, Norwood Russell (1958), *Patterns of Discovery: An Inquiry into the Conceptual Foundations of Science*. Cambridge, UK: Cambridge University Press.

Hesse, Mary (1966), *Models and Analogies in Science*. Notre Dame, IN: University of Notre Dame Press.

Holton, Gerald (1978), "Subelectrons, Presuppositions, and the Millikan-Ehrenhaft Debate," *Historical Studies in the Physical Sciences* 9: 161–224.

Horwich, Paul (ed.) (1993), *World Changes: Thomas Kuhn and the Nature of Science*. Cambridge, MA: MIT Press.

Hoyningen-Huene, Paul (1993), *Reconstructing Scientific Revolutions: Thomas S. Kuhn's Philosophy of Science*. Translated by Alexander T. Levine. Chicago: University of Chicago Press.

Kitcher, Philip (1993), *The Advancement of Science: Science Without Legend, Objectivity Without Illusions*. New York: Oxford University Press.

Kuhn, T. S. (1957), *The Copernican Revolution: Planetary Astronomy in the Development of Western Thought*. Cambridge, MA: Harvard University Press.

——— (1977), *The Essential Tension: Selected Essays in Scientific Tradition and Change*. Chicago: University of Chicago Press.

——— (1978), *Black-Body Theory and the Quantum Discontinuity, 1894–1912*. Oxford: Oxford University Press.

——— (1996), *The Structure of Scientific Revolutions*. Chicago: University of Chicago Press.

——— (2000), *The Road since Structure: Philosophical Essays, 1970–1993 (with an autobiographical interview)*. Edited by James Conant and John Haugeland. Chicago: University of Chicago Press.

Lakatos, Imre, and Alan Musgrave (eds.) (1974), *Criticism and the Growth of Knowledge*. Cambridge: Cambridge University Press.

Longino, Helen E. (1990), *Science as Social Knowledge: Values and Objectivity in Scientific Inquiry*. Princeton, NJ: Princeton University Press.

Masterman, Margaret (1974), "The Nature of a Paradigm," in Imre Lakatos and Alan Musgrave (eds), *Criticism and the Growth of Knowledge*. Cambridge, UK: Cambridge University Press, 59–89.

Millikan, Robert A. (1911), "The Isolation of an Ion, a Precision Measurement of its Charge, and the Correction of Stokes's Law," *Physical Review* 32: 350–397.

——— (1965), "The Electron and the Light-Quantum from the Experimental Point of View," *Nobel Lectures—Physics 1922–41*. Amsterdam: Elsevier.

Nersessian, Nancy J. (1984), *Faraday to Einstein: Constructing Meaning in Scientific Theories*. Boston: Kluwer Academic Publishers.

Nickles, Thomas (ed.) (2003), *Thomas Kuhn*. New York: Cambridge University Press.

Pickering, Andrew (ed.) (1992), *Science as Practice and Culture*. Chicago: University of Chicago Press.

Reichenbach, Hans (1938), *Experience and Prediction: An Analysis of the Foundations and the Structure of Knowledge*. Chicago: University of Chicago Press.

Salmon, Wesley, (1990), "Rationality and Objectivity in Science, or Tom Kuhn meets Tom Bayes." *Scientific Theories, V. 14, Minnesota Studies in Philosophy of Science*. Minneapolis: University of Minnesota Press, 175–204.

Shapere, Dudley (1964), "The Structure of Scientific Revolutions," *Philosophical Review* LXXIII: 383–394.

Suppe, Fredrick (ed.) (1977), *The Structure of Scientific Theories*. Urbana: University of Illinois Press.

Suppes, Patrick (1969), "Models of Data," *Studies in the Methodology and Foundations of Science: Selected Papers from 1951 to 1969*. Dordrecht, Netherlands: D. Reidel, 24–35.

Toulmin, Stephen (1961), *Foresight and Understanding*. New York: Harper and Row.

Traweek, Sharon (1988), *Beamtimes and Lifetimes: The World of High Energy Physicists*. Cambridge, MA: Harvard University Press.

See also **Carnap Rudolf; Feyerabend, Paul; Hanson, Norwood Russell; Hempel, Carl Gustav; Lakatos, Imre; Incommensurability, Observation; Popper, Karl Raimund; Protocol Sentences; Scientific Change; Scientific Progress; Scientific Revolutions**

IMRE LAKATOS

(5 November 1922–2 February 1974)

Lakatos was born Imre Lipsitz to Jewish parents in Budapest on November 5, 1922. He was not sent to a Jewish school, and when his family moved to Debrecen in eastern Hungary (in 1932) he attended the local *realgymnasium* (a secondary school with an emphasis on the sciences). Having excelled at school Lakatos went on to study mathematics, physics, and philosophy at the University of Debrecen, from which he graduated in 1944. During the war years he gravitated to the Marxist left. When the Nazi occupation of 1944 placed Hungarian Jews in mortal danger, he used a false identity to escape the labor gangs and the deportations, unlike many of his family and friends, including his mother and grandmother, who died in Auschwitz. During the occupation he belonged to an underground Marxist group, and late in 1944 he adopted the name "Lakatos" ("Locksmith"). After the war he continued his education, now at the prestigious Eötvös College in Budapest. At this time he wrote his first published works, on politics and its relation with science, written from a Marxist perspective. In 1947, he was awarded a Ph.D. for his dissertation *On the Sociology of Concept Formation in Natural Science.* Meanwhile, the Moscow-backed communist government in Hungary suffered from a shortage of dedicated Marxists to fill its offices. Thus it was that in 1947, the young Lakatos was attached to the Ministry of Education with responsibility for the "democratic reform of higher education." In practice this meant a rapid expansion in student numbers, together with brutal measures to bring independent intellectual centers under Party control—including Eötvös College, where Lakatos is still remembered for his role as denouncer-in-chief. At this time he was also a research student of the Hegelian-Marxist philosopher György Lukács, and he traveled to Moscow University in 1949. On his return to Hungary in the spring of 1950, he was arrested, charged with "revisionism," and imprisoned for almost four years (including a period in solitary confinement), one of many Party members caught up in the Stalinist purges. On his release after Stalin's death Lakatos returned to academia. Between 1954 and 1956, he worked on probability-and-measure theory under the mathematician Alfred Rényi. Crucially, one of his tasks was to translate into Hungarian György

Pólya's *How to Solve It*, a book on mathematical heuristics (Pólya, working in the United States, wrote *How to Solve It* for his undergraduates). At the same time, Lakatos began to question the entire edifice of Marxist thought. When demonstrations against the government erupted in October 1956, Lakatos was at the forefront of the student movement demanding academic and intellectual freedom. His conviction that science and philosophy should suffer no external control was by now firmly established. The uprising was quashed by Soviet troops, and Lakatos left Hungary for Vienna in late November 1956.

In 1957, he secured a Rockefeller fellowship to King's College, Cambridge, England, where (supervised by R. B. Braithwaite) he wrote a Ph.D. thesis, *Essays in the Logic of Mathematical Discovery,* a later version of which was eventually published as *Proofs and Refutations.* On taking his doctorate in 1960 he joined the London School of Economics (LSE), where he remained until his untimely death. In 1969, he was appointed professor of logic. During this time, his philosophical interests broadened to include physical science, and he developed his Methodology of Scientific Research Program. On the political front, the LSE was the British center of the student uprising around 1968. Remembering the damage done to Hungarian intellectual life by political interference both from the state and from student activists, Lakatos urged the university to resist demands for a student role in policymaking. In public he insisted on clear distinctions: between logic and psychology; between science and pseudo-science; and in this case, between the "constructive" student demand for the right to criticize the university and the "destructive" demand to take part in its decision-making processes. In private his tone was playful and his logic supple. In his publications he sharply distinguished his position from that of "irrationalists" such as Feyerabend (see Feyerabend, Paul); in his lectures he recommended Feyerabend to his students; and, in letters, he and Feyerabend argued for so long and took each other so seriously that in the end their positions were in danger of collapsing into one another. This combination of public dogmatism and private openness was motivated by an acute sense of the fragility of intellectual liberty. The enemies of free inquiry, whether they be totalitarian governments or ideologically blinkered students, cannot be resisted by the force of argument alone. Defenders of freedom, Lakatos thought, must stand ready to use whatever combination of rhetoric and military force the occasion demands.

In the early 1970s, Lakatos was full of plans for books and papers on mathematics and science, but he was dogged by ill health. He died after a heart attack on February 2, 1974, aged fifty-one.

Proofs and Refutations

Lakatos' earliest published works are Hungarian book reviews written in 1946–1947 when he was finishing his first Ph.D., *On the Sociology of Concept Formation in Natural Science.* These book reviews range over politics, science, and literature, but they invariably criticize their subjects from a Marxist perspective. For example, one author claims that as the result of scientific progress, modern man is alienated from nature—but, contends Lakatos, this is the case only under capitalism. In another review Lakatos complains that undialectical thinking makes a mystery of the fact that yesterday's progressive social class can become tomorrow's empty husk. Lakatos' concern in these reviews and in his work for the Ministry of Education was to place the resources and achievements of bourgeois intellectual life at the disposal of the new, postwar, communist reality. No doubt he reasoned that this was the only hope for their survival and growth. If intellectual culture remained bourgeois and "idealist," it would die for lack of relevance.

Lakatos had abandoned this Marxist framework by the time he began his Cambridge Ph.D. a decade later. Nevertheless, he retained a dialectically oriented study of the emergence of novel concepts. Indeed, two themes—the growth of knowledge and the relation between intellectual life and politics—dominated his writing from his first postgraduate publications of 1946 to his death in 1974. At the beginning of his Cambridge thesis, he declared: "The three major—apparently incompatible—'ideological' sources of [this] thesis are Pólya's mathematical heuristic, Hegel's dialectic and Popper's critical philosophy" (Lakatos [1962] 1978, 70n). The thesis is a case study in the growth of informal mathematics, that is, mathematics that have not been translated into the language of a system of formal logic. The case in hand is the Descartes–Euler formula, $V - E + F = 2$, where V is the number of vertices of a polyhedron, E the number of edges, and F the number of faces. A cube, for example, has 8 vertices, 12 edges, and 6 faces: $8 - 12 + 6 = 2$. The formula holds for the Platonic solids and many other polyhedra. Obvious questions arise: Does it hold for all polyhedra, or only for a special class of them? What exactly is a polyhedron anyway? Formal logic (that is, the logic expressed in such systems as the predicate calculus) offers no answers to these questions, though it may help to clarify answers found by some other means. If, therefore, one supposes that the rationality of mathematics lies entirely in its use of formal logic, then there can be no rational means of addressing these questions. One can only hope to stumble on an answer by a lucky guess, intuitive leap, or stroke of genius. This is where Pólya enters the picture.

Pólya's books on mathematical heuristics began as teaching aids. Some of his tips on mathematical research are altogether general (e.g., Do you know how to solve a problem similar or related to the one in hand?), while others are topic specific (e.g., If you want to discover new theorems about solids, remember that theorems in plane geometry often have three-dimensional analogues). These research strategies are fallible, but they do offer students and researchers in mathematics a more productive approach than simply tinkering with the material and hoping that luck or inspiration will supply an insight. Pólya's heuristics also suggest a philosophical account of the growth of mathematical knowledge (psychologistic talk of "inspiration" or "genius" is no account at all). Pólya suggested that Lakatos should use the Descartes–Euler conjecture as a case study. Philosophically, Pólya contributed three further thoughts to Lakatos' thesis:

1. Natural science and mathematics have many heuristic patterns in common. Pólya's work provides examples in pure mathematics of enumerative induction, inference to the best explanation, and the testing of general hypotheses by checking that their logical consequences are true.
2. The mathematical operations used to test a conjecture may eventually form the kernel of a proof. For example, one wishes to test the Descartes–Euler formula. One could try lots of different polyhedra in turn, but this would be hopelessly slow. One can, however, generate a vast collection of polyhedra out of the Platonic solids by "roofing" (building a pyramid taking one side of an existing polyhedron as its base) and "slicing" (cutting off a corner). It is easy to check that roofing and slicing do not change the alternating sum $V - E + F$. One has, therefore, checked the hypothesis for all polyhedra generated from the Platonic solids by roofing and slicing. But, by rearranging the elements, one also has the means to prove it for this class of solids.
3. Proofs and tests may suggest definitions and theorems. Pólya provides cases in which, rather than first seizing on definitions and conjectures by luck or insight and then later finding a suitable proof, mathematicians tailor their definitions and theorems to suit a promising proof idea. For example, the roofing-and-slicing idea suggests a definition: Let polyhedra constructed from the Platonic solids by finite iterations of roofing and slicing be called "P-constructable." The earlier argument is now the proof of a theorem: For all P-constructable polyhedra, $V - E + F = 2$.

The introduction of new definitions leads to Hegel. From Hegel, Lakatos learned that the arrival of new concepts in science is not usually announced by an explicit definition. Rather, existing concepts are developed in use. This may be a surreptitious expansion occasioned by consideration of a new sort of object. For example, in Lakatos' case study, the concept 'polyhedron' is quietly stretched as mathematicians contemplate solids with holes and solids formed by joining simple polyhedra at a vertex or along an edge. Lakatos also shows how a new proof can reinvent an entire field of study. Cauchy suggested a proof of the Descartes–Euler formula that involved removing one side of a polyhedron and flattening the remaining figure onto a plane. This is to treat a polyhedron as a closed surface rather than as a solid, and thereby to shift the problem from geometry to topology. Such changes, from minor concept-stretching to revolutions in the very subject matter, may not be apparent when they happen. They may come to light only later, when the growth of knowledge is rationally reconstructed. This explains the unusual literary structure of Lakatos' essay. The main text is a dialogue in a fictional mathematics class, while the historical sources are supplied in footnotes. Philosophical rational reconstruction is not concerned with accuracy in historical detail. Rather, the point is to make known the subterranean conceptual shifts masked by the use of old words for new ideas.

A consequence of Pólya's heuristic is that one may end up proving something deeper and more interesting than first intended. Lakatos' "dialectical" interest in conceptual change shows that as a result of trying to prove a theorem, one may end up with a whole new theoretical language. Thus, Lakatos' second Hegelian inheritance was a distinction between formal logic, which analyzes and rearranges the conceptual resources already available, and a heuristic rationality that develops new concepts out of old. A movement to a new language L_2 may introduce contradictions and refutations that were not present in the conceptual order associated with the old language L_1. Formal logic would recommend sticking with L_1, but this may be overridden if there are good heuristic reasons to prefer L_2. Indeed, the contradictions and refutations thus introduced may require improvements to the mathematical theory that might not otherwise have been made. Hegel expressed this distinction in psychologistic terms inherited from Kant: Formal logic is the natural tool of the Understanding, while "dialectical" or "speculative" thinking is the business of Reason. Lakatos preferred to distinguish between "language statics" and "language dynamics," but the point is the same. The logical analysis of concepts alone can only clarify and entrench the present conceptual order, when the real philosophical task is to understand the progress from one stage of conceptual development to the next.

Lakatos' third Hegelian lesson is that there is no general method for the development of concepts. Hegel is associated with the thesis–antithesis–synthesis schema, but one ought not to expect to apply this formula mechanically. In Hegel's work the development of any concept comes in three "moments," the third being in some sense a return to or rediscovery of the first, suitably modified by passage through the second. However, the details in any given case cannot be anticipated. There are no laws of dialectical growth as there are laws of formal logic. "Formalism," for Hegel, is the false supposition that there is a universal scientific method that may be grasped abstractly in advance of any particular inquiry. In his Cambridge thesis, Lakatos explicitly denied the possibility of a general system of heuristic rules, though this passage vanished from published versions of the text.

After Pólya and Hegel, Lakatos' third source was Popper and his critical philosophy (see Popper, Karl Raimund). Popper argued that empirical science is not built up by establishing true laws of nature. Rather it develops by the refutation of conjectures. Popper's view depends on the asymmetry between proof and disproof: No argument can conclusively prove a general statement because no argument can rule out altogether the possibility that an exception to the proposed law will be discovered in the future. On the other hand, a single counterexample is sufficient to refute a general statement. Because Popper put refutation rather than confirmation at the core of his model of science, he had no need for a nondeductive logic of induction. Pólya's thought that mathematics shares heuristic patterns with natural science enabled Lakatos to import Popperian ideas about science into his thinking about mathematics. Specifically, he developed Pólya's observation that a theorem and its proof may evolve together so that the final theorem is carefully tailored to capture the results of an insightful proof strategy. Lakatos gave this a Popperian gloss: Mathematics evolves through a process of conjecture and refutation. What is more (following Pólya), the same thought experiment can be both proof and test. This is most obvious when all the steps in a proof are reversible. However, a proof with nonreversible steps in it can also function as a source of counterexamples because it decomposes a theorem into its logical dependencies. One theorem may depend on many lemmas, and by finding a counterexample to a lemma, one may refute the theorem itself (alternatively one may learn that the lemma is stronger than it need be, and so one can improve the proof). Mathematics, in this view, is not about proving theorems from self-evident axioms. It is a matter of offering conjectures for refutation—but the refutations are found in proofs and other thought experiments rather than in empirical experiments and observations. This, then, is Lakatos' first Popperian thesis.

The second Popperian lesson is that formal logic is indispensable to the development of concepts. In Hegelian dialectic, no conception is simply false. A concept is found to be one-sided, partial, or otherwise inadequate. This inadequacy is exposed by contemplation of its dialectical twin. Finally, a new conception supersedes them both, repairing the inadequacy but preserving what was true in the original. Hegel thought that this process of conceptual evolution had to be separate from the work of formal logic. Formal logic requires that terms keep their meanings unchanged from start to finish, otherwise one commits the fallacy of equivocation. On the other hand, the very point of dialectic is to develop the meanings of terms. Hegel attempted to exhibit a dialectical logic that made no appeal to the notion of contradiction found in formal logic precisely because he understood the nature of formal rigor. However, this separation is artificial. Mathematical practice does not divide into formal and dialectical phases. Mathematical thought experiments, whether they function as proofs or tests, are structured by formal logic. The criterion for a successful proof is still given by the formal definition of a valid argument.

Lakatos' third Popperian claim is that mathematical heuristic does not begin inductively from facts, but from guesses at solutions to problems and questions. Neither science nor mathematics can start from bare facts because facts do not spontaneously order themselves into inductive tables. They can be so ordered only in the light of a conjecture, a problem, or some other governing idea. This is as true of mathematical facts about the parts of polyhedra as it is of empirical data. Indeed (still following Popper), there are no bare facts, in either science or mathematics. In particular, there are no self-evident mathematical axioms. Hence Lakatos is anti-foundationalist and fallibilist in his philosophy of mathematics.

Some of the fictional dialogue in Lakatos' Ph.D. thesis concerning the Descartes–Euler formula was published in four parts in the *British Journal for the Philosophy of Science* (Lakatos 1963–1964). This treatment is confined to the early history of the formula, during which the proofs and refutations made appeal to geometric intuition. It stops short of the absorption of the formula into the relatively abstract systems of modern algebra. A common criticism was that Lakatos' view did not apply to these more abstract parts of modern mathematics. Lakatos' plans to publish a more comprehensive account were cut short by his death in 1974, but selections from his thesis were published posthumously as *Proofs and Refutations* (Lakatos 1976). This material extended the Descartes–Euler story to include algebraic treatments, and included studies of topics from analysis and set theory. Lakatos was in no doubt that his heuristic view applied to the whole of mathematics. In the introduction to *Proofs and Refutations*, Lakatos described the book as an argument against "formalism," which he defined as the tendency to identify mathematics with its formal axiomatic abstraction. This, he argued, caused the philosophical neglect of everything about mathematics not captured in fully formalized axiom systems of meta-mathematics. In particular, the history and heuristics of mathematical practice vanish from sight. When one comes to assess Lakatos' philosophy of mathematics, it is well to distinguish this anti-formalism from the claim that the growth of mathematics is marked by dramatic refutations of a Popperian sort—for, in the case of advanced mathematics, the former claim is rather more plausible than the latter.

Transition to Research Programmes

Lakatos observed that his three "ideological" sources appeared to be inconsistent. In fact, the inconsistency is more than apparent. Popper held that epistemology can have no interest in the process of conjecture production and that philosophers ought to confine their attention to the logic of conjecture evaluation. Pólya, on the other hand, showed that there can be a fallible logic of conjecture production (his heuristic). Moreover, though he separated them in thought, Pólya's studies showed that the "contexts of discovery and justification" overlap in practice, because the same thought experiment can function as a test and then later as a proof. Unsurprisingly, the deepest contradictions are between the Popperian and Hegelian elements. Though Lakatos argued that theorems develop by responding to counterexamples, he quickly introduced a distinction between "logical" counterexamples (that is, counterexamples in the ordinary Popperian sense) and "heuristic" counterexamples (cases that, while not strictly inconsistent with the theorem in hand, show it to be conceptually deficient in some respect). Logical counterexamples belong to language statics (in Hegelian terms, the rigid, formal logic of the Understanding). Heuristic counterexamples belong to language dynamics (in Hegelian jargon, the dialectic of Reason). Popper was contemptuous of any philosophy that paid attention to fine distinctions of meaning. In his view, scientists use explicit, stipulative definitions to establish their terms with as much precision as required for the task in hand, and philosophers ought to do likewise. He had no sympathy with the Hegelian project of revealing the subtle shifts of meaning hidden in the evolving use of a single word. This, though, is precisely what Lakatos did with the central terms of his dialogue ('polyhedron,' 'proof,' etc.). Therefore, much of Lakatos' achievement in *Proofs and Refutations* could never be absorbed into Popperian philosophy.

The final contradiction in *Proofs and Refutations* between Popper and Hegel concerns the possibility of a general method. Hegel, though he organized some of his works into triples within triples, insisted that there is no general logic of scientific or philosophical progress. Each episode in the intellectual history of humanity has its own character, which must be traduced if it is forced into some all-purpose mould. Popper, for his part, believed himself to have discovered a general logic of science. Science, in his view, has a characteristic logical process (philosophically articulated as critical rationalism) that distinguishes it from other activities in general and from pseudo-science in particular. *Proofs and Refutations* vacillates between these extremes. On the Popperian side, Lakatos offers highly general heuristic patterns that might find application outside mathematics. There is no reason in principle why lemma incorporation or "monster adjustment" (to take two examples) should not be found in the development of legal arguments (for example) (*monster adjustment* is the redescription of an outlandish counterexample in such a way that it ceases to be outlandish and thereby ceases to be a counterexample.) On the Hegelian side, Lakatos offers no *general* solution to the problem of identifying ad hoc-ness and degeneration, preferring instead to appeal to the judgment of persons with refined mathematical taste. Indeed, he argues that any of his heuristic patterns can lead to triviality and degeneration if pursued mindlessly (in this sense *Proofs and Refutations* is an anarchist tract in the sense of Feyerabend). These tensions between the Hegelian and Popperian elements of Lakatos' early views became acute as he tried to move beyond the piecemeal studies of his Ph.D. thesis.

As it turned out, Lakatos tended to land on the Popperian side of dilemma. In 1965 (the year after he published part of his thesis in the *British Journal for the Philosophy of Science*), he wrote a brief paper titled "A Renaissance of Empiricism in the Recent Philosophy of Mathematics?" (Lakatos 1978b). In this paper he developed a contrast between "Euclidean" and "quasi-empirical" theories. In a so-called Euclidean theory, truth flows down from self-evident axioms to derived theorems. In a quasi-empirical theory, falsehood is transmitted up from theorems to axioms. Such theories are *quasi-empirical* because the falsifying criticism need not come from empirical observation or experiment. He went on to claim that mathematics is quasi-empirical in this sense. In other words, he stripped the Hegelian elements from his earlier work, leaving a straightforwardly Popperian philosophy of mathematics that required him to claim that the growth of mathematics is marked by Popperian refutations.

The Methodology of Scientific Research Programs

While Lakatos was working on his philosophy of mathematics, the Popperian school had identified the alleged relativism and irrationalism of Thomas Kuhn as the principal philosophical threat to free enquiry (see Kuhn, Thomas). Lakatos agreed that Kuhn's philosophy had to be resisted, but by the late 1960s, he was convinced that Popper's account of scientific method was not adequate. To meet the need, he developed his own Methodology of Scientific Research Programs (Lakatos 1978a) (see Research Programs). In his Cambridge Ph.D. dissertation, he had suggested that theorems and proofs ought not to be regarded as separate entities. A theorem and its proof evolve together, and this common evolution makes sense of the terms employed in stating the theorem and the lemmas deployed in the proof. The natural unit of philosophical appraisal is thus not the theorem, but the theorem–proof pair. Similarly, in *Methodology of Scientific Research Programmes*, he replaced the single theory as the unit of appraisal with the "program," a temporal sequence of theories. In both cases the point was to allow philosophers to understand the growth of knowledge over time.

Two features distinguish the Methodology of Scientific Research Programs from Lakatos' earlier work on mathematics. One is the almost total absence of any discussion of shifts in meaning. The distinction between language statics and language dynamics that motivated the search for hidden conceptual changes was quietly dropped. The other is the specification of criteria by which progress and degeneration might be judged. The early Lakatos agreed with Hegel (and Feyerabend) that there can be no general rule for judging scientific progress, because each moment in the growth of knowledge has its own inner logic, which it is the task of philosophy to exhibit. In the absence of a general rule, it is a matter of judgment whether this or that development represents progress or degeneration. Feeling the threat of Kuhn's "irrationalism," the later Lakatos tried to articulate explicit criteria to do the job that he had previously left to good scientific taste. Indeed, he dismissed the exercise of taste as "elitism," which is ironic given his determination to exclude students from university decision making. At the same time, Feyerabend, in conversation and correspondence, persuaded Lakatos that a rigidly mechanical rule would fail to respect the special merits of individual programs. Therefore, he found himself arguing that his Methodology of Scientific Research Programs is *both* sufficiently exacting to distinguish progressive from degenerative programs *and* sufficiently supple to account for particular cases.

After Lakatos' death, there were sporadic attempts to develop his Methodology of Scientific Research Programs and to apply it more widely than Lakatos was able to in his lifetime (e.g., Howson 1976). Some philosophers attempted to develop his papers on mathematics into a Methodology of Mathematical Research Programs (e.g., Hallett 1979). None of these efforts succeeded in rebutting the charges of rigidity and arbitrariness leveled against Lakatos' later work, and the supply of would-be heirs eventually dried up. On the other hand, there has lately been a rising tide of historically oriented philosophy of mathematics, for which *Proofs and Refutations* is often cited as an inspiration. *Proofs and Refutations*, with its three ideological sources, is subtle and rich, while *Methodology of Scientific Research Programme*s is relatively rigid and ideologically uniform. Moreover, Kuhnian relativism was a temporary threat, while formalism is a permanent menace. For these reasons, one may expect *Proofs and Refutations* to be the most durable part of Lakatos' contribution to philosophy.

BRENDAN LARVOR

References

Hallett, Michael (1979), "Towards a Theory of Mathematical Research Programmes," *British Journal for the Philosophy of Science* 30: 1–25, 135–159.

Howson, Colin (ed.) (1976), *Method and Appraisal in the Physical Sciences*. Cambridge: Cambridge University Press.

Lakatos, I. ([1962] 1978) "Essays in the Logic of Mathematical Discovery," in Worrall and Currie (eds), *Mathematics, Science and Epistemology (Philosophical Papers volume 2)*. Cambridge: Cambridge University Press.

——— (1963–4), "Proofs and Refutations," *British Journal for the Philosophy of Science* [BJPS] 14: 1–25, 120–139, 221–245, 296–342.

——— (1976), *Proofs and Refutations*. Edited by Worrall and Zahar. Cambridge: Cambridge University Press. (Consisting of the BJPS article plus additional material from the Ph.D. thesis)

——— (1978a), *The Methodology of Scientific Research Programmes (Philosophical Papers volume 1)*. Edited by Worrall and Currie. Cambridge: Cambridge University Press.

——— (1978b), "A Renaissance of Empiricism in the Recent Philosophy of Mathematics?" in Worrall and Currie (eds), *Mathematics, Science and Epistemology (Philosophical Papers volume 2)*. Cambridge: Cambridge University Press, 24–42.

See also **Feyerabend, Paul; Kuhn, Thomas; Popper, Karl Raimund; Scientific Change; Scientific Progress**

LAWS OF NATURE

The main difficulty in developing an account of laws is to distinguish laws from accidental truths without leaving mysterious how scientists could ever gain knowledge of laws. Those accounts that adequately distinguish laws from accidental truths fail to make clear how scientists could ever determine which general truths are laws; and those that make knowledge of laws feasible seem unable to adequately distinguish laws from accidental generalizations.

Contrast the following generalizations:

All spheres of gold have a mass of less than 100,000 kg.
All spheres of U^{235} have a mass of less than 100,000 kg.

Assuming both are true, the former is merely accidentally true, whereas the latter expresses a lawful relation; the critical mass of U^{235} makes it impossible for there to be spheres of U^{235} of such a large mass. The question is what makes for the difference. A simple regularity view of laws is inadequate, since both generalizations describe regularities. Moreover, the distinction cannot be drawn based on features of the generalizations themselves, as Hempel and Oppenheim (1948) attempt to do; both generalizations are universal in form, are of unlimited scope, make no essential reference to particulars, and involve purely qualitative predicates. Therefore, either there must be something outside the regularity (and the generalization used to describe it) that makes one a law and the other not, or law claims must be in some way stronger than universal generalizations. The former option is the route taken by sophisticated regularity accounts, which Dretske (1977) labels "universal truth + X." According to such views, laws are simply regularities (leaving aside probabilistic laws for the moment), but some regularities are not laws. There is nothing about the regularities themselves that distinguishes laws from accidental regularities. Law statements are universal generalizations that satisfy some additional external criterion. There are as many such accounts as there are functions for laws. For example, laws might be those generalizations that are used to make predictions (e.g., Goodman 1954), are resilient (Skyrms 1980), function in explanations (e.g., Braithwaite 1953), are integrated into the best systematization of the facts (e.g., Lewis 1973), and so on. Alternatively, many have argued that laws are not mere regularities, but are ontologically stronger. There are two main accounts that follow this route: those that conceive of laws as nomically necessary regularities (e.g., Pargetter 1984) and those that conceive of laws as relations between universals (Dretske 1977; Armstrong 1983; Tooley 1987).

Possible-Worlds Accounts

Perhaps the most natural way to distinguish laws from accidental regularities is to conceive of them as nomically necessary truths, since laws seem to involve some sort of natural (as opposed to logical) necessity. "It is a law that α" is equivalent to "It is nomically necessary that α"; and the latter is true if and only if α is true in all nomically accessible worlds. Accidental generalizations, as distinguished from laws, are not true in all nomically accessible worlds.

Four problems arise for such accounts. First, they are committed to metaphysically dubious entities. Bigelow and Pargetter (1990) have attempted to address this problem by conceiving of possible worlds as complex structural universals. However, one might still find this objectionable, since it requires a commitment to uninstantiated structural universals, else there would be only one possible world—the complex structural universal that is instantiated by the actual world. In fact, in order to use possible worlds to make sense of logical necessity, such an account requires the existence of nomically impossible universals. Otherwise, "It is a law that α" would be equivalent to "It is logically necessary that α," since all possible worlds would be nomically accessible.

This leads to a second difficulty, that of providing an account of nomic accessibility that does not rely on a prior understanding of lawfulness. It would be circular, for example, to analyze "It is a law that α" is true if and only if α is true in all nomically accessible worlds and then require an understanding of "It is a law that α" to make sense of nomic accessibility. Pargetter's response is analogous to Lewis's (1986a) argument that his analysis of possible worlds is not circular. Lewis argues that possible worlds exist and understanding the notion of a possible world does not rely on a previous understanding of possibility, since possible worlds are like the actual world, differing only in what occurs in them. Pargetter (1984) argues likewise that the nomic accessibility relation exists and that one need not have a prior understanding of lawfulness to understand this accessibility relation. Nomic accessibility does not require that all accessible worlds have the same laws; it merely requires that the appropriate generalizations be true in these worlds. Moreover, what this involves is easily understood, since it is analogous to a generalization being true in different parts of the universe. The central problem with this response is that it is uninformative, since it simply treats the accessibility relation as a primitive. (For alternative accounts, see Vallentyne 1988 and Mormann 1994.)

The third difficulty is whether such accounts can make sense of probabilistic laws. Tritium has a half-life of 12.26 years, which means that tritium atoms have a probability of $\frac{1}{2}$ of decaying in 12.26 years. There are typically two ways to understand this probability in terms of possible worlds: One might consider one particular tritium atom and understand the probability for this atom as the proportion of nomically possible worlds (with the same history until now) in which that atom (or its counterpart) decays in the next 12.26 years; alternatively, one might consider *all* tritium atoms in all nomically possible worlds and understand the probability as the proportion of all nomically possible tritium atoms that decay in 12.26 years. In order for such accounts to even get off the ground, an assumption must be made about the likelihood of each possible world—typically that each is equally probable. Of course, if probabilities are grounded in possible worlds, it is unclear what could ground this probability assignment. It might be taken as a primitive, but that would not answer which primitive probability assignment should be used—should they be treated as equally probable, or should some be treated as more likely than others? Moreover, as van Fraassen (1989) has argued, if there are a countable infinity of possible worlds, they cannot be treated as equally likely, and "if they are infinitely many and form a continuum (surely the most plausible idea) then it literally makes no sense to say: the objective chance is *the* measure that treats them all as equally likely" (79).

Perhaps the most serious difficulty faced by possible-worlds accounts is how scientists could ever gain knowledge of which generalizations are laws. It would require knowing not only what goes on in other possible worlds, but also which possibilities are nomically accessible. Knowledge of logic might yield answers to the former, but knowledge of the latter is more problematic. Consider Bigelow and Pargetter's (1990) account that treats possible worlds as complex structural universals. The question is how one could know which complex structural universals are nomically possible and which are not. Perhaps scientists could run experiments to determine which lower-level structural universals are possible. Experiments allow scientists to instantiate possibilities that might otherwise not have existed; and since actuality implies possibility, this would provide some access to which structural universals are possible. However, since it is impossible to instantiate more than one most complex structural universal, that is, the one that is instantiated by the entire actual world, there is at least one level at which it is in fact impossible to instantiate other possibilities. Moreover, to determine which generalizations are nomically necessary, scientists would have to know what all the nomic possibilities are, or which structural universals are nomically impossible, and it is unclear how instantiating some of the possibilities yields knowledge of which are impossible.

The problem for probabilistic laws seems even worse, since it is unclear what the proportion of possible worlds with a certain type of event implies about the proportion of those events in the actual world (van Fraassen 1989), much less what the proportion of a type of event in the actual world implies about the proportion of events in other possible worlds, which is what knowledge of probabilistic laws would need to rely on. To resolve these difficulties, one might argue that scientists can simply use the methods of inference already used to figure out which generalizations are laws. However, such a response is inadequate, since the question is precisely how these methods could yield reliable knowledge laws, *if* laws are understood as nomically necessary truths. One could argue that this simply shows that laws ought not to be understood as nomically necessary truths, since it is possible for scientists to gain knowledge of which generalizations are laws, and they could not have such knowledge were laws nomically necessary truths.

Universals Accounts

The three main proponents of the universals account (Dretske 1977; Armstrong 1983 and 1997; Tooley 1987) have slightly different formulations, but the general idea is to conceive of laws as relations between universals. Law statements, rather than being universally quantified statements ranging over individuals, are singular statements about the relations between universals. However, not all relations between universals are laws, since accidental generalizations can be described as second-relations of extensional inclusion between universals. Tooley (1987) and Armstrong (1997) differ in their accounts of what makes laws distinct. According to Tooley, statements expressing nomological relations are contingent, irreducibly and purely of order two or higher, and logically entail the appropriate first-order generalization. Probabilistic laws are contingent, irreducible *probabilification* relations between universals, where the relation of probabilification between universals F and G makes it the case (due to logical probability) that, given that x has F, it is probable to degree k that x has G. According to Armstrong, laws are irreducible relations of necessitation (1983) or causation (1997) between universals, and probabilistic laws are probabilistic relations between universals that specify the probability of an instance of one universal necessitating or causing an instance of a second.

Both accounts successfully distinguish laws and accidental generalizations, since the relation of extensional inclusion is reducible to a first-order generalization and is neither a necessitation nor a causal relation between universals. This comes, however, at a price. In particular, it is unclear how such claims could entail anything about what happens in particular instances. Moreover, this problem infects accounts of both deterministic and probabilistic laws. (See van Fraassen 1989 for a discussion of this and other problems, and Armstrong 1997 for a reply.)

Even if this problem can be solved, another difficulty remains. If laws are relations between universals, how could scientists gain knowledge of which generalizations correspond to lawful relations and which are merely accidental? Armstrong argues that scientists can observe causal relations in single instances, draw causal generalizations from these, and then use inference to the best explanation (IBE) to infer that the causal regularity holds due to a lawful relation between the universals instantiated in the causal instances. Leaving aside problems with observing causation in single instances (see Causality), this would not allow scientists to discriminate laws and accidental regularities. For Armstrong, there are two kinds of accidental regularities: causal and noncausal regularities. Assuming scientists could observe causation in single instances, the latter could be excluded. However, there might be accidental causal regularities. The question is how IBE could discriminate an accidental causal regularity from a lawful one.

Similar problems arise for Tooley. Tooley argues that when $(x)(Px \rightarrow Qx)$ survives potential falsification, there is good reason to infer that there is a nomological relation between universals P and Q. In fact, it is only if laws are relations between universals that surviving potential falsification can provide support for the lawful status of a generalization. Otherwise, assuming that the universe is potentially infinite, assigning a nonzero probability to a universal generalization would be unjustifiable. Of course, since generalizations are supposed to follow from law claims, the law claim cannot be better confirmed than the generalization (Woodward 1992). Moreover, scientists would be unable to discriminate a generalization that is accidental from one that has a nomological relation. Either no accidental generalization can survive falsification (but then Tooley needs to clarify why) or scientists need to determine, prior to (or independently of) the attempted falsification, which generalizations are nomological and therefore ought to be assigned nonzero prior probabilities.

An alternative account that falls broadly into the universals category treats laws as metaphysically necessary truths, much like "Water is H_2O," rather than as contingent relations between distinct universals. The central motivation for such views is the idea that properties are individuated, or perhaps even constituted, by the causal powers they have and, therefore, by the lawful relations in which they stand (e.g., Shoemaker 1980; Swoyer 1982).

Sophisticated Regularity Accounts

There have been numerous attempts to develop sophisticated regularity accounts that conceive of laws as universal truths satisfying some additional functional requirement. The most fully developed of these is the best systems account, according to which laws are those generalizations that function in the appropriate way in the best systematization of the facts. There are various versions of this account that differ in significant respects (Ramsey 1928; Kitcher 1986 and 1989; Lewis 1973) (see also Ramsey, Frank Plumpton). However, the focus here is on Lewis's later (1994) version, according to which laws are those generalizations that are axioms or theorems in the true deductive system that achieves the best balance of simplicity and strength. Simplicity and strength often conflict, so the best system must balance the two. The use of simplicity also requires that the predicates used in the axioms refer to natural kinds. Otherwise, one could use gerrymandered predicates to artificially alter the simplicity of the system. Moreover, there might be different criteria used to measure strength, simplicity, and the balance between them. Lewis's hope is that nature will be kind and there will be only one system that will be best on any standards of strength, simplicity, and balance. If nature is unkind, then there may be nothing deserving the title of 'laws.'

To extend this account to probabilistic laws (or even deterministic laws in a chancy world) the systems must be limited to those that never had any chance of being false (Lewis 1986b). Probabilistic laws are those generalizations about chance that are theorems in the best system, where the best system balances simplicity, strength, and fit, and fit is understood in terms of the degree to which the probabilistic laws conform to the actual course of history. In other words, the chance of the actual history occurring will be higher according to some systems than others; the higher the chance, the better the fit. Since fit is not the only criterion the best system must satisfy, probabilities will not in general be equivalent to actual frequencies. Lewis was initially doubtful that this account of laws would work, since it seemed to lead to a contradiction when combined with his *principal principle* (Lewis 1986b) and ultimately led him to revise it (Lewis 1994). Others have argued that the contradiction need not have arisen in the first place (e.g., Roberts 2001).

This account of laws appears to avoid the epistemic difficulties faced by other accounts. Scientists can justify law claims by determining which universal generalizations are axioms or theorems in the best deductive systems. It is even reasonable to think that scientists already use strength and simplicity as a guide to theory choice. Nevertheless, epistemic problems do arise. A central difficulty is accounting for how scientists could distinguish natural from nonnatural kinds. One common answer to how this is done—that natural kinds are those picked out by the best theories—is not open to Lewis, assuming simplicity plays a role in theory selection. If simplicity guides theory choice (and it must if science is to discover laws, according to Lewis), then science's best theories might fail to pick out natural kinds, since it will be possible to use nonnatural predicates to achieve gains in simplicity. This is precisely why Lewis added this requirement in the first place. (For a discussion of other potential epistemic difficulties, see van Fraassen 1989, 55–59.) Nevertheless, the epistemic problems faced by this account seem, at least on the surface, to be less daunting than those of other accounts. Lewis's account also has the advantage of making sense of the connection between laws and modality, without relying on modal notions to define laws. Nomic necessity is defined in terms of laws, rather than the other way around. A proposition is nomically necessary if and only if it is entailed by the laws of nature. In terms of possible worlds, a world W' is nomically possible relative to another world W if and only if the laws of W are true in W', though they may not be laws in W'. A proposition is nomically necessary if and only if it is true in all nomically possible worlds. Lewis is also able to distinguish between laws and accidental generalizations, since not all generalizations will be axioms or theorems in the best system. This will presumably apply to such claims as "All pieces of gold have a mass of less than 10,000 kg."

This leads to the fundamental problem faced by such Humean positions: They fail to adequately account for the necessity involved in laws and therefore incorrectly distinguish between laws and accidental generalizations. While Lewis's account does provide for a distinction between laws and accidental generalizations, the question is whether it draws the distinction correctly. Criticisms have generally taken the form of counterexamples that attempt to show that Lewis's account yields the wrong answer about which generalizations are laws (Armstrong 1983; Tooley 1987; Carroll 1994; van Fraassen 1989). While these counterexamples are decisive against Lewis only if one shares the intuitions about which generalizations ought to count as laws or about how laws and modal notions are connected (Loewer 1996), they nevertheless make clear why many find Lewis's account inadequate. One counterintuitive consequence of Lewis's account is that there will be some nomic possibilities compatible with the laws but not compatible with their being the laws. This leads to a revision of the distinction between initial conditions and laws. As a result, Lewis's account carves the distinction between laws and accidental generalizations in a way many find counterintuitive.

Other Approaches and Issues

In response to these difficulties, Carroll (1994) and Lange (2000) argue that it is impossible to give a reductive analysis of laws that does not rely on nomic notions, while van Fraassen (1989) and Giere (1999) argue that there are no laws. Others have argued that there are no strict laws, even in fundamental physics (Cartwright 1983 and 1989). Instead, law statements require *ceteris paribus* clauses or perhaps might be better understood as claims about capacities. Earman and Roberts (1999) disagree with Cartwright about fundamental physics, but, building on an insight of Hempel's (1988), argue that the special sciences can have no strict laws, at least not formulated purely in the language of that science. (For related discussion about psychological laws, see Davidson 1970; for biological laws, see Beatty 1995; for economic laws, see Hausman 1992; and for social science laws, see Kincaid 1990.) The use and nature of laws has also played an integral part in debates about numerous other philosophical issues, such as causation, explanation, reductionism, determinism, confirmation, and induction.

JESSICA PFEIFER

References

Armstrong, David (1983), *What Is a Law of Nature?* Cambridge: Cambridge University Press.
––––– (1997), *A World of States of Affairs*. Cambridge: Cambridge University Press.
Beatty, John (1995), "The Evolutionary Contingency Thesis," in Gereon Wolters and James Lennox (eds.), *Concepts, Theories and Rationality in the Biological Sciences*. Pittsburgh: Pittsburgh University Press, 45–81.
Bigelow, John, and Robert Pargetter (1990), *Science and Necessity*. Cambridge: Cambridge University Press.
Braithwaite, R. (1953), *Scientific Explanation*. Cambridge: Cambridge University Press.
Carroll, John (1994), *Laws of Nature*. Cambridge: Cambridge University Press.
Cartwright, Nancy (1983), *How the Laws of Physics Lie*. Oxford: Clarendon Press.
––––– (1989), *Nature's Capacities and Their Measurement*. Oxford: Oxford University Press.
Davidson, Donald (1970), "Mental Events," in *Essays on Actions and Events*. Oxford: Clarendon Press, 207–225.
Dretske, Fred (1977), "Laws of Nature," *Philosophy of Science* 44: 248–268.
Earman, John, and John Roberts (1999), "Ceteris Paribus, There Is No Problem of Provisos," *Synthese* 118: 439–478.
Giere, Ronald (1999), *Science without Laws*. Chicago: University of Chicago Press.
Goodman, Nelson (1954), *Fact, Fiction, and Forecast*. London: Althone Press.
Hausman, Daniel (1992), *The Inexact and Separate Science of Economics*. Cambridge: Cambridge University Press.
Hempel, Carl (1988), "Provisos: A Problem Concerning the Inferential Function of Scientific Laws," in Adolf Grunbaum and Wesley Salmon (eds.), *The Limits of Deductivism*. Berkeley and Los Angeles: University of California Press, 19–36.
Hempel, Carl, and Paul Oppenheim (1948), "Studies in the Logic of Explanation," *Philosophy of Science* 15: 135–175.
Kincaid, Harold (1990), "Defending Laws in the Social Sciences," *Philosophy of the Social Sciences* 20: 56–83.
Kitcher, Philip (1986), "Projecting the Order of Nature," in Robert Butts (ed.), *Kant's Philosophy of Physical Science*. Dordrecht, Netherlands: D. Reidel, 201–235.
––––– (1989), "Explanatory Unification and the Causal Structure of the World," in Philip Kitcher and Wesley Salmon (eds.), *Scientific Explanation*. Minneapolis: University of Minnesota Press, 410–505.
Lange, Marc (2000), *Natural Laws in Scientific Practice*. Oxford: Oxford University Press.
Lewis, David (1973), *Counterfactuals*. Cambridge, MA: Harvard University Press.
––––– (1986a), *On the Plurality of Worlds*. Cambridge: Basil Blackwell.

─── (1986b), *Philosophical Papers: Volume II*. New York: Oxford University Press.

─── (1994), "Humean Supervenience Debugged," *Mind* 103: 473–490.

Loewer, Barry (1996), "Humean Supervenience," *Philosophical Topics* 24: 101–127.

Mormann, Thomas (1994), "Accessibility, Kinds, and Laws: A Structural Explication," *Philosophy of Science* 61: 389–406.

Pargetter, Robert (1984), "Laws and Modal Realism," *Philosophical Studies* 46: 335–347.

Ramsey, Frank (1928), "Universals of Law and of Fact," in David Mellor (ed.), *Philosophical Papers*. Cambridge: Cambridge University Press.

Roberts, John (2001), "Undermining Undermined: Why Humean Supervenience Never Needed to Be Debugged (Even if It's a Necessary Truth)," *Philosophy of Science* 68 (Proceedings): S98–S108.

Shoemaker, S. (1980), "Causality and Properties," in van Inwagen, P. (ed.), *Time and Cause*. Dordrecht: Reidel, 109–136.

Skyrms, Brian (1980), *Causal Neccessity*. New Haven, CT: Yale University Press.

Swoyer, C. (1982), "The Nature of Natural Laws," *Australian Journal of Philosophy* 60: 203–223.

Tooley, Michael (1987), *Causation*. Oxford: Clarendon Press.

Vallentyne, Peter (1988), "Explicating Lawhood," *Philosophy of Science* 55: 598–613.

van Fraassen, Bas (1989), *Laws and Symmetry*. Oxford: Clarendon Press.

Woodward, James (1992), "Realism about Laws," *Erkenntnis* 36: 181–218.

See also **Biology, Philosophy of; Causality; Chemistry, Philosophy of; Confirmation Theory; Determinism; Economics, Philosophy of; Explanation; Hempel, Carl Gustav; Induction, Problem of; Mechanism; Physicalism; Reductionism; Scientific Models; Theories**

PHILOSOPHY OF LINGUISTICS

Linguists study various topics, ranging from politeness register to the etymology of the word "dude"; but what has caught the attention of philosophers (and what has given rise to interesting questions in the philosophy of science) has been work in generative linguistics as developed by Noam Chomsky (see Chomsky, Noam) and several generations of his students. Generative linguistics per se has a tradition that dates back further than Chomsky (e.g., to work in phonology by Roman Jakobson), but Chomsky was the first to bring serious formal methods to the field and the first to pursue ways of embedding linguistic theory in other more basic sciences, such as biology. The field that has emerged in the wake of this effort has seen its share of empirical successes and continues to show promise, but it has also given rise to a series of interesting debates about the nature of language and scientific practice.

Among the issues that have emerged with the development of generative linguistics are questions about the nature of language and the object of inquiry in linguistics, about the role of reference and language/world relations, and about the plausibility and nature of rules and representations in linguistic theorizing, as well as a cluster of issues surrounding methodology in linguistics (among other things, for example, generative linguistics incorporates an unabashed appeal to intuitions as part of its data). The most radical issue, however, concerns how to understand the object of study itself.

The Object of Study

Most current work in generative linguistics holds that the object of study is not an external abstract language that one comes to acquire, but rather the study of the faculty that underwrites one's linguistic competence. The motivation for this is clear enough: The commonsense notion of a language has been under pressure for some time. Consider Weinreich's contention that a language is a dialect with an army and a navy: Why is Italian a language and Veneto a dialect? It seems to be a purely political decision, not driven by any facts about the linguistic forms themselves. Individuating dialects (or identifying a natural group of "same-language speakers") is no more feasible than separating dialects from languages. Exactly when do two people speak the same dialect? There are differences between the way any two people speak—does that make them speakers of different dialects? In the end, people say that they "speak the same dialect" when they identify with each other enough. Once again, linguistic identity recapitulates political identity. Even trying to say that a particular agent speaks a single idiolect (his/her own individual language), which might be identified by a series of rules, does not stand up to serious scrutiny. Which set of rules is appropriate for describing one's idiolect? The way one speaks shifts radically depending upon one's age, discourse partners, and context. Is the conversation in a classroom? in a bar? with an interlocutor from a foreign country? What sentences one produces will vary markedly from situation to situation. Then, too, there is the question of errors. One may stutter and stammer or hiccup during a conversation. Are those noises to be counted as part of one's idiolect? If not, then, why not?

The considerations just reviewed provide good reason for being suspicious of language as an abstract external object, whether construed nationally, locally, or individually. But what is to replace this conception? Generative linguists hold that one should focus on the faculty that underwrites linguistic competence. That is, they are not interested in languages so much as the tacit theory that the agent deploys in the production and comprehension of linguistic behavior (or at least in judging the acceptability of certain linguistic forms).

This tacit theory is also enlisted to account for the fact that the data the language learner is exposed to are not sufficient (by themselves) to explain the linguistic competence that the agent comes to have. The language learner faces the problem, familiar in the philosophy of science, that the available evidence radically underdetermines the theory. The sentences that a child hears—and even the explicit corrections and affirmations it receives—when first acquiring a language are compatible with infinitely many, wildly divergent grammars. Nonetheless, the learner apparently manages to select a grammar. And, by and large, learners in similar environments select similar grammars. Infinitely many grammars, perfectly compatible with the evidence the child has access to, are simply ignored. Understanding the nature of the tacit theory that language users employ is one way of illuminating just how it is that agents are able to acquire the linguistic competence they do. Current linguistic and psychological theories suggest that the tacit theory employed by a linguistic agent may be part of human biological endowment and, as a corollary, that the mechanisms under study in linguistic theory will be embedded in human cognitive psychology and ultimately in human biology (see Pinker 1994 for a readable account and defense of this view).

This view about the object of study is not shared by all linguists. For example, Katz (1985) endorsed a Platonist view that takes the object of study in linguistics to be an abstract mathematical object outside of space and time—this would contrast with a position like Chomsky's, in which the object of study is a mental object of some form. The Platonist view has been advanced most visibly by Katz (1981), and it has been at least endorsed by Gazdar et al. (1985). It may be, however, that the position rests on a confusion. Higginbotham (1983) has observed that even if grammars are abstract objects, there is still the empirical question of which grammar a particular agent is employing. George (1989) has further clarified the issue, holding that one needs to distinguish between (i) a grammar, which is the abstract object that an agent knows, (ii) a psycho-grammar, which is the cognitive state that constitutes the agent's knowledge of the grammar, and (iii) a physio-grammar, which is the physical manifestation of the psycho-grammar in the brain. If this picture is right, then the Platonist position in linguistics may be trading on a failure to distinguish between grammars and psycho-grammars.

Perhaps more pressing is the dispute between what Chomsky (1986) has characterized as conceptions of language of *E*-language and *I*-language. From the *E*-language perspective, a natural language is a kind of social object the structure of which is purported to be established by convention (Lewis 1975) (see Conventionalism), and persons may acquire varying degrees of competence in their knowledge and use of that social object. In Chomsky's view, such objects would be of little scientific interest if they did exist (since they would not be "natural" objects), but in any case such objects *do not* exist. Alternatively, an *I*-language is not an external object, but is rather a state of an internal system that is part of the agent's biological endowment. An agent might have *I*-language representations of English sentences, but those internal representations are not to be confused with spoken or written English sentences. They are rather data structures in a kind of internal computational system.

Chomsky understands the *I*-language computational system to be individualistic (see Methodological Individualism). That means that the properties of the system can be specified independently of the environment that the agent is embedded in. Thus, it involves properties like the agent's rest mass and genetic makeup, but not relational properties like the agent's weight and IQ. The difference can be described by analogy to the difference between primate physiology and primate ecology. The former study is "narrow" in that it is concerned with the properties and structure of the primate in isolation (e.g., with bone and muscle architecture). The latter study is "wide" in that it is concerned with primate/environment relations (e.g., with the role the primate's musculature might play in its interactions with its environment—allowing it to swing from tree to tree, say). In Chomsky's view, linguistic theory is much more analogous to physiology, and is narrow in precisely that sense. Both the claim that *I*-language is individualistic and the claim that it is computational have led to a number of philosophical skirmishes.

The Language Faculty and the External World

One of the immediate questions raised by the idea that *I*-language is individualistic has to do with how semantics could be possible—in particular, how *referential* semantics could be possible, where referential semantics is a theory of the relation between linguistic forms and aspects of the external world. The worry is this: If generative linguistics is a chapter of narrow (individualistic) psychology, and semantics (and indeed meaning) is concerned with relations of language/world (or at least mind/world), then how are the two enterprises to be squared? There are two parts to the question. Is it really the case that meaning involves language/world relations? And, if so, then how can semantics (and the theory of meaning) be reconciled with generative linguistics?

The case for meaning involving language/world relations was put most vividly by Putnam (1975), who offered a number of thought experiments intended to show that the meanings of linguistic utterances (and, indeed, of tokens of mentalese, should there be any) are not determined solely by the internal psychological states of the speaker, but by those states along with the speaker's local environment and social milieu. Consider the example of an agent, Hilary, who has in his lexicon the words *elm* and *beech*. However, all Hilary knows about elms is that they are trees called 'elm,' and all he knows of beeches is that they are trees called 'beech.' He has no knowledge, linguistic or otherwise, that would allow him to identify elms or beeches, or to tell them apart; in particular, there is nothing in his concept *elm* (*beech*) that makes it true of all and only elms (beeches). (The fact that elms are called 'elm' while beeches are called 'beech' will not do: For this to work, Hillary would need to know what '*elm*' and '*beech*' pick out—but this, of course, is exactly what is in question. Kripke [1980] first made this point with respect to the meanings of proper names.) What, then, allows Hilary's uses of '*elm*' to pick out exactly those things that are, in fact, elms? (And likewise for his uses of '*beech*.') Putnam's answer is that the tokens get their reference, in part, from Hilary being a member of a linguistic community that contains experts who *can* identify elms and beeches. And it is these experts who determine the reference of '*elm*' and '*beech*'; Hilary's use of the words is normatively constrained by the expert's use—Hilary means by '*elm*' whatever the experts mean by '*elm*.' So the reference of Hilary's tokens of '*elm*' and '*beech*' is partly determined by knowledge about how to identify elms and beeches. But this knowledge is not in Hilary's head, it is distributed across Hilary's linguistic community.

Next, consider Oscar. Oscar is a normal English speaker, with the word '*water*' in his lexicon. Oscar's uses of '*water*' refer to the substance with the chemical formula H_2O (this being the substance that the relevant experts identify as water). Now somewhere in the universe, there is a planet, Twin-Earth, that is qualitatively very similar to Earth. In particular, it has "lakes," "rivers," and "oceans" filled with a colorless, odorless liquid that often falls from the sky and is necessary for the life on Twin-Earth. The intelligent inhabitants of Twin-Earth—twin-earthlings—drink large quantities of this liquid, wash their dishes in it, have bubbling fountains of it; in fact, they use it in all the ways Oscar and other earthlings use water. Despite its superficial resemblance to water, however, this substance does not have the chemical formula H_2O, but the formula XYZ. Among the twin-earthlings is Twin-Oscar, a normal Twin–English speaker, with the word '*water*' in his lexicon. Twin-Oscar and the other twin-earthlings use this word to refer to the liquid that plays the same role on Twin-Earth as water does on Earth. But this substance, on Twin-Earth, is not H_2O, but XYZ. So despite having qualitatively identical mental states—and perhaps even qualitatively very similar linguistic communities—Oscar's uses of '*water*' and Twin-Oscar's uses of '*water*' refer to different substances. What Oscar's and Twin-Oscar's uses of '*water*' are about depends crucially on their local environments.

So Hilary and an arborist, despite having different concepts of elms, refer to the same things by tokens of '*elm*'. And Oscar and Twin-Oscar, despite having the same concept associated with '*water*', refer to different things by tokens of that word. Thus, Putnam's examples apparently show that the meanings of words cannot depend entirely on the mental states of the speaker: Meaning depends crucially on facts external to the speaker. Call such an approach to meaning *referential semantics*.

Chomsky (e.g., 1995a and 2000) has launched numerous attacks on referential semantics. The initial worry is that referential semantics conflicts with the internalist, individualist nature of *I*-language. *I*-language is to be characterized in terms of nonrelational properties of the linguistic agent: It is a "narrow" property (or system) of the agent's mind/brain. Referential semantics, however, deals with relational properties: It is in the business of detailing relations between linguistic items and parts of the world external to the agent.

Chomsky admits that the naturalistic study of *I*-language is not restricted a priori to internalist investigations. The relations that bits of *I*-language (lexemes, phrases, and sentences, conceived of as elements of a mental computation system) bear to the extra-mental world could, *in principle*, be examined naturalistically. Chomsky believes, however, that these relations are, in fact, naturalistically intractable—the relations that linguistic items bear to the external world are asystematic enough that no serious, naturalistically acceptable explanatory theory can be given for them. Chomsky (1995) illustrates this point with the word *water*, which Putnam took to pick out all and only those things with the property of being H$_2$O (give or take impurities). Suppose Peter fills a cup out of the tap; it is a cup of water. But after a tea bag is dipped in it, it is no longer water—it is tea. Suppose further that, across town, Noam fills another cup from the tap; Noam's tap draws water from a reservoir into which a large quantity of tea leaves has been dumped as part of a purification process. Noam's cup contains water (albeit contaminated), even if what it contains is chemically indistinguishable from what is in Peter's cup. Thus, whether a substance counts as tea or water (containing tea only as an impurity) depends crucially on the particular interests and intentions of the speaker in the context. There simply is no single substance that infallibly serves as the reference of '*water*'.

Any object that could serve as a referent in referential semantics, argues Chomsky, is bound to be so gerrymandered and ill-behaved that it could not plausibly be a real constituent of the world. Is it to be believed that there is some single substance, an "object" in the world, that is what is in Noam's cup—as well as the River Thames, the Pacific Ocean, and bottles of Evian—but not what is in Peter's cup—nor bottles of Windex or cans of Coca Cola? And the complexity that attends '*water*' is entirely typical of natural language. Consider that, during World War II, Dresden was burned to the ground; but it was rebuilt, and it deserves to be proud of its many new buildings. Or that a bank, which raised its interest rates because it hoped to bring in new business, might be destroyed in an earthquake and have to move across the street. What are these things, picked out by '*Dresden*' and '*bank*', that are at once concrete and abstract, apparently intentional, and can survive destruction? Consider, too, that there is likely a flaw somewhere in this paper, or that the average family has 2.3 children. In the simplest referential picture, '*flaw*' refers to flaws and '*the average family*' refers to the average family. But, surely, flaws and average families are naturalistically dubious; it is difficult to imagine a naturalistic inquiry into the nature of flaws. Of course, the surface form of sentences containing '*flaw*' and '*average family*' might be misleading; at the level of logical form, these linguistic items might not be referential—they could instead be adjectival modifiers or adverbials. But natural language is replete with apparent reference to naturalistically suspicious entities; to pass naturalistic muster, referential semantics must contend with each and every case.

Chomsky concludes that there is no relation of reference holding between linguistic items and objects in the world, at least not one about which anything interesting and general can be said. Any account of reference must ultimately advert to intentionality, a subject forever out of reach of naturalistic inquiry (see Intentionality).

Ludlow (2003) mounts a defense of referential semantics for *I*-languages. He suggests that semanticists bite the metaphysical bullet and accept that—perhaps in addition to the substances, objects, and properties catalogued by physical science—there are things exactly like those needed for referential semantics. There are cities and banks that survive destruction and act intentionally; there is a substance, water, the nature of which depends sensitively on its origin and the uses to which it is put; there are, perhaps, even such things as flaws and average families. In this view, metaphysical intuitions—about, say, whether or not a particular substance in a particular context is water—are underwritten by the structure of *I*-language. And if metaphysical intuitions—at least of the sort probed in Putnam-like examples—are by and large correct, semanticists and philosophers are in a position to reason from the structure of *I*-language to the structure of the world and vice versa.

This Kantian view of the nature of referential semantics presents a dilemma, however. If the semanticist insists that the entities and substances invoked as referents really exist, but agrees with Chomsky that such things are not fit for naturalistic study, then referential semantics is nonnaturalistic. It then lies outside the *scientific* study of language and is instead a philosophical epicycle on naturalistic linguistics. If, on the other hand, the entities and substances of referential semantics are natural objects, it is a serious question as to what place they hold in the vast array of objects posited by the other sciences. How, for example, is the referent of '*water*', water, related to atoms and molecules (objects of chemistry and physics) and to the representational/computational systems of the brain (objects of linguistics, psychology, and neurobiology)? (If Chomsky is correct, any adequate answer will be more complicated than "The referent of '*water*' is identical with H_2O.") This seems tantamount to asking questions about the relation between *I*-language and the world that Chomsky thinks is naturalistically legitimate but also naturalistically intractable.

Questions About Rules and Representations

The idea that linguistic theory involves the investigation of rules and representations (or principles and parameters) of an internal computational system has also led to philosophical questions about the nature of these rules and representations. For example, Quine (1970) has argued that, since many possible grammars may successfully describe an agent's linguistic behavior, there is no way in principle to determine which grammar an agent is using (see Quine, Willard Van). For his part, Chomsky (1980) has argued that, if one considers the *explanatory adequacy* of a grammar in addition to its *descriptive adequacy*, then the question of which grammar is correct is answerable in principle. That is, because the theory of grammar must be consistent with the theory of language acquisition, acquired language deficits, and, more generally, cognitive psychology, then there are many constraints available to rule out competing grammatical theories. To illustrate, two descriptively adequate theories of tense may differ in their assumptions about whether tenses like past and future are more basic, or whether a grasp of terms of temporal order (like 'before' and 'after') are more basic. Acquisition data might shed light on such a standoff if it could be shown that children acquire the use of one set of linguistic items markedly before the other.

Another set of worries about rule following have stemmed from Kripke's (1982) reconstruction of arguments in Wittgenstein (1953 and 1956). The idea is that there can be no brute fact about what rules and representations a system is using apart from the intentions of the designer of the system. Since, when studying humans, there is no access to the intentions of the designer, there can be no fact of the matter about what rules and representations underlie linguistic abilities. The conclusion drawn by Kripke (1982) is that "it would seem that the use of the idea of rules and of competence in linguistics needs serious reconsideration, even if these notions are not rendered meaningless" (1983, 31n, 22).

Chomsky (1986) appears to argue that one can know certain facts about computers in isolation, but Chomsky's current position (1995) is that computers, unlike the human language faculty, are artifacts and hence the product of human intentions. The language faculty is a natural object and embedded within human biology, so the facts about its structure are no more grounded in human intentions than are facts about the structure of human biology.

Kripke's argument is often stated in the form of a problem about justification: Speakers believe they know what they mean by their words; what justifies that belief? Chomsky (1986) responds to this worry by observing that, in the syntactic realm, there is no reason to suppose that speakers have first-person authority with respect to the rules they follow. Speakers produce grammatical sentences effortlessly, but the procedures they use in producing them are highly abstract and can remain obscure even under careful scrutiny. Justification is not available, or necessary, for successful linguistic communication. Scientific inquiry into the nature of those rules, by contrast, is no more or less justified than inquiry into the guiding principles of any system whose operation cannot be directly observed.

Such considerations address the epistemic side of Kripke's argument, but they do not address the metaphysical problem: What is it about someone that makes him or her a follower of rule *R*? One possible response may come from Chomsky's (1965) suggestion that the object of inquiry for syntactic theorizing is "an ideal speaker-listener, in a completely homogeneous speech-community, who knows its language perfectly and is unaffected by such grammatically irrelevant conditions as memory limitations, distractions, shifts of attention and interest, and errors (random or characteristic) in applying his knowledge in actual performance" (3). Such an idealized speaker could in principle produce evidence capable of distinguishing between any two substantively different rule systems. The apparent reliance on the relevant problematic notions ("grammatical," "ungrammatical") in linguistic theory might thus be seen as an artifact of considerations that are appropriately idealized away. This line of response may, however, be susceptible to Kripke's critique of solutions that depend on *ceteris paribus* clauses. In that critique, Kripke argues that there is no non-question-begging way to decide which features of the system should be idealized away from and which should not.

Perhaps a more promising solution can be found in an argument, due to Soames (1998), that an inability to know what fact about someone makes that person a follower of rule *R* in no way undermines the metaphysical possibility that there is such a brute fact. Soames suggests that Kripke, of all people, should have seen the error here—a conflation of metaphysical and epistemic possibility.

Methodological Issues

If the language faculty is an internal computational/representational system, a number of questions arise about how to best go about investigating and describing it. For example, there has been considerable attention paid to the role of formal rigor in linguistic theory. On this score, a number of theorists (e.g., Gazdar et al. 1985; Bresnan and Kaplan 1982; Pullum 1989) have argued that the formal rigor of their approaches—in particular their use of well-defined recursive procedures—count in their favor. However, Ludlow (1992) has argued that this sort of approach to rigorization would be out of synch with the development of other sciences (and indeed, of branches of mathematics), where formalization follows in the wake of the advancing theory.

Another methodological issue concerns the nature of evidence available to investigations of the language faculty. For generative linguists, evidence from a written or spoken corpus is at best twice removed from the actual object of investigation, and given the possibility of performance errors, is notoriously unreliable at that. Much of the evidence adduced in linguistic theory has therefore been from speakers' intuitions of acceptability, as well as intuitions about possible interpretations. This raises a number of interesting questions about the reliability of introspective data and the kind of training required to have reliable judgments. There is also the question of why one should have introspective access to the language faculty at all. It is fair to say that these questions have not been adequately explored to date (except in a critical vein; see Devitt 1995; Devitt and Sterelny 1987).

A third methodological issue relates to the use of parsimony and simplicity in the choice between linguistic theories (see Parsimony). While tight definitions of simplicity within a linguistic theory seem to be possible (see Halle 1961; Chomsky and Halle 1968; Chomsky 1975), finding a notion of simplicity that allows one to choose between two competing theoretical frameworks is another matter. Some writers (e.g., Postal 1972; Hornstein 1995) have argued that generative semantics and the most recent version of generative linguistics, minimalism (discussed in the next section), are simpler than their immediate competitors because they admit fewer levels of representation. In response, Ludlow (1998) has maintained that there is no objective criterion for evaluating the relative amount of theoretical machinery across linguistic theories. Ludlow offers that the only plausible definition of simplicity would be one that appealed to "simplicity of use," suggesting that simplicity in linguistics may not be a feature of the object of study itself, but rather an ability to easily grasp and utilize certain kinds of theories.

An alternative approach would be to take a leaf from Sober (1975) and argue for a view of simplicity according to which a theory is simpler than another if it is more easily embedded within more basic sciences. In such a view the simpler linguistic theory would be the one that could more naturally be embedded into cognitive psychology or even, following recent work in the minimalist program, into low-level biophysical and mathematical principles.

Issues Raised by the Minimalist Program

Although generative linguistics has gone through a number of permutations over the last 50 years, perhaps the most interesting has been the recent development of the minimalist program, outlined in Chomsky (1995a). Setting aside the technical details of the project, the headline idea is that the core language faculty did not evolve slowly over an extended period of time but was, rather, the result of a sudden mutation that, in effect, wired together two discrete cognitive systems—the conceptual/intentional (C/I), involving meaning and thought; and the perceptual/articulatory (P/A), responsible for speech production and perception. The working hypothesis is that the wiring solution was "optimal" and governed by basic low-level biological and mathematical constraints (such as those that account for the prevalence of recursive and fractal patterns in nature). This speculative hypothesis, if correct, would suggest that the linguistic theory should be looking for very specific kinds of properties and principles linking the C/I and P/A systems—properties that might naturally emerge from low-level biophysical principles (much as the Fibonacci pattern in a sunflower does). While speculative in the extreme, this new research program has shown some surprising successes and clearly calls for further investigation of its basic guiding assumptions. Although the project has yet to be explored in a formal way by philosophers of linguistics, it should prove a fascinating domain for future investigations.

Conclusion

Although the philosophy of linguistics is not as well explored as the philosophy of physics or of biology, it is certainly no less rich a domain of inquiry. Not only does it involve the usual concerns of scientific methodology (simplicity, the nature and trustworthiness of the data, etc.), but it also deals with kinds of entities (rules and representations) that are not routinely found in the basic sciences and are not well understood. Furthermore, the philosophy of linguistics is concerned with questions of the embeddability of linguistics into more basic sciences (possibly even into low-level biophysical and mathematical systems), as well as which parts of linguistics (if any) involve agent/environment relations and which parts are purely individualistic. For all these reasons, this subdiscipline of the philosophy of science promises to be a fertile area of investigation, plausibly able to illuminate some of the deeper cognate questions being explored in the philosophy of other sciences, as well as the philosophy of science generally.

JOSHUA BROWN
PETER LUDLOW
TIM SUNDELL

References

Bresnan, J., and R. Kaplan (1982), "Introduction: Grammars as Mental Representations of Language," in Bresnan (ed.), *The Mental Representation of Grammatical Relations*. Cambridge, MA: MIT Press, xvii–lii.

Chomsky, N. (1965), *Aspects of the Theory of Syntax*. Cambridge, MA: MIT Press.

——— (1975), *The Logical Structure of Linguistic Theory*. New York: Plenum.

——— (1980), *Rules and Representations*. New York: Columbia University Press.

——— (1986), *Knowledge of Language*. New York: Praeger.

——— (1993), "Explaining Language Use," in J. Tomberlin (ed.), *Philosophical Topics 20*, 205–231.

——— (1995a), "Language and Nature," *Mind* 104: 1–61.

——— (1995b), *The Minimalist Program*. Cambridge, MA: MIT Press.

Chomsky, N., and M. Halle (1968), *The Sound Pattern of English*. New York: Harper and Row.

Devitt, M. (1995), *Coming to Our Senses: A Naturalistic Program for Semantic Localism*. Cambridge: Cambridge University Press.

Devitt, M., and K. Sterelny (1987), *Language and Reality: An Introduction to the Philosophy of Language*. Cambridge, MA: MIT Press.

Gazdar, G., E. Klein, G. Pullum, and I. Sag (1985), *Generalized Phrase Structure Grammar*. Cambridge, MA: Harvard UP.

George, A. (1989), "How Not to Become Confused about Linguistics," in George (ed.), *Reflections on Chomsky*. Oxford: Basil Blackwell, 90–110.

Halle, M. (1961), "On the Role of Simplicity in Linguistic Description," in *Proceedings of Symposia in Applied Mathematics* 12 (Structure of Language and Its Mathematical Aspects). Providence, RI: American Mathematical Society, 89–94.

Higginbotham, J. (1983), "Is Grammar Psychological?" in L. Cauman, I. Levi, C. Parsons, and R. Schwartz (eds.), *How Many Questions: Essays in Honor of Sydney Morgenbesser*. Indianapolis: Hackett.

Hornstein, N. (1984), *Logic as Grammar*. Cambridge, MA: MIT Press.

——— (1995), *Logical Form: From GB to Minimalism*. Oxford: Blackwell.

Katz, J. (ed.) (1985), *The Philosophy of Linguistics*. Oxford: Oxford University Press.

——— (1981), *Language and Other Abstract Objects*. Totowa, NJ: Rowman and Littlefield.

Kripke, S. (1980), *Naming and Necessity*. Cambridge, MA: Harvard University Press.

——— (1982), *Wittgenstein on Rules and Private Language*. Cambridge: Harvard University Press.

Lewis, D. (1975), "Language and Languages," in K. Gunderson (ed.), *Language, Mind and Knowledge*. Minneapolis: University of Minnesota Press, 3–35.

Ludlow, P. (forthcoming), "Referential Semantics for *I*-Languages?" in N. Hornstein and L. Antony (eds.), *Chomsky and His Critics*. Oxford: Blackwell.

——— (1998), "Simplicity and Generative Grammar," in R. Stainton and K. Murasugi (eds.), *Philosophy and Linguistics*. Boulder, CO: Westview Press.

——— (1992), "Formal Rigor and Linguistic Theory," *Natural Language and Linguistic Theory* 10: 335–344.

Pinker, S. (1994), *The Language Instinct: How the Mind Creates Language*. New York: William Morrow and Company.

Postal, P. (1972), "The Best Theory," in S. Peters (ed.), *Goals of Linguistic Theory*. Englewood Cliffs, NJ: Prentice-Hall, 131–179.

Pullum, Geoffrey (1989), "Formal Linguistics Meets the Boojum," *Natural Language and Linguistic Theory* 7: 137–143.

Putnam, H. (1975), "The Meaning of Meaning," in Gunderson (ed.), *Language, Mind and Knowledge. Minnesota Studies in the Philosophy of Science* (Vol. 7). Minneapolis: University of Minnesota Press, 131–193.

Quine, Willard Van (1970), "Methodological Reflections on Current Linguistic Theory," *Synthese* 21: 368–398.

Soames, S. (1998), "Skepticism about Meaning: Indeterminacy, Normativity, and the Rule-Following Paradox," in A. Kazmi (ed.), *Meaning and Reference. Canadian Journal of Philosophy* 23(Suppl).

Sober, E. (1975), *Simplicity*. Oxford: Oxford University Press.

Wittgenstein, L. (1953), *Philosophical Investigations*. Translated by G. E. M. Anscombe. New York: Macmillan.

——— (1956), *Remarks on the Foundations of Mathematics*. Translated by G. E. M. Anscombe. Cambridge, MA: MIT Press.

See also **Chomsky, Noam; Innate-Acquired Distinction; Intentionality; Quine, Willard Van; Social Sciences, Philosophy of**

LEVELS OF SELECTION

See **Biology, Philosophy of; Natural Selection**

LOCALITY

The principle of local action has played an important role in the development of modern physics, and has been taken by many philosophers to be a necessary condition for intelligible causal explanations. However, recent evidence from quantum mechanics and quantum field theory seems to point toward fundamental limitations on the ability to provide locally causal explanations of physical phenomena (see Quantum Mechanics; Quantum Field Theory).

Locality in the History of Philosophy

According to a popular view, the most primitive cause/effect relation is that which holds between two physical objects that make contact in space and time, known as *contact action*. Furthermore, it is supposed that between any cause and effect, there must be a continuous chain of primitive causes by contact action; and if there is no continuous chain in space and time between two events, neither can be a cause of the other. This view has been advocated in one form or another by a number of philosophers of diverse persuasions. For example, Aristotle claims that "it is evident, therefore, that in all locomotion there is nothing intermediate between mover and moved" (Aristotle 1941). Similarly, in establishing the foundations for his new physics, Descartes takes it as an a priori principle that causation occurs only by local contact (see Suppes 1954). Moreover, Hume (1978)—the arch-critic of a priori knowledge of causal relations—claims that the concept of causation includes the concept of contiguity:

> [W]hatever objects are consider'd as causes or effects, are *contiguous*; and . . . nothing can operate in a time or a place, which is the ever so little remov'd from those of its existence. (§I.3.2)

Einstein (1948) claims that the principle of local action is a necessary presupposition for the existence of empirically testable natural laws:

> For the relative independence of spatially distinct things (A and B), this idea is characteristic: an external influence on A has no immediate effect on B; this is known as the "principle of local action" . . . . The complete suspension of this basic principle would make impossible . . . the establishment of empirically testable laws in the sense familiar to us. (321)

Finally, a number of influential contemporary accounts of causation tie the notion of causal connectedness to a space-time picture (see, e.g., Salmon 1984).

Locality in Modern Physics

The principle of local action was a cornerstone of the mechanical philosophy of Cartesian and neo-Cartesian physics. However, in Newton's theory of gravitation, the inverse square law seems to entail that some causes are spatially separated from their effects. Although numerous attempts, both physical and philosophical, were made to explain nonlocal gravitational forces (see McMullin 1989), a satisfactory resolution was not reached until Einstein supplied a field-theoretic formulation of gravity. Einstein's general theory of relativity was the culmination of a line of development that had begun in the early nineteenth century with Michael Faraday's introduction of the concept of "lines of force" emanating from a magnet. Faraday's idea was incorporated into Maxwell's dynamical theory of the electromagnetic field, in which electromagnetic field quantities are associated with each point of space, and disturbances in the field propagate through space via wave motion. Problems arising from Maxwell's theory ultimately led to Einstein's special theory of relativity, which grounds the principle of local action in the assumption of the constancy of the speed of light in all reference frames.

According to textbook presentations, special relativity is based on the limit principle: No physical process can propagate faster than the speed of light. If two events cannot be connected by a light signal, then they are said to be *spacelike separated* (i.e., they are simultaneous in some inertial reference frame). Thus, there can be no cause/effect relation between two spacelike separated events. However, the status of the principle of locality in special relativity continues to be a subject of dispute among philosophers. For example, it has been claimed that special relativity is not premised on the limit principle and probably does not entail it (Nerlich 1982). It has also been claimed that the limit principle is a statistical generalization that need not hold for individual processes (Cushing 1996; Maudlin 1994 provides an extended discussion of the role of locality in special relativity) (see Space-Time).

Entanglement and the Einstein–Podolsky–Rosen Result

In quantum mechanics, a pair of spatially separated systems can occupy an "entangled" state in which the values of their dynamical variables are perfectly correlated. In particular, if S_1 and S_2 are spatially separated systems with state spaces H_1 and H_2, then the state space for the composite system $S_1 + S_2$ is the tensor product $H_1 \otimes H_2$. For any $u_2 \in H_1$ and $v_2 \in H_2$, there is a vector $u \otimes v \in H_1 \otimes H_2$ called the "product" of u and v. Product states can be thought of as describing conjunctive states of affairs: $S_1 + S_2$ is in state $u \otimes v$ just in case S_1 is in state u and S_2 is in state v. However, since $S_1 + S_2$ is a quantum system, it also has states that are superpositions of product states. In particular, if u_1, u_2 are distinct states of S_1 and v_1, v_2 are distinct states of S_2, then

$$\psi = \frac{1}{\sqrt{2}}(u_1 \otimes v_1) + \frac{1}{\sqrt{2}}(u_2 \otimes v_2),$$

is a state of $S_1 + S_2$ that cannot be decomposed as a simple product. Such states are said to be *entangled*.

Composite systems in classical physics also have correlated states. But unlike the classical case, an entangled quantum state cannot be taken to represent an ensemble of composite systems each of which is in some definite product state. Indeed, unlike any correlated state in classical physics, entangled states are "pure" (i.e., statistically irreducible) states of the composite system. However, entangled states look mixed to local observers at S_1 or S_2. In fact, there is no pure (vector) state v of S_1 that agrees with ψ on the probabilities assigned to the various propositions about S_1, and similarly for S_2. In general, when $S_1 + S_2$ is in an entangled state, it is impossible to think of S_1 and S_2 as having their own (pure) quantum states.

The entangled state ψ predicts perfect correlations between measurements on the component systems. In particular, there is a measurement M_1 on S_1 that discriminates between the states u_1 and u_2, and there is a measurement M_2 on S_2 that discriminates between the states v_1 and v_2. If $S_1 + S_2$ is in the state ψ, then the two outcomes of M_1 are equally likely, and the two outcomes of M_2 are equally likely. However, outcomes of M_1 and M_2 are perfectly correlated: If M_1 yields an outcome corresponding to u_1, then M_2 will yield an outcome corresponding to v_1; and if M_1 yields an outcome corresponding to u_2, then M_2 will yield an outcome corresponding to v_2.

In their argument against the completeness of quantum mechanics, Einstein, Podolsky, and Rosen [EPR] (1935) made use of an entangled state that predicts perfect correlations between both the positions and the momenta of a pair of particles. They note that if the position of the first particle is ascertained, then the position of the second particle can be predicted with certainty. Similarly, if the momentum of the first particle is ascertained, then the momentum of the second particle can be predicted with certainty. Now, if one assumes (as EPR did) that the principle of local action holds, then a measurement on the first particle can neither alter nor bring into being properties of the second particle. Thus, a *position* measurement on the first particle should be thought of as a means of discovering the preexisting position of the second particle; and a *momentum* measurement on the first particle should be thought of as a means of discovering the preexisting momentum of the second particle. But then the second particle must have had a definite position and momentum before any measurement was performed. Since, however, a quantum-mechanical state never assigns a definite position and momentum to any object, EPR concluded that the quantum-mechanical state does not provide a complete description of the properties of the second particle.

EPR claimed to have shown that each particle must have a "hidden" state that determines the values of all of its dynamical variables—in other words, there are hidden variables. EPR also assumed that the hidden state of one system cannot be instantaneously influenced by events in distant locations. Nonetheless, neither EPR nor any other similarly inclined physicists were able to find a hidden variable theory that obeys the principle of local action. This failure, it is now known, was inevitable: Bell's theorem (Bell 1964) shows that no local hidden variable theory can explain the correlations described in the EPR experiment.

Bell's Theorem

According to hidden variable theories, quantum states merely provide statistical information about ensembles of systems, each of which has its own definite state (which includes a specification of the values of all relevant dynamical variables). A local hidden variable theory attributes a definite state to each local system and requires that changes in the state of one system cannot instantaneously bring about changes in the state of a distant system. Bell's theorem shows that no such local hidden variable theory can reproduce the predictions of quantum mechanics.

In the thirty years prior to the proof of Bell's theorem, the question of hidden variables had been largely pushed aside, in particular since von Neumann (1932) had supposedly shown that hidden variables are inconsistent with the empirical predictions of quantum mechanics. Few physicists took notice at the time when, in 1952, David Bohm constructed a hidden variable theory that is empirically equivalent to quantum mechanics. As Bell (1982) points out, Bohm's theory is not ruled out by the (mathematically valid) no-go theorems of von Neumann and Kochen-Specker because Bohm's hidden variables are contextual; that is, the hidden state of a system cannot be specified without taking into account its context, including the setting of measurement devices in distant locations. In fact, Bohm's hidden variables are patently nonlocal. Bell's theorem shows that this feature of Bohm's theory holds for any hidden variable theory that reproduces the predictions of quantum mechanics.

Consider the most simple correlation experiment, in which there is a pair of measurement devices situated in distant wings of a laboratory and each measurement device has (at least) two distinct settings. Let L_a denote the event that the left device is in setting a, and let R_b denote the event that the right device is in setting b. Suppose that each experiment has two possible outcomes, denoted by − and +. Let $L_a^\pm$ denote the event that the device on the left registers a ± outcome when in setting a, and let $R_b^\pm$ denote the event that the device on the right registers a ± outcome when in setting b. A quantum-mechanical realization of such an experiment is given by a pair of spin-$\frac{1}{2}$ particles in the singlet state:

$$\psi = \frac{1}{\sqrt{2}}(x_1 \otimes y_1 - x_2 \otimes y_2). \quad (1)$$

The measuring devices can be taken to be a pair of Stern-Gerlach magnets, each of which can be oriented at various angles in a plane from a common (arbitrarily chosen) axis. For each possible measurement on each particle, there are two possible outcomes, spin up and spin down. In this case, quantum mechanics supplies the following probabilities:

$$P_{QM}(L_a^+) = P_{QM}(R_b^+) = \frac{1}{2}, \quad (2)$$

and

$$P_{QM}(L_a^+ \cap R_b^+) = \frac{1}{2}\cos^2\theta_{ab}, \quad (3)$$

where θ_{ab} is the difference between the angles of orientation of the magnets on the left and right. If θ_{ab} is not an integer multiple of $\pi/4$, then the left and right measurement outcomes are statistically correlated:

$$P_{QM}(L_a^+ \cap R_b^+) \neq P_{QM}(L_a^+) \times P_{QM}(R+b). \quad (4)$$

Of course, the existence of a correlation between spatially separated events does not necessarily indicate a nonlocal connection, because the two events might have a common cause in the intersection of their past light cones. Suppose then that the quantum state corresponds to a probability distribution P_{HV} over a space Λ of hidden variables. Suppose for simplicity that Λ is finite. Suppose also that the domain of the probability function P_{HV} includes the events L_a, R_b^+ etc.). Let

$$P_{ab} = \sum_{\lambda \in \Lambda}[P_{HV}(L_a^+ \cap R_b^+ \mid L_a \cap R_b \cap \lambda) \times P_{HV}(\lambda)], \quad (5)$$

for all a, b, and let

$$P_1 = \sum_{\lambda \in \Lambda}[P_{HV}(L_1^+ \mid L_1 \cap \lambda) \times P_{HV}(\lambda)], \quad (6)$$

$$P_3 = \sum_{\lambda \in \Lambda} [P_{\text{HV}}(R_3^+ \mid R_3 \cap \lambda) \times P_{\text{HV}}(\lambda)]. \qquad (7)$$

Thus, this hidden variable model reproduces the quantum mechanical probabilities in state ψ just in case $P_a = P_{\text{QM}}(L_a^+)$, $P_b = P_{\text{QM}}(R_b^+)$, and $P_{ab} = P_{\text{QM}}(L_a^+ \cap R_b^+)$.

The hidden variable λ is local just in case it determines the outcomes of measurements on S_1 independently of what is occurring at S_2, and vice versa. That is, once the value of the hidden variable λ and the setting of the measurement apparatus at S_1 is fixed, then the outcomes of measurements on S_1 are determined; and similarly for S_2. (In the more general case of stochastic hidden variables, λ and the setting at S_1 will fix the probabilities for various outcomes at S_1.) This assumption is captured succinctly by Bell's locality condition:

$$\begin{aligned} &P_{\text{HV}}(L_a^+ \cap R_b^+ \mid L_a \cap R_b^+ \cap \lambda) \\ &= P_{\text{HV}}(L_a^+ \mid L_a \cap \lambda) \times P_{\text{HV}}(R_b^+ \mid R_b \cap \lambda)]. \end{aligned} \qquad (8)$$

If Bell's locality condition is conjoined with a "no conspiracy" condition (*viz.*, the event that a certain measurement occurs is probabilistically independent of λ), then Bell's inequality follows:

$$0 \leq P_1 + P_3 + P_{24} - P_{14} - P_{23} - P_{13} \leq 1. \qquad (9)$$

Thus, Bell's inequality is satisfied by the statistical predictions of any "reasonable" local hidden variable model of this experiment.

For specific choices of angles for the measurement devices, the quantum mechanical predictions violate Bell's inequality. For example, for the settings $\theta_{24} = \pi/2$, $\theta_{13} = \theta_{14} = \theta_{23} = \pi/6$, the sum of the quantum-mechanical probabilities equals $-\frac{1}{8}$. Thus, the predictions of quantum mechanics cannot be reproduced by any local hidden variable model. Moreover, these predictions have now been verified in a number of different experiments (for a review, see Redhead 1994, 107ff). Thus, the phenomena cannot be explained by a local hidden variable model.

Interpretations of Bell's Theorem

Many philosophers and physicists think that the violation of Bell's inequality points toward some form of nonlocality, whether or not quantum mechanics is a complete theory. However, a small group of dissenters claim that the violation of Bell's inequality has nothing to do with locality but should be seen as a consequence of the use by quantum mechanics of a nonclassical probability theory. Moreover, even among those who think that the violation of Bell's inequality entails nonlocality, there is still widespread disagreement about what exactly this means. Some claim that the violation of Bell's inequality shows that the world is thoroughly interconnected and holistic (or "nonseparable"), but not that the principle of local action is false. Others claim that the violation of Bell's inequality shows that causes can be spatially separated from their effects. The following section examines arguments for these three positions.

Quantum Mechanics Is Local

In a minimalist interpretation of Bell's theorem, the violation of Bell's inequality is due to the fact that local systems have incompatible observables, that is, observables for which there are no joint probabilities. The primary support for this interpretation comes from a theorem by Arthur Fine (1982a and 1982b) that shows that Bell's inequality is satisfied if and only if all joint probabilities are well defined. More precisely, suppose that there are joint probabilities $P(L_a^+, L_b^\pm)$ and $P(R_c^+, R_d^\pm)$ that return the already-given marginal probabilities:

$$P_{\text{QM}}(L_a^+) = P(L_a^+, L_b^+) + P(L_a^+, L_b^-), \qquad (10)$$

$$P_{\text{QM}}(R_c^+) = P(L_c^+, L_d^+) + P(R_c^+, R_d^-). \qquad (11)$$

Note that these joint probabilities are not supplied by quantum mechanics.

If such joint probabilities exist, then the marginal probabilities must satisfy Bell's inequality (de Muynck 1986). The minimalist will then point out that this derivation of Bell's inequality does not use any locality condition, and so the violation of Bell's inequality does not entail nonlocality. Contrapositively, the minimalist claims that since the existence of joint probabilities entails Bell's inequality, the violation of Bell's inequality shows that joint probabilities do not exist. The minimalist interpretation of Bell's theorem has also been defended within the context of particular interpretations of quantum mechanics. For example, advocates of the *consistent histories* interpretation have argued that "locality is not the only assumption that goes into the proof of the Bell inequalities, and thus their violation by quantum theory is not a proof of nonlocality" (Brun and Griffiths 2000). Furthermore, it has recently been claimed that if quantum mechanics is approached from an information-theoretic perspective, then the theory is "essentially local" (Fuchs and Peres 2000).

However, the minimalist interpretation of Bell's theorem has been criticized on the grounds that it ignores the issue of contextuality (van Fraassen 1991, 102; Shimony 1993, II.9). In particular, in a contextual hidden variable theory, unconditional probabilities such as $P_{\text{HV}}(L_a^+ \mid \lambda)$ are not physically significant. Rather, the physically significant probabilities are those conditionalized on all relevant measurement settings, for example, $P_{\text{HV}}(L_a^+ \mid L_a \cap R_b \cap \lambda)$.

However, Bell's inequality cannot be derived for the latter conditional probabilities. In other words, the existence of joint conditional probabilities does not entail Bell's inequality.

Holism and Nonseparability

Some philosophers have argued that the violation of Bell's inequality may entail nonlocality but it does not entail that there is superluminal causation. The main supporting argument for this position draws on Jarrett's (1984) analysis of Bell's locality condition. Jarrett shows that Bell's locality condition is equivalent to the conjunction of two conditions (the labels here are due to Shimony):

1. Outcome Independence:

$$P_{\text{HV}}(L_a^+ \cap R_b^+ \mid L_a \cap R_b \cap \lambda)$$
$$= P_{\text{HV}}(L_a^+ \mid L_a \cap R_b \cap \lambda)$$
$$\times P_{\text{HV}}(R_b^+ \mid L_a \cap R_b \cap \lambda). \quad (12)$$

2. Parameter Independence:

$$P_{\text{HV}}(L_a^+ \mid L_a \cap R_b \cap \lambda) = P_{\text{HV}}(L_a^+ \mid L_a \cap \lambda), \quad (13)$$

$$P_{\text{HV}}(R_b^+ \mid L_a \cap R_b \cap \lambda) = P_{\text{HV}}(R_b^+ \mid R_b \cap \lambda). \quad (14)$$

According to the orthodox interpretation of quantum mechanics, the quantum state ψ gives maximal information about the system. Thus, the orthodox interpretation can be thought of as the trivial hidden variable theory in which λ supplies no information beyond that supplied by the quantum state. Since the probabilities assigned by the quantum state are insensitive to which measurements are being performed on distant systems, the orthodox interpretation satisfies parameter independence. On the other hand, since distant measurement outcomes are correlated (see equation 4), outcome independence is violated. Moreover, it has been claimed that since outcomes are not determined by the quantum state—and hence cannot be controlled—this nonlocality could not be exploited to send a signal faster than the speed of light. Thus, orthodox quantum mechanics is consistent with special relativity (Shimony 1993, II.10). Proponents of the orthodox interpretation have also claimed that hidden variable theories violate parameter independence and that such a violation allows for superluminal signaling, and therefore hidden variable theories are inconsistent with special relativity. However, advocates of hidden variables have replied by pointing out that superluminal signaling is possible only if the hidden variables could be controlled, and this is not generally possible (e.g., in Bohm's theory).

Action at a Distance

Bell's inequality follows from the conjunction of outcome independence and parameter independence. So, either outcome independence or parameter independence (or both) is false. Bell's theorem by itself says no more. Nonetheless, there have been many attempts over the years to narrow down the interpretive options by deriving Bell's inequality from one of Jarrett's conditions. On the one hand, some argue that Bell's inequality can be derived from the assumption of hidden variables (*viz.*, the existence of joint probabilities), and therefore its violation supplies good evidence against "realism." Others, however, argue that Bell's inequality follows from locality alone, and so its violation entails nonlocality.

Maudlin (1994) argues that quantum-mechanical correlations can be explained only on the supposition of nonlocal causes. For example, he parodies Jarrett's analysis of Bell's locality condition by showing that it is equivalent to the conjunction of two conditions:

$$P_{\text{HV}}(L_a^+ \mid L_a \cap R_b^+ \cap \lambda) = P_{\text{HV}}(L_a^+ \mid L_a \cap \lambda), \quad (15)$$

$$P_{\text{HV}}(R_b^+ \mid R_b \cap L_a^+ \cap \lambda) = P_{\text{HV}}(R_b^+ \mid R_b \cap \lambda), \quad (16)$$

and,

$$P_{\text{HV}}(L_a^+ \mid L_a \cap R_b \cap R_b^+ \cap \lambda) = P_{\text{HV}}(L_a^+ \mid L_a \cap R_b^+ \cap \lambda), \quad (17)$$

$$P_{\text{HV}}(R_b^+ \mid L_a \cap R_b \cap L_a^+ \cap \lambda) = P_{\text{HV}}(R_b^+ \mid R_b \cap L_a^+ \cap \lambda). \quad (18)$$

Maudlin (1994) then points out that that it would be appropriate to call the first condition "outcome independence" and the second condition "parameter independence" (95). But now orthodox quantum mechanics violates "parameter independence" but not "outcome independence." The conclusion that should be drawn, claims Maudlin, is that Jarrett's analysis does nothing to show that the nonlocality found in orthodox quantum mechanics is more benign than the nonlocality found in hidden variable theories. (For another argument that quantum mechanics by itself—i.e., without additional interpretive assumptions—entails nonlocality, see Stapp 1997.)

The Subtleties of Nonlocality

While philosophers have been mainly concerned with investigating the consequences of the violation of Bell's inequality, physicists have also been trying to find ways to use nonlocality as a physical resource (e.g., to speed up computation). In the course of these investigations, it has been discovered that the violation of Bell's inequality is just one of many manifestations of nonlocality in quantum mechanics. Each vector state for a composite system is either a product state, or it is entangled. If a vector state is entangled, then it violates Bell's inequality (Gisin and Peres 1992), and therefore its correlations cannot be reproduced by a local hidden variable model. More generally, an arbitrary (possibly mixed) state ρ of a composite system is said to be separable just in case it is a mixture of product vector states; otherwise it is said to be nonseparable.

It is not difficult to see that separable states satisfy Bell's inequality. (Indeed, the "hidden variables" can be taken to be the quantum product states that are mixed together to form the separable state.) Werner (1989), however, shows that not all nonseparable states violate Bell's inequality. In particular, consider the mixture

$$W_n = \frac{1}{2} M_n + \frac{1}{2} P_s, \qquad (19)$$

where $M_n = (1/n^2)(I \otimes I)$ is the maximally mixed state of $\mathbf{C}^n \otimes \mathbf{C}^n$, and P_s is any maximally entangled, symmetric, pure state of $\mathbf{C}^n \otimes \mathbf{C}^n$. (For example, when $n = 2$, P_s could be the singlet state.) Werner uses an ingenious argument to show that W_n is nonseparable. However, he then goes on to construct a local hidden variable model for the correlations of W_n in Bell-type experiments. So, is W_n local or nonlocal?

There are at least two good reasons for thinking that the Werner state is nonlocal. First, a local hidden variable model, in Bell's sense, need account for the outcomes of only a certain special class of measurements; there might be other measurements whose statistics in W_n cannot be reproduced by such a model. In fact, Popescu (1995) shows that (for $n \geq 5$) a local observer can select a subensemble from W_n that violates Bell's inequality. That is, after an initial preparatory measurement on W_n is performed, then a Bell-type measurement can be performed that yields manifestly nonlocal results.

Since the initial preparatory measurement is purely local, it cannot create entanglement where none already existed. Therefore, it seems plausible to say that the original state W_n was already nonlocal (although its nonlocality was "hidden").

Second, the Werner state permits a teleportation scheme with higher fidelity than any classical communication channel (see Popescu 1994). Suppose that an observer O_1 has a particle P_1 in some unknown quantum state ψ, and O_1 wants to supply enough information to a second observer O_2 so that O_2 can prepare a particle in an identical state. On the one hand, if O_1 has access only to classical means of communication, the best O_1 can do is to make a measurement on P_1 (which can supply only partial information about its state), and then report the outcome to O_2. On the other hand, suppose that O_1 and O_2 share a pair (P_2, P_3) of particles in the Werner state. Suppose also that O_1 makes a measurement on the pair (P_1, P_2) and reports the outcome of this measurement to O_2. It can then be shown that O_2 is more likely to infer correctly the initial state ψ of P_1 than would be possible if O_2 had access to only classical communication from O_1. Thus, the Werner state allows for the transmission of more information than any classical procedure. Whether this improved communication ability amounts to superluminal information transfer is a matter of dispute.

Nonlocality in Relativistic Quantum Field Theory

The special theory of relativity (at least according to its most popular interpretation) prohibits action at a distance, while quantum mechanics seems to require it. Surely this poses a serious problem of consistency: How can two theories be true (or at least approximately true) when their most basic principles contradict each other? Philosophers have often confronted this apparent contradiction by looking for creative ways to reinterpret relativity and quantum mechanics; some have even concluded that special relativity must be false. And yet, there already is a theory, viz., relativistic quantum field theory, that is both relativistic and quantum mechanical. Although relativistic quantum field theory has been immensely successful in applications (in fact, it forms the basis for all of contemporary particle physics), philosophers have hardly begun to investigate how it manages to combine relativistic causality and quantum nonlocality (see Quantum Field Theory).

There are two structural features of relativistic quantum field models that mark them as distinctively relativistic. First, if *A* and *B* are spacelike separated regions, then any observable that can be measured in *A* is compatible with any observable that can be measured in *B*. This compatibility relation ensures that (nonselective) measurements performed in *A* cannot influence the statistics of measurements performed in *B*, and vice versa. Second, in relativistic quantum field models the spectrum of the four-momentum observable (i.e., the set of its possible measurement outcomes) is contained in the forward light cone. Thus, any measurement of the four-momentum will yield a result consistent with the predictions of special relativity; in particular, there can be no detectable energy-momentum transfer faster than light.

However, recent investigations have shown that these relativistic features of quantum field models do not preclude them from having nonlocal states. In fact, it has been shown (roughly speaking) that the percentage of nonlocal states grows in proportion to the dimension of the state space of the system. Thus, while systems with low-dimensional state spaces (e.g., spin-$\frac{1}{2}$ particles) have relatively few nonlocal states, systems with infinite-dimensional state spaces (e.g., field theories) have a very high percentage of nonlocal states. More specifically, for any two spacelike separated regions *A, B*, the set of field states that are Bell correlated across *A* and *B* is everywhere dense in the state space (Halvorson and Clifton 2000). Furthermore, every field state is maximally Bell correlated across unbounded tangent space-time regions, that is, "Rindler wedges" (Summers and Werner 1988). Finally, the vacuum state is non-separable across any pair of space-like separated regions, no matter how distant (Halvorson and Clifton 2000).

HANS HALVORSON

References

Aristotle (1941), "Physics," in Richard McKeon (ed.), *The Basic Works of Aristotle*. New York: Random House, VII, 2; 244b 16.
Bell, John S. (1964), "On the Einstein-Podolsky-Rosen Paradox," *Physics* 1: 195–200.
——— (1982), "On the Impossible Pilot Wave," *Foundations of Physics* 12: 989–999.
Brun, Todd, and Robert Griffiths (2000), Letter to the Editor, *Physics Today* 53.
Cushing, James (1996), "What Measurement Problem?" in Rob Clifton (ed.), *Perspectives on Quantum Reality*. Dordrecht, Holland: Kluwer, 167–181.
de Muynck, Willem (1986), "The Bell Inequalities and their Irrelevance to the Problem of Locality in Quantum Mechanics," *Physics Letters A* 114: 65–67.
Einstein, Albert (1948), "Quantenmechanik und Wirklichkeit," *Dialectica* 2: 320–324.
Einstein, Albert, Boris Podolsky, and Nathan Rosen (1935), "Can Quantum-Mechanical Description of Physical Reality Be Considered Complete?" *Physical Review* 17: 777–780.
Fine, Arthur (1982a), "Hidden Variables, Joint Probability, and the Bell Inequalities," *Physical Review Letters* 48: 291–294.
——— (1982b), "Joint Distributions, Quantum Correlations, and Commuting Observables," *Journal of Mathematical Physics* 23: 1306–1310.
Fuchs, Christopher, and Asher Peres (2000), Letter to the Editor, *Physics Today* 53.
Gisin, Nicolas, and Asher Peres (1992), "Maximal Violation of Bell's Inequality for Arbitrarily Large Spin," *Physics Letters A* 162: 15–17.
Halvorson, Hans, and Rob Clifton (2000), "Generic Bell Correlation Between Arbitrary Local Algebras in Quantum Field Theory," *Journal of Mathematical Physics* 41: 1711–1717.
Hume, David (1978), *A Treatise of Human Nature*. Edited by L. A. Selby-Bigge. New York: Oxford University Press.
Jarrett, Jon (1984), "On the Physical Significance of the Locality Conditions in the Bell Arguments," *Noûs* 18: 569–589.
Maudlin, Tim (1994), *Quantum Non-Locality and Relativity*. New York: Blackwell.
McMullin, Ernan (1989), "The Explanation of Distant Action: Historical Notes," in McMullin and James Cushing (eds.), *Philosophical Consequences of Quantum Theory: Reflections on Bell's Theorem*. South Bend, IN: University of Notre Dame Press, 272–302.
Nerlich, Graham (1982), "Special Relativity Is Not Based on Causality," *British Journal for the Philosophy of Science* 33: 361–382.
Popescu, Sandu (1994), "Bell's Inequalities Versus Teleportation: What is Non-Locality?" *Physical Review Letters* 72 (1994): 797–799.
——— (1995), "Bell's Inequalities and Density Matrices: Revealing 'Hidden' Nonlocality," *Physical Review Letters* 74: 2619–2622.
Redhead, M. L. G. (1994), *Incompleteness, Nonlocality, and Realism: A Prolegomenon to the Philosophy of Quantum Mechanics*. New York: Oxford University Press.
Salmon, Wesley (1984), *Scientific Explanation and the Causal Structure of the World*. Princeton, NJ: Princeton University Press.
Shimony, Abner (1993), *Search for a Naturalistic World View*. New York: Cambridge University Press.
Stapp, Henry (1997), "Nonlocal Character of Quantum Theory," *American Journal of Physics* 65: 300–304.
Summers, Stephen, and Reinhard Werner (1988), "Maximal Violation of Bell's Inequalities for Algebras of Observables in Tangent Spacetime Regions," *Annales de l'Institut Henri Poincaré* 49: 215–243.
Suppes, Patrick (1954), "Descartes and the Problem of Action at a Distance," *Journal of the History of Ideas* 15: 146–152.
van Fraassen, Bas (1991), *Quantum Mechanics: An Empiricist View*. New York: Oxford University Press.
von Neumann, John (1932), *Mathematische Grundlagen der Quantenmechanik*. Berlin: Springer.
Werner, R. F. (1989), "Quantum States with Einstein-Rosen-Podolsky Correlations Admitting a Hidden-Variable Model." *Physical Review A* 40: 4277–4281.

See also **Emergence; Quantum Field Theory; Quantum Mechanics; Reductionism**

LOGICAL EMPIRICISM

Logical empiricism dominated philosophical thinking about science from the late 1930s through the 1970s, so much so that nearly all philosophical writing about science in this period located itself, quite consciously, either within the fold of or in opposition to logical empiricism. Middle ground was, for a time, not easily found, and to ignore logical empiricism was to betray a profound ignorance of professional philosophy of science. Indeed, for many, the distinctly professional philosophy of science that emerged in the 1950s and 1960s had been made possible by—and was, perhaps, even identical to—logical empiricism. It is the task of this article to convey the main ideas and development of this profoundly influential philosophical movement.

Logical Positivism Versus Logical Empiricism

'Logical empiricism' is often used to refer to a philosophical school thought to be developed from *logical positivism*, whose more stringent *verifiability* criterion of cognitive meaning was replaced by a looser *confirmability* criterion; and an endorsement of scientific realism replaced logical positivism's rejection of the very question of whether the terms of a scientific theory refer to or of whether a successful theory's main claims should be taken to be, approximately true (see Cognitive Significance; Scientific Realism; Verifiability). (For the distinction between logical positivism and logical empiricism in this form, see Salmon 1999, 334.) But the notion that 'logical empiricism' suggests an identifiable, discrete, and conscious departure from something called logical positivism is tantamount to historical falsehood. In fact, logical empiricism, logical positivism, and *wissenschaftliche Philosophie* (scientific philosophy) had no fixed referents even in their heyday, and for good reason: What was usually behind these terms was an *attitude* or *approach,* rather than a theory or doctrine (a study of the terms used by participants themselves might suggest *wissenschaftliche Philosophie* as the most apt general term for the movement). And for this reason, current historical work should resist any urge to fix these terms' referents retrospectively (Hardcastle and Richardson 2003); logical positivism thus is not best described as having turned into, or given rise to, any particular successor movement, let alone one that embraced relaxed criteria of cognitive significance or scientific realism. It is worth noting that in contrast to 'logical positivism,' a term introduced in Blumberg and Feigl (1931), the provenance of 'logical empiricism' is murky.

On the other hand, insofar as logical empiricism denotes something like an intellectual development from and reaction to the main texts of logical positivism, the term is both useful and historically appropriate. Accordingly, this article on logical empiricism will present the tenets of logical positivism and the subsequent intellectual efforts that related themselves, directly and, on occasion, in opposition to these tenets; logical empiricism is really the story of the *development* of themes articulated within logical positivism. Following a discussion of five of logical positivism's main themes, then, logical empiricism will be described as it was reflected in ensuing philosophical work on analyticity, cognitive significance, holism, explanation, and the proper attitude toward scientific theories (see Analyticity; Cognitive Significance; Explanation; Theories).

Logical empiricism developed alongside intellectual projects with which it conflicted (socially and politically, as well as intellectually). The account of logical empiricism given here describes these conflicts without the typical "mortality" metaphor, in which philosophical views are born, mature, age, and die, becoming then objects of historical study. Instead, logical positivism is presented as the *distillation* of a particular scientific/philosophical ethos, and logical empiricism as the gradual, but detectable, *dilution* of that ethos over several decades and across several realms. This alternative metaphor presents logical empiricism as not a dead philosophical idea of mere historical interest, but as a set of identifiable traces left by the problems, approaches, and thinking associated with logical empiricism—problems, approaches, and thinking that animate and explain philosophy of science in the twenty-first century.

Distillation: Central Themes of Logical Positivism

The central themes of logical positivism are:

1. Antimetaphysics
2. A close relation to the natural sciences
3. Logic

At logical positivism's heart is antimetaphysics, or, to put it positively, a "spirit of enlightenment and anti-metaphysical factual research" (*geist der Aufklärung und der antimetaphysischen Tatsachenforschung*) (Hahn, Carnap, and Neurath [1929] 1973, 301). Indeed, it is important to express this central theme positively, since the image of logical positivism (and logical empiricism) as a negative, even destructive, movement is an egregious misrepresentation. The clearest expression of logical positivism's spirit of enlightenment is the 1929 pamphlet *The Scientific Conception of the World: The Vienna Circle* [*Wissenschaftliche Weltauffassung: Der Wiener Kreis*] (hereafter SCW), the 64-page "manifesto" of the Vienna Circle, the group of scientists, mathematicians, and like-minded thinkers that gathered around Moritz Schlick in Vienna in the 1920s, including Rudolf Carnap, Herbert Feigl, Philipp Frank, Hans Hahn, Otto Neurath, and Friedrich Waismann (see the listings for many of these figures in this volume, along with Vienna Circle). The pamphlet's author was listed simply as the "Wiener Kreis," but Neurath in fact wrote the bulk of the text, with contributions from Hahn and Carnap (Neurath 1973). The document announced to the European intellectual world the Circle's "scientific world-conception," which, it made clear, was characterized not so much by theses of its own, but rather by its basic attitude, its points of view, and direction of research:

> Neatness and clarity are striven for, and dark distances and unfathomable depths rejected. In science there are no 'depths'; there is surface everywhere: all experience forms a complex network, which cannot always be surveyed and can often be grasped only in parts. Everything is accessible to man; and man is the measure of all things.... The scientific world-conception knows no *unsolvable riddle*. Clarification of the traditional philosophical problems leads us partly to unmask them as pseudo-problems, and partly to transform them into empirical problems and thereby subject them to the judgment of empirical science. The task of philosophical work lies in this clarification of problems and assertions, not in the propounding of special 'philosophical' pronouncements. (Hahn et al. [1929] 1973, 305–306; cf. Carnap 1963, 20–34)

To what was this "spirit of enlightenment" directed? In practice, logical positivism directed itself toward questions concerning the logical structure of the sciences—the "clarification of problems and assertions" in them. The SCW summarized, for example, various "fields of problems" still to be addressed at "the foundations" of arithmetic, physics, geometry, biology, psychology, and the social sciences. Significantly, though, the scientific world-conception also addressed "questions of life" (304–305) in an enlightened manner, and thereby linked the Vienna Circle and logical positivism with political and social sensitivity, if not activism. As the SCW declared:

> [E]ndeavors toward a new organization of economic and social relations, toward the unification of mankind, toward a reform of school and education all show an inner link with the scientific world-conception; it appears that these endeavors are welcomed and regarded with sympathy by the members of the Circle, some of whom indeed actively further them. (Hahn et al. [1929] 1973, 305; cf. Carnap 1963, 20–26)

The later impression that logical positivism, and thus logical empiricism, consisted mainly in the systematic *rejection* of various philosophical traditions and methods as (worthless) "metaphysics" is owed largely to A. J. Ayer's *Language, Truth, and Logic*, an early, popular, and polarizing report on the Vienna Circle (see Ayer, Alfred Jules). The young Ayer visited the Circle from late 1932 through March of 1933 and managed to grasp its opposition to metaphysics, but little of its positive program. Nevertheless, *Language, Truth and Logic*'s infamous first line—"The traditional disputes of philosophers are, for the most part, as unwarranted as they are unfruitful" (Ayer 1936, 33)—fixed for the next three decades an image of logical positivism and empiricism. (In fairness, Blumberg and Feigl's [1931] earlier but less influential English presentation of logical positivism also failed to convey logical positivism's enlightenment theme.) Ayer's employment of logical positivism's verifiability criterion as a test for cognitive meaningfulness, and thus as a means to attack metaphysical claims, will be taken up below.

The second theme of logical positivism, one subsequently prominent in logical empiricism, is its close relationship to the natural sciences. Indeed, the distance between Ayer's understanding of logical positivism and logical positivism itself is clearest at the very end of *Language, Truth, and Logic*, where Ayer calls for the "philosopher to become a scientist... if he is to make any substantial contribution towards the growth of human knowledge" (Ayer 1936, 153). Ayer took himself to be calling for an unrealized philosophical future, but in fact logical positivism was already steeped in science. Its adherents aspired in their philosophical work to the sobriety and clarity of science, as represented particularly in relativistic physics (see Conventionalism; Space-Time). And its practitioners in Vienna, Berlin, and Prague were themselves adept at relativistic and quantum physics; the Circle's Philipp Frank, in fact, succeeded Einstein in Prague in 1912 when Einstein went to Berlin. Yet Ayer's insulation from actual science, combined with his authorship of *Language, Truth, and Logic*, led to the unfortunate, mistaken, and ironic impression that logical positivism was actually *detached* from contemporary science.

Steeped in science, logical positivism nevertheless distinguished itself from science, although what sort of project it *was* exactly would be subject to continued debate. The distinction between logical positivism and science itself was reflected most clearly in the fact that the former attended not to particular domains of experience (that was the focus of the various separate sciences) but to experience, and its *logical* structure, *as a whole*. Logic—the study of the *form* of scientific statements and the experience they describe—emerges as a third theme of logical positivism. And it is in this context that logical positivism's verifiability criterion is best appreciated. In SCW the criterion can be glimpsed as the claim that "for us, *something is 'real' through being incorporated into the total structure of experience*" (Hahn et al. [1929] 1973, 308, emphasis in original; see also Schlick [1930–1931] 1959). A claim or entity that could *not* be incorporated into the "total structure of experience" was therefore not real, in the strongest sense, and thus not even describable. Attempts to express such a claim or describe such an entity were at best confused and at worst disingenuous: metaphysics in the most damning sense (see, e.g., Blumberg and Feigl 1931; Carnap [1932] 1959, esp. §7).

Logic, particularly the quantificational logic articulated by Gottlob Frege and developed in 1910 by Whitehead and Russell (1963) and in 1921 by Wittgenstein (1961), was thus of enormous use to logical positivism. Indeed, the Circle read Wittgenstein's *Tractatus* painstakingly (Feigl 1969, 634). Moreover, the logical positivists' view of logic allowed for a statement to be true in a language purely by virtue of the meaning of its terms, that is, to be *analytically* true; such sentences provided an essential component of their epistemological account of mathematics and the formal sciences generally (see Analyticity). If mathematical statements were ultimately analytic statements of logic, and if logical statements were truths that could be produced at will, as it were, by the articulation of a language (perhaps an artificial language) in which they always came out true, then the truth of a mathematical sentence could be explained by reference to nothing more puzzling than a simple convention—a decision, freely taken, to render the sentence in question true (see Conventionalism). Knowledge of the truths of mathematics, *qua* sentences of logic, would then be as transparent or obvious as the language itself. Such a result embodied the enlightenment ethos of the scientific world-conception, and, correspondingly, challenges to it (as well as to, more generally, the view of logic at the heart of logical positivism) carried particular weight.

It remains to recognize two further important themes of logical positivism. One, implicit in the *generality* of logic, is the unity of science (*Einheitswissenshaft*) (see Unity and Disunity of Science; Unity of Science Movement). The SCW duly sounded the unity-of-science theme, which would become central to logical empiricism in the United States: "The goal ahead," the SCW stated, "is *unified science*. . . .The endeavor is to link and harmonize the achievements of individual investigators in their various fields of science" (Hahn et al. [1929] 1973, 306). Sciences, that is, were not to be distinguished on the basis of different methods, subject matters, or attitudes; their attitudes and methods did not differ, and differences in subject matter reflected merely pragmatic divisions of labor. Just beneath the surface of the unity-of-science theme was the denial of a division between the natural and social sciences, that is, between *Natur-* and *Geisteswissenschaften*; this denial had substantial political import for logical empiricism (Reisch 2004).

Finally, emphasis on the unity of science served to underscore a final theme implicit in the other three, *viz.,* the question of logical positivism's own place in the realm of inquiry or knowledge. In the 1920s and 1930s, the logical positivists struggled to describe their work in its own terms, that is, to describe their perspective as a philosophy informed by but not identical to science and its world-conception. Some, such as Schlick, subscribed to Wittgenstein's apparent view that the propositions of philosophy, properly understood, were themselves meaningless (Wittgenstein 1961; Schlick [1930–1931] 1959). Against Schlick's view, Carnap (1934) argued that a sufficiently rich metalanguage allowed for the expression of truths about the logical structure of science, including truths about that metalanguage itself. This was a matter within logical positivism that was not resolved and, indeed, became a central question for logical empiricism.

Dilution of Logical Empiricism: Analyticity and Cognitive Significance

World War II's significance to the development of logical empiricism is enormous, and at present only partly understood and appreciated (Reisch 2004). The war interrupted work within scientific philosophy in the obvious, material, way, but significantly also by virtue of logical positivism's perceived peripheral significance compared with the war effort. After World War II, scientific philosophy—now a distinctively North American endeavor, with Carnap, Hempel, Feigl, Frank, and Hans Reichenbach having emigrated to the United States—resumed work put aside in the late 1930s and early 1940s, although it did so in a very different cultural and political climate (see Carnap, Rudolf; Hempel, Carl Gustav; Reichenbach, Hans).

The result, by the early 1950s, consisted of three papers that would each have, for decades to come, a profound effect on philosophical thinking about science, challenging many (but, importantly, not all) of the themes associated with logical positivism. These papers were Quine's (1951) *Two Dogmas of Empiricism* and two essays published by Hempel in the early 1950s (Hempel 1965) (see Hempel, Carl Gustav; Quine, Willard Van). The former argued against the widely accepted view that certain truths were to be accounted for as true by virtue of the meanings of their terms (and, indeed, the article rejected as dogma the doctrine that there *were* analytic truths in this sense), while Hempel's two papers taken together reviewed and eventually rejected the notion that statements on their own had cognitive significance, by which was meant empirical meaning (see Analyticity; Cognitive Significance). Together the line of argument in these works pointed scientific philosophers sympathetic to logical positivism toward a holism about the meaning of statements and a pragmatism about any separation between science and metaphysics. The result, ultimately, was a nearly complete dilution, by the 1960s, of logical empiricism. An examination of these papers (each of which, incidentally, summarized discussions and work of several previous years) will illustrate this development.

Two Dogmas of Empiricism ostensibly takes up the following question: In virtue of what, precisely, are certain "analytic" sentences, the truth of which seems to be unavoidable, true? (Quine's example is "All bachelors are unmarried men.") Recognizing that the logical positivists' answer appealed to the meaning of the nonlogical terms in such sentences ('bachelor' and 'unmarried men'), Quine presses the question of how, precisely, the meaning of these terms manages to accomplish such a feat (see Analyticity). A good portion of *Two Dogmas of Empiricism* is given over, then, to showing how successive versions of the putative semantic relations between terms, and between sentences and a language, fail to provide the needed explanation of the analytic truths in question, typically because the purported explanation rests on concepts as much in need of clarification as meaning itself (Quine considers, specifically, definition, interchangeability *salva veritate*, semantic rules, and verificationism). Having presented a comprehensive, if not exhaustive, list of potential accounts and found them all wanting, Quine proceeds to doubt the presumption of the question, doubting, that is, that so-called analytic sentences are true by virtue of meaning. He then offers an alternative view. The single exception Quine recognizes to his list of failed accounts of analyticity, though, is telling. Quine allows that truths arising from the "explicitly conventional introduction of novel notations for sheer abbreviation" *does* suffice to account for some analytic truths, for here the definiendum becomes synonymous with the definiens simply because it has been created expressly for this purpose: "Here we have a really transparent case of synonymy created by definition; would that all species of synonymy were as intelligible" (26). Such "explicitly conventional" definition is rejected simply because it in fact is at the root of very few, if any, of the *actual existing* instances of analytic statements *in the present language*. Radically revising, or even discarding, present language, or more generally present theories, is not a live option.

Quine's rejection of explicitly conventional definition in *Two Dogmas of Empiricism* reflects not just his own conservatism but, significantly, that of the philosophy of science from the early 1950s on (see Quine, Willard Van). The enlightenment optimism of the SCW was no longer seriously considered; a departure from the past by means of the adoption of a new, modern, scientific attitude and the creation of a new, transparent language suitable to modern needs was simply dismissed.

Quine's critical treatment/rejection of various explanations of analytic truth is followed by his own account of those truths. His alternative is at once novel and conservative:

> The totality of our so-called knowledge or beliefs, from the most casual matters of geography and history to the

profoundest laws of atomic physics or even pure mathematics and logic, is a man-made fabric which impinges on experience only along the edges.... A conflict with experience at the periphery occasions readjustments in the interior of the field.... But the total field is so underdetermined by... experience that there is much latitude of choice as to what statements to reevaluate in the light of any single contrary experience.... If this view is right... it becomes folly to seek a boundary between synthetic statements, which hold contingently on experience, and analytic statements, which hold come what may. Any statement can be held true come what may, if we make drastic enough adjustments elsewhere in the system.... Conversely... no statement is immune to revision. (42–43)

Analytic statements, then, are simply those abandoned last, if ever, in the face of experience. Contained in this influential metaphor, which would come to be known as the web of belief, is a powerful challenge to several of logical positivism's central themes (see Duhem Thesis). Quine's holism—reflected particularly in his insistence that present and future language and theory be continuous with past theory and language—is in considerable tension with the progressiveness of the Vienna Circle (a tension embodied, indeed, in the work of Neurath (1973) (see Neurath, Otto). Further, Quine's (1951) pragmatism, displayed above in his recognition of several different but acceptable reactions to "contrary experience," "blurs," as he puts it, the "supposed boundary between speculative metaphysics and natural science" (10).

The effect of *Two Dogmas of Empiricism* was dramatic, and was only heightened when an analysis from a somewhat different starting point led to a very similar set of recommendations. Hempel's *Empiricist Criteria of Cognitive Significance: Problems and Changes*, which in many ways continued work begun by Carnap (1936–1937), reviewed and ultimately abandoned the thesis that there existed a "sharp dividing line ... between those sentences which do have cognitive significance and those which do not," or, for that matter, between significant and insignificant, or metaphysical, *systems* or *theories* (see Cognitive Significance; Hempel, Carl Gustav). Posing the problem in terms of a search for a formal relation between putatively cognitively significant statements and "observation sentences" (which contain only observational terms and are thus empirically unobjectionable), Hempel examined and rejected both verificationism (construed as the thesis that a cognitively significant sentence must be entailed by a consistent finite set of observation sentences) and falsificationism (that its negation be so entailed), as well as closely related proposals, on the grounds that each criterion either admitted clearly insignificant claims or, alternatively, barred clearly significant ones.

Pursuing the different tack of isolating the cognitive significance of *terms* (on the basis of "observation" terms), Hempel (1965) established first the inadequacy of the reductionist strategy Carnap (1936–1937) had outlined for theoretical terms, and then endorsed the semantic holism Quine also forwarded:

It is not correct to speak ... of the "experiential meaning" of a term or a sentence in isolation. [A] single statement usually has no experiential implications.... [T]he occurrence of certain observable phenomena can be derived from it only by conjoining it with a set of other, subsidiary, hypotheses ... [that] will usually be observation sentences [and] accepted theoretical statements. (112)

More significantly, Hempel was led from this holism to philosophical morals that, again, echo Quine's. After a failed search for criteria to separate cognitively significant from cognitively insignificant systems or theories (by way of barring those containing "isolated" sentences, that is, those without "experiential bearing"), Hempel (1965) cautiously suggests that "it is not possible to formulate ... criteria which would separate those ... systems whose isolated sentences ... have a significant function from those in which the isolated sentences are ... mere useless appendages. (117)

Rather,

[C]ognitive significance ... is a matter of degree: Significant systems range from those whose entire extralogical vocabulary consists of observation terms, through theories whose formulation relies heavily on theoretical constructs, on to systems with hardly any bearing on potential empirical findings. (117)

As with Quine's conclusion in *Two Dogmas of Empiricism*, such a claim constitutes (and was understood at the time to constitute) an abandonment of the *spirit* of the verifiability criterion, and of logical positivism. Moreover, here, as again with Quine (1951), logical empiricism took direction from Hempel's conclusion, which (like Quine's) revealed a certain conservatism in opposition to logical positivism's "spirit of enlightenment and anti-metaphysical factual research." For example, Hempel's critique of cognitive significance relied upon the appeal to the value of theory, even (indeed, especially) theory disconnected from experience or practice. As Hempel (1965) put it:

The history of scientific endeavor shows that if we wish to arrive at precise, comprehensive, and well-confirmed general laws, we have to rise above the level of direct observation .... In following ... a narrowly phenomenalistic or positivistic course, we ... deprive ourselves of the tremendous fertility of theoretical constructs, and we ... often render the formal structure of the expurgated theory clumsy and inefficient. (116)

Hempel's brief for empirically isolated theory contrasts deeply with the antimetaphysical spirit of logical positivism, and would shape logical empiricism in many ways, for decades.

Dissolution of Logical Empiricism: Unity, Realism, and the Philosophy of Science

It is important to recognize that the challenges to logical empiricism posed by Quine and Hempel, among others, as deep as they were, remained in a broadly scientific philosophical context. In other respects, logical empiricist themes were endorsed and elaborated. This is the case, for example, regarding the unity of science, which figured in logical empiricism in two significant ways.

Hempel's studies of cognitive significance were undertaken at nearly the same time as studies of what Hempel called the "logic of explanation," the attempt to characterize, as with cognitive significance, the formal relation obtaining between an event or a regularity to be explained, an *explanandum,* and that which explains the explanandum, the *explanans* (see Explanation). Hempel was motivated to pursue a logic of explanation in order to counter the view that methodology within the social sciences, notably history, differed in kind from what was found in the natural sciences. Both, Hempel urged, accomplish explanation by showing how the explanandum was to be expected given the laws inevitably cited in the explanans. In his first discussion of explanation, *The Function of General Laws in History,* Hempel (1942) argued that historical explanation "aims at showing that the event in question was not 'a matter of chance,' but was to be expected in view of certain ... conditions. The expectation referred to is not prophecy or divination, but rational scientific anticipation which rests on the assumption of general laws" (39).

Hempel later attempted to extend the fundamental idea contained in his account of explanation—that explanation consisted in subsuming the explanandum under general laws—to both the use of statistical laws to explain particular facts and the explanation of the laws themselves. The explication of scientific explanation in all its guises emerged as a research project of great fecundity in the 1950s and 1960s, as philosophers such as Braithwaite (1953) and Nagel (1961) pursued the essentially Hempelian project of providing a single explication of scientific explanation in all its guises, including, notably, explanations that made use of functions, particularly within a biological context. In this respect, Nagel's (1961) account of functional explanation came to exercise particular influence (see Function). This research project remained both popular and fruitful through the 1980s, as exemplified by Kitcher and Salmon (1989). Its roots, however, lay in the unity-of-science thesis. (For an authoritative overview of research on explanation, see Salmon 1990 and 1998.)

Hempel also pursued a parallel, and equally fecund, project with respect to confirmation; the effort here was to express the single relationship between evidence and hypothesis as it was to be found across all the sciences (Hempel 1965) (see Confirmation Theory). Hempel's efforts to capture the confirmation relation syntactically were dealt a fatal blow by Goodman (1955) (see Induction, Problem of), although valuable and influential work in this direction continued under the heading of inductive logic in the hands of Carnap, who made extensive use of the logical (as opposed particularly to a frequentist) interpretation of probability (Carnap 1950 and 1952) (see Inductive Logic; Probability).

In another, different guise, the unity of science proved far less successful. The Unity of Science movement, an official collaboration of Neurath, Carnap, and Charles Morris, had begun in Europe in 1934 with enormous ambition and promise (see Unity of Science Movement). It included international congresses, the *Journal of Unified Science* (a reincarnated version of the earlier logical positivists' organ *Erkenntnis*), and a separate Library of Unified Science, containing, in one vision, two hundred separate monographs (Neurath, Carnap, and Morris 1971). But World War II, a series of personal disputes between the collaborators, and, possibly, increasing (and justified) association of the Unity of Science movement with socialism and communism during the onset of the Cold War in the United States resulted in the nearly complete disintegration of the enterprise by the early 1950s. Thus the idea of scientific unity survived, even thrived, in the search for universal formal accounts of explanation and confirmation, while it languished in the cultural forum that had previously identified it with political aims—specifically, progressive socialism (Reisch 2004).

Hempel's welcoming of theoretical terms and entities disconnected from observation, as well as Quine's appeal to purely pragmatic criteria for preferring certain ontologies to others in an account of the world, contributed significantly to the consideration of the question of under what conditions, if any, the success of a scientific theory warranted belief that its central claims were in fact true or that the entities it mentioned in fact existed. Scientific realism—the position that success did warrant actual belief in, rather than mere acceptance of, a theory—was subsequently the focus of much debate from the late 1950s on, leading in turn to careful attention to the distinction between observable and nonobservable entities and the nature of success for scientific theories (see Scientific Realism). The debate itself, quite apart from any resolution it reached, demonstrated the intellectual distance logical empiricism had come from the attitude heralded in the SCW.

Thus by the late 1950s, logical empiricism embodied a tension between the historically oriented holism suggested by Quine and (less so) Hempel and the pursuit of general, unified accounts of science by way of specifying formal relations definitive of explanation or confirmation. This tension provides one (but hardly the only) means to understand the reaction to Thomas Kuhn's (1962/1970) profoundly influential *The Structure of Scientific Revolutions,* an essay that pitted an image of science that Kuhn understood to be logical empiricism against another image garnered from a close reading of scientific changes and the texts surrounding them (see Kuhn, Thomas; Scientific Revolutions). The latter image, Kuhn argued, engaged on its own terms, would transform the former and possibly lead to a rejection of both the unity-of-science thesis and the notion that science itself was a social institution with epistemic authority and privilege that it had earned. Yet, while *The Structure of Scientific Revolutions* was offered (and is now perceived) as an attack on logical empiricism, in fact several logical empiricists (most notably, Carnap) endorsed it. Such were the tensions within logical empiricism by the early 1960s.

Logical empiricism, understood as the reexamination, modification, and (alternatively) rejection and endorsement of the themes of logical positivism, is perhaps no more detectable within philosophy of science in the early twenty-first century than in current discussions of the nature and place of the philosophical examination of science, a topic, as mentioned above, that exercised the Vienna Circle. A good number of logical empiricists or their heirs subscribe to some version of Carnap's understanding of philosophy as the analysis of the logical structure of the concepts of science; work on confirmation, explanation, and other general concepts proceeds on several fronts. Others take up Quine's development of his own holism in his call for continuity between science and philosophical thought *about* science; Quine's "naturalized" account of epistemology describes philosophy of science as science itself. So motivated, philosopher-scientists have contributed to a number of scientific fields since the late twentieth century, most notably biology and physics. Finally, a felt need to connect philosophy of science in either guise to social and political matters, as well as study of the traditional aims and history of logical positivism and empiricism, has reopened discussion of the social and political dimensions of the philosophy of science and informed, for example, the *Institut Wiener Kreis,* an especially active institute of the University of Vienna dedicated to promoting historical study and understanding of the Vienna Circle and its aims. The perspective of logical empiricism thus informs the best philosophy of science done today.

Gary Hardcastle

References

Ayer, Alfred Jules (1936), *Language, Truth, and Logic.* London: Victor Gollancz.

——— (ed.) (1959), *Logical Positivism.* New York: Free Press.

Blumberg, Albert E., and Herbert Feigl (1931), "Logical Positivism: A New Movement in European Philosophy," *Journal of Philosophy* 28: 281–296.

Braithwaite, R. B. (1953), *Scientific Explanation.* Cambridge: Cambridge University Press.

Carnap, Rudolf ([1932] 1959), "The Elimination of Metaphysics through Logical Analysis of Language" in A. J. Ayer (ed.), *Logical Positivism.* New York: Free Press, 60–81. Originally published as "Überwindung der Metaphysik durch der Logische Analyse der Sprache," *Erkenntnis* 2: 219–241.

——— (1934), *Logische Syntax der Sprache.* Wien: Springer. Translated as *Logical Syntax of Language.* London: Kegan Paul, 1937.

——— (1936–1937), "Testability and Meaning," *Philosophy of Science* 3 (1936): 419–471, and 4 (1937): 1–40.

——— (1950), *Logical Foundations of Probability.* Chicago: University of Chicago Press.

——— (1952), *The Continuum of Inductive Methods.* Chicago: University of Chicago Press.

——— (1963), "Intellectual Autobiography," In Paul A. Schilpp (ed.), *The Philosophy of Rudolf Carnap.* LaSalle, IL: Open Court, 3–84.

Feigl, Herbert (1969), "The Wiener Kreis in America," in D. Fleming and B. Bailyn (eds.), *The Intellectual Migration: Europe and America, 1930–1960.* Cambridge, MA: Belknap, 630–673.

Goodman, Nelson (1955), *Fact, Fiction, and Forecast.* Cambridge, MA: Harvard University Press.

Hahn, Hans, Rudolf Carnap, and Otto Neurath ([1929] 1973), "The Scientific Conception of the World: The Vienna Circle," in Marie Neurath and Robert S. Cohen (eds.), *Otto Neurath: Empiricism and Sociology.* Dordrecht, Holland: Reidel, 299–318. Originally published as "Wissenshaftliche Weltauffassung: Der Wiener Kreis," *Veröffentlichungen des Vereines Ernest Mach*. Wien: Artur Wolf Verlag.

Hardcastle, Gary L., and Alan Richardson (eds.) (2003), *Logical Empiricism in North America*. Minneapolis: University of Minnesota Press, xiv–xvi.

Hempel, Carl G. (1942), "The Function of General Laws In History," *Journal of Philosophy* 39: 35–48.

——— (1965), "Empiricist Criteria of Cognitive Significance: Problems and Changes," in Hempel, *Aspects of Scientific Explanation*. New York: Free Press, 101–122.

Kitcher, Philip S., and Wesley Salmon (1989), *Scientific Explanation*. Minneapolis: University of Minnesota Press.

Kuhn, Thomas (1962), *Structure of Scientific Revolutions*. Chicago: University of Chicago Press.

Nagel, Ernest (1961), *The Structure of Science: Problems in the Logic of Scientific Explanation*. New York: Harcourt, Brace & World.

Neurath, Otto (1973). *Empiricism and Sociology*. Edited by Marie Neurath and Robert S. Cohen. Dordrecht, Holland: Reidel.

Neurath, Otto, Rudolf Carnap, and Charles C. Morris (1971), *Foundations of the Unity of Science: Toward an International Encyclopedia of Unified Science*, 2 vols. Chicago: University of Chicago Press.

Quine, Willard Van (1951), "Two Dogmas of Empiricism," *Philosophical Review* 60: 20–43.

——— (1953), *From a Logical Point of View*. Cambridge, MA: Harvard University Press.

Reisch, George (2004), *How the Cold War Transformed Philosophy of Science*. Cambridge: Cambridge University Press.

Salmon, Wesley C. (1990), *Four Decades of Scientific Explanation*. Minneapolis: University of Minnesota Press.

——— (1998), "Scientific Explanation: How We Got from There to Here," in *Causality and Explanation*. Oxford: Oxford University Press.

——— (1999), "The Spirit of Logical Empiricism: Carl G. Hempel's Role in Twentieth-Century Philosophy of Science," *Philosophy of Science* 66: 333–350.

Schlick, Moritz ([1930–31] 1959), "The Turning Point in Philosophy," in A. J. Ayer (ed.), *Logical Positivism*. New York: Free Press, 53–59.

Whitehead, Alfred North, and Bertrand Russell (1963), *Principia Mathematica*, 2nd ed. Cambridge: Cambridge University Press.

Wittgenstein, Ludwig (1961), *Tractatus Logico-Philosophicus*. Translated by D. F. Pears and B. F. McGuinness. London: Routledge and Kegan Paul.

See also **Analyticity; Ayer, Alfred Jules; Carnap, Rudolf; Cognitive Significance; Confirmation Theory; Explanation; Explication; Feigl, Herbert; Hahn, Hans; Hempel, Carl Gustav; Induction, Problem of; Inductive Logic; Kuhn, Thomas; Nagel, Ernest; Neurath, Otto; Phenomenalism; Physicalism; Popper, Karl Raimund; Probability; Protocol Sentences; Quine, Willard Van; Rational Reconstruction; Reichenbach, Hans; Scientific Realism; Schlick, Moritz; Theories; Unity and Disunity of Science; Unity of Science Movement; Verifiability; Vienna Circle**

M

ERNEST MACH

(18 February 1838–19 February 1916)

Mach was a physicist, psychologist, philosopher, and historian of science, as well as a political figure (for biographical detail, see Blackmore 1972). He studied physics at the University of Vienna from 1855 to 1861, continuing there as a lecturer until 1864. After spending three years as professor of mathematics at Graz, he received a chair at Prague, where he stayed until 1895. For the next six years, Mach occupied a chair in the History and Philosophy of the Inductive Sciences at Vienna. He suffered a stroke in 1898 and retired in 1901.

Mach's Influence

As a political figure, Mach served in the Austrian parliament and was so influential amongst the Austrian and Russian left that Lenin wrote *Materialism and Empirico-Criticism* as a criticism of Mach's anti-materialism. In physics, he was the first to understand supersonic shock waves and was a major influence upon a generation of physicists, including Einstein (who credited Mach for being a philosophical forerunner of relativity theory), Schrödinger, Planck, and Heisenberg. In psychology, Mach was a founder of Gestalt theory and made numerous contributions to sense physiology (see Psychology, Philosophy of). His research on Mach bands anticipated the modern understanding that the senses have neural nets that preprocess information before sending it to the brain. In philosophy, he was a major influence on the Vienna Circle (especially Frank, Hahn, and Carnap) and remains an inspiration to empiricist conceptions of science (see Vienna Circle). He was one of the most influential intellectuals in Vienna at a time when Vienna was the center of Western intellectual activity.

Mach's Psychology and Biology

Although he received his degree in physics, Mach attended many classes in physiology, and from the beginning of his career the majority of his research was not in physics but in the physiology of the senses. His central project, developed most extensively in his *Analysis of Sensations*, was to understand the relationship between sensations (the actual phenomenal experience) and the physical stimuli that trigger them (Mach [1914] 1984).

How, for instance, do eyes convert light particles into three-dimensional visual fields, when those particles themselves contain no information about their point of origin?

Although an empiricist in many regards, Mach held a strong notion of the a priori, maintaining that three-dimensional space is biologically innate (Banks 2003). Mach differed from the Kantian tradition by holding that the a priori is formed through evolution and development. Thus, there is nothing necessary about the current human spatial intuition; had humans evolved differently, they would perceive space differently. A similar account is given for other aspects of cognition, such as the understanding of matter, time, color, and even mathematics. The current human condition is historically contingent upon the particular evolutionary pathway humans accidentally took, and thus the world known through human senses should not be confused with the actual world.

Mach was strongly influenced by Darwin and by a variety of evolutionary ideas that are rejected today but were prominent within nineteenth-century German culture (see Evolution). Borrowing from the latter, Mach saw humans as nature's way of understanding itself, almost a waking up of nature. Science was a continuation of this process in that it was the "waking up" of humans. Evolution in prescientific times was an unconscious adaptation of organisms to their physical environments. The selection pressure was on survival and not on truth or higher social ideals. Humans thus adapted to convenient local minima and have come to misinterpret the human view of the world for the world itself. Thus there is a confusion between biological programming and reality. It is the biological purpose of science to lift humans out of this condition by understanding the nature of our psychological programming, and to provide us with a stable environment for future cognitive growth. It was an optimistic outlook; Mach's brand of positivism held out the hope that psychology embedded within an evolutionary framework would change the way humans saw the world, and thus lead not only to knowledge but, more importantly, to sociopolitical harmony.

Mach's Philosophy of Physics

Two central areas dominate Mach's philosophy of physics: his opposition to atomism and his opposition to Newton's conception of absolute space and time. Both arise from his antimetaphysical attitude, which in turn derives from his placing physics within a biopsychological framework.

His opposition to atomism arose from his biological conception of the purpose of science. For Mach, the social-political value of a scientific worldview is that it is nonmetaphysical and stable, and thus can provide a basis for positive scientific and social progress. Science should be nonmetaphysical so that all humans can agree to it. Furthermore, a nonmetaphysical science is stable in that it is not grounded in speculation but in description. This led him to favor a view of physics that emphasized description of the phenomena over the positing of ontologies and theories. Descriptions are simply more stable and less likely to be overthrown by future science, and thus provide a stable environment for cognitive change.

Turning to Mach's theories of space and time, it is important to note that Mach writes far more on the *psychology* of space than on the physics of space. In particular, his research on Mach bands develops the idea that even at a sensory level, one experiences only relations between things, and not the things themselves (Ratliff 1965). Mach's critique of Newton's theories of absolute space and time (which in turn influenced Einstein) was thus derivative of this psychological outlook. The history of physics thus owes a small but important debt to psychology. This is still controversial. What is agreed upon is that Mach rejected the mechanical view of physics and developed an influential alternative in which space, time, and matter were redefined as relations. For instance, he defined matter in terms of its relational interactions with other matter, that is, how much acceleration two objects impart to each other. With space and time, Mach similarly argued for a relational view. Humans have no psychological access to what space and time "are," so physics should simply mathematically describe the relations between objects.

In his influential *Science of Mechanics*, he argues that space and time cannot be used to measure the absolute changes of objects, as the very concepts of space and time are arrived at by observing the changes in objects (Mach [1893] 1960). That is, an intuition of space and time is prerequisite to measuring motion. Thus one cannot claim after measuring a motion that one has measured 'space' and 'time.' All one can do is give mathematical accounts of the spatial–temporal relationships of things. Mach, then, was a radical naturalistic epistemologist who turned to biology and psychology to give a naturalistic account of the entire human condition, including physics.

PAUL POJMAN

References

Banks, Erik (2003), *Ernest Mach's World Elements*. Dordrecht and Boston: Kluwer Academic.
Blackmore, John T. (1972), *Ernest Mach: His Work, Life, and Influence*. Berkeley and LA: University of California Press.
Mach, Ernest ([1914] 1984), *The Analysis of Sensations and the Relation of the Physical to the Psychical*. Translated by C. M. Williams. LaSalle, IL: Open Court.
—— ([1893] 1960). *The Science of Mechanics: A Critical and Historical Exposition of its Principles*. Chicago: Open Court Publishing.
Ratliff, Floyd (1965), *Mach Bands: Quantitative Studies on Neural Networks in the Retina*. San Francisco: Holden-Day.

See also **Logical Empiricism; Phenomenalism; Space–Time; Verifiability; Vienna Circle**

MATERIALISM

See **Physicalism**

MECHANISM

Interest in mechanisms has experienced a recent upsurge in the philosophy of science generally (e.g., Salmon 1984; Glennan 1996) and in the philosophy of biology and neuroscience in particular (see e.g., Bechtel and Richardson 1993; Craver and Darden 2001; Machamer, Darden, and Craver 2000). Scientific explanation often involves identifying the mechanism responsible for a phenomenon of interest. This entry provides a generic account of what mechanisms are and how they are appealed to in explanations and then turns to the question of how scientists discover them. What are mechanisms?

Four Aspects of Mechanisms

The notion of mechanism has four aspects: (i) a phenomenal aspect, (ii) a componential aspect, (iii) a causal aspect, and (iv) an organizational aspect. Mechanisms can differ from one another in each of these aspects. Consider them in turn, first using a common mousetrap as an example and then considering the more complicated mechanism of action potential generation in neurons:

The Phenomenal Aspect

Mechanisms do things; they are the mechanisms *of* the things that they *do*. A mousetrap traps mice, and the mechanism for generating action potentials generates action potentials. These tasks performed by the mechanism as a whole are the *phenomena* explained by the working of the mechanism. There are no mechanisms *simpliciter*—only mechanisms *for* phenomena. A mechanism's phenomenon partially determines the mechanism's boundaries (i.e., what is "in" the mechanism and what is not). As Kauffman (1971) clearly emphasized, an item is considered "part of" the mechanism only if it is relevant to a mechanism's phenomenon.

The Componential Aspect

Mechanisms have components, or working parts. Mechanisms all have at least two components. The old-fashioned mousetrap has six: a platform, a trigger, a latch, a catch, a spring, and an impact bar (see Figure 1). Trivially, the components are proper parts of the mechanism as a

MECHANISM

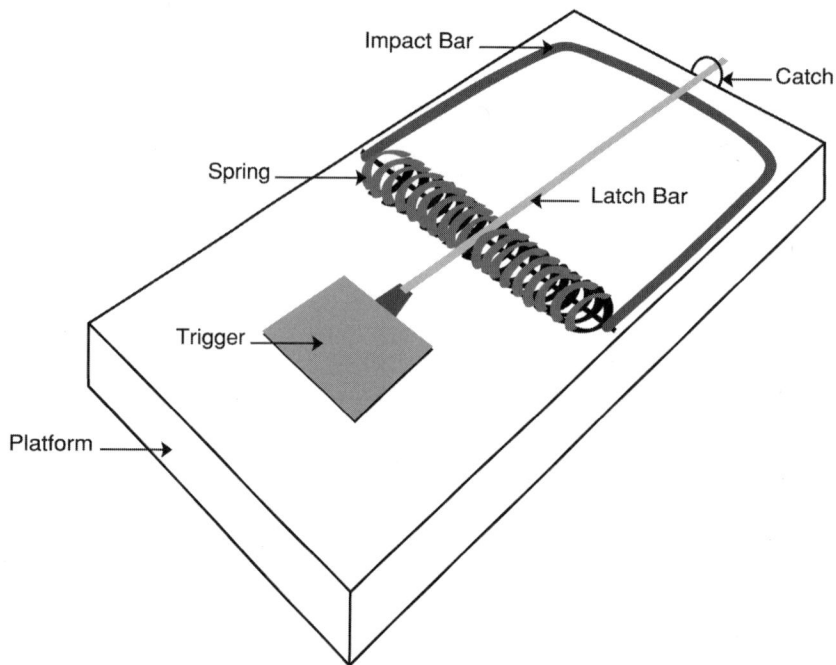

Fig. 1. The Mousetrap.

whole. More restrictively, as just noted, the parts of a mechanism are those that are relevant to the phenomenon explained by the mechanism. The parts are relevant to the phenomenon by virtue of certain of their properties (and not others). But for the rigidity of the bar and the tension on the spring, a mousetrap would catch no mice. The buoyancy of the platform, in contrast, is not properly included in the mechanism for catching mice.

The Causal Aspect

The components of mechanisms act and interact with one another. If they did not, they would not do anything. *Pressing* the trigger *releases* the catch, *allowing* the spring to *launch* the impact bar. The verbs in this description of the mousetrap refer to the relevant causal relations among the component parts. Talk of causal relations is a schematic placeholder to be filled in with one or more appropriate accounts of the kinds of causing exhibited in a given case. Philosophical attempts to develop univocal analyses of such causal relationships have yet to garner widespread acceptance. Yet the intransigence of the causal relation to a single uniform philosophical analysis should not distract attention from the central role that causal relations play in mechanistic explanations.

The Organizational Aspect

The components of mechanisms and their causal relations are organized spatially and temporally in the production of the phenomenon. The *spatial organization* of a mechanism includes the relative locations, shapes, sizes, orientations, connections, and boundaries of the mechanism's components. In the mousetrap, the trigger and the catch have to be so located with respect to one another that a small amount of pressure on the trigger moves the trigger bar enough to dislodge from the catch. The catch is circular and accommodates the size of the trigger bar. When the mechanism is loaded, the parts are connected to one another: The trigger bar restrains the blunt bar because it is stuck in the catch. As the mousetrap "fires," temporal organization takes center stage. The temporal organization of a mechanism includes the order, rates, durations, and frequencies of the activities in the mechanism. If a mousetrap is to work, it should work quickly, it should not discharge until there is pressure on the trigger, and there should not be significant delays between the steps of its working. Spatial and temporal organization are two important varieties of mechanistic organization. There are familiar patterns of mechanistic organization that can be found in different mechanisms for different phenomena. Some mechanisms are feed-forward, with each step following upon its predecessor without forks, joins, or cycles (like the common mousetrap); others may work in parallel or have significant feedback connections.

A Neurobiological Example: The Mechanism of the Action Potential

Mousetraps fire, and so do neurons. The firing of a neuron is known as an action potential. Action potentials are changes in the electrical potential difference across the cell membrane that propagate along the length of the neuron. This difference, known as the membrane potential (V_m), consists of the separation of charged ions on either side of the membrane. In the neuron's resting state, positive ions line up against the membrane's extracellular surface, and negative ions line up on the intracellular side, producing a polarized resting potential (V_{rest}) of roughly −60 mV. The action potential (as indicated in Figure 2) consists of (I) a rapid rise in V_m (reaching a maximum value of roughly +20 mV), followed by (II) an equally rapid decline in V_m to values below V_{rest}, and then (III) an extended hyperpolarized afterpotential during which the neuron is less excitable. These three features characterize the phenomenon to be explained by the action potential mechanism.

The components of this mechanism include the cell membrane, positively charged sodium (Na^+) ions, positively charged potassium (K^+) ions, and two types of voltage-sensitive ion channels that selectively allow, respectively, Na^+ or K^+ ions to diffuse through the membrane. It is the temporally organized activities of these channels that produce the action potential phenomenon.

The mechanism of the action potential starts with a cumulative depolarization of the cell body (i.e., V_m becomes greater than V_{rest}), typically through the effect of neurotransmitters on ion channels in the cell's dendrites (the "receiving" ends of the neuron). Action potentials are generated in the axon hillock, an ion-channel-dense region of membrane at the interface of the cell body and the axon (the "sending" end of the neuron). Depolarization of the cell body opens voltage-sensitive Na^+ channels (increasing membrane conductance to Na^+), allowing Na^+ to diffuse down its concentration gradient from the Na^+-rich extracellular fluid into the relatively Na^+-poor intracellular fluid (illustrated by the membrane conductance curve for Na^+ in Figure 2). The resulting flood of Na^+ drives the voltage of the cell toward the Na^+ equilibrium potential (E_{Na}; roughly +55 mV), accounting for the rapid rising phase of the action potential (I).

This rapid depolarization of the membrane has two consequences that account for the declining phase of the action potential (II). The first is the inactivation of the Na^+ channel, which slows and eventually stops the ascent of V_m toward E_{Na}. The second is the delayed activation of voltage-sensitive K^+ channels, increasing the K^+ conductance of the membrane and allowing K^+ to diffuse down its concentration gradient from the K^+-rich intracellular fluid into the K^+-poor extracellular fluid. This diffusion of K^+ drives the membrane potential back down toward the K^+ equilibrium potential (E_K; roughly −75 mV) and even below the resting potential of the membrane.

Thus begins the final, afterpotential phase of the action potential (III), which is characterized by both the hyperpolarization of the membrane (i.e., V_m is

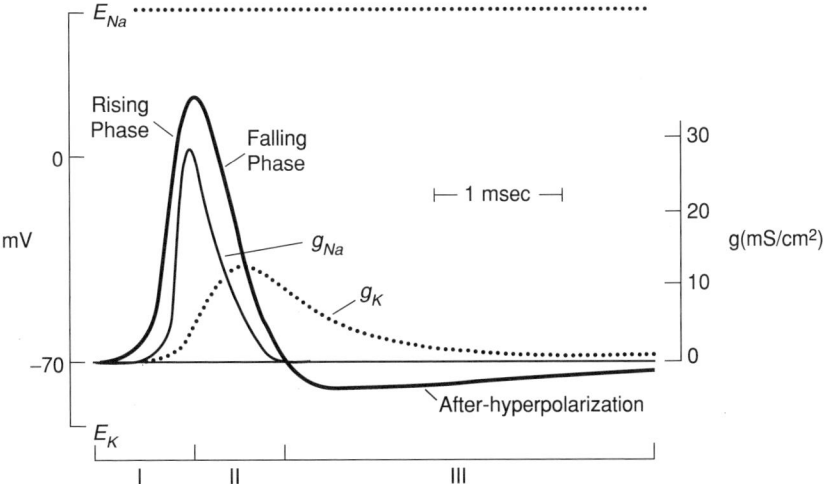

Fig. 2. The Action Potential.

lower than V_{rest}) and a period of reduced excitability. The membrane hyperpolarizes after the action potential because K^+ channels are slow to return to their resting closed state. The residual K^+ conductance tugs V_m away from V_{rest} and toward E_K.

The parts in the mechanism for generating action potentials are the membrane, the ions, and the ion channels. These parts are causally connected; they act and interact in regular ways to produce the action potential. These activities depend crucially upon the spatial organization of the components; ion channels *span* the membrane, allowing ion *movement* between the intracellular and extracellular fluids. Spatial organization is also fundamental to understanding the molecular mechanisms of channel activation and inactivation and for understanding the propagation of action potentials along axons. Yet, it is temporal organization that is most evident in the mechanism of the action potential; it is the relative orders and durations of the activation and inactivation of Na^+ and K^+ channels that explain the characteristic waveform (I, II, and III) of the action potential.

Levels of Mechanisms

Often mechanisms are nested within mechanisms. In such cases, some phenomenon (ψ) of a mechanism (M) is explained by the organized activities (ϕ) of lower-level components (X) that can themselves be taken as phenomena to be explained by the activities (ρ) of still lower level components (Z). Thinking about mechanisms provides a straightforward way to think about levels (see Craver 2001b). In this case, the relationship between lower and higher mechanistic levels is a compositional relationship with the additional restriction that the lower-level parts are components of (and hence organized within) the mechanism at the higher level. The requirement that lower-level parts be organized (at least spatially and temporally) within the higher-level mechanism distinguishes mechanistic levels from mere aggregates, such as piles of sand (Wimsatt 1986); from mere collections of improper parts, such as the cubes into which a television might be arbitrarily sliced (Haugeland 1998); and from mere inclusive sets, such as the collected songs of the Ramones. Lower mechanistic levels are entities and activities organized to exhibit the behavior of the mechanism as a whole.

Mechanistic levels should not be confused with intuitive ontic levels (e.g., Oppenheim and Putnam 1958), which map out a monolithic stratigraphy of levels across theories, sciences, and types of entities. Just as there are no mechanisms *simpliciter*, there are no mechanistic levels *simpliciter*. Mechanistic levels, instead, are defined only with respect to some highest-level mechanism M and its phenomenon ψ (pronounced "psi"). This, however, does not mean that the investigator cannot move upward, treating M as part of a yet higher level mechanism that generates its own phenomenon. Different levels of a mechanism involve different entities and activities. Accordingly, different vocabularies are typically used to describe mechanisms at different levels (Bechtel 1995). Exactly how many levels there are and how they are to be individuated are empirical questions that are answered differently for different phenomena.

Representing Mechanisms

There are many conventions for describing and representing mechanisms. Verbal accounts are generally insufficient to convey an understanding of a mechanism, especially if there are any nonlinearities in its behavior. Accordingly, verbal descriptions are often accompanied by diagrams representing the components, their activities (often depicted with arrows), and the relevant features of their organization (see Figure 1 above). Temporal relations are often represented spatially, either with labeled events conjoined by arrows or in separate frames. Diagrams afford the viewer the opportunity to follow through the parallel sequences of activities within the mechanism in one glance. With increasing frequency, the working of a mechanism may be represented in animated shorts. Extremely complicated mechanisms, however, frequently require the viewing time afforded by static two-dimensional representations so that aspects of the mechanism can be taken in piecemeal.

Descriptions of mechanisms, whether verbal or pictorial, may be more or less gappy, with holes or question marks to be filled in as details of the mechanism are discovered. Sometimes these are appreciated by the person portraying the mechanism, but many times the gaps are not even recognized until, for example, another component is discovered and researchers try to figure out what it contributes. Descriptions of mechanisms may also be more or less abstracted from the details of the operation of any particular mechanism, highlighting broad patterns of organization (e.g., with equations) or exhibiting precisely the spatial, temporal, and hierarchical organization of the components and activities of the mechanism.

Often the activities within a mechanism are characterized mathematically. For example, in describing the action potential mechanism, equations are advanced describing the changes in magnitude of Na^+ concentrations over time. Once such equations are developed, mathematical models of the overall operation of the mechanism can be advanced.

Mechanistic Explanations

Since mechanisms are often responsible for generating phenomena for which explanations are sought, it is not surprising that scientists frequently advance accounts of mechanisms as explanations. That is, to explain an action potential, they proceed much as in the example above—identifying the components of the responsible mechanism, describing the activities performed by the components, and showing how these components and activities are organized. They frequently present this information in diagrams, and often the account offered is gappy. Presenting a mechanism as an explanation, however, does not fit the standard deductive-nomological account of explanation, according to which explanation involves deriving a statement of the phenomenon to be explained from laws and relevant initial conditions. It is not laws that do the explanatory work but the account of the operation of the mechanism.

One might try to reconcile the two accounts of explanations by insisting that there is a law characterizing each mechanism. Typically, however, there is too much variability in a given mechanism (e.g., in the generation of action potentials in different neurons) for this to be plausible. It is better to recognize mechanistic explanation as an alternative model of explanation. Its prevalence in a variety of sciences such as physiology and neuroscience may account for the fact that these sciences have not been the primary source of examples of deductive-nomological explanation and have been relatively neglected by philosophers of science. Once mechanisms are recognized for their explanatory role in these sciences, though, we can also identify a number of philosophical issues to be pursued. One of these concerns their discovery.

How Are Mechanisms Discovered?

Characterizing the Phenomenon

One of the first tasks in discovering mechanisms is identifying the phenomenon—determining what it is that the mechanism does. The world does not come obviously prepackaged in terms of phenomena. How one characterizes the phenomenon critically affects how one goes about trying to discover the responsible mechanism and whether that quest will prove successful. Accordingly, characterizations of phenomena prove controversial and frequently are revised in the course of inquiry as one discovers that the mechanism does something different than one thought.

Phenomena are often subdivided, consolidated, or reconceptualized entirely as the discovery process proceeds. Researchers may recognize the need to subdivide a phenomenon into many distinct phenomena, as when learning and memory researchers were forced to recognize that there were many different kinds of memory requiring more or less distinct mechanisms to explain them. Alternatively, researchers may be forced to consolidate many different phenomena into a single phenomenon, as when it became understood that burning, respiring, and rusting were all due to a common mechanism and thus are examples of one phenomenon, oxidation. Finally, investigators may need to reconceptualize the phenomenon to be explained entirely. For example, early physiologists focused on the fact that animals burn foodstuffs and release heat. But after further investigation, researchers recharacterized this phenomenon as transforming energy into usable forms (e.g., ATP bonds).

Identifying Components

The discovery of mechanisms also involves identifying the components of the mechanism and their activities. Bechtel and Richardson (1993) used the term *decomposition* to describe analysis of a phenomenon into activities that, when properly organized, exhibit the phenomenon. In one of their main examples, they describe how the biological process of fermentation, over three decades of research, was decomposed into a set of more basic chemical reactions (oxidations, reductions, phosphorylations, etc.). This is *functional decomposition*. But frequently the process of decomposition begins by breaking the mechanism apart into component entities and only then investigating what the components do. This is *structural decomposition*. Ultimately, one measure of the adequacy of either form of decomposition is that it maps onto the other so that specific components are related to particular activities. Bechtel and Richardson call this identification of activities with components *localization*.

Often the search for the components of a mechanism is guided by an accepted store of components and activities that are reasonably well understood by a science at a particular time and that are available for use in thinking about how a mechanism works (Craver and Darden 2001). In the early stages of mechanism discovery, there may be no such store; there is either no idea of, or considerable controversy over, what the components and activities might be. The brain provides a useful example (Mundale 1998). There has been considerable controversy, for example, about what counts as a brain region, with different investigators using different criteria to divide the brain into parts at different times. Early attempts to map brain areas focused on the sulci and gyri resulting from the folding of cortex. Although prominent features of the brain, these tend not to be closely linked to component activities. With the identification of different types of neurons and the existence of cortical layers of varying thicknesses, numerous early-twentieth-century scientists, including Korbinian Brodmann, used these cytoarchitectural features to demarcate brain areas. Brodmann explicitly thought that different areas were likely to perform different operations, but he lacked any means for linking the regions he differentiated with function. More recent brain mappers have invoked yet additional criteria such as connectivity to other regions to identify brain areas. A major reason for controversy over these is that researchers are interested in components that perform the activities that generate the relevant phenomena. As the sulci and gyri of the brain illustrate, it is possible to differentiate structures within a mechanism that are not working components, that is, parts that carry out the relevant activities. In the relevant sense, these are not components of the mechanism. Similar challenges arise in functional decomposition—one may propose a decomposition into activities but not ones performed by any of the mechanism's components. Moreover, as hinted above, the search for components and for activities is interdeterminate—conceptions of the activities thought to be performed guide the identification of components, and vice versa.

How do scientists arrive at satisfactory decompositions that describe mechanisms in terms of their organized parts and activities? Often scientists begin the discovery process by proposing that there is a single component in the mechanism that alone is responsible for the phenomenon (e.g., attributing pleasure to activation of the brain's pleasure center). Sometimes this claim is correct, but even when it is, the task of identifying the mechanism that generates the phenomenon awaits decomposition of that component itself.

True decomposition is frequently guided either by the available store of components or by the available tools for investigating these components. Often scientists, functioning much like engineers, attempt to organize known components and activities in such a way that they might possibly produce the phenomenon. This process may involve reasoning analogically from other mechanisms (discovered in nature or human artifacts) and the activities performed in them. Such "how possibly" reasoning is, of course, fallible, since even two phenomena that are very similar may be generated by two very different mechanisms. In fact, sometimes the discovery process is slowed dramatically by pursuit of false leads generated by this engineering heuristic. On the other hand, even an erroneous proposal often advances the inquiry, since now experimental evidence can be generated that points to a more adequate decomposition. Experimental strategies for decomposing a mechanism are discussed further below.

Discovering the Organization of a Mechanism

Beyond delineating the phenomenon and revealing the components, a third major goal in the discovery of a mechanism is to determine how these components and activities are organized in the mechanism. Typically there are both spatial and temporal aspects to the organization of a mechanism. For example, the rate and duration of the phenomenon places time constraints on the activities of the components, and uncovering the order, rate, and duration of the steps in a mechanism often provides important clues into how the mechanism works. Likewise, discovering aspects of the spatial organization of a mechanism (the size, shape, position, orientation, etc., of the components) is often crucial for suggesting possible mechanisms and for ruling out others (see Craver and Darden 2001).

The relative importance of spatial and temporal organization varies from mechanism to mechanism. Spatial organization is of fundamental importance in, for example, the mechanisms of enzyme degradation because enzymes that can break down cellular substances need to be kept separate from other cellular substances that are not to be broken down. Spatial organization also helps provide efficiency in production mechanisms in which intermediate products are literally passed from one activity to the next (as in the Kreb's cycle).

If a phenomenon involves a change from one state or set of conditions to another (e.g., from glucose to alcohol, from sensory stimulus to recognition), it is common to think of that change as being executed by a linear sequence of steps. In part this is common because human conscious cognitive activities are serial—humans proceed from thinking of one thing to thinking of another. But, for very good reasons, such as ensuring proper regulation of a process, many natural mechanisms are not organized linearly. As a result, they are difficult for humans to conceptualize, at least without the aid of external representations such as diagrams in which one can represent backward as well as forward linkages.

The naturalness of linear organization means that in trying to fit multiple parts and activities together into a coherent description of a mechanism, researchers often begin by trying to organize them linearly. Often researchers begin to appreciate more complex modes of organization only when these attempts fail to account for the phenomenon. In modeling a chemical process, for example, one may find that there is no way to link together known basic reactions to get from the initial input to the product. This often leads to the exploration of more complex modes of organization such as a cycle. Thus, one common pattern in the process of discovering mechanisms is to begin with linear organization and then add complexity as required.

Experiments in Mechanism Discovery

Typically, the components, activities, and organization of a mechanism cannot be understood without the aid of well-designed experiments. Experimentation figures not just in the testing of models of mechanisms that have been hypothesized independently, but in the very process of discovering the mechanism.

Experimentation requires some means of intervening in the operation of the mechanism as well as a means of recording the effects of those interventions. Sometimes interventions into a mechanism are performed "by nature," through accidental damage, disease, or genetic mutation or variation. Other times the interventions are intentional and designed by the researcher to perturb some isolated aspect of the phenomenon or some component or activity in the mechanism.

A taxonomy of experimental approaches to developing and testing descriptions of mechanisms can be developed by focusing on where the intervention and recording techniques are applied (Bechtel and Richardson 1993; Craver 2001a). In the sense discussed above, a phenomenon and the mechanism that produces it are at two mechanistic levels, the phenomenal level (L_P) and the level of the mechanism (L_M) (see Figure 3). As illustrated in Figure 3, experiments may intervene and record entirely at the phenomenal level, bridge phenomenal and mechanistic levels, or intervene and record entirely within the mechanistic level.

First, both the intervention and the recording may be conducted at L_P without going down to L_M (see Figure 4). For example, one can intervene to vary the inputs to a mechanism or the conditions under which it operates (e.g., temperature) and record variations in the phenomenon. Much experimentation in cognitive psychology (e.g., requiring subjects to perform a task under varying conditions, such as cognitive load, and using reaction time as the measure of the effect) is of this sort and, when done well, can provide abundant information about the internal design of the mechanism. For example, evidence that two tasks interfere with each other provides further evidence that some component or components may be involved in both tasks. A great deal can also be learned about a mechanism by determining the range of input conditions under which it works properly and under which it fails or malfunctions.

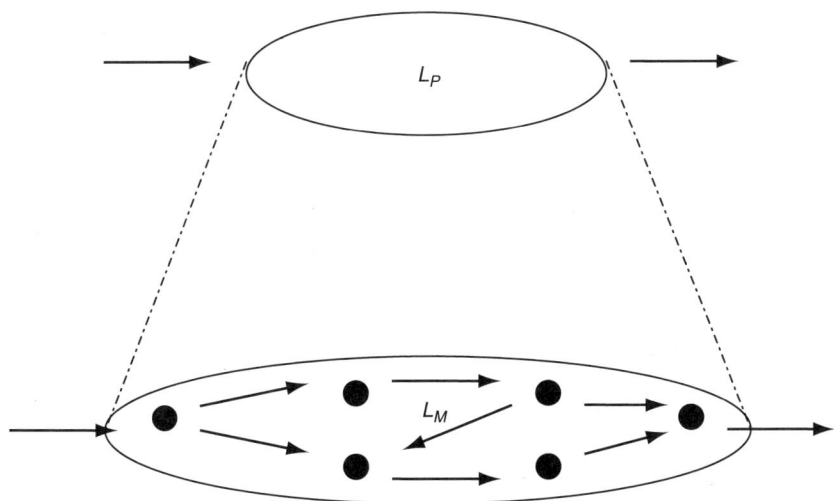

Fig. 3. Phenomenal Level (top) and Mechanism Level (bottom).

Second, experiments may bridge L_P and L_M. (Many experiments bridge several such levels at once.) Such experiments may be top-down (intervening at L_P and recording at L_M) or bottom-up (intervening at L_M and recording at L_P), and the experimental intervention may be either excitatory (somehow stimulating the target of the intervention) or inhibitory (somehow removing or impairing the target of the intervention). Top-down excitatory experiments are prevalent in cognitive neuroscience, where researchers intervene to engage an organism in some cognitive task while recording the activities of component brain regions, neurons, or molecules. Bottom-up excitatory experiments are also common. Neural stimulation studies, for example, use electrodes to excite individual neurons, and the effects are recorded for the cognitive phenomenon in which those neurons are involved. Additionally, bottom-up inhibitory experiments are a staple of most sciences that search for mechanisms. In neuroscience, for example, one may intervene to remove a brain region, a receptor molecule, or a neurotransmitter and record the effects on the phenomena in which those components are putatively involved. It is not uncommon for researchers to find a way to impair the activity before they figure out what the relevant components are or which are being affected. For example, one can discover a chemical poison that impairs a metabolic process but not know what component of the mechanism the poison is acting upon.

Third, inhibitory and excitatory techniques can also be applied within L_M. In this case, one intervenes to excite or inhibit some component or activity in the mechanism and then records the results of that intervention elsewhere in the mechanism. This form of experiment is especially important for determining how the components of the mechanisms are organized together in the production of the phenomenon.

There are significant epistemological challenges in interpreting the results of excitatory and inhibitory interventions into the working of the mechanism. Bottom-up inhibitory experiments may be foiled by redundancy, reorganization, and failures of specificity in the intervention. Intervention to remove or inhibit a component or activity may result in little or no change to the phenomenon if the removed or inhibited component is redundant (like the human kidney). Likewise, the mechanism may reorganize in the face of a loss of its component, leaving the phenomenon intact or only mildly transformed. In general, in removing a part of a mechanism and observing the behavior of the mechanism as a whole, researchers learn not what the removed part does but rather what the rest of the mechanism can do in its absence. Finally, the intervention may have nonspecific effects on other components in the mechanism, thereby indirectly altering the phenomenon and foiling the inference from the recorded changes to the function of the inhibited part. This problem is often exacerbated in "natural

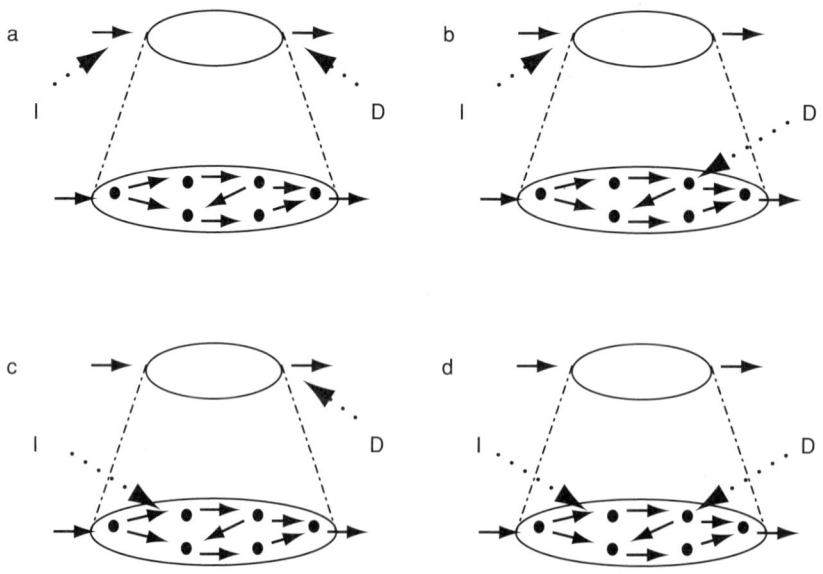

Fig. 4. Points of intervention and recording in experiments. Experiments may (a) both intervene and record at the phenomenal level, (b) intervene at the phenomenal level and record at the mechanistic level, (c) intervene at the mechanistic level and record at the phenomenal level, or (d) both intervene and record at the mechanistic level.

experiments" in which the intervention has not been tightly controlled by an investigator and so may have had a rather nonlocal impact on the components of the mechanism.

Similar epistemological difficulties attend the use of top-down excitatory experimental strategies. One example of such an experiment is to provide a stimulus to an organism and record from individual neurons in its brain or to use neuroimaging to record where there is increased blood flow in the brain. The epistemic challenges here are no less than when the intervention is within the mechanism. Activity in a part of a mechanism when the whole mechanism has been stimulated shows only that the component in question does respond to the stimulation. It does not yet show what activity it performs. Many neurons in the brain, for example, will respond when the organism is presented with a visual stimulus. One can gain more of a clue as to what a component is contributing by varying the intervention and determining the range of interventions to which the component is responsive (e.g., that it is only responsive to visual stimuli moving to the left). Even so, a given active neuron may perform an activity that is largely incidental to the phenomenon one is investigating (e.g., how objects are identified).

One way investigators begin to acquire confidence in their physical and functional decompositions is by drawing upon multiple modes of investigation, especially by invoking both inhibition and excitatory interventions. If lesioning a component eliminates the phenomenon of interest and exciting it produces the phenomenon, compelling evidence is provided that the component figures in generating that phenomenon. But just what does it contribute? Often answering that question depends on formulating a hypothesis about what many different components are contributing, and developing an account of how the components together produce the phenomenon. For example, researchers working on how the brain recognizes objects identified different brain regions in which individual cells would respond to different aspects of a stimulus—some responded whenever a given color was present, another when a given shape was present, and yet others when a particular object was present. By also knowing how these various brain regions were connected to each other, researchers began to piece together an account of the overall mechanism (Bechtel 2001).

As researchers reach the stage of reasonably worked out hypotheses about what different components contribute, additional tools can be invoked to help figure out the mechanism. For example, researchers often begin to build models, including computational ones, that characterize what each component is thought to contribute and to simulate their interaction. To the degree that the model predicts the phenomenon, one acquires confidence that one's account is at least close to correct. (The fit between a model and the phenomenon is often a matter of degree, and the degree of fit deemed sufficient often changes as research on the mechanism proceeds.) But failures are equally informative, since they often lead researchers to posit yet unidentified components and activities and begin to seek evidence for them.

Not surprisingly, there is no foolproof procedure for discovering mechanisms. But there are a range of strategies that can be identified by careful examination of actual science.

Conclusion

Four aspects of mechanisms have been identified: (i) the phenomenal, (ii) the componential, (iii) the causal, and (iv) the organizational. The generation of a phenomenon is often the product of a mechanism, and describing the mechanism provides an explanation of the phenomenon. The sciences concerned with identifying mechanisms have developed a variety of conceptual and experimental tools for this purpose. The philosophical analysis of mechanisms and their discovery is still in a relatively early stage but has advanced far enough that it is safe to predict that careful attention to mechanisms and mechanistic explanation is likely to yield significant advance in the philosophical understanding of science.

CARL CRAVER
WILLIAM BECHTEL

References

Bechtel, William (1995), "Biological and Social Constraints on Cognitive Processes: The Need for Dynamical Interactions Between Levels of Organization," *Canadian Journal of Philosophy* 20 (supplement): 133–164.

——— (2001), "Decomposing and Localizing Vision: An Exemplar for Cognitive Neuroscience," in Bechtel, Pete Mandik, Jennifer Mundale, and Robert S. Stufflebeam (eds.), *Philosophy and the Neurosciences: A Reader*. Oxford: Basil Blackwell, 225–249.

Bechtel, William, and Robert C. Richardson (1993), *Discovering Complexity: Decomposition and Localization as Strategies in Scientific Research*. Princeton, NJ: Princeton University Press.

Craver, Carl (2001a), "Interlevel Experiments and Multilevel Mechanisms in the Neuroscience of Memory," *Philosophy of Science* 68 (supplement): S83–S97.

——— (2001b), "Role Functions, Mechanisms, and Hierarchy," *Philosophy of Science* 68: 53–74.

Craver, Carl, and Lindley Darden (2001), "Discovering Mechanisms in Neuroscience: The Case of Spatial Memory," in Peter K. Machamer, Rick Gush, and Peter

McLaughlin (eds.), *Theory and Method in Neuroscience*. Pittsburgh: University of Pittsburgh Press.
Glennan, Stuart (1996), "Mechanisms and the Nature of Causation," *Erkenntnis* 44: 49–71.
Haugeland, John (1998), *Having Thought*. Cambridge, MA: Harvard University Press, chap. 10.
Kauffman, Stuart A. (1971), "Articulation of Parts Explanation in Biology and the Rational Search for Them," in R. C. Bluck and R. S. Cohen (eds.), *PSA 1970*. Dordrecht, Netherlands: Reidel.
Machamer, Peter, Lindley Darden, and Carl Craver (2000), "Thinking about Mechanisms," *Philosophy of Science* 67: 1–25.
Mundale, Jennifer (1998), "Brain Mapping," in William Bechtel and George Graham (eds.), *A Companion to Cognitive Science*. Oxford: Basil Blackwell, 129–139.
Oppenheim, Paul, and Hilary Putnam (1958), "Unity of Science as a Working Hypothesis," in Herbert Feigl, Grover Maxwell, and Michael Scriven (eds.), *Minnesota Studies in the Philosophy of Science*. Minneapolis: University of Minnesota Press, 3–36.
Salmon, Wesley (1984), *Scientific Explanation and the Causal Structure of the World*. Princeton, NJ: Princeton University Press.
Wimsatt, William (1986), "Forms of Aggregativity," in A. Donagan, A. Perovich, and M. Wedin (eds.), *Human Nature and Natural Knowledge*. Dordrecht, Netherlands: Reidel, 259–291.

See also **Explanation; Cognitive Science; Neurobiology; Reductionism**

METHODOLOGICAL INDIVIDUALISM

Methodological individualism (MI) is a set of related but distinct theses about how the social sciences should proceed. Drawing inspiration from the atomist program in natural science, MI holds that all social explanation should be in terms of individuals. Classic figures in the history of the social sciences such as Weber advocated some version of the doctrine, which is espoused by major schools of thought in economics and elsewhere in the social sciences (Blaug 1992; Gordon 1991). It is opposed by the holist tradition that began with the founders of sociology (Durkheim 1965) and continues to this day throughout the social sciences. While it has been usually treated as a conceptual or philosophical thesis, MI is probably best thought of as a series of more or less empirical theses.

Most individualist claims can be classified under one of the following four types (Kincaid 1996 and 1997):

1. Ontological claims:
    (i) Societies are composed of individuals.
    (ii) Societies do not act independently of individuals.
    (iii) Social entities do not exist.
2. Claims about theory reduction:
    (iv) Any social theory is in principle reducible to a theory referring entirely to individuals. (see Reductionism)
3. Claims about explanation:
    (v) Theories referring only to individuals can fully explain all social phenomena.
    (vi) Individualist mechanisms are a necessary condition for social explanation. (see Explanation)
4. Claims about confirmation:
    (vii) No social theory without individualist mechanisms can be well confirmed; and
    (viii) Searching for individualist theories is the best route to successful social science (see Confirmation Theory)

There are, of course, various real and alleged interconnections among these claims.

Perhaps the most common and logically central claim is the reductionist one. To reduce one theory to another is to show that the reducing theory can do all the explanatory work of the reduced theory, which is only a special case of the reducing theory. Since different theories have different vocabularies, theory reduction requires systematic linkages between the categories of the two theories. Individualism in its reductionist guise thus claims that the concepts of the social sciences can be equated with descriptions of individual behavior in such a way that all social science explanations can be put in individualist terms.

Many have claimed that this reductionist thesis follows from the ontological truism that society is made of and does not act independently of individual people (Watkins 1973). That conclusion does not follow, however. Chairs are made of molecules, but there are so many ways of making a chair that it is unlikely that a systematic connection will be found that would allow the replacement of the category 'chair' with some molecular description, however complex. Social entities such as corporations or states may have a similarly loose relation to the individual behaviors constituting them. Whether that is the case is an empirical issue that is unlikely to have any general answer across all domains of social research.

While the ontological claim that society does not exist and act independently of individuals is quite plausible, the presumed failure of theory reduction mentioned above argues against the further ontological claim that societies do not exist. If there are successful theories that make essential reference to social entities, that is one good reason to countenance their existence.

Of course the methodological individualist might deny the assumption that social explanations—explanations in terms of social entities—ever succeed in the first place. Two versions of this claim were identified above: No explanation is adequate unless it is entirely in terms of individuals (claim [v]) and some reference to individuals is necessary (claim [vi]). The latter, logically weaker claim is usually put as a claim about mechanisms: explanations in terms of social entities are only acceptable if the individualist mechanisms bringing them about are provided (see Mechanism).

To be interesting, these theses should be independent of assertions about theory reduction. It is unclear, however, that the stronger of the two is independent. If an individualist theory can fully explain everything that a nonindividualist theory can, then it can state those explanations only if it can relate the categories of the nonindividualist theory in every case to its own. But this goes back to the requirements for reduction.

The claim that individualist mechanisms are needed for explanation is widespread (Elster 1985; Little 1989). It is often defended as an instance of the general truth that all explanations require citing the mechanism involved. The general principle is implausible and of dubious value. It is implausible because one seemingly explains one macroscopic physical event by a previous one without any idea of the mechanism—"the crash of the plane caused the collapse of the building" is explanatory even without describing how it did so. The general principle is of dubious value because 'mechanism' is ill-defined. There can be mechanisms at many different levels of detail, so a demand for mechanisms is ambiguous. Individualists need to show that the mechanism cannot be social—as when competition between firms is cited to explain prices—and that it must be about individuals instead of neurons or genes.

Even if individualist mechanisms are needed to explain, this may not be much of a victory for methodological individualism. The mechanisms in question are likely to invoke the role individuals play and the constraints they face in institutions. Many game-theoretic and rational choice accounts provide individualist explanations, but they take norms, rules of the game, institutional constraints, and other such nonindividualist explainers as givens (see Game Theory).

Another motivation for individualist mechanisms is confirmation. The basic idea is that mechanisms are needed to rule out confounding cases. Aside from the question of levels raised above, this rationale suffers from the obvious fact that scientific claims can be confirmed without mechanisms. Newton's laws of motion were predictively very successful for a long period of time without any account of the underlying cause of gravity; Darwin had a similar success in describing evolution with an incorrect mechanism of genetic transmission. No doubt mechanisms play a useful role in confirming and explaining. But when, where, and at what level of detail they are useful is an empirical question that depends on the context.

HAROLD KINCAID

References

Blaug, M. (1992), *The Methodology of Economics*. Cambridge: Cambridge University Press.
Durkheim, Emile (1965), *The Rules of the Sociological Method*. New York: Free Press.
Elster, Jon (1985), *Making Sense of Marx*. Cambridge: Cambridge University Press.
Gordon, S. (1991), *The History and Philosophy of the Social Sciences*. London: Routledge.
Kincaid, Harold (1996), *Philosophical Foundations of the Social Sciences: Analyzing Controversies in Social Research*. Cambridge: Cambridge University Press.
——— (1997), *Individualism and the Unity of Science: Essays on Reduction, Explanation, and the Special Sciences*. Lanham, MD: Rowman and Littlefield.
Little, Daniel (1989), *Understanding Peasant China*. New Haven, CT: Yale University Press.
Watkins, John (1973), "Methodological Individualism: A Reply," in John O'Neill (ed.), *Modes of Individualism and Collectivism*. London: Heineman, 179–185.

See also **Confirmation Theory; Decision Theory; Economics, Philosophy of; Explanation; Game Theory; Mechanism; Reductionism; Social Sciences, Philosophy of**

MIND-BODY PROBLEM

See **Consciousness; Intentionality; Physicalism; Supervenience**

MOLECULAR BIOLOGY

The term 'molecular biology' was introduced by Warren Weaver in 1938 in an internal report of the Rockefeller Foundation: "And gradually there is coming into being a new branch of science—molecular biology . . . in which delicate modern techniques are being used to investigate ever more minute details of certain life processes" (as quoted in Olby 1974, 442). What Weaver may have only dimly foreseen is that these new techniques would ultimately transform the practice of biology in a way comparable only to the emergence of the theory of evolution in the nineteenth century. At the beginning of the twenty-first century, molecular biology has become most of biology, either *constitutively,* insofar as biological structures are characterized at the molecular level as a prelude for further study, or at least *methodologically,* as molecular techniques have become a preferred mode of experimental investigation of a domain. Recent biological work at the organismic and lower levels of organization (cytology, development, neurobiology, physiology, etc.) increasingly fall under the former rubric. Work in demography, epidemiology, and ecology falls under the latter, with ecology perhaps being the subdiscipline within biology that has most resisted molecularization. Work in evolution falls under both: constitutively, when the evolution of molecules and molecular structures forming organisms is studied for its own sake, and methodologically, when molecular techniques (most notably, DNA sequencing) are used to reconstruct evolutionary history. This article will be largely restricted to the constitutive aspect of molecular biology, since that is what has so far (perhaps deservedly) commanded most philosophical attention.

The decade following Weaver's introduction of molecular biology saw the steady increase in the use of "delicate" molecular techniques, in particular, x-ray crystallography, to study biological macromolecules "minutely," increasingly with an emphasis on proteins. The central problem was the elucidation of the three-dimensional structures (the relative positions of the atoms) of biological macromolecules. The structure of proteins was supposed to explain their behavior. Proteins were singled out because they were believed to be the most important of these macromolecules. In particular, since the establishment of biochemistry as a discipline in the 1920s, enzymes and their interactions had been held to be the key to understanding *metabolism* (the catchall term for the complex chemical reaction systems that characterize life). All enzymes are proteins. Until the early 1940s, it was believed that the hereditary material (the genes) was also likely to be proteins. The nucleic acids, constructed out of only four nucleotide base types (adenosine [A], cytosine [C], guanine [G], and thymine [T]), were believed to be insufficiently complex to be able to specify the immense variety of known genes.

However, experimental work starting in the early 1940s showed that the hereditary substance—specifying 'genes' (see Genetics)—was deoxyribonucleic acid (DNA). Attention then shifted to deciphering the physical structure of DNA, a problem that was solved by Watson and Crick (1953) with their double-helix model. The construction of this model and its subsequent confirmation marks a development of signal importance for modern biology (Sarkar 2005, Ch. 1). It ushered in the "classical" age of molecular biology (see the next two sections) with an intriguing informational interpretation of biology (see Biological Information). Important conceptual innovation also came from Monod and Jacob in the early 1960s, who constructed the allostery model to explain cooperative behavior in proteins and the operon model of gene regulation (Monod 1971; Jacob 1973; see below). Genes were interpreted as DNA sequences either specifying proteins (the *structural* genes) or controlling the action of other genes (the *regulatory* genes). Perhaps the most important development in classical molecular biology was the establishment of a genetic "code" delineating the relation of DNA sequences to amino acid residue sequences in proteins. (Both DNA and protein are linear molecules in the sense that they consist of units connected in a chain through strong [covalent] chemical bonds.) Gene *expression* took place by the *transcription* of DNA to ribonucleic acid, RNA, at the chromosomes (in the nucleus), and the *translation* of these transcripts into protein at the ribosomes (in the cytoplasm). The one gene–one enzyme credo of classical genetics was transformed into the one DNA segment–one protein chain credo of molecular biology (see Genetics).

Crucial to the program of molecularizing biology was the expectation—first explicitly stated by Waddington (1962)—that gene regulation explained tissue differentiation and, ultimately, morphogenesis in complex organisms. Genetic reductionism, the thesis that genes alone can explain organismic features, long predates molecular biology (Sarkar 1998). However, the molecular interpretation of the gene allowed the general explanatory success of molecular biology to be co-opted as a success of molecular genetics. In such a context, Waddington's thesis was positively received and helped usher in an era dominated by *developmental genetics,* according to which organismic development was to be understood through the action of genes. Mayr (1961) and others introduced the metaphor of the genetic program to characterize the putative relation between genomic DNA and organismic features. As molecular genetics began to dominate the research agenda of molecular biology in the 1970s, the emergence of organismic features came to be viewed as determined by "master control genes" (Gehring 1998). This view was initially supported by the demonstration that some DNA sequences (such as the homeobox) were conserved across a wide variety of species. DNA came to be viewed as the molecule "defining" life, a view that helped initiate the massive genome sequencing projects of the 1990s, which were supposed to produce a gene-based complete biology that delivered on all the promises of molecular developmental genetics. In general, because of the presumed primacy of DNA in influencing organismic features, starting in the early 1960s, molecular genetics began to dominate research in molecular biology.

Genetics and development were the earliest biological subdisciplines to be redefined by molecular biology. In the case of evolutionary biology, as early as the 1950s, Crick (1958) pointed out that the genotype/phenotype relation could be reinterpreted as the relation between DNA and protein, with proteins constituting the subtlest form of the expression of a phenotype of an organism. Consequently, the evolution of proteins (and, later, DNA sequences), especially the question of what maintained their diversity within a population, became a topic of investigation—in the 1960s, these studies led to the neutralist challenge to the received view of evolution (see Evolution). More importantly, changes at the level of DNA sequences, provided that these were selectively neutral, permitted the construction of a "molecular clock" that could be used to reconstruct evolutionary history more accurately than could be achieved by traditional morphological methods (see Population Genetics).

Meanwhile, biochemistry and immunology were reconstituted by the new molecular biology in ways that were not unexpected. That enzyme interactions and specificity would be explained in molecular terms was no surprise (see "Classical Molecular Biology" below). However, immunological specificity was also believed to be explainable by the same mechanism. This model of immune action was coupled to a selectionist theory of cell proliferation to generate the clonal theory of antibody formation, which combined molecular and cellular mechanisms in a novel fashion (see Immunology). In both biochemistry and immunology, what was largely at stake was the development of models that could explain the observed specificity of interactions: Enzymes reacted with only very few substrates; antibodies were highly specific to their antigens.

By the late 1970s, it became clear that the simplicity of the picture of genetics inherited from the 1960s was being lost. The initial picture was generated from an exploration of the genomes of prokaryotes (single-celled organisms without a nucleus), especially the bacterium *Escherichia coli*. In prokaryotes, every piece of DNA has a structural or regulatory function. In the 1970s, it was discovered that the genetics of eukaryotes (organisms with cells with nuclei) turned out to have an unexpected complexity. In particular, large parts of the genomic DNA sequences apparently had no function: These segments of "junk" DNA were interspersed between genes on chromosomes and also within genes. After RNA transcription, noncoding segments within genes were *spliced out* before translation. Gene regulation in eukaryotes was qualitatively different and more complicated than in prokaryotes. Some organisms used nonstandard genetic codes and other alternatives (see Sarkar 1996 for a detailed account.)

Subsequent work in molecular biology has only added to this picture of complexity, so much so that it is reasonable to suspect that the classical picture is breaking down. RNA transcripts are subject to *alternative splicing,* with the same DNA gene corresponding to several proteins. RNA is edited, with bases added and removed, before translation at the ribosome, to such an extent that it is sometimes difficult to maintain that some gene actually does code for a given protein. There is no obvious relation between the number of genes in an organism and its morphological or behavioral complexity. Most importantly, it now appears that a fair amount of the DNA thought to be junk is transcribed into RNA though not translated. Thus, presumably, much of the so-called junk DNA is functional, though the nature of these functions remains controversial.

This article will concern both classical molecular biology and the postgenomic molecular biology of the modern era. It will not only discuss issues in the philosophical interpretation of the classical era, which are fairly well characterized, but also include more speculative discussions of issues raised by recent developments.

Classical Molecular Biology

Classical molecular biology can be viewed in continuity with both the genetics and the biochemistry of the era that preceded it. From biochemistry—in particular, the study of enzymes in the 1920s and 1930s—early molecular biology inherited the mechanistic proposal that the function or behavior of biological molecules was "determined" by its structure, an idea that went back to Ehrlich's "sidechain" theory in the late nineteenth century. In the 1950s, structural modeling of biological macromolecules, especially proteins, was pioneered by Pauling and his collaborators using data from x-ray crystallography (see, e.g., Pauling and Corey 1950). By the early 1960s, a handful of such structures were fully solved. These structures, along with the structure of DNA, seemed to confirm the hypothesis that structure explains behavior. Perhaps more surprisingly, it was found that structural interactions seemed to be mediated entirely by the shape of active sites on molecules and that the sensitive details of structure and shape were maintained by very weak interactions.

These experimental observations led to four seemingly innocuous rules about the behavior of biological macromolecules, which in the 1960s and 1970s formed the theoretical core of molecular biology (Sarkar 1998, 149–150):

1. The weak interactions rule: The interactions that are critical in molecular processes are very weak.
2. The structure-function rule: The behavior of biological macromolecules can be explained from their structures, as determined by techniques such as crystallography.
3. The molecular shape rule: These structures, in turn, can be characterized entirely by molecular size, external shape (especially), and some general properties (such as hydrophobicity) of the different regions of the surfaces;
4. The lock-and-key fit rule: In molecular interactions, molecules interact only when there is a lock-and-key fit between the two molecular surfaces. There is no interaction when these fits are destroyed.

A lock-and-key-fit thus based on shape is an obvious way of achieving stereospecific capacity, thus resolving the critical problem for classical molecular biology. Because they are most intimately involved in the explanation of specificity, the molecular shape and lock-and-key-fit rules are the most important in this respect. In what follows, these will be called the rules of classical molecular biology.

In the 1960s and 1970s, these rules were deployed with remarkable success. As noted earlier, enzymatic and immunological interactions were among those that were immediately brought under the aegis of the new molecular biology. Two other cases are even more philosophically interesting:

- The allostery model explains why some molecules such as hemoglobin show *cooperative* behavior. In the case of hemoglobin, there is a nonlinear increase in the binding of oxygen after binding is first initiated. This is explained by conformational—shape—changes in the molecular subunits of hemoglobin; and

- The operon model explains *feedback*-mediated gene regulation in prokaryotes: The presence of a substrate activates the production of a protein that interacts with it, and its absence inhibits that production (see Monod 1971 for an accessible accurate account of these two examples and a conceptual summary of theoretical reasoning in early molecular biology).

Both cooperativity and feedback phenomena formed part of the traditional repertoire of holists in biology (see the next section, which will discuss the philosophical significance of the success of such structural explanation in molecular biology).

However, the 1950s also saw the elaboration of a radically different model of biological specificity, based on the concept of *information*, which was introduced into genetics only in 1953 (Sarkar 1996). This concept soon came to play a foundational role in molecular genetics. DNA was supposed to be the repository of biological information, a genetic "program" was supposed to convert this information into the adult organism, and new information was supposed to result from random mutation (and be maintained by selection) and never incorporated into the genome from the environment. Crick (1958) enshrined these assumptions in what he called the central dogma of molecular biology:

> This states that once "information" has passed into protein *it cannot get out again.* In more detail, the transfer of information from nucleic acid to nucleic acid, or from nucleic acid to protein may be possible, but transfer from protein to protein, or from protein to nucleic acid is impossible. (153) (emphasis in original)

Information, according to Crick, was the sequence of nucleotide bases in DNA or the sequence of amino acid residue in protein molecules. Note the contrast here with the stereospecific physical model of specificity. The dogma has continued to be an important regulative principle of molecular biology in the sense that it is presumed for further theoretical reasoning: Whether it survives recent developments will be discussed later in this essay.

However, the complexities of eukaryotic genetics, as discovered in the 1970s and 1980s, already began to challenge the central dogma (but see Thiéffry and Sarkar 1998). Much of this work was made possible by the development of technologies based on the polymerase chain reaction in the 1980s. There were five salient discoveries that challenged the simple picture inherited from prokaryotic genetics (Sarkar 2005, Ch. 8) (see Genetics):

1. The genetic code is not fully universal, the most extensive variation being found in mitochondrial DNA in eukaryotes. However, there is also some variation across taxa (see Fox 1987 for a review).
2. DNA sequences are not always read sequentially in blocks. There are overlapping genes, genes within genes, and so on (Barrell, Air, and Hutchison 1976). Thus, two or more different proteins could be specified by the same gene.
3. As noted earlier, not all DNA in the genome is functional. Intervening sequences—within and between structural genes—must be spliced out from transcripts (Berget, Moore, and Sharp 1977; Chow et al. 1977). This discovery helped resolve the so-called C-value paradox (Cavalier-Smith 1978), that is, the absence of any obvious correlation between the size of the genome and the morphological and behavioral complexity of an organism.
4. The same transcript may be spliced in different ways (Berk and Sharp 1978). One consequence of such alternative splicing is that, as with overlapping genes, two or more different proteins could be specified by the same gene.
5. Besides splicing, RNA is sometimes subject to extensive editing before translation at the genome (Cattaneo 1991).

These developments have led to skepticism of the relevance of the coding model of the DNA/protein relationship and of the informational model of specificity (see the next section). Though philosophers (and some biologists) have been slow to recognize this, the credo of one DNA segment–one protein chain has long become irrelevant in molecular biology. The modern era presents even more significant challenges, as later sections of this essay will underscore.

Philosophical Interpretations

Philosophy of biology only emerged as a recognizable part of philosophy of science only in the late 1960s. In the early years, considerable attention was paid to molecular biology, especially with respect to the issue of reductionism, but starting in the late 1970s, attention within philosophy of biology began to be concentrated solely on evolutionary theory, much to the detriment of the field. Attention shifted back to molecular biology in the 1990s, with some work now being done on the question of biological information besides reductionism. Since then, classical molecular biology has been increasingly scrutinized by philosophers, though not as much as it deserves. This section will focus on reduction and information. However, important philosophical work has also been done on other forms of conceptual change in molecular biology and, lately, experimentation in the field (Culp 1995; Rheinberger 1997).

Reduction

The first question about molecular biology that interested philosophers was whether it could be interpreted as a reductionist enterprise in the same way as the kinetic theory of matter was reductionist within classical physics (see Reductionism). The model of reduction then in vogue was due to Nagel (1961) with some modification by Schaffner (1967): It viewed reduction as a deductive-nomological explanation but with the reduced laws as the explananda (see Explanation; Nagel, Ernest). The debate soon centered on the question of whether molecular genetics was reducing or replacing Mendelian genetics. While Schaffner (1967) made the case for successful reduction, this position was attacked by Hull (1972) on the grounds that molecular biology did not have laws and theories (as logical empiricists envisioned those entities). Subsequently, an antireductionist consensus developed (also influenced by Kitcher [1984]).

This consensus was subsequently challenged by Sarkar (1989 and 1998), Waters (1990), and others, but only by rejecting the Nagel-Schaffner formal model as being relevant to substantive questions about reduction. (Even earlier, Wimsatt [1976] had argued against the relevance of the Nagel-Schaffner model.) In these analyses, what is at stake is that properties of wholes are being explained by properties of parts interacting locally. The allostery and operon models are philosophically critical exemplars of this approach because the former explains cooperativity and the latter feedback, both of which formed part of the conceptual repertoire of traditional holists (see Emergence). Enzymatic and immunological specificity provide more mundane examples. However, most of these cases are much simpler than that of providing fully reductionist explanations of quintessentially Mendelian genetic phenomena such as the segregation or assortment of alleles. In these cases—central to the question of reducing Mendelian genetics to molecular genetics—reductionist explanation remains piecemeal and, in many ways, incomplete. However, there is every reason to believe that the relevant lacunae will be filled without requiring new conceptual or theoretical resources.

Nevertheless, even during the classical era, a few anomalies remained, though none serious enough to call into question the viability of the reductionist project. In particular, there has never been a successful parts-whole account of dominance (that is, the dominance of one trait or allele over another) (see Genetics). There is some reason to believe that explaining dominance at the molecular level will require appeal to topological properties of networks, but such a move would take explanation beyond the reductionist realm (see "Philosophical Speculations" below).

Information

Though it is commonplace to talk of biological information, no successful formal definition of the concept in the context of molecular biology has ever been given. Because of difficulties that the concept of information encountered in the late 1980s and 1990s, this failure led Sarkar (1996) to suggest that information in molecular biology was a metaphor masquerading as a theoretical concept (Griffiths 2001) (see Scientific Metaphors). For Crick (1958), information consisted of sequences, of DNA or protein. Informally, this is what 'information' is probably taken to mean in most contexts. The first point to note is that any such definition would require that the concept of information being used *not* be Shannon's (1948) communication-theoretic notion of information, which requires the estimation of the frequency of symbols drawn from a set. Thus, mathematical information theory based on Shannon's concept simply becomes irrelevant in this context (for a contrary position, see Yockey 1992). At the very least, any usable concept of biological information must refer to individual sequences and be symbolic, semantic, or semiotic, in the sense that it must capture the idea that the sequence is a "sign" for something else (Sarkar 2005, Ch. 10). As such, it must account for biological specificity.

The concept was central to two related theoretical interpretations within molecular biology:

(a) that the DNA/protein relation is a genetic code, typically extended to suggest that all phenotypic traits are encoded in the DNA of the genome; and
(b) that the genome constitutes a genetic program for the organism.

As discussed earlier, developments within eukaryotic genetics began to limit the scope of the genetic code in the 1970s. Any claim of the existence of a genetic program at the very least constitutes a claim of genetic reductionism, and at the very worst a claim of genetic determinism. Genetic reductionism must be clearly distinguished from physical reductionism (the physical explanation of properties of wholes from properties of parts, which was discussed earlier). Genetic reductionism is the claim that organismic features are satisfactorily explained by appeal to properties of genes or DNA (without recourse to properties of other molecules). It is, for instance, central to the project of developmental genetics. Such a reductionism was never very plausible; consequently, the metaphor of a genetic "program" was always troubled (Keller 2000). Nevertheless, the "program" metaphor was quite influential during the heyday of developmental genetics. As the next two sections will underscore, it does not survive even in a mitigated form in the postgenomic era. The failure of genetic reductionism makes any stronger claim of genetic determinism irrelevant.

Viewing information as sequence, Crick (1958) also proposed the *sequence hypothesis:* that the sequence of amino acid residues in a protein (also called its *primary* structure) determines its three-dimensional conformation (also called its *tertiary* structure). Attempting to show how this comes about came to be called the protein folding problem. It has never been successfully solved, and not for lack of effort (Sarkar 1998). Moreover, for many proteins, it is known that sequence alone is insufficient for specifying three-dimensional conformation. It may even be the case that the same sequence can lead to several different conformations. This failure casts additional doubts on the utility of the concept of biological information stored in the genome, at least in the sense Crick intended it. Even if the genetic code were as exceptionless and predictively successful as was believed in the 1960s, all that it would allow is the inference of an amino acid residue sequence from the DNA. If the protein sequence does not determine its conformation, ipso facto, the DNA sequence cannot. It follows that the information within the DNA cannot specify phenotypes even further removed from the genome.

To the extent that the genetic code still remains useful, a proper explication of the concept of biological information remains an unaccomplished philosophical task of some importance (see Biological Information).

The Modern Era

By the "modern era" of molecular biology is meant the period beginning with the production of large genomic sequences in the 1990s. It is also referred to as the Genomic, Postgenomic, or, less accurately, Proteomic era ("proteomic" is less accurate because, to date, there has been limited progress in proteomics; see below). What marks this era is the study of large genomic sequences, and not individual alleles that had been previously identified by their phenotypic effects.

Genomics and Postgenomics

Genomics was ushered in by the decision to sequence the entire human genome as an organized project (the Human Genome Project [HGP]), involving a large number of laboratories in the late 1980s. Subsequently, similar projects were established to sequence the genome of many other species. To date, genomes of over 150 species have been sequenced. Almost every month sees the announcement of the completion of sequencing for a new species. The sheer volume of sequence information that has been produced has spawned a new discipline of "bioinformatics" dedicated to the computerized analyses of biological data.

When the HGP was first proposed, there was considerable controversy among biologists about its wisdom (Tauber and Sarkar 1992; Cook-Deegan 1994). There were:

(i) doubts about its ability to deliver on the bloated promises made by proponents of its scientific and, especially, medical benefits;
(ii) questions whether such organized "Big Biology" projects were wise science policy because of their potential effect on the ethos of biological research; and
(iii) worries that society would be legally and medically ill-prepared to cope with the results of sequencing that came too rapidly, in contrast to the normal slower accumulation of human genomic sequence information. It was feared that legislation protecting genetic privacy and preventing genetic discrimination would not be in place; there would be a shortage of genetic counselors; and so on.

In one important respect, the critics were correct: There have been few immediate medical benefits from the HGP, and no significant such innovation seems forthcoming. Instead, recent work underscores the importance of gene/environment interactions that critics had routinely invoked to criticize the claims of the HGP (see Heredity and Heritability). However, in another sense, even the most acerbic critics should now accept that the scientific results of the sequencing projects, taken together, have been breathtaking.

MOLECULAR BIOLOGY

Contrary to the expectations of the HGP's proponents, few successful predictions about organismic development have come from sequence information alone (Stephens 1998). However, genomic research is persistently throwing up surprises:

1. The most important surprise from the HGP is that there are probably only about 30,000 genes in the human genome, compared with an estimate of 140,000 as late as 1994 (Hahn and Wray 2002). In general, plant genomes are expected to contain many more genes than the human genome. Morphological or behavioral complexity is not correlated with the number of genes that an organism has. This has been called the G-value paradox (ibid).
2. The number of genes is also not correlated with the size of the genome, as measured by the number of base pairs. The fruit fly *Drosophila melanogaster* has 120 million base pairs but only 14,000 genes; the worm *Caenorhabditis elegans* has 97 million base pairs but 19,000 genes; the mustard weed *Arabidopsis thaliana* has only 125 million base pairs and 26,000 genes; while humans have 2,900,000,000 base pairs and 30,000 genes (Hahn and Wray 2002).
3. At least in humans, the distribution of genes on chromosomes is highly uneven. Most of the genes occur in highly clustered sites. Most of these genes are expressed in many tissues—the so-called "housekeeping" genes (Lercher, Urrutia, and Hurst 2002). However, the spatial distribution of cluster sites appears to be random across the chromosomes. (Cluster sites tend to be rich in C and G, whereas gene-poor regions are rich in A and T.) In contrast, the genomes of arguably less complex organisms, including *D. melanogaster*, *C. elegans*, and *A. thaliana,* do not have such pronounced clustering.
4. Only 2% of the human genome codes for proteins, while 50 % of the genome is composed of repeated units. Coding regions are interspersed by large areas of noncoding DNA. However, some functional regions, such as *HOX* gene clusters, do not contain such intervening sequences.
5. Scores of genes appear to have been horizontally transferred from bacteria to humans and other vertebrates, though apparently not to other eukaryotes. However, this issue remains highly controversial.
6. Once attention shifts from the genome to the proteome, or the protein complement of a cell (see below for more detail), a strikingly different pattern emerges. The human proteome is far more complex than the proteomes of the other organisms for which the genomes have so far been sequenced. According to some estimates, about 59% of the human genes undergo alternative splicing, and there are at least 69,000 distinct protein sequences in the human proteome. In contrast, the proteome of *C. elegans* has at most 25,000 protein sequences (Hahn and Wray 2002).
7. It now appears that noncoding DNA is routinely transcribed into RNA but not translated in complex organisms (Mattick 2003). It seems that these RNA transcripts form regulatory networks that are critical to development. Interestingly, the amount of noncoding DNA sequences in organisms appears to grow monotonically with the morphological complexity of organisms.
8. At least in *A. thaliana*, there is evidence of genome-wide non-Mendelian inheritance during which specifications from the grand-parental, rather than parental, generation are transmitted to organisms (Lolle et al. 2005).

An important task of modern molecular biology is to make sense of these disparate unexpected discoveries. One conclusion seems unavoidable: Any concept of the gene reasonably close to that in classical genetics will be irrelevant to the molecular biology of the future (see Genetics).

Proteomics

The term "proteome" was introduced only in 1994 to describe the total protein content of a cell produced from its genome (Williams and Hochstrasser 1997). Unlike the genome, the proteome is not even approximately a fixed feature of a cell (let alone an organism), but changes over time during development. Deciphering the proteome, and following its temporal development during the life cycle of each tissue of an organism, has emerged as the major challenge for molecular biology in the postgenomic era. This project has been encouraged by the discovery of unexpected universality of developmental processes at the level of cells and proteins (Gerhart and Kirschner 1997). For instance, even though hundreds of genes are known to specify molecules involved in transport across cellular membranes, there are only about twenty transport mechanisms in all living systems. The emergence of proteomics in the wake of the various sequencing projects signals an acceptance of the position that studying processes largely at the DNA level will not suffice to explain phenomena at the cellular and higher levels of organization. Even genomics did not go far enough; a sharper break with the past will be necessary.

Nevertheless, in one very important sense, the emergence of proteomics recaptures the spirit of early molecular biology, when all molecular types, but especially proteins, were the foci of interest, and the deification of DNA had not replaced a pluralist vision of the molecular basis for life. In the late 1960s, Brenner and Crick proposed "Project K, the complete solution of *E. coli.*" *E. coli* (strain K-12) was selected as a model organism because of its simplicity (as a unicellular prokaryote) and ease of laboratory manipulation. Project K included: (i) a "detailed test-tube study of the structure and chemical action of biological molecules (especially proteins)"; (ii) completion of the models of protein synthesis; (iii) work on the structure and function of cell membranes; (iv) the study of control mechanisms at every level of organization; and (v) the study of the behavior of natural populations, including population genetics. Once *E. coli* was solved, and biology was supposed to move on to more complex organisms (Crick 1973, 67).

Notice that DNA receives no preferential attention at the expense of other molecular components in Project K and that the centrality of proteins as the most important active molecules in a cell is recognized. Project K accepts that there is much more to the cell than DNA; it accepts that no simple solution of the cell's behavior can be read from the genomic sequence. After a generation of infatuation with DNA and genetic reductionism, the aims of proteomics return in part to the vision of biology incorporated in Project K. However, at least in one important way, that project went even beyond proteomics as currently understood: It emphasized all levels of organization, whereas the explicit aims of proteomics are limited to the protein level. The future will probably require further expansion.

Meanwhile, work on proteins has also generated unexpected challenges. In particular, the four rules of classical molecular biology have not survived intact, and at least the last three will require some modification. It now appears—though the essential idea goes back to the 1960s—that the fit between interacting sites of protein molecules is more dynamic than in the classical model, with the active site often "inducing" an appropriate fit (see e.g., Koshland and Hamadani 2002). It also appears that a more complicated model than the original allostery model will be required to account for many cases of cooperativity. A systematic philosophical appraisal of these developments is yet to be undertaken.

Philosophical Speculations

The developments described in the last section are so recent that any attempt to interpret their philosophical significance must remain partly speculative. Some of the empirical generalizations noted will undoubtedly be challenged by further work in the near future—if the recent past of molecular biology is any guide to its future. Moreover, there has been very little philosophical attention to these developments.

Beyond Reduction?

That the four rules of classical molecular biology are being challenged, at least to some extent, is not reason enough to generate any new skepticism about the reductionist interpretation of explanation in molecular biology. They do not bring the physical explanation of wholes by parts into question. However, if an RNA-based (or other) regulatory network turns out to be crucial to explaining development (and evolution, as Mattick 2003 argues), the reductionist interpretation may be in trouble. If network-based explanations are ubiquitous, it is quite likely that what will often bear the explanatory weight in such explanations is the topology of the network. As noted earlier, some classical phenomena such as dominance have already been known to resist straightforward reductionist explanation (Sarkar 1998).

Topological explanations have not received the kind of attention from philosophers they deserve, even though networks have lately entered the center stage of scientific attention (Mattick and Gagen 2005). Here "topology" refers to the connectivity properties of systems such as networks, which, without loss of generality, can be modeled as directed graphs. The vertices of such a graph represent components of a system, and edges (between vertices), with appropriate directionality and weights, represent interactions between such vertices. How topological an explanation is becomes a matter of degree: The more an explanation depends on individual properties of a vertex, the closer an explanation comes to traditional reduction (the components matter more than the structure) (see Reductionism). Conversely, the more an explanation is independent of individual properties of a vertex, the less reductionist it becomes. In the latter case, if explanations invoke properties of a graph that measure its connectivity, then these are topological explanations. Such connectivity measures include the number of edges in the graph, the distribution of edge degree between vertices (the "degree" of a vertex being the number of edges incident on it), and so on. (For a review of network theory, see Newman 2003). If topological explanations become necessary in molecular biology, it will mark a serious philosophical break with the reductionist classical era.

Beyond DNA Information?

As noted earlier, there is as yet no fully satisfactory account of biological information that is appropriate for molecular biology. However, the developments within eukaryotic genetics and, especially, genomics strongly suggest that the view that DNA is the sole carrier of information, however it is characterized, cannot be sustained at least for organisms more complicated than prokaryotes and perhaps not even for them. Most of the critical interactions that determine the future behavior of a cell seem to occur at the level of RNA: splicing, RNA editing, and so on. Because of this feature of cellular interactions, Sarkar (2005, Ch. 14) has speculated that the DNA genome consists of a relatively static set of sequestered modular templates (resulting in the "SMT model" of the cell), far from the classical view of the genome coding a program for development. The failure of the sequence hypothesis for many proteins only increases skepticism about the classical picture.

The routine generation of untranslated RNA transcripts from the genome also suggests that should cellular processes be viewed informationally, RNA networks form a parallel information-processing system partly independent from the genomic DNA (Mattick 2003). At present, it is unclear whether such information must also be viewed semiotically, though it seems likely, since the simplest way in which RNA sequences can be viewed as carriers of information is by the specification of information by the RNA sequences.

Similarly, the discovery of ubiquitous non-Mendelian genetic specification in *A. thaliana* (Lolle et al. 2005) also suggests that there is yet another parallel system of heredity that can also potentially be viewed informationally and, once again, is not specified through DNA. It is also possible that all such phenomena are best interpreted not informationally but using the more traditional—generally structural—conceptual apparatus of physics and chemistry. However, the distinction between the two frameworks becomes blurred in the case of RNA because the relation between the sequence and three-dimensional conformation seems to be relatively straightforward, at least much more so than in the case of proteins.

Finally, in these discussions of biological information, two issues should be distinguished:

- whether an informational framework for molecular biology is of any use; and
- whether, within any such framework, DNA (or, more restrictively, genomic DNA) is the sole repository of that information.

The problems mentioned here provide a forceful argument against the second claim, leaving open the status of the first.

Toward a Dynamic Account of the Organism

One problem with informational interpretations of molecular biology is that they have always been static: Time does not enter explicitly into accounts of biology based on the transfer of information, though, implicitly, such transfer must take place during some time interval. Recall that the proteome is not a static feature of the organism, let alone the cell: Proteomics requires a commitment to the characterization of cellular and organismic change over time. Moreover, the recent discoveries of potentially ubiquitous RNA network-based regulation also underscore the importance of dynamic accounts explicitly taking time into consideration. Moreover, new microarray techniques and their extensions are increasingly making temporal stages of cellular changes empirically accessible. The challenge remains to develop a theoretical framework to interpret the empirical information.

Any such framework can begin with either a physicalist or an informational characterization of cellular processes or a mixture of both, though prospects for a physicalist account do not seem particularly promising because of the sheer complexity of the molecular networks involved (Sarkar 2005, Ch. 10). But a dynamic informational account also leads to uncharted territory. In retrospect, what seems surprising is how successful the static framework for classical molecular biology has been, given that organisms are obviously dynamic entities undergoing development over time.

Conclusions: An Invitation

Molecular biology has not received the extent of philosophical attention it deserves, and the little it has received has been limited to the classical period (see Darden and Tabery 2005 for a more detailed summary than what has been presented here).

There are at least two reasons why philosophers should invest more work on the subject:

- Without at least a partial methodological commitment to molecular concepts and techniques, any subdiscipline within biology will likely soon be relegated to irrelevance. Philosophy of biology that does not take molecular biology into account will remain incomplete.
- Modern molecular biology raises fundamentally new epistemological questions, especially about the relevance of physical and semiotic informational accounts that have dominated discussions of biology for the last century. The deployment of philosophical (particularly formal) techniques may contribute significantly to the advancement of the field.

The most important task in the philosophy of biology for the next few decades will be to conceptualize the functional role of DNA within the cell so as to explain the surprising organization and other properties of the genome that were discussed earlier. Physical and informational accounts will probably have to interact in order to create a consistent satisfactory picture. As the last section indicates, any such attempt must necessarily begin with a clearer account than what is currently available of what 'information' means in a biological context. This is probably where philosophers have the most to contribute to the future of molecular biology (see Biological Information). Perhaps techniques from formal epistemology or semantics will enable progress where traditional biological tools have largely failed.

Sahotra Sarkar

References

Barrell, B. G., G. M. Air, and C. Hutchison III (1976), "Overlapping Genes in Bacteriophage PhiX174," *Nature* 264: 34–41.

Berget, S., C. Moore, and P. Sharp (1977), "Spliced Segments at the 5' Terminus of Adenovirus 2 Late mRNA," *Proceedings of the National Academy of Sciences (USA)* 74: 3171–3175.

Berk, A., and P. Sharp (1978), "Structure of the Adenovirus 2 Early mRNAs," *Cell* 14: 695–711.

Cattaneo, R. (1991), "Different Types of Messenger RNA Editing," *Annual Review of Genetics* 25: 71–88.

Cavalier-Smith, T. (1978), "Nuclear Volume Control by Nucleoskeletal DNA, Selection for Cell Volume and Cell Growth Rate, and the Solution of the DNA C-Value Paradox," *Journal of Cell Science* 34: 247–278.

Chow, L., R. Gelinas, T. Broker, and R. Roberts (1977), "An Amazing Sequence Arrangement at the 5' Ends of Adenovirus 2 Messenger RNA," *Cell* 12: 1–18.

Cook-Deegan, R. (1994), *The Gene Wars*. New York: Norton.

Crick, F. H. C. (1958), "On Protein Synthesis," *Symposia of the Society for Experimental Biology* 12: 138–163.

—— (1973), "Project K: "The Complete Solution of *E. coli*," *Perspectives in Biology and Medicine* 17: 67–70.

Culp, S. (1995), "Objectivity in Experimental Inquiry: Breaking Data-Technique Circles," *Philosophy of Science* 62: 438–458.

Darden, L., and J. Tabery (2005), "Molecular Biology," in E. N. Zalta (ed.), *The Stanford Encyclopedia of Philosophy*. http://plato.stanford.edu/archives/spr2005/entries/molecular-biology

Fox, T. D. (1987), "Natural Variation in the Genetic Code," *Annual Review of Genetics* 21: 67–91.

Gehring, W. (1998), *Master Control Genes in Development and Evolution: The Homeobox Story*. New Haven, CT: Yale University Press.

Gerhart, J., and M. Kirschner (1997), *Cells, Embryos, and Evolution*. Oxford: Blackwell Science.

Griffiths, P. (2001), "Genetic Information: A Metaphor in Search of a Theory," *Philosophy of Science* 67: 26–44.

Hahn, M. W., and G. A. Wray (2002), "The G-Value Paradox," *Evolution and Development* 4: 73–75.

Hull, D. (1972), "Reduction in Genetics—Biology or Philosophy?" *Philosophy of Science* 39: 491–499.

Jacob, F. (1973), *The Logic of Life: A History of Heredity*. New York: Pantheon.

Keller, E. F. (2000), *The Century of the Gene*. Cambridge, MA: Harvard University Press.

Kitcher, P. (1984), "1953 and All That: A Tale of Two Sciences," *Philosophical Review* 93: 335–373.

Koshland, D. E., Jr. and K. Hamadani (2002), "Proteomics and Models for Enzyme Cooperativity," *Journal of Biological Chemistry* 277: 46841–46844.

Lercher, M. J., A. O. Urrutia, and L. D. Hurst (2002), "Clustering of Housekeeping Genes Provides a Unified Model of Gene Order in the Human Genome," *Nature Genetics* 31: 180–183.

Lolle, S. J., J. L. Victor, J. M. Young, and R. H. Pruitt (2005), "Genome-Wide Non-Mendelian Inheritance of Extra-Genomic Information in *Arabidopsis*," *Nature* 434: 505–509.

Mattick, J. (2003), "Challenging the Dogma: The Hidden Layer of Non-Protein-Coding RNAs in Complex Organisms," *BioEssays* 25: 930–939.

Mattick, J., and M. J. Gagen (2005), "Accelerating Networks," *Science* 307: 856–857.

Mayr, E. (1961), "Cause and Effect in Biology," *Science* 134: 1501–1506.

Monod, J. (1971), *Chance and Necessity: An Essay on the Natural Philosophy of Modern Biology*. New York: Knopf.

Nagel, E. (1961), *The Structure of Science*. New York: Harcourt, Brace and World.

Newman, M. E. J. (2003), "The Structure and Function of Complex Networks," *SIAM Review* 45: 167–256.

Olby, R. C. (1974), *The Path to the Double Helix*. Seattle: University of Washington Press.

Pauling, L., and R. B. Corey (1950), "Two Hydrogen-Bonded Spiral Configurations of the Polypeptide Chains," *Journal of the American Chemical Society* 71: 5349.

Rheinberger, H. J. (1997), *Toward a History of Epistemic Things: Synthesizing Proteins in the Test Tube*. Stanford, CA: Stanford University Press.

Sarkar, Sahotra (1989), "Reductionism and Molecular Biology: A Reappraisal," Ph. D. Dissertation, Department of Philosophy, University of Chicago.

——— (1996), "Biological Information: A Skeptical Look at Some Central Dogmas of Molecular Biology," in Sarkar (ed.), *The Philosophy and History of Molecular Biology: New Perspectives*. Dordrecht, Netherlands: Kluwer, 187–231.

——— (1998), *Genetics and Reductionism*. New York: Cambridge University Press.

——— (2005), *Molecular Models of Life: Philosophical Papers on Molecular Biology*. Cambridge, MA: MIT Press.

Schaffner, K. F. (1967), "Approaches to Reduction," *Philosophy of Science* 34: 137–147.

Shannon, C. E. (1948), "A Mathematical Theory of Information," *Bell System Technical Journal* 27: 379–423, 623–656.

Stephens, C. (1998), "Bacterial Sporulation: A Question of Commitment?" *Current Biology* 8: R45–R48.

Tauber, A. I., and S. Sarkar (1992), "The Human Genome Project: Has Blind Reductionism Gone Too Far?" *Perspectives on Biology and Medicine* 35: 220–235.

Thiéffry, D., and S. Sarkar (1998), "Forty Years under the Central Dogma," *Trends in Biochemical Sciences* 32: 312–316.

Waddington, C. H. (1962), *New Patterns in Genetics and Development*. New York: Columbia University Press.

Waters, C. K. (1990), "Why the Anti-Reductionist Consensus Won't Survive the Case of Classical Mendelian Genetics," in A. Fine, M. Forbes, and L. Wessels (eds.), *PSA 1990: Proceedings of the 1990 Biennial Meeting of the Philosophy of Science Association*. East Lansing, MI: Philosophy of Science Association, 125–139.

Watson, J. D., and F. H. Crick (1953), "Molecular Structure of Nucleic Acids—A Structure for Deoxyribose Nucleic Acid," *Nature* 171: 737–738.

Williams, K. L., and D. F. Hochstrasser (1997), "Introduction to the Proteome," in M. R. Wilkins, K. L. Williams, R. D. Appel, and D. F. Hochstrasser (eds.), *Proteome Research: New Frontiers in Functional Genomics*. Berlin: Springer, 1–12.

Wimsatt, W. C. (1976), "Reductive Explanations: A Functional Account," *Boston Studies in the Philosophy of Science* 32: 671–710.

Yockey, H. P. (1992), *Information Theory and Molecular Biology*. Cambridge, UK: Cambridge University Press.

See also **Biological Information; Biology; Explanation; Function; Genetics; Heredity and Heritability; Holism; Mechanisms; Neurobiology, Philosophy of; Genetics; Physicalism; Reductionism**